T0300112

An Introductory Course of

PARTICLE
PHYSICS

AN INTRODUCTORY COURSE OF
PARTICLE
PHYSICS

PALASH B. PAL

SAHA INSTITUTE OF NUCLEAR PHYSICS

KOLKATA, INDIA

CRC Press
Taylor & Francis Group
Boca Raton   London   New York

CRC Press is an imprint of the
Taylor & Francis Group, an **informa** business

A CHAPMAN & HALL BOOK

CRC Press
Taylor & Francis Group
6000 Broken Sound Parkway NW, Suite 300
Boca Raton, FL 33487-2742

International Standard Book Number-13: 978-1-4822-1698-1 (Hardback)

Visit the Taylor & Francis Web site at
http://www.taylorandfrancis.com

and the CRC Press Web site at
http://www.crcpress.com

To

Lincoln Wolfenstein

Whose inspiration has guided me
not only through my PhD thesis
but throughout my life

Contents

Contents

List of Figures

List of Tables

Preface

Modern theory of particle interactions is based on gauge theories. Among the four fundamental interactions, all but the gravitational one have been moulded in the form of gauge theories. There are two important ingredients in a gauge theory: first, it has to be a quantum field theory, and second, it must have some internal symmetry that governs its dynamics.

In giving the reader a taste of fundamental interactions among elementary particles, I have not assumed any prior knowledge of quantum field theory. I have presented the necessary tools of quantum field theory in Ch. 4 of this book. Of course, such a brief introduction can never be complete, and many results have been stated without proof. I have taken care to ensure that the number of such results is kept to a minimum without seriously getting into the nitty-gritty of quantum field theory, and the nature of these results is more didactic than conceptual. But if the reader wants to understand the reasons behind these formulas, he or she will have to supplement this book with a beginner-level textbook on quantum field theory.

I do not take the same attitude regarding some advanced topics of quantum field theory. These have been dealt with in more detail. Such discussions start from Ch. 11 of the book and appear in different sections of the chapters that follow. In other words, I assume only a beginner's knowledge of quantum field theory, not an expert's knowledge.

In regard to group theory, I have assumed a somewhat different level of preparation from the reader. I assumed that the reader is familiar with the basic definition and properties of a group, and is even familiar with the group SU(2) and its representations. This seemed like a fair assumption since these topics are covered by any textbook on quantum mechanics in the context of angular momentum operators. For continuous groups bigger than SU(2), I do not assume any prior knowledge. Their representations have been discussed in detail in the text.

The idea of writing such a book grew when I taught a one-semester course on particle physics for several years during the last decade at my home institute, the Saha Institute of Nuclear Physics. However, the book contains much more than what could be covered in that one-semester course. Students enter our institute after completing their M.Sc. programs, so most of them already come with some background of particle physics. Some preliminary material presented in the book can therefore be omitted for them, but I have included it for the sake of completeness. The book also contains some advanced material that could not be covered in the course due to lack of time. I think finally I have been able to cover the subject at a level where an M.Sc. student can start reading the book and understand a good part of it, and also beginning Ph.D. students can learn something from it. Anyone teaching a course from the book will have to probably choose from the topics, depending on the level of students in the course. But the book is not only about teaching courses: I hope established researchers will also find the book useful.

I have put in a lot of exercises for the benefit of the students. I have placed them not at the end of the chapters but within the text, in places where I think performing the exercise would be most helpful for the students. Even if a student does not feel like solving a particular exercise while reading through the text, it is advised that the student reads the statement of the exercise, because some exercises complement the text in a useful way. Some exercises have been marked with a '♩' sign, signifying that they have in fact been worked out elsewhere in the book, although I don't mention where.

It took me a lot of time to finish the book, much more than what I would have liked. I am very fortunate that Lincoln Wolfenstein, who introduced me to the world of particle physics, has kindly read earlier drafts of several chapters of the book and has made illuminating and helpful remarks. I am also deeply indebted to a long list of students and colleagues, who have either helped me out when I was stuck, or suggested modifications to earlier drafts, or read some chapters and made extensive comments. The list includes Biswajit Adhikary, Sunanda Banerjee, Bireswar Basu-Mallick, Gautam Bhattacharyya, Francisco Botella, Gustavo Branco, Sayan Chakrabarti, Somdeb Chakraborty, Subhasis Chattopadhyay, Abhishek Chowdhury, Dipankar Das, Asit Kumar De, Jadunath De, Amit Ghosh, Kumar Sankar Gupta, Avaroth Harindranath, Aminul Islam, Anirban Kundu, Amitabha Lahiri, Debasish Majumdar, Parthasarathi Mitra, Kuntal Mondal, Santanu Mondal, Swagata Mukherjee, Munshi Golam Mustafa, José F. Nieves, Shoili Pal, Jorge Romão, Shibaji Roy, Satyajit Saha, Santosh Samaddar, Subir Sarkar, Ashoke Sen, João Silva — and I apologize to anyone whose name may have been omitted. I also received help and advice from Michael Barnett, Kunio Inoue, Serguey Petcov, Hitoshi Murayama, Abdelhak Djouadi, Sven-Olaf Moch, Kevin McFarland, Heinrich Päs, Subir Sarkar, Horst Wahl and Graeme Watt regarding obtaining permissions for reproducing several illustrations used in the book. Publishers of various journals have also been very kind in granting permissions. I have acknowledged them in proper places. I cannot overstate my indebtedness to my home institute, the Saha Institute of Nuclear Physics, for providing me with wonderful infrastructural facilities. For some extended period during the writing of this book, I was a visitor at the Instituto Superior Técnico in Lisbon, and I thank my hosts for their hospitality. Peripheral help from Kamakshya Prasad Modak and Souvik Priyam Adhya is also appreciated. And finally, I want to thank the wonderful support staff at the CRC

Press, who have helped me in a lot of ways in preparing the manuscript in its final form.

Despite all the help I have received, I am sure that some mistakes must have remained in the book. I am responsible for all of them. Since I have submitted a camera-ready version of the book to the publishers, I am even responsible for any typographic errors. I will appreciate if any reader points out any mistake and will keep a list on the internet. For now, the errata will be located at `http://www.saha.ac.in/theory/palashbaran.pal/books/partphys/errata.html`. In case I have to change this web address, please search on the internet with `palash+particle+course`.

Palash B. Pal

Notations

This is not an exposition of all symbols used in the book. It is an explanation of why such a list could not be made.

It would have been nice if a one-to-one correspondence could be maintained between the physical quantities and the symbols that represent them. Alas, it was not possible to do that without creating notations which were unnecessarily non-transparent, or uncomfortably different from accepted ones, or both. Take, for example, the lowercase letter 's'. It is used to represent spin, for obvious reasons. But it is also used customarily to denote one of the Mandelstam variables. One might think of using the uppercase letter for one of the two, but then it will be confused with the symbol for the strangeness quantum number. Similarly, the letter C is used for charge conjugation as well as for the charm quantum number. Maybe the worst confusion occurs with B, which can be the baryon number or the bottom quantum number, as well as the symbol for the branching ratio of a decay channel.

In the end, I stopped worrying about these things by making two adjustments with my conscience. First, I assumed that the reader would be mature enough to understand the meaning of a symbol from its immediate context. Second, in some cases, I used different font shapes of the same letter, sometimes with shades of gray, to stand for different things. Here are some (and definitely not all) examples of uses of this sort:

---

s Mandelstam variable	π Irrational number	Γ Decay rate
s Spin vector	π Pion	Γ Gamma function

In some cases, I have pointed out such shape differences within the text. If the reader carefully notices them, the differences should be apparent. Conversely, if the reader feels that looking at font shapes is too much of a bother, at least he or she is no worse off than looking at identical symbols for different quantities.

In a very few cases, I have used notation that is definitely non-standard. It is customary to denote antiparticles by a bar above the symbol for denoting the corresponding particle, e.g., $\bar{\nu}$ for antineutrinos. I never liked this idea because the overbar is used for constructing bilinears of fermion fields with definite transformation properties under the Lorentz group, e.g., $\overline{\psi}\psi$ denotes a scalar. So I have used hats to denote antiparticles. Thus, $\hat{q}$ stands for antiquark and $\hat{\nu}$ for antineutrinos. I am not claiming that the hat has not been used for any other purpose. For example, unit 3-vectors have been denoted by $\hat{n}$. My only claim is that such other uses occur in very few places, so that they do not conflict with my use of symbols for antiparticles in a large way.

There is a bound state of the charm quark and its antiquark that is referred to as J/ψ in the literature, and called *j-psi*. I do not like using the slash in the name. Using the flexibilities of LaTeX typesetting, I have superposed the shapes of the two letters involved to create the notation Ψ. Some of my friends find it outrageous. I obviously don't.

Chapter 1

Scope of particle physics

1.1   What are elementary particles?

Particle physics is the branch of physics which studies elementary particles and their properties.

It is easy to clarify what we mean by "elementary particles": these are particles which do not have any substructure. In other words, these are objects which are not made up of smaller objects. Turning the tables around, we can say that the elementary particles are the fundamental constituents of all objects in the universe.

It is not so easy to tell which particles are elementary. It is possible to say which particles are *not* elementary, if one knows of some experiment that shows substructure of the object. But no experiment can guarantee that a given object does *not* have any substructure. A new experiment might show substructure in an object that was earlier considered elementary. Atoms used to be thought of as indivisible until a series of experiments performed in the late 19th and early 20th century confirmed the existence of electrons and the atomic nucleus. It was soon realized that for all atoms except the lightest one, viz., hydrogen, the nucleus has a substructure. Protons and neutrons came to be collectively called *nucleons* because they are the constituents of nuclei, and treated as elementary particles. A few decades later, new experiments indicated that the nucleons themselves have substructure, so they cannot be called elementary particles any longer. We now believe that they are made up of *quarks*, which are elementary particles like the electrons.

Thus the list of elementary particles changes with time. There is no guarantee that today's elementary particles would not turn out to be composite objects tomorrow. But this cannot also stop us from having a discussion on elementary particles as they are known today. If things change tomorrow, the discussion will change, no doubt. We can only employ the best knowledge that is available at the moment to decide which objects are elementary. Needless to say, that will be our approach in this book. In §1.2 and §1.4, we are going to provide an inventory of all particles which are believed to be elementary. We will discuss the properties of these particles in this book.

An elementary particle does not have to be subatomic, in the sense that it does not have to be a constituent of atoms in some way or other. Naïvely, when we think about elementary particles, we have one of the two following propositions in our mind, or maybe some mixture of the two:

1. Elementary particles are constituents of larger conglomerates.

2. Elementary particles do not have any constituents.

It is important, therefore, to be absolutely clear at the outset that when we talk about the field of "elementary particle physics", we use the word "elementary" *only* in the second sense. The existence of the first property might be crucial in detecting a particular elementary particle, but is not necessary for deciding whether a particular particle should be included in the list of elementary particles. In other words, the capability of forming larger conglomerates is a property that an elementary particle may or may not have. In fact, most elementary particles do not have this property for one reason or other. For example, if the particle is very unstable, it would not have the chance of coming in contact with other particles to form a big, macroscopic chunk of matter. Also, not all particles can bind with others. One of the best-known elementary particles, the photon, cannot bind to other particles and is not a constituent of any bigger structure.

1.2   Inventory of elementary fermions

In 1897, the electron was discovered by J. J. Thomson, who suggested that electrons were constituents of atoms. Before that, it was believed that the atoms were indivisible. In search for all constituents of atoms, it was soon discovered that the negative electric charge of the electrons is counterbalanced in atoms by the positive charge of the nucleus. The nucleus, though small, contains protons and neutrons, which are much heavier than the electrons. The electrons, protons and neutrons thus constitute atoms, a picture that was established finally by the discovery of the neutron in 1932 by James Chadwick.

The electron, the proton and the neutron are all examples of *fermions*, which have an intrinsic angular momentum, or *spin*, that is a half-integral multiple of the fundamental constant $\hbar$. In particular, all three of these have spin equal to $\frac{1}{2}\hbar$. The other alternative is to have spin in integral multiples of $\hbar$, and particles carrying such spin are called *bosons*. In this section, we will consider elementary fermions only.

The proton and the neutron are not elementary particles, as we know now. But the electron is. The electron thus goes back longer into history than any other particle which is considered elementary even now.

With the first observation of the muon (μ^-) in 1937 and the subsequent study of its properties, it was clear that there are other elementary particles which are very much similar in their properties to the electron, except that they are heavier. Rabi asked his famous question, "Who ordered that?" expressing his surprise and disgust at the fact that Nature seems to be repeating things. We don't know the answer to Rabi's question, but we know that the repetition does exist. Of course there are some differences in the properties of the muon and the electron. For example, the muon, being heavier, can decay into lighter particles, which the electron cannot. But, as this example shows, the apparent differences can all be traced to the difference in mass. At a basic level, the electron and the muon have the same interactions. In modern terminology, we call the muon a particle of the second *generation*. Much later, in 1975, another such particle has been discovered, which is called the tau particle (τ), or simply the tau. It belongs to the third generation.

In 1930, in order to explain the continuous spectrum of electrons in nuclear beta decay, Pauli proposed that a neutral fermion was produced in such processes. Neutral fermions are collectively called *neutrinos*. Pauli's conjecture was proved in 1956, and thus the existence of one kind of neutrinos was confirmed. In the modern terminology, these are called the *electron-neutrinos*(ν_e) since they are related to beta decay, where an electron is emitted. In the 1960s, it was realized that the muon is also accompanied by its own neutrino, which is now called the *muon-neutrino* (ν_μ). The tau particle is also believed to have its own neutrino which is different from ν_e and ν_μ, and is denoted by ν_τ. The neutrinos, therefore, also conform to the generational structure.

These neutrinos, along with the electron, the muon and the tau, form a class of elementary particles which are called *leptons*. There is another class of elementary particles which are called *quarks*. Unlike the leptons, no one has seen quarks in their free state. Quarks always appear in bound states, examples of which are protons and neutrons that occur in atomic nuclei. Bound states involving quarks are called *hadrons*.

Properties of quarks must be inferred from the properties of hadrons. Six types of quarks are necessary to understand the properties of all known hadrons. These quarks are given the names 'up' (u), 'down' (d), 'charm' (c), 'strange' (s), 'top' (t) and 'bottom' (b). They also seem to follow the generational structure, in the sense that the properties of u, c and t are similar, and so are the properties of d, s and b. Thus the six quarks come in three generations. These, along with the three generations of leptons mentioned earlier, exhaust the list of elementary fermions as it is known today.

A cautionary remark should accompany this list. It is known that at least any particle carrying a non-zero electric charge must have an antiparticle which carries an opposite charge. Neutrinos are electrically neutral. It is not clear whether they have antiparticles which are different from themselves. But all other particles listed above are charged. The electron and its likes are negatively charged, so they have their own antiparticles as well. The antiparticle of the electron is called the *positron*, whereas those for the muon

Table 1.1: Elementary fermions. All of them have spin-$\frac{1}{2}$. The antiparticles have not been included in the list.

	Name	Symbol
Leptons	Electron	e
	Electron-neutrino	ν_e
	Muon	μ
	Muon-neutrino	ν_μ
	Tau	τ
	Tau-neutrino	ν_τ
Quarks	Up	u
	Down	d
	Charm	c
	Strange	s
	Top	t
	Bottom	b

and the tau do not have any such special names: they are called just antimuon and antitau. All quarks have electric charge, so there must be six different types of antiquarks. Their names are also quite mundane, antiup, antidown etc.

It may be a matter of taste to decide whether to count the antiparticles as separate entries while preparing a list of elementary particles. On the one hand, it can be argued that any antiparticle is indeed different from the corresponding particle, at least in the electric charge and often in other respects as well, so they ought to be counted separately. On the other hand, it can be argued that the properties of any antiparticle follow from the properties of the corresponding particle, and therefore the antiparticle is not an independent object.

In compiling a list of all fundamental fermions in Table 1.1, we have taken the latter viewpoint. The names appearing in this table, along with their antiparticles, complete the list of elementary fermions to the best of our knowledge today. There are also elementary particles which are bosons. We will get acquainted with them in §1.4.

1.3   Which properties?

We have announced our list of elementary fermions. Let us suppose that we will also agree on our list of elementary bosons. Even after that, it remains to be decided which properties of these particles should fall within the purview of particle physics. It seems that if we successfully identify all elementary particles and know all of their properties, we will know their behavior in all sorts of circumstances and in any sort of conglomeration. This would mean that we would know the properties of any system. Thus, in principle, it might seem that this branch of physics should cover all aspects of physics, and indeed everything else that there is to know. Following the properties of single particles, we should be able to know everything about any system.

But this would be a worthless pursuit. We really cannot discuss "all" properties of all systems starting from elementary particles. The problems are twofold. First, if we have a large number of particles in a system, there is enormous practical difficulty in discussing the properties of that system starting from those of individual particles. Often, while discussing a physical process, it is beneficial to proceed by taking larger units comprised of many particles, whose identities are not disturbed in the process. For example, while discussing a football match, we talk about the players and the ball, treating them as units. It would not make much sense to talk about the atoms in the ball, or the electrons in the ball.

The second kind of problem is that, in discussing large systems, one often encounters useful parameters which appear as co-operative effects, i.e., because of the presence of a large number of particles. Things like temperature and entropy cannot be defined for a single particle, yet they are undeniably very helpful concepts while discussing thermal properties of a system.

It is therefore customary to treat physical problems in a step-by-step manner, not going beyond the substructure more than is necessary to understand a particular problem. For the football match, we do not need to get into the anatomy of the players. If a player is injured and has to be operated upon, then we stop considering the player as a unit and discuss his or her anatomy. But even then, we think in terms of bones and organs rather than molecules and atoms. For discussing the properties of a system, it is usually not necessary to go beyond one level of substructure.

Turning the table around, we can say that the properties of any elementary particle are hardly discernible in systems which have more than one level of structure. This will be our guiding principle in determining the scope of the subject. We will discuss elementary particles, and the next level of structures formed by them.

For example, we will study interactions of the electron and its antiparticle, the positron. We can also discuss the positronium system, which is a bound state of an electron and a positron. We will study protons, because they are the first level structures formed from quarks. In fact, the proton is just an example of the class of structures known, as mentioned earlier, as hadrons. There are many hadrons, which can be broadly divided into two classes. Hadrons which have half-integral spin, e.g., the proton and the neutron, fall into the class called *baryons*. Hadrons with integral spin, like the pion, are called *mesons*. In other words, baryons are fermions, mesons are bosons. We will encounter many kinds of hadrons and discuss their properties in this book.

But a hydrogen atom will not fall within the purview of particle physics, because that is a bound state of the electron and the proton, and the latter itself is a bound state of elementary quarks. Even nuclei are not included in discussions of particle physics, because they are bound states of protons and neutrons, neither of which is elementary. We will make exception to this rule only in a few places in the book when deuterium, tritium and helium nuclei will make their appearances in our discussions. However, even in such places, the aim will not be so much to understand the properties of those bound states, but rather to illustrate some general principles that apply to elementary particles as well.

To summarize then, we will study the properties of elementary particles as well as the first level of structures formed by them. There are two kinds of properties associated with any system. The first kind can be called static properties, which the system possesses even if it is left alone. The second kind of properties can be classified as dynamical, and involve the interactions of the system with other systems. For example, the mass, the spin, the charge and the magnetic moment of a particle are static properties. On the other hand, if we ask how an electron behaves when it collides with a pion, we are essentially inquiring about a dynamic property of the electron.

In practice, the difference between these two kinds of properties is less pronounced than what their definitions might suggest. The reason is that even a static property is revealed to us only through interactions. Even if we agree that the electron has a charge when it is left alone, we do not know about it until the electron undergoes some sort of interaction with other things, or even with another electron. Thus, whether static or dynamic, a property of a particle is known through interactions. It is therefore good to take an inventory of interactions between particles.

1.4   Fundamental interactions

There are four kinds of fundamental interactions, as far as we know. The oldest known one is the gravitational interaction, known since the time of Newton in the 17th century. In the 19th century, electricity and magnetism were unified into the electromagnetic theory, and this is the second kind of interaction that we recognize to be fundamental. In the early part of the 20th century, with the discovery of the atomic nucleus, it was necessary to introduce two more kinds of fundamental interactions. One of them is the strong interaction, required to explain the stability of the atomic nuclei despite the fact that the protons in the nuclei exert repulsive Coulomb forces on one another. The other is the weak interaction, needed to explain the phenomenon of beta radioactivity, in which an electron or a positron comes out of a nucleus.

In this book, gravitational interaction will not be discussed at all. The reason is twofold. First, the gravitational interaction between elementary particles is much feebler than all other interactions. For example, the grav-

Table 1.2: Elementary bosons.

Spin	Name	Number	Symbol
1	Photon	1	γ
	W bosons	2	W^+, W^-
	Z boson	1	Z
	Gluons	8	g
0	Higgs boson	1	H

itational potential between two protons is roughly 10^{-39} times weaker than the Coulomb potential between them. Thus, gravitational interactions can be neglected to a very good approximation. The second reason is that there is no universally accepted quantum theory of gravitation. In this book, we will discuss the other three kinds of interactions, using quantum theory as the theoretical basis.

In quantum theory, interactions are also described in terms of exchange of particles. For example, electromagnetic interactions are described by the mediation of the photons, which are seen as the quanta of the electromagnetic field. Photons have spin 1 (in units of $\hbar$) and are therefore bosons. Other interactions are also mediated by spin-1 bosons. The weak interactions are mediated by the particles called W^+ and W^- which have electric charges shown as the superscripts and an uncharged boson Z. The strong interactions are mediated by eight particles, which are collectively called gluons. Like the quarks, the gluons have also never been seen as free particles. Their existence is inferred from theoretical calculations regarding the properties of hadrons.

In addition, the standard theory of electroweak interactions postulates that there is a spinless boson. It is called the Higgs boson. Since the inception of the standard model, there were strong theoretical reasons to believe that it existed, although there were alternatives to the standard model which avoid this prediction. In 2012, an announcement was made about the existence of a boson that conforms with the Higgs boson of the standard model, although there is not enough data to deduce various properties of the particle and therefore to confirm that it is indeed the Higgs boson proposed in the standard model. We will assume that it is so and stick to the standard model, according to which the elementary bosons are those listed in Table 1.2.

Unlike the list of fermions, the names of all antiparticles have been included in the list of bosons. The reason is that many of the elementary bosons are their own antiparticles. The photon, for example, is its own antiparticle. So is the Z boson, and the Higgs boson according to the standard model. The W^+ and the W^- are antiparticles of each other. Each gluon has its antiparticle included in the set of eight.

1.5   High energy physics

The branch of physics that is called *particle physics* is also called *high energy physics*. The two names suggest that the study of the most minute objects is somehow related to the use of highest energies.

It is easy to argue that there has to be a connection between these two extremes. If we shine an object with some kind of waves and want to see substructures of the object at a length scale smaller than a certain value ℓ, we need waves whose wavelength is smaller than, or at most comparable to, ℓ:

$$\lambda \lesssim \ell. \tag{1.1}$$

The relation between the wave and particle natures is provided by the de-Broglie relation:

$$p = \frac{2\pi\hbar}{\lambda}, \tag{1.2}$$

where p denotes the momentum of the particles that correspond to the waves, or in other words are the quanta of the waves. Eq. (1.1) can then be written as

$$p \gtrsim \frac{\hbar}{\ell}. \tag{1.3}$$

This is the crucial equation. It tells us that, in order to study length scales or order ℓ, we must use probes whose quanta should have a minimum momentum whose magnitude is inversely proportional to ℓ. Thus, in order to know about smaller and smaller structures, we need to use more and more energetic probes.

We can look at the question in another way. Suppose we have some objects whose lengths are of order ℓ, and we want to know whether they are made up of some building blocks. Well, if there are such building blocks, or particles, they are no doubt confined in some manner within a distance of order of ℓ, which defines the size of the supposedly composite object. By the uncertainty principle, we conclude that the momentum uncertainty of these objects satisfies the relation

$$\Delta p \gtrsim \frac{\hbar}{\ell}, \tag{1.4}$$

Now, if the momentum uncertainty is that big, the momentum itself must at least be of that order. This sets up a minimum value for the kinetic energy of the particles that constitute the composite object. Since kinetic energy is a monotonically increasing function of momentum, it follows that the expression for the minimum possible kinetic energy should be of the form $\ell^{-\alpha}$, where α is positive, and its value depends on the relation connecting momentum and kinetic energy. For example, for non-relativistic particles the kinetic energy is a quadratic function of the momentum, so that $\alpha = 2$. For massless particles, $\alpha = 1$.

Obviously, the particles must also possess an overall attractive potential energy. In order that the composite system is bound, the total energy must be negative, i.e., the magnitude of the potential energy has to be larger than that of the kinetic energy. However, the two magnitudes do not differ by huge factors, as exemplified in Ex. 1.1 below. Thus, the magnitude of the total energy, called the *binding energy* of the composite object, is usually comparable with the individual contributions of the kinetic and the potential energies. We need to overcome this binding energy if we want to break up the object and see its constituent particles. For this, we need probes which are more energetic than the the the binding energy. The minimum energy of the probes will thus have to be proportional to $\ell^{-\alpha}$, which shows that we need higher energies in order to gain access to smaller length scales.

□ **Exercise 1.1** Consider the electron in the hydrogen atom. Treating it as a classical system and using Newton's laws of motion, show that the magnitude of its potential energy is twice that of the kinetic energy.

□ **Exercise 1.2** Take the size of a typical atom to be of order 10^{-8} cm. Use the arguments above to show that its binding energy has to be at least of order of a few eV. [**Note :** $1\,\mathrm{eV} = 1.6 \times 10^{-12}$ erg.]

□ **Exercise 1.3** For nuclei which have typical dimensions of order 10^{-13} cm, show that the binding energy has to be at least of the order of tens of MeVs.

□ **Exercise 1.4** Electrons have been bombarded at energies of the order of 100 GeV, and those experiments have not revealed any substructure of the electron. Even if the electron has any substructure, argue that its size must be smaller than of order 10^{-16} cm.

Indeed, the history of discovery of smaller and smaller substructures is intimately connected to the ability of obtaining and controlling probes of higher and higher energies. In the second half of the 19th century, there were immense advancements in the science and technology of electricity, which made it possible to produce moderately high voltages. Such voltages were applied to electrodes to impart energies of the order of tens of eVs to atomic electrons and strip them out of the atoms, activities which led to the discovery of the electron. Radioactivity was discovered at the turn of the century. Soon after, Rutherford realized that the products of radioactive decay are very energetic, have kinetic energies of the order of a few MeVs. He used these as probes and shot them at atoms, discovering the nuclear structure of atoms.

A big revolution in this field was ushered by Lawrence, who in the 1930s discovered a technique by which velocities of particles could be increased through periodic kicks. Cyclotrons were built to utilize this technique, and the age of particle accelerators began. At first, the accelerators were designed to impart a few, or at most a few tens of, MeVs of energy on the particles circulating within them. In the pursuit of smaller and smaller distances, the higher and higher energies were sought. This required modifications of the Lawrencian cyclotron, which worked perfectly only if the particles did not have velocities or energies in the extreme relativistic regime. In the 1960s, when one could produce probes having kinetic energies of the order of a GeV, substructures of protons were revealed in some epoch-making experiments. By the end of the 20th century, the experimentalists could achieve energies that were a couple of orders of magnitude higher than GeV.

It has to be appreciated that these are very high energies. The mass of the electron is 9.1×10^{-28} g. The energy that an object possesses due to its mass, to be called the *rest energy*, was given in the famous formula by Einstein:

$$E = mc^2 . \tag{1.5}$$

For the electron, this rest energy (sometimes also called the rest-mass energy) is 0.511 MeV. Thus, even with tens of MeVs of kinetic energy, an electron is extremely relativistic.

The proton is about 2000 times heavier: its rest energy is 938 MeV, or roughly about 1 GeV. The neutron is very slightly heavier, by only about 1.3 MeV. But the energies available in a modern day state-of-the-art accelerator machine are much larger than the mass energies of the protons or neutrons. Thus, even these particles will be extremely relativistic in these machines.

☐ **Exercise 1.5** *Calculate the rest energy of the electron in the MeV unit to verify the number given in the text.*

☐ **Exercise 1.6** *Calculate the speed of an electron if its kinetic energy is 10 MeV. [**Note :** Do not use the Newtonian formula for kinetic energy. For a particle not in a bound state, the kinetic energy is the total energy minus the rest energy.]*

1.6   Relativity and quantum theory

Since we must deal with high energies, or equivalently high momenta, we cannot discuss these matters non-relativistically: it is imperative that we have a relativistic theory of particle interactions. Moreover, since we are talking about very small objects, clearly we cannot talk in classical terms: quantum nature of the particles must be taken into account in understanding their properties.

Relativistic quantum theory is not a trivial extension of the non-relativistic version. In other words, one cannot merely use relativistically correct expressions for the Hamiltonian of any given system and use them to perform the same kind of calculations as one performs in non-relativistic quantum mechanics. One needs a completely different starting point. The reason is simple. In non-relativistic quantum mechanics, we take time as the parameter in which things evolve. On the other hand, spatial co-ordinates are treated as operators, as is obvious from their commutators with components of momentum:

$$\left[x_i, p_j\right] = i\hbar\delta_{ij}\,. \tag{1.6}$$

We calculate the expectation value of the co-ordinate operator for a particle. We also state that the uncertainties in a position co-ordinate and its canonical momentum must satisfy the uncertainty relation:

$$\Delta x \, \Delta p_x \geq \hbar\,, \tag{1.7}$$

and similar relations for other components. There is also a time-energy uncertainty relation,

$$\Delta t \, \Delta E \geq \hbar\,, \tag{1.8}$$

but that has a different interpretation in non-relativistic quantum mechanics, since time is not an operator and one cannot talk about the measurement uncertainty of time. Rather, we say that if one wants to make a measurement of energy to an accuracy better than ΔE, then the time needed to make the measurement, Δt, satisfies the inequality of Eq. (1.8).

In a relativistic theory, space and time are intimately connected, and they form what is called the spacetime. The Lorentz transformation equations, for example, involve linear combinations of spatial co-ordinates and time. In such a setting, therefore, it is imperative that space and time must be treated on the same footing. One cannot have spatial co-ordinates as operators while time is a parameter in the theory. So one takes both space and time as parameters, and the physical objects depend on them.

This changes the whole approach. At a conceptual level, this change is best exemplified by the position-momentum uncertainty principle of Eq. (1.7). It no longer has the operator interpretation of non-relativistic quantum theory. Rather, it should now be understood as something similar to the time-energy uncertainty relation of Eq. (1.8). Thus, we will take Eq. (1.7) to mean that a momentum measurement with precision better than Δp_x is possible only if the linear extent of the system, Δx, satisfies that inequality. We point out that this is the meaning in which uncertainty relation has been talked about in §1.5.

At an operational level, the change of approach implies that, rather than talking of the position operators and other operators which depend on time, we should base our physical theories on some objects which depend on both space and time. The simplest such objects are *fields*, which in classical physics mean functions of space and time. For example, when one specifies the electromagnetic field, one gives the functional dependence of electric and magnetic fields on the spatial co-ordinates and time. At any spacetime point, the value of a classical field is thus a number, and the collection of such numbers over the entire spacetime constitutes the state in which the field is.

However, the structure of classical fields does not work. It produces systems which have no ground state, and are therefore completely unstable and pathological. To cure this problem, one needs quantum fields. The basic difference with classical fields is that a quantum field at a given spacetime point is not a number. It is an operator which acts on states, creating or annihilating particles. These statements will be explained in some more detail in Ch. 4, where we will also discuss how this idea cures the problem of ground state.

And that brings in the point that quantum fields are associated with creation and annihilation of particles. This is a very big bonus, because in high energy processes, particles can of course be created from the available energy. For example, suppose an electron collides with its antiparticle, a positron. The end product of the collision might be two such pairs of the particle and its antiparticle:

$$e^- + e^+ \longrightarrow e^- + e^+ + e^- + e^+ \,, \tag{1.9}$$

provided the total kinetic energy of the initial electron and the positron exceeded the mass energy of the new pair. It is impossible to discuss such processes in the framework of non-relativistic quantum mechanics.

And that is not just because the energies are in the relativistic regime for the example mentioned above. It is not necessary that the energy for producing the final particles should come from kinetic energies of the initial ones. Mass energy can also do the job. For example, there are many unstable particles which decay into other particles, and such decays are possible because the total mass of the decay products is less than the mass of the original particle. In such cases, decays are possible even in the rest frame of the parent particle where it has no kinetic energy. One needs quantum field theory to discuss such phenomena, because, as we said, quantum field theory is armed with a mechanism of particle creation and annihilation, whereas non-relativistic quantum mechanics is not.

In summary, it is imperative that we use quantum field theory. So we discuss rudiments of this theory in Ch. 4 before embarking on properties of elementary particles. But before ending this chapter, we want to make some concluding remarks.

1.7   Natural units

Usually, we employ a system of units where there are three independent kinds of units, and we conveniently take them as the units of mass, length and time. Units of these three fundamental quantities are unrelated, and we choose them independently of one another. Other units, like those of force or energy, can be derived from these basic units.

Although there is nothing wrong with this program, one often curbs the freedom of choosing units while discussing relativistic quantum theories, be-

Table 1.3: Dimensions of important kinematic quantities in natural units.

Quantity	Dimension in	
	conventional units	natural units
Length	$[L]$	$[M]^{-1}$
Time	$[T]$	$[M]^{-1}$
Velocity	$[LT^{-1}]$	$[M]^0$
Momentum	$[MLT^{-1}]$	$[M]^1$
Force	$[MLT^{-2}]$	$[M]^2$
Angular momentum	$[ML^2T^{-1}]$	$[M]^0$
Energy	$[ML^2T^{-2}]$	$[M]$
Energy density	$[ML^{-1}T^{-2}]$	$[M]^4$
Action	$[ML^2T^{-1}]$	$[M]^0$

cause it makes calculations much simpler. In a relativistic theory, the constant speed of light c must appear in physical formulas. We can choose the units of length and time in such a way that this speed has the value 1 in our units. Effectively, it means that if we choose 'second' as the unit of time, we choose the unit of length as the distance that light travels in 1 second. This does not affect the unit of mass, which can be chosen in such a way that the value of $\hbar$ becomes 1. That will also simplify some formulas, since factors of $\hbar$ appear everywhere in quantum theory. If we choose the units by these restrictions,

$$c = \hbar = 1, \tag{1.10}$$

they are called *natural units*. We will use them throughout in this book.

Effectively, this means that we will have only one independent unit: the other two will be determined by the conditions in Eq. (1.10). We will take the unit of mass to be the independent one, and denote it by $[M]$. Remembering Einstein's formula for the rest energy, $E_0 = mc^2$, we conclude that energy must also have the dimension of mass since $c = 1$ in the natural system of units. From Planck's formula, $E = h\nu$ or $E = \hbar\omega$, we have to then infer that frequency or ω has the dimension of $[M]$, which means time has the dimension of $[M]^{-1}$. Since $c = 1$, length must also have the same dimension. And the list can be easily extended to many other things, including the examples that have been presented in Table 1.3.

It is to be noted that we are not losing any information by omitting factors of c and h in our formulas. We can always put them back by performing simple dimensional analysis. In other words, if we want to go back to conventional units, we put factors of the form $c^a \hbar^b$ in each term of an expression, and then equate powers of length and time dimensions to find a and b. The procedure is simple and short. This is why Feynman commented that keeping track of the c's and the $\hbar$'s is a complete waste of time.

☐ **Exercise 1.7** The Planck mass (M_P) is defined in the natural units by the relation

$$M_P = \frac{1}{\sqrt{G_N}}, \tag{1.11}$$

where G_N is Newton's gravitational constant. Rewrite this relation in conventional system of units, reinstating factors of c and $\hbar$ into the definition. Hence, find the value of M_P.

☐ **Exercise 1.8** Repeat Ex. 1.2 (p 9) to Ex. 1.4 (p 9) in the natural units.

1.8   Plan of the book

The structure of the rest of the book is as follows. In Ch. 2 and Ch. 3, I cover some introductory material. Ch. 4 gives a brief introduction to the basics of quantum field theory. This introduction has to be quick as well as dirty, because many of the assertions made in the chapter have not been proved. It has been written more or less in a recipe style. More advanced topics of quantum field theory, which are not expected to be covered in a first course of the subject, have been discussed in detail in some of the later chapters, as we will describe here.

In Ch. 5, I discuss quantum electrodynamics (QED for short), the field theory of electromagnetic interactions. In Ch. 6 and Ch. 7, I discuss various discrete symmetries that are crucial in the study of particle physics.

Then, in Ch. 8 and Ch. 10, I discuss various global symmetries of strong interactions. This is followed by an introduction to Yang-Mills gauge theories in Ch. 11, and then quantum chromodynamics (QCD) in Ch. 12. Both perturbative and non-perturbative aspects of QCD have been discussed. In Ch. 13, I discuss the structure of hadrons that is understood from consideration of various deep inelastic scattering experiments.

There is an intervening chapter, Ch. 9, which I have not cited in the previous paragraph. This chapter contains a brief discussion of various experimental techniques. An experimentalist would be highly dissatisfied with this chapter. It can at best be called a summary of experimental questions that a theorist should be aware of, and a history of development of experimental techniques related to the study of elementary particles.

The next few chapters discuss weak interactions, starting with the Fermi theory in Ch. 14. Spontaneously broken quantum field theories have been introduced in Ch. 15, and the standard electroweak model introduced in Ch. 16 with only the leptons, and extended to quarks in Ch. 17. In Ch. 18, I discuss various global symmetries of the standard model, including chiral symmetries and chiral anomalies. Ch. 19 is devoted to the discussion of properties of gauge bosons of the standard model, as well as the Higgs boson.

In Ch. 20, I go back to quarks again, this time introducing the heavy quarks, discussing properties of hadrons which contain at least one heavy quark. In Ch. 21, I discuss CP violation, with detailed account of the neutral kaon system, as well as the B mesons.

Ch. 22 contains discussion on signatures of neutrino mass, so far the only solid evidence for physics beyond the standard model. Ch. 23 is a summary of various theories beyond the standard model. The presentation in this chapter is only qualitative. No details of the theories are presented. Since none of these theories have any experimental signature, I consider these theories to be outside the scope of the book.

Chapter 2

Relativistic kinematics

Einstein published his ideas on relativity in two major instalments. The first instalment 1905 examined how the perception of time, space and energy might differ for observers in uniform relative motion. This constituted the basis of the *special theory of relativity*. When the speed of any particle is very high, close to the speed of light, the kinematical consequences of this theory were found to be drastically different from those derived from Newton's laws of motion and associated ideas of space and time. Since the study of particle physics is intimately connected to high energies and therefore large speeds, we must take relativistic effects into account in discussing kinematics of particle motion, which is what we do in this chapter. We will be brief, because we will assume that the reader is familiar with the basic tenets of the special theory of relativity. For fuller treatments, the reader is advised to consult textbooks on the subject.

Einstein's second instalment of relativity came around 1915, where he discussed observers in accelerated motion with respect to one another. This *general theory of relativity* turned out to be a theory of gravitation. Since we decided to neglect all gravitational effects, the ramifications of this theory will be ignored completely.

2.1   Lorentz transformation equations

The special theory of relativity is based on two axioms. One of them is the principle of relativity, which states that all physical laws should have the same form to all observers who might be moving in uniform relative motion with respect to one another. This principle was first advocated by Galileo, and was inherent in the Newtonian formulation of dynamics.

Later, when Maxwell formulated his theory of electromagnetism, it was found that the equations involve a constant which has the dimension of velocity. Further, it was realized that the constant equalled the speed of propagation of electromagnetic waves in the vacuum. This created an apparent contradiction with Newtonian dynamics. To resolve the problem, Einstein took a second axiom, viz., that the speed of light in the vacuum is the same for all inertial observers.

It is this second axiom which necessitated reformulation of the laws of dynamics. For two observers in relative motion, the notion of time and space had to be modified in order that the speed of light remains constant for both of them. The consequences of these axioms are summarized in the *Lorentz transformation equations*. Consider a frame of reference S, and another one called S' which moves with a uniform speed v along a direction which is called the common x-axis in both frames. Suppose S' has a co-ordinate system whose origin coincides with that of S at time $t = 0$. Then the relation between the location and time of any event from the two frames will be given by

$$
\begin{aligned}
x' &= \frac{x - vt}{\sqrt{1 - v^2}} \\
y' &= y \\
z' &= z \\
t' &= \frac{t - vx}{\sqrt{1 - v^2}} \, .
\end{aligned}
\tag{2.1}
$$

☐ **Exercise 2.1** Rewrite the Lorentz transformation equations using conventional units where the magnitudes of c and $\hbar$ are not unity. From them, recover the Galilean transformation equations by taking the limit $c \to \infty$.

Note that Eq. (2.1) is a set of homogeneous linear equations. They can therefore be written in a matrix notation. Introducing a shorthand

$$
x^\mu \equiv \left(t, x, y, z \right),
\tag{2.2}
$$

we can write the Lorentz transformation equations of Eq. (2.1) in the compact form

$$
x'^\mu = \Lambda^\mu{}_\nu x^\nu,
\tag{2.3}
$$

where $\Lambda^\mu{}_\nu$ are elements of the matrix

$$
\Lambda = \frac{1}{\sqrt{1 - v^2}}
\begin{pmatrix}
1 & -v & 0 & 0 \\
-v & 1 & 0 & 0 \\
0 & 0 & 1 & 0 \\
0 & 0 & 0 & 1
\end{pmatrix} .
\tag{2.4}
$$

In Eq. (2.3), the index ν is summed over, a fact that we have not explicitly indicated. Indeed, that will be our notation from now on: whenever an index is repeated, we will assume that it is summed over, unless otherwise noted. This *summation convention* would apply not only for vector indices such as ν that appears in Eq. (2.3), but also on matrix element indices, symmetry group indices and so on. A summation for repeated indices will not be assumed when the indices run over values for which there is no obvious symmetry or structure. For example, if we talk of similar kinds of terms in an equation coming from different particles, the sum will be explicitly indicated.

Whenever we assume summation over repeated indices, the range of the values of the indices should be understood from the context, and will not be indicated explicitly. For example, for the vector index ν appearing in Eq. (2.3), the sum is over all four possible values which are indicated in writing Eq. (2.2).

The transformation equations can be written in another useful form by introducing a parameter ϑ defined by the relation

$$\sinh \vartheta = \frac{v}{\sqrt{1-v^2}},$$ (2.5)

or equivalently by

$$\tanh \vartheta = v.$$ (2.6)

The parameter ϑ is called *rapidity*. Using this parameter, the matrix of Eq. (2.4) can be written in the form

$$\Lambda = \begin{pmatrix} \cosh\vartheta & -\sinh\vartheta & 0 & 0 \\ -\sinh\vartheta & \cosh\vartheta & 0 & 0 \\ 0 & 0 & 1 & 0 \\ 0 & 0 & 0 & 1 \end{pmatrix}.$$ (2.7)

Of course, Eq. (2.4) or Eq. (2.7) gives the appropriate transformation matrix if we consider the relative motion between the two frames to be along the common x-axis. If the relative motion is in some other direction, the form of the matrix, as well as the form of the Lorentz transformation equations in Eq. (2.1), will change accordingly.

Consider now another frame of reference S'' which is moving with a speed u with respect to the frame S', the direction of the relative velocity being along the common x-axis. Obviously, the co-ordinates of any event adjudged from the S'' and the S' frames will be related by the equation

$$x''^\mu = \Lambda'^\mu{}_\nu x'^\nu,$$ (2.8)

where the matrix Λ' is given by

$$\Lambda' = \frac{1}{\sqrt{1-u^2}} \begin{pmatrix} 1 & -u & 0 & 0 \\ -u & 1 & 0 & 0 \\ 0 & 0 & 1 & 0 \\ 0 & 0 & 0 & 1 \end{pmatrix}.$$ (2.9)

Combining Eqs. (2.4) and (2.9), we obtain that the relation between the co-ordinates assigned by the frames S'' and S is given by

$$x''^\mu = \Lambda'^\mu{}_\nu \Lambda^\nu{}_\rho x^\rho.$$ (2.10)

Taking the matrix product of Λ' and Λ, we find that it can be expressed in the form of the matrix in Eq. (2.4) with v replaced by

$$w = \frac{u+v}{1+uv}.$$ (2.11)

This expresses the *velocity addition law*. If an object is moving with a velocity v compared to an observer, and something is moving in the same direction with a velocity u compared to that object, the velocity of the second object with respect to the original observer is given by w given in Eq. (2.11).

☐ **Exercise 2.2** Supply the missing steps in arriving at Eq. (2.11).

☐ **Exercise 2.3** Derive the velocity addition law by using the representation of the Lorentz transformation matrix that appears in Eq. (2.7). Hence, show that while velocities add by the involved rule of Eq. (2.11), rapidities add arithmetically.

☐ **Exercise 2.4** Any component of velocity of any moving object, or a frame of reference, must satisfy the relation

$$-1 < v < +1 \tag{2.12}$$

in order that the Lorentz transformation equations are meaningful. Show that, when two such velocities are added through Eq. (2.11), the resulting velocity also satisfies this constraint.

2.2   Vectors and tensors on spacetime

The transformation law of Eq. (2.1) has a great deal of similarity with the laws of rotation. Consider two sets of co-ordinate axes in ordinary 3-dimensional space, with a common origin. If the co-ordinates of a point are given by x^i in one set of axes and by x'^i in the other, then there exists a relation of the form

$$x'^i = R^i{}_j x^j . \tag{2.13}$$

The matrix R is called the rotation matrix. This is a homogeneous linear relation, just like the relations in Eq. (2.1).

It is easy to see that rotations can be made part of the more general notation that appears in Eq. (2.3). Consider a matrix Λ of the form

$$\Lambda = \begin{pmatrix} 1 & 0 & 0 & 0 \\ 0 & & & \\ 0 & & R & \\ 0 & & & \end{pmatrix} . \tag{2.14}$$

Obviously, this changes the spatial co-ordinates by the rule of Eq. (2.13) while keeping time unchanged. On the other hand, a matrix like that given in Eq. (2.4) affects both time and space. From now on, we will use the name "Lorentz transformation" for both kinds. Transformations of the form given in Eq. (2.14) will be called *rotations*, and those which affect time will be called *boosts*.

In a rotation, the length of the co-ordinate vector remains invariant. Thus, for the vectors $\mathbf{x}'$ and $\mathbf{x}$, we can write

$$\mathbf{x}' \cdot \mathbf{x}' = \mathbf{x} \cdot \mathbf{x} . \tag{2.15}$$

This can be generalized for Lorentz transformations, including boosts and rotations, by saying that in such transformations, we obtain

$$t'^2 - \mathbf{x}' \cdot \mathbf{x}' = t^2 - \mathbf{x} \cdot \mathbf{x} . \tag{2.16}$$

For rotations, this trivially reduces to Eq. (2.15) since $t' = t$ under such transformations. For boosts, only this more general form holds, and it can be checked by using Eq. (2.1) directly. We use this analogy to call the foursome appearing in Eq. (2.2) as a 4-vector. It is a vector on the 4-dimensional *spacetime*. Time t and the spatial co-ordinates constitute co-ordinates on spacetime. Time t is usually called the 0th component of the spacetime 4-vector x^μ, i.e., the component with $\mu = 0$. The spatial components go from 1 to 3. If somewhere we want to mention only the spatial components, as in Eq. (2.13), we will use Roman letters. Indices of this kind are supposed to take values from 1 to 3, and repeated indices would imply a sum over these values.

The invariance of Eq. (2.16) can be expressed in a more compact form by writing

$$g_{\mu\nu}x'^\mu x'^\nu = g_{\mu\nu}x^\mu x^\nu \,, \tag{2.17}$$

where $g_{\mu\nu}$, called the *metric tensor*, is given by

$$g_{\mu\nu} = \mathrm{diag}(1,-1,-1,-1)\,. \tag{2.18}$$

A more compact form is obtained by defining

$$x_\mu = g_{\mu\nu}x^\nu \tag{2.19}$$

and similarly x'_μ, so that we can write the invariance of Eq. (2.17) as

$$x'^\mu x'_\mu = x^\mu x_\mu\,. \tag{2.20}$$

The object x_μ, with the lower index, is called a *covector*, or more elaborately a *covariant vector*. In order to distinguish it from it, the object x^μ defined in Eq. (2.2) is also called a *contravariant vector*. When no confusion can arise, we will use the word "vector" for either of the two types.

The notion of a vector can now be extended trivially. Any object whose components transform like the components of x^μ will be called a vector. Thus, in a transformation given by Eq. (2.3), the components of a vector A^μ will change to A'^μ, where

$$A'^\mu = \Lambda^\mu{}_\nu A^\nu\,. \tag{2.21}$$

We can also define tensors in the usual manner. For example, a rank-2 tensor will be defined by the following transformation rule:

$$F'^{\mu\nu} = \Lambda^\mu{}_\alpha \Lambda^\nu{}_\beta F^{\alpha\beta}\,. \tag{2.22}$$

□ **Exercise 2.5** Using the transformation rule for x^μ from Eq. (2.3), show the transformation rule for the covector x_μ is given by

$$x'_\mu = \Lambda_{\mu\nu}x^\nu = \Lambda_\mu{}^\nu x_\nu\,. \tag{2.23}$$

□ **Exercise 2.6** Use the transformation rules of Eqs. (2.3) and (2.23), and the invariance of Eq. (2.20), to show that the Lorentz transformation matrices satisfy the relation

$$\Lambda^\mu{}_\nu \Lambda_\mu{}^\rho = \delta^\rho_\nu . \tag{2.24}$$

Or equivalently, use Eqs. (2.3) and (2.17) to show that

$$g_{\mu\nu}\Lambda^\mu{}_\alpha \Lambda^\nu{}_\beta = g_{\alpha\beta} . \tag{2.25}$$

□ **Exercise 2.7** Using Eq. (2.24) or otherwise, show that the inverse transformation of Eq. (2.3) (i.e., expressing the unprimed co-ordinates in terms of the primed ones) can be written as

$$x^\mu = \Lambda_\nu{}^\mu x'^\nu . \tag{2.26}$$

We now find the transformation properties of derivatives with respect to the co-ordinates. Consider a scalar function ϕ of the co-ordinates, and consider the transformation property of its gradient, $\partial\phi/\partial x^\mu$. In the transformed co-ordinates, the components of the gradient would be

$$\frac{\partial\phi}{\partial x'^\mu} = \frac{\partial\phi}{\partial x^\nu}\frac{\partial x^\nu}{\partial x'^\mu} , \tag{2.27}$$

by use of the chain rule. Using Eq. (2.26) now, we can write it as

$$\frac{\partial\phi}{\partial x'^\mu} = \Lambda_\mu{}^\nu \frac{\partial\phi}{\partial x^\nu} . \tag{2.28}$$

Comparing with the transformation rule given in Eq. (2.23), we see that the gradient transforms as a covector. We have denoted covectors by lower indices. Inspired by that, we will often use the notation

$$\partial_\mu \equiv \frac{\partial}{\partial x^\mu} \tag{2.29}$$

when there cannot be any confusion about which co-ordinates have been used in the derivatives. It should be remembered that although the covector has components

$$x_\mu \equiv \left(t, -x, -y, -z\right), \tag{2.30}$$

where the minus signs appear from the metric tensor when we use Eq. (2.19), the components of ∂_μ are given by

$$\partial_\mu \equiv \left(\frac{\partial}{\partial t}, \frac{\partial}{\partial x}, \frac{\partial}{\partial y}, \frac{\partial}{\partial z}\right), \tag{2.31}$$

since these are derivatives with respect to the components of the contravariant vector.

2.3   Velocity, momentum and energy

We wrote Lorentz transformation equations in Eq. (2.1), assuming the co-ordinate origins of two frames coincide at $t = 0$. In the more general situation when the origins do not necessarily coincide, the transformation equations will have inhomogeneous terms. However, even in that case, we can write the difference between two nearby events in a homogeneous form, i.e., we can write

$$dx'^{\mu} = \Lambda^{\mu}{}_{\nu} dx^{\nu} . \tag{2.32}$$

Following the arguments of §2.2, we can then construct the invariant

$$ds^2 = dx^{\mu} dx_{\mu} . \tag{2.33}$$

The *velocity 4-vector* can then be defined as

$$u^{\mu} = \frac{dx^{\mu}}{ds} , \tag{2.34}$$

and the *momentum 4-vector* of a particle of mass m by

$$p^{\mu} = m u^{\mu} = m \frac{dx^{\mu}}{ds} . \tag{2.35}$$

☐ **Exercise 2.8** From Eq. (2.33), show that

$$u^{\mu} u_{\mu} = 1 , \tag{2.36}$$
$$p^{\mu} p_{\mu} = m^2 . \tag{2.37}$$

Note that, for a particle moving along the line $x^{\mu}(s)$, we can write

$$ds^2 = dt^2 - d\boldsymbol{x}^2 = dt^2 (1 - \boldsymbol{v}^2) , \tag{2.38}$$

where

$$\boldsymbol{v} = \frac{d\boldsymbol{x}}{dt} \tag{2.39}$$

is the 3-velocity, i.e., the velocity in 3-dimensional space. Thus

$$ds = dt \sqrt{1 - \boldsymbol{v}^2} , \tag{2.40}$$

and therefore the components of the velocity 4-vector are given by

$$u^{\mu} = \left(\frac{1}{\sqrt{1 - \boldsymbol{v}^2}}, \frac{\boldsymbol{v}}{\sqrt{1 - \boldsymbol{v}^2}} \right) . \tag{2.41}$$

Obviously, components of the momentum 4-vector of a particle are given by

$$p^{\mu} = \left(\frac{m}{\sqrt{1 - \boldsymbol{v}^2}}, \frac{m\boldsymbol{v}}{\sqrt{1 - \boldsymbol{v}^2}} \right) . \tag{2.42}$$

The spatial components reduce to the Newtonian definition $\boldsymbol{p} = m\boldsymbol{v}$ when $v \ll 1$. The temporal component is the energy of the particle:

$$E = \frac{m}{\sqrt{1 - \boldsymbol{v}^2}} \, . \tag{2.43}$$

If we expand it as a power series in v, the first term, i.e., the velocity-independent term, is the rest energy, m. The next term is the Newtonian expression for kinetic energy.

Many elementary textbooks of special relativity write Eq. (2.43) as $E = Mc^2$ (the factor of c^2 is hidden in Eq. (2.43) by our choice of natural units) by defining $M = m/\sqrt{1 - v^2/c^2}$ and call M the 'kinetic mass'. With this definition, M depends on the velocity. We will follow the more modern practice of using the word *mass* to mean the Lorentz invariant quantity m that first appeared in Eq. (2.35). With this definition, m (or mc^2, if natural units are not used) is the rest energy of a particle. Additional energy acquired because of a non-zero velocity is the particle's kinetic energy.

☐ **Exercise 2.9** *Show that Eq. (2.37) can be written as*

$$E^2 = \boldsymbol{p}^2 + m^2 \, , \tag{2.44}$$

which is the relativistic energy-momentum relation of a free particle. The relation between energy and momentum for any particular object is often called the dispersion relation of the object. Thus, Eq. (2.44) is the dispersion relation of a free particle.

☐ **Exercise 2.10** *Expand the expression for energy for small values of v. Show that the first non-trivial dependence on velocity agrees with the Newtonian definition of kinetic energy.*

2.4   Covariance

Once the 4-vector notation is put into place, it is easy to construct physical equations which are consistent with the principle of relativity. We only have to make sure that all terms in an equation transform the same way under Lorentz transformations. For example, if both sides of an equation are vectors, both will transform by the rule given in Eq. (2.21), so that the equality will hold for the transformed vectors in any other frame. Once this criterion is met, an equation is called *covariant*, i.e., all its terms have the same transformation properties.

Indeed, we have used this consideration in giving definitions of some quantities earlier. See, for example, the definition of Eq. (2.34). The quantity dx^μ transforms as a vector, whereas ds is a scalar. Therefore, the derivative dx^μ/ds transforms like a vector, and we called it the velocity 4-vector. If this is multiplied by another scalar m, the result will be another 4-vector, which is the momentum 4-vector introduced in Eq. (2.35).

We can ask what would be the derivative dp^μ/ds. Of course it will be a 4-vector, so let us call

$$\frac{dp^\mu}{ds} = F^\mu \, . \tag{2.45}$$

When the velocity is very small, the spatial components of this equation should give the rate of change of 3-momentum per unit time. In Newtonian mechanics, that is the definition of force. Thus, F^μ can be called the *force 4-vector*, and Eq. (2.45) the classical equation of motion that describes the motion of a particle in a relativistically consistent manner.

□ **Exercise 2.11** *Show that the components of the force 4-vector can be written as*

$$F^\mu = \left(\frac{\boldsymbol{F} \cdot \boldsymbol{v}}{\sqrt{1 - \boldsymbol{v}^2}}, \frac{\boldsymbol{F}}{\sqrt{1 - \boldsymbol{v}^2}} \right), \tag{2.46}$$

where $\boldsymbol{F} = d\boldsymbol{p}/dt$, as defined in Newtonian mechanics. For small values of $\mathrm{v} \equiv |\boldsymbol{v}|$, show that Eq. (2.45) has the same form as Newton's second law of motion.

2.5   Invariances and conservation laws

In a given physical process, if we know all forces acting on all participating particles, we can in principle solve the classical equation of motion, Eq. (2.45), to determine outcome of the process. This constitutes the program of *dynamics*.

In practice, this is often a cumbersome procedure since the forces might be complicated, and the equations of motion might not be analytically solvable. It is therefore natural to ask how much information can be obtained in a given situation without knowing the forces exactly.

Conservation laws can be of much help in this regard. We already mentioned that Lorentz transformations, in the generalized sense in which we use the term, contain rotations. Lorentz invariance, as a result, contains rotational invariance, which implies the conservation of angular momentum. Invariance under boosts also implies some conservation laws, but they will not be very useful in our discussion. We will only make a passing comment about these conservation laws in §3.6. In addition, we assume that all theories are invariant under the redefinition of the origins of space and time, i.e., homogeneity of space and time is assumed. Energy and momentum, or 4-momentum in short, is conserved as a consequence of this assumption.

We seem to be implying that an invariance is equivalent to a conservation law. Indeed, it is. If there is a transformation induced by a continuous parameter under which the action of a system is invariant, it implies a conservation law. This statement is famously known as *Noether's theorem*.

We will not give a general proof of this theorem here. Instead, we show an example to corroborate the theorem. Consider a system with only one co-ordinate x, whose Lagrangian is given by $L(x, \dot{x})$. The action, calculated between the times t_1 and t_2, is given by

$$\mathscr{A} = \int_{t_1}^{t_2} dt \, L(x, \dot{x}) \,. \tag{2.47}$$

Now suppose that the action is invariant if we replace x by $x + \epsilon$, where ϵ is an arbitrary constant. Obviously, $\dot{x}$ is unaffected by this transformation. Thus, we can write

$$\mathscr{A} = \int_{t_1}^{t_2} dt \, L(x + \epsilon, \dot{x}) \,. \tag{2.48}$$

Taking the difference of these two equations, we obtain

$$0 = \int_{t_1}^{t_2} dt \ [L(x + \epsilon, \dot{x}) - L(x, \dot{x})] \ . \tag{2.49}$$

The parameter ϵ is continuous, i.e., it can take arbitrary values. In particular, it is possible to take infinitesimally small values of ϵ, and in this limit Eq. (2.49) implies

$$0 = \epsilon \int_{t_1}^{t_2} dt \ \frac{\partial L}{\partial x} \ . \tag{2.50}$$

By Lagrange's equation of motion,

$$\frac{\partial L}{\partial x} = \frac{d}{dt} \left(\frac{\partial L}{\partial \dot{x}} \right) \equiv \frac{dp}{dt} \ , \tag{2.51}$$

where $p = \partial L / \partial \dot{x}$ is the momentum which is canonically conjugate to the co-ordinate x. Putting this back into Eq. (2.50), we obtain

$$p(t_2) - p(t_1) = 0 \ , \tag{2.52}$$

which means that the momentum does not change with time. This shows that invariance with respect to the translation of a co-ordinate implies the conservation of its conjugate momentum.

□ **Exercise 2.12** Consider a Lagrangian, involving two co-ordinates x and y, which is invariant under the rotation

$$\begin{aligned} x' &= x \cos \theta - y \sin \theta \ , \\ y' &= x \sin \theta + y \cos \theta \ , \end{aligned} \tag{2.53}$$

for any value of θ. Show that the quantity $xp_y - yp_x$ is conserved, where p_x and p_y are momenta that are canonically conjugate to x and y.

An advantage of using these conservation laws is that they are obeyed by quantum mechanics as well. Thus, any consequence that we might deduce from them will be valid irrespective of whether the system interacts classically, or whether quantum mechanical effects are important. In §2.6 and §2.7, we will discuss some simple consequences of these conservation laws.

It should be kept in mind that these are not the only conservation laws that we can use. We will encounter many other conservation laws and invariances in later chapters and will discuss their consequences. Some of them are exact, some are valid approximately in some specific situations. In a sense, the goal of particle physics is to discover conservation laws and invariances in particle interactions. They help us in guessing the natures of fundamental interactions. Also, the basic dynamics of particle interactions is based on some invariances called *gauge invariances*, which will be discussed later in the book. For the moment, we discuss only the consequences of 4-momentum conservation, which follows from spacetime symmetries.

2.6   Kinematics of decays

2.6.1   Lifetime and time dilation

An unstable particle decays into several other particles. This is a statistical phenomenon, which means that if we have a number N_0 of a certain kind of unstable particles at time $t = 0$, not all of them will decay at the same time. At time t, if we look at the collection and find the number $N(t)$ of the original particles which have escaped decay, this number behaves as

$$N(t) = N_0 e^{-t/\tau} \tag{2.54}$$

for some τ, which is called the *lifetime* of the unstable particle. Alternatively, we also write the relation as

$$N(t) = N_0 e^{-\Gamma t}, \tag{2.55}$$

so that

$$\Gamma = \frac{1}{\tau}. \tag{2.56}$$

The quantity Γ is called the *decay rate*.

The definitions of lifetime and decay rate, as given above, apply to the rest frame of the decaying particle. If the particle is moving in a frame, the number of surviving particles obeys a different relation, and the lifetime comes out to be different. The reason can be seen from the Lorentz transformation equations given in Eq. (2.1). In the rest frame, suppose a particle was created at time t_1 and it decayed at time t_2. Both events occurred at the same spatial point in this frame, i.e., the creation of the particle took place at (x, t_1) and the decay at (x, t_2), suppressing the other co-ordinates. Consider now the same two events from the point of view of another frame which is moving along the x-direction with a speed v with respect to the rest frame. From Eq. (2.1), the instants of time when these two events took place in this frame would be given by

$$t'_1 = \frac{t_1 - vx}{\sqrt{1 - v^2}}, \qquad t'_2 = \frac{t_2 - vx}{\sqrt{1 - v^2}}, \tag{2.57}$$

so that

$$t'_2 - t'_1 = \frac{t_2 - t_1}{\sqrt{1 - v^2}}. \tag{2.58}$$

This shows that the interval between creation and decay of the particle would be lengthened by the factor $1/\sqrt{1 - v^2}$. So the lifetime is enhanced by the same factor. Equivalently, we can say that the decay rate would be diminished the factor $\sqrt{1 - v^2}$. This phenomenon is called *time dilation*.

2.6.2   Two-body decays

Consider a process where a particle called a decays into two particles, called a_1 and a_2. Symbolically, we write the process as:

$$a(p) \rightarrow a_1(p_1) + a_2(p_2), \tag{2.59}$$

where in parentheses, we have put in the notation for the 4-momenta of the particles. The mass of the decaying particle is m, and those of the products are m_1 and m_2.

Conservation of 4-momentum means

$$p^\mu = p_1^\mu + p_2^\mu \tag{2.60}$$

for this process. Therefore,

$$p^\mu p_\mu = (p_1 + p_2)^\mu (p_1 + p_2)_\mu . \tag{2.61}$$

Using Eq. (2.37) for all particles, we obtain

$$m^2 = m_1^2 + m_2^2 + 2(E_1 E_2 - \boldsymbol{p}_1 \cdot \boldsymbol{p}_2) , \tag{2.62}$$

where E_1 and E_2 are the energies of the decay products.

The energies E_1 and E_2 can be individually determined only in a specific frame. For example, it is convenient to consider the rest frame of the decaying particle. In this frame, the decay products come out back to back, i.e., in opposite directions along a straight line:

$$\boldsymbol{p}_1 = -\boldsymbol{p}_2 . \tag{2.63}$$

Thus, Eq. (2.62) can be rewritten as

$$\frac{1}{2}(m^2 - m_1^2 - m_2^2) = \sqrt{(\mathbf{p}^2 + m_1^2)(\mathbf{p}^2 + m_2^2)} + \mathbf{p}^2 , \tag{2.64}$$

where $\mathbf{p} = |\boldsymbol{p}_1| = |\boldsymbol{p}_2|$. Solving p from here and putting into the energy-momentum relation, we obtain

$$E_1 = \frac{m^2 + m_1^2 - m_2^2}{2m} , \qquad E_2 = \frac{m^2 - m_1^2 + m_2^2}{2m} . \tag{2.65}$$

☐ **Exercise 2.13** Supply the missing steps between Eqs. (2.64) and (2.65).

☐ **Exercise 2.14** In a frame where the initial particle has a non-zero 3-momentum, show that the magnitude of momentum of the decay product a_1 is given by

$$p_1 = \frac{\mu^2 \mathbf{p} \cos\theta + \sqrt{\mu^4 \mathbf{p}^2 \cos^2\theta - (m_1^2 E^2 - \mu^4)(E^2 - \mathbf{p}^2 \cos^2\theta)}}{E^2 - \mathbf{p}^2 \cos^2\theta} , \tag{2.66}$$

where θ is the angle between the vectors $\mathbf{p}$ and $\boldsymbol{p}_1$, and $\mu^2 = \frac{1}{2}(m^2 + m_1^2 - m_2^2)$. If $m_1 = 0$, show that this gives

$$E_1 = \frac{m^2 - m_2^2}{2(E - \mathbf{p}\cos\theta)} . \tag{2.67}$$

2.6.3 Three-body decays

We now consider the case of one particle decaying into three particles. In the notation of Eq. (2.59), the process can be written as

$$a(p) \rightarrow a_1(p_1) + a_2(p_2) + a_3(p_3) \,. \tag{2.68}$$

Unlike the case of two-body decays, the energies of the final particles are not predictable even in the rest frame of the decaying particle. Depending on the relative orientations of the 3-momenta of the decay products, energy is shared in different ways among the particles.

We can find the maximum and the minimum possible energy that any of the three decay products might possess. For example, consider the particle called a_3 in Eq. (2.68). It must carry at least its mass energy, m_3. Thus,

$$\left(E_3\right)_{\min} = m_3 \,, \tag{2.69}$$

which corresponds to the situation when this particle is produced at rest while the other two particles take away all the kinetic energy. On the contrary, maximum energy will be obtained when the magnitude of its 3-momentum will be maximum, i.e., when the other two particles will be produced in the same direction, opposite to that of p_3. In this case, we can think of the other two particles as one single particle with mass $m_1 + m_2$, and treat the problem like a two-body decay process. Borrowing Eq. (2.65) with obvious changes in notations, we can write

$$\left(E_3\right)_{\max} = \frac{m^2 + m_3^2 - (m_1 + m_2)^2}{2m} \,. \tag{2.70}$$

☐ **Exercise 2.15** *Consider nuclear beta decay, in which a nucleus X turns into another nucleus Y with the emission of an electron and an antineutrino:*

$$X \rightarrow Y + e + \widehat{\nu}_e \,. \tag{2.71}$$

The masses of the two nuclei are much larger compared to their difference Q, as well as to the masses of the other particles. In such a situation, show that the maximum energy attainable by the electron is given by

$$\left(E_e\right)_{\max} = Q - m_{\nu_e} \,. \tag{2.72}$$

All the energy values discussed in this context refer to the rest frame of the decaying particle. In other frames, of course, the ranges of energies of each decay product will be different. In order to obtain frame-independent conclusions, one often uses combinations like

$$m_{12}^2 \equiv (p_1 + p_2)^2 = (p - p_3)^2 \,. \tag{2.73}$$

Note that if we evaluate this quantity in the rest frame of the decaying particle, we obtain

$$m_{12}^2 = (p - p_3)^2 = m^2 + m_3^2 - 2mE_3 . \qquad (2.74)$$

Thus, the maximum and minimum possible values of m_{12}^2 can be obtained from the expressions for minimum and maximum values of E_3 in the rest frame of the decaying particle.

$$\left(m_{12}^2\right)_{\max} = (m - m_3)^2 ,$$
$$\left(m_{12}^2\right)_{\min} = (m_1 + m_2)^2 . \qquad (2.75)$$

The range of values of m_{23}^2 and m_{31}^2 are given by similar expressions.

We can also find correlated bounds on the three quantities m_{12}^2, m_{23}^2 and m_{31}^2. First, note that these three quantities are not independent:

$$m_{12}^2 + m_{23}^2 + m_{31}^2 = m^2 + m_1^2 + m_2^2 + m_3^2 , \qquad (2.76)$$

a constant for a given decay channel, where the word *channel*, in this context, means a specified collection of particles in the end product. We can therefore take any two of these quantities, e.g. m_{23}^2 and m_{12}^2, as independent and ask ourselves what is the range of values that one of these quantities can attain for a given value of the other quantity. For this, note that in any frame,

$$m_{23}^2 = m_2^2 + m_3^2 + 2(E_2 E_3 - \boldsymbol{p}_2 \cdot \boldsymbol{p}_3) . \qquad (2.77)$$

Since the left hand side is a Lorentz invariant, we can evaluate it by taking the energies and momenta of particles 2 and 3 in any frame. It is most convenient to work in the *center-of-mass (CM) frame* of particles 1 and 2, i.e., in a frame where the 3-momenta of these two particles are equal and opposite. From the definition of Eq. (2.73), we then obtain that

$$m_{12} = E_1^\star + E_2^\star , \qquad (2.78)$$

where the stars are indicative of the particular frame used. Now note that

$$m^2 = p^2 = (p_1 + p_2 + p_3)^2 = (p_1 + p_2)^2 + p_3^2 + 2(p_1 + p_2)_\mu \, p_3^\mu$$
$$= m_{12}^2 + m_3^2 + 2m_{12}E_3^\star ,$$
$$m_1^2 = p_1^2 = (p_1 + p_2 - p_2)^2 = (p_1 + p_2)^2 + p_2^2 - 2(p_1 + p_2)_\mu \, p_2^\mu$$
$$= m_{12}^2 + m_2^2 - 2m_{12}E_2^\star , \qquad (2.79)$$

so that $E_2^\star$ and $E_3^\star$ are functions of m_{12} only, since the individual masses are all fixed. The same thing can be said about the magnitudes of the 3-momenta of these two particles:

$$\boldsymbol{p}_2^\star = \frac{\sqrt{(m_{12}^2 + m_2^2 - m_1^2)^2 - 4m_2^2 m_{12}^2}}{2m_{12}} ,$$
$$\boldsymbol{p}_3^\star = \frac{\sqrt{(m^2 - m_{12}^2 - m_3^2)^2 - 4m_3^2 m_{12}^2}}{2m_{12}} . \qquad (2.80)$$

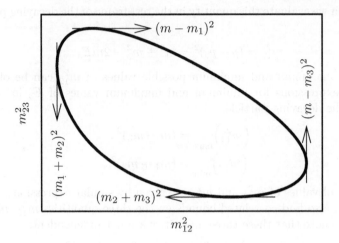

Figure 2.1: Kinematically allowed region in a three-body decay. We have used some hypothetical values for various masses to illustrate the nature of the allowed region.

But the angle between p_2 and p_3 is not determinable from the conservation laws, even for a fixed value of m_{12}^2. Therefore, referring to Eq. (2.77), we see that for a given value of m_{12}^2, there is a range of possible values of m_{23}^2 depending on the angle between the vectors $p_2^\star$ and $p_3^\star$. The minimum and maximum values of m_{23}^2 will correspond to the situations when p_2 and p_3 will be in the same or in opposite directions, i.e.,

$$\left(m_{23}^2\right)_{\substack{\max \\ \min}} = m_2^2 + m_3^2 + 2(E_2^\star E_3^\star \pm \mathbf{p}_2^\star \mathbf{p}_3^\star) \,. \qquad (2.81)$$

Using the momenta and energies from Eqs. (2.79) and (2.80), we can chalk out the allowed values of m_{23}^2 for any value of m_{12}^2. This gives an allowed region in a plot of m_{12}^2 versus m_{23}^2, whose shape has been shown schematically in Fig. 2.1. It means that, if we take the three products from the decay of a single particle, measure the momenta of all three, form the invariants m_{12}^2 versus m_{23}^2, the values would fall somewhere within a curve of the form shown. If we take the data for many such events, we will obtain a point corresponding to each event. A plot of such points for a particular decay channel is called a *Dalitz plot*. The distribution of the points within the kinematical boundary of Fig. 2.1 provides important information about the mechanism of the decay, as we will see later in §4.11.

☐ **Exercise 2.16** *Notice that the allowed region in a Dalitz plot is a bounded region. This means that the values of $\left(m_{23}^2\right)_{\max}$ and $\left(m_{23}^2\right)_{\min}$ must be equal for the maximum and minimum possible values of m_{12}^2. Verify that this is indeed the case because $\mathbf{p}_2^\star \mathbf{p}_3^\star = 0$ in these cases.*

2.7 Kinematics of scattering processes

2.7.1 Center-of-mass frame

Decay processes have one particle in the initial state. Scattering processes have more. If the scattering is *elastic*, the final state contains the same particles as in the initial state. Of course scatterings can also be *inelastic*, where the particle contents of the initial state and the final state are not the same. The difference might be in the type of particles present, or in the number, or in both.

Earlier, we have seen that the kinematical formulas for decay processes look very simple in the rest frame of the decaying particle. Similarly, for scattering processes also there is one frame in which the formulas look simple. This is the frame in which the total initial 3-momentum is zero, which is called the *CM frame*. Let us analyze scattering processes in this frame.

We take the simplest case of *2-to-2* scatterings, i.e., processes in which there are two particles in the initial state as well as in the final state. Let us denote the process symbolically by

$$a_1(p_1) + a_2(p_2) \rightarrow a_1'(p_1') + a_2'(p_2') , \tag{2.82}$$

where the 4-momenta of the particles, as in Eq. (2.59), have been shown in the parentheses. Note that we have not assumed that the scattering is elastic. Each of the four particles appearing in the process might be different. Their masses will be denoted by m_1, m_2, m_1' and m_2', in the order that they appear in Eq. (2.82).

By definition, we must have

$$\boldsymbol{p}_1 + \boldsymbol{p}_2 = 0 \tag{2.83}$$

in the CM frame. Momentum conservation then tells us that, in the final state also, total 3-momentum must vanish, i.e.,

$$\boldsymbol{p}_1' + \boldsymbol{p}_2' = 0 . \tag{2.84}$$

Also, the energy in the initial state is

$$E_{\text{tot}} \equiv E_1 + E_2 , \tag{2.85}$$

and we must then have

$$E_1' + E_2' = E_{\text{tot}} . \tag{2.86}$$

Using the Einstein relation between energy and momentum, we can write this equation as

$$\sqrt{\boldsymbol{p}'^2 + m_1'^2} + \sqrt{\boldsymbol{p}'^2 + m_2'^2} = E_{\text{tot}} , \tag{2.87}$$

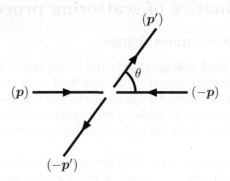

Figure 2.2: Kinematics of two-particle scattering in the CM frame.

where p' is the magnitude of p'_1, and therefore also of p'_2, through Eq. (2.84). Solving this, we obtain the value of p', and use this to obtain

$$E'_1 = \frac{E^2_{\text{tot}} + m'^2_1 - m'^2_2}{2E_{\text{tot}}} , \qquad E'_2 = \frac{E^2_{\text{tot}} - m'^2_1 + m'^2_2}{2E_{\text{tot}}} . \qquad (2.88)$$

This shows that, given the total energy of the particles in the initial state, the energies of the particles in the final state are uniquely determined. It is easy to see that the same does not apply to 3-momentum. The magnitude, of course, is determined. But the directions are not. Because of Eq. (2.84), we need to specify the direction of any one of the two final particles. This requires two parameters, like the polar and azimuthal angles on a sphere. In Fig. 2.2, we show one of these, which we call θ. This is called the *scattering angle*. The other is the orientation of the plane on which all four momenta should lie. Here, they are shown to lie on the plane of the paper, but they could lie on any other plane containing the momentum vectors of the initial particles.

2.7.2   Fixed-target frame

Some experiments are done by hitting a fixed target with a beam of particles. For the analysis of such experiments, it is often convenient to use the laboratory frame itself, i.e., the frame in which one of the initial particles is at rest. Commonly, it is called the *lab frame*, an obvious legacy of the times when all scattering experiments were done this way, i.e., by hitting a fixed target with a beam of particles. However, these days many experiments are done in which neither of the initial particles is at rest in the laboratory, including some experiments in which the lab frame, taken in the literal sense, is really the CM frame. To avoid confusion, it seems better to use the name *fixed-target frame*, or *FT frame* for short.

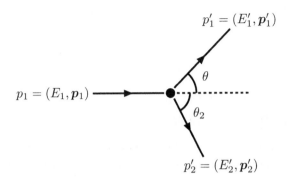

$p_1' = (E_1', \boldsymbol{p}_1')$

$p_1 = (E_1, \boldsymbol{p}_1)$

θ

θ_2

$p_2' = (E_2', \boldsymbol{p}_2')$

Figure 2.3: Kinematics of two-particle scattering in the FT frame. Particle 2 is at rest in this frame, shown as a blob.

The process is still given by Eq. (2.82), with the stipulation that, in the frame in question,

$$\boldsymbol{p}_2 = 0. \qquad (2.89)$$

Momentum conservation then implies that

$$\boldsymbol{p}_1 = \boldsymbol{p}_1' + \boldsymbol{p}_2', \qquad (2.90)$$

which means that the vectors $\boldsymbol{p}_1$, $\boldsymbol{p}_1'$ and $\boldsymbol{p}_2'$ lie on the same plane. In Fig. 2.3, this plane has been taken as the plane of the paper. Our notations for different angles have been shown in this figure.

The general formulas are quite cumbersome, so we give the results for the case when one of the initial particles and one of the final particles are massless. In particular, we take $m_1 = m_1' = 0$. Then, with the geometry shown in Fig. 2.3, we can write the components of the 4-momenta of different particles in the following manner:

$$
\begin{aligned}
p_1^\mu &= (E_1, E_1, 0, 0)\,, \\
p_2^\mu &= (m_2, 0, 0, 0)\,, \\
p_1'^\mu &= (E_1', E_1' \cos\theta, E_1' \sin\theta, 0)\,, \\
p_2'^\mu &= (E_2', \mathrm{p}_2' \cos\theta_2, -\mathrm{p}_2' \sin\theta_2, 0)\,.
\end{aligned} \qquad (2.91)
$$

Among the component equations implied by Eq. (2.90), one is trivial. The other two are:

$$
\begin{aligned}
E_1' \sin\theta - \mathrm{p}_2' \sin\theta_2 &= 0\,, \\
E_1' \cos\theta + \mathrm{p}_2' \cos\theta_2 &= E_1\,,
\end{aligned} \qquad (2.92)
$$

In addition, there is the energy conservation law, which gives:

$$E_1' + E_2' = E_1 + m_2\,. \qquad (2.93)$$

From these equations, one can determine the energies of the final particles as a function of the scattering angle:

$$E_1' = \frac{m_2^2 + 2m_2 E_1 - m_2'^2}{2(m_2 + E_1 - E_1 \cos\theta)} \,, \tag{2.94}$$

and E_2' is obtained through Eq. (2.93).

□ **Exercise 2.17** Derive Eq. (2.94).

An interesting piece of information comes out of Eq. (2.94). Consider an endergonic reaction, i.e., a reaction for which the total mass of the final particles exceeds that of the initial particles. In the simplified case that we are considering here with $m_1 = m_1' = 0$, it means $m_2 < m_2'$. Since E_1' must be positive, Eq. (2.94) implies that we must have

$$E_1 > \frac{m_2'^2 - m_2^2}{2m_2} \,. \tag{2.95}$$

In other words, particles in the beam hitting the target must possess a threshold energy in order that the reaction can happen.

□ **Exercise 2.18** Notice, in Eq. (2.95), that the threshold energy is larger than the mass difference between the final and the initial states. Explain qualitatively why supplying just the difference of energy, i.e., having $E_1 = m_2' - m_2$, is not enough.

□ **Exercise 2.19** For an endergonic reaction, what is the threshold energy in the CM frame?

Chapter 3

Symmetries and groups

3.1   The role of symmetries

Quantum field theory provides a set of rules with which, given the Lagrangian of a system, one can calculate rates of physical processes like decay or scattering. These rules will be discussed in Ch. 4.

One might ask, if quantum field theory can do everything from a Lagrangian, what does the subject of particle physics do then? Does it do anything that does not fall within the purview of quantum field theory?

The answer to the last question is 'yes', and the tasks are twofold. First, by considering interactions among particles, we will have to guess what the Lagrangian is. Of course there are some requirements on the Lagrangian from the basic properties of spacetime. These will be discussed in Ch. 4. But in addition to that, there are other symmetries between particles which play a crucial role in the structure of the Lagrangian. By discovering these symmetries, particle physics provides the input to quantum field theory.

Second, it is difficult to discuss bound states in quantum field theory. For properties of bound states, it is therefore much easier to be guided by the symmetries alone.

It is important to note that symmetries take the center stage in both these tasks. And group theory provides the mathematical structure for dealing with symmetries. For this reason, we devote this chapter on group theory, giving special attention to groups which will be useful for us in discussions later in the book.

3.2   Group theory

A set $\mathcal{G}$ is called a *group* if there exists a binary composition rule between its elements, denoted by ∘, that possesses the properties listed below.

1. For any elements a and b belonging to $\mathcal{G}$, the result of the binary composition, $a \circ b$, must also be an element of $\mathcal{G}$. This is called the closure property.

2. The composition rule is associative, i.e., $a \circ (b \circ c) = (a \circ b) \circ c$ for any elements a, b and c.

3. There exists an identity element e in the sense that, for each $a \in \mathcal{G}$, we obtain $a \circ e = e \circ a = a$.

4. For each element $a \in \mathcal{G}$, there exists an element $a^{-1} \in \mathcal{G}$ such that $a \circ a^{-1} = a^{-1} \circ a = e$. The element a^{-1} is called the inverse of the element a.

In short, the composition rule must be closed, associative, should have an identity, and an inverse for each element. And we can see why this mathematical structure is appropriate for discussing symmetries of any system.

By a system, we might mean a physical object, or a mathematical equation, or anything else. Symmetry operations are operations which leave the system unchanged. Consider a set whose elements are these symmetry operations. And the composition rule, denoted earlier by $\circ$, means just applying one operation after another. In what follows, we will often omit the symbol $\circ$ for the composition rule and denote $a \circ b$ simply by ab. The composition rule, accordingly, will often be called group multiplication.

We apologize that the name *group multiplication* can be confusing for certain groups. We warn that the group composition rule may not have anything to do with ordinary multiplication of numbers. For example, in Ex. 3.1, we present a group in which the group composition rule is in fact the addition of integers. However, this usage of the term is not at all unusual, so we will use it: the reader should be careful.

Clearly, if we apply two such operations in succession, the result does not change the system either, which means that the composition rule is closed. If we apply three specific symmetry operations in succession, it certainly does not matter how we might envisage them as being grouped, because each operation leaves the system unchanged. So the result will be associative. There is also an obvious symmetry operation for any system, viz., doing nothing to the system. This is the identity element. And if we apply some operation which does not change the system, applying it in the opposite direction must also not change the system. This defines the inverse of an element. Thus we conclude that all symmetry operations on any system form a group.

In passing, in a manner of warning, let us observe some of the things that are *not* essential in the definition of a group:

1. The element a^{-1} need not be different from a. In fact, it is obvious that the inverse of the identity element is the identity element itself.

2. A group operation need not be commutative, i.e., $a \circ b \neq b \circ a$ in general. Of course if either a or b is the identity element, an equality sign will

hold. Also, if $b = a^{-1}$, an equality sign will hold. But the equality sign need not hold for all pairs of elements a and b of the group. If in case it does so for a particular group, the group is called a *commutative group* or an *Abelian group*.

☐ **Exercise 3.1** Verify that the set of all integers forms a group under the operation of ordinary addition of numbers. What is the identity element? What is the inverse of the integer 2? Is this group Abelian? [**Note :** *This group is usually called* ℤ, *the first letter of the German word 'Zahl' which means 'number' ('Zahlen' is plural).*]

☐ **Exercise 3.2** The set of all integers does not form a group under the operation of ordinary multiplication of numbers. Why?

☐ **Exercise 3.3** Verify that the set of all positive real numbers forms a group under the operation of ordinary multiplication of numbers.

☐ **Exercise 3.4** Verify that the set of all real numbers forms a group under the operation of ordinary addition of numbers.

3.3 Examples and classification

3.3.1 Examples

Since this is not a textbook on group theory, it is not expected that we give examples of all different kinds of groups that occur in different contexts. As examples, we present here only some groups which will be useful for us in the subsequent discussion in this book.

a) The group $\mathbb{Z}_2$

Consider a set of two elements which will be called a and b, and a composition rule that can be represented in the form of the table below:

$$
\begin{array}{c|cc}
 & a & b \\
\hline
a & a & b \\
b & b & a
\end{array}
\tag{3.1}
$$

The row head denotes the first element in the composition and the column head the second, and the result of the composition is given in the matrix. Clearly, it shows that a is the identity element. The operation b, applied twice, gives the identity.

What kind of symmetry operations might b represent? An easy example that comes to mind is reflection from a fixed plane. Twice reflected, the result is the same as that of doing nothing. If we consider a geometrical figure that is invariant under reflection, its symmetry group would be given by two elements, whose composition would be governed by the table in Eq. (3.1).

We can also think of the invariance of a mathematical expression. Consider the quantity x^2 where x is a real number, and ask ourselves which operations

keep it invariant for arbitrary values of x. Of course, the identity transformation does that. And then there is also the operation

$$x \rightarrow -x \,, \qquad (3.2)$$

which can be thought of as a reflection about the point zero on the real line.
If this operation is called b and the identity operation a, the composition rule
is still governed by Eq. (3.1).

   Another way to envisage the group is by the solutions of the equation
$x^2 = 1$, i.e., by the square roots of unity, where the group composition rule is
simple multiplication of numbers. Because the power on x is 2 in the equation,
the group is called $\mathbb{Z}_2$.

   □ **Exercise 3.5** *In analogy with $\mathbb{Z}_2$, one can define the groups $\mathbb{Z}_N$ for
   any positive integral value of N. The group $\mathbb{Z}_N$ has N elements which
   correspond to the roots of the equation $x^N = 1$. Write down the
   elements of the group $\mathbb{Z}_3$ and construct the multiplication table.*

   □ **Exercise 3.6** *Define a group whose elements are integers from 0 to
   $N-1$, and the composition rule is 'addition mod N' (i.e., perform
   a simple addition, and then take the remainder of the result when
   divided by N). Show that the group composition table for this group
   is exactly the same as that obtained in Ex. 3.5 for the group $\mathbb{Z}_N$.
   [**Note** : This is also an equivalent definition of the group $\mathbb{Z}_N$. We explained why
   the letter $\mathbb{Z}$ is used to denote such groups; you can guess the reason for the subscript
   N.]*

b)   Group U(1)

Let us now consider the solutions of the equation $|z|^2 = 1$ for complex values
of z. The solutions are of the form

$$z = e^{-i\theta} \qquad (3.3)$$

for real θ. It is straightforward to check that such numbers form a group,
where the group composition rule is just ordinary multiplication of complex
numbers.

   A complex number $e^{-i\theta}$ can be thought of as a unitary 1×1 matrix. For
this reason, the aforementioned group is called U(1): the letter U denotes
unitary, and the number 1 indicates the size of the matrix.

Numbers of the form e^θ, for all real values of θ, form a group as well if ordinary multiplication
is taken as the group composition rule. But that is a different group: not the U(1) group. In
fact, the elements of this group can be put in one-to-one correspondence with the elements
of the group defined in Ex. 3.4. The factor of i in the exponent of Eq. (3.3) is crucial in the
definition of a U(1) group. It ensures unit absolute value for the elements of the group. The
minus sign in the definition is just a matter of convention: we will use the same convention for
other groups as well.

c)   Unitary groups

Taking a cue from the U(1) group, we can think of larger groups with matrices,
the group composition rule being the ordinary rule of matrix multiplication.

The set of all $N \times N$ matrices does not form a group, because it includes matrices which have zero determinants, and therefore no inverses. However, if we think of all unitary $N \times N$ matrices, they do form a group. This group is called U(N). Let us use the symbol U to denote any element of the group U(N). By definition, U obeys the relation

$$U^\dagger U = 1. \tag{3.4}$$

Taking the determinant of both sides we obtain

$$\left| \det U \right|^2 = 1, \tag{3.5}$$

which means that the determinant of such a matrix will be of the form

$$\det U = e^{i\alpha} \tag{3.6}$$

for some real α. Thus, the determinant can never be zero, and each unitary matrix must have an inverse. In fact, Eq. (3.4) tells us that the inverse is given by the relation

$$U^{-1} = U^\dagger. \tag{3.7}$$

Another interesting property of the U(N) group follows from Eq. (3.6). Let us write an arbitrary element U in the form

$$U = U'D, \tag{3.8}$$

where D is a multiple of the unit matrix, with each diagonal element equal to $e^{i\alpha/N}$ so that its determinant is equal to that of U, and U' is a matrix whose determinant is unity. Clearly, if we follow this prescription and write down the matrices U' for all elements of the U(N) group, the matrices U' will form a group by themselves. This group is called SU(N), where the letter S stands for *special*, in the sense that the determinant of any element of this group has a special value, viz., unity. The matrix D, acting on any column matrix, will change the phase of each element by the same amount. This is like a U(1) transformation on the elements of the column matrix. Moreover, the matrices U' will commute with the matrices D, since the latter, by construction, is a multiple of the unit matrix. Thus, in short, we can write the matrix U(N) in the form

$$U(N) = SU(N) \times U(1). \tag{3.9}$$

If $N = 1$, this statement is vacuous, since SU(1) is a trivial group with just the unit element. But for all integral $N > 1$, the statement is non-trivial. It says that we can talk about the SU(N) part and the U(1) part independently of each other.

☐ **Exercise 3.7** What is so special about the determinant being equal to unity? In other words, can we form a group with unitary matrices whose determinant has some other value?

We can think of the group U(N) in terms of some invariance. For this, suppose an $N \times N$ matrix U acts on a column matrix ψ with N elements. The product $U\psi$ will be a column matrix, and the action of the matrix can be represented by writing

$$\psi' = U\psi. \tag{3.10}$$

Then, it is easy to see that

$$\psi'^\dagger \psi' = \psi^\dagger \psi \tag{3.11}$$

by using Eq. (3.4). Thus, unitary transformations keep the norm of a state invariant. If we denote the components of ψ by $\psi_1, \psi_2, \cdots \psi_N$, we can say that, under the action of any element of U(N), the expression

$$\sum_{i=1}^{N} |\psi_i|^2 \tag{3.12}$$

is invariant.

d) Orthogonal groups

We can also consider symmetries of an expression like

$$\sum_{i=1}^{N} x_i^2 \,, \tag{3.13}$$

where the x_i are real variables. The group corresponding to this symmetry is called O(N), where the letter O is for orthogonal. It is easy to see that the elements of this group are orthogonal matrices, i.e., any element O satisfies the condition

$$O^\top O = 1. \tag{3.14}$$

An important orthogonal group is O(3). Any element of this group, by definition, is a linear transformation on x, y and z that keeps the expression $x^2 + y^2 + z^2$ constant. Among these, there are elements which can be represented by a matrix with unit determinant, and they form a group by themselves. This group is called SO(3). It represents rotations around a fixed origin in 3-dimensional space.

☐ **Exercise 3.8** *A general rotation in 2 dimensions can be represented by the co-ordinate transformations given in Eq. (2.53, p 25). Define $\xi = x + iy$ and show that the transformations are equivalent to a phase rotation on the complex parameter ξ. This means that the group SO(2) is the same as the group U(1).*

3.3.2   Classifications

We have given examples of a lot of groups, many of which will be encountered in later parts of the book. It will be useful, at this point, to classify the groups in terms of the symmetry operations they represent.

a)   Discrete versus continuous

One kind of classification can be made by considering whether one element of the group can be changed, without going outside the elements of the set that define the group, to another element. Consider, e.g., elements of the group U(1). As said earlier, the elements are of the form $e^{-i\theta}$ with real θ. We can change the value of θ continuously and access different elements of the group. Such groups are called *continuous groups*.

In contrast, consider the group $\mathbb{Z}_2$. It has two elements, as noted earlier. The elements can be thought of as the square roots of unity, i.e., as the numbers ± 1 and the group composition rule as ordinary multiplication of numbers. There is no way we can continuously go from $+1$ to -1, or vice versa, since the intermediate numbers do not belong to the group. The group is characterized by discrete elements, and would therefore fall in the category of *discrete groups*.

There can also be groups which have several connected parts, but the parts are separate from each other in the sense that one cannot change the elements of one part continuously to obtain an element of another part. Consider the group O(3) mentioned a little earlier. The determinant of a 3×3 orthogonal matrix must be either $+1$ or -1. The elements of any orthogonal matrix of determinant $+1$ can be continuously changed to obtain any other orthogonal matrix of the same determinant, but one cannot obtain the other determinant by making continuous changes. Therefore we can say that the O(3) group has two disconnected parts.

b)   Spacetime versus internal

A group is called a *spacetime group* if its elements inflict a change of spatial co-ordinates and/or time. For example, the rotation group falls in this category, because when a rotation is applied on an object, the co-ordinates of individual points change.

In contrast, we can think of transformations between quantities defined at the same spacetime point. For example, we can change the phase of a wavefunction at each point, or imagine a transformation that changes one kind of particle to another. Such transformations are called *internal*, and their symmetry group also can be designated by the same adjective.

c)   Local versus global

Transformations of any group are characterized by one or more parameters. An example is the parameter θ that was used to represent a U(1) transformation. Bigger groups will need more than one parameter, as we will presently see. If these parameters do not depend on the spacetime co-ordinates, the transformation, as well as the symmetry, is called *global*. If the contrary is true, the transformation and the symmetry group are called *local*.

It should be appreciated that among the kinds of distinctions mentioned here, only the first one is an inherent property of the mathematical structure of the group. The others depend on the context in which the group structure is invoked, or the manner in which it is done. The same mathematical group can describe either a spacetime symmetry or an internal symmetry, depending on the symmetry operations that the elements represent and the states on which they act. For example, we discussed that a $\mathbb{Z}_2$ symmetry might be related to mirror reflections or of changing the signs of wavefunctions of some particles. In the first case, it would be a spacetime symmetry, whereas in the second case, it would be an internal symmetry. Similarly, the same mathematical group can be either global or local. But the same group cannot be both discrete and continuous: that distinction is something inherent in the group.

3.4   Generators

3.4.1   Parameters and generators

A discrete group can have a finite number of elements. It is therefore possible to write down explicitly the result of the composition for any pair of elements. Such groups are called *finite groups*. Discrete groups may also have infinite number of elements, as exemplified in Ex. 3.1 *(p 37)*. And for continuous groups, the number of elements is necessarily infinite. For *infinite groups*, it is not possible to write down the result of the composition for each pair of elements, since the number of such pairs is infinite.

One therefore has to devise some strategy for expressing the result of group composition in the case of infinite groups. The strategy employed is to represent the infinite number of elements by a *finite* number of parameters, each of which can take infinite number of values. We have already used this strategy while talking about the U(1) group, whose elements were written as $e^{-i\theta}$. To write down all elements, we thus need only a single parameter θ. The number of group elements is infinite because θ can take infinite number of values.

For larger groups, we need more than one parameter. To include such larger groups, we employ a generalization of the strategy employed for U(1) and write a group element G in the form

$$G = \exp(-iT_a\theta_a), \tag{3.15}$$

where θ_a are the parameters, and there is of course an implied sum over the subscript a. The objects T_a are called *generators* of the group. The factor of $-i$ in the definition is just a matter of convention, as mentioned earlier.

A few simple observations can be made here. First, there is no loss of generality in taking all θ_a's to be real, since a complex number can always be written using two real numbers. Second, when all parameters are equal to zero, we obtain the identity element of the group. Third, the function on the right hand side of Eq. (3.15) is continuous in the parameters θ_a. Thus, a group element can be written in that form only if it is connected to the identity through continuous changes of parameters. If there are discrete operations in the group, those cannot be represented in this form. Fourth, we can use Eq. (3.15) to define the generators by the relation

$$T_a = i \left. \frac{\partial G}{\partial \theta_a} \right|_{\theta=0} . \qquad (3.16)$$

The derivative has to be evaluated at the point where all parameters vanish, i.e., at the identity element.

Corresponding to each parameter, we therefore need one generator. The group U(1) has one parameter and therefore one generator. Using Eq. (3.16) and the representation in the form shown in Eq. (3.3), we find that this generator is just 1.

For counting the number of parameters necessary for representing the elements of U(N), we start with the most general $N \times N$ matrix. It has N^2 elements which can all be complex. Thus, if there is no other restriction on the elements, an $N \times N$ matrix will be characterized by $2N^2$ real parameters. A unitary matrix has fewer parameters since, by definition, it must satisfy Eq. (3.4). We can equate both sides, element by element. Notice that $U^\dagger U$ is hermitian for any matrix U. Its diagonal elements are then automatically real, and we get N constraints by equating each of them to 1. Among the off-diagonal elements of $U^\dagger U$, the elements in the lower half are complex conjugates of the elements in the upper half. The elements in the upper half can be taken to be the independent ones, and there are $\frac{1}{2}N(N-1)$ of them. Each of them may be complex, so we should obtain two equations corresponding to each off-diagonal element: one for the real part and one for the imaginary part. The number of constraints from the off-diagonal elements is thus $N(N-1)$. Adding this number with the number of constraints from the diagonal elements, we find that in all there are N^2 constraints on the elements of the matrix U. This means that we will be left with N^2 independent parameters. In addition, for SU(N) group elements, the determinant is unity, which imposes an extra condition. So the number of independent parameters required to designate an SU(N) matrix is $N^2 - 1$. Accordingly, there are $N^2 - 1$ generators.

Unitary matrices also satisfy the condition

$$UU^\dagger = 1 . \qquad (3.17)$$

One might wonder whether the elements of this equation put extra constraints on the elements of U. However, these are not independent conditions. In fact, Eqs. (3.4) and (3.17) are equivalent, since both can be derived from Eq. (3.7).

☐ **Exercise 3.9** Show that the number of generators of an O(N) group is $\frac{1}{2}N(N-1)$. The number is same if the group is SO(N). Why?

3.4.2   Algebra

The properties of a group are manifest in the multiplication rule of all pairs
of elements of the group. Since the elements of a continuous group can be
expressed in the exponential form as given in Eq. (3.15), we need to use the
Baker–Campbell–Hausdorff formula for their multiplication. This formula
says that if A and B do not necessarily commute,

$$e^A e^B = \exp\left(A + B + \frac{1}{2}[A,B] + \cdots\right), \tag{3.18}$$

where there are further terms involving multiple commutators. Clearly then,
in order to multiply two group elements of the form given in Eq. (3.15),
we need to know the commutators between the generators. We know that
group multiplication is closed. It means that, if A and B both are linear
superpositions of generators like what appears in the exponent of the right
hand side of Eq. (3.15), the exponent on the right hand side of Eq. (3.18)
must also be a linear superposition of the generators. For that to happen, the
commutators must be in the form of superposition of generators. Thus, the
collection of all commutators can be written in the form

$$\left[T_a, T_b\right] = if_{abc}T_c. \tag{3.19}$$

This is called the *algebra* of the group. The quantities f_{abc} are called the
structure constants of the group. Obviously, they obey the conditions

$$f_{abc} = -f_{bac}. \tag{3.20}$$

Symmetry properties of the structure constants for interchanges involving the
third index will be discussed in §11.2.

As a digression, one can note that the generators defined through Eq. (3.15) span a vector
space, i.e., form the basis of a *vector space*. Elements of a vector space have the property that
any two elements can be added to obtain an element, and also any element can be multiplied
by a number to obtain another element. Generators of an algebra definitely satisfy both these
properties because any linear superposition of the generators can be used as a generator. For
example, if one has two generators T_1 and T_2, the exponent of Eq. (3.15) contains the expression
$\theta_1 T_1 + \theta_2 T_2$. But this can also be written as $\frac{1}{2}(\theta_1+\theta_2)(T_1+T_2) + \frac{1}{2}(\theta_1-\theta_2)(T_1-T_2)$, which
means that we can use $T_1 \pm T_2$ also as generators, using different parameters for expressing the
same group element. Any particular set of generators chosen for a group can be seen as a basis
in the vector space of generators.

   □ **Exercise 3.10** Verify that the structure constants depend on the
   choice of generators. Consider the SU(2) algebra, which can be written in terms of the hermitian generators J_i (for $i = 1, 2, 3$) as

$$[J_i, J_j] = i\epsilon_{ijk}J_k, \tag{3.21}$$

   where ϵ_{ijk} is completely antisymmetric in its indices, with

$$\epsilon_{123} = +1. \tag{3.22}$$

   Now take a different set of generators of SU(2), e.g., $J_\pm \equiv J_1 \pm iJ_2$
   and $J_0 \equiv J_3$. Find the structure constants for this choice.

3.5 Representations

3.5.1 Matrix and differential representations

A *representation* of a group $\mathcal{G}$ is obtained by assigning an operator R_G to each element G of the group in such a way that

$$R_{G_1} R_{G_2} = R_{G_1 G_2}, \tag{3.23}$$

where G_1 and G_2 are arbitrary elements of the group. Similarly, we can also talk about the representation of an algebra by assigning an operator R_a to each generator T_a such that

$$R_a R_b - R_b R_a = i f_{abc} R_c. \tag{3.24}$$

Needless to say, a representation of the algebra automatically defines a representation of the group whose algebra it is.

We mentioned the word *operator* while giving the definition of a representation. Operators are defined on a vector space. We have talked about vector spaces earlier. The elements of a vector space are most commonly and most obviously called *vectors*. The name sounds fine at first, because ordinary vectors that we are familiar with, through our knowledge of dynamics and electromagnetic theory, are indeed elements of a vector space in the 3-dimensional space. But this name creates some confusion, because the tensors of rank 2 (or of any other rank) are also elements of a vector space. To avoid this trouble of terminology, we will use the word *states* to denote the elements of a vector space. An operator, in our terminology, would transform one state to another (not necessarily different) state.

Thus, a representation is a way of visualizing group elements by considering what they do to a particular choice of states. Consider the rotation group, whose elements are rotations. If a rotation is applied to a vector, the components of the vector change. Since a vector in ordinary space has three components, the result of application of rotation can be represented by a 3×3 matrix. And this matrix can be used to represent the rotation itself. Such a representation is called a *matrix representation*. It should perhaps be said here that, although the operators form the representation of the group, sometimes we loosely say that the vector space itself, or the states, form a representation. Call it abuse of language if you like, but it is quite common. Thus, it is common to say that vectors form a representation of the rotation group.

A particular explicit representation of an algebra therefore involves the choice of basis in two different vector spaces. First, the generators of a group form a vector space, and we need to make a choice for the independent generators. The dimension of this vector space is equal to the number of generators. This statement has nothing to do with the theory of representations that we are talking about in this section. Second, a matrix representation of the generators is based on a vector space of some finite number of dimensions, and we need to choose a basis in this vector space in order to arrive at an explicit representation. As we will see now, this number of dimensions can start from 1 for any group, irrespective of the number of generators it might have.

It is easy to see that at least one matrix representation exists for any group. This is defined by

$$R_G = 1 \qquad (3.25)$$

for all group elements. Here we assign a number to each group element. Obviously, Eq. (3.23) is satisfied if such an association is made. Since a number can be thought as a 1×1 matrix, this can be called a 1-dimensional representation of the group. Of course this representation is not *faithful*, i.e., it does not assign a different matrix to each different group element, but that is not a necessary property for a representation.

☐ **Exercise 3.11** *What is the corresponding representation of the algebra of a group?*

There may be other 1-dimensional representations as well. In addition, there may be matrix representations of other dimensions. For SU(2), there are representations of all dimensions, which readers should have seen in ordinary quantum mechanics, when considering rotation matrices for states of different spins. For other groups, this is not the case. Representations of other groups will be discussed as and when the need arises.

☐ **Exercise 3.12** *Find a faithful 1-dimensional representation for the group $\mathbb{Z}_2$.*

We can have different kinds of representations altogether. Suppose we apply a rotation on a scalar function, $\varphi(\boldsymbol{x})$. Note that scalar functions constitute a vector space as well, since we can add two such functions, or multiply one function by a number, and we end up with a scalar function. By definition, a scalar is something that remains invariant under a rotation, even though the argument of the function, i.e., the co-ordinate vector, changes under it. In other words, suppose there is a point P at which we want to evaluate the value of the scalar. The co-ordinate vector of this point is $\boldsymbol{x}$. We apply a certain rule, encoded in $\varphi(\boldsymbol{x})$, to obtain the value of the scalar at the point P. Now we make a rotation of the co-ordinates, so that the co-ordinate vector of P changes to $\boldsymbol{x}'$. We now have to apply a new rule, say $\varphi'(\boldsymbol{x}')$, such that we obtain the same value for the function at P. So we want

$$\varphi'(\boldsymbol{x}') = \varphi(\boldsymbol{x}). \qquad (3.26)$$

For the sake of definiteness, let us suppose that the rotation performed was about the z-axis, so that the z co-ordinate remained unchanged, and the other co-ordinates changed by Eq. (2.53, *p 25*). For an infinitesimal rotation, these changes can be written as

$$x' = x - y\theta, \qquad y' = y + x\theta. \qquad (3.27)$$

Thus, ignoring higher order terms in θ, we should have

$$\varphi(x, y, z) = \varphi'(x - y\theta, y + x\theta, z), \qquad (3.28)$$

or equivalently

$$\varphi'(x, y, z) = \varphi(x + y\theta, y - x\theta, z)$$
$$= \varphi(x, y, z) + \theta\left(y\frac{\partial}{\partial x} - x\frac{\partial}{\partial y}\right)\varphi(x, y, z). \qquad (3.29)$$

The functional form changes by the action of a rotation. We can write

$$\varphi' = D_z\varphi, \qquad (3.30)$$

using the notation D_z to denote a rotation about the z-axis. We can write this rotation in the form prescribed in Eq. (3.15), denoting the relevant generator by L_z and keeping only up to the first order term in the parameter θ:

$$D_z = 1 - iL_z\theta. \qquad (3.31)$$

Comparing this form with Eq. (3.29), we obtain

$$L_z = -i\left(x\frac{\partial}{\partial y} - y\frac{\partial}{\partial x}\right), \qquad (3.32)$$

meaning that this is how L_z works when it acts on a function. This is a *differential representation* of the generator for rotation around the z-axis. Similar representations for L_x and L_y are obvious.

It is clear why both types of representations are necessary. A matrix has a finite number of rows and columns, and can act on something that has a finite number of elements. Thus, e.g., if we want to see how the three components of a spatial vector mix under a rotation, we need a matrix representation of order 3. On the other hand, co-ordinates are represented by continuous parameters, so we need differential representation of generators to find out how a function of the co-ordinates changes.

If we are talking about a vector field, e.g., then both kinds of changes take place. Corresponding to Eq. (3.30), we can now symbolically write

$$\boldsymbol{A}'(\boldsymbol{x}) = D\mathbb{R}\boldsymbol{A}(\boldsymbol{x}). \qquad (3.33)$$

Here D contains differential operators which ensure that we obtain the same vector at the same point P in space. This is the same thing we had to do for scalar functions as well, and the recipe for this change must be the same as that for a scalar function. For a vector field, in addition, the co-ordinate axes have changed after rotation, so the components of this vector along the new axes will involve a superposition of the components along the old axes. These changes in the components can be represented by a 3-dimensional matrix representation of rotation, which mixes the components.

Using generators, we can write

$$D = \exp(-i\boldsymbol{L}\cdot\boldsymbol{\theta}), \qquad (3.34)$$

where L has been defined through Eq. (3.32) and its cyclic permutations. Similarly, we can define

$$\mathbb{R} = \exp(-i\boldsymbol{S} \cdot \boldsymbol{\theta}) \,, \tag{3.35}$$

where $\boldsymbol{S}$ denotes a set of three matrices. For infinitesimal θ, we can keep only up to first order terms and obtain

$$\boldsymbol{A}'(\boldsymbol{x}) = \left(1 - i\boldsymbol{L} \cdot \boldsymbol{\theta} - i\boldsymbol{S} \cdot \boldsymbol{\theta}\right)\boldsymbol{A}(\boldsymbol{x}) \,. \tag{3.36}$$

Compared to the scalar function, there is an extra contribution coming from the mixing of the components. This represents the matrix representation of the object whose rotation is being considered.

Since both L and S generate effects of rotation, there must be some common properties that they share. This is the algebra, which determines the group multiplication property between any two elements of the group. The rotation group has the algebra of SU(2), and therefore the components of both L and S should satisfy the same kind of relations as shown in Eq. (3.21). This means that the differential generators should satisfy the relations

$$[L_i, L_j] = i\epsilon_{ijk}L_k \tag{3.37}$$

when operating on any function. And the matrix generators should satisfy the relations

$$[S_i, S_j] = i\epsilon_{ijk}S_k \,, \tag{3.38}$$

where the multiplication implied in the definition of the commutator bracket is just matrix multiplication. If, instead of vectors, we take some tensor which acts as a representation of the rotation group, we need to use different dimensional matrices to represent its transformation property. Even those matrices would obey the same commutation relation. Said another way, if we can find three matrices obeying Eq. (3.38), they constitute a representation of the SU(2) algebra.

□ **Exercise 3.13** *Check the validity of Eq. (3.37) by letting the left hand side operate on an arbitrary function of co-ordinates.*

3.5.2   Irreducible representations

Vectors constitute a representation of the rotation group because if a rotation is applied, one vector turns into another vector, with different components along some predetermined axes. The important point is that, by the operation of a rotation, vectors turn into vectors, and vectors only: we do not need anything other than vectors in order to define the effect of rotation. Contrast it with vectors along the x-axis only. They do not constitute a representation because, after an arbitrary rotation is applied, we do not obtain a vector along the x-axis. Therefore, the x-components of vectors do not form a representation: the vectors do. Objects which transform like a representation of a group should be self-contained under the action of the group.

Now consider scalars and vectors together. Certainly this is also a representation because the effect of a rotation would not produce, say, a rank-2 tensor. However, it is also true that under rotation, a vector does not become a scalar or vice versa. The scalars and the vectors are independently self-contained. In such cases, we say that the representation obtained by considering them together is a *reducible representation*. On the other hand, if we consider the three components of a vector, we need all of them to write down the effect of a general rotation. Such a representation is called an *irreducible representation*.

The same thing can be said in a different manner. Suppose we start with a state in the vector space that has only one non-zero component. Let us call it $X_{(i)}$, where the subscript indicates that only the ith component is non-zero. Now we start operating this state by the operator representation of all group elements, one by one. The question is: will the vectors resulting from these operations include the vector $X_{(j)}$ for *any* $j \neq i$? If the answer is 'yes', it implies that the representation is irreducible. If not, the representation is reducible.

Clearly, any reducible representation must be a direct sum of several irreducible ones. In our example above, the reducible representation contained vectors and scalars, both of which constitute irreducible representations. In order to discuss representations, it is therefore sufficient to discuss irreducible representations only. Properties of any reducible representation can be inferred from the properties of the irreducible representations it contains. In view of this, from now on we will mostly talk about irreducible representations and drop the adjective *irreducible* wherever no confusion may arise.

A special property of Abelian groups should be mentioned here. For such groups, all pairs of group elements commute, which means that all generators commute among themselves. In any matrix representation, we can therefore choose the basis so that all generators are diagonal matrices. When exponentiated, all group elements will also be diagonal in this basis. Acting on states, they would not mix any state with any other, which means all states will be 1-dimensional irreducible representations. This does not necessarily mean that the states remain invariant under the group operations. A state can still transform by a phase under a group operation. If it does, we say that the state has a non-zero charge under the Abelian group. If it doesn't, i.e., if it really remains invariant, the charge is zero.

3.5.3 Kronecker product of representations

What happens if we take the product of components of two states in two (not necessarily different) representations? For example, consider the components of a vector, A_i, and those of another vector, B_j. The question is: how do the products $A_i B_j$ transform under rotation?

There are, of course, nine such products for ordinary 3-dimensional space. Since the components of each vector mix among themselves under a rotation, the product itself does the same. Thus all nine products, taken together, should constitute a representation. It is not clear though whether this representation is irreducible. If we consider the general case of products of two arbitrary representations, the result in general is not irreducible. Showing the Kronecker product of two representations means showing which irreducible representations occur in product terms.

Let us get back to the example of products of the components of two vectors, i.e., terms of the form $A_i B_j$. Under a rotation, the components of A and B transform as

$$A'_i = \mathbb{R}_{ik} A_k , \quad B'_j = \mathbb{R}_{jl} B_l . \tag{3.39}$$

Here, while discussing rotations, we do not distinguish between upper and lower indices. Thus,

$$A'_i B'_j = \mathbb{R}_{ik} \mathbb{R}_{jl} A_k B_l . \tag{3.40}$$

The rotation group is the orthogonal SO(3) group, so the matrices $\mathbb{R}$ representing rotation must be orthogonal matrices. This means that they should satisfy the relations

$$\mathbb{R}_{ik} \mathbb{R}_{il} = \delta_{kl} , \quad \mathbb{R}_{ik} \mathbb{R}_{jk} = \delta_{ij} . \tag{3.41}$$

Using these relations in Eq. (3.40), we find that

$$A'_i B'_i = A_i B_i . \tag{3.42}$$

In other words, the combination of the form $A_i B_i$, also called the dot product of the two vectors, is invariant under rotations. This constitutes a 1-dimensional representation. Symbolically, we can write this result as

$$\mathbf{3} \times \mathbf{3} = \mathbf{1} + \cdots \quad \text{in SO(3)} , \tag{3.43}$$

signifying that the Kronecker product contains a singlet, or 1-dimensional representation. The dots in this equation imply that we have not yet explored which other irreducible representations occur in this product.

To finish this task, we note that Eq. (3.40) implies the relation

$$A'_i B'_j \pm A'_j B'_i = \left(\mathbb{R}_{ik} \mathbb{R}_{jl} \pm \mathbb{R}_{jk} \mathbb{R}_{il} \right) A_k B_l$$
$$= \mathbb{R}_{ik} \mathbb{R}_{jl} \left(A_k B_l \pm A_l B_k \right) , \tag{3.44}$$

where the last step has been obtained by redefining the dummy indices in one of the terms. This equation shows that the symmetric and the antisymmetric combinations do not mix under rotations: the symmetric ones remain symmetric and the antisymmetric ones remain antisymmetric. It implies that the antisymmetric combinations transform like an irreducible representation. Since there are three possible antisymmetric combinations, this representation will be 3-dimensional representation, i.e., it will act on states which are triplets. Similarly, there are six possible symmetric combinations, but one of them has already been accounted for in Eq. (3.42). So we are left with a 5-dimensional representation. The complete form of Eq. (3.43) is therefore

$$3 \times 3 = 1 + 3 + 5 \qquad \text{in SO(3)}. \tag{3.45}$$

For the rotation group, the dimensionality d of the representation and the total angular momentum quantum number j are related through the definition

$$d = 2j + 1. \tag{3.46}$$

In terms of the angular momentum quantum number, Eq. (3.45) means that two angular momenta, 1 each, combine to form angular momentum 0, 1 or 2. The reader must be familiar with this result, and its generalization, from basic quantum mechanics courses. There was not much need of repeating it here except for introducing the method which will be useful in finding representations of other groups when we need them.

3.5.4 Decomposition under a subgroup

A *subgroup* is a subset of elements of a group which satisfy all properties of being a group. In other words, $\mathcal{G}'$ will be a subgroup of the group $\mathcal{G}$ provided

1. The elements of $\mathcal{G}'$ form a subset of the elements of $\mathcal{G}$.

2. $\mathcal{G}'$ itself is a group under the same group operation that defines $\mathcal{G}$.

In a representation of $\mathcal{G}$, if we collect all operators corresponding to the elements of a subgroup $\mathcal{G}'$, they obviously constitute a representation of $\mathcal{G}'$. If the states corresponding to the representation of $\mathcal{G}$ are self-contained by the action of the elements of $\mathcal{G}$, obviously they are self-contained by the action of a subset of elements of $\mathcal{G}$. But an irreducible representation of $\mathcal{G}$ does not necessarily remain an irreducible representation of $\mathcal{G}'$. Suppose we start with a state $X_{(i)}$ which has a non-zero element only in the ith position. Because the representation is irreducible, it means that, for any $j \neq i$, we can find a group element G such that $GX_{(i)} = X_{(j)}$. But there is no guarantee that this element will be part of $\mathcal{G}'$. If it is not, then it means that the ith and the jth components do not fall in the vector space of an irreducible representation of $\mathcal{G}'$. In general then, an irreducible representation of $\mathcal{G}$ decomposes under a subgroup $\mathcal{G}'$.

Consider an example of the group SU(4). According to the definition given in Eq. (3.12), this group is defined by matrices of unit determinant which, acting on four complex elements z_1 through z_4, ensure that

$$|z_1|^2 + |z_2|^2 + |z_3|^2 + |z_4|^2 = \text{invariant}. \tag{3.47}$$

Take the subset of these matrices which keep only $|z_1|^2 + |z_2|^2$ invariant, without affecting z_3 and z_4 at all. This will define an SU(2) subgroup of the SU(4) group. However, as this description shows, the elements of this SU(2) mix the components z_1 and z_2 among themselves, but not with z_3 and z_4. This means that the 4-dimensional representation of SU(4) is not an irreducible representation of the SU(2) subgroup. The first two components are now singled out and they span a 2-dimensional representation of the SU(2).

As another example, consider a vector A in the ordinary co-ordinate space. If we start with A_x and apply arbitrary rotations, the result cannot be expressed without using A_y and A_z. This is why we say that the three components of a vector transform in a 3-dimensional representation, and it is irreducible because all three are needed in order to write down the result of a general rotation. But now consider only rotations around the z-axis. It is easy to see that these rotations, by themselves, form a group. This is the group SO(2), because these are rotations in the x-y plane. This is a subgroup of the rotation group SO(3). If we apply the operations in this subgroup, the x and y components mix, but the z-component remains unaffected. This means that the z-component is a singlet of SO(2). From the discussion above, it might seem that the x and y components transform like a doublet or a 2-dimensional representation. But even that is not true. The reason for this was hinted in Ex. 3.8 *(p 40)*. If we consider the combinations $A_x \pm iA_y$, we will see that neither of them mixes with the other: each combination changes only by a phase. Thus, $A_x \pm iA_y$ and A_z are singlets of the subgroup SO(2) which has only L_z as its generator. We can summarize this statement by saying that

$$3 \longrightarrow 1 + 1 + 1 \qquad \text{for SO(3)} \supset \text{SO(2)}. \qquad (3.48)$$

The + sign here, as well as in Eq. (3.45), does not really mean summation in any sense: it is just an indication that the 3-dimensional representation of SO(3) decomposes into three 1-dimensional representations of SO(2). In order to emphasize this non-arithmetical meaning of the signs, we have not used a simple equality sign between the left hand side and the right hand side of the equation, but rather used the symbol $\longrightarrow$.

In a sense, the result of Eq. (3.48) was inevitable. The group SO(2) has only one generator, so it must be an Abelian group. We mentioned earlier that Abelian groups can have only 1-dimensional irreducible representations.

3.6   Lorentz group

3.6.1   Generators and algebra

Lorentz transformations, i.e., boosts and rotations together, form a group. One can take each of the properties mentioned in §3.2 that defines a group, and verify that it is satisfied by Lorentz transformations.

There are shortcuts for this procedure. Recall Eq. (2.25, *p 21*), which must be satisfied by any Lorentz transformation. In matrix notation, this can be written as

$$\Lambda^\top g \Lambda = g \,. \tag{3.49}$$

It is then easy to see that the product of two Lorentz transformations will also satisfy this equation. Associativity is guaranteed for matrix multiplication. And, taking the determinant of both sides of Eq. (3.49), we obtain

$$\left(\det \Lambda\right)^2 = 1 \,, \tag{3.50}$$

which implies that a Lorentz transformation matrix must be non-singular, and hence its inverse must exist. It is easy to see, by multiplying both sides of Eq. (3.49) by Λ^{-1} from the right and by its transpose from the left, that the inverses are also Lorentz transformations. Thus, all group properties are satisfied.

The group elements can be generated by six parameters. There can be three independent boost parameters, one along each orthogonal axis in the 3-dimensional space. Any rotation is by an angle around an axis. The angle is one parameter, and it takes two parameters to specify an axis in 3-dimensional space. Thus, there are three parameters required to specify any rotation. Taking boosts and rotations together, we need six parameters.

There is another way to realize that there are six parameters. The Lorentz transformation equations were given in Eq. (2.3, *p 17*). We can write this equation in the form

$$x'^\mu = \left(\delta^\mu{}_\alpha + \omega^\mu{}_\alpha\right) x^\alpha \,, \tag{3.51}$$

where $\omega^\mu{}_\alpha$ contain the transformation parameters. However, any Lorentz transformation keeps the quantity $g_{\mu\nu} x^\mu x^\nu$ invariant, as mentioned in Eq. (2.17, *p 20*). Thus,

$$g_{\mu\nu} x^\mu x^\nu = g_{\mu\nu} x'^\mu x'^\nu = g_{\mu\nu} \left(\delta^\mu{}_\alpha + \omega^\mu{}_\alpha\right) \left(\delta^\nu{}_\beta + \omega^\nu{}_\beta\right) x^\alpha x^\beta \,. \tag{3.52}$$

For infinitesimal $\omega^\mu{}_\alpha$'s, keeping only up to first order terms in the small parameters, we obtain

$$g_{\mu\nu} \left(\omega^\mu{}_\alpha \delta^\nu{}_\beta + \delta^\mu{}_\alpha \omega^\nu{}_\beta\right) = 0 \,, \tag{3.53}$$

or,

$$\omega_{\beta\alpha} + \omega_{\alpha\beta} = 0 \,. \tag{3.54}$$

The parameters $\omega_{\alpha\beta}$ are therefore antisymmetric in the indices. Since each index can take four values, there are $\binom{4}{2} = 6$ independent combinations, which represent six parameters.

For a six-parameter group, there are six generators. Earlier, we have shown that the differential generators of rotation are given by Eq. (3.32) and its cyclic permutations. The extension to Lorentz group should be obvious, and we can write the differential representation of the generators as

$$\mathscr{J}_{\mu\nu} = i(x_\mu \partial_\nu - x_\nu \partial_\mu). \tag{3.55}$$

For rotation generators, both indices will have to be spatial. For example, L_z, defined in Eq. (3.32), is nothing but $\mathscr{J}_{12}$. (While comparing the two expressions, we need to remember the presence of the minus signs in Eq. (2.30, *p 21*) in the definition of x_μ, and the absence of them in Eq. (2.31, *p 21*) in the definition of ∂_μ.) Boost generators, on the other hand, have one time index and one space index.

> □ **Exercise 3.14** Show that $\mathscr{J}_{01}$ can generate Lorentz boosts in the x-direction, yielding the co-ordinate transformation rules of Eq. (2.1, *p 17*). [**Hint :** *First show it for an infinitesimal boost, then try to exponentiate it and obtain the result in the form implied in Eq. (2.7, p 18).*]

Note that $\mathscr{J}_{\mu\nu} = -\mathscr{J}_{\nu\mu}$, so that Eq. (3.55) defines only six independent non-zero combinations, which is exactly what we need. From these differential representations, it is straightforward to show that

$$\left[\mathscr{J}_{\mu\nu}, \mathscr{J}_{\lambda\rho}\right] = i(g_{\mu\rho}\mathscr{J}_{\nu\lambda} + g_{\nu\lambda}\mathscr{J}_{\mu\rho} - g_{\mu\lambda}\mathscr{J}_{\nu\rho} - g_{\nu\rho}\mathscr{J}_{\mu\lambda}). \tag{3.56}$$

This equation contains the commutation relations of all pairs of generators of the group, and is therefore the *algebra* of the group.

3.6.2   Representations

We have derived the algebra by identifying differential operators for the generators of the Lorentz group in Eq. (3.55). As discussed in §3.5.1, the algebra might be satisfied by some finite dimensional matrices as well. These matrices will then constitute a finite dimensional representation of the Lorentz algebra. Already, we have found a 4-dimensional representation of the Lorentz group. This is the representation that follows from the definition of the group, and is called the *fundamental representation*. Some of the group elements in this representation have been given in Eqs. (2.4) and (2.14).

> □ **Exercise 3.15** Find the representation of the Lorentz group generators that follows from the definition of the group elements given in Eqs. (2.4) and (2.14). Show that the matrix elements of the generator $J_{\mu\nu}$ can be written as
> $$\left(J_{\mu\nu}\right)_{\lambda\rho} = i(g_{\mu\lambda}g_{\nu\rho} - g_{\nu\lambda}g_{\mu\rho}). \tag{3.57}$$
> Verify that they obey the algebra given in Eq. (3.56).

In order to find all representations of the Lorentz algebra, it is better to rewrite the algebra in a different way. We define

$$J_i \equiv \frac{1}{2}\epsilon_{ijk}\mathscr{J}_{jk}, \tag{3.58a}$$

$$K_i \equiv \mathscr{J}_{0i}. \tag{3.58b}$$

The commutation relations between the operators introduced in Eq. (3.58) can be easily read from Eq. (3.56):

$$\left[J_i, J_j\right] = i\epsilon_{ijk}J_k \,,$$
$$\left[K_i, K_j\right] = -i\epsilon_{ijk}J_k \,,$$
$$\left[J_i, K_j\right] = i\epsilon_{ijk}K_k \,. \tag{3.59}$$

We see that the commutation relations of the J-operators are self-contained in the sense that they do not involve the K-operators. This means that the J-operators, by themselves, generate a group. Indeed, it is the rotation group, or SU(2). The operators K do not have these properties, and the commutation relations involving K mix up the two kinds of operators.

To avoid such mixing up, we define a different set of operators by the relations

$$N_i^{(\pm)} \equiv \frac{1}{2}\left(J_i \pm iK_i\right) \,. \tag{3.60}$$

Then it is straightforward to deduce the following commutation relations:

$$\left[N_i^{(+)}, N_j^{(+)}\right] = i\epsilon_{ijk}N_k^{(+)} \,,$$
$$\left[N_i^{(-)}, N_j^{(-)}\right] = i\epsilon_{ijk}N_k^{(-)} \,, \tag{3.61}$$

so that the three operators $N_i^{(+)}$ generate an SU(2), and so do the three operators $N_i^{(-)}$. Moreover, it is also seen that

$$\left[N_i^{(+)}, N_j^{(-)}\right] = 0 \,, \tag{3.62}$$

which means that all generators of one SU(2) commute with all generators of the other SU(2). Mutually commuting algebras are denoted with $\times$, so we can conclude that the Lorentz algebra is an SU(2) $\times$ SU(2) algebra. The generators of each SU(2) factor, i.e., $N_i^{(+)}$ and $N_i^{(-)}$, can also be shown to be hermitian, as SU(2) generators should be.

□ **Exercise 3.16** *If we write a unitary matrix U in the form $U = e^{-iA}$, show that A must be hermitian.*

□ **Exercise 3.17** *Consider a Lorentz boost in the x-direction with a velocity v. For infinitesimal v, how do the components of a 4-vector change? From this, find the 4-dimensional matrix representation of the boost generator K_x and show that it is anti-hermitian. From this and similar results, argue that the generators $N_i^{(\pm)}$ defined in Eq. (3.60) are hermitian.*

Once this is established, we can write down all finite dimensional representations of the Lorentz group using the knowledge of SU(2) representations learned while studying the theory of angular momentum in quantum mechanics courses. We know that each SU(2) representation is characterized by a number which can be an integer or a half-integer, representing the highest eigenvalue of any of the hermitian generators in that representation. For SU(2) × SU(2), each finite dimensional representation will be characterized by two numbers, n_+ and n_-, specifying its transformation property under the two SU(2) subgroups. Moreover, note that Eq. (3.60) implies that

$$J_3 = N_3^{(+)} + N_3^{(-)} , \qquad (3.63)$$

so that the angular momentum of the representation will be the sum $n_+ + n_-$. Some examples of finite dimensional irreducible representations are given below:

Representation	Spin
$(0,0)$	0
$(0,\frac{1}{2})$	$\frac{1}{2}$
$(\frac{1}{2},0)$	$\frac{1}{2}$
$(\frac{1}{2},\frac{1}{2})$	1

$$\qquad (3.64)$$

The list goes on, with higher and higher values of n_+ and n_-.

In §2.5, we mentioned that the conservation laws implied by invariance under boosts will not be very useful for us. Let us digress a bit and see what kind of conservation laws they are.

Let us consider a collection of point particles. From Eqs. (3.58b) and (3.55), we see that the boost operators are given by

$$K^i = \sum_{\text{particles}} \left(t p^i - x^i E \right). \qquad (3.65)$$

Let us define the *center of energy* co-ordinate of this collection of particles by the relation

$$\sum_{\text{particles}} x^i E = x^i_{\text{CE}} E_{\text{tot}} , \qquad (3.66)$$

where the subscript 'tot' means a sum over all particles. Then Eq. (3.65) can be written as

$$x^i_{\text{CE}} = \left(\frac{p^i_{\text{tot}}}{E_{\text{tot}}} \right) t - \frac{K^i}{E_{\text{tot}}} . \qquad (3.67)$$

Since p^i_{tot} and E_{tot} are conserved, this equation implies that the center of energy moves with a constant velocity.

3.6.3   Extended Lorentz group and its representations

We remarked that Lorentz transformations keep the combination

$$t^2 - \mathbf{x} \cdot \mathbf{x} \qquad (3.68)$$

invariant, a fact that we wrote in the form of Eq. (2.16, *p 19*) earlier. As shown in §3.2, the set of all transformations that keeps a particular mathematical expression invariant always forms a group. We can therefore define a group by the set of all transformations on t and $\boldsymbol{x}$ which keep the expression given in Eq. (3.68) invariant. This group goes by the name O(3,1).

The reason behind the name can be seen from the expression in Eq. (3.68). Transformations which keep the combination $x^2 + y^2 + z^2$ invariant form a group called O(3), where the number in the parentheses is the number of variables whose sum of squares is maintained invariant. The invariant in Eq. (3.68) also contains superposition of squares of several variables. However, it is not a sum of squares. One of the squares appears with a different sign compared to the other three. This fact is indicated by calling the group O(3,1). It can be called O(1,3) as well; it makes no difference.

Lorentz transformations are certainly members of the group O(3,1). However, a little thought reveals that the group O(3,1) contains other transformations which are neither boosts nor rotations. For example, consider the transformation

$$t \to -t, \qquad \boldsymbol{x} \to \boldsymbol{x}. \tag{3.69}$$

Obviously it keeps the expression of Eq. (3.68) invariant. So does the transformation

$$t \to t, \qquad \boldsymbol{x} \to -\boldsymbol{x}. \tag{3.70}$$

The first of these operations is called *time reversal*, and the second *parity transformation*. The name *space-inversion* is also used alternatively for the second. Although the latter name is more descriptive of the operation involved, we will mostly use just the single word *parity*, a curtailed form of the former name.

It is easy to see, and in fact it will be shown shortly, that the transformations of Eqs. (3.69) and (3.70) cannot be the result of a boost or of a rotation in 3-dimensional space. A rotation, of whatever magnitude, can be built by adding smaller rotations. The same goes for boosts. In this sense, they are continuous transformations. The operations shown in Eqs. (3.69) and (3.70), on the other hand, are discrete.

The possibility for the existence of such transformations can also be seen from Eq. (3.49), which is equivalent to the statement that the expression in Eq. (3.68) is invariant by the transformation of Eq. (2.3, *p17*). As we have shown, Eq. (3.49) implies Eq. (3.50), which says that

$$\det \Lambda = \pm 1. \tag{3.71}$$

Continuous transformations cannot make the determinant jump from the value +1 to the value −1 or vice versa. Discrete transformations can.

□ **Exercise 3.18** Use the formulas for rotations in a 2-dimensional space, given in Eq. (2.53, *p25*), to show that the inversion of both co-ordinates can be seen as the result of a rotation in two dimensions, where the determinant of the transformation matrix is +1.

There is another thing which can be discontinuous. For this, we go back to Eq. (2.25, *p 21*), which is nothing but Eq. (3.49) in an indexed notation. This equation, for $\alpha = \beta = 0$, reads

$$\left(\Lambda^0{}_0\right)^2 - \left(\Lambda^1{}_0\right)^2 - \left(\Lambda^2{}_0\right)^2 - \left(\Lambda^3{}_0\right)^2 = 1. \tag{3.72}$$

Clearly, this implies

$$\left(\Lambda^0{}_0\right)^2 \geq 1, \tag{3.73}$$

which means that $\Lambda^0{}_0$ can either be equal to or larger than 1, or equal to or smaller than -1. Once again, only discrete transformations can take the value from one branch to another.

The discussion makes it clear that the group O(3,1) has four disconnected branches, as listed below:

Branch	$\det \Lambda$	$\Lambda^0{}_0$
1	$+1$	$\geq +1$
2	$+1$	≤ -1
3	-1	$\geq +1$
4	-1	≤ -1

(3.74)

Obviously, all rotations and boosts fall in the first branch. The entire O(3,1) group, containing all branches, is called the *extended Lorentz group*. The first branch listed in Eq. (3.74), which contains all rotations and boosts, is sometimes called the *proper Lorentz group* when the distinction with the extended group has to be made clear.

□ **Exercise 3.19** *Give examples of transformations which belong to the branches 2, 3 and 4.*

Earlier, we have identified finite dimensional irreducible representations of the Lorentz group. Now, we try to do the same for the extended Lorentz group. In §3.5.4, we argued that an irreducible representation of a group $\mathcal{G}$ does not necessarily remain irreducible under a subgroup $\mathcal{G}'$. Said another way, an irreducible representation of a smaller group $\mathcal{G}'$ does not automatically qualify as a representation for a bigger group $\mathcal{G}$ containing it.

Indeed, this is the case for many of the irreducible representations of the SU(2) × SU(2) Lorentz group. This can be seen from the way the generators of this group behave under discrete transformations like parity or time reversal. Recall the differential generators of the Lorentz group given in Eq. (3.55). Clearly, under time reversal, the generators of the form $\mathscr{J}_{ij}$ remain unchanged, but those of the form $\mathscr{J}_{0i}$ change by a sign. The same is true under parity transformation. Looking at Eq. (3.58), we then conclude that under either of these discrete transformations,

$$J_i \longrightarrow J_i, \qquad K_i \longrightarrow -K_i. \tag{3.75}$$

The definition in Eq. (3.60) now implies that, under the same discrete transformations,

$$N_i^{(+)} \longleftrightarrow N_i^{(-)}, \tag{3.76}$$

i.e., generators of the two SU(2) factors get interchanged. Thus an element of the representation (n_1, n_2) of the Lorentz group becomes an element of the representation (n_2, n_1) by the action of parity or time reversal. If a representation has $n_1 = n_2$, it is closed under the discrete symmetries, and therefore qualifies as representation of the extended Lorentz group. For others, we must take the combination $(n_1, n_2) + (n_2, n_1)$ if we want to obtain an irreducible representation of the extended group. For example, among the examples that appear in Eq. (3.64), the spin-0 representation, $(0, 0)$, and the spin-1 representation, $(\frac{1}{2}, \frac{1}{2})$, are representations of the extended Lorentz group. But a fermion representation like $(\frac{1}{2}, 0)$ is not. An irreducible representation of the extended Lorentz group is $(\frac{1}{2}, 0) + (0, \frac{1}{2})$, and this is in fact the Dirac representation that we will encounter in Ch. 4. Representations like $(\frac{1}{2}, 0)$ or $(0, \frac{1}{2})$ can be useful if parity or time reversal symmetries are not relevant, and we will see them in the context of weak interaction theories in Ch. 14.

Physical theories, we believe, should be invariant under proper Lorentz transformations. Whether they are invariant under the extended Lorentz group is a question that we will address in later chapters.

3.7 Poincaré group

In addition to Lorentz transformation, we believe that redefinitions of the origin of spacetime should also have no effect on physical theories. This means that we should also have the symmetry

$$x^\mu \rightarrow x^\mu + a^\mu, \tag{3.77}$$

where a^μ is an arbitrary constant 4-vector. This statement can be combined with the invariant of Lorentz transformations given in Eq. (3.68) to say that, physical theories should be invariant under all transformations which keep the quantity

$$\Delta x^\mu \, \Delta x_\mu \tag{3.78}$$

invariant, where Δx^μ is the difference of co-ordinates of two points in spacetime. Transformations which keep this combination invariant is called the *Poincaré group*.

Poincaré group therefore has 10 generators. Six of them belong to the Lorentz group which is a subgroup of the Poincaré group. The other four generate spacetime translations, whose parameters are the components of a^μ Eq. (3.77). The generators of these translations are the components of the 4-momentum.

It should be noted that the expression shown in Eq. (3.78) is also invariant under discrete transformations like parity and time reversal. When we want to indicate the group including these discrete operations, we can call it the *extended Poincaré group*. If we discuss only the connected part of the group, we can use the name *proper Poincaré group*. The distinction will be made only where it is important to do so.

Chapter 4

A brisk tour of quantum field theory

We have mentioned in §1.6 that a relativistic quantum theory naturally leads us to field theory. It was also commented there that the structure of classical field theory has some problems, and we should use quantum field theory. Since such theories describe all three fundamental interactions that we are going to discuss, it is imperative that we have an idea of the basic tenets and the main working tools of quantum field theory. In this chapter, we present the important concepts and tools in capsule form. If the reader is already familiar with the methods of quantum field theory, this chapter may easily be skipped. Other readers will feel unhappy at many places in this chapter, where results will appear without proofs. Unfortunately, there is no alternative: if we have to do justice to the discussion of quantum field theory, we would require so many pages that the reader would feel even more unhappy. So, a reader unfamiliar with quantum field theory will have either to accept some results and understand how a result can be applied without understanding how it is derived, or consult a quantum field theory book to obtain more details.

4.1 Motivating quantum fields

Already in Ch. 1, we have emphasized the necessity of the use of fields in relativistic quantum theory. Anything that depends on space and time is called a *field* in physics. We assume that the reader is familiar with classical fields, which are number-valued functions of space and time. When we say 'number-valued', we mean that, corresponding to a given point at a given time, the value of the field can be expressed as one or more numbers, either real or complex. For example, when we talk about the electromagnetic field in classical physics, we mean a collection of six numbers (three components of the electric field and three of the magnetic field) corresponding to each spacetime point. Taken for all points together, they describe the state of the electromagnetic field.

Relativistic quantum fields cannot be number-valued, for reasons that we now try to explain. As we know, the relativistic relation between energy and momentum of a free particle, viz.,

$$E^2 = \boldsymbol{p}^2 + m^2 , \tag{4.1}$$

is quadratic in energy. This is in contrast with the non-relativistic formula for the kinetic energy:

$$E = \frac{1}{2m} \boldsymbol{p}^2 , \tag{4.2}$$

and it is problematic on two counts.

To appreciate the first problem, let us recall how one sets up a quantum mechanical equation of motion. The expression for energy is taken as the expression for the Hamiltonian operator H, where $\boldsymbol{p}$ is understood to be the gradient operator $-i\boldsymbol{\nabla}$ acting on the wavefunction Ψ. Then one writes the time-evolution equation through the Schrödinger equation:

$$i\frac{\partial}{\partial t}\Psi = H\Psi . \tag{4.3}$$

For the non-relativistic formula of Eq. (4.2), this procedure makes perfect sense. However, if we want to write down the Hamiltonian operator starting from the relativistic formula of Eq. (4.1), it would look like

$$H = \sqrt{-\boldsymbol{\nabla}^2 + m^2} , \tag{4.4}$$

and the square root would make no sense at all.

This may be just a matter of convention, but it is worth stating. By the phrase *Schrödinger equation*, some people mean only the non-relativistic Schrödinger equation, where a non-relativistic Hamiltonian appears in place of H in Eq. (4.3). We will adopt a more general use of the term where, irrespective of the form and content of the Hamiltonian, an equation of the form of Eq. (4.3) will be called the Schrödinger equation.

To avoid the problem with the square root, we can decide to give up the Schrödinger equation and write a different relativistic equation. After all, the Schrödinger equation implies the equivalence of the operators $i\partial/\partial t$ and H. So, rather than forming the equation by letting these operators act on the wave function, we can let the squares of these operators act on the wave function. That would give the equation

$$-\frac{\partial^2}{\partial t^2}\Psi = H^2\Psi . \tag{4.5}$$

This can be called the *Klein–Gordon equation* in a generalized sense, and it puts us face to face with the second problem. Suppose we put the free Hamiltonian of Eq. (4.1) into this equation. No matter what we do, we will find not only the positive energy eigenvalues given by

$$E_p = +\sqrt{\boldsymbol{p}^2 + m^2} \tag{4.6}$$

in the eigenvalue spectrum, but also their negative counterparts, $-E_p$. Since the magnitude of the 3-momentum can take all positive values, the quantities $-E_p$ will contain negative energy eigenvalues with arbitrarily large magnitude, extending all the way up to negative infinity. This means that there is no ground state for this system. Such pathological systems cannot be physical.

If Ψ is interpreted not as a wavefunction but rather as a classical field, the same problem remains with the plane wave solutions of this field. As we said earlier, the solution to the wave equation gives the state of the system in this case, and we would again encounter a situation that the field has no ground state. Despite this problem, it is worthwhile to construct the plane wave solutions. For one thing, these solutions form a complete set of functions, and any solution can be expanded in a Fourier series in terms of these plane waves. So, in §4.2, we discuss various field equations and their plane wave solutions and write fields as superpositions of these plane waves. And, following a build-up in the intermediate sections, in §4.7 we discuss how the problem with negative energies disappears with an operator interpretation of the fields.

4.2 Plane wave solutions

Our aim would be to construct an action functional involving fields, from which all dynamics can be obtained. We will want the action to be Poincaré invariant. Lorentz invariance forms a subset of Poincaré invariance, so the action should definitely be Lorentz invariant. In order to achieve this, we need to consider fields which have definite Lorentz transformation properties. The representations of the Lorentz group were discussed in §3.6, and the simplest ones were identified in Eq. (3.64, *p 56*). In fact, we will need only the ones mentioned there, viz., the scalar representation, the Dirac representation and the spin-1 or vector representation. We will discuss the plane wave solutions of these kinds of fields here.

4.2.1 Scalar fields

By plane wave solutions of any kind of field, we mean solutions whose dependence on spacetime co-ordinates is of the form $e^{ip\cdot x}$. This factor, by itself, is a scalar under Lorentz transformations. For scalar fields, the plane wave solutions can therefore be just of the form

$$ e^{-ip\cdot x} \qquad \text{or} \qquad e^{+ip\cdot x} . \tag{4.7} $$

One question: why did we write two kinds of terms in Eq. (4.7)? With arbitrary p^μ in the exponent, should we not be able to obtain all plane waves with either of the terms written in Eq. (4.7)? The answer to this second question is obviously 'yes', but we do need both kinds of terms because the p^μ's that we want to use are not entirely arbitrary. We want the scalar field to satisfy the Klein–Gordon equation, Eq. (4.5). When we substitute the expression on the right-hand side of Eq. (4.1) for H^2 and use the co-ordinate space representation of the momentum operator, the Klein–Gordon equation can be written as

$$\left(\Box + m^2\right)\phi(x) = 0\,,\tag{4.8}$$

where

$$\Box = \partial^\mu \partial_\mu \,.\tag{4.9}$$

If we substitute the solutions of Eq. (4.7) into Eq. (4.8), we obtain

$$p^2 = m^2\,.\tag{4.10}$$

Thus, all components of p^μ are not really arbitrary. We will use only the spatial components of p^μ as independent variables. Moreover, Eq. (4.10) yields two solutions for the temporal component, $\pm E_p$, where E_p is given in Eq. (4.6). As we said before, there is a problem with the negative solution. Thus, we will always take the temporal component to be positive, i.e., take the solution given in Eq. (4.6).

Now, any function can be expanded as a Fourier transform by using plane waves. But such an expansion will have both positive and negative frequencies. Here, p_0 takes the role of the frequency, and we have decided to use only the positive solution, E_p, in its place. This is why we need both the solutions of Eq. (4.7). Writing

$$p \cdot x = E_p t - \boldsymbol{p} \cdot \boldsymbol{x}\,,\tag{4.11}$$

we see that the first solution represents positive frequency, or positive energy, whereas the second one represents negative energies. Between the two, all energies are covered, and we can use them to perform a Fourier transform and write

$$\phi(x) = \int D^3 p \left(a(\boldsymbol{p})e^{-ip\cdot x} + a^\dagger(\boldsymbol{p})e^{+ip\cdot x} \right).\tag{4.12}$$

The integration measure has been denoted by $D^3 p$. We will take

$$D^3 p \equiv \frac{d^3 p}{\sqrt{(2\pi)^3 2E_p}}\,.\tag{4.13}$$

Compared to the usual factor used for Fourier transforms, there is an extra $\sqrt{2E_p}$ in the denominator. This is purely conventional. It should be realized that any overall x-independent factor with the plane waves can be absorbed in the definition of a and $a^\dagger$. Our choice of the normalization simplifies the relation between the objects $a(\boldsymbol{p})$ and $a^\dagger(\boldsymbol{p})$, as we will see in §4.7.

Notice, for the moment, that the two terms in Eq. (4.12) are conjugates of each other, which ensures that

$$\phi(x) = \phi^\dagger(x) \, . \tag{4.14}$$

In other words, $\phi(x)$ represents a hermitian field, or a real field. If we want to deal with a scalar field that is complex, the most general Fourier transform would be

$$\phi(x) = \int D^3p \left(a(\boldsymbol{p})e^{-ip\cdot x} + \widehat{a}^\dagger(\boldsymbol{p})e^{+ip\cdot x} \right) . \tag{4.15}$$

With $a \neq \widehat{a}$, the two terms are no more conjugates of each other, so this field is complex.

4.2.2   Vector fields

We now discuss fields which transform like vectors under Lorentz transformations. We start with the photon field for two reasons. First, the classical theory of electromagnetic fields is well-known, and second, the photon field serves as a prototype for other massless fields that will come up in the discussion later in the book. After the photon field, we have a short discussion on massive vector fields.

a)   Photon field

Maxwell's electromagnetic equations are written in terms of the electric field $\boldsymbol{E}$ and the magnetic field $\boldsymbol{B}$. In 4-dimensional language, they are components of an antisymmetric rank-2 tensor $F_{\mu\nu}$, called the field-strength tensor. In quantum theory, we have to deal with the 4-vector A_μ, which is related to $F_{\mu\nu}$ through the definition

$$F_{\mu\nu} = \partial_\mu A_\nu - \partial_\nu A_\mu \, . \tag{4.16}$$

Components of the 4-vector A_μ are called potentials in classical physics. This is what we take as a field in quantum theory. We will call it the *photon field*.

The plane wave solutions for the photon field must also be of the form shown in Eq. (4.7). But the photon field transforms like a vector under Lorentz transformations. So the exponential factors must be multiplied by an x-independent factor which transforms like a vector. Thus the positive energy solutions can be written as

$$\epsilon^\mu(\boldsymbol{p})e^{-ip\cdot x} \, . \tag{4.17}$$

The negative energy solutions will be similar, with opposite sign in the exponent.

The vector $\epsilon_\mu(\boldsymbol{p})$ is called the *polarization vector*. It seems that, in order to span the 4-dimensional spacetime, we can have four independent such vectors.

But the classical Maxwell field equations show that the electromagnetic field
has only two independent degrees of freedom, because the components of E
and B, six altogether in number, are constrained by the equations

$$\nabla \cdot B = 0, \qquad \nabla \times E = -\frac{\partial B}{\partial t}, \tag{4.18}$$

which are called he homogeneous Maxwell equations, because they are indeed
homogeneous functions of degree 1 of the electric and the magnetic fields.
The divergence equation gives one constraint condition. The curl equation
is a vector equation, which contains three components, and therefore three
constraints. Thus only two of the six components of E and B are independent,
leaving two independent polarization vectors for each momentum.

The other Maxwell equations, which express the relation between the field
strengths and the sources, can be written as

$$\partial_\mu F^{\mu\nu} = j^\nu, \tag{4.19}$$

where j^ν is the 4-vector for current density. Using Eq. (4.16), it can be written
as

$$\Box A^\nu - \partial_\mu \partial^\nu A^\mu = 0 \tag{4.20}$$

in the absence of sources. To make it look like the Klein–Gordon equation,
we can use the constraint

$$\partial_\mu A^\mu = 0. \tag{4.21}$$

Thankfully, this does not result in any loss of generality since Eq. (4.16)
implies that A_μ is not uniquely defined: a redefinition of the form

$$A_\mu \to A_\mu + \partial_\mu f \tag{4.22}$$

leaves physics unchanged for any function f. This arbitrariness can be used to
ensure Eq. (4.21). The freedom, or the arbitrariness, mentioned in Eq. (4.22)
is called *gauge invariance*. Conditions like Eq. (4.21) which curb this freedom
are called *gauge conditions*.

In passing, we should make a comment about the choice of units in writing the inhomogeneous
Maxwell equations, Eq. (4.19). There are many different systems of units for writing electromagnetic quantities. Depending on the convention, an extra constant factor might appear on
the right hand side of this equation. Such constants need not be just numerical factors like
4π: they might even have dimensions, depending on our definition of what j^ν is. We have
decided to put all such possible factors into the definition of j^ν. This system is called the
Heaviside–Lorentz units.

☐ **Exercise 4.1** Show that, in the Heaviside–Lorentz units, the electric
field for a point particle of charge q located at the origin of co-
ordinates is given by

$$E(r) = \frac{q}{4\pi r^3}\, r. \tag{4.23}$$

[**Hint :** For a point charge at rest at the origin of co-ordinates, $j^0(r) = q\delta^3(r)$
whereas $j_i = 0$.]

Looking back at the plane wave solution of Eq. (4.17), we see that Eq. (4.21) implies that the polarization vectors should satisfy the condition

$$\epsilon^{\mu}(\boldsymbol{p})p_{\mu} = 0 \,. \tag{4.24}$$

This means that only three of the four components of $\epsilon^{\mu}(\boldsymbol{p})$ are independent. Moreover, we can make one component zero by using the freedom denoted in Eq. (4.22). Thus, only two components of the polarization vector are physical, and consequently there can be only two independent polarization vectors.

This is how it works. Consider a photon moving in the z-direction, so that its 4-momentum can be written as

$$p^{\mu} = (E, 0, 0, E) \,, \tag{4.25}$$

consistent with the fact that the photon is massless. Eq. (4.24) then tells us that the components of the polarization vector must satisfy the relation

$$E\epsilon^{0} - E\epsilon^{3} = 0 \,, \tag{4.26}$$

so that $\epsilon^{0} = \epsilon^{3}$. Further, Eq. (4.22) says that, under transformations of the form

$$\epsilon^{\mu}(\boldsymbol{p}) \to \epsilon^{\mu}(\boldsymbol{p}) + Kp^{\mu} \tag{4.27}$$

for arbitrary K, physics remains unchanged. These are the gauge transformations. For the momentum 4-vector that we have chosen, this means that we can change ϵ^{0} and ϵ^{3} at our will: these components therefore cannot be physical. The other two components are unaffected by the gauge transformation, and are therefore the physical components for the photon momentum given in Eq. (4.25). For a photon moving in an arbitrary direction, we can take the physical polarization vectors to be perpendicular to the direction of motion. In other words, the spatial part of the polarization vectors should satisfy the condition

$$\boldsymbol{\epsilon}(\boldsymbol{p}) \cdot \boldsymbol{p} = 0 \,. \tag{4.28}$$

□ **Exercise 4.2** *Show that the gauge choice of Eq. (4.24) remains unaffected under the gauge transformations given in Eq. (4.27) since the photon is massless.*

Denoting the different polarization vectors by an index r, we can write down the plane wave expansion of the photon field as

$$A^{\mu}(x) = \sum_{r} \int D^{3}p \Big(a_{r}(\boldsymbol{p})\epsilon_{r}^{\mu}(\boldsymbol{p})e^{-ip\cdot x} + a_{r}^{\dagger}(\boldsymbol{p})\epsilon_{r}^{\mu*}(\boldsymbol{p})e^{ip\cdot x} \Big) \,. \tag{4.29}$$

Note that we have taken the Fourier components in a form which makes the photon field real, as it should be. Also, it should be mentioned that the sum on the index r can run over four independent polarization vectors as well: there is a well-defined formalism that ensures that the characteristics of a physical photon depend only on two polarization vectors, both of which satisfy Eq. (4.21).

Following the argument in §4.2.1, we have the freedom to set the normalization of the polarization vectors arbitrarily. Since the physical polarizations are space-like, we can take them of unit length, i.e.,

$$\boldsymbol{\epsilon} \cdot \boldsymbol{\epsilon} = 1 \,. \tag{4.30}$$

If we deal with four independent polarization vectors out of which two will
be unphysical, one of these vectors must be time-like. So we can take the
normalization to be

$$\epsilon_r^\mu \epsilon_{\mu r'}^* = -\zeta_r \delta_{rr'} , \tag{4.31}$$

with

$$\zeta_0 = -1, \qquad \zeta_1 = \zeta_2 = \zeta_3 = +1 . \tag{4.32}$$

Note that the expression on the right hand side of Eq. (4.31) does not involve
a sum over the repeated index r since this is just a counting index. According
to our summation convention stated in §2.1, any summation over this index
will be explicitly indicated, as in the following equation:

$$\sum_{r=0}^{3} \zeta_r \epsilon_r^\mu \epsilon_r^{\nu *} = -g^{\mu\nu} , \tag{4.33}$$

which is the completeness relation satisfied by the polarization vectors.

As mentioned earlier, the sum involves unphysical polarization states. In real problems, we will
encounter such sums over physical states only. Consider a photon moving in the z-direction, so
that we can use Eq. (4.25) as its 4-vector. Subject to the conditions given in Eqs. (4.24) and
(4.28), the physical polarization vectors will be given by

$$\epsilon_{(1)}^\mu = (0, 1, 0, 0) , \qquad \epsilon_{(2)}^\mu = (0, 0, 1, 0) , \tag{4.34}$$

or linear combinations of these two. We can then write

$$\sum_{\substack{\text{pol} \\ \text{(physical)}}} \epsilon_r^\mu \epsilon_r^{\nu *} = -g^{\mu\nu} - \frac{p^\mu p^\nu}{(p \cdot n)^2} + \frac{p^\mu n^\nu + p^\nu n^\mu}{p \cdot n} , \tag{4.35}$$

where $n^\mu = (1, 0, 0, 0)$. The result in this form is valid for any photon momentum p^μ, not
necessarily in the z-direction. We will show later that the terms containing p^μ and/or p^ν do
not contribute to any physical amplitude, and so one can effectively use the formula

$$\sum_{\substack{\text{pol} \\ \text{(physical)}}} \epsilon_r^\mu \epsilon_r^{\nu *} = -g^{\mu\nu} . \tag{4.36}$$

☐ **Exercise 4.3** Verify that, for the photon momentum in the z-
direction, the expression given in Eq. (4.35) has non-zero values only
if $\mu = \nu = 1$ or $\mu = \nu = 2$, which is what is expected from Eq. (4.34).

b)   Proca field

Eq. (4.19) suggests an easy and straightforward way of generalization for
massive vector fields. In absence of sources, we can write the equation as

$$\partial_\mu F^{\mu\nu} + M^2 A^\nu = 0 , \tag{4.37}$$

where $F^{\mu\nu}$ and A^ν are related through Eq. (4.16). This is called the *Proca
equation*.

Since $F^{\mu\nu}$ is antisymmetric in its indices, $\partial_\nu \partial_\mu F^{\mu\nu} = 0$. So Eq. (4.37) implies

$$\partial_\nu A^\nu = 0 \,, \tag{4.38}$$

an equation that the Proca field must satisfy. The polarization vectors of the Proca field therefore satisfy Eq. (4.24) just like the photons do. However, the freedom described in Eq. (4.27) is not available for the Proca field. Because of the presence of the A^ν term, the Proca equation is not invariant under gauge transformations of the form given in Eq. (4.22).

Because of the constraint of Eq. (4.24), there can be three independent polarization states of a massive vector boson. We can take the polarization vectors to be orthogonal to one another, i.e.,

$$\epsilon_\mu^{(r)}(\boldsymbol{p}) \left(\epsilon_{(s)}^\mu(\boldsymbol{p}) \right)^* = -\delta_s^r \,. \tag{4.39}$$

Suppose a massive vector boson has the 4-momentum $p^\mu = (E_p, \mathrm{p}\widehat{\boldsymbol{n}})$, where $\widehat{\boldsymbol{n}}$ is a unit 3-vector. As for the case of the photon, we can take two polarization vectors of the form $(0, \widehat{\boldsymbol{n}}')$ with $\widehat{\boldsymbol{n}} \cdot \widehat{\boldsymbol{n}}' = 0$. These are transverse degrees of polarization. For a massive vector boson, we can obtain one more, viz.,

$$\epsilon_l^\mu(p) = \frac{1}{M}(\mathrm{p}, E_p\widehat{\boldsymbol{n}}) \,. \tag{4.40}$$

This is a longitudinal polarization state since the spatial part is along the direction of the 3-momentum of the vector boson. With these three states of polarization, it is easy to see that

$$\sum_r \epsilon_\mu^{(r)}(\boldsymbol{p}) \left(\epsilon_\nu^{(r)}(\boldsymbol{p}) \right)^* = -g_{\mu\nu} + \frac{p_\mu p_\nu}{M^2} \,. \tag{4.41}$$

If we are talking about a real Proca field, the plane wave expansion will be given by Eq. (4.29), with the polarization vectors appropriate for the Proca field. For a complex Proca field, $a_r^\dagger(\boldsymbol{p})$ appearing on the right side of Eq. (4.29) will have to be replaced by $\widehat{a}_r^\dagger(\boldsymbol{p})$ so that the two terms are not hermitian conjugates of each other. However, there are problems with dealing with Proca fields which will be alluded to in §14.9.

4.2.3   Dirac fields

Dirac suggested that, for spin-$\frac{1}{2}$ fermions, one should not use the Klein–Gordon equation. Instead, he introduced the Hamiltonian

$$H = \boldsymbol{\alpha} \cdot \boldsymbol{p} + \beta m \,. \tag{4.42}$$

The Schrödinger equation with the Dirac Hamiltonian is often called the Dirac equation:

$$i\frac{\partial \psi}{\partial t} = -i\boldsymbol{\alpha} \cdot \boldsymbol{\nabla}\psi + \beta m\psi \,. \tag{4.43}$$

Multiplying both side from the left by β, this equation can also be written in the form

$$i\gamma^\mu \partial_\mu \psi - m\psi = 0 , \qquad (4.44)$$

where we introduce the notation

$$\gamma^\mu = \{\beta, \beta\alpha^i\} . \qquad (4.45)$$

We will also define the corresponding matrices with lower indices in the usual way:

$$\gamma_\mu = g_{\mu\nu}\gamma^\nu . \qquad (4.46)$$

It should be noted that we have put a Lorentz index on these matrices, despite the fact that these are constant matrices. They are the same in all frames. The reason for equipping these matrices with Lorentz indices will be explained later, in §4.4. In any case, even at this point it should be obvious that the objects presented in Eq. (4.45) are different from Lorentz vectors: for a particular value of μ, the object A^μ is a number, whereas γ^μ is a matrix.

In order that the solutions to the Dirac equation reproduces the relativistic relation between energy and momentum, Eq. (4.1), the objects γ^μ should satisfy the relations

$$\left[\gamma_\mu, \gamma_\nu\right]_+ = 2g_{\mu\nu} , \qquad (4.47)$$

where the square brackets with a subscripted plus sign denote the anticommutator,

$$[A, B]_+ \equiv AB + BA . \qquad (4.48)$$

Hermiticity of the Hamiltonian of Eq. (4.42) imposes the following property:

$$\gamma_\mu^\dagger = \gamma_0 \gamma_\mu \gamma_0 . \qquad (4.49)$$

□ **Exercise 4.4** *Take the square of the Hamiltonian given in Eq. (4.42), assuming the objects α and β commute with all components of momentum, but nothing about their mutual commutation properties. Show that the result gives $H^2 = p^2 + m^2$ if Eq. (4.47) is satisfied.*

The anticommutation property cannot be satisfied if the objects γ^μ are numbers. We have to take them as matrices. In Appendix F, we argue that they have to be 4×4 matrices. Eq. (4.44) then says that any solution for ψ must be a 4-component column vector. Let us therefore try solutions of the form

$$\psi(x) = u_p e^{-ip \cdot x} , \qquad (4.50)$$

where u_p is a 4-component column vector. This is called a *spinor* solution. According to the discussion of §4.2.1 following Eq. (4.7), the u_p appearing in

Eq. (4.50) must be a positive energy spinor. Similarly, there will be negative energy spinors, defined by solutions of the form

$$\psi(x) = v_{\boldsymbol{p}}e^{+ip\cdot x} \,. \tag{4.51}$$

There will be four linearly independent solutions altogether, two of the first kind and two of the second. We will denote these different solutions by an extra subscript index, i.e., by writing $u_{\boldsymbol{p},s}$ and $v_{\boldsymbol{p},s}$. When we write any mathematical formula with only one subscript on the spinors, the subscript should always be understood to be the 3-momentum appearing in the plane wave solution, and the formula to be valid irrespective of which of the two different solutions we take.

To obtain the spinors, we insert Eqs. (4.50) and (4.51) into the Dirac equation, Eq. (4.44). This gives the following equations for the spinors:

$$\left(\not{p} - m\right)u_{\boldsymbol{p}} = 0 \,, \tag{4.52}$$

$$\left(\not{p} + m\right)v_{\boldsymbol{p}} = 0 \,, \tag{4.53}$$

where we have introduced the useful notation

$$\not{p} \equiv \gamma^{\mu}p_{\mu} \,. \tag{4.54}$$

Of course, the explicit solutions of Eqs. (4.52) and (4.53) depend on the explicit forms of the Dirac matrices. These are representation dependent, and therefore are not of direct interest to us. Independent of the representation, we can make a few observations that will be helpful for future manipulations with the spinors. First, note how Eq. (4.52) looks for $\boldsymbol{p} = 0$. In this case $p_0 = m$, so that the equation reduces to

$$\left(\gamma_0 - 1\right)u_{\boldsymbol{0}} = 0 \,, \tag{4.55}$$

which means that $u_{\boldsymbol{0}}$ is an eigenvector of γ_0 with eigenvalue $+1$. Similarly, Eq. (4.53) shows that $v_{\boldsymbol{0}}$ should be an eigenvector of γ_0 with eigenvalue -1. Note that there will be two solutions of each type, since there are four eigenvectors of the 4×4 matrix γ_0.

We now define normalized eigenvectors of γ_0 by the equations

$$\gamma_0\xi_s = \xi_s \,, \qquad \gamma_0\chi_s = -\chi_s \,, \tag{4.56}$$

with

$$\xi_s^{\dagger}\xi_{s'} = \chi_s^{\dagger}\chi_{s'} = \delta_{ss'} \,, \qquad \xi_s^{\dagger}\chi_{s'} = \chi_s^{\dagger}\xi_{s'} = 0 \,. \tag{4.57}$$

The index on the eigenvectors can take two values, giving two different eigenvectors.

Clearly, u_0 and v_0 must be given, apart from some optional normalization constant, by ξ_s and χ_s respectively. And the solutions for arbitrary momentum can be constructed in the following way:

$$u_{\boldsymbol{p},s} = N_{\boldsymbol{p}}(\not{p} + m)\xi_s \,, \tag{4.58}$$

$$v_{\boldsymbol{p},s} = N_{\boldsymbol{p}}(-\not{p} + m)\chi_{-s} \,, \tag{4.59}$$

where $N_{\boldsymbol{p}}$ is a normalizing factor, to be defined shortly. To show that these are indeed solutions, we need to check that they do indeed satisfy Eqs. (4.52) and (4.53) respectively. Substitution of these forms into the left hand sides of the equations yields the combination

$$(\not{p} - m)(\not{p} + m) = \not{p}\not{p} - m^2 \,. \tag{4.60}$$

This is zero since

$$\not{p}\not{p} = \gamma_\mu \gamma_\nu p^\mu p^\nu = \gamma_\mu \gamma_\nu p^\nu p^\mu = \gamma_\nu \gamma_\mu p^\mu p^\nu$$

$$= \frac{1}{2}\left[\gamma_\mu, \gamma_\nu\right]_+ p^\mu p^\nu = g_{\mu\nu} p^\mu p^\nu = p^2 \,, \tag{4.61}$$

which equals m^2, by Eq. (4.10).

As for the normalization factor $N_{\boldsymbol{p}}$, we choose

$$N_{\boldsymbol{p}} = \frac{1}{\sqrt{E_{\boldsymbol{p}} + m}} \,. \tag{4.62}$$

In Appendix F, we show that this choice implies the following normalization conditions on the spinors:

$$u_{\boldsymbol{p},s}^\dagger u_{\boldsymbol{p},s'} = v_{\boldsymbol{p},s}^\dagger v_{\boldsymbol{p},s'} = 2E_{\boldsymbol{p}}\delta_{s,s'} \tag{4.63}$$

and

$$u_{\boldsymbol{p},s}^\dagger v_{-\boldsymbol{p},s'} = v_{\boldsymbol{p},s}^\dagger u_{-\boldsymbol{p},s'} = 0 \,. \tag{4.64}$$

Using these spinor solutions, we can write

$$\psi(x) = \sum_s \int D^3p \left(d_s(\boldsymbol{p}) u_s(\boldsymbol{p}) e^{-ip\cdot x} + \widehat{d}_s^\dagger(\boldsymbol{p}) v_s(\boldsymbol{p}) e^{+ip\cdot x} \right) . \tag{4.65}$$

Note that we use the notations $u(\boldsymbol{p})$ and $u_{\boldsymbol{p}}$ interchangeably, here and elsewhere in the book.

□ **Exercise 4.5** Express u_0 and v_0 in terms of the normalized eigenvectors of γ_0.

4.3   Lagrangian

Before proceeding further with fields, let us discuss what we want to do with them.

Of course, we want to study the evolution of fields. This can be done by setting up an action for the fields and deriving the equations of motion from the action. In particle mechanics, the action is defined as the time integral of the Lagrangian:

$$\mathscr{A} = \int dt \, L \, . \tag{4.66}$$

Since fields are functions of both time and space, it is more useful to write the action as

$$\mathscr{A} = \int dt \int d^3x \, \mathscr{L} = \int d^4x \, \mathscr{L} \, , \tag{4.67}$$

where $\mathscr{L}$ is the Lagrangian density. Since this is the object we will use always, and not the object L that appears in Eq. (4.66), we will drop the word 'density' and refer to it simply as *Lagrangian*. In case we want to refer to L, the spatial integral of the Lagrangian, we will call it the *total Lagrangian*.

What will the Lagrangian consist of? It will contain fields, of course. In addition, there can be first derivatives of the fields. The Lagrangian formulation does not allow the presence of any higher order derivatives in the Lagrangian.

For the purpose of a general discussion, let us denote a general field by $\Phi(x)$. Thus

$$\mathscr{L} = \mathscr{L}(\Phi, \partial_\mu \Phi) \, . \tag{4.68}$$

Of course, Φ need not be only one field. All relevant fields are summarized in that notation.

The action principle says that the action must be extremum for the classical solution. The fields must satisfy some condition for this to be realized. Omitting details, we write the result:

$$\partial_\mu \left(\frac{\partial \mathscr{L}}{\partial(\partial_\mu \Phi^A)} \right) = \frac{\partial \mathscr{L}}{\partial \Phi^A} \, , \tag{4.69}$$

where the index A selects out any one field among those present in the Lagrangian. This equation is called the *Euler–Lagrange equation* for the field Φ^A.

Since all physics comes from the action and the physics is Poincaré invariant, the action itself must be Poincaré invariant. Looking back at Eq. (4.67) and appreciating the fact that the integration measure d^4x is Poincaré invariant, we conclude that the Lagrangian has to be Poincaré invariant.

This is the advantage of proceeding through a Lagrangian formulation. Quantum mechanics is often introduced through the Hamiltonian formulation. We can of course define the Hamiltonian (which is, truly speaking, the Hamiltonian density) through the Lagrangian by

$$\mathcal{H} = \sum_A \Pi_A \partial_0 \Phi^A - \mathcal{L} \,, \qquad (4.70)$$

where Π_A is called the *canonical momentum* corresponding to the field Φ^A, defined as

$$\Pi_A = \frac{\partial L}{\partial(\partial_0 \Phi^A)} \,. \qquad (4.71)$$

But this Hamiltonian will not be Poincaré invariant, and therefore will not be as convenient as the Lagrangian.

With Lagrangians, the task is simple. Take the fields, and try making Poincaré invariants with them. There are two parts of Poincaré invariance, as discussed in Ch. 3. First, the Lagrangian will have to be translation invariant. This can be easily ensured by not putting any explicit dependence of x^μ in the Lagrangian. Derivatives are of course allowed, as indicated in Eq. (4.68), because they are not affected by a constant shift of the co-ordinates.

The second part of ensuring Poincaré invariance is ensuring Lorentz invariance. The Lagrangian should contain only Lorentz invariant combinations of fields. This is the task that we take up in §4.4. Once this is done, we will be able to see examples of Lagrangians that we can work with.

The Lagrangian may have internal symmetries apart from the spacetime symmetries that we discussed above. We will talk about such symmetries briefly in §4.6, but will take them up seriously beginning with Ch. 5.

Another property of Lagrangians should be kept in mind. In order that the Hamiltonian is hermitian, the Lagrangian should be hermitian. That means that any given term in the Lagrangian must either be hermitian by itself, or if it is not, the Lagrangian must contain the hermitian conjugate term as well.

There is another important point. It is the action which really governs the dynamics. Suppose we have a term which is a total divergence, of the form $\partial_\mu S^\mu$ for some collection of fields S^μ. We can use the Gauss theorem to turn the integral of this term into a surface integral. Assuming all fields vanish at the spacetime surface at infinity, the surface integral will be zero. The lesson is that a total divergence term is irrelevant in the Lagrangian.

One final constraint should be discussed here. In conventional units, action has the dimension of $\hbar$. In natural units, since we have taken $\hbar = 1$, action should be dimensionless, as has been noted in Table 1.3 *(p 13)*. Eq. (4.67) then implies that the mass dimension of the Lagrangian is given by

$$\dim \mathcal{L} = 4 \,. \qquad (4.72)$$

In a term in the Lagrangian, any combination of fields must therefore be multiplied by a constant such that the overall combination has mass-dimension equal to 4. The constants, in general, are called *coupling constants*. In case only the fields themselves make up for the dimension 4, the coupling constant would be dimensionless. If the dimension of the fields in a Lagrangian term is less than 4, the coupling constant would have a positive mass dimension. The remaining possibility, where the field combination has dimension greater than 4 and consequently the coupling constant has negative mass-dimension, suffers from a big problem. With the presence of any such coupling constant, the entire theory becomes sick. One obtains infinite results for amplitudes of physical processes. We will therefore consider theories in which such constants do not appear, theories which are called *renormalizable*. Exceptions can be made only if the theory is supposed to be an approximation, valid up to a certain maximum energy. These kind of theories are called *effective field theories*.

4.4   Making Lorentz invariants with fields

4.4.1   Invariants with scalar and vector fields

Making Lorentz invariants with scalar fields is easy. The scalar field itself is invariant under the Lorentz group, so any polynomial made out of the scalar field must also be invariant. For example, if we consider a scalar field $\phi(x)$, there can be terms like ϕ^2 or ϕ^3 or any higher power of ϕ in the Lagrangian. There may also be a term like $(\partial_\mu \phi)(\partial^\mu \phi)$, or the square or cube or higher powers of this expression. There might be some restrictions coming from hermiticity, renormalizability, and other symmetries, which we will encounter as we go along.

We now consider a vector field V^μ. It is easy to construct invariants using this field: one merely has to notice that all Lorentz indices are contracted. For example, we can write a term $V^\mu V_\mu$. One can also have terms like $(\partial_\mu V_\nu)(\partial^\mu V^\nu)$, or $(\partial_\mu V_\nu)(\partial^\nu V^\mu)$. The combination $\partial_\mu V^\mu$ cannot occur. Or, rather, it is useless to employ this combination, since it is set to zero for vector fields, as indicated in Eq. (4.21).

4.4.2   Invariants involving Dirac fields

There cannot be any term in the Lagrangian which would have only one Dirac field. The reason is that the Dirac field carries an angular momentum of $\frac{1}{2}$. Any term in the Lagrangian has to be Lorentz invariant, which implies that the total angular momentum should be zero. So, fermion fields must occur in even numbers. It is therefore instructive to see how different combinations of two fermion fields behave under Lorentz transformations.

a)   Basic invariant

Under Lorentz transformations, a fermion field transforms as follows:

$$\psi(x) \longrightarrow \psi'(x') = \exp\left(-\frac{i}{4}\omega^{\mu\nu}\sigma_{\mu\nu}\right)\psi(x), \tag{4.73}$$

where

$$\sigma_{\mu\nu} = \frac{i}{2}\left[\gamma_\mu, \gamma_\nu\right], \tag{4.74}$$

and $\omega^{\mu\nu}$'s are the parameters of transformation. This is a representation in the form advocated in Eq. (3.15, *p42*), because when all $\omega^{\mu\nu}$'s vanish, we obtain the identity transformation. Thus $\omega^{\mu\nu}$'s represent the departure from identity transformation. They can be defined by writing the co-ordinate transformation rule, Eq. (2.3, *p17*), in the form

$$x'^\mu = \Lambda^\mu{}_\nu x^\nu = \left(\delta^\mu{}_\nu + \omega^\mu{}_\nu\right)x^\nu = \left(g^{\mu\nu} + \omega^{\mu\nu}\right)x_\nu. \tag{4.75}$$

When we defined generators in Eq. (3.15, *p42*), there was only a factor of $-i$ in the exponent. Compared to that, it might seem that Eq. (4.73) has an extra factor of $\frac{1}{4}$. Actually, there has been no change of convention between Eqs. (3.15) and (4.73). To appreciate this statement, first it has to be realized that $\frac{1}{2}\sigma_{\mu\nu}$, and not $\sigma_{\mu\nu}$, are the generators of the Lorentz group. This fact, elaborated in Ex. 4.6, accounts for a factor of $\frac{1}{2}$ present in the exponent of Eq. (4.73). Another factor of $\frac{1}{2}$ comes from the fact that the Lorentz group has six generators, as elucidated in §3.6. These can be taken as $\frac{1}{2}\sigma_{\mu\nu}$ with $\mu < \nu$. The exponent of Eq. (4.73) has an unrestricted sum over the indices so that each independent $\sigma_{\mu\nu}$ appears twice in the sum.

□ **Exercise 4.6** *Starting from the definition of Eq. (4.74), show that the matrices $\frac{1}{2}\sigma_{\mu\nu}$ satisfy the same commutation commutation relation as the $\mathscr{J}_{\mu\nu}$'s, shown in Eq. (3.56, p54). [**Hint :** Use Eq. (4.47), and the identities*

$$[A, BC] = [A, B]C + B[A, C], \tag{4.76a}$$
$$[A, BC] = [A, B]_+ C - B[A, C]_+, \tag{4.76b}$$

which hold for any objects A, B and C whose multiplication is associative.]

To keep the notation compact in relations such as the one in Eq. (4.73), we will omit the co-ordinate as the argument of the field. It will be understood that the primed field should have the primed co-ordinate as its argument, whereas the unprimed field should have the unprimed co-ordinate. Taking the hermitian conjugate of Eq. (4.73), we obtain

$$\psi^\dagger \longrightarrow \psi'^\dagger = \psi^\dagger \exp\left(+\frac{i}{4}\omega^{\mu\nu}\sigma^\dagger_{\mu\nu}\right). \tag{4.77}$$

This shows that

$$\psi'^\dagger\psi' = \psi^\dagger \exp\left(+\frac{i}{4}\omega^{\mu\nu}\sigma^\dagger_{\mu\nu}\right)\exp\left(-\frac{i}{4}\omega^{\mu\nu}\sigma_{\mu\nu}\right)\psi. \tag{4.78}$$

But the two exponential factors in the middle do not cancel each other, because $\sigma_{\mu\nu}^{\dagger} \neq \sigma_{\mu\nu}$. In fact, from Eq. (4.49), it is easy to show that

$$\sigma_{\mu\nu}^{\dagger} = \gamma_0 \sigma_{\mu\nu} \gamma_0 \,, \tag{4.79}$$

which means that the right hand side of Eq. (4.78) cannot be written as $\psi^{\dagger}\psi$, or in other words, the combination $\psi^{\dagger}\psi$ is not invariant under a Lorentz transformation.

□ **Exercise 4.7** Verify Eq. (4.79).

Let us then try to find whether there is an invariant of the form $\psi^{\dagger} M \psi$, where M is some fixed matrix. Since the transformation property of ψ has been shown in Eq. (4.73), the given combination will be invariant if $\psi^{\dagger} M$ has the following transformation property:

$$\psi^{\dagger} M \longrightarrow \psi'^{\dagger} M = \psi^{\dagger} M \exp\left(+\frac{i}{4}\omega^{\mu\nu}\sigma_{\mu\nu}\right). \tag{4.80}$$

Using Eq. (4.77), we see that this implies

$$\psi^{\dagger} \exp\left(+\frac{i}{4}\omega^{\mu\nu}\sigma_{\mu\nu}^{\dagger}\right) M = \psi^{\dagger} M \exp\left(+\frac{i}{4}\omega^{\mu\nu}\sigma_{\mu\nu}\right). \tag{4.81}$$

Since we want this equation to be satisfied for arbitrary ψ and arbitrary values of $\omega^{\mu\nu}$, we require

$$\sigma_{\mu\nu}^{\dagger} M = M \sigma_{\mu\nu} \,. \tag{4.82}$$

Using Eq. (4.79) and $\left(\gamma_0\right)^2 = 1$ which follows from Eq. (4.47), this can be written as

$$\sigma_{\mu\nu}\gamma_0 M = \gamma_0 M \sigma_{\mu\nu} \,. \tag{4.83}$$

In other words, $\gamma_0 M$ commutes with all $\sigma_{\mu\nu}$'s. This is easily ensured if $\gamma_0 M$ is the unit matrix, or if

$$M = \gamma_0 \,. \tag{4.84}$$

For future purposes, it is therefore convenient to introduce the notation

$$\overline{\psi} \equiv \psi^{\dagger}\gamma_0 \,, \tag{4.85}$$

and summarize our exercise by saying that under Lorentz transformations, $\overline{\psi}$ transforms as

$$\overline{\psi} \longrightarrow \overline{\psi}' = \overline{\psi} \exp\left(+\frac{i}{4}\omega^{\mu\nu}\sigma_{\mu\nu}\right), \tag{4.86}$$

so that the combination $\overline{\psi}\psi$ is an invariant.

The choice of Eq. (4.84) is not unique. There exists a matrix that anti-commutes with all gamma matrices, i.e.,

$$\left[\gamma^{\mu}, \gamma_5\right]_{+} = 0. \tag{4.87}$$

This matrix, called γ_5, is given by

$$\gamma_5 \equiv i\gamma^0\gamma^1\gamma^2\gamma^3. \tag{4.88}$$

The matrices $\sigma_{\mu\nu}$ contain a pair of gamma matrices, and therefore commute with γ_5:

$$\left[\sigma_{\mu\nu}, \gamma_5\right] = 0. \tag{4.89}$$

Thus, we can also use γ_5 in place of $\gamma_0 M$. In this case, we obtain the result that $\overline{\psi}\gamma_5\psi$ is also an invariant under Lorentz transformations.

There is an important mathematical theorem, called *Schur's lemma*, which says that a matrix can commute with all generators of an irreducible representation if and only if it is a multiple of the unit matrix. Existence of the matrix γ_5 that commutes with every $\sigma_{\mu\nu}$ clearly indicates that the representation we are dealing with is not irreducible. Indeed, it is very easy to see that if we define

$$\sigma_{\mu\nu}^{(\pm)} = \frac{1}{2}(1 \pm \gamma_5)\sigma_{\mu\nu}, \tag{4.90}$$

then the matrices $\frac{1}{2}\sigma_{\mu\nu}^{(+)}$ satisfy the same commutation commutation relation as the $\mathscr{J}_{\mu\nu}$, and so do the matrices $\frac{1}{2}\sigma_{\mu\nu}^{(-)}$. This already shows that the set $\sigma_{\mu\nu}$ is not irreducible: it can be decomposed into two different sets, each of which satisfies the same algebra. These two sets are in fact the matrix representation of the generators in the two different spin-$\frac{1}{2}$ representations that we mentioned in Eq. (3.64, *p 56*). The Dirac field contains these two irreducible representations, and can be said to belong to the $(\frac{1}{2},0) + (0,\frac{1}{2})$ representation. This is obviously reducible, as the plus sign between the two irreducible representations indicates.

It can be easily seen that the component representations defined by the generators appearing in Eq. (4.90) are 2-dimensional. By the choice of the overall constant in the definition of γ_5, we have ensured that

$$\left(\gamma_5\right)^2 = 1. \tag{4.91}$$

The eigenvalues of γ_5 are therefore ± 1, each one being doubly degenerate. We can take a basis for the Dirac matrices such that γ_5 is diagonal, with two positive eigenvalues in the first two diagonal positions, and the two negative ones in the last two positions. Then, the matrices $\frac{1}{2}(1 \pm \gamma_5)$ will be of the form

$$\frac{1}{2}(1 + \gamma_5) = \text{diag}(1,1,0,0), \qquad \frac{1}{2}(1 - \gamma_5) = \text{diag}(0,0,1,1). \tag{4.92}$$

Consequently, $\sigma_{\mu\nu}^{(+)}$ will be a set of matrices with non-zero elements only in the upper left 2×2 block, and will inflict non-trivial transformations only among the two upper components of any column vector. Similarly, $\sigma_{\mu\nu}^{(-)}$ will inflict non-trivial transformation only among the two lower components. In this sense, each set constitutes a 2-dimensional representation.

□ **Exercise 4.8** *Following the example of γ_5 discussed here, construct a general proof of Schur's lemma. In other words, show that if there exists a matrix Y that commutes with all generators T_a in a representation and is not a multiple of the unit matrix, then one can construct two or more (depending on the number of eigenvalues that Y has) sets of matrices, in the fashion shown in Eq. (4.90), which would satisfy the same algebra as the T_a's.*

b)   General bilinears of fermion fields

How do general combinations of the form $\overline{\psi}F\psi$ transform? We have already encountered two such combinations, and found that they are scalars, or invariants. Before exploring other possibilities for F, let it first be realized that it is not necessary to deal with arbitrary matrices in place of F. Since ψ has four components, the matrix F must be a 4×4 matrix. It will thus be enough if we identify 16 independent basis matrices and derive the transformation rules of the resulting bilinears. The conventional choice groups them conveniently into several categories, listed below.

Combination name	F	Number of matrices
S	1	1
V	γ_μ	4
T	$\sigma_{\mu\nu}$	6
A	$\gamma_\mu\gamma_5$	4
P	γ_5	1

$$(4.93)$$

   In fact, the names of the combinations contain hints to their Lorentz transformation properties. The combination $\overline{\psi}\psi$ is called S because it is a scalar, as we have seen. Of course we have also seen that $\overline{\psi}\gamma_5\psi$ behaves the same way. It turns out that under the extended Lorentz group, the behaviors of these two combinations are different. In Ch. 6, we will see that the combination $\overline{\psi}\gamma_5\psi$ is odd under parity. For such scalars, the name *pseudoscalar* is often used, which explains why this combination has been dubbed P in Eq. (4.93).

   Among the rest, we first consider the behavior of $\overline{\psi}\gamma^\lambda\psi$ under a Lorentz transformation. Using Eqs. (4.73) and (4.86), we find

$$\overline{\psi}'\gamma^\lambda\psi' = \overline{\psi}\left(1 + \frac{i}{4}\omega_{\mu\nu}\sigma^{\mu\nu}\right)\gamma^\lambda\left(1 - \frac{i}{4}\omega_{\mu\nu}\sigma^{\mu\nu}\right)\psi + \mathcal{O}\left(\omega^2\right)$$
$$= \overline{\psi}\gamma^\lambda\psi + \frac{i}{4}\omega_{\mu\nu}\overline{\psi}\left[\sigma^{\mu\nu}, \gamma^\lambda\right]\psi + \mathcal{O}\left(\omega^2\right) . \qquad (4.94)$$

For evaluating the commutator, we use Eq. (4.76b). Using the anticommutators of Eq. (4.47), we obtain

$$\left[\gamma^\mu\gamma^\nu, \gamma^\lambda\right] = 2g^{\nu\lambda}\gamma^\mu - 2g^{\mu\lambda}\gamma^\nu . \qquad (4.95)$$

Using the definition of $\sigma^{\mu\nu}$ from Eq. (4.74), we then obtain

$$\left[\sigma^{\mu\nu}, \gamma^\lambda\right] = 2i\left(g^{\nu\lambda}\gamma^\mu - g^{\mu\lambda}\gamma^\nu\right) . \qquad (4.96)$$

Thus

$$\omega_{\mu\nu}\left[\sigma^{\mu\nu}, \gamma^\lambda\right] = 2i\left(\omega_\mu{}^\lambda\gamma^\mu - \omega^\lambda{}_\nu\gamma^\nu\right)$$
$$= 2i\left(\omega^{\mu\lambda}\gamma_\mu - \omega^{\lambda\nu}\gamma_\nu\right) = -4i\omega^{\lambda\mu}\gamma_\mu , \qquad (4.97)$$

where in writing the last step, we have used the property that $\omega_{\mu\nu}$ is anti-symmetric in its indices, Eq. (3.54, *p 53*). Putting this back into Eq. (4.94), we obtain

$$\overline{\psi'}\gamma^\lambda\psi' = \overline{\psi}\gamma^\lambda\psi + \omega^{\lambda\mu}\overline{\psi}\gamma_\mu\psi. \tag{4.98}$$

Comparing Eqs. (4.98) and (4.75), we conclude that the combination $\overline{\psi}\gamma^\lambda\psi$ transforms like a 4-vector under a Lorentz transformation. This is the reason we called this combination V in Eq. (4.93), and this is also why we put a Lorentz index on the set of four constant matrices presented in Eq. (4.45).

Since γ_5 commutes with all $\sigma_{\mu\nu}$'s, the transformation property of the combination $\overline{\psi}\gamma^\lambda\gamma_5\psi$ should be the same as that of $\overline{\psi}\gamma^\lambda\psi$ so far as the proper Lorentz transformations are concerned. Under parity, however, the transformation is different, as we show in Ch. 6. To emphasize this feature, the combination $\overline{\psi}\gamma^\lambda\gamma_5\psi$ is called the *axial vector*, and is denoted by A, as in Eq. (4.93). Finally, the bilinear where $\sigma_{\mu\nu}$ is sandwiched between spinors is called T because it transforms like a rank-2 tensor.

□ **Exercise 4.9** Show that $\overline{\psi}\sigma_{\mu\nu}\psi$ transforms like a rank-2 tensor under proper Lorentz transformations.

4.4.3   Final recipe

Writing down Lorentz invariants can now be reduced to an easy recipe: contract all Lorentz indices that appear anywhere, either with fields, or with derivatives, or in fermion bilinears. This can be used even when we want to write down the Lagrangian including more than one kind of field. For example, if there are two scalar fields ϕ_1 and ϕ_2, any combination of them, like $\phi_1^2\phi_2^2$, should be invariant. With a vector field A^μ, a combination like $A^\mu\phi^\dagger\partial_\mu\phi$ will be invariant. So will be the combination $\overline{\psi}\gamma_\mu\psi A^\mu$, or $\overline{\psi}\sigma_{\mu\nu}\psi F^{\mu\nu}$. Of course, before writing any term in the Lagrangian, we should check its hermiticity. If it is not hermitian, we must add the hermitian conjugate. We should also check whether the term is renormalizable. Explicit examples will be shown shortly.

4.5   Lagrangians for free fields

We have already discussed the equations of motion for free fields. For scalar fields, we should have the Klein–Gordon equation, given in Eq. (4.8). For a Dirac field, we should obtain the Dirac equation, Eq. (4.44). And, for the photon field, the relevant equation is Eq. (4.19), with the source term on the right hand side set to zero.

Notice that all these equations are linear in the fields. If we want to derive them from a Lagrangian via the Euler–Lagrange equations, Eq. (4.69), then the corresponding Lagrangian must be quadratic in the fields. We can therefore think of the possible quadratic terms that one can construct with a given kind of field in order to arrive at the free Lagrangian for the field.

Consider a scalar field ϕ first. The obvious quadratic term is ϕ^2, and this can be one term in the Lagrangian. The Lagrangian can also contain derivatives. A quadratic term involving the derivative is $(\partial_\mu \phi)(\partial^\mu \phi)$. Combining the two terms, we can write the free Lagrangian as

$$\mathscr{L}_0 = \frac{1}{2}(\partial_\mu \phi)(\partial^\mu \phi) - \frac{1}{2}m^2\phi^2 \,. \tag{4.99}$$

If ϕ is a complex scalar field, the free Lagrangian is

$$\mathscr{L}_0 = (\partial_\mu \phi)^\dagger (\partial^\mu \phi) - m^2\phi^\dagger \phi \,. \tag{4.100}$$

In either case, it is straightforward to see that the resulting Euler–Lagrange equation is indeed the Klein–Gordon equation.

☐ **Exercise 4.10** *Verify this statement by deducing the Euler–Lagrange equations from the Lagrangians of Eqs. (4.99) and (4.100).* [**Note :** *For the complex scalar field, there should be two Euler–Lagrange equations, one for ϕ and one for $\phi^\dagger$. Each should be equivalent to the other.*]

☐ **Exercise 4.11** *What are the mass dimensions of the field ϕ and the constant m that appear in the Lagrangian?*

Note that we have put factors of $\frac{1}{2}$ in both terms of Eq. (4.99) but not in Eq. (4.100). An overall factor in the Lagrangian does not affect the Euler–Lagrange equation. Then why was it introduced? The answer is to maintain a consistency with the Lagrangian for the complex scalar field. The complex field can be broken up into its real part and imaginary part in the form

$$\phi = \frac{1}{\sqrt{2}}\left(\phi_1 + i\phi_2\right), \tag{4.101}$$

and here the numerical factor ensures that ϕ is normalized the same way that ϕ_1 and ϕ_2 are. If we put this into Eq. (4.100), the terms involving ϕ_1 and ϕ_2 come with factors of $\frac{1}{2}$, as seen in Eq. (4.99).

For the Dirac field, consider the Lagrangian

$$\mathscr{L}_0 = i\overline{\psi}\gamma^\mu \partial_\mu \psi - m\overline{\psi}\psi \,. \tag{4.102}$$

It is easy to see that it gives the Dirac equation: take $\overline{\psi}$ as the Φ^A that appears in Eq. (4.69) and it follows immediately. But there is an apparent problem, viz., the the first term of Eq. (4.102) is not hermitian. We can make the Lagrangian hermitian by replacing the first term by the average of the term and its hermitian conjugate. That would give

$$\mathscr{L}'_0 = \frac{i}{2}\left(\overline{\psi}\gamma^\mu \partial_\mu \psi - (\partial_\mu \overline{\psi})\gamma^\mu \psi\right) - m\overline{\psi}\psi \,. \tag{4.103}$$

According to the rules we set up earlier, this is the Lagrangian that we should use. However, we should notice that

$$\mathscr{L}_0 - \mathscr{L}'_0 = \frac{i}{2}\left(\overline{\psi}\gamma^\mu \partial_\mu \psi + (\partial_\mu \overline{\psi})\gamma^\mu \psi\right)$$
$$= \partial_\mu\left(\frac{i}{2}\overline{\psi}\gamma^\mu \psi\right), \tag{4.104}$$

i.e., the two forms of the Lagrangians differ by a total derivative. As we said earlier, total derivative terms are irrelevant, so we can use either of them. We will use the Lagrangian of Eq. (4.102), which is simpler.

□ **Exercise 4.12** Find the Euler–Lagrange equation corresponding to the field ψ from the Lagrangian of Eq. (4.102), and show that it is just the hermitian conjugate of the Dirac equation.

□ **Exercise 4.13** Show that Eq. (4.103) also gives the Dirac equation.

We now discuss the photon field. If we find the Lagrangian of an electro-magnetic field classically, we obtain

$$\mathscr{L}_0 = -\frac{1}{4} F_{\mu\nu} F^{\mu\nu} . \tag{4.105}$$

However, in quantum field theory, this Lagrangian gives problems, which will be alluded to in §4.10. So one adds another term to it and writes

$$\mathscr{L}_0 = -\frac{1}{4} F_{\mu\nu} F^{\mu\nu} - \frac{1}{2\xi} (\partial_\mu A^\mu)^2 . \tag{4.106}$$

The extra term is called a *gauge-fixing term*. It sounds like an arbitrary insertion and an insult to the classical wisdom, but actually it is neither, if we insist on the subsidiary condition on the photon field in the form of Eq. (4.21). At the classical level, the added term is zero. At the quantum level, it makes calculation possible. The Lagrangian contains a parameter ξ, which can have arbitrary values. All physical amplitudes should be independent of this parameter.

4.6   Noether currents and charges

We mentioned Noether's theorem earlier in §2.5, and demonstrated how to derive conservation laws from symmetries of the action. We now give a more complete recipe for deriving the conserved quantities, and give some examples from the free Lagrangians encountered in §4.5. We will keep the present discussion confined to internal symmetries only, and will not give details. Details of the deductions, as well as more general discussions including spacetime symmetries, can be obtained in books on quantum field theory.

Internal symmetries involve transformation between fields at a given space-time point. Suppose we have an infinitesimal transformation of the form

$$\Phi^A(x) \longrightarrow \Phi'^A(x) = \Phi^A(x) + \delta\Phi^A(x) . \tag{4.107}$$

Suppose these changes in the fields are inflicted through small changes of a number of parameters, which we denote by θ_r. It can then be shown that the change in the action is given by

$$\delta\mathscr{A} = \int d^4x \sum_r \delta\theta_r \partial_\mu J_r^\mu , \tag{4.108}$$

where

$$J_r^\mu = \sum_A \frac{\partial\mathscr{L}}{\partial(\partial_\mu \Phi^A)} \frac{\delta\Phi^A}{\delta\theta_r} . \tag{4.109}$$

For a symmetry transformation, the action should not change, so that we should obtain

$$\partial_\mu J_r^\mu = 0 \,. \tag{4.110}$$

For each value of the index r, the 4-vector J_r^μ, whose divergence vanishes, is called the *Noether current*. More explicitly, the equation can be written as

$$\frac{\partial J_r^0}{\partial t} + \boldsymbol{\nabla} \cdot \boldsymbol{J}_r = 0 \,. \tag{4.111}$$

Integrating this equation on the entire space and assuming that the fields, and therefore the currents, vanish at infinity, we obtain

$$\frac{dQ_r}{dt} = 0 \,, \tag{4.112}$$

where Q_r is the *Noether charge*, defined as

$$Q_r = \int d^3x \, J_r^0 \,. \tag{4.113}$$

Eq. (4.112) then tells us that the Noether charge is conserved.

As an example, first consider the free Lagrangian of a complex scalar field, given in Eq. (4.100). Clearly, it is invariant under a phase rotation

$$\phi \longrightarrow \phi' = e^{-i\theta}\phi \,, \tag{4.114}$$

where θ is a parameter that is independent of spacetime co-ordinates. The transformation of $\phi^\dagger$ follows from here, but we have to remember that ϕ and $\phi^\dagger$ (or alternatively, ϕ_1 and ϕ_2 introduced in Eq. (4.101)) are really two independent fields in the Lagrangian. For infinitesimal values of the phase parameter, denoted by $\delta\theta$, the changes of these two fields can be written as

$$\delta\phi = -i\delta\theta\,\phi \,, \qquad \delta\phi^\dagger = +i\delta\theta\,\phi^\dagger \,. \tag{4.115}$$

Following Eq. (4.109) now, we can write the Noether current for this transformation as

$$\begin{aligned}
J^\mu &= \frac{\partial \mathscr{L}}{\partial(\partial_\mu\phi)}\frac{\delta\phi}{\delta\theta} + \frac{\partial \mathscr{L}}{\partial(\partial_\mu\phi^\dagger)}\frac{\delta\phi^\dagger}{\delta\theta}\\
&= -i(\partial^\mu\phi^\dagger)\phi + i(\partial^\mu\phi)\phi^\dagger \,.
\end{aligned} \tag{4.116}$$

□ **Exercise 4.14** *The Dirac Lagrangian, given in Eq. (4.102), is invariant under phase rotations of ψ. Show that the corresponding Noether current is given by*

$$J^\mu = \overline{\psi}\gamma^\mu\psi \,. \tag{4.117}$$

4.7   Quantum fields as operators

We mentioned in §4.1 that it is imperative that we have an operator interpretation of quantum fields. What we have done so far is that we have expressed fields as superpositions of plane waves: the scalar field in Eqs. (4.12) and (4.15), the Dirac field in Eq. (4.65), and the photon field in Eq. (4.29). If these expressions should behave as operators, it means that the Fourier components, like $a(\boldsymbol{p})$ of Eq. (4.12) or $d(\boldsymbol{p})$ of Eq. (4.65), should be operators. In fact, this is the reason why we added daggers at different places, where just a complex conjugation sign was expected if we were talking of number-valued functions.

To understand what these operator do, we consider the real scalar field in some detail. From the Lagrangian of Eq. (4.99), we can first find the canonical momenta corresponding to the field ϕ. This will be

$$\Pi = \partial_0 \phi \,. \tag{4.118}$$

Thus the Hamiltonian is given by

$$\mathscr{H} = \Pi \partial_0 \phi - \mathscr{L} = \frac{1}{2}\left[\left(\frac{\partial \phi}{\partial t}\right)^2 + \left(\boldsymbol{\nabla}\phi\right)^2 + m^2\phi^2 \right] \,. \tag{4.119}$$

If we now integrate it over all space to find the total Hamiltonian, and substitute the plane wave expansion of the field ϕ from Eq. (4.12), we obtain

$$H = \int d^3x \, \mathscr{H} = \frac{1}{2}\int d^3p \, E_p\Big(a^\dagger(\boldsymbol{p})a(\boldsymbol{p}) + a(\boldsymbol{p})a^\dagger(\boldsymbol{p}) \Big)\,, \tag{4.120}$$

without assuming anything about the commutation properties of $a(\boldsymbol{p})$ and $a^\dagger(\boldsymbol{p})$. The similarity of this expression with the Hamiltonian of a collection of harmonic oscillators suggests that we declare the operator nature of $a(\boldsymbol{p})$ and $a^\dagger(\boldsymbol{p})$ through the relations

$$\left[a(\boldsymbol{p}), a^\dagger(\boldsymbol{p}') \right] = \delta^3(\boldsymbol{p} - \boldsymbol{p}')\,,$$
$$\left[a(\boldsymbol{p}), a(\boldsymbol{p}') \right] = 0\,,$$
$$\left[a^\dagger(\boldsymbol{p}), a^\dagger(\boldsymbol{p}') \right] = 0\,. \tag{4.121}$$

It is now easy to check the following commutation relations:

$$\left[H, a(\boldsymbol{p}) \right] = -E_p a(\boldsymbol{p})\,, \tag{4.122a}$$
$$\left[H, a^\dagger(\boldsymbol{p}) \right] = +E_p a^\dagger(\boldsymbol{p})\,. \tag{4.122b}$$

Consider now an eigenstate $|\Psi\rangle$ of the total Hamiltonian with some eigenvalue, say $\mathcal{E}$. The relations in Eq. (4.122) imply that

$$H a(\boldsymbol{p}) |\Psi\rangle = (\mathcal{E} - E_p)a(\boldsymbol{p}) |\Psi\rangle \,, \tag{4.123a}$$
$$H a^\dagger(\boldsymbol{p}) |\Psi\rangle = (\mathcal{E} + E_p)a^\dagger(\boldsymbol{p}) |\Psi\rangle \,. \tag{4.123b}$$

In other words, $a(\boldsymbol{p})|\Psi\rangle$ and $a^\dagger(\boldsymbol{p})|\Psi\rangle$ are also eigenstates of H. The action of $a(\boldsymbol{p})$ reduces the eigenvalue by E_p, whereas the action of $a^\dagger(\boldsymbol{p})$ increases the eigenvalue by the same amount. This is interpreted by saying that, acting on a state consisting of a number of particles, the operator $a(\boldsymbol{p})$ annihilates a particle of 3-momentum $\boldsymbol{p}$ from the state, whereas the operator $a^\dagger(\boldsymbol{p})$ creates a particle of 3-momentum $\boldsymbol{p}$ in the state. These particles are the quanta of the field. So $a(\boldsymbol{p})$ is called the *annihilation operator* and $a^\dagger(\boldsymbol{p})$ the *creation operator*.

The problem of negative energy solutions, discussed in §4.1, disappears in this new interpretation. In Eq. (4.12), $a(\boldsymbol{p})$ is associated with the positive energy plane wave solution, and it annihilates a particle with positive energy. In the same equation, $a^\dagger(\boldsymbol{p})$ was associated with the negative energy plane wave solution. In the operator interpretation, it means that it annihilates a negative amount of energy from a state, or in other words, it creates energy into a state. It is the energy *difference* before and after operating with the operator that is positive or negative, not the energy of the state.

For the photon field, the annihilation and creation operators come with a polarization index. The task of the operators is to annihilate or create a photon in the polarization state indicated by the index.

For the complex scalar field, the operator $a(\boldsymbol{p})$ and $a^\dagger(\boldsymbol{p})$ still satisfy the commutation relations of Eq. (4.121). So do their hatted counterparts. And either of the hatted operators commutes with either of the unhatted ones. This means that both daggered operators are creation operators and both undaggered operators are annihilation operators.

Further relation between the two creation operators is obtained by constructing the Noether charge from the conserved current of Eq. (4.116). Once the plane wave expansion of Eq. (4.15) is inserted into this expression, we obtain

$$Q = \int d^3p \left(a^\dagger(\boldsymbol{p})a(\boldsymbol{p}) - \widehat{a}^\dagger(\boldsymbol{p})\widehat{a}(\boldsymbol{p}) \right). \tag{4.124}$$

The combination $a^\dagger a$ is the number operator. So, this equation shows that the quanta created by $\widehat{a}^\dagger(\boldsymbol{p})$ has the opposite conserve charge compared to the quanta created by $a^\dagger(\boldsymbol{p})$. The former kind of quanta are called the *antiparticles* of the latter kind. The field operator ϕ for a complex field then either annihilates a particle, or creates an antiparticle. Exactly the opposite is done by the operator $\phi^\dagger$.

The interpretation is the same for the Dirac field operator, for which the operator $d(\boldsymbol{p})$ annihilates a particle of momentum $\boldsymbol{p}$, whereas $d^\dagger(\boldsymbol{p})$ creates an antiparticle of the same momentum. The opposite is done by $\overline{\psi}$, because from Eq. (4.65) we get

$$\overline{\psi}(x) = \sum_s \int D^3p \left(d_s^\dagger(\boldsymbol{p})\overline{u}_s(\boldsymbol{p})e^{+ip\cdot x} + \widehat{d}_s(\boldsymbol{p})\overline{v}_s(\boldsymbol{p})e^{-ip\cdot x} \right), \tag{4.125}$$

which involves the operators $d_s^\dagger(\boldsymbol{p})$ and $\widehat{d}_s(\boldsymbol{p})$. The difference, for the case of fermions, is that these operators do not obey any commutation relation. Rather, they obey the anticommutation relations

$$\left[d_s(\boldsymbol{p}), d_s^\dagger(\boldsymbol{p}')\right]_+ = \delta^3(\boldsymbol{p} - \boldsymbol{p}'), \qquad (4.126a)$$

$$\left[\widehat{d}_s(\boldsymbol{p}), \widehat{d}_s^\dagger(\boldsymbol{p}')\right]_+ = \delta^3(\boldsymbol{p} - \boldsymbol{p}'), \qquad (4.126b)$$

all other anticommutators being zero.

□ **Exercise 4.15** Show that the commutation relations of Eq. (4.121) imply the following equal-time commutation relations on the field $\phi(x)$:

$$\left[\phi(x), \partial_0\phi(y)\right]\bigg|_{x_0=y_0} = i\delta^3(\boldsymbol{x} - \boldsymbol{y}). \qquad (4.127)$$

[**Note :** *From the standpoint of quantum field theory, this relation is more fundamental. The right hand side of the commutator of creation and annihilation operators depend on the choice of our integration measure given in Eq. (4.13).*]

□ **Exercise 4.16** For a fermion field, show that the anticommutation relations of Eq. (4.126) imply the following anticommutation relation for the field operator:

$$\left[\psi(x), \psi^\dagger(y)\right]_+\bigg|_{x_0=y_0} = \delta^3(\boldsymbol{x} - \boldsymbol{y}), \qquad (4.128)$$

whereas the anticommutator between $\psi(x)$ and $\psi(y)$ vanishes, and so does the anticommutator between two daggered field operators.

4.8   States

The fields are operators. What do they operate on?

Of course they act on states. States consist of particles: any number of them, any kind of them. The simplest state is the one which contains no particle. This is called the *vacuum state*, defined by

$$a(\boldsymbol{p})\,|0\rangle = 0, \qquad (4.129)$$

where in this equation, $a(\boldsymbol{p})$ is a generic notation for the annihilation operator of any field corresponding to any momentum $\boldsymbol{p}$.

A one-particle state should be defined by the action of a creation operator on the vacuum, i.e., should be of the form

$$|A(\boldsymbol{p})\rangle = N_{\boldsymbol{p}} a^\dagger(\boldsymbol{p})\,|0\rangle, \qquad (4.130)$$

where $a^\dagger$ denotes the creation operator for the particle called A, and $N_{\boldsymbol{p}}$ is a normalization constant. The definition of states with more particles, including particles of different types, should be obvious.

The normalization of states with one or more particles has to be dealt with care. The vacuum state can be simply normalized by imposing the condition

$$\langle 0 | 0 \rangle = 1 \,. \tag{4.131}$$

The same cannot be done for states with particles. To appreciate the point, consider the inner product of two one-particle states. The definition of Eq. (4.130) gives

$$\langle A(\boldsymbol{p}') | A(\boldsymbol{p}) \rangle = N_{\boldsymbol{p}'}^* N_{\boldsymbol{p}} \langle 0 | a(\boldsymbol{p}') a^\dagger(\boldsymbol{p}) | 0 \rangle \,. \tag{4.132}$$

Using the commutation or anticommutation relation between the creation and annihilation operators depending on whether the particle in question is a boson or a fermion, and then using Eqs. (4.129) and (4.131), we obtain

$$\langle A(\boldsymbol{p}) | A(\boldsymbol{p}') \rangle = N_{\boldsymbol{p}'}^* N_{\boldsymbol{p}} \delta^3(\boldsymbol{p} - \boldsymbol{p}') \,. \tag{4.133}$$

This is problematic when we look at the normalization of a state, i.e., when $\boldsymbol{p} = \boldsymbol{p}'$. The right-hand side of Eq. (4.133) then contains a delta function with vanishing argument, which is not defined. However, note that

$$\delta^3(\boldsymbol{q}) = \int d^3x \, e^{i \boldsymbol{q} \cdot \boldsymbol{x}} \,. \tag{4.134}$$

So, for $\boldsymbol{q} = 0$, the result of the integral on the right hand side is the volume of the space. Thus, the problem with normalization can be avoided if we define our theory within a finite (but very large) volume $\mathscr{V}$. Choosing

$$N_{\boldsymbol{p}} = \sqrt{2E_p} \,, \tag{4.135}$$

we find that Eq. (4.133) gives the normalization condition

$$\langle A(\boldsymbol{p}) | A(\boldsymbol{p}) \rangle = 2E_p \mathscr{V} \,. \tag{4.136}$$

Note that this implies that, unlike the vacuum state which is dimensionless, any one-particle state has a non-trivial mass dimension:

$$\dim | A(\boldsymbol{p}) \rangle = -1 \,. \tag{4.137}$$

The arbitrary volume $\mathscr{V}$ taken for the purpose of normalization should drop out of the calculation of any physically measurable quantity.

4.9 Interactions

Let us look back at the free Lagrangians presented in §4.5. As mentioned there, these Lagrangians are all quadratic in fields. The operator interpretation shows why this is so. Consider, e.g., the Lagrangian of real scalar fields in Eq. (4.99). Both terms contain the field operator combination $\phi\phi$,

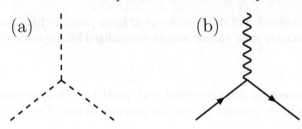

Figure 4.1: Interactions in diagrammatic form. Diagram (a) corresponds to the ϕ^3 interaction, diagram (b) to the interaction $\overline{\psi}\gamma^\mu\psi A_\mu$.

with derivatives or with constants which are not important for this discussion. Each field operator ϕ either creates or annihilates a particle, as discussed. If it acts on a one-particle state, the combination will annihilate the particle and create it again. Thus, in effect, nothing happens at any point: the particle just moves on.

If we consider any term in the Lagrangian which has three or more field operators, the situation is different. Consider, for the sake of concreteness, a term of the form ϕ^3 with a real scalar field ϕ. If it acts on a state with two particles, it can annihilate them and create a new one. Or it can act on a state of one particle and the final result will be a state with two particles. In any case, it signifies a non-trivial event. Such terms are therefore interaction terms.

Or consider a term $\overline{\psi}\gamma^\mu\psi A_\mu$ in the Lagrangian, where ψ is a Dirac field and A_μ is the photon field. Let us recall, in a tabular form, what the different field operators can do.

Operator	Can annihilate	Can create
ψ	a fermion	an antifermion
$\overline{\psi}$	an antifermion	a fermion
A_μ	a photon	a photon

Overall, there will be eight possibilities, corresponding to one entry from each row. For example, the interaction might annihilate a fermion and a photon, and create a fermion. Or may be it can annihilate a fermion-antifermion pair into a photon. And so on.

We can summarize all such information in a diagrammatic language. Fig. 4.1 shows the basic building blocks of this language. For example, the diagram to the left shows three lines meeting at a point, implying that the interaction at that point would involve creation and annihilation of three particles. Dashed lines conventionally denote spin-0 particles. In the diagram on the right hand side, the wavy line signifies a photon being created or annihilated. The solid line, with an outgoing arrow, stands for the operator $\overline{\psi}$: a fermion created or an antifermion annihilated. And the solid line with an

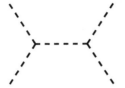

Figure 4.2: Two vertices of ϕ^3 interaction making a Feynman diagram for a two-particle elastic scattering.

incoming arrow represents the work of ψ: annihilation of a fermion or creation of an antifermion.

It should be pointed out that these representations of basic interaction vertices in Fig. 4.1 cannot correspond to physical processes. Consider, e.g., the diagram in Fig. 4.1a. Each leg represents a particle being created or annihilated. It would be obviously impossible to have all three legs correspond to annihilation, because then the final state would be the vacuum. The vacuum has zero energy, whereas each initial particle must carry some positive energy, so energy cannot be conserved. For the same reason, one cannot produce three particles from the vacuum. A little more effort shows that $1 \to 2$ or $2 \to 1$ processes are also impossible.

☐ **Exercise 4.17** *Show that the vertex of Fig. 4.1a cannot give a physical process of one particle decaying into two.* [**Hint :** *Consider energy conservation in the rest frame of the decaying particle.*]

☐ **Exercise 4.18** ♩ For the vertex of Fig. 4.1b, we mentioned eight possibilities. Show that none of them can correspond to physical processes. [**Hint :** *It is energy conservation once again.*]

But these diagrams are not totally useless for this reason. Problems in quantum field theory are usually addressed through perturbation theory. The free Lagrangian, i.e., the part of the Lagrangian that is quadratic in the field, is taken as the unperturbed Lagrangian, and the coupling terms are treated as perturbations on it. When a proposed diagram has only one vertex, that can be seen as a first order perturbation result. What we have said above can be summarized by saying that in the first order of perturbation theory, the interactions that we considered do not give rise to any physical effect.

We can then try second order in perturbation theory. The result can be represented in the form of a diagram with two interaction vertices. In higher and higher orders of perturbation theory, we can use these basic interaction vertices as units to build up more complicated diagrams which can represent physical processes. Such diagrams are called *Feynman diagrams*. For example, consider Fig. 4.2. It contains two vertices of ϕ^3 interaction. If we consider that lines to the left correspond to incoming particles and the lines to the right outgoing particles, the diagram represents a scattering process where there are two particles in the initial as well as in the final state. This is an elastic scattering since all lines correspond to the same particle, so that the initial and final masses are the same.

The problem about energy conservation encountered with Fig. 4.1 is evaded here because at each vertex there is one line which does not correspond to any of the initial or final particles, and some rules that apply for physical particles need not apply for them. As a concrete example, consider the left vertex of Fig. 4.2. Let the two left lines correspond to incoming particles, with 4-momenta p^μ and p'^μ, which satisfy the relations $p^2 = p'^2 = m^2$. The particle line in the middle will then have a 4-momentum $(p + p')^\mu$, but

$$(p + p')^2 = p^2 + p'^2 + 2p \cdot p' = 2(m^2 + p \cdot p'), \qquad (4.138)$$

which is not equal to m^2. Thus, the particle in the intermediate line does not satisfy the energy-momentum relation. Then so be it! We cannot perform measurements on that particle, so there is no reason to assume that it satisfies the energy-momentum relation. Of course, this can happen only for intermediate lines, and the particles in these lines are called *virtual particles* or *off-shell particles*. In contrast, particles which satisfy the energy-momentum relation are called either *physical* or *on-shell*.

4.10   From Lagrangian to Feynman rules

This entire journey that we are undertaking, from a Lagrangian to the Feynman diagrams, is intended for calculating amplitudes and rates for physical processes. A Feynman diagram for a certain physical process corresponds to a contribution to the amplitude for that process. And this contribution can be written down just by looking at the diagram if we remember a set of rules, or a recipe for doing so. These rules are called *Feynman rules*. There are different kinds of elements in a diagram: external lines, internal lines, vertices, loops and so on. Each of them contributes a factor to the amplitude. We discuss the rules by such categories.

4.10.1   External lines

Roughly speaking, the Feynman rule for an external line is obtained by the following steps. First, we need to look at the plane wave expansion of the corresponding field and identify the relevant term. For example, if the line corresponds to an incoming line, the relevant term is the one which contains the annihilation operator for the particle. Once this is done, we discard the exponential factor, the creation or annihilation operator and the integration measure that we called D^3p. Whatever is left in the term is the Feynman rule for the external line.

For example, look at the complex scalar field in Eq. (4.15). Once we discard the parts mentioned just above, nothing is left in either term. Thus,

Table 4.1: Feynman rules for external lines.

Type of particle	Feynman rule for	
	incoming	outgoing
scalar	1	1
fermion	$u_{\boldsymbol{p},s}$	$\bar{u}_{\boldsymbol{p},s}$
antifermion	$\bar{v}_{\boldsymbol{p},s}$	$v_{\boldsymbol{p},s}$
photon	$\epsilon_r^\mu(\boldsymbol{p})$	$\epsilon_r^{\mu*}(\boldsymbol{p})$

the Feynman rule is nothing, or just a factor of 1. For the Dirac field, however, the spinors will remain, and for the photon field, the polarization vector. We summarize all these results in Table 4.1.

4.10.2 Internal lines

The Feynman rule for an internal line contains the *propagator* for that line. We describe how to obtain this for a generic field Φ. Take the free Lagrangian for a field and write it in the form

$$\mathscr{L}_0 = \frac{1}{2}\Phi\mathcal{O}\Phi + \text{total derivatives} \qquad (4.139)$$

if Φ is real (i.e., hermitian), or as

$$\mathscr{L}_0 = \overline{\Phi}\mathcal{O}\Phi + \text{total derivatives} \qquad (4.140)$$

if Φ is a complex field, and $\overline{\Phi}$ is the conjugation of Φ defined in a way that $\overline{\Phi}\Phi$ is Lorentz invariant. Thus, for example, for scalar fields $\overline{\Phi}$ is simply $\Phi^\dagger$, whereas for a Dirac field, it is the combination defined in Eq. (4.85).

$\mathcal{O}$ will invariably contain differential operators, coming from the derivatives in the Lagrangian. Replace all derivatives ∂_μ in it by $-ip_\mu$ (since $p_\mu = i\partial_\mu$), and take the inverse of $\mathcal{O}$. That is the propagator. The Feynman rule is i times the propagator.

As an example, consider the free Lagrangian of a complex scalar field, given in Eq. (4.100). We can write it as

$$\mathscr{L}_0 = \partial_\mu\left(\phi^\dagger\partial^\mu\phi\right) - \phi^\dagger\Box\phi - m^2\phi^\dagger\phi\,, \qquad (4.141)$$

so that we identify

$$\mathcal{O} = -\Box - m^2\,. \qquad (4.142)$$

The Feynman rule for a scalar internal line will then be given by

$$i\Delta_F(p) = \frac{i}{p^2 - m^2}\,. \qquad (4.143)$$

For a Dirac field, Eq. (4.102) shows that

$$\mathcal{O} = i\gamma^\mu \partial_\mu - m\,, \tag{4.144}$$

and therefore the propagator should be the inverse of $(\not p - m)$. It means that the propagator is a matrix which, when multiplied by $(\not p - m)$, gives the unit matrix. A little reflection and the use of Eq. (4.61) show that the Feynman rule for an internal fermion line is

$$iS_F(p) = i\,\frac{\not p + m}{p^2 - m^2}\,. \tag{4.145}$$

For the photon field, we start from the Lagrangian of Eq. (4.106) and rewrite it as

$$\mathcal{L}_0 = -\frac{1}{2}\partial_\mu A_\nu(\partial^\mu A^\nu - \partial^\nu A^\mu) - \frac{1}{2\xi}(\partial_\nu A^\nu)(\partial_\mu A^\mu)$$

$$= \frac{1}{2}A_\nu\left(g^{\mu\nu}\Box - (1-\frac{1}{\xi})\partial^\mu\partial^\nu\right)A_\mu + \partial_\mu(\cdots)\,. \tag{4.146}$$

The propagator is then given by $D_{\mu\lambda}$, where

$$\left(-g^{\mu\nu}p^2 + (1-\frac{1}{\xi})p^\mu p^\nu\right)D_{\mu\lambda} = \delta^\nu_\lambda\,. \tag{4.147}$$

Writing $D_{\mu\lambda}(p) = ag_{\mu\lambda} + bp_\mu p_\lambda$, we can solve the co-efficients a and b and obtain the Feynman rule for an internal photon line as

$$iD_{\mu\lambda}(p) = -\frac{i}{p^2}\left(g_{\mu\lambda} - (1-\xi)\frac{p_\mu p_\lambda}{p^2}\right)\,. \tag{4.148}$$

Note what happens if we take $\xi \to \infty$: the propagator cannot be defined. This is the reason why we had to introduce the gauge-fixing term in Eq. (4.106). We cannot do without this term, i.e., we must use some finite value of ξ so that the gauge-fixing term does not vanish. Since the final results do not depend on the value of ξ, we will take $\xi = 1$ so that the photon propagator is given by

$$iD_{\mu\lambda}(p) = -\frac{ig_{\mu\lambda}}{p^2}\,. \tag{4.149}$$

This is called the 't Hooft–Feynman gauge.

□ **Exercise 4.19** Go through the same procedure without the gauge-fixing term, and show that the procedure of finding the inverse of $\mathcal{O}$ fails.

□ **Exercise 4.20** Consider a Proca field, for which the free Lagrangian can be written as

$$\mathcal{L}_0 = -\frac{1}{4}F_{\mu\nu}F^{\mu\nu} + \frac{1}{2}M^2 A_\mu A^\mu\,. \tag{4.150}$$

Show that in this case, the propagator can be obtained without adding any gauge-fixing term, and the propagator is given by

$$D_{\mu\nu}(p) = \frac{1}{p^2 - M^2}\left(-g_{\mu\nu} + \frac{p_\mu p_\nu}{M^2}\right)\,. \tag{4.151}$$

4.10.3 Vertices

The Feynman rule for a vertex depends on the interaction term in the Lagrangian that the vertex comes from. Take the term, strip it of all field operators, and multiply by i to obtain the Feynman rule for the vertex. If there are n factors of any of the fields for $n > 1$, multiply by a factor $n!$. The following examples will make the algorithm clear:

$$
\begin{array}{ll}
\text{Interaction term} & \text{Feynman rule} \\
\text{in Lagrangian} & \text{for vertex} \\
\hline
h\bar{\psi}\psi\phi & ih \\
h\bar{\psi}\gamma_5\psi\phi & ih\gamma_5 \\
-\lambda\phi^4 & -4!i\lambda \\
\hline
\end{array}
\qquad (4.152)
$$

If an interaction term contains the derivative of any field, the corresponding Feynman rule will contain a factor of momentum. To see how this rule follows from the prescription given above for the vertices, suppose there is an interaction term that contains $\partial_\mu\phi$, where ϕ is a complex scalar field. From the plane wave expansion of the field given in Eq. (4.15), we obtain

$$
\partial_\mu\phi(x) = \int D^3p \left(-ip_\mu a(\boldsymbol{p})e^{-ip\cdot x} + ip_\mu \hat{a}^\dagger(\boldsymbol{p})e^{+ip\cdot x} \right). \qquad (4.153)
$$

This means that, apart from other factors that would come from other field operators in this interaction term, there should be a factor $-ip_\mu$ for a particle annihilated, or a factor $+ip_\mu$ for an antiparticle created. Said another way, the factor will be $-ip_\mu$ if p_μ is the 4-momentum in the direction of the charge carried by the particle.

4.10.4 Other factors

There are various kinds of other factors that can enter the amplitude. Here is a list.

Loop factors : For every independent loop in a Feynman diagram, there must be one 4-momentum that is not determined by the momenta of external particles. So there will be one arbitrary momentum for each loop, and we should integrate over all possible values of this momentum, with a factor of

$$
\int \frac{d^4q}{(2\pi)^4}. \qquad (4.154)
$$

Antisymmetry factors : Fermion creation and annihilation operators anticommute, as we have seen in Eq. (4.126). There are two consequences of this fact in Feynman rules.

1. If there is a closed fermion loop in the diagram, one should put an overall factor of -1 for it and take a trace of the Dirac matrices occurring in the expression.

2. If two diagrams differ by the interchange of two fermion lines only, there must be an extra negative sign between the two diagrams.

Symmetry factors : If, in a diagram, there is a set of n internal lines corresponding to the same field which begin and end on the same two vertices, there should be an extra factor of $1/n!$ for the amplitude of the diagram. For lines with complex fields, it means that the arrows all point from one vertex to another. For real fields, arrows are irrelevant. The reason for this factor will be explained with an example in §4.12.2.

Another passing comment. The exponential co-ordinate dependent factors of the plane wave expansion seem not to appear anywhere in this scheme. Actually, they do, in a quiet way. All such factors corresponding to the lines meeting at a vertex ensure that the 4-momentum is conserved at the vertex.

4.10.5   Feynman amplitude

When we calculate the rate of any process, we first draw the Feynman diagrams corresponding to the process. Then we look for the Feynman rules for the geometrical elements that constitute any diagram: external lines, internal lines, vertices, and so on. We string together the Feynman rules corresponding to all such geometrical elements to form the *Feynman amplitude* of a process. To be more precise, the product of the Feynman rules of all geometrical elements gives i times the Feynman amplitude $\mathscr{M}$ for the diagram. If a physical process has many diagrams, the Feynman amplitudes of different diagrams should be added.

For example, consider the diagram of Fig. 4.2 *(p 89)*. The geometrical elements that we talked about are of the following nature for this diagram:

- Four external lines: according to the rules given in Table 4.1 *(p 91)*, there is a factor of 1 for each such line.

- Two vertices: if the interaction term in the Lagrangian is $-\mu\phi^3$, the Feynman rule for each vertex should be $-6i\mu$.

- One internal line: its Feynman rule should be its propagator.

Combining all these factors, we find that the Feynman amplitude for the diagram is given by

$$i\mathscr{M} = (-6i\mu)^2 \, \frac{i}{k^2 - m^2}, \qquad (4.155)$$

where m is the mass of the particle. The 4-momentum k of the internal line is determined by those of the external lines. For example, if the two

external lines to the left of the diagram represent two particles coming in
with 4-momenta p and q, we should write $k = p + q$.

It should be noted that the Feynman amplitude is not the transition amplitude in the quantum mechanical sense, whose modulus squared gives the probability of transition. Transition probability must be dimensionless, whereas
Feynman amplitudes are not so in general (see Ex. 4.23 *(p 100)* below). The
quantum mechanical amplitude can be obtained by multiplying the Feynman
amplitude by some standard factors. These factors do not depend on the interactions, so they can be easily separated out, and the formulas for rates of
physical processes can be written directly in terms of the Feynman amplitude,
which is what we do in some of the remaining sections of this chapter.

4.11   Calculation of decay rates

Armed with the Feynman rules, we can write down the Feynman amplitude
of a diagram. The amplitude for a process will be the sum of amplitudes of
all diagrams that contribute to the process. Once this amplitude is obtained,
we can use it to find rates of various physical processes.

4.11.1   General formula

The simplest kind of processes that we consider are the ones where the initial
state has just one particle. In the final state, there can be any number of
them. The process would then correspond to the decay of the initial particle
into the final ones. The rate of the decay, i.e., the inverse of the lifetime, is
given by

$$\Gamma = \frac{1}{2E} \left(\prod_a \int \frac{d^3 p_a'}{(2\pi)^3 2E_a'} \right) (2\pi)^4 \delta^4 \left(p - \sum_a p_a' \right) |\mathscr{M}|^2, \qquad (4.156)$$

where $\mathscr{M}$ is the Feynman amplitude. We have denoted the properties of final
state particles with primed parameters. The initial particle has an energy E.

We want to note an important feature of this formula. The Feynman
amplitude for any process is Lorentz invariant. The δ-function, of course, is a
Lorentz invariant function because if two momenta, say p^μ and q^μ, are equal
in one frame of reference, they will be equal in every frame. Moreover, the
integration measures are also Lorentz invariant, as indicated in Ex. 4.21.

This implies two things. First, we can perform the integration in a frame
that is most convenient for us. To obtain the decay rate in a particular frame,
one merely has to put in the value of the energy of the initial particle that
is appropriate for that frame. Second, suppose we consider the decay rate in
the rest frame of the decaying particle and call it Γ_0. In any other frame, the
decay rate will be given by

$$\frac{\Gamma_0}{\Gamma} = \frac{E}{m}. \qquad (4.157)$$

Since lifetime is the inverse of Γ, this shows the time dilation of lifetime.

□ **Exercise 4.21** *Convince yourself that the integration measure appearing in Eq. (4.156) is Lorentz invariant by showing that*

$$\int d^4p\, \delta(p^2 - m^2)\Theta(p_0) = \int \frac{d^3p}{2E_p}, \qquad (4.158)$$

where Θ is the step function which is zero for negative arguments and 1 for positive arguments. Next, show that $\Theta(p_0)$ is Lorentz invariant for an on-shell particle satisfying $p^2 = m^2$, so that the entire left hand side is Lorentz invariant.

□ **Exercise 4.22** *Verify that, for one particle decaying into N particles, the mass dimension of the Feynman amplitude $\mathcal{M}$ is $3 - N$.*

4.11.2   Two-body decays

We will often discuss two-body decays, so it would be convenient to have a more processed form of the result for such cases. Of course, without the knowledge of the amplitude, we cannot really perform the integral that appears in Eq. (4.156). But we can at least try to see how much progress we can make without knowing the amplitude.

We will find out the decay rate in the rest frame of the parent particle, a quantity which we have called Γ_0. Suppose the mass of the parent particle is m, and that of the product particles are m_1' and m_2'. Then,

$$\begin{aligned}
\Gamma_0 &= \frac{1}{2m} \int \frac{d^3p_1'}{(2\pi)^3 2E_1'} \int \frac{d^3p_2'}{(2\pi)^3 2E_2'} (2\pi)^4 \delta^4(p - p_1' - p_2') \big|\mathcal{M}\big|^2 \\
&= \frac{1}{32\pi^2 m} \int \frac{d^3p_1'}{E_1'} \int \frac{d^3p_2'}{E_2'} \delta^4(p - p_1' - p_2') \big|\mathcal{M}\big|^2 .
\end{aligned} \qquad (4.159)$$

We can easily perform the integration over p_2' by using the delta function and obtain

$$\Gamma_0 = \frac{1}{32\pi^2 m} \int \frac{d^3p_1'}{E_1' E_2'} \delta(m - E_1' - E_2') \big|\mathcal{M}\big|^2 . \qquad (4.160)$$

We can write

$$d^3p_1' = d\Omega\, dp'\, p'^2 , \qquad (4.161)$$

showing explicitly the angular and the magnitude variables in the integration measure. Here p' denotes the magnitude of either of the outgoing 3-momenta, both being equal in the rest frame of the decaying particle. We can then take the differential of the energy-momentum relation for a free particle to obtain

$$p'dp' = E'dE' , \qquad (4.162)$$

where E' can be either E_1' or E_2'. Putting these things, we obtain

$$\frac{d\Gamma_0}{d\Omega} = \frac{1}{32\pi^2 m} \int \frac{dE_1'\, p'}{E_2'} \delta(m - E_1' - E_2') \big|\mathcal{M}\big|^2 . \qquad (4.163)$$

Here, E_2' should be treated as a function of E_1' through the relation

$$E_2'^2 = E_1'^2 - m_1'^2 + m_2'^2, \tag{4.164}$$

which follows from the energy-momentum relation of free particles. Thus, the argument of the delta function is a function of E_1'. We can simplify it by using the formula for the delta function of a function of a variable,

$$\delta\big(f(x)\big) = \sum_a \frac{\delta(x - x_a)}{|df/dx|} \tag{4.165}$$

where the quantities x_a are the solutions of the equation $f(x_a) = 0$. Solution of E_1', subject to the condition imposed by the delta function, was derived in Eq. (2.65, *p 27*). Using it, we can write

$$
\begin{aligned}
\delta(m - E_1' - E_2') &= \frac{1}{\left|1 + \frac{dE_2'}{dE_1'}\right|} \delta\left(E_1' - \frac{m^2 + m_1^2 - m_2^2}{2m}\right) \\
&= \frac{E_2'}{m}\, \delta\left(E_1' - \frac{m^2 + m_1^2 - m_2^2}{2m}\right).
\end{aligned}
\tag{4.166}
$$

Putting this back and performing the integration on E_1', we obtain

$$\frac{d\Gamma_0}{d\Omega} = \frac{1}{32\pi^2 m}\frac{\mathbf{p}'}{m}\,\overline{|\mathcal{M}|^2}. \tag{4.167}$$

Substituting the solution for the common 3-momentum from Eq. (2.64, *p 27*), we can write this as

$$\frac{d\Gamma_0}{d\Omega} = \frac{\overline{|\mathcal{M}|^2}}{64\pi^2 m}\sqrt{\left[1 - \left(\frac{m_1' + m_2'}{m}\right)^2\right]\left[1 - \left(\frac{m_1' - m_2'}{m}\right)^2\right]}. \tag{4.168}$$

Once we know the amplitude, we can perform the integration over the angles and find the decay rate. However, it should be remembered that the quantity $\overline{|\mathcal{M}|^2}$ that appears in Eq. (4.168) has been obtained after performing all delta function integrations which imply energy-momentum conservation. Thus, in the expression for $\overline{|\mathcal{M}|^2}$, we should put in all constraints that come from energy-momentum conservation before we attempt to perform the angular integrations.

4.11.3 Three-body decays

Let us now look for some insights from three-body decays. Starting from the general formula of Eq. (4.156), we can write

$$\Gamma_0 = \frac{1}{(2\pi)^5}\frac{1}{16m}\int \frac{d^3 p_1'}{E_1'}\int \frac{d^3 p_2'}{E_2'}\,\delta(m - E_1' - E_2' - E_3')\frac{\overline{|\mathcal{M}|^2}}{E_3'} \tag{4.169}$$

in the rest frame of the decaying particle. In writing this form, we have already integrated over the 3-momentum of the third particle, so that we can impose

$$\boldsymbol{p}_1' + \boldsymbol{p}_2' + \boldsymbol{p}_3' = 0 \,. \tag{4.170}$$

Clearly, for fixed energies of the decay products, the angles between the momenta of the products are determined. Thus, there are only two independent kinematical parameters in the final state, which can be taken to be E_1' and E_2'. The other energy, E_3', is determined by energy conservation, i.e., from the delta function that remains in the above expression for the decay rate.

We can use Eq. (4.170) to write the argument of the delta function appearing in Eq. (4.169) as

$$m - E_1' - E_2' - \sqrt{\boldsymbol{p}_1'^2 + \boldsymbol{p}_2'^2 + 2\boldsymbol{p}_1'\boldsymbol{p}_2'\cos\theta_{12} + m_3^2} \,, \tag{4.171}$$

where θ_{12} is the angle between $\boldsymbol{p}_1'$ and $\boldsymbol{p}_2'$. Using Eq. (4.165) now, we can write the delta function as

$$\frac{E_3'}{\boldsymbol{p}_1'\boldsymbol{p}_2'} \, \delta\left(\cos\theta_{12} - \frac{\boldsymbol{p}_3'^2 - \boldsymbol{p}_1'^2 - \boldsymbol{p}_2'^2}{2\boldsymbol{p}_1'\boldsymbol{p}_2'}\right) \,. \tag{4.172}$$

Using Eq. (4.162), we can then write the decay rate in the form

$$\Gamma_0 = \frac{1}{(2\pi)^5} \frac{1}{16m} \int dE_1' d\Omega_1' \int dE_2' d\Omega_2'$$

$$\times \delta\left(\cos\theta_{12} - \frac{\boldsymbol{p}_3'^2 - \boldsymbol{p}_1'^2 - \boldsymbol{p}_2'^2}{2\boldsymbol{p}_1'\boldsymbol{p}_2'}\right) \overline{|\mathcal{M}|^2} \,. \tag{4.173}$$

The integration over the angular variables can be performed trivially. As we said earlier, only the angle θ_{12} has any kinematical significance. Its integration gives 1 because of the delta function. The other three angles present in $d\Omega_1'$ and $d\Omega_2'$ merely tell us the orientation of the plane containing the final state momenta $\boldsymbol{p}_1'$, $\boldsymbol{p}_2'$ and $\boldsymbol{p}_3'$. Their integration gives a factor $2(2\pi)^2$, so that

$$\Gamma_0 = \frac{1}{(2\pi)^3} \frac{1}{8m} \int dE_1' \int dE_2' \, \overline{|\mathcal{M}|^2} \,. \tag{4.174}$$

This expression can be written in an alternative form by using, instead of E_1' and E_2' as independent variables, the variables like m_{12}^2 which were defined in §2.6.3. Using the expression of m_{23}^2 similar to that given in Eq. (2.74, *p 29*), we see that in the rest frame of the decaying particle, one obtains

$$dm_{23}^2 = -2m \, dE_1' \tag{4.175}$$

etc. Thus, Eq. (4.174) can also be written as

$$\Gamma_0 = \frac{1}{(2\pi)^3} \frac{1}{32m^3} \int dm_{23}^2 \int dm_{13}^2 \, \overline{|\mathcal{M}|^2} \,. \tag{4.176}$$

In fact, it is inconsequential which two quantities of the form m_{AB}^2 we use as integration variables, since the sum of the three is a constant, as indicated in Eq. (2.76, *p 29*).

Eq. (4.176) shows something quite interesting. In §2.6.3, we discussed Dalitz plots. We found that, if we find out the values of dm_{23}^2 and dm_{13}^2 for a particular decay event and plot them as points in a graph using those two variables as the two axes, the points would always lie within a certain region, schematically shown in Fig. 2.1 *(p 30)*, whose boundaries depend on the masses of the decaying particle and its products. We now see from Eq. (4.176) that the density of the points in a certain region of the graph is proportional to the value of $\overline{|\mathcal{M}|^2}$ for the decay. For example, if the Feynman amplitude is a constant, independent of the momenta of the decay products, then the points would be evenly spread on the Dalitz plot. Deviation from the uniform distribution provides information about momentum dependence of the Feynman amplitude.

4.12 Calculation of cross-sections

4.12.1 General formula

Decay processes contain one particle in the initial state. The next natural issues to consider are processes which contain two particles in the initial state. These are scattering processes. The rate of occurrence of such a process per unit volume is given by

$$\frac{\Gamma}{V} = n_1 n_2 v_{\text{rel}} \sigma, \tag{4.177}$$

where the two reactants have number densities n_1 and n_2, and v_{rel} is the relative velocity between them. The characteristics of the interaction are buried in the quantity σ, called *scattering cross-section*. It is given by

$$\sigma = \frac{1}{4\sqrt{(p_1 \cdot p_2)^2 - m_1^2 m_2^2}} \left(\prod_a \int \frac{d^3 p_a'}{(2\pi)^3 2E_a'} \right)$$
$$\times (2\pi)^4 \delta^4 (p_1 + p_2 - \sum_a p_a') |\mathcal{M}|^2, \tag{4.178}$$

where p_1^μ and p_2^μ represent the 4-momenta of the incoming particles, and the index a counts particles in the final state. Recalling the discussion following Eq. (4.156), we can conclude that the integration in these formulas gives a Lorentz invariant result. Thus, the cross-section itself is Lorentz invariant.

The first factor on the right hand side of Eq. (4.178) depends only on the properties of initial state particles, and can therefore be called the *initial state factor*. Calculations of cross-sections are most often done in the CM frame introduced in §2.7. In this frame, $\boldsymbol{p}_1$ and $\boldsymbol{p}_2$ are collinear. The collinearity is

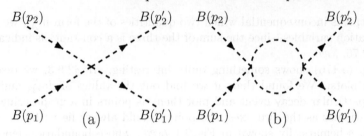

Figure 4.3: Simplest diagrams for elastic scattering of scalar particles having a ϕ^4 interaction.

also trivially true in the fixed target (FT) frame, where one of the 3-momenta is zero. For such frames, it is easy to see that the prefactor of Eq. (4.178) can be written as

$$4\sqrt{(p_1 \cdot p_2)^2 - m_1^2 m_2^2} = 4E_1 E_2 \left| \frac{\boldsymbol{p}_1}{E_1} - \frac{\boldsymbol{p}_2}{E_2} \right| = 4E_1 E_2 v_{\text{rel}}. \qquad (4.179)$$

We will often use this latter form in calculations.

Most of the times, we will be concerned with the case where the final state contains two particles as well, so it is worthwhile to write down the formula for that case explicitly. It is

$$\sigma = \frac{1}{v_{\text{rel}}} \frac{1}{2E_1} \frac{1}{2E_2} \int \frac{d^3 p_1'}{(2\pi)^3 2E_1'} \int \frac{d^3 p_2'}{(2\pi)^3 2E_2'}$$
$$\times (2\pi)^4 \delta^4(p_1 + p_2 - p_1' - p_2') \left| \mathscr{M} \right|^2. \qquad (4.180)$$

☐ **Exercise 4.23** As a generalization of Ex. 4.22 (p 96), show that if a process involves N_i particles in the initial state and N_f in the final state, the Feynman amplitude for the process will have mass dimension $4 - (N_i + N_f)$.

4.12.2   Illustrative example

As an example, consider the elastic scattering process

$$B(p_1) + B(p_2) \to B(p_1') + B(p_2'), \qquad (4.181)$$

for some spinless particle B, where the quantities in parentheses stand for the 4-momenta of the particles. Suppose the particle B is its own antiparticle, so that it is described by a real scalar field $\phi(x)$, whose Lagrangian has only one interaction term:

$$\mathscr{L}_{\text{int}} = -\frac{\lambda}{4!} \phi^4. \qquad (4.182)$$

In Fig. 4.3, we show two diagrams of how this process might happen. Fig. 4.3a uses the interaction only once, so that it represents a contribution from first order in perturbation theory. Fig. 4.3b is second order. Note that diagram 4.3b has a loop in it, because of which it falls into the category of *loop diagrams*, or more specifically a *1-loop diagram*. Fig. 4.3a, on the other hand, has no loop of lines. Such diagrams are called *tree diagrams*.

According to the prescriptions laid down in §4.10, the Feynman amplitude of the diagram in Fig. 4.3a is given by

$$i\mathcal{M}_a = -i\lambda \,, \tag{4.183}$$

without any other numerical factor. In fact, this is the reason why we had put the factor of $1/4!$ in defining the constant accompanying the accompanying the field operators ϕ^4 while writing down the interaction Lagrangian. The Feynman amplitude for the loop diagram will involve two factors of λ, and is expected to be smaller than the tree diagram, assuming λ is small because we are using perturbation theory. In the calculation of the cross-section, we neglect this diagram from now on.

> □ **Exercise 4.24** At the 1-loop level, Fig. 4.3b is not the only possible diagram. There can be two more diagrams. Try to draw them. [**Note :** *We did not draw them because we are not taking the 1-loop contribution in the calculation of the cross-section anyway.*]

The Feynman amplitude is independent of the momenta of the particles involved, so that we can take the absolute square of its magnitude outside the integrals appearing in Eq. (4.180) very easily and write

$$
\begin{aligned}
\sigma &= \frac{\lambda^2}{4E_1 E_2 v_{\rm rel}} \int \frac{d^3 p_1'}{(2\pi)^3 2E_1'} \int \frac{d^3 p_2'}{(2\pi)^3 2E_2'} (2\pi)^4 \delta^4(p_1 + p_2 - p_1' - p_2') \\
&= \frac{\lambda^2}{64\pi^2 E_1 E_2 v_{\rm rel}} \int \frac{d^3 p_1'}{E_1'} \int \frac{d^3 p_2'}{E_2'} \delta^4(p_1 + p_2 - p_1' - p_2') \,. \tag{4.184}
\end{aligned}
$$

We first integrate over p_2'. This is easy. The δ-function appearing in the integrand is a 4-dimensional one, which contains the three spatial components. If we integrate over p_2', the result will be

$$
\sigma = \frac{\lambda^2}{64\pi^2 E_1 E_2 v_{\rm rel}} \int \frac{d^3 p_1'}{E_1' E_2'} \delta(E_1 + E_2 - E_1' - E_2') \,. \tag{4.185}
$$

Of course, the time component of the δ-function still remains. In addition, it will have to be remembered that the spatial components p_2' are not independent variables any longer. Rather, after the integration over the spatial parts of the δ-function, p_2' has been fixed to have the value given by

$$
p_1' + p_2' = p_1 + p_2 \,. \tag{4.186}
$$

Keeping this in mind, let us now look at the remaining integration in the evaluation of the cross-section.

Before we can perform the integration, we need to decide on a frame of reference. A very convenient choice is the center-of-mass or the CM frame introduced in §2.7. In this frame, the total 3-momentum of the initial particles is zero, i.e.,

$$\boldsymbol{p}_1 + \boldsymbol{p}_2 = 0. \tag{4.187}$$

The magnitudes of $\boldsymbol{p}_1$ and $\boldsymbol{p}_2$ are thus equal. Since in the present case we are talking of a situation where the two initial state particles have the same mass as well, we conclude that their energies are also equal, so we can denote both by the same symbol:

$$E_1 = E_2 \equiv E. \tag{4.188}$$

Moreover, Eqs. (4.186) and (4.187) imply that, for a 2-to-2 scattering, we must also have

$$\boldsymbol{p}_1' + \boldsymbol{p}_2' = 0 \tag{4.189}$$

in the center-of-mass frame. By the same argument, we now conclude that

$$E_1' = E_2' \equiv E'. \tag{4.190}$$

The expression of Eq. (4.185) can then be written as

$$\sigma = \frac{\lambda^2}{64\pi^2 E^2 v_{\text{rel}}} \int \frac{d^3 p'}{E'^2} \delta(2E - 2E'). \tag{4.191}$$

The angular integrations can be performed trivially, and we can use the property $\delta(2x) = \frac{1}{2}\delta(x)$, and obtain

$$\sigma = \frac{\lambda^2}{32\pi E^2 v_{\text{rel}}} \int \frac{dp' p'^2}{E'^2} \delta(E - E'), \tag{4.192}$$

where $\text{p}' = |\boldsymbol{p}'|$. Utilizing Eq. (4.162) and performing the remaining integration, we obtain

$$\sigma = \frac{\lambda^2}{32\pi E^2 v_{\text{rel}}} \frac{\text{p}}{E}. \tag{4.193}$$

Notice that in the CM frame $v_{\text{rel}} = 2\text{p}/E$, so that we can write the final result as

$$\sigma = \frac{\lambda^2}{64\pi E^2}. \tag{4.194}$$

Having obtained the result, let us discuss which aspects of the result could have been guessed without going through the detailed calculations. Looking at the interaction Lagrangian, it is clear that a vertex should obtain a factor of λ in the amplitude. Since the cross-section involves the square of the amplitude, it must contain a factor λ^2. Next, the cross-section has the dimension of area, i.e., of inverse mass-squared in natural units. Thus, there must be some factor with dimensions of inverse mass-squared in the cross-section, and this can come only from the mass and the energy (or equivalently, energy and momentum) of the particles. At very high energies where the mass can be neglected, the relevant factor can only be $1/E^2$. In the other extreme, when the 3-momenta are negligible, the relevant factor can only be $1/m^2$. There is no way of telling, without doing the calculations, that the $1/E^2$ dependence is valid for all energies. This is a speciality of this particular process, and will not be valid for a general cross-section. So, for this process, even before starting the calculation, we could have said that the cross-section would be proportional to λ^2/E^2 at high energies and λ^2/m^2 at low energies.

In fact, once we knew that the Feynman amplitude in this case has no angular dependence, we could have even predicted the factors of π in the final formula. From the formula in Eq. (4.180), we see that there is, naively speaking, an overall factor of $1/\pi^2$. Because the final amplitude is independent of the scattering angle, the angular integration would produce a factor of 4π, so we could guess that we would be left with a factor of π in the denominator, and indeed that is the case in Eq. (4.194). In more complicated cases where the Feynman amplitude would depend on the scattering angle, this simple-minded expectation may not hold.

□ **Exercise 4.25** *Suppose, in addition to the quartic interaction shown in Eq. (4.182), there is also a cubic interaction term involving the ϕ fields. Show that now there are three more diagrams which contribute to the scattering process of Eq. (4.181) at the tree level.*

It might be worthwhile at this point to discuss how the permutation factors come into the Feynman rules for vertices. Notice that in the interaction of Eq. (4.182), each factor of ϕ is of the form given in Eq. (4.12), so that each factor of ϕ can either create a particle or annihilate a particle. Recall also that a field operator ϕ can create or annihilate a particle of any momentum since the expression for the field contains an integral over all momenta.

Consider first the particle in the initial state with momentum p_1. It needs to be annihilated from the initial state. This annihilation operator, with momentum p_1, can come from any of the four factors of ϕ present in the interaction Lagrangian. Hence we already obtain a factor of 4 in the amplitude. For any of these choices made, we can annihilate the particle of momentum p_2 from any of the remaining three field operators. Once that is done, the final state particle with momentum p_1' can be created from the creation operator present in any of the two remaining factors of ϕ. And for the last one, we will have no choice left, so that we obtain an overall factor of $4 \times 3 \times 2 = 4!$. This factor, along with the factor $\lambda/4!$ that appears in the Lagrangian, implies that the Feynman rule for the vertex should be $-i\lambda$.

We can perform a similar counting to understand how the symmetry factors come in. For this, we look at Fig. 4.3b *(p 100)*. Notice that the 4-momentum of both internal lines cannot be determined by momentum conservation. If the 4-momentum flowing on one of them is denoted by l, the momentum on the other can be written as $p_1 + p_2 - l$. Thus, the Feynman amplitude of this diagram will be of the form

$$i\mathcal{M}_b = \left(\frac{-i\lambda}{4!}\right)^2 K \int \frac{d^4 l}{(2\pi)^4} \frac{i}{l^2 - m^2} \frac{i}{(p_1 + p_2 - l)^2 - m^2}, \qquad (4.195)$$

where K is a numerical factor which comes from different assignments of the fields to the lines in the diagram. To see what this factor is, first consider any one of the vertices. Among the four factors of ϕ, we will need to decide which two go as internal lines. This can be done in $\binom{4}{2}$ or six ways. The remaining two can be assigned to the external lines in two different ways, as described earlier for the tree diagram. On the other vertex, the same argument is repeated, and we obtain the same factors. Finally, the two lines from each vertex selected to be internal lines can be joined in two different ways to complete the loop. Thus,

$$K = (6 \times 2)^2 \times 2 = \frac{(4!)^2}{2!},$$ (4.196)

so that

$$i\mathcal{M}_b = \frac{(-i\lambda)^2}{2!} \int \frac{d^4 l}{(2\pi)^4} \frac{i}{l^2 - m^2} \frac{i}{(p_1 + p_2 - l)^2 - m^2}.$$ (4.197)

This is exactly what comes out of the prescriptions of §4.10: factors of $(-i\lambda)$ for both vertices, and a symmetry factor of $1/2!$ for two internal lines which have identical end points.

4.12.3   Mandelstam variables

From the discussion that follows Eq. (4.156) about the decay rates, it is clear that the cross-section of a scattering process given in Eq. (4.178) is a Lorentz invariant quantity. This piece of information is not conveyed in an obvious manner if we look at the cross-section calculated in a given frame in terms of energies and momenta values that apply to that frame, e.g., the formula given in Eq. (4.194).

It is much more elegant and useful if instead we express the cross-section of any process in terms of Lorentz invariant variables only. Such invariants can be constructed by taking dot products of of the 4-momenta of incoming and outgoing particles. Note that for any external momentum p, an invariant of the form p^2 is no good: it is not a variable, just the mass squared of the corresponding particle. Only dot products of *different* momenta qualify as variables.

For a 2-to-2 scattering, let us denote the 4-momenta of the two initial-state particles by p_1 and p_2, and of the final-state particles by p_1' and p_2'. We can construct the following Lorentz invariants:

$$\mathfrak{s} = (p_1 + p_2)^2,$$
$$\mathfrak{t} = (p_1 - p_1')^2,$$
$$\mathfrak{u} = (p_1 - p_2')^2.$$ (4.198)

These are called the *Mandelstam variables*. The definitions contain the dot products of different 4-momenta in the external states.

In the CM frame, where Eq. (4.187) holds, the Mandelstam variable $\mathfrak{s}$ equals $(E_1 + E_2)^2$. Thus the quantity $\sqrt{\mathfrak{s}}$ can be interpreted as the total energy of the incoming particles in the CM frame, called E_{tot} in §2.7.1. The interpretation of the other two Mandelstam variables is clumsy, or even unclear. The reason is that, given that there are two particles in the initial state, it is not clear which one we call particle 1 and assign the 4-momentum called p_1. The same problem applies to the particles in the final state. Thus, what might be called the invariant $\mathfrak{t}$ by one person might be the invariant $\mathfrak{u}$ in another person's way of defining things. Some conventions need to be set up before one names the invariants.

Suppose now we have set up some convention such that in the CM frame, we call the angle between $\boldsymbol{p}_1$ and $\boldsymbol{p}'_1$ the scattering angle, as was shown in Fig. 2.2 *(p 32)*. Then

$$\mathsf{t} = m_1^2 + m_1'^2 - 2(E_1 E'_1 - \mathsf{p}\mathsf{p}' \cos\theta), \qquad (4.199)$$

where m_1 and m'_1 are the masses of the relevant particles, $\mathsf{p} = |\boldsymbol{p}_1|$ and $\mathsf{p}' = |\boldsymbol{p}'_1|$. We showed in Eq. (2.88, *p 32*) how E'_1 in the CM frame can be expressed in terms of s. A similar expression can be derived for E_1 as well. Once we use these expressions, we can express the scattering angle in terms of the Mandelstam variables s and t.

☐ **Exercise 4.26** *Show that the three Mandelstam variables defined in Eq. (4.198) are not independent by deriving the relation*

$$\mathsf{s} + \mathsf{t} + \mathsf{u} = m_1^2 + m_2^2 + m_1'^2 + m_2'^2. \qquad (4.200)$$

☐ **Exercise 4.27** *For a scattering where the initial state has two particles but the final state has N, show that the number of independent Lorentz invariant variables is $\frac{1}{2}(N-1)(N+2)$.*

4.13 Differential decay rates and cross-sections

From an experimental point of view, it is useful considering the quantities obtained by not performing the full integrations that appear in the definitions of decay rate or cross-sections. If we do not perform the integration with respect to a certain kinematical variable, the result that we obtain gives us the differential rate of change of the total quantity with respect to that kinematical variable. Such quantities are called *differential rates* or *differential cross-sections*, depending on which total quantity would have been obtained if we finished the integration.

4.13.1 Angular distribution in the CM frame

The most commonly studied quantity is the angular distribution. For 2-to-2 scattering viewed in the CM frame, this can be written in a compact form, as we now show. We start from Eq. (4.180) and perform the integration over $\boldsymbol{p}'_2$. This gives

$$\sigma = \frac{1}{64\pi^2 E_1 E_2 v_{\text{rel}}} \int \frac{d^3 p'_1}{E'_1 E'_2} \delta(E_1 + E_2 - E'_1 - E'_2) |\mathcal{M}|^2. \qquad (4.201)$$

We can now use Eq. (4.161), where p' is, as before, the magnitude of momentum of either of the final particles, which are equal in the CM frame. For the differential cross-section, we do not perform the integration over the solid angle Ω and write

$$\frac{d\sigma}{d\Omega} = \frac{1}{64\pi^2 E_1 E_2 v_{\text{rel}}} \int \frac{d\mathsf{p}' \, \mathsf{p}'^2}{E'_1 E'_2} \delta(\sqrt{\mathsf{s}} - E'_1 - E'_2) |\mathcal{M}|^2, \qquad (4.202)$$

using the Mandelstam variable s which is equal to the total initial energy in the CM frame.

The solutions for the final energies that come out as a result of energy-momentum conservation were deduced in Eq. (2.88, *p 32*), which show that

$$E_2' = \sqrt{E_1'^2 - m_1'^2 + m_2'^2}\,. \tag{4.203}$$

Putting this in the expression for the differential cross-section, we encounter the delta function of a function of E_1'. Using Eq. (4.165), we obtain

$$\delta(\sqrt{s} - E_1' - E_2') = \frac{E_2'}{\sqrt{s}}\,\delta\left(E_1' - \frac{s^2 + m_1'^2 - m_2'^2}{2s}\right), \tag{4.204}$$

using the solution of Eq. (2.88, *p 32*). The integration variable can also be changed to E_1' by using Eq. (4.162) with minimal notational change. Plugging all this back into Eq. (4.202) and performing the integration over E_1', we obtain

$$\frac{d\sigma}{d\Omega} = \frac{p'}{64\pi^2\sqrt{s}E_1 E_2 v_{\text{rel}}}\left|\mathcal{M}\right|^2, \tag{4.205}$$

where now p', as well as the factor containing the amplitude, must be evaluated by using the conservation laws, i.e., should be consistent with Eq. (2.88, *p 32*).

In the CM frame,

$$v_{\text{rel}} = \left(\frac{p}{E_1} + \frac{p}{E_2}\right) = \frac{p\sqrt{s}}{E_1 E_2}, \tag{4.206}$$

where we write the common magnitude of the 3-momentum of initial-state particles as p. Thus, the expression for the differential cross-section can be further simplified to

$$\frac{d\sigma}{d\Omega} = \frac{1}{64\pi^2 s}\frac{p'}{p}\left|\mathcal{M}\right|^2. \tag{4.207}$$

For elastic scattering $p = p'$, so that the formula is further simplified. More generally, the magnitudes p and p' can be determined in terms of the Mandelstam variable s and the masses of the particles, and one finally obtains

$$\frac{d\sigma}{d\Omega} = \frac{1}{64\pi^2 s}\left(\frac{[s - (m_1' + m_2')^2][s - (m_1' - m_2')^2]}{[s - (m_1 + m_2)^2][s - (m_1 - m_2)^2]}\right)^{1/2}\left|\mathcal{M}\right|^2. \tag{4.208}$$

Remember that this angular distribution is obtained in the CM frame, i.e., the quantity $d\Omega$ is an element of solid angle in the CM frame.

□ **Exercise 4.28** Show that in the CM frame, p can be expressed in terms of s and the masses:

$$p = \frac{\sqrt{[s - (m_1 + m_2)^2][s - (m_1 - m_2)^2]}}{2\sqrt{s}}, \tag{4.209}$$

and p' is given by an exactly similar expression with the masses of the final-state particles instead of the initial-state ones.

Hence, verify the differential cross-section formula Eq. (4.208).

As an example, consider the scattering problem discussed in §4.12.2. Using the amplitude from Eq. (4.183), we obtain

$$\frac{d\sigma}{d\Omega} = \frac{\lambda^2}{256\pi^2 E^2} \tag{4.210}$$

from Eq. (4.208). Integration over the solid angle would produce a factor of 4π, giving the total cross-section of Eq. (4.194). Of course, this is a trivial case because the differential cross-section is the same in all directions. This will not be true in general, because the Feynman amplitude can depend on the angular variables. Non-trivial differential cross-section would result from such dependence, and the nature of its variation would provide important information about the nature of the interaction.

For 2-to-2 scattering, Eq. (4.200) implies that there are only two Lorentz invariant Mandelstam variables, which can be taken as s and t. In the CM frame, the scattering process can be fully described by the center of mass energy and the scattering angle. We said that the center of mass energy is a function of s only, and that the scattering angle can be expressed as a function of s and t. Thus the differential cross-section, which contains the scattering angle, can depend on both s and t. The total cross-section is obtained by integrating over all angles, and therefore can depend only on s.

4.13.2 Invariant form of angular distribution

It is also instructive to derive an expression for angular distribution of scattering cross-section in a Lorentz invariant manner. For this, we first use the fact that, for isotropic interactions, the cross-section can never depend on the azimuthal angle, so that

$$d\Omega = 2\pi \, d(\cos\theta) \tag{4.211}$$

in any frame, where θ is the scattering angle in that frame. The Mandelstam variable s does not depend on the scattering angle, but t and u do. Hence we can replace the θ-variation in terms of the variation with respect to t or u. In what follows, we take t, which is defined as

$$t = (p_1 - p_1')^2 = m_1^2 + m_1'^2 - 2(E_1 E_1' - \mathsf{p}_1 \mathsf{p}_1' \cos\theta). \tag{4.212}$$

In the CM frame, the magnitudes of the momenta of the outgoing particles do not depend on the scattering angle. So we can write

$$dt = 2\mathsf{p}\mathsf{p}' d(\cos\theta_{\text{CM}}), \tag{4.213}$$

where $\mathbf{p}$ and $\mathbf{p}'$ denote the magnitudes of the 3-momentum of any particle in the incoming and outgoing states. Combining this equation with Eq. (4.211), we can write

$$\frac{d\sigma}{dt} = \frac{\pi}{\mathrm{p p}'} \frac{d\sigma}{d\Omega_{\mathrm{CM}}}. \tag{4.214}$$

Substituting the expression for the differential cross-section in the CM frame that appears in Eq. (4.207), we find

$$\frac{d\sigma}{dt} = \frac{1}{64\pi \mathrm{sp}^2} \left| \mathcal{M} \right|^2$$
$$= \frac{1}{16\pi} \frac{1}{\left[\mathrm{s} - (m_1 + m_2)^2 \right] \left[\mathrm{s} - (m_1 - m_2)^2 \right]} \left| \mathcal{M} \right|^2, \tag{4.215}$$

using Eq. (4.209) for $\mathbf{p}$. Notice that the right hand side involves s, the masses and the Feynman amplitude, all of which are Lorentz invariant quantities. Thus, this is a completely invariant form of the differential cross-section.

4.13.3   Angular distribution in FT frame

From the invariant form of the differential cross-section, one can derive the angular dependence in any frame. Of particular interest is the fixed-target or FT frame. Using the independence on azimuthal angle from Eq. (4.211), we can write

$$\frac{d\sigma}{d\Omega_{\mathrm{FT}}} = \frac{1}{2\pi} \frac{dt}{d(\cos\theta_{\mathrm{FT}})} \frac{d\sigma}{dt}, \tag{4.216}$$

where the scattering angle is defined to be the angle between $\mathbf{p}_1$ and $\mathbf{p}'_1$ in the frame in which $\mathbf{p}_2 = 0$. For the Mandelstam variable t, we have already given an expression in Eq. (4.212). Alternatively, we can also write

$$t = (p_2 - p'_2)^2 = m_2^2 + m_2'^2 - 2m_2 E'_2$$
$$= m_2^2 + m_2'^2 - 2m_2(E_1 + m_2 - E'_1). \tag{4.217}$$

Therefore,

$$\frac{dt}{d(\cos\theta_{\mathrm{FT}})} = 2m_2 \frac{dE'_1}{d(\cos\theta_{\mathrm{FT}})}. \tag{4.218}$$

Equating the two expressions for t from Eqs. (4.212) and (4.217) and taking derivatives with respect to $\cos\theta$, we obtain

$$(E_1 + m_2)\frac{dE'_1}{d(\cos\theta_{\mathrm{FT}})} = \mathrm{p}_1 \left(\mathrm{p}'_1 + \cos\theta_{\mathrm{FT}} \frac{d\mathrm{p}'_1}{d(\cos\theta_{\mathrm{FT}})} \right). \tag{4.219}$$

The derivative of $\mathbf{p}'_1$ can easily be changed to the derivative of E'_1 by using the energy-momentum relation, as shown in Eq. (4.162). Once that is done, one easily obtains

$$\frac{dE'_1}{d(\cos\theta_{\mathrm{FT}})} = \frac{\mathrm{p}_1 \mathrm{p}_1'^2}{\mathrm{p}'_1(E_1 + m_2) - \mathrm{p}_1 E'_1 \cos\theta_{\mathrm{FT}}}. \tag{4.220}$$

Further, in this frame, $\mathsf{s} = m_1^2 + m_2^2 + 2E_1 m_2$. Putting this into Eq. (4.215) and using Eqs. (4.216) and (4.218), we obtain

$$\frac{d\sigma}{d\Omega_{\text{FT}}} = \frac{1}{64\pi^2 m_2 \mathbf{p}_1} \frac{\mathbf{p}_1'^2}{\mathbf{p}_1' (E_1 + m_2) - \mathbf{p}_1 E_1' \cos\theta_{\text{FT}}} \left|\mathcal{M}\right|^2. \qquad (4.221)$$

The expression is compact but misleading, because the angular dependence does not come only through the term $\cos\theta_{\text{FT}}$ occurring in the denominator and also possibly through the Feynman amplitude: $\mathbf{p}_1'$ and therefore E_1' also depend on the scattering angle. The expressions for the angular dependence of $\mathbf{p}_1'$ and E_1' are quite cumbersome.

□ **Exercise 4.29** Show that

$$E_1' = \frac{\mu^2 (E_1 + m_2) + \mathbf{p}_1 \cos\theta_{\text{FT}} \sqrt{\mu^4 - 4m_1'^2 M_\theta^2}}{2M_\theta^2}, \qquad (4.222)$$

where

$$\mu^2 = m_1^2 + m_2^2 + m_1'^2 - m_2'^2 + 2E_1 m_2,$$
$$M_\theta^2 = (E_1 + m_2)^2 - \mathbf{p}_1^2 \cos^2\theta_{\text{FT}}. \qquad (4.223)$$

For $m_1' = 0$, the expression simplifies considerably. In this case, one obtains

$$\frac{d\sigma}{d\Omega_{\text{FT}}} = \frac{m_1^2 + m_2^2 - m_2'^2 + 2E_1 m_2}{128\pi^2 m_2 \mathbf{p}_1 \left(E_1 + m_2 - \mathbf{p}_1 \cos\theta_{\text{FT}}\right)^2} \left|\mathcal{M}\right|^2. \qquad (4.224)$$

If we consider elastic scattering with $m_1 = m_1' = 0$ and $m_2 = m_2'$, the formula reduces to something that looks even simpler:

$$\frac{d\sigma}{d\Omega_{\text{FT}}} = \frac{1}{64\pi^2} \frac{1}{(m_2 + E_1 - E_1 \cos\theta)^2} \left|\mathcal{M}\right|^2. \qquad (4.225)$$

□ **Exercise 4.30** Consider the case $m_1 = m_1' = 0$. Starting from the basic formula for 2-to-2 scattering cross-section, Eq. (4.180), or from the form obtained in Eq. (4.201) after the integration over $\mathbf{p}_2'$, perform the integration over E_1' and arrive at Eq. (4.224) with m_1 set to zero.

□ **Exercise 4.31** Show that, for elastic scattering with $m_1 = 0$, Eq. (4.222) reduces to

$$E_1' = \frac{m_2 E_1}{m_2 + E_1(1 - \cos\theta)}. \qquad (4.226)$$

4.13.4 Other differentials

We can use the differentials for decay rates as well. One can measure how the decay products are distributed angularly, or the differential decay rate. Besides, one can also consider other kinds of differential cross-sections or decay

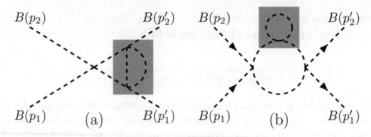

Figure 4.4: Some loop diagrams for elastic scattering of scalar particles that were not shown in Fig. 4.3 *(p 100)*. We have omitted the arrows on the external lines. Some regions of the diagrams have been shaded. The reason for this has been explained in the text.

rates, where the distribution of final particles is seen not as a function of angles, but of the energy of a certain particle. Of course, this question does not arise if the final state contains only two particles, because then the energy of both of them will be determined by the properties of the initial state, and one cannot observe the product particles at various energies. But it can be done for any final state with three or more particles.

□ **Exercise 4.32** *Suppose a particle of mass m is moving in a certain frame with a speed v. It decays to another particle of mass m', and a massless particle. Show that the differential decay rate is given by*

$$\frac{d\Gamma}{d\Omega} = \left(\frac{m^2 - m'^2}{64\pi^2 m^3}\right) \frac{(1-v^2)^{3/2}}{(1 - v\cos\theta)^2}\left|\mathscr{M}\right|^2, \tag{4.227}$$

where θ is the angle between the 3-momenta of the initial particle and the massless particle in the final state.

4.14 Feynman diagrams that do not represent physical amplitudes

So far we have used Feynman diagrams to represent physical amplitudes. Sometimes, it is useful to draw a Feynman diagram that does not correspond to a physical amplitude, but is very useful nevertheless.

We will make the point with the help of an example. Consider the elastic scattering between two spinless bosons, as discussed in §4.12.2. In Fig. 4.3 *(p 100)*, we have shown two diagrams that contribute to this process. With only a ϕ^4 interaction present in the theory, there is only one tree diagram, which has been shown. In addition, we have shown a 1-loop diagram. Of course, there can be other loop diagrams. Some samples have been shown in Fig. 4.4.

Obviously, such samples suggest other diagrams. For example, in Fig. 4.4a, we see two external lines have been joined by other lines. The two external

Figure 4.5: Representation of 4-point and 2-point functions in a theory with $\lambda\phi^4$ interaction. The 4-momenta on the external lines need not satisfy the energy-momentum relation. The dark blobs represent any combinations of lines and vertices consistent with the interactions of the theory.

lines joined in this fashion need not be the two outgoing lines. They can be any two of the four external lines. Similarly, the extra blob appearing in Fig. 4.4b can appear on the other internal line, or even on both internal lines. Instead of considering all these diagrams from scratch, it would be much helpful if we have a ready-made factor for the kinds of objects shown in the shaded regions of Fig. 4.4. In the diagram on the left, the shaded region has four legs. For the diagram on the right, the shaded region has two legs. The factors to be put in place of the shaded objects are called, accordingly, contributions to the *4-point function* and the *2-point function*, respectively. More generally, the 4-point function can be denoted symbolically by Fig. 4.5a, and the 2-point function by Fig. 4.5b, where the blobs can represent any collection of lines and vertices consistent with the Lagrangian of the underlying theory. In general, a diagram with n external legs will contribute a factor that will be called an *n-point function*. A more elaborate name would be n-point *Green function*. They are functions of the momenta of the legs that go into or come out of these diagrams.

One cannot fail to notice that, apart from shapes and orientations of the lines which are irrelevant, the shaded region of Fig. 4.4a is exactly the same as the diagram of Fig. 4.3b *(p 100)*. In other words, the two have the same topology. There is a difference which does not show in the diagram, however. In Fig. 4.3b *(p 100)*, the lines going out of the loop are all external lines, which represent physical particles. The 4-momentum of any these lines, therefore, satisfies the energy-momentum relation, Eq. (2.37, *p 22*), with the appropriate value of the mass. For the shaded region of Fig. 4.4a, this is not the case. Two of the lines coming out of this shaded region are internal lines of the diagram, which represent virtual particles. In general, for an n-point function, none of the external lines need to represent an on-shell particle. The momenta can be kept general, except for the fact that total 4-momenta must be conserved. We will see in later chapters that such n-point functions are quite useful for evaluating loop diagrams.

Chapter 5

Quantum electrodynamics

As noted earlier in this book we will discuss three of the four fundamental interactions: the strong, the weak and the electromagnetic. Among the three, there are many reasons why the electromagnetic interaction should be discussed first. First, this is the only one among the three for which there was a classical theory, and the insight from the classical theory might be helpful in discussing the quantum theory. Second, the quantum theory of the electromagnetic interactions, called *quantum electrodynamics* or *QED* for short, was developed earlier than the quantum theories of strong and weak interactions, so historically also this interaction comes first. And lastly, the theory is based on a local internal symmetry. This idea was taken over from QED and generalized to develop the present theories of weak and strong interactions. Because of this reason, and because the internal symmetry is much simpler for QED than for the theories strong and the weak interactions, it would be appropriate to study electromagnetic interactions first. This is what we do in this chapter.

5.1 Gauge invariance

5.1.1 Global phase symmetry

We mentioned an internal symmetry in the preface to this chapter. Let us start by exposing the nature of this symmetry.

Consider the free Dirac Lagrangian, given earlier in Eq. (4.102, *p 81*), and repeated here for the sake of convenience:

$$\mathscr{L}_0 = i\overline{\psi}\gamma^\mu\partial_\mu\psi - m\overline{\psi}\psi. \tag{5.1}$$

Obviously, this Lagrangian is invariant under a change of phase of the field ψ. Explicitly, suppose we change over to a new field

$$\psi'(x) = \exp(-ieQ\theta)\psi(x), \tag{5.2}$$

where e, Q and θ are all real numbers, with different interpretations. The quantity e represents a universal constant which sets up the scale of the phase, Q is a characteristic of the field ψ and may therefore vary from one kind of field to another, and θ is a variable which determines how large the phase is. Note that we have not changed the spacetime co-ordinates at all: the new field is defined in terms of the old field at the same spacetime point. Such symmetries are called *internal symmetries*, as was first mentioned in §3.3.2.

Since $\overline{\psi}$ involves the hermitian conjugate of ψ, Eq. (5.2) implies the following transformation on it:

$$\overline{\psi}'(x) = \overline{\psi}(x)\exp(+ieQ\theta) \,. \tag{5.3}$$

Eqs. (5.2) and (5.3) show that

$$i\overline{\psi}'\gamma^\mu\partial_\mu\psi' - m\overline{\psi}'\psi' = i\overline{\psi}\gamma^\mu\partial_\mu\psi - m\overline{\psi}\psi \,, \tag{5.4}$$

because the phase cancels between the ψ and the $\overline{\psi}$. Thus, the Lagrangian of Eq. (5.1) is invariant under the transformation given in Eq. (5.2).

The statement just made is true provided θ is independent of the spacetime co-ordinates, i.e., is a constant over spacetime. Such symmetries are called *global symmetries*, as mentioned in §3.3.2. It can be easily seen that such a global symmetry exists also in the free Lagrangian of a complex scalar field, Eq. (4.100, *p 81*).

What is the symmetry group? Note that the transformation of the fields involve a complex number of modulus unity, i.e., a phase transformation. The relevant group is called U(1), which was explained in §3.3.1(b).

In Eq. (5.2), we seem to have defined e, Q and θ all through the same expression. Certainly this is not possible. θ is a U(1) parameter, so this can be defined independent of Eq. (5.2). But then it needs to be multiplied by eQ, which depends on the kind of field on which we apply the U(1) transformation. So it seems that only the product of these two quantities is determined, leaving an arbitrariness of specifying either of these two quantities. Indeed that is true: the definition of any U(1) charge has a multiplicative arbitrariness.

5.1.2 Local symmetry

Suppose we now want to see what happens if the parameter θ, appearing in Eq. (5.2), depends on spacetime co-ordinates. In this case, we obtain

$$\partial_\mu\psi'(x) = \exp(-ieQ\theta)\left[\partial_\mu\psi(x) - ieQ(\partial_\mu\theta)\psi(x)\right] \,. \tag{5.5}$$

There is an extra term, involving the derivatives of θ. Because of this, the Lagrangian will not be invariant under the transformation. In fact, if we change ψ to ψ', the Lagrangian $\mathscr{L}_0$ changes to $\mathscr{L}_0'$, where

$$\mathscr{L}_0' - \mathscr{L}_0 = eQ(\partial_\mu\theta)\overline{\psi}(x)\gamma^\mu\psi(x) \,. \tag{5.6}$$

If we want the local symmetry, we would need to work with a modified Lagrangian. So, instead of the Lagrangian of Eq. (5.1), let us try

$$\mathscr{L} = i\overline{\psi}\gamma^\mu D_\mu\psi - m\overline{\psi}\psi \,, \tag{5.7}$$

where

$$D_\mu = \partial_\mu + ieQA_\mu \,, \tag{5.8}$$

bringing in a new A_μ. This D_μ is usually called the *covariant derivative*. The prescription of replacing ordinary derivatives by covariant derivatives is called *minimal substitution*.

Let us now demand that this A_μ also transforms non-trivially when the phase of the fermion field is changed, in such a way that the new Lagrangian is invariant under the local symmetry. That would mean that, under the local symmetry, A_μ changes to A'_μ such that

$$\mathscr{L}'_0 - eQ\overline{\psi}'\gamma^\mu\psi' A'_\mu = \mathscr{L}_0 - eQ\overline{\psi}\gamma^\mu\psi A_\mu \,, \tag{5.9}$$

or

$$A'_\mu = A_\mu + \partial_\mu\theta \,. \tag{5.10}$$

But this is exactly the redefinition that we talked about in Eq. (4.22, *p 66*). We can thus identify A_μ with the photon field and conclude that, in the urge for making the global phase symmetry local, we have introduced the photon field in a natural manner.

Once the photon field has entered into our discussion, it is imperative that we add terms in the Lagrangian that involve the photon field only. Such terms were shown in Eq. (4.106, *p 82*). Adding these terms, we can now write down the Lagrangian involving the fields $\psi(x)$ and $A_\mu(x)$ in the form

$$\mathscr{L} = i\overline{\psi}\gamma^\mu D_\mu\psi - m\overline{\psi}\psi - \frac{1}{4}F_{\mu\nu}F^{\mu\nu} - \frac{1}{2\xi}(\partial_\mu A^\mu)^2 \,. \tag{5.11}$$

This is the Lagrangian of quantum electrodynamics, or QED.

□ **Exercise 5.1** In the Lagrangian for the complex scalar field, Eq. (*4.100, p 81*), if we replace the derivatives by D_μ defined in Eq. (5.8), show that the resulting Lagrangian is invariant under a local phase symmetry of the field ϕ. This is the Lagrangian of the so-called *scalar QED*.

Local symmetries are also called *gauge symmetries*. Theories incorporating gauge symmetries are called *gauge theories*. The spin-1 particles which are necessary to keep the gauge invariance are called *gauge bosons*. QED is therefore a gauge theory, based on the gauge group U(1) whose elements give transformations of fields $\psi(x)$. The gauge boson for QED is the photon. We will see later that the theories of strong and weak interactions are also gauge theories, based on other gauge groups.

5.1.3 Charge conservation

Noether's theorem tells us that a continuous symmetry implies a divergence-less current, which in turn implies a conserved quantity. The recipe for finding the conserved current or the Noether current was given in §4.6. Using the technique elaborated there, we can calculate the Noether current for the Lagrangian of Eq. (5.1) and obtain

$$j^\mu = eQ\overline{\psi}\gamma^\mu\psi\,. \tag{5.12}$$

It is purely a matter of convention whether we define the Noether current with or without the overall factor of e, or even with an extra overall numerical factor. Since the expression given on the right hand side of Eq. (5.12) has zero divergence and since the basic unit of electric charge, e, is a universal constant, the quantity without the factor of e is also divergenceless.

What is the conservation law that follows from this symmetry? If we have just one fermion field as shown in §5.1.1 and §5.1.2, we can write down the conserved current without the factor eQ that appears in Eq. (5.12). The conserved quantity corresponding to this current is the number of particles minus the number of antiparticles associated with the field ψ.

In the real world, more than one fermion field interacts with the photon. The Lagrangian containing all fermion fields and their electromagnetic interactions is then of the form

$$\mathscr{L} = \sum_A \left(i\overline{\psi}_A\gamma^\mu D_\mu\psi_A - m_A\overline{\psi}_A\psi_A\right) - \frac{1}{4}F_{\mu\nu}F^{\mu\nu} \tag{5.13}$$

plus the gauge-fixing term. The Lagrangian is invariant under transformations of the photon field shown in Eq. (5.10), and of the fermion fields which are of the type shown in Eq. (5.2). The infinitesimal form of these transformations can be written as

$$\delta\psi_A = -ieQ_A\psi_A\,. \tag{5.14}$$

The Noether current is then given by

$$j_\mu = e\sum_A Q_A\overline{\psi}_A\gamma^\mu\psi_A\,. \tag{5.15}$$

This sum of current densities of individual fields, weighted by their electric charge, is called the electric current density. The corresponding conserved quantity is the electric charge.

We commented that the Lagrangian of Eq. (5.13) is invariant under the transformations shown in Eqs. (5.2) and (5.10). One might wonder, since a gauge-fixing term has to be introduced in the Lagrangian, whether this statement, as well as the conclusions about Noether charges and currents derived from it, can be trusted.

Without the gauge-fixing term, any differentiable function of spacetime co-ordinates $\theta(x)$ can appear in Eq. (5.10). Once we introduce the gauge-fixing term, we will have to make sure that this term is also invariant under the symmetry. Taking the 4-divergence of each side of Eq. (5.10), we see that this is achieved if

$$\Box\theta(x) = 0\,. \tag{5.16}$$

So the gauge symmetry of Eqs. (5.2) and (5.10) is still there, although in a somewhat restricted form in the sense that not any arbitrary function $\theta(x)$ can be used in defining the gauge transformations: one has to use the functions which satisfy Eq. (5.16). Accordingly, the gauge-fixing term does not change any of our conclusion about Noether charges and currents. In fact, one of the solutions of Eq. (5.16) is a constant θ for all x, which gives the global U(1) symmetry mentioned in §5.1.1. This global symmetry is enough to draw conclusions about Noether charges and currents.

5.2   Interaction vertex

Let us now look closely at the Lagrangian of Eq. (5.11). Of course it contains the quadratic terms for the Dirac field and the photon field. Apart from those, there is only one more term, which is

$$\mathscr{L}_{\text{int}} = -eQ\overline{\psi}\gamma^{\mu}\psi A_{\mu}\,. \tag{5.17}$$

This term involves three field operators and must therefore be an interaction term. We have put a subscript to the Lagrangian to remind us of that. The universal quantity e thus appears in all electromagnetic interactions, and can be called the coupling constant of the U(1) group itself, or the *gauge coupling constant*. Q, on the other hand, depends on the particular field ψ, and will be called the electric charge of the field. As mentioned earlier, there is a multiplicative arbitrariness in the definitions of e and Q. In what follows, we will use a convention in which the proton has $Q = +1$ and the electron has $Q = -1$.

We said that the fields are operators in quantum field theory. Any term in the Lagrangian, being a combination of operators, must be an operator itself. In particular, the interaction term of Eq. (5.17) is an operator, and we can ask what it might do. The field operator ψ can annihilate a particle or create an antiparticle. The operator $\overline{\psi}$ can do just the opposite. And the photon field operator A_{μ} can either create or annihilate a photon. Taking everything together, we see that there are eight possibilities of events at an interaction vertex, as described in §4.9. We summarize all these possibilities in graphical form in Fig. 5.1, taking the fermion to be an electron for the sake of definiteness.

Let us explain the convention of arrows used in Fig. 5.1. There exist many conventions regarding arrows. Some people always put an inward-going arrow on the leg that corresponds to the operator ψ and an outward-going arrow on the leg that corresponds to $\overline{\psi}$. Thus, if an electron is going into the vertex, it is annihilated by the operator ψ, so this obtains an inward arrow. The same arrow is given if a positron is going out of the vertex. Some people find this convention confusing, because they want an outward arrow to represent anything, i.e., particle or antiparticle, coming out of the vertex, and an inward arrow for anything going into the vertex. We have made a compromise here by using the first kind of arrows on the line, but in addition, in the legend for the line, putting a negative sign if the direction of momentum is opposite to the direction of the arrow. For example, in the diagram of Fig. 5.1b, the upper left line represents a positron, but then the momentum of the positron is $-\boldsymbol{p}'$

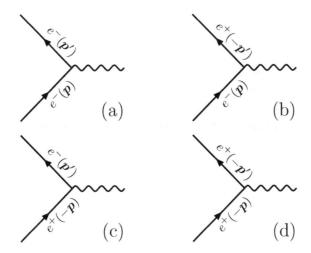

Figure 5.1: Possibilities of creation and annihilation of particles at the basic QED vertex. See text for the arrow convention on the fermion lines. No arrow is drawn on the photon line. For each of the four possibilities shown here, the photon can either be emitted or be absorbed at the vertex, making a total of eight possibilities.

in the direction of the arrow shown, so it is really a positron coming into the vertex with a momentum p'. This diagram then represents a vertex where an electron and a positron are annihilated. In Fig. 5.1c, an electron as well as a positron are created at the vertex. And finally, in Fig. 5.1d, the creation of a positron is shown by the lower line, whereas the upper line represents the annihilation of a positron. In each case, the photon might be created or annihilated at the vertex, because the photon field A_μ contains both creation and annihilation operators. No matter which of these eight possibilities is realized at a vertex, the Feynman rule for the vertex is the same, and it has been shown in Fig. 5.2 *(p 118)*.

The enumeration of the possibilities should not be taken to signify that there are physical processes whose initial and final state particles have been represented by the diagrams of Fig. 5.1. The possibilities shown in Fig. 5.1 only show the combination of creation and annihilation that can occur at a vertex. In order for this combination to be a physical process, other constraints need to be satisfied, like energy-momentum conservation.

For example, let us consider whether we can have the process

$$e(p) \to e(p') + \gamma(q), \tag{5.18}$$

where in parentheses we have put our notation for the 4-vector corresponding to each particle. Let us use the on-shell conditions for the initial and the final

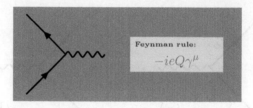

Figure 5.2: Feynman rule for the QED vertex.

electrons,

$$p^2 = m^2, \qquad p'^2 = m^2, \tag{5.19}$$

and try to see what conditions will the photon have to satisfy in order that such a process might be possible.

Energy-momentum conservation would imply

$$p^\mu - q^\mu = p'^\mu. \tag{5.20}$$

Squaring both sides and using Eq. (5.19), we obtain

$$q^2 = 2p \cdot q. \tag{5.21}$$

Writing

$$p^\mu = (E, \mathrm{p}\widehat{p}), \qquad q^\mu = (\omega, \mathrm{q}\widehat{q}), \tag{5.22}$$

where $\widehat{p}$ and $\widehat{q}$ are unit vectors in the directions of p and q, we obtain

$$\omega^2 - \mathrm{q}^2 = 2(E\omega - \mathrm{pq}\widehat{p} \cdot \widehat{q}). \tag{5.23}$$

The quantity $\widehat{p} \cdot \widehat{q}$ is just the cosine of the angle between p and q, whose value must be within -1 and $+1$. Thus we obtain the inequalities

$$E\omega - \mathrm{pq} \le \frac{1}{2}(\omega^2 - \mathrm{q}^2) \le E\omega + \mathrm{pq}. \tag{5.24}$$

For a photon, we expect $\omega = \mathrm{q}$, so that the expression in the middle is zero. However, on the left side we now have $(E-\mathrm{p})\mathrm{q}$, which is positive since $E > \mathrm{p}$ by dint of the on-shell conditions in Eq. (5.19). So at least the left part of the inequalities in Eq. (5.24) cannot be satisfied, which means that the electron cannot emit a photon.

The statement above applies for free electrons. If, e.g., the electron is bound within an atom, the conclusion would be different because the on-shell conditions of Eq. (5.19) would not apply. Alternatively, even with free electrons, we can check whether the inequalities of Eq. (5.24) can be satisfied for any combination of photon energy ω and photon momentum q, allowing for the possibility that, for the photon, they are not necessarily related by a free particle dispersion relation of the type given in Eq. (2.44, *p 23*). Since $E > p$, $E\omega - pq > E(\omega - q)$. So we can strengthen the left inequality by writing $E(\omega - q) < \frac{1}{2}(\omega^2 - q^2)$. Now, if $\omega > q$, i.e., $\omega - q$ is positive, this newly found inequality would translate to $2E < \omega + q$. We can again strengthen this inequality by replacing the q on the right hand side by ω, which gives $E < \omega$. This is certainly not possible because of energy conservation.

We are thus left with the option $\omega < q$. If this is the case, then the quantity in the middle of Eq. (5.24) is negative. Since the rightmost quantity is obviously positive, the right side of the inequality is automatically satisfied. The left side then produces a subsidiary condition

$$E\omega - pq < 0\,, \tag{5.25}$$

or $p/E > \omega/q$. The left hand side of this inequality is the speed of the initial electron, whereas the right hand side is the phase velocity of light. Thus the process shown in Eq. (5.18) is possible, i.e., an electron can spontaneously emit a photon, if the process takes place in a medium in which the speed of light is less than the speed of the initial electron. Such a process is called *Čerenkov radiation*. In the vacuum, it cannot take place.

Proceeding in similar fashion, we can show that none of the vertices shown in Fig. 5.1 can give rise to a physical process in the vacuum. It implies that in the first order in the QED interaction, no physical process is possible.

□ **Exercise 5.2** *There is a much simpler way of showing that processes like that shown in Eq. (5.18) cannot take place. Consider a frame in which the initial electron is at rest. Its energy must then be m. Show that the energy of the final particles is definitely greater than m, so that energy conservation cannot be satisfied.*

5.3   Elastic scattering at second order

We therefore are compelled to look for physical processes in the second order of interaction, which correspond to Feynman diagrams containing two of the vertices shown in Fig. 5.1. For example, consider two copies of the possibility shown in Fig. 5.1a. Take the photon to be outwardly going at one vertex and inwardly coming at the other. We can then join the two vertices so that the same photon that leaves one vertex arrives at the other. This gives us the diagram of Fig. 5.3. There are many such diagrams possible at the second order, giving rise to a variety of physical phenomena. In this section, we will consider elastic scattering processes only.

5.3.1   Electron–electron scattering

Looking at Fig. 5.3, we see that it represents the process

$$e^-(\boldsymbol{p}_1) + e^-(\boldsymbol{p}_2) \rightarrow e^-(\boldsymbol{p}'_1) + e^-(\boldsymbol{p}'_2)\,, \tag{5.26}$$

which is elastic scattering between two electrons. Here the subscripts on the momentum vectors do not represent any component of those vectors. Rather, a letter along with a numerical subscript stands for the name of a 4-momentum. Following the recipe given in Ch. 4, let us try to write down the amplitude for this diagram.

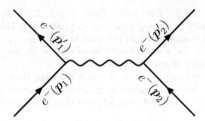

Figure 5.3: A Feynman diagram for electron–electron scattering at the second order in perturbation theory. There is a different diagram at the same order, which has been described in the text.

First, follow one solid line in the diagram, say the one on the left. Moving against the arrows, we encounter first an outgoing electron line, then a vertex, and finally an incoming electron line. Combining the Feynman rules corresponding to each of these elements of the picture, we obtain the factors

$$\overline{u}_{p'_1, s'_1} \, ie\gamma^\mu \, u_{p_1, s_1} . \qquad (5.27)$$

Notice that since we are talking about the electron field, we have used $Q = -1$ in the Feynman rule for the vertex. The subscripts s_1 and s'_1 correspond to the particular form of the positive energy solution that have been created and annihilated at the vertex.

The factors coming from the other solid line are similar, except that the momenta are different. And finally, there is the propagator of the photon field, which we use in the 't Hooft–Feynman gauge. Combining all these factors, we can write the amplitude for the diagram in Fig. 5.3 as

$$\left[\overline{u}_{p'_1, s'_1} \, ie\gamma^\mu \, u_{p_1, s_1} \right] \left(\frac{-ig_{\mu\nu}}{(p_1 - p'_1)^2} \right) \left[\overline{u}_{p'_2, s'_2} \, ie\gamma^\nu \, u_{p_2, s_2} \right] . \qquad (5.28)$$

This is not really the full amplitude at the second order in perturbation theory. The reason is that another diagram contributes to the same process at this order. In the diagram we have shown, the electron with momentum p'_1 is created at the same vertex with the electron with momentum p_1 is annihilated. But that need not be the case. It is also possible that, at the vertex where the electron with momentum p_1 is annihilated, the electron created has the momentum p'_2. This gives another diagram which looks much the same as that in Fig. 5.3, but with the outgoing lines exchanged. Invoking the antisymmetry rule stated in §4.10.4, we can then write down the full amplitude at the second order as

$$i\mathcal{M} = ie^2 \left(\frac{\left[\overline{u}_{p'_1, s'_1} \, \gamma^\mu \, u_{p_1, s_1} \right] \left[\overline{u}_{p'_2, s'_2} \, \gamma_\mu \, u_{p_2, s_2} \right]}{(p_1 - p'_1)^2} \right.$$
$$\left. - \frac{\left[\overline{u}_{p'_2, s'_2} \, \gamma^\mu \, u_{p_1, s_1} \right] \left[\overline{u}_{p'_1, s'_1} \, \gamma_\mu \, u_{p_2, s_2} \right]}{(p_1 - p'_2)^2} \right)$$

$$\equiv ie^2 \left(\frac{T_1}{\mathfrak{t}} - \frac{T_2}{\mathfrak{u}} \right). \tag{5.29}$$

In the last step, we have introduced a shorthand notation for the numerators. The denominators, it should be noted, are precisely the Mandelstam variables $\mathfrak{t}$ and $\mathfrak{u}$.

In calculating the absolute square of the modulus and hence the cross-section, we will neglect the mass of the electron, assuming that the energies involved are much higher than the electron mass. We will also assume that we do not have any preference about the spin orientations of the incoming as well as the outgoing electrons. It means that, for the initial electrons, we have an equal mixture of both kinds of spin, so we will average over them. For the final electrons, we don't care what their spins are, so we sum over them. Since the different spin channels are incoherent, meaning that they can, in principle, be separated out, the contributions from different spin combinations in the external lines should add in the probability and not in the amplitude. Thus, the differential cross-section will be given by

$$\frac{d\sigma}{d\Omega} = \frac{1}{64\pi^2 \mathfrak{s}} \overline{|\mathscr{M}|^2}, \tag{5.30}$$

where

$$\overline{|\mathscr{M}|^2} = \frac{1}{4} \sum_{\text{spins}} \left| \mathscr{M} \right|^2, \tag{5.31}$$

the factor of $\frac{1}{4}$ coming from the averaging of spins in the initial state.

The complex conjugation of the amplitude can be performed by using the rules shown in §F.3.2 of Appendix F, utilizing the rules for spin sums given in Eq. (F.104, *p 749*). This gives

$$\sum_{\text{spins}} T_1 T_1^* = \text{Tr} \left(\not{p}_1 \gamma^\nu \not{p}_1' \gamma^\mu \right) \times \text{Tr} \left(\not{p}_2 \gamma_\nu \not{p}_2' \gamma_\mu \right)$$

$$= 16 \left(p_1^\mu p_1'^\nu + p_1^\nu p_1'^\mu - p_1 \cdot p_1' g^{\mu\nu} \right) \left(p_{2\mu} p_{2\nu}' + p_{2\nu} p_{2\mu}' - p_2 \cdot p_2' g_{\mu\nu} \right)$$

$$= 32 \left((p_1 \cdot p_2)^2 + (p_1 \cdot p_2')^2 \right). \tag{5.32}$$

In writing the last step, we have used kinematic relations like

$$p_1 \cdot p_2 = p_1' \cdot p_2', \qquad p_1' \cdot p_2 = p_1 \cdot p_2', \tag{5.33}$$

which follow from 4-momentum conservation and the equality of the masses of all external particles. Note that in this limit where the electron mass is neglected, the Mandelstam variables are given by

$$\mathfrak{s} = 2p_1 \cdot p_2, \qquad \mathfrak{t} = -2p_1 \cdot p_1', \qquad \mathfrak{u} = -2p_1 \cdot p_2', \tag{5.34}$$

so that we can also write

$$\sum_{\text{spins}} T_1 T_1^* = 8(\mathfrak{s}^2 + \mathfrak{u}^2). \tag{5.35}$$

Then we note that T_2 is obtained from T_1 by interchanging the momenta p_1' and p_2', so that

$$\sum_{\text{spins}} T_2 T_2^* = 32\left((p_1 \cdot p_2)^2 + (p_1 \cdot p_1')^2\right) = 8(\mathsf{s}^2 + \mathsf{t}^2). \tag{5.36}$$

There are also the cross terms. Employing the same methods, we obtain

$$\sum_{\text{spins}} T_1 T_2^* = \text{Tr}\left(\slashed{p}_1 \gamma^\nu \slashed{p}_2' \gamma_\mu \slashed{p}_2 \gamma_\nu \slashed{p}_1' \gamma^\mu\right). \tag{5.37}$$

These can be simplified by using the Dirac matrix contraction formulas shown in §F.1.3 of Appendix F. Thus,

$$\sum_{\text{spins}} T_1 T_2^* = -2\,\text{Tr}\left(\slashed{p}_1 \slashed{p}_2 \gamma_\mu \slashed{p}_2' \slashed{p}_1' \gamma^\mu\right)$$

$$= -8\,\text{Tr}\left(\slashed{p}_1 \slashed{p}_2 p_2' \cdot p_1'\right) = -32(p_1 \cdot p_2)^2 = -8\mathsf{s}^2. \tag{5.38}$$

The other cross term is the complex conjugate of this term. Since this term turned out to be real, the other term would be equal to it. Organizing all the terms obtained, we can now write

$$\overline{|\mathcal{M}|^2} = 2e^4\left(\frac{\mathsf{s}^2 + \mathsf{u}^2}{\mathsf{t}^2} + \frac{\mathsf{s}^2 + \mathsf{t}^2}{\mathsf{u}^2} + \frac{2\mathsf{s}^2}{\mathsf{t}\mathsf{u}}\right). \tag{5.39}$$

Plugging this into Eq. (5.30), we obtain the differential cross-section as

$$\frac{d\sigma}{d\Omega} = \frac{\alpha^2}{2s}\left(\frac{\mathsf{s}^2 + \mathsf{u}^2}{\mathsf{t}^2} + \frac{\mathsf{s}^2 + \mathsf{t}^2}{\mathsf{u}^2} + \frac{2\mathsf{s}^2}{\mathsf{t}\mathsf{u}}\right), \tag{5.40}$$

where, instead of using e for which there are many conventions, we have used the *fine-structure constant*. In the unit of electric charge that we have been using, the relation between them is

$$\alpha = \frac{e^2}{4\pi}. \tag{5.41}$$

Earlier, we commented that the total cross-section is Lorentz invariant. There is no reason why the differential cross-section should be Lorentz invariant. A formula like Eq. (5.40), expressed completely in terms of Mandelstam variables, should not be taken to imply that the differential cross-section is Lorentz invariant. The absolute square of the amplitude is of course Lorentz invariant, as seen from Eq. (5.39). But, in writing the differential cross-section, we have used Eq. (4.202, *p 105*), which is a formula valid only in the CM frame. In any other frame, the square of the matrix element would be multiplied by different factors, thus making the differential cross-section a frame-dependent quantity.

The formula given in Eq. (5.40) can be rewritten by using the scattering angle. In the CM frame that we have been using, the different 4-momenta can be written as follows if we neglect electron mass:

$$p_1^\mu = (E, E\hat{n}), \qquad p_2^\mu = (E, -E\hat{n}),$$
$$p_1'^\mu = (E, E\hat{n}'), \qquad p_2'^\mu = (E, -E\hat{n}'), \tag{5.42}$$

where $\hat{\boldsymbol{n}}$ and $\hat{\boldsymbol{n}}'$ are unit 3-vectors, the angle between them being the scattering angle θ. Then

$$\mathsf{s} = 4E^2, \tag{5.43a}$$

$$\mathsf{t} = -2E^2(1 - \cos\theta) = -4E^2 \sin^2\frac{\theta}{2}, \tag{5.43b}$$

$$\mathsf{u} = -2E^2(1 + \cos\theta) = -4E^2 \cos^2\frac{\theta}{2}. \tag{5.43c}$$

Plugging these back into Eq. (5.40), we obtain

$$\frac{d\sigma}{d\Omega} = \frac{\alpha^2}{2\mathsf{s}} \left(\frac{1 + \cos^4\frac{\theta}{2}}{\sin^4\frac{\theta}{2}} + \frac{1 + \sin^4\frac{\theta}{2}}{\cos^4\frac{\theta}{2}} + \frac{2}{\sin^2\frac{\theta}{2}\cos^2\frac{\theta}{2}} \right). \tag{5.44}$$

Using standard trigonometric identities, this can be put in a form which looks a lot simpler:

$$\frac{d\sigma}{d\Omega} = \frac{\alpha^2}{\mathsf{s}} \left(\frac{1}{\sin^4\frac{\theta}{2}} + \frac{1}{\cos^4\frac{\theta}{2}} + 1 \right). \tag{5.45}$$

The total cross-section can be obtained by integrating this expression over the angles. However, there is an important point that needs to be remembered while performing this integration. The expression for cross-section given in Eq. (4.178, *p 99*) contains integrations over the phase spaces of the final state particles. In the process under discussion here, the final state contains two identical particles. We cannot treat their phase spaces independently. The delta function that appears in the formula ensures that the two electrons will come out back to back in the CM frame. However, one electron going in the direction marked by the unit vector $\hat{\boldsymbol{n}}'$ and the second one going in the direction $-\hat{\boldsymbol{n}}'$ is indistinguishable from the first electron going along $-\hat{\boldsymbol{n}}'$ and the second along $\hat{\boldsymbol{n}}'$, and these two situations cannot be considered different. So we can take $\hat{\boldsymbol{n}}'$ to be in one hemisphere only. While integrating Eq. (5.45) over the solid angle, we should let θ run between 0 and $\pi/2$. Equivalently, we can integrate over the full solid angle but then divide the result by 2 at the end.

☐ **Exercise 5.3** *Verify that Eq. (5.45) indeed follows from Eq. (5.44).*

In passing, we should comment on which features of this result should have been obvious from the beginning. Once we see the diagrams, we know that there are two vertices. The Feynman rule for each vertex has a factor of e, so that the amplitude must contain a factor e^2. In the cross-section, the square of the amplitude is involved, which must therefore contain a factor e^4, i.e., must be proportional to α^2. Then comes the question of dimension. Cross-section has the dimension of the square of length, which is M^{-2} in natural units. So we must have something in the denominator that has the dimension of square of energy. Since we are working in the energy range where the electron mass can be neglected, electron energy is the only independent variable having the dimension of energy. Hence the formula for the cross-section must contain a factor of $1/E^2$, i.e., a factor of $1/\mathsf{s}$. Combining this with the vertex factors, we find that the cross-section must be of the form

$$\sigma = \frac{\alpha^2}{\mathsf{s}} \times (\text{numerical factors}), \qquad (5.46)$$

where the numerical factors cannot be determined without detailed calculation. In the case of differential cross-section, these numerical factors can depend on the scattering angle.

We can use similar arguments to figure out the momentum dependence of the cross-section in the very low energy limit, where the electrons can be considered non-relativistic. Although this limit is not very useful in the context of particle physics where the focus is on high energy, we discuss it in order to gain some experience in dimensional arguments. We see in Eq. (5.29) that the denominators of the Feynman amplitude contain the Mandelstam variables t and u, and this feature is independent of the electron energy: it comes from the photon propagator. It is easy to see that, in the CM frame, the values for these two Mandelstam variables are given by

$$\mathsf{t} = -4\mathbf{p}^2 \sin^2 \frac{\theta}{2}, \qquad \mathsf{u} = -4\mathbf{p}^2 \cos^2 \frac{\theta}{2}, \qquad (5.47)$$

without again any assumption about the magnitude of the electron energy. Thus the Feynman amplitude has an overall factor of $1/\mathbf{p}^2$. In the expression for the cross-section, we will have to square the Feynman amplitude, which will therefore have a factor of $1/\mathbf{p}^4$. In order to obtain the right dimension, we must have a numerator which has the dimension of $(\text{mass})^2$. We can use the electron mass and $\mathbf{p}$ to obtain this: the energy will not give an independent term. Since $\mathbf{p} \ll m$ in the non-relativistic limit, the mass terms will dominate, and so we should obtain

$$\sigma = \frac{\alpha^2 m^2}{\mathbf{p}^4} \times (\text{numerical factors}). \qquad (5.48)$$

It is interesting to note that the momentum dependence in the non-relativistic limit is the same as that obtained by a classical treatment of Rutherford scattering.

☐ **Exercise 5.4** *Verify Eq. (5.47). Remember not to make any assumption about the magnitude of the electron energy.*

☐ **Exercise 5.5** *We have used the 't Hooft–Feynman gauge for the photon propagator in the calculation. More generally, the photon propagator is given by Eq. (4.148, p 92), which contains some extra terms. With the help of the Dirac equation for the spinors, show that these extra terms do not contribute to the amplitude.*

5.3.2   Electron–positron scattering

For an experimental particle physicist, this is a much more important process than electron–electron scattering. The reason is that it is much easier to

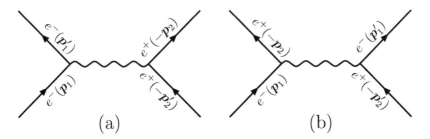

Figure 5.4: Diagrams for electron–positron scattering at the second order in perturbation theory.

arrange high energy collision between an electron and a positron. Because they have opposite electric charges, the electron and the positron can be accelerated in opposite directions by using the same machinery, and then can be made to collide. The initial state has no electric charge, so that in the final state, any particle-antiparticle pair can be produced provided there is enough energy. Thus, these collisions can be used very effectively for new particle searches.

Production of new particles involves inelastic collisions, a typical example of which will be seen in some detail in §5.4.2. Here, we want to discuss the elastic scattering part, which would provide a background to all inelastic processes. Like the electron–electron scattering, there are two diagrams for this elastic scattering process,

$$e^-(\boldsymbol{p}_1) + e^+(\boldsymbol{p}_2) \to e^-(\boldsymbol{p}_1') + e^+(\boldsymbol{p}_2'),\tag{5.49}$$

at the second order of perturbation. These have been shown in Fig. 5.4.

The amplitude for each of these diagrams can be written down easily, following the Feynman rules described earlier. We obtain

$$i\mathcal{M}_a = ie^2\frac{[\overline{u}_{\boldsymbol{p}_1',s_1'}\gamma^\mu u_{\boldsymbol{p}_1,s_1}][\overline{v}_{\boldsymbol{p}_2,s_2}\gamma_\mu v_{\boldsymbol{p}_2',s_2'}]}{(p_1-p_1')^2} \equiv ie^2 T_1/\mathfrak{t},\tag{5.50a}$$

$$i\mathcal{M}_b = ie^2\frac{[\overline{v}_{\boldsymbol{p}_2,s_2}\gamma^\mu u_{\boldsymbol{p}_1,s_1}][\overline{u}_{\boldsymbol{p}_1',s_1'}\gamma_\mu v_{\boldsymbol{p}_2',s_2'}]}{(p_1+p_2)^2} \equiv ie^2 T_2/\mathfrak{s},\tag{5.50b}$$

introducing the shorthands T_1 and T_2, and using the Mandelstam variables $\mathfrak{s}$ and $\mathfrak{t}$. The total amplitude is given by

$$\mathcal{M} = \mathcal{M}_a - \mathcal{M}_b,\tag{5.51}$$

since the two diagrams differ by the exchange of two of the external lines which are both fermionic lines.

At this point, we want to introduce a little bit of jargon. Since the amplitude of Eq. (5.50b) contains the Mandelstam variable $\mathfrak{s}$ in the denominator, the corresponding diagram, i.e., the one shown in Fig. 5.4b, is often called an $\mathfrak{s}$-*channel diagram*. By the same token, the diagram in Fig. 5.4a can be called a $\mathfrak{t}$-*channel diagram*.

We are interested in calculating the cross-section for the case where the initial state has unpolarized particles, and the final state spins are summed over. It means that we need the quantity

$$\overline{|\mathcal{M}|^2} = \frac{1}{4} \sum_{\text{spins}} |\mathcal{M}|^2$$

$$= \frac{e^4}{4} \sum_{\text{spins}} \left(\frac{T_1 T_1^*}{t^2} + \frac{T_2 T_2^*}{s^2} - \frac{T_1 T_2^* + T_2 T_1^*}{st} \right). \tag{5.52}$$

Here,

$$\sum_{\text{spins}} T_1 T_1^* = \text{Tr} \left(\not{p}_1 \gamma^\nu \not{p}_1' \gamma^\mu \right) \times \text{Tr} \left(\not{p}_2 \gamma_\mu \not{p}_2' \gamma_\nu \right)$$

$$= 8(s^2 + u^2), \tag{5.53}$$

neglecting electron mass and following some the same steps that we had gone through for the electron–electron scattering. Similarly,

$$\sum_{\text{spins}} T_2 T_2^* = \text{Tr} \left(\not{p}_1 \gamma^\nu \not{p}_2 \gamma^\mu \right) \text{Tr} \left(\not{p}_2' \gamma^\nu \not{p}_1' \gamma^\mu \right)$$

$$= 32 \left((p_1 \cdot p_2')^2 + (p_1 \cdot p_1')^2 \right) = 8(u^2 + t^2). \tag{5.54}$$

The cross terms can also be evaluated in a similar manner, yielding

$$\sum_{\text{spins}} T_1 T_2^* = \sum_{\text{spins}} T_2 T_1^* = -32(p_1 \cdot p_2')^2 = -8u^2. \tag{5.55}$$

Finally, then, we obtain

$$\frac{d\sigma}{d\Omega} = \frac{\alpha^2}{2s} \left(\frac{s^2 + u^2}{t^2} + \frac{u^2 + t^2}{s^2} + \frac{2u^2}{st} \right). \tag{5.56}$$

This can be easily written in terms of the scattering angle by using Eq. (5.43):

$$\frac{d\sigma}{d\Omega} = \frac{\alpha^2}{2s} \left(\frac{1 + \cos^4 \frac{\theta}{2}}{\sin^4 \frac{\theta}{2}} + (\cos^4 \frac{\theta}{2} + \sin^4 \frac{\theta}{2}) - \frac{2 \cos^4 \frac{\theta}{2}}{\sin^2 \frac{\theta}{2}} \right). \tag{5.57}$$

It looks a little less intimidating if we use the trigonometric identity

$$\cos^4 \frac{\theta}{2} + \sin^4 \frac{\theta}{2} = \frac{1}{2}(1 + \cos^2 \theta). \tag{5.58}$$

5.3.3   Compton scattering

Compton scattering is the name given to the elastic scattering of a free electron with a photon:

$$e^-(p) + \gamma(k) \to e^-(p') + \gamma(k'). \tag{5.59}$$

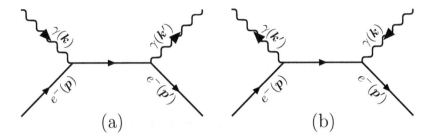

Figure 5.5: Diagrams for Compton scattering at the second order in perturbation theory.

There are two diagrams corresponding to this process at the second order, shown in Fig. 5.5.

Let us write the amplitude in the form

$$i\mathcal{M} = e^2 \epsilon_\mu(\boldsymbol{k}) \epsilon_{\mu'}^*(\boldsymbol{k}') \overline{u}(\boldsymbol{p}') \, i\Gamma^{\mu\mu'} \, u(\boldsymbol{p}), \tag{5.60}$$

separating out the factors which will be the same for both diagrams. The difference between the diagrams lies in $\Gamma^{\mu\mu'}$, which receives the following contribution from the two diagrams:

$$i\Gamma_a^{\mu\mu'} = i\gamma^{\mu'} \frac{i(\not{p} + \not{k} + m)}{(p+k)^2 - m^2} \, i\gamma^\mu,$$

$$i\Gamma_b^{\mu\mu'} = i\gamma^\mu \frac{i(\not{p} - \not{k}' + m)}{(p-k')^2 - m^2} \, i\gamma^{\mu'}. \tag{5.61}$$

The total $\Gamma^{\mu\mu'}$ is equal to $\Gamma_a^{\mu\mu'} + \Gamma_b^{\mu\mu'}$.

The absolute square of the matrix element, summed over final as well as averaged over initial spin and polarization, is given by

$$\overline{|\mathcal{M}|^2} = \frac{e^4}{4} \left(\sum_{\text{pol}} \epsilon_\mu(\boldsymbol{k}) \epsilon_\nu(\boldsymbol{k})^* \right) \left(\sum_{\text{pol}} \epsilon_{\mu'}(\boldsymbol{k}')^* \epsilon_{\nu'}(\boldsymbol{k}') \right)$$

$$\times \sum_{\text{spins}} \left[\overline{u}(\boldsymbol{p}') \, \Gamma^{\mu\mu'} \, u(\boldsymbol{p}) \right] \left[\overline{u}(\boldsymbol{p}') \, \Gamma^{\nu\nu'} \, u(\boldsymbol{p}) \right]^*. \tag{5.62}$$

The polarization sums, of course, are on the physical states of polarization. To perform the sums, we use Eq. (4.36, *p 68*) and obtain

$$\overline{|\mathcal{M}|^2} = \frac{e^4}{4} \sum_{\text{spins}} \left[\overline{u}(\boldsymbol{p}') \, \Gamma^{\mu\mu'} \, u(\boldsymbol{p}) \right] \left[\overline{u}(\boldsymbol{p}') \, \Gamma_{\mu\mu'} \, u(\boldsymbol{p}) \right]^*$$

$$= \frac{e^4}{4} \, \text{Tr} \left[(\not{p} + m) \Gamma_{\mu\mu'}^\ddagger (\not{p}' + m) \Gamma^{\mu\mu'} \right], \tag{5.63}$$

where, for any matrix F that is either a Dirac matrix, or some combination of them including the unit matrix, we use the notation

$$F^\ddagger = \gamma_0 F^\dagger \gamma_0. \tag{5.64}$$

We digress here to talk about a property of Feynman amplitudes involving photons as external lines. If a process has n external photons, we can write the amplitude as

$$\mathcal{M} = \epsilon_{\mu_1}(k_1) \cdots \epsilon_{\mu_n}(k_n) \mathcal{M}^{\mu_1 \cdots \mu_n}, \tag{5.65}$$

separating out the polarization vectors. For the photons in the final state, the polarization vectors will be complex conjugated, something which is not shown in the expression above. Now, Eq. (5.10) shows the effect of a gauge transformation. If we write the same formula in terms of the Fourier expansion, it will show that a gauge transformation involves the change

$$\epsilon_\mu \rightarrow \epsilon'_\mu = \epsilon_\mu + k_\mu \times \text{(arbitrary number)} \tag{5.66}$$

for any photon. Since the theory is gauge invariant, such changes can leave amplitudes unchanged only if

$$(k_1)_{\mu_1} \mathcal{M}^{\mu_1 \cdots \mu_n} = 0, \tag{5.67}$$

and similar conditions with other momenta are also met. For the amplitude of Compton scattering, it means that we should have

$$k_\mu \bar{u}(p') \Gamma^{\mu\mu'} u(p) = 0, \qquad k'_{\mu'} \bar{u}(p') \Gamma^{\mu\mu'} u(p) = 0, \tag{5.68}$$

which can be verified in a straightforward manner.

This feature has another important implication. In deriving the polarization sum formula in §4.2, we first showed that the sum over physical polarization vectors is given by Eq. (4.35, *p 68*), and then commented that in this sum, the terms containing the momentum vector of the photon do not contribute to physical amplitude. The reason for this comment is now clear: if there is a term in the polarization sum containing the factor k_μ, this term does not contribute because of the gauge invariance condition, Eq. (5.68).

□ **Exercise 5.6** *Take the expressions given in Eq. (5.61) and verify Eq. (5.68).* [**Note :** *Don't forget to use the Dirac equation for the spinors, and also the 4-momentum conservation equation $p + k = p' + k'$.*]

For the rest of the calculation, we will assume that the energies are high so that the electron mass can be neglected in $\overline{|\mathcal{M}|^2}$. Using the expression of $\Gamma^{\mu\mu'}$ from Eq. (5.61), we obtain

$$\overline{|\mathcal{M}|^2} = \frac{e^4}{4} \operatorname{Tr}\left[\not{p} \left(\frac{\gamma_\mu(\not{p} + \not{k})\gamma_{\mu'}}{\mathfrak{s}} + \frac{\gamma_{\mu'}(\not{p} - \not{k'})\gamma_\mu}{\mathfrak{u}} \right) \right.$$
$$\left. \not{p'} \left(\frac{\gamma^{\mu'}(\not{p} + \not{k})\gamma^\mu}{\mathfrak{s}} + \frac{\gamma^\mu(\not{p} - \not{k'})\gamma^{\mu'}}{\mathfrak{u}} \right) \right]. \tag{5.69}$$

The traces can be evaluated in a straightforward manner. For example, the term containing $1/\mathfrak{s}^2$ in the trace is simplified as follows, with the help of the cyclicity of the trace operation and the contraction formulas given in §F.1.3 of Appendix F:

$$\frac{1}{\mathfrak{s}^2} \operatorname{Tr}\left[\gamma^\mu \not{p} \gamma_\mu (\not{p} + \not{k}) \gamma_{\mu'} \not{p'} \gamma^{\mu'} (\not{p} + \not{k}) \right] = \frac{4}{\mathfrak{s}^2} \operatorname{Tr}\left[\not{p}(\not{p} + \not{k})\not{p'}(\not{p} + \not{k}) \right]. \tag{5.70}$$

We can now use the trace formulas, along with the expressions for the Mandelstam variables,

$$\mathfrak{s} = 2p \cdot k, \qquad \mathfrak{t} = -2p \cdot p', \qquad \mathfrak{u} = -2p \cdot k', \tag{5.71}$$

which are valid when the electron mass is neglected. These relations allow us
to write

$$\frac{4}{s^2}\,\mathrm{Tr}\left[\slashed{p}(\slashed{p}+\slashed{k})\slashed{p}'(\slashed{p}+\slashed{k})\right] = -\frac{8u}{s}\,. \tag{5.72}$$

The term with u in both denominators will similarly give a contribution of
$-8s/u$. Finally, there are two interference terms. One of them will produce a
contribution

$$\begin{aligned}\frac{1}{su}\,\mathrm{Tr}\left[\slashed{p}\gamma_\mu(\slashed{p}+\slashed{k})\gamma_{\mu'}\slashed{p}'\gamma^\mu(\slashed{p}-\slashed{k}')\gamma^{\mu'}\right] &= -\frac{2}{su}\,\mathrm{Tr}\left[\slashed{p}\slashed{p}'\gamma_{\mu'}(\slashed{p}+\slashed{k})(\slashed{p}-\slashed{k}')\gamma^{\mu'}\right]\\ &= -\frac{32}{su}(p+k)\cdot(p-k')p\cdot p'\\ &= \frac{8(s+u+t)t}{su}\,,\end{aligned} \tag{5.73}$$

which vanishes if the electron mass is neglected, as can be seen from Eq.
(4.200, *p 105*). Similarly, the other interference term vanishes in this limit,
and we obtain

$$\overline{|\mathscr{M}|^2} = -2e^4\,\frac{s^2+u^2}{su}\,. \tag{5.74}$$

The differential cross-section in the CM frame can then be written down easily
by using Eq. (4.207, *p 106*). We obtain

$$\frac{d\sigma}{d\Omega} = -\frac{\alpha^2}{2s}\,\frac{s^2+u^2}{su}\,. \tag{5.75}$$

The overall minus sign might look perplexing, but we need to remember that
the Mandelstam variable u is negative.

☐ **Exercise 5.7** Obtain the leading contribution to the differential cross-
section in the FT frame where the initial electron is at rest. Express
the answer in terms of the initial photon energy ω and the scattering
angle θ.

☐ **Exercise 5.8** Redo the calculation in the opposite limit, i.e., for $\omega \ll m$. Show that the differential cross-section in this limit, in the FT
frame of the electron, is given by

$$\frac{d\sigma}{d\Omega} = \frac{\alpha^2}{2m^2}(1+\cos^2\theta)\,, \tag{5.76}$$

where θ is the angle between the directions of k and k'.

☐ **Exercise 5.9** On integrating Eq. (5.76) over the scattering angle, one
obtains

$$\sigma = \frac{8\pi\alpha^2}{3m^2}\,, \tag{5.77}$$

an expression obtained from classical electromagnetic theory for the
cross-section of scattering of electromagnetic radiation by a free electron. Find this cross-section in the unit cm². [**Note :** *This cross-section
is called the* Thomson cross-section, *and is often used as a benchmark value for electromagnetic cross-sections.*]

☐ **Exercise 5.10** Suppose that the initial photon is virtual, with an invariant 4-momentum squared equal to Ω^2. Show that in this case one obtains

$$\overline{|\mathscr{M}|^2} = -2e^4 \, \frac{s^2 + u^2 - 2t\Omega^2}{su}.$$

(5.78)

5.4   Inelastic scattering at second order

A scattering is inelastic when the final state particles and the initial state particles are not the same. We discuss a few examples of such processes.

5.4.1   Pair creation and pair annihilation

By 'pair creation', we mean the creation of an electron–positron pair from the annihilation of two photons, i.e.,

$$\gamma(k_1) + \gamma(k_2) \to e^-(p_1) + e^+(p_2).$$

(5.79)

'Pair annihilation' means the opposite process, in which an electron–positron pair annihilates into two photons:

$$e^-(p_1) + e^+(p_2) \to \gamma(k_1) + \gamma(k_2).$$

(5.80)

The diagrams for these processes are very similar to those responsible for Compton scattering. In fact, they are so similar that we need not display them separately. Just take a diagram for the Compton process, reinterpret an outgoing photon line as an incoming photon line, and an incoming electron line as an outgoing positron line, and the result is a pair creation process. Note that the changes that we just mentioned do not need any different combination of field operators. The same operator that can create a photon can also annihilate a photon. Similarly, the same field operator that annihilates an electron can create a positron. Processes which are related by renaming of this sort are said to be related by *crossing symmetry*.

When two processes are related like this, the amplitude squared of one of them can be easily found from the amplitude squared of the other. To see how, let us write down the renamings or redefinitions needed for going from the Compton scattering process of Eq. (5.59) to the pair creation process of Eq. (5.79). As described a bit earlier, these would be

$$k \to k_1 , \quad k' \to -k_2 , \quad p \to -p_2 , \quad p' \to p_1 .$$

(5.81)

The negative sign over an entire 4-vector changes an incoming line to an outgoing line, and vice versa. Now look at the Mandelstam invariants for the two processes, and make a little chart of what happens to each of them as we make the transformations shown in Eq. (5.81):

Compton process		Pair creation	
$\mathfrak{s} \equiv (p+k)^2$	$\Longrightarrow$	$(-p_2 + k_1)^2 \equiv$	$\mathfrak{t}$
$\mathfrak{t} \equiv (p-p')^2$	$\Longrightarrow$	$(-p_2 - p_1)^2 \equiv$	$\mathfrak{s}$
$\mathfrak{u} \equiv (p-k')^2$	$\Longrightarrow$	$(-p_2 + k_2)^2 \equiv$	$\mathfrak{u}$

(5.82)

It means that the roles of the Mandelstam variables $\mathfrak{s}$ and $\mathfrak{t}$ have inter-changed. We should make the same adjustments while writing the matrix element squared for the process at hand. In addition, we should introduce an extra minus sign for each fermion that has been crossed from the initial state to the final state, or vice versa. In this case, there is only one such crossing, so the square of the matrix element can be obtained from Eq. (5.74) by making this replacement:

$$\overline{|\mathcal{M}|^2} = 2e^4 \, \frac{\mathfrak{t}^2 + \mathfrak{u}^2}{\mathfrak{t}\mathfrak{u}} \,. \tag{5.83}$$

The differential cross-section for pair creation is therefore given by

$$\frac{d\sigma}{d\Omega} = \frac{\alpha^2}{2\mathfrak{s}} \frac{\mathfrak{t}^2 + \mathfrak{u}^2}{\mathfrak{t}\mathfrak{u}} \,. \tag{5.84}$$

□ **Exercise 5.11** Find the differential cross-section for the pair creation process using crossing symmetry.

□ **Exercise 5.12** Verify that $\overline{|\mathcal{M}|^2}$ for electron–positron scattering can be deduced from Eq. (5.39) by using crossing symmetry.

5.4.2 Muon pair production

In §5.1, we showed how the requirement of local phase invariance leads to the introduction of the photon field. It of course does not matter which field ψ we started with: we would have reached the same conclusion. Moreover, notice that the transformation property of the photon field, shown in Eq. (5.10), does not depend on any characteristic of the fermion field ψ. For example, it does not depend on the parameter Q which denotes the charge of the particle corresponding to the field ψ. Thus, no matter which field we start with, or how many of them, it is the same photon field that can ensure local gauge invariance.

So suppose we want to discuss some process where both electrons and muons are involved. We will write down the free Lagrangians for both fields. Then we will consider phase rotations of these fields. Both fields will have $Q = -1$, since muons have the same charge as the electrons. Once we make the symmetry local, the photon field will come in and the muon field will have the same kind of interaction term with the photon field. With this added interaction, we will be able to deal with many more kinds of scattering processes. One such process, which is very important, is

$$e^-(\boldsymbol{p}_1) + e^+(\boldsymbol{p}_2) \to \mu^-(\boldsymbol{p}_1') + \mu^+(\boldsymbol{p}_2') \,. \tag{5.85}$$

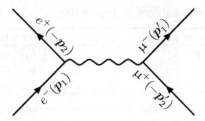

Figure 5.6: Diagram for muon–antimuon production at the second order in perturbation theory.

Here, a muon–antimuon pair is produced from the electron–positron pair. Obviously, since the masses in the initial and final states are different, it is an inelastic process.

The process has similarities with electron–positron elastic scattering that was discussed in §5.3.2. As for diagrams, we will have a diagram that looks like Fig. 5.4b *(p 125)*, except that the final state lines will now correspond to a muon and an antimuon. It has been shown in Fig. 5.6, but the similarity ends there, because there is no diagram that would look like Fig. 5.4a *(p 125)*. The reason is that, the photon field couples to only one kind of field: it does not have any vertex that involves both the electron and the muon fields. This is obvious from the derivation given in §5.1.

Given the diagram, we can write down the amplitude easily. We get

$$i\mathcal{M} = [\bar{v}_{\boldsymbol{p}_2} ie\gamma^\mu u_{\boldsymbol{p}_1}] \left(\frac{-ig_{\mu\nu}}{(p_1 + p_2)^2} \right) [\bar{u}_{\boldsymbol{p}'_1} ie\gamma^\nu v_{\boldsymbol{p}'_2}], \qquad (5.86)$$

or

$$\mathcal{M} = \frac{e^2}{\mathfrak{s}} [\bar{v}_{\boldsymbol{p}_2} \gamma^\mu u_{\boldsymbol{p}_1}][\bar{u}_{\boldsymbol{p}'_1} \gamma_\mu v_{\boldsymbol{p}'_2}]. \qquad (5.87)$$

We have not shown the other indices on the spinors: they are implied. Also, we are using the same notation for the electron spinors and muon spinors: the primes on the momenta will remind us that the corresponding spinor belongs to the muon field.

The muon is more than 200 times heavier than the electron. So, in order that the given process is possible, the electron–positron pair in the initial state must possess a huge energy, at least equal to the rest energy of the muon–antimuon pair. At such high energies, the electron mass can be neglected. So we obtain

$$\overline{|\mathcal{M}|^2} = \frac{1}{4} \sum_{\text{spins}} \left| \mathcal{M} \right|^2$$

$$= \frac{e^4}{4\mathfrak{s}^2} \text{Tr} \left(\not{p}_1 \gamma^\nu \not{p}_2 \gamma^\mu \right) \text{Tr} \left((\not{p}'_2 - M)\gamma_\nu (\not{p}'_1 + M)\gamma_\mu \right), \qquad (5.88)$$

where M is the mass of the muon. Note that we have averaged over the initial spins and summed over the final ones. Evaluating the traces in the usual way, we obtain

$$\overline{|\mathcal{M}|^2} = \frac{4e^4}{s^2} \left(p_1^\mu p_2^\nu + p_1^\nu p_2^\mu - p_1 \cdot p_2 g^{\mu\nu} \right)$$
$$\times \left(p'_{1\mu} p'_{2\nu} + p'_{1\nu} p'_{2\mu} - (p'_1 \cdot p'_2 + M^2) g_{\mu\nu} \right)$$
$$= \frac{8e^4}{s^2} \left((p_1 \cdot p'_1)^2 + (p_1 \cdot p'_2)^2 + M^2 p_1 \cdot p_2 \right). \tag{5.89}$$

In writing the last step, we have used kinematical relations like

$$p_1 \cdot p'_1 = p_2 \cdot p'_2, \qquad p_1 \cdot p'_2 = p_2 \cdot p'_1, \tag{5.90}$$

which follow from momentum conservation.

In the CM frame, we can write the different 4-momenta as

$$p_1^\mu = (E, \mathbf{p}\hat{\mathbf{n}}), \qquad p_2^\mu = (E, -\mathbf{p}\hat{\mathbf{n}}),$$
$$p_1'^\mu = (E, \mathbf{p}'\hat{\mathbf{n}}'), \qquad p_2'^\mu = (E, -\mathbf{p}'\hat{\mathbf{n}}'), \tag{5.91}$$

where $\hat{\mathbf{n}}$ and $\hat{\mathbf{n}}'$ are unit 3-vectors, with $\hat{\mathbf{n}} \cdot \hat{\mathbf{n}}' = \cos\theta$, θ being the scattering angle. Since we have neglected the electron mass, we have $\mathbf{p} = E = \frac{1}{2}\sqrt{s}$. This gives

$$\mathbf{p}' = \sqrt{E^2 - M^2} = \frac{1}{2}\sqrt{s - 4M^2}. \tag{5.92}$$

Putting these in, we obtain

$$\overline{|\mathcal{M}|^2} = e^4 \left(1 + \frac{1}{E^2} (\mathbf{p}'^2 \cos^2\theta + M^2) \right)$$
$$= e^4 \left((1 + \cos^2\theta) + \frac{M^2}{E^2} (1 - \cos^2\theta) \right). \tag{5.93}$$

Putting this result into Eq. (4.207, *p 106*) and using Eq. (5.92), we obtain

$$\frac{d\sigma}{d\Omega} = \frac{\alpha^2}{4s} \sqrt{1 - \frac{4M^2}{s}} \left((1 + \cos^2\theta) + \frac{4M^2}{s} (1 - \cos^2\theta) \right). \tag{5.94}$$

The total cross-section is given by

$$\sigma = \int_{-1}^{+1} d(\cos\theta) \int_0^{2\pi} d\varphi \, \frac{d\sigma}{d\Omega}$$
$$= \frac{4\pi\alpha^2}{3s} \sqrt{1 - \frac{4M^2}{s}} \left(1 + \frac{2M^2}{s} \right). \tag{5.95}$$

□ **Exercise 5.13** ♪ *Express the angular cross-section in terms of the Mandelstam variables. At high energies where the muon mass can be neglected, show that*

$$\frac{d\sigma}{dt} = \frac{2\pi\alpha^2}{s^2} \frac{t^2 + u^2}{s^2} \tag{5.96}$$

using Eq. (4.215, p 108).

□ **Exercise 5.14** What will be the differential cross-section for the elastic scattering $e\mu \to e\mu$ at energies when both electron and muon masses can be neglected? [**Hint** : *Use crossing symmetry.*]

There are many interesting aspects of these results that we have deduced. As a sample, let us take a look at the differential cross-section in Eq. (5.94). We notice that its angular dependence comes only through $\cos^2 \theta$, which implies that the probability for a muon coming out at a certain angle θ with the original electron direction is the same as that for coming out at the angle $\pi - \theta$. When this happens, we say that the process is *forward-backward symmetric*, or alternatively that the *forward-backward asymmetry* of the process is zero. This was also the case with the electron–electron scattering, as seen from Eq. (5.45). But in that case, the reason is easy to understand. There are two electrons produced in the final state. If one of them goes at an angle θ with the original beam direction, the other must go at an angle $\pi - \theta$. Thus, the number of scattered particles at angles θ and $\pi - \theta$ must be the same. But when a muon–antimuon pair is produced, this argument does not hold. We can ask, what is the reason that this scattering turns out to be forward-backward symmetric? The answer depends on matters that we will discuss later, and will appear in §16.7.

Let us now look at another aspect. Suppose we are performing an experiment where electrons and positrons are collided, and we are measuring the total cross-section, irrespective of the final state. In other words, we are measuring the cross-section of electron and positron going to anything in the final state. Also suppose that we are ignorant about any particle of higher mass, so that we are expecting only the elastic scattering. The elastic cross-section goes down like $1/s$ for large CM energy, a comment made in the context of electron–electron collision in Eq. (5.46) but is obviously also valid in this case. So, if we plot the quantity σs versus s itself, we will obtain a horizontal straight line if elastic scattering is the only contributing process. As we keep increasing s and go past a certain value, a new channel opens in the form of muon–antimuon, and there is an extra contribution to the cross-section. The value of σs then rises to a higher value. The higher value then indicates that the CM energy of the electron or the positron is now higher than the mass of another particle. Thus, by looking at electron–positron cross-section, we can discover new particles. Said another way, electron–positron collision is a very good probe for discovering new particles. Many high energy machines use this technique.

Why are we making such a big deal about electrons and positrons? Can we not do the same thing by colliding, say, muons with antimuons, or protons with antiprotons? The problem is that muons are unstable, with a short lifetime. So it is not easy to create and maintain a beam of them. Protons do not have a problem with lifetimes, and they can of course be used. They are used in some of the biggest machines in the world. There are advantages and disadvantages of using proton-antiproton machines compared to electron–positron machines, some of which will be discussed in Ch. 9.

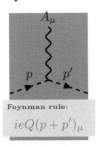

Figure 5.7: Feynman rules for scalar QED vertices.

5.5 Scalar QED

Electromagnetic interactions of scalar particles can also be described through minimal substitution, as has been hinted in Ex. 5.1 *(p 114)*. The Lagrangian for this theory is

$$\mathscr{L} = \left(D_\mu \phi\right)^\dagger \left(D^\mu \phi\right) - m^2 \phi^\dagger \phi - \frac{1}{4} F_{\mu\nu} F^{\mu\nu} - \frac{1}{2\xi}(\partial_\mu A^\mu)^2 \,, \quad (5.97)$$

where $\phi(x)$ is a complex scalar field and D_μ, as before, is given by Eq. (5.8). Note that this Lagrangian contains two different kinds of vertices involving the photon field: a cubic vertex with one photon field, and a quartic vertex with two photon fields. The Feynman rules for these vertices are shown in Fig. 5.7.

The Noether current for the free Lagrangian of a scalar field was given in Eq. (4.116, *p 83*). It should be remembered that Noether currents come from terms in the Lagrangian that contain derivatives of fields, as seen in Eq. (4.109, *p 82*). With the introduction of photon field in the Lagrangian through the prescription of minimal substitution, there are now new terms involving $\partial_\mu \phi$ and $\partial_\mu \phi^\dagger$ which were not present in the free Lagrangian of the scalar field. As a result, the expression for the Noether current will change:

$$J_\mu = -ieQ(\partial_\mu \phi^\dagger)\phi + ieQ(\partial_\mu \phi)\phi^\dagger - 2(eQ)^2 \phi^\dagger \phi A_\mu \,. \quad (5.98)$$

☐ **Exercise 5.15** Derive the expression for the Noether current given in Eq. (5.98) and show that it is gauge invariant.

☐ **Exercise 5.16** Derive the Euler–Lagrange equation for the photon field from Eq. (5.97) and verify that it is of the standard form given in Eq. (4.19, *p 66*), where the current density 4-vector is given by the expression on the right hand side of Eq. (5.98).

5.6   Multi-photon states

5.6.1   Generalities

A state containing n photons is obviously defined as

$$|k_1, \epsilon_1; k_2, \epsilon_2; \cdots k_n, \epsilon_n\rangle = a^\dagger(k_n, \epsilon_n) \cdots a^\dagger(k_2, \epsilon_2) a^\dagger(k_1, \epsilon_1) |0\rangle , \quad (5.99)$$

where $a^\dagger(k_1, \epsilon_1)$ denotes the creation operator for a photon with 4-momentum k_1 and polarization vector ϵ_1. There will in general be a normalizing factor on the right hand side as well, but that is not relevant for our discussion here.

If we write down the wavefunction of the state in Eq. (5.99) in the momentum representation, it will of course contain the momenta and the polarization vectors of the photons. Polarization vector of a photon is not uniquely defined. It depends on the choice of gauge, as remarked earlier in Eq. (4.27, *p 67*). For the present purpose, it is most convenient to use a gauge in which the time components of all polarization vectors are zero. As for the momentum 4-vector, we notice that its time component, i.e., the energy, is not independent of the 3-momentum. Thus, we can use only the 3-vectors of polarization and momentum for each photon in the expression for the states. Also, we cannot use the dot product of the momentum and polarization of the same photon, since

$$\epsilon_1 \cdot \boldsymbol{k}_1 = 0 \qquad\qquad (5.100)$$

in this gauge, along with similar equations for all other photons.

How will a multi-photon state be created in a physical process? It will of course be created through interactions, and each photon will be created from a factor of the photon field operator A^μ in the Lagrangian. While explaining the Feynman rules for photons, we noted in Table 4.1 *(p 91)* that each photon creation should provide a factor of ϵ_μ^* to the amplitude. Thus, the n-photon state should have one factor — no more, no less — of the polarization vector for each of the photons. In what follows, we will assume that we have taken linear polarization vectors which have real components, and therefore do not write the complex conjugation sign.

The rotational properties of the wavefunction will depend on the angular momentum of the state. For example, if we want to construct a state with zero angular momentum, $J = 0$, the wavefunction should be a scalar. For a $J = 1$ state, the wavefunction should be a vector.

Finally, since photons are bosons, all multi-photon states must be Bose-symmetric, i.e., they must be completely symmetric under the exchange of any two photons. With these things in mind, let us now explicitly see the wavefunctions with two and three photons. We will proceed as follows. First, we will disregard Bose symmetry and construct the wavefunction consistent with all other requirements. Then we will invoke Bose symmetry and pick up only the symmetric wavefunctions.

5.6.2   Two-photon states

a) $J = 0$ states

Let us first consider $J = 0$ states of two photons. We can discuss the problem in the center-of-mass frame of the two photons, so that the momenta of the photons can be written as

$$\boldsymbol{k}_1 = \boldsymbol{k}\,, \qquad \boldsymbol{k}_2 = -\boldsymbol{k}\,. \tag{5.101}$$

This $\boldsymbol{k}$, and the polarization vectors of the two photons should be used to construct the wavefunction.

As stated earlier, the wavefunction must have one power of each polarization. We can construct the scalar

$$\epsilon_1 \cdot \epsilon_2 f(\mathrm{k})\,, \tag{5.102}$$

where $\mathrm{k} = |\boldsymbol{k}|$, and $f(\mathrm{k})$ denotes an arbitrary function of the argument. This is Bose symmetric, and is therefore an acceptable state. Another example of a scalar is

$$\epsilon_1 \times \epsilon_2 \cdot \boldsymbol{k} f(\mathrm{k})\,. \tag{5.103}$$

The cross product is antisymmetric under the interchange of the two photons, but notice that $\boldsymbol{k}$ changes sign under such an interchange, so that overall the expression is symmetric. It is therefore a different $J = 0$ state of two photons.

b) $J = 1$ states

We now try to construct $J = 1$ states with two photons. These wave functions should behave like vectors. Without involving the momentum, we can construct the combination

$$\epsilon_1 \times \epsilon_2 f(\mathrm{k})\,. \tag{5.104}$$

But this is antisymmetric under the interchange of the two photons, and therefore is not allowed. For the same reason,

$$(\epsilon_1 \cdot \epsilon_2)\boldsymbol{k} f(\mathrm{k}) \tag{5.105}$$

is not allowed. How about

$$(\epsilon_1 \times \epsilon_2) \times \boldsymbol{k} f(\mathrm{k})? \tag{5.106}$$

It passes the interchange test because it is symmetric, but it is of no use because the triple cross product is in fact zero. This can be easily seen from the formula

$$(\boldsymbol{A} \times \boldsymbol{B}) \times \boldsymbol{C} = (\boldsymbol{A} \cdot \boldsymbol{C})\boldsymbol{B} - (\boldsymbol{B} \cdot \boldsymbol{C})\boldsymbol{A} \tag{5.107}$$

that applies for any three vectors $\boldsymbol{A}$, $\boldsymbol{B}$ and $\boldsymbol{C}$.

Here we run out of possibilities. There does not seem to be any vector combination of polarizations and momenta of two photons that is Bose symmetric. The conclusion is that two photons cannot be in a $J = 1$ state. This is called Yang's theorem.

☐ **Exercise 5.17** Prove Eq. (5.107) by using the identity given in Eq. (D.22a, *p 731*) of Appendix D.

5.6.3   Three-photon states

a) $J = 0$ states

Consider the following wavefunctions:

$$\Psi_1 = (\epsilon_1 \times \epsilon_2) \cdot \epsilon_3 f(k_1, k_2, k_3) \tag{5.108}$$
$$\Psi_2 = \epsilon_1 \cdot \boldsymbol{k}_{23}\, \epsilon_2 \cdot \boldsymbol{k}_{31}\, \epsilon_3 \cdot \boldsymbol{k}_{12}\, f(k_1, k_2, k_3) . \tag{5.109}$$

where

$$\boldsymbol{k}_{12} \equiv \boldsymbol{k}_1 - \boldsymbol{k}_2 \tag{5.110}$$

etc. For $J = 0$ states, the wavefunction should be a scalar under rotation. Thus, the function $f(k_1, k_2, k_3)$ should be a scalar. As long as this scalar is antisymmetric under the interchange of any two photons, the overall wavefunction would be symmetric and would therefore be acceptable. An example of a completely antisymmetric function is

$$f(k_1, k_2, k_3) = (\omega_2 - \omega_3)(\omega_3 - \omega_1)(\omega_1 - \omega_2) , \tag{5.111}$$

where $\omega = |\boldsymbol{k}|$ for any photon.

b) $J = 1$ states

For $J = 1$, we need vector combinations. Here are some:

$$\Psi_1 = \Big[(\epsilon_1 \cdot \epsilon_2)\epsilon_3 + (\epsilon_2 \cdot \epsilon_3)\epsilon_1 + (\epsilon_3 \cdot \epsilon_1)\epsilon_2\Big] F(k_1, k_2, k_3) \tag{5.112}$$
$$\Psi_2 = \Big[(\epsilon_1 \cdot \epsilon_2)(\boldsymbol{k}_{12} \times \epsilon_3) + \text{cyclic permutations}\Big] f(k_1, k_2, k_3) . \tag{5.113}$$

Here, $F(k_1, k_2, k_3)$ must be completely symmetric and $f(k_1, k_2, k_3)$ completely antisymmetric in order that the states are symmetric under the interchange of any two photons.

5.7   Higher-order effects

5.7.1   Electromagnetic form factors

The lowest order contribution to the QED vertex was shown in Fig. 5.2 *(p 118)*, where also the Feynman rule for this vertex was presented. This is the basic

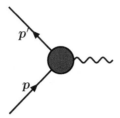

Figure 5.8: General form for the coupling between a fermion and the photon. The momenta carried by the fermions are shown.

vertex. But, in an experiment, we may not always probe the basic vertex. If we see the coupling of a fermion with a single photon, all we can say is that we are seeing the effect of something that has been schematically shown in Fig. 5.8, where the blob in the middle can contain any number of lines of any kind. Together, they will give rise to an effective vertex, which is what we will probe in an experiment. We can ask the question, what will be the characteristics of this effective vertex?

In the classical Lagrangian density, the interaction of particles with the electromagnetic field is contained in a term $-j^\mu(x)A_\mu(x)$, where $j^\mu(x)$ is the 4-vector for current density. In quantum field theory, of course, $j^\mu(x)$ is an operator that involves creation and/or annihilation of particles and antiparticles. In a situation depicted by Fig. 5.8, we are effectively probing the matrix element of $j^\mu(x)$ between two 1-particle states $\big|f(\boldsymbol{p})\big\rangle$ and $\big|f(\boldsymbol{p}')\big\rangle$, where the symbols in the parentheses denote the 4-momentum eigenvalues of the states. We will use the shorthand notation

$$\langle j_\mu(x)\rangle \equiv \frac{1}{\sqrt{2E_p\mathscr{V}}\,\sqrt{2E_{p'}\mathscr{V}}}\,\langle f(\boldsymbol{p}')\,|j_\mu(x)|\,f(\boldsymbol{p})\rangle\;. \tag{5.114}$$

Here, $\mathscr{V}$ is the volume of the entire system in which we consider all physical processes to take place. The necessity for using this volume was first noted in §4.8 where we discussed the normalization of one-particle states. In fact, the two square roots in the denominator of the right hand side of Eq. (5.114) occur precisely so that the quantity defined on the left side has the same dimension as j^μ, and becomes the expectation value of the current density in the limit $\boldsymbol{p} = \boldsymbol{p}'$. Now note that, by using the momentum operator P, we can write

$$j_\mu(x) = e^{iP\cdot x}j_\mu(0)e^{-iP\cdot x}\;. \tag{5.115}$$

Acting on momentum eigenstates, the exponentials involving the momentum operator would produce exponentials involving the momentum eigenvalues of the states. Therefore,

$$\langle f(\boldsymbol{p}')\,|j_\mu(x)|\,f(\boldsymbol{p})\rangle = e^{-iq\cdot x}\,\langle f(\boldsymbol{p}')\,|j_\mu(0)|\,f(\boldsymbol{p})\rangle\;, \tag{5.116}$$

where

$$q = p - p' \tag{5.117}$$

is the momentum carried away by the photon. The remaining matrix element on the right hand side of Eq. (5.116) does not have any co-ordinate dependence. Let us write it as

$$\langle f(\boldsymbol{p}') \, | j_\mu(0)| \, f(\boldsymbol{p}) \rangle = e \, \bar{u}(\boldsymbol{p}') \Gamma_\mu(p, p') u(\boldsymbol{p}) \,, \tag{5.118}$$

which defines the quantity $\Gamma_\mu(p, p')$, called the *vertex function*. We have put in an explicit factor of e in the definition, because somewhere or other within the blob, there must be an electromagnetic vertex involving the photon, and all such vertices have a factor of e. The spinors $\bar{u}(\boldsymbol{p}')$ and $u(\boldsymbol{p})$ occur because of the Feynman rules for creating and annihilating on-shell fermions. It should be noted that, since we have taken the fermion lines to be on-shell, the photon cannot be on-shell, as argued in §5.2. Thus, the vertex function is not the amplitude of a physical process. It is the type of Green function discussed in §4.14.

□ **Exercise 5.18** *Show that the definition of Eq. (5.118) implies that the vertex function is dimensionless.* [**Note :** *Recall the dimension of one-particle states from Eq. (4.137, p 87).*]

The vertex function carries a Lorentz index. This can come from the momenta involved, and also from the Dirac matrices. It is to be remembered that there might be interrelations between different such terms. For example, the Gordon identity, presented in Eq. (F.123, *p 752*) of Appendix F, shows that it is not necessary to consider a term with $(p + p')_\mu$ in the expression for the vertex function. In addition, the electromagnetic current is conserved,

$$\partial_\mu j^\mu(x) = 0 \,. \tag{5.119}$$

So, taking the divergence of Eq. (5.116), we obtain the condition

$$q^\mu \, \bar{u}(\boldsymbol{p}') \Gamma_\mu(p, p') u(\boldsymbol{p}) = 0 \,, \tag{5.120}$$

which must be satisfied by the vertex function. Subject to this constraint, the most general form of the vertex function is given by

$$\Gamma_\mu = F_1 \gamma_\mu + (iF_2 + \widetilde{F}_2 \gamma_5) \sigma_{\mu\nu} q^\nu + \widetilde{F}_3 (\slashed{q} q_\mu - q^2 \gamma_\mu) \gamma_5 \,, \tag{5.121}$$

where F_1, F_2, $\widetilde{F}_2$ and $\widetilde{F}_3$ are Lorentz invariant quantities, which are called *form factors*.

It should be noticed that we have used nothing except Lorentz invariance and gauge invariance in order to arrive at the expression for the vertex function given in Eq. (5.121). Thus, this form of the vertex can be used even for fermions which are not elementary particles, e.g., for the proton and the neutron. And although the neutron has no electric charge, the vertex function shows that it can still have interactions with a photon because of its substructure. Our subsequent discussion will show that some of the effects of these interactions can be interpreted as a magnetic dipole moment of the neutron.

The form factors are not necessarily constants. They can be functions of Lorentz invariant combinations of the various momenta involved. Two such combinations are p^2 and p'^2, but they are irrelevant because both of them are equal to the square of the mass m of the relevant fermion:

$$p^2 = p'^2 = m^2 \,. \tag{5.122}$$

The dot product $p \cdot p'$ is a variable, and it can be traded for q^2, because they are related through Eq. (5.117). Thus, all form factors can be treated as functions of q^2.

☐ **Exercise 5.19** In Eq. (5.121), there are terms with γ_μ and $\sigma_{\mu\nu}$, as well as with a factor of γ_5 attached to both. In addition, we see a term with $\not{q}\gamma_5$, but no counterpart without the γ_5. Why?

☐ **Exercise 5.20** In Eq. (5.121), we have put a factor of i with the term having $\sigma_{\mu\nu}$, but no such factor for the term having $\sigma_{\mu\nu}\gamma_5$. This has been done to ensure that the relevant form factors are real. Prove this statement. [**Hint :** $j_\mu(x)$ must be hermitian since the interaction will have to be hermitian, and A^μ itself is hermitian.]

☐ **Exercise 5.21** Consider the possibility of a massive spin-1 particle decaying into two photons. The Feynman amplitude can be written in the form

$$\mathscr{M} = \mathscr{E}^\alpha(q)\epsilon^\mu(k)\epsilon^\nu(k')T_{\alpha\mu\nu} \,, \tag{5.123}$$

where $\mathscr{E}$ denotes the polarization vector of the decaying spin-1 particle, and $q = k + k'$. Write down the most general form for $T_{\alpha\mu\nu}$ involving form factors, using only the momenta k and k', as well as the metric tensor $g_{\mu\nu}$ and the completely antisymmetric tensor $\varepsilon_{\mu\nu\lambda\rho}$ which are properties of spacetime. Now write down the conditions for electromagnetic current conservation which are analogous to Eq. (5.120) for the present case. Next, use Bose symmetry, which implies

$$T_{\alpha\mu\nu}(k, k') = T_{\alpha\nu\mu}(k', k) \,. \tag{5.124}$$

In addition, each polarization 4-vector is orthogonal to the respective momentum 4-vector. After imposing all these conditions, show that $T_{\alpha\mu\nu}$ vanishes, in compliance with Yang's theorem.

5.7.2   Form factors in non-relativistic limit

Some physical insight into these form factors can be obtained by considering their non-relativistic limits. Let us consider them one by one.

a)   Form factor F_1

The form factor $F_1(q^2)$ appears with the bilinear involving γ_μ. We first use the Gordon identity, Eq. (F.123, *p 752*), to write

$$\overline{u}(p')F_1(q^2)\gamma_\mu u(p) = \frac{F_1(q^2)}{2m}\overline{u}(p')\Big[(p+p')_\mu - i\sigma_{\mu\nu}q^\nu\Big]u(p) \,. \tag{5.125}$$

Now, first consider the limit where the external fermion lines are both at rest, i.e., $q^\mu = 0$ and $p^\mu = p'^\mu = (m, \mathbf{0})$. In this case, the effective coupling to the electromagnetic field will be

$$- \langle j_\mu(x) \rangle \, A^\mu(x) = -\frac{eF_1(0)}{2m\mathscr{V}} \overline{u}(\mathbf{0}) u(\mathbf{0}) A^0(x)$$

$$= -\frac{eF_1(0)}{\mathscr{V}} A^0(x), \tag{5.126}$$

where we have made use of the normalization condition in the form given in Eq. (F.128, $p\,753$). In the effective Hamiltonian density, the term would be present with an opposite sign. Since A^0 is the scalar potential, the quantity multiplying it in the Hamiltonian density should be the charge density. Multiplying by the volume $\mathscr{V}$, we conclude that $eF_1(0)$ is the charge of the fermion whose coupling we are considering, i.e.,

$$F_1(0) = Q, \tag{5.127}$$

in the notation that we had introduced in Eq. (5.2).

When we start considering non-zero values of the photon momentum q, the second term in the right hand side of Eq. (5.125) will also contribute. Notice, however, that the structure of this term is exactly the same as that of the F_2-term. Let us then clump this part with F_2 and consider its interpretation.

b)   Form factor F_2

Once the Gordon identity of Eq. (5.125) is employed, we find that the terms involving the $\sigma_{\mu\nu}$ matrices are of the form

$$F_m(q^2) \overline{u}(p') i \sigma_{\mu\nu} q^\nu u(p), \tag{5.128}$$

where

$$F_m(q^2) = -\frac{F_1(q^2)}{2m} + F_2(q^2). \tag{5.129}$$

Now consider expanding $F_m(q^2)$ in a power series and take only the first term, i.e., the constant term $F_m(0)$. We also consider the limit in which the initial and the final fermions are both non-relativistic, so that we can use the zero-momentum spinors, and also put the energy equal to the mass as a first approximation. In the effective interaction, we can identify this contribution as

$$- \langle j_\mu(x) \rangle \, A^\mu(x) = -\frac{eF_m(0)e^{-iq\cdot x}}{2m\mathscr{V}} \overline{u}(\mathbf{0}) i \sigma_{\mu\nu} q^\nu u(\mathbf{0}) A^\mu(x). \tag{5.130}$$

This can be rewritten as

$$- \langle j_\mu(x) \rangle \, A^\mu(x) = \frac{eF_m(0) \partial^\nu (e^{-iq\cdot x})}{2m\mathscr{V}} \overline{u}(\mathbf{0}) \sigma_{\mu\nu} u(\mathbf{0}) A^\mu(x)$$

$$= -\frac{eF_m(0)e^{-iq\cdot x}}{2m\mathscr{V}} \overline{u}(\mathbf{0}) \sigma_{\mu\nu} u(\mathbf{0}) \partial^\nu A^\mu(x) + \cdots, \tag{5.131}$$

where the dots now contain a total derivative term. This should be of no consequence, and so we will omit it in subsequent steps. Using the antisymmetry of the σ-matrices in the Lorentz indices, this can be written as

$$- \langle j_\mu(x) \rangle \, A^\mu(x) = \frac{e F_m(0) e^{-iq \cdot x}}{4m \mathcal{V}} \overline{u}(\mathbf{0}) \sigma_{\mu\nu} u(\mathbf{0}) F^{\mu\nu}(x). \qquad (5.132)$$

In §F.3.3 of Appendix F, we demonstrate that the matrices σ_{0i}, sandwiched between rest solutions of spinors, vanish. Thus the contributions come from σ_{ij} terms only. It was pointed out in §4.4 that $\frac{1}{2}\sigma_{\mu\nu}$ are the representations of the Lorentz group generators. The subset $\frac{1}{2}\sigma_{ij}$ form the representation of the rotation group generators, i.e,. of spin. The spin vector can be denoted by $\boldsymbol{S}$:

$$\frac{1}{2}\sigma_{ij} = \varepsilon_{ijk} S^k. \qquad (5.133)$$

Then

$$\sigma_{ij} F^{ij} = 2\varepsilon_{ijk} F^{ij} \Sigma^k = 4 B_k S^k, \qquad (5.134)$$

where B^k are the components of the magnetic field. In the Hamiltonian density, we will therefore have the negative of this, which is $4\boldsymbol{B} \cdot \boldsymbol{S}$. The term is then an interaction with the magnetic field, which is written as $-\boldsymbol{\mu} \cdot \boldsymbol{B}$ in classical electrodynamics. Using this analogy, we conclude that the magnetic moment of the particle is given by

$$\boldsymbol{\mu} = -\frac{e F_m(0)}{m} \overline{u}(\mathbf{0}) \boldsymbol{S} u(\mathbf{0}). \qquad (5.135)$$

The expectation value of spin is given by

$$\langle \boldsymbol{S} \rangle = \frac{u^\dagger(\mathbf{0}) \boldsymbol{S} u(\mathbf{0})}{u^\dagger(\mathbf{0}) u(\mathbf{0})}. \qquad (5.136)$$

In §4.2, we showed that the rest-frame solution for spinors are eigenstates of γ_0, so we can replace $u^\dagger$ by $\overline{u}$ in the equation above. Then, imposing the normalization condition in the form given in Eq. (F.128, *p 753*), we can write

$$\boldsymbol{\mu} = \frac{e}{m}\left(Q - 2m F_2(0)\right) \langle \boldsymbol{S} \rangle, \qquad (5.137)$$

where we have also reinserted the definition of F_m from Eq. (5.129) and used the interpretation of $F_1(0)$ from Eq. (5.127). Thus we see that the magnetic moment of a fermion can have a contribution that is proportional to the charge eQ of the particle. This is called the *Dirac contribution* to the magnetic moment. But in addition, there can be other contributions involving $F_2(0)$, which have nothing to do with the charge of the particle and therefore can exist even for uncharged particles. This part is called the *anomalous magnetic moment*.

Customarily, the value of the magnetic moment for a charged particle is expressed in terms of the Landé g-factor, defined for a particle of charge eQ by the relation

$$\boldsymbol{\mu} = \frac{eQ}{2m}\, g \,\langle \boldsymbol{S} \rangle \,. \tag{5.138}$$

Comparing Eqs. (5.137) and (5.138), we see that Landé g-factor for the Dirac part of the magnetic moment is given by

$$g_D = 2\,, \tag{5.139}$$

whereas the anomalous magnetic moment gives a contribution

$$g_A = -\frac{4m}{Q}\, F_2(0)\,. \tag{5.140}$$

For any uncharged fermion, of course, the Landé g-factor is undefined, and one only has the anomalous contribution to the magnetic moment, the second term in the parenthesis of Eq. (5.137).

c) Form factor $\widetilde{F}_2$

The interpretation of this form factor is trivial, once we note that

$$\sigma_{\mu\nu}\gamma_5 = -\frac{i}{2}\varepsilon_{\mu\nu\alpha\beta}\sigma^{\alpha\beta}\,, \tag{5.141}$$

an identity which has been discussed in §F.1. We can now follow exactly the steps that we went through for the magnetic moment, and obtain, in place of Eq. (5.132), the expression

$$-\frac{1}{2}\frac{e\widetilde{F}_2(0)e^{-iq\cdot x}}{4m\mathscr{V}}\varepsilon_{\mu\nu\alpha\beta}\overline{u}(\mathbf{0})\sigma^{\alpha\beta}u(\mathbf{0})F^{\mu\nu}(x)\,. \tag{5.142}$$

This looks very much like Eq. (5.132), except that it involves a different form factor, and instead of the electromagnetic field strength tensor $F^{\mu\nu}$ we now have its dual,

$$\frac{1}{2}\varepsilon_{\mu\nu\alpha\beta}F^{\mu\nu} \equiv \widetilde{F}_{\alpha\beta}\,. \tag{5.143}$$

The components of the dual field-strength tensor can be obtained from the components of the field-strength tensor by making the following replacements:

$$\boldsymbol{E} \to \boldsymbol{B}\,, \qquad \boldsymbol{B} \to -\boldsymbol{E}\,. \tag{5.144}$$

This shows that the effective interaction Hamiltonian arising out of $\widetilde{F}_2(0)$ has the form

$$-\boldsymbol{d} \cdot \boldsymbol{E} \tag{5.145}$$

in the non-relativistic limit, where

$$d = 2e\widetilde{F}_2(0)S. \tag{5.146}$$

In classical electromagnetic theory, the interaction energy proportional to the electric field comes from the electric dipole moment. We thus see that a non-zero $\widetilde{F}_2(0)$ signifies an *electric dipole moment* of a fermion.

d)   Form factor $\widetilde{F}_3$

In this case, similar analysis shows that the interaction energy contains the term

$$2e\widetilde{F}_3(0)S \cdot j. \tag{5.147}$$

This has no analogy in classical electrodynamics. The quantity $\widetilde{F}_3(0)$ is called the *anapole moment* of the fermion.

> □ **Exercise 5.22** Show that the contribution proportional to $\widetilde{F}_3(0)$ is indeed of the form shown in Eq. (5.147). [**Hint :** You will need the non-relativistic reduction of $\bar{u}\gamma_5 u$, which has been done in §F.3.]

It should *not* be understood, from the statements made above, that only $\widetilde{F}_3$ is a purely quantum effect. In fact, all the effects discussed here, other than the electric charge, cannot exist in classical physics. Magnetic or electric dipole moments can exist for extended charge distributions in classical theory. But here we are talking about point particles, and in quantum theory, they can also have interactions that mimic classical interactions of magnetic and electric dipoles. In addition, the momentum dependence of all these form factors is also a hallmark of quantum physics.

5.7.3   Vertex function at one-loop

Let us calculate the contribution to the QED vertex function coming from one-loop diagrams. The diagram is shown in Fig. 5.9. For the sake of concreteness, we take the fermion in the diagram to be electron, so that $Q = -1$. We denote the one-loop contribution by $\Gamma_\mu^{(1)}$. Looking at the diagram, we can write

$$-ie\Gamma_\mu^{(1)} = \int \frac{d^4k}{(2\pi)^4} ie\gamma_\alpha iS_F(p'+k) ie\gamma_\mu iS_F(p+k) ie\gamma_\beta \frac{-ig^{\alpha\beta}}{k^2}, \tag{5.148}$$

where $iS_F(p)$ denotes the Feynman rule for the electron propagator at momentum p. Using the form of the propagator from Eq. (4.145, p 92), we obtain

$$\Gamma_\mu^{(1)} = ie^2 \int \frac{d^4k}{(2\pi)^4} \frac{\gamma_\alpha(\not{p}'+\not{k}+m)\gamma_\mu(\not{p}+\not{k}+m)\gamma^\alpha}{[(p'+k)^2-m^2][(p+k)^2-m^2]k^2}. \tag{5.149}$$

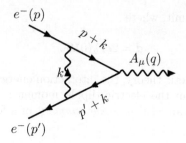

Figure 5.9: One-loop correction to the vertex function in QED.

The method for evaluating such loop integrals has been described in Appendix G. Using Eq. (G.2, *p 756*), we can rewrite this integral as

$$\Gamma_\mu^{(1)} = 2ie^2 \int_0^1 d\zeta_1 \int_0^{1-\zeta_1} d\zeta_2 \int \frac{d^4k}{(2\pi)^4} \frac{N_\mu(k)}{\left[k^2 + 2k \cdot (\zeta_1 p' + \zeta_2 p)\right]^3} , \quad (5.150)$$

where ζ_1 and ζ_2 are Feynman parameters, and $N_\mu(k)$ is the same numerator that appears in Eq. (5.149). In writing this form, we have used Eq. (5.122) for the electron lines in the outer legs. Making now a shift in the integration momentum k, we obtain

$$\Gamma_\mu^{(1)} = 2ie^2 \int_0^1 d\zeta_1 \int_0^{1-\zeta_1} d\zeta_2 \int \frac{d^4k}{(2\pi)^4} \frac{N_\mu(k - \zeta_1 p' - \zeta_2 p)}{\left[k^2 - (\zeta_1 p' + \zeta_2 p)^2\right]^3} . \quad (5.151)$$

Note that, using Eq. (5.122) as well as the momentum-conservation equation, Eq. (5.117), we can write

$$(\zeta_1 p' + \zeta_2 p)^2 = (\zeta_1 + \zeta_2)^2 m^2 - \zeta_1 \zeta_2 q^2 \qquad (5.152)$$

in the denominator.

Let us now turn our attention to the numerator. Notice that the indices on the first and the last Dirac matrices are contracted, so we can use the contraction formulas of §F.1.3 and write

$$N_\mu(k) = -2(\not p + \not k)\gamma_\mu(\not p' + \not k) + 4m\left[2k_\mu + p_\mu + p'_\mu\right] - 2m^2\gamma_\mu . \quad (5.153)$$

In the quantity appearing in the numerator of Eq. (5.151), we can separate out terms depending on the power of k they contain. The term linear in k would vanish on integration, so we need not consider it. There is only one term containing two powers of k, which is

$$-2\not k\gamma_\mu\not k = -2(2k_\mu - \gamma_\mu\not k)\not k = -4k_\mu k_\nu\gamma^\nu + 2\gamma_\mu k^2 . \qquad (5.154)$$

It has been argued in Appendix G that if the integrand of a momentum integration is of the form $k_\mu k_\nu f(k^2)$, it can be transformed into the form $\frac{1}{4} g_{\mu\nu} k^2 f(k^2)$. Using this, we see that the quadratic term in k appearing in the numerator can be written as

$$\gamma_\mu k^2 . \tag{5.155}$$

The terms in the numerator which are independent in k can be written, by using Eq. (5.117), as

$$- 2\Big((1 - \zeta_1)\slashed{q} + (1 - \zeta_1 - \zeta_2)\slashed{p}'\Big)\gamma_\mu\Big((1 - \zeta_1 - \zeta_2)\slashed{p} - (1 - \zeta_2)\slashed{q}\Big)$$
$$+ 4m(1 - \zeta_1 - \zeta_2)(p_\mu + p'_\mu) - 4m(\zeta_1 - \zeta_2)q_\mu . \tag{5.156}$$

Since the vertex function makes sense only within the spinors, as shown in Eq. (5.118), we might as well use some properties of the spinors in order to simplify the expression. We can use the basic equations for the spinors, Eq. (4.52, p 71), to turn the occurrences of $\slashed{p}$ and $\slashed{p}'$ into masses. We can also use the Gordon identity, proved in §F.3.1, for the term containing $p_\mu + p'_\mu$. In addition, we can use Dirac matrix identities like Eq. (F.45, p 741) to write

$$\gamma_\mu \slashed{q} = q_\mu - i\sigma_{\mu\nu}q^\nu , \qquad \slashed{q}\gamma_\mu = q_\mu + i\sigma_{\mu\nu}q^\nu , \tag{5.157}$$

so that finally the k-independent part of the numerator will have three kinds of terms. There will be terms proportional to q_μ, which will vanish once the integrations of the Feynman parameters are performed. The terms containing γ_μ will join hands with the quadratic term obtained in Eq. (5.155) to define the form factor F_1:

$$F_1(q^2) = 2ie^2 \int_0^1 d\zeta_1 \int_0^{1-\zeta_1} d\zeta_2 \int \frac{d^4k}{(2\pi)^4}$$
$$\frac{k^2 - 2m^2[(\zeta_1 + \zeta_2)^2 - 2(1 - \zeta_1 - \zeta_2)] - 2q^2(1 - \zeta_1)(1 - \zeta_2)}{\Big[k^2 - (\zeta_1 + \zeta_2)^2 m^2 + \zeta_1\zeta_2 q^2\Big]^3} . \tag{5.158}$$

And the co-efficient of $i\sigma_{\mu\nu}q^\nu$ will define the form factor F_2:

$$F_2(q^2) = 4ime^2 \int_0^1 d\zeta_1 \int_0^{1-\zeta_1} d\zeta_2 \int \frac{d^4k}{(2\pi)^4} \frac{(\zeta_1 + \zeta_2)(1 - \zeta_1 - \zeta_2)}{\Big[k^2 - (\zeta_1 + \zeta_2)^2 m^2 + \zeta_1\zeta_2 q^2\Big]^3} . \tag{5.159}$$

These equations give the form factors at the one-loop level. Notice that the electric dipole moment and the anapole moment form factors are zero in this case. The reason for this will be explained in Ch. 6.

□ **Exercise 5.23** *Show explicitly that the co-efficient of q_μ vanishes in the one-loop vertex function.*

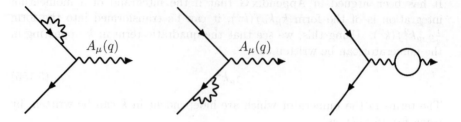

Figure 5.10: More one-loop corrections to the vertex function in QED.

It should be pointed out that the diagram of Fig. 5.9 is not the only diagram at the one-loop level that contributes to the effective vertex. There are other contributions, as shown in Fig. 5.10. It is easy to see that these diagrams contribute only to $F_1(q^2)$. Thus, the expression for $F_1(q^2)$ given in Eq. (5.158) is not complete even so far as only the one-loop contributions are concerned. What is worse, the integration diverges, so that we will obtain an infinite result for $F_1(q^2)$ if we take Eq. (5.158) literally. Infinities like this occur in many loop diagrams, and we will discuss in §12.2 how to deal with them. Right now, we focus on the other form factor. The expression given in Eq. (5.159) is indeed the correct expression for $F_2(q^2)$ at the one-loop level, and let us try to understand its implication.

5.7.4   Anomalous magnetic moment

Let us discuss the form factor F_2. Its value at $q^2 = 0$ is related to the anomalous magnetic moment of the electron, as discussed earlier. From Eq. (5.159), we obtain

$$F_2(0) = 4ime^2 \int_0^1 d\zeta_1 \int_0^{1-\zeta_1} d\zeta_2 \int \frac{d^4k}{(2\pi)^4} \frac{(\zeta_1 + \zeta_2)(1 - \zeta_1 - \zeta_2)}{\left[k^2 - (\zeta_1 + \zeta_2)^2 m^2\right]^3} .$$

$$(5.160)$$

In order to evaluate this, we first perform the integration over the loop momentum k. This can be done by performing Wick rotation, as described in Appendix G. In particular, applying Eq. (G.35, p 762), we obtain

$$\int \frac{d^4k}{(2\pi)^4} \frac{1}{\left[k^2 - A^2\right]^3} = -\frac{i}{32\pi^2 A^2} .$$

$$(5.161)$$

Putting this result of integration, we obtain

$$F_2(0) = \frac{\alpha}{2\pi m} \int_0^1 d\zeta_1 \int_0^{1-\zeta_1} d\zeta_2 \frac{1 - \zeta_1 - \zeta_2}{\zeta_1 + \zeta_2} .$$

$$(5.162)$$

The remaining integration is straightforward, and the final result is

$$F_2(0) = \frac{\alpha}{4\pi m} \, ,$$

(5.163)

which means that the anomalous contribution to the Landé g-factor of the electron is given by

$$g_A = \frac{\alpha}{\pi} \, ,$$

(5.164)

as suggested by Eq. (5.140).

Chapter 6

Parity and charge conjugation

6.1 Discrete symmetries in classical electrodynamics

Consider a charge q' held fixed at a point r' and another charge q which moves in its electric field. In Newtonian mechanics, the equation of motion will be given by

$$m\frac{d^2 r}{dt^2} = \frac{qq'}{4\pi} \frac{r - r'}{|r - r'|^3}, \qquad (6.1)$$

where m is the mass of the charge q and r denotes its instantaneous position. Looking at this equation, we immediately notice that there are certain discrete symmetries that it possesses. These are listed below.

1. If we change all co-ordinate vectors by a sign, each side of Eq. (6.1) changes by a sign so that the equality of the two sides remains unaffected.

2. If we change the sign of time, the equation is unchanged because it involves only the second derivative with respect to time.

3. If all charges are changed by a sign, the equation is unaffected.

We have no doubt considered a very simple situation where there are only two charges, and one of them is static so that we can use the Coulomb law. If we have more charges, and some of them are even moving, the equations of motion will be more complicated. However, the discrete symmetries mentioned above still remain.

The first discrete symmetry mention above is called *space inversion* or *parity*. The second one is called *time reversal*. Both of these were mentioned in §3.6, where it was also said that these two operations are part of the extended Lorentz group. The third operation is called *charge conjugation*, and its nature will be discussed in detail in §6.3.

In this chapter, we will discuss how parity and charge conjugation symmetries manifest themselves in quantum theories. Since our motivation comes from classical electrodynamics, we will of course show that these symmetries are respected by quantum electrodynamics. But our discussion will not be restricted to QED only. Rather, we will lay down a general framework which will enable us to discuss different kinds of theories, including the ones where these symmetries are not respected. Time reversal will not be treated in this chapter. Because of some intricacies relating to this symmetry, it will be treated separately in Ch. 7.

□ **Exercise 6.1** *The equation of motion remains unchanged even if we change some, but not all, components of co-ordinate vectors by a sign. Convince yourself, by following the outline given below, that this does not amount to an independent discrete symmetry.*

a) *Suppose we change the signs of two spatial co-ordinates only, leaving the third unchanged. Show that this is not a discrete operation at all. In fact, this is the result of a rotation.*

b) *Suppose we now consider changing the sign of one spatial co-ordinate only, leaving the other two unchanged. Show that this can be seen as the parity operation compounded with a rotation.*

6.2 Parity transformation of fields

6.2.1 Parity invariance of a Lagrangian

Our definition of the parity operation, given in §6.1, shows that a theory will be parity invariant if any of its consequences do not change when we change the spatial co-ordinates by a sign, i.e., change the spacetime co-ordinates from x^μ to

$$\widetilde{x}^\mu \equiv (t, -\boldsymbol{x}) \,. \tag{6.2}$$

We are discussing quantum field theory through its Lagrangian. The Lagrangian is a function of fields and their first derivatives. The fields, in turn, are functions of spacetime. The derivatives of the fields are also, in general, functions of spacetime. Thus, the Lagrangian itself is a function of spacetime, and we can denote this functional dependence by writing $\mathscr{L}(x)$ for the Lagrangian.

The fields will transform under the parity operation in general. Since, as explained in §3.6.3, parity is part of the extended Lorentz group, these transformations will be linear, like any other Lorentz transformation. These transformations will affect the Lagrangian. The changed Lagrangian will have the same physical consequences if its form in the new co-ordinates is the same as that of the original Lagrangian in the original co-ordinates, i.e., if

$$\mathscr{P}\mathscr{L}(x)\mathscr{P}^{-1} = \mathscr{L}(\widetilde{x}) \,. \tag{6.3}$$

If this happens, the action remains unchanged by the parity transformation, and the physical consequences remain unaffected.

We now show that the free Lagrangians of scalar, Dirac field and vector fields are automatically invariant under parity. When interactions are introduced, the Lagrangian may or may not obey parity invariance, depending on the nature of the interaction.

6.2.2   Free scalar fields

Let us first consider a real scalar field. The parity operation should turn the field $\phi(x)$ into $\phi(\tilde{x})$. The most general linear relation of this sort can be written as

$$\phi_P(x) \equiv \mathscr{P}\phi(x)\mathscr{P}^{-1} = \eta_P\phi(\tilde{x}), \tag{6.4}$$

where η_P is a numerical constant. The Lagrangian of a free scalar field was given earlier in Eq. (4.99, p 81). We reproduce it here, separating the time and space derivatives:

$$\mathscr{L}(x) = \frac{1}{2}\left[\left(\partial_t\phi(x)\right)^2 - \left(\boldsymbol{\nabla}_{\boldsymbol{x}}\phi(x)\right)^2 - m^2\left(\phi(x)\right)^2\right]. \tag{6.5}$$

We have put a subscript on the gradient operator to remind us that it involves derivatives with respect to the spatial co-ordinates $\boldsymbol{x}$.

Under parity operation, $\phi(x)$ changes to $\phi_P(x)$, as defined in Eq. (6.4). So the Lagrangian changes to

$$
\begin{aligned}
\mathscr{P}\mathscr{L}(x)\mathscr{P}^{-1} &= \frac{1}{2}\left[\left(\partial_t\phi_P(x)\right)^2 - \left(\boldsymbol{\nabla}_{\boldsymbol{x}}\phi_P(x)\right)^2 - m^2\left(\phi_P(x)\right)^2\right] \\
&= \frac{1}{2}\eta_P^2\left[\left(\partial_t\phi(\tilde{x})\right)^2 - \left(\boldsymbol{\nabla}_{\boldsymbol{x}}\phi(\tilde{x})\right)^2 - m^2\left(\phi(\tilde{x})\right)^2\right],
\end{aligned} \tag{6.6}
$$

making use of Eq. (6.4). On the other hand,

$$\mathscr{L}(\tilde{x}) = \frac{1}{2}\left[\left(\partial_t\phi(\tilde{x})\right)^2 - \left(\boldsymbol{\nabla}_{-\boldsymbol{x}}\phi(\tilde{x})\right)^2 - m^2\left(\phi(\tilde{x})\right)^2\right]. \tag{6.7}$$

Here, we have the gradient operator with respect to the spatial co-ordinates of $\tilde{x}^\mu$, but it really does not matter, because the gradient is squared. Comparing Eqs. (6.6) and (6.7), we see that the requirement of parity invariance, Eq. (6.3), is satisfied provided

$$\eta_P = \pm 1. \tag{6.8}$$

Parity invariance thus constrains possible values of η_P. The factor is called the *intrinsic parity* for the field ϕ.

If, instead of a real scalar field, we consider a complex scalar field, the situation does not change very much. We can still start with the definition of Eq. (6.4). Note that, if we take the hermitian conjugate of this equation, we obtain

$$\mathscr{P}\phi^\dagger(x)\mathscr{P}^{-1} = \eta_P^*\phi^\dagger(\tilde{x}) \equiv \phi_P^\dagger(x). \tag{6.9}$$

Putting this into the free Lagrangian of the complex scalar field and going through similar steps, it is straightforward to show that the free Lagrangian is invariant under parity provided $|\eta_P| = 1$, i.e., η_P is a phase factor.

Further restriction on the value of η_P comes from the fact that parity, applied twice, is just the identity operation:

$$\mathscr{P}^2 = 1. \tag{6.10}$$

To see the consequence of this fact, consider a further parity operation on Eq. (6.4). This gives

$$\mathscr{P}^2\phi(x)\mathscr{P}^{-2} = \eta_P\,\mathscr{P}\phi(\tilde{x})\mathscr{P}^{-1} = \eta_P^2\phi(x). \tag{6.11}$$

Using Eq. (6.10) now, we conclude that $\eta_P^2 = 1$, which means that even for complex scalar fields, the possible values of intrinsic parity are given by Eq. (6.8).

Spin-0 fields with negative intrinsic parity are sometimes called *pseudoscalar fields*. In contrast, the name *scalar fields* is reserved for spin-0 fields with positive intrinsic parity. The latter use sometimes creates confusion, because spin-0 fields, irrespective of their parity property, are collectively called *scalar fields* as well. We will use the term mostly in the latter sense, and will sometimes use the phrase *intrinsic scalar* when a spin-0 field with positive intrinsic parity will be implied.

6.2.3 Free photon field

We write the Lagrangian for a free photon field as

$$\mathscr{L}(x) = -\frac{1}{4}\Big(\partial_\mu A_\nu(x) - \partial_\nu A_\mu(x)\Big)\Big(\partial^\mu A^\nu(x) - \partial^\nu A^\mu(x)\Big)$$
$$= -\frac{1}{2}\partial_\mu A_\nu(x)\Big(\partial^\mu A^\nu(x) - \partial^\nu A^\mu(x)\Big). \tag{6.12}$$

As for the scalar case, it is better to rewrite this where the time and space derivatives are shown separately. In addition, since the field A^μ has temporal as well as spatial components, we separate them as well and write

$$\mathscr{L}(x) = -\frac{1}{2}\Big[\partial_0 A_i\Big(\partial^0 A^i - \partial^i A^0\Big) + \partial_i A_0\Big(\partial^i A^0 - \partial^0 A^i\Big)$$
$$+ \partial_i A_j\Big(\partial^i A^j - \partial^j A^i\Big)\Big], \tag{6.13}$$

not mentioning the dependence on the spacetime point x for each field. We can now retrace the steps followed for scalar fields to find that this Lagrangian is parity invariant if we define

$$\mathscr{P}A_0(x)\mathscr{P}^{-1} = -\eta_P A_0(\tilde{x}),$$
$$\mathscr{P}\boldsymbol{A}(x)\mathscr{P}^{-1} = +\eta_P \boldsymbol{A}(\tilde{x}), \tag{6.14}$$

and $\eta_P = \pm 1$, as obtained for scalars in Eq. (6.8). This η_P is called the intrinsic parity of the field A^μ. If $\eta_P = +1$, the field is often called an *axial vector field*. On the other hand, if $\eta_P = -1$, the unadorned *vector field* is often used. As for the scalar field, this is confusing, because the phrase *vector field* is used for all spin-1 fields, irrespective of their intrinsic parity. To avoid this confusion, we will use the name *polar vector field* for spin-1 fields with negative intrinsic parity when it will be necessary to distinguish them from axial vector fields.

From the free Lagrangian only, we cannot tell whether the intrinsic parity of the photon field — or any other field which has the same kind of kinetic terms — is positive or negative. We will see later in §6.2.5 that interactions help to fix the intrinsic parity for the photon.

Some authors write Eq. (6.14) in a supposedly compact form:

$$\mathscr{P} A_\mu(x) \mathscr{P}^{-1} = -\eta_P A^\mu(\tilde{x}) . \tag{6.15}$$

In my opinion, this is an abuse of notation. True, that numerically it contains the same rules as those given in Eq. (6.14) with our choice of metric that was announced in Eq. (2.18, p 20). But parity transformation rules should not depend on the metric. Someone might choose to work with a metric with the opposite sign. With this choice, Eq. (6.14) would still remain the correct parity transformation rules, although Eq. (6.15) will not be satisfied.

6.2.4   Free fermion fields

For a free Dirac field, the Lagrangian is

$$\mathscr{L}(x) = \overline{\psi}(x) \left[i\gamma_0 \partial_t + i\boldsymbol{\gamma} \cdot \boldsymbol{\nabla}_{\boldsymbol{x}} - m \right] \psi(x) , \tag{6.16}$$

so that

$$\begin{aligned}
\mathscr{L}(\tilde{x}) &= \overline{\psi}(\tilde{x}) \left[i\gamma_0 \partial_t + i\boldsymbol{\gamma} \cdot \boldsymbol{\nabla}_{-\boldsymbol{x}} - m \right] \psi(\tilde{x}) \\
&= \overline{\psi}(\tilde{x}) \left[i\gamma_0 \partial_t - i\boldsymbol{\gamma} \cdot \boldsymbol{\nabla}_{\boldsymbol{x}} - m \right] \psi(\tilde{x}) .
\end{aligned} \tag{6.17}$$

To obtain the left hand side of Eq. (6.3), we need the transformation property of the field $\psi(x)$ under parity. Again, the relationship should be linear. Since $\psi(x)$ is a 4-component object, the most general relation in this case would be of the form

$$\psi_P(x) \equiv \mathscr{P} \psi(x) \mathscr{P}^{-1} = \mathbb{P} \psi(\tilde{x}) , \tag{6.18}$$

where $\mathbb{P}$ is some 4×4 matrix. This also means

$$\overline{\psi}_P(x) = \psi_P^\dagger(x)\gamma_0 = \psi^\dagger(\tilde{x})\mathbb{P}^\dagger\gamma_0 = \overline{\psi}(\tilde{x})\gamma_0\mathbb{P}^\dagger\gamma_0 . \tag{6.19}$$

So

$$\mathscr{P}\mathscr{L}(x)\mathscr{P}^{-1} = \overline{\psi}(\tilde{x})\gamma_0\mathbb{P}^\dagger\gamma_0 \left[i\gamma_0 \partial_t + i\boldsymbol{\gamma} \cdot \boldsymbol{\nabla}_{\boldsymbol{x}} - m \right] \mathbb{P}\psi(\tilde{x}) . \tag{6.20}$$

This will equal the expression of Eq. (6.17) provided the following relations are satisfied:

$$\gamma_0 \mathbb{P}^\dagger \gamma_0 \gamma_0 \mathbb{P} = \gamma_0 \,,$$
$$\gamma_0 \mathbb{P}^\dagger \gamma_0 \boldsymbol{\gamma} \mathbb{P} = -\boldsymbol{\gamma} \,,$$
$$\gamma_0 \mathbb{P}^\dagger \gamma_0 \mathbb{P} = 1 \,. \tag{6.21}$$

It is possible to find a $\mathbb{P}$ that satisfies these equations, which means that the free Dirac Lagrangian is parity invariant. The solution, in fact, is of the form

$$\mathbb{P} = \eta_P \gamma_0 \,, \tag{6.22}$$

where η_P can be any phase as far as the relations in Eq. (6.21) are concerned. So the parity transformation property of the Dirac field is given by

$$\mathscr{P}\psi(x)\mathscr{P}^{-1} = \eta_P \gamma_0 \psi(\widetilde{x}) \,. \tag{6.23}$$

The quantity η_P is the intrinsic parity of the Dirac field. If we further utilize Eq. (6.10), we find that the possible values of η_P are ± 1. Thankfully, no one has come up with separate names for spin-$\frac{1}{2}$ fields with positive and negative parities and created the same kind of confusion in the terminology that exists for spin-0 and spin-1 fields.

6.2.5 Interacting fields

So far, we have discussed free Lagrangians only, and found that they cannot determine the intrinsic parities of the fields. Physical processes involve interactions. Interaction terms in a Lagrangian can impose further restrictions on the parity properties. As long as one can find a consistent assignment of intrinsic parities of all fields appearing in a Lagrangian, we can say that parity is a good symmetry. If we cannot find any such solution, we will have to conclude that the Lagrangian does not respect parity symmetry. We illustrate these statements with some examples.

Suppose we have a theory with just one real scalar field, and the interaction terms in the Lagrangian are

$$\mathscr{L}_{\text{int}} = -\mu\phi^3 - \lambda\phi^4 \,. \tag{6.24}$$

If we take Eq. (6.4) and follow the procedure by which we found the intrinsic parity of the free field, we would find that parity invariance with these interaction terms would require $\eta_P^3 = 1$ and $\eta_P^4 = 1$. These conditions are satisfied by $\eta_P = +1$ but not by $\eta_P = -1$. This means that, with these interactions, parity is a conserved symmetry, and the intrinsic parity of the field ϕ is positive. This is one example of how interactions fix the parity of a field.

In order to discuss interactions involving fermion fields, let us first determine how fermion bilinears transform under parity. Let us denote a general bilinear by $\overline{\psi}F\psi$, where F is a constant 4×4 matrix. Then,

$$\mathscr{P}\overline{\psi}(x)F\psi(x)\mathscr{P}^{-1} = \mathscr{P}\overline{\psi}(x)\mathscr{P}^{-1}F\mathscr{P}\psi(x)\mathscr{P}^{-1}$$
$$= \overline{\psi}(\widetilde{x})\gamma_0 F\gamma_0\psi(\widetilde{x}), \tag{6.25}$$

using Eqs. (6.18) and (6.19), with the solution of $\mathbb{P}$ taken from Eq. (6.22). For different choices of F, the corresponding matrix that appears in the parity transformed bilinear, to be called F_P, are tabulated below:

F	$\mathbb{1}$	γ_0	γ_i	σ_{0i}	σ_{ij}	$\gamma_0\gamma_5$	$\gamma_i\gamma_5$	γ_5
F_P	$\mathbb{1}$	γ_0	$-\gamma_i$	$-\sigma_{0i}$	σ_{ij}	$-\gamma_0\gamma_5$	$\gamma_i\gamma_5$	$-\gamma_5$

$$\tag{6.26}$$

So now let us look at the interaction involving scalars and fermions. Suppose we have an interaction term

$$\mathscr{L}_{\text{int}} = -h\overline{\psi}\psi\phi. \tag{6.27}$$

Using Eqs. (6.4) and (6.26), we obtain

$$\mathscr{P}\mathscr{L}_{\text{int}}(x)\mathscr{P}^{-1} = -h\eta_P^{(\phi)}\overline{\psi}(\widetilde{x})\psi(\widetilde{x})\phi(\widetilde{x}), \tag{6.28}$$

where we have used the symbol $\eta_P^{(\phi)}$ to denote the intrinsic parity of the scalar field, in order to distinguish it from the same property of the Dirac field which does not appear in the parity-transformed Lagrangian. Comparing the expression with $\mathscr{L}_{\text{int}}(\widetilde{x})$, we conclude that the interaction is parity invariant provided the intrinsic parity of the scalar field is positive. By a similar argument, we can conclude that if the only interaction is of the form

$$\mathscr{L}_{\text{int}} = -h'\,\overline{\psi}\gamma_5\psi\phi, \tag{6.29}$$

parity would still be a symmetry of the Lagrangian, and the intrinsic parity of the field ϕ is negative, i.e., it is a pseudoscalar field.

Now consider that we have an interaction of the form

$$\mathscr{L}_{\text{int}} = -\overline{\psi}(h + h'\gamma_5)\psi\phi. \tag{6.30}$$

In this case, one term will require $\eta_P^{(\phi)} = +1$ in order to conserve parity, while the other would require $\eta_P^{(\phi)} = -1$. Both terms taken together, we cannot find any solution for $\eta_P^{(\phi)}$ that will conserve parity. Thus, such an interaction will be parity violating.

We can think of other ways in which such an impasse can be reached. For example, suppose the interaction Lagrangian is of the form

$$\mathscr{L}_{\text{int}} = -\mu\phi^3 - \lambda\phi^4 - h'\,\overline{\psi}\gamma_5\psi\phi. \tag{6.31}$$

We argued earlier that if we had only the first two terms, we would have obtained parity invariance, with $\eta_P^{(\phi)} = +1$. If we had only the last two terms, we would have also obtained parity invariance, with $\eta_P^{(\phi)} = -1$. But with all three terms present, the interactions are parity violating.

Table 6.1: Behavior of common electromagnetic quantities under parity. Behavior under time reversal has also been included for later use.

Quantity	Usual Symbol	Behavior under Parity	Time reversal
Scalar potential	A_0 or ϕ	+	+
Vector potential	$\mathbf{A}$	−	−
Electric field	$\mathbf{E}$	−	+
Magnetic field	$\mathbf{B}$	+	−
Polarization vector	ϵ	(same as $\mathbf{A}$)	

□ **Exercise 6.2** Consider a massless fermion field $\psi(x)$.

a) Show that one can define the parity transformation in the following way:

$$\mathscr{P}\psi(x)\mathscr{P}^{-1} = \eta_P \gamma_0 \gamma_5 \psi(\widetilde{x})\,, \qquad (6.32)$$

where $\widetilde{x}^\mu = (t, -\boldsymbol{x})$.

b) Show that, in this case, if a field $\phi(x)$ has an interaction of the form $\overline{\psi}\psi\phi$ with the fermion, it is a pseudoscalar field. On the other hand, if the interaction is $\overline{\psi}\gamma_5\psi\phi$, the field is an intrinsic scalar.

c) Show that the simultaneous presence of $\overline{\psi}\psi\phi$ and $\overline{\psi}\gamma_5\psi\phi$ interactions cannot be parity invariant.

□ **Exercise 6.3** Prepare a more general table than that in Eq. (6.26), for the case when the two fermion fields are different. In other words, if

$$\mathscr{P}\overline{\psi}_2(x)F\psi_1(x)\mathscr{P}^{-1} = \overline{\psi}_2(\widetilde{x})F_P\psi_1(\widetilde{x})\,, \qquad (6.33)$$

find F_P for the choices of F shown in Eq. (6.26).

□ **Exercise 6.4** Argue that all parity violating amplitudes arising out of the Lagrangian of Eq. (6.31) must contain factors of both μ and h'.

As a last example, we show that the QED Lagrangian, discussed at length in Ch. 5, is parity invariant. Of course the free parts are invariant for positive or negative intrinsic parity for the fermion and the photon, as we have already demonstrated. The interaction part is given by

$$\mathscr{L}_{\text{int}}(x) = -eQ\overline{\psi}(x)\gamma^\mu\psi(x)A_\mu(x)$$
$$= -eQ\left(\overline{\psi}(x)\gamma_0\psi(x)A_0(x) - \overline{\psi}(x)\gamma_i\psi(x)A_i(x)\right). \qquad (6.34)$$

Taking help of parity transformation properties of fermion bilinears and the photon field, we can write

$$\mathscr{P}\mathscr{L}_{\text{int}}(x)\mathscr{P}^{-1} = eQ\eta_P^{(A)}\left(\overline{\psi}(\widetilde{x})\gamma_0\psi(\widetilde{x})A_0(\widetilde{x}) - \overline{\psi}(\widetilde{x})\gamma_i\psi(\widetilde{x})A_i(\widetilde{x})\right)\,, (6.35)$$

where $\eta_P^{(A)}$ is the intrinsic parity of the photon, the quantity denoted simply by η_P in Eq. (6.14). Comparing this with the expression for $\mathscr{L}_{\text{int}}(\tilde{x})$ through Eq. (6.3), we find that the interaction is indeed parity invariant, and it forces the intrinsic parity of the photon to be negative:

$$\eta_P^{(A)} = -1 \,. \tag{6.36}$$

This fixes the parity transformation properties of the electric and magnetic fields, which we summarize in Table 6.1. But note that the procedure says nothing about the intrinsic parity of the fermion since this quantity cancels in the expression for the parity transformed Lagrangian.

☐ **Exercise 6.5** *Show that the Lagrangian of scalar QED is also invariant under parity, and that it also implies Eq. (6.36).*

☐ **Exercise 6.6** *Show that, if for some spin-1 field B_μ, the interaction with fermions is given by*

$$\mathscr{L}_{\text{int}} = a\,\overline{\psi}\gamma^\mu\gamma_5\psi B_\mu \tag{6.37}$$

where a is a constant, the Lagrangian is parity invariant provided B_μ is an axial vector field.

☐ **Exercise 6.7** *Show that, if for some spin-1 field Z_μ, the interaction with fermions contains a mixture of polar vector and axial vector currents, i.e., is of the form*

$$\mathscr{L}_{\text{int}} = \overline{\psi}\gamma^\mu(a + b\gamma_5)\psi Z_\mu \tag{6.38}$$

where a and b are constants, the Lagrangian cannot be parity invariant. More generally, show that if a spin-1 field W^μ has the interaction

$$\mathscr{L}_{\text{int}} = \overline{\psi}_1\gamma^\mu(a + b\gamma_5)\psi_2 W_\mu + \text{h.c.} \tag{6.39}$$

with two fermion fields ψ_1 and ψ_2, ("h.c." stands for hermitian conjugate) the Lagrangian cannot be parity invariant.

6.3   Charge conjugation

6.3.1   Nature of the transformation

When we talked about charge conjugation in §6.1, we identified it as a symmetry of the Coulomb force formula under the change of signs of electric charges of all particles. As discussed in §4.7, a particle and its antiparticle have opposite charges. Thus, changing the sign of the charge of all particles can also be seen as exchanging each particle with its antiparticle.

In field theory, the basic objects that constitute the action are fields, not particles. If a field operator contains the annihilitor of a certain particle, the hermitian conjugate of the field operator would contain the annihilator of its antiparticle. Thus, in the language of fields, changing particles into antiparticles can be interpreted as changing any field with its hermitian conjugate.

This is a discrete operation, and is usually called *charge conjugation*. Here, we consider the possibility of this operation being a symmetry of the action, and the consequences of such a symmetry if it exists.

A few comments should be made here. First, if we take the hermitian conjugate of a field, that not only changes the electric charge, but all other charges that might be associated with it. Indeed, a particle and its antiparticle not only differ in the signs of their electric charges, but of all charges. For example, baryons carry a *baryon number*, particles like the electron and the muon carry a *lepton number*. There are other examples of such quantities which we will encounter later in the book. All of these will be reversed in the operation of charge conjugation.

The second comment is that this discrete symmetry has a different character than parity transformation or time reversal. The latter are operations on the spacetime variables. Charge conjugation, on the other hand, is an operation on the fields: it does not affect the spacetime co-ordinates directly.

6.3.2 Free bosonic fields

Let us start with a free scalar field ϕ. From the discussion above, we understand that the charge conjugation operation will change it to $\phi^\dagger$. But like in the case of parity, there might be a numerical constant appearing in the transformation equation, so that the most general definition of the operation of charge conjugation on a scalar field would be

$$\mathscr{C}\phi(x)\mathscr{C}^{-1} = \eta_C\phi^\dagger(x)\,. \tag{6.40}$$

The numerical co-efficient η_C, called the *intrinsic charge conjugation*, is an intrinsic property of the field. To see the constraints on this co-efficient, let us take the hermitian conjugate of Eq. (6.40). Remembering that $\mathscr{C}$ is a unitary operator so that

$$\mathscr{C}^{-1} = \mathscr{C}^\dagger\,, \tag{6.41}$$

we obtain

$$\mathscr{C}\phi^\dagger(x)\mathscr{C}^{-1} = \eta_C^*\phi(x)\,. \tag{6.42}$$

We now apply the charge conjugation operator one more time on both sides of Eq. (6.40). This gives

$$\mathscr{C}^2\phi(x)\mathscr{C}^{-2} = \eta_C\mathscr{C}\phi^\dagger(x)\mathscr{C}^{-1} = |\eta_C|^2\phi(x)\,. \tag{6.43}$$

Since, like parity, $\mathscr{C}$ is a toggle operator, i.e.,

$$\mathscr{C}^2 = 1\,, \tag{6.44}$$

we conclude that η_C can be an arbitrary phase, i.e., a complex number with unit modulus. However, for real scalar fields, Eqs. (6.40) and (6.42) are identical, which tells us that η_C must be real, i.e., $\eta_C = \pm 1$.

For a general vector field V_μ, the analog of Eq. (6.40) would be

$$\mathscr{C} V_\mu(x) \mathscr{C}^{-1} = \eta_C V_\mu^\dagger(x). \tag{6.45}$$

The constraints on the intrinsic factor η_C are similar: in general, η_C can be a phase. For real vector fields like the photon, the only allowed values are $\eta_C = \pm 1$.

□ **Exercise 6.8** *Show that Eq. (6.40) implies*

$$\mathscr{C} a(\boldsymbol{p}) \mathscr{C}^{-1} = \eta_C \widehat{a}(\boldsymbol{p}), \tag{6.46}$$

which clearly shows the interchange of the particle and the antiparticle under charge conjugation.

6.3.3   Free fermion fields

One might be tempted to think that the analog of Eqs. (6.40) and (6.45) for a Dirac field should be written as

$$\mathscr{C} \psi(x) \mathscr{C}^{-1} = \eta_C \psi^\dagger(x). \tag{6.47}$$

But this would be wrong. The reason is contained in Eqs. (4.73) and (4.77), which show that $\psi(x)$ and $\psi^\dagger(x)$ do not transform the same way under Lorentz transformations. So, even if this relation is imposed in one frame, it would not hold in another frame which is rotated or boosted with respect to the former one. Thus, the correct relation must be of the form

$$\mathscr{C} \psi(x) \mathscr{C}^{-1} = \eta_C \widehat{\psi}(x), \tag{6.48}$$

where $\widehat{\psi}(x)$ involves the elements of $\psi^\dagger(x)$ but nevertheless transforms like $\psi(x)$ under Lorentz transformations.

Let us then assume that $\widehat{\psi}(x)$ is defined by the relation

$$\widehat{\psi}(x) = \gamma_0 \mathbb{C} \psi^*(x) \tag{6.49}$$

for some matrix $\mathbb{C}$ whose components are numbers. Here, the object ψ^* has the same components as $\psi^\dagger$, except that it is thought as a column matrix, just like ψ. The matrix $\mathbb{C}$ in Eq. (6.49) must have appropriate properties such that the Lorentz transformation properties of $\widehat{\psi}(x)$ and $\psi(x)$ are identical. The question now is: can we find such a matrix $\mathbb{C}$?

Obviously, since charge conjugation is a unitary operation, $\gamma_0 \mathbb{C}$ has to be a unitary matrix. Since γ_0 itself is unitary, it implies that $\mathbb{C}$ should also be unitary, i.e.,

$$\mathbb{C}^\dagger = \mathbb{C}^{-1}. \tag{6.50}$$

The transformation property of $\psi(x)$, given in Eq. (4.73, *p 76*), implies the following transformation property of $\psi^*(x)$:

$$\psi^*(x) \longrightarrow \psi'^*(x') = \exp\left(+\frac{i}{4} \omega^{\mu\nu} \sigma_{\mu\nu}^*\right) \psi^*(x), \tag{6.51}$$

so that

$$\widehat{\psi}(x) \longrightarrow \widehat{\psi}'(x') = \gamma_0 \mathbb{C} \psi^*(x) = \gamma_0 \mathbb{C} \exp\left(+\frac{i}{4}\omega^{\mu\nu}\sigma^*_{\mu\nu}\right)\psi^*(x). \quad (6.52)$$

On the other hand, we want this transformation to look exactly like that of $\psi(x)$, i.e., we want

$$\widehat{\psi}(x) \longrightarrow \widehat{\psi}'(x') = \exp\left(-\frac{i}{4}\omega^{\mu\nu}\sigma_{\mu\nu}\right)\widehat{\psi}(x)$$

$$= \exp\left(-\frac{i}{4}\omega^{\mu\nu}\sigma_{\mu\nu}\right)\gamma_0 \mathbb{C} \psi^*(x). \quad (6.53)$$

Comparing these two equations, we see that we want the matrix $\mathbb{C}$ to satisfy the relation

$$\gamma_0 \mathbb{C} \sigma^*_{\mu\nu} = -\sigma_{\mu\nu}\gamma_0 \mathbb{C}, \quad (6.54)$$

or

$$\sigma^*_{\mu\nu} = -\mathbb{C}^{-1}\gamma_0 \sigma_{\mu\nu}\gamma_0 \mathbb{C} = -\mathbb{C}^{-1}\sigma^\dagger_{\mu\nu}\mathbb{C}, \quad (6.55)$$

where in the last step we have used Eq. (4.79, *p 77*). Then, taking the hermitian conjugate, we obtain

$$\sigma^\top_{\mu\nu} = -\mathbb{C}^{-1}\sigma_{\mu\nu}\mathbb{C}, \quad (6.56)$$

where we have used the unitary nature of the matrix $\mathbb{C}$. Clearly, it will be sufficient to search for a matrix $\mathbb{C}$ that satisfies the relation

$$\mathbb{C}^{-1}\gamma_\mu \mathbb{C} = -\gamma^\top_\mu, \quad (6.57)$$

because this relation guarantees Eq. (6.56) through the definition of the sigma matrices. No matter what choice we make about the Dirac matrices, it is always possible to find a matrix $\mathbb{C}$ that satisfies this equation, as we discuss in Appendix F. We will not need the explicit form for the matrix. We will only need its properties given in Eqs. (6.50) and (6.57), and the important property given in Eq. (6.60) below.

One might wonder why we have put a factor of γ_0 in the definition of the Lorentz covariant conjugate in Eq. (6.49). It must be emphasized that this is purely a matter of convention, and no physics depends on it. What is important is that $\psi^*(x)$, component by component, does not transform like $\psi(x)$ under Lorentz transformation. So we must put in a matrix to mix the components. We could just well have put a matrix $\mathbb{C}'$ in place of $\gamma_0\mathbb{C}$ in Eq. (6.49), and proceeded to find relations equivalent to Eq. (6.57) or Eq. (6.60) in terms of this $\mathbb{C}'$.

There are of course different conventions in this regard. Many people define

$$\widehat{\psi}(x) = \mathbb{C}\gamma_0^\top \psi^*(x). \quad (6.58)$$

It is easy to see, using Eq. (6.57), that this definition differs from that in Eq. (6.49) by an overall sign. This sign will not affect any physical implication because fermion fields have to occur in even numbers in any Lagrangian term in order to conserve angular momentum.

Some other texts use the definition

$$\widehat{\psi}(x) = \mathbb{C}\gamma_0\psi^*(x).\tag{6.59}$$

This is wrong in general. It can be easily shown that this equation cannot hold in arbitrary representations of the Dirac matrices. It holds only in the representations in which γ_0 is either a symmetric or an antisymmetric matrix, in which case it coincides with the definition in Eq. (6.58) or Eq. (6.49).

☐ **Exercise 6.9** *Use the toggle property of charge conjugation, given in Eq. (6.44), to show that*

$$\mathbb{C}^\top = -\,\mathbb{C},\tag{6.60}$$

i.e., the matrix $\mathbb{C}$ must be antisymmetric.

6.3.4   Interacting fields

As in the case of parity, we find that the free Lagrangians are invariant under charge conjugation symmetry, but they leave a lot of arbitrariness in the definition of the intrinsic charge conjugation property of any field. Once interactions are introduced, this situation is modified. In some cases, interactions fix the intrinsic charge conjugation properties of different fields. In some other cases, interactions do not allow any consistent definition of intrinsic charge conjugation properties of the participating fields, in which case we say that charge conjugation symmetry is violated by the interactions. We will discuss some examples of both kinds here.

Consider first the interaction shown in Eq. (6.24). It is easy to see that it is invariant under charge conjugation, and it dictates that the phase η_C should be $+1$ for the field ϕ. Next, consider fermion bilinears. First, we note that the hermitian conjugate of Eq. (6.48) gives

$$\mathscr{C}\overline{\psi}(x)\mathscr{C}^{-1} = \eta_C^*\psi^\top\mathbb{C}^\dagger.\tag{6.61}$$

Using this, we obtain

$$\begin{aligned}\mathscr{C}\overline{\psi}F\psi\mathscr{C}^{-1} &= \mathscr{C}\overline{\psi}\mathscr{C}^{-1}F\mathscr{C}\psi\mathscr{C}^{-1}\\ &= \psi^\top\mathbb{C}^\dagger F\gamma_0\mathbb{C}\psi^*,\end{aligned}\tag{6.62}$$

where F is any 4×4 matrix, including the unit matrix. In writing this equation, we have suppressed the spacetime dependence because it is the same for all fields in this equation.

The right hand side of Eq. (6.62) looks very different from the usual expressions for bilinears that we have used so far, where the field at the left had a bar on it and the field on the right has nothing. However, the expression obtained can be put into the usual form. For this, we consider the expression $\psi^\top M\psi^*$ for any numerical matrix M, and write it showing the sum involved in it explicitly, i.e.,

$$\psi^\top M\psi^* = \psi_a M_{ab}\psi_b^* = M_{ab}\psi_a\psi_b^*.\tag{6.63}$$

We have taken the elements of the matrix M to the front because it is a matrix whose elements are numbers, which commute with everything. But now, let us consider what will happen if we want to bring the ψ^* in front of the ψ. Fermion field operators anticommute, so there will be a negative sign. Thus, we can write

$$M_{ab}\psi_a\psi_b^* = -M_{ab}\psi_b^*\psi_a = -\psi_b^* M_{ab}\psi_a \,. \tag{6.64}$$

Reverting back to the matrix notation now, we can summarize our result as

$$\psi^\top M\psi^* = -\psi^\dagger M^\top \psi = -\overline{\psi}\gamma_0 M^\top \psi \,. \tag{6.65}$$

Applying this identity on the combination of matrices that appear in Eq. (6.62), we can write the charge conjugation property of a fermion bilinear in the form

$$\mathscr{C}\overline{\psi}F\psi\mathscr{C}^{-1} = \overline{\psi}F_C\psi \,, \tag{6.66}$$

where

$$F_C = -\gamma_0(\mathbb{C}^{-1}F\gamma_0\mathbb{C})^\top = \mathbb{C}F^\top\mathbb{C}^{-1} \,, \tag{6.67}$$

using Eq. (6.60). Using this and the definition of the matrix $\mathbb{C}$ in Eq. (6.57), we can now summarize the transformation properties of bilinears in a tabular form:

$$\begin{array}{c||c|c|c|c|c} F & \mathbb{1} & \gamma_\mu & \sigma_{\mu\nu} & \gamma_\mu\gamma_5 & \gamma_5 \\ \hline F_C & \mathbb{1} & -\gamma_\mu & -\sigma_{\mu\nu} & \gamma_\mu\gamma_5 & \gamma_5 \end{array} \,. \tag{6.68}$$

Using this table, we can now analyze interactions involving fermions. For example, consider the interaction term of QED, $\overline{\psi}\gamma_\mu\psi A^\mu$. Since the vector current is odd under charge conjugation, we conclude that the photon must also be odd:

$$\eta_C^{(A)} = -1 \,. \tag{6.69}$$

On the other hand, if some vector boson couples only to the axial vector current of fermions, charge conjugation symmetry is still obeyed, the boson being even under charge conjugation. But if the same vector boson couples to both vector and axial vector currents of fermions, there is no consistent assignment of η_C, so charge conjugation symmetry must be violated.

☐ **Exercise 6.10** Show that the interactions of scalar QED are invariant under charge conjugation.

☐ **Exercise 6.11** Derive charge conjugation rules for fermion bilinears involving two different fermion fields.

☐ **Exercise 6.12** Consider the interaction

$$\mathscr{L}_{\rm int} = h\,\overline{\psi}_1\gamma_5\psi_2\phi + \text{h.c.} \,, \tag{6.70}$$

where ϕ is a complex scalar field.

    a) Write the hermitian conjugate term explicitly.

    b) Show that this interaction respects charge conjugation symmetry and find the resulting relation between h and the intrinsic charge conjugation properties of various fields.

☐ **Exercise 6.13** Show that interactions of the form shown in Eq. (6.39) are not invariant under charge conjugation unless either a or b is zero.

6.4   Parity properties of particle states

So far, we have talked about parity and charge conjugation properties of quantum fields. In experiments, however, we observe not the fields but rather the particles and antiparticles, which are quanta of the fields. It is therefore necessary to know how the parity properties of the particles are related to those of the fields.

6.4.1   Intrinsic parity for bosons

Let us start with a real scalar field. The parity transformation property of such a field has been given in Eq. (6.4). We let both sides act on the vacuum state and obtain

$$\mathscr{P}\phi(x)\left|0\right\rangle = \eta_P \phi(\widetilde{x})\left|0\right\rangle , \tag{6.71}$$

where the vacuum is parity symmetric, i.e.,

$$\mathscr{P}\left|0\right\rangle = \mathscr{P}^{-1}\left|0\right\rangle = \left|0\right\rangle . \tag{6.72}$$

We now take the plane wave expansion of the scalar field from Eq. (4.12, *p 64*). Since

$$p \cdot \widetilde{x} = E_p t - \boldsymbol{p} \cdot \widetilde{\boldsymbol{x}} = E_p t + \boldsymbol{p} \cdot \boldsymbol{x} . \tag{6.73}$$

we obtain

$$\phi(\widetilde{x}) = \int D^3 p \left(e^{-iE_p t - i\boldsymbol{p} \cdot \boldsymbol{x}} a(\boldsymbol{p}) + e^{iE_p t + i\boldsymbol{p} \cdot \boldsymbol{x}} a^\dagger(\boldsymbol{p}) \right) . \tag{6.74}$$

Changing the integration variable from $\boldsymbol{p}$ to $-\boldsymbol{p}$, we can rewrite it as

$$\phi(\widetilde{x}) = \int D^3 p \left(e^{-ip \cdot x} a(-\boldsymbol{p}) + e^{+ip \cdot x} a^\dagger(-\boldsymbol{p}) \right) . \tag{6.75}$$

We now put Eqs. (4.12) and (6.75) into Eq. (6.71). Recalling that the annihilation operator annihilates the vacuum, we obtain

$$\mathscr{P}a^\dagger(\boldsymbol{p})\left|0\right\rangle = \eta_P a^\dagger(-\boldsymbol{p})\left|0\right\rangle , \tag{6.76}$$

using the independence of different Fourier components. Now, $a^\dagger(\boldsymbol{p})|0\rangle$ is a one particle state, which we denote by $|B(\boldsymbol{p})\rangle$. Then the last equation can be rewritten as

$$\mathscr{P}|B(\boldsymbol{p})\rangle = \eta_P |B(-\boldsymbol{p})\rangle\,, \tag{6.77}$$

implying that the intrinsic parity of a one-particle state is equal to the intrinsic parity of the field in this case. It also implies that the parity operation changes the direction of the 3-momentum of the particle, something that should be obvious from the definition of momentum.

The result remains the same if we have a complex scalar field, and it also works for antiparticle states. In fact, the result for antiparticle states comes directly from following the argument given above. For the particle states, we need to follow the same path, but we need to start from the hermitian conjugate of Eq. (6.4), which is:

$$\mathscr{P}\phi^\dagger(x)\mathscr{P}^{-1} = \eta_P\phi^\dagger(\widetilde{x})\,. \tag{6.78}$$

For spin-1 particles like the photon as well, the result is the same: viz., the intrinsic parity of a particle state is the same as that of the field.

6.4.2 Intrinsic parity for fermions and antifermions

We can apply the parity operator directly on the plane wave expansion of the field $\psi(x)$. Parity is a linear operator, a property that we discuss in more detail in §7.1 (see, in particular, Eq. (7.13, *p 190*)). It means that when such an operator acts on something times a number, the number is unaffected by the operation. In the expression for $\psi(x)$ in Eq. (4.65, *p 72*), the exponentials as well as the spinors are numerical factors, which should remain unaffected. Thus, the action of parity gives

$$\mathscr{P}\psi(x)\mathscr{P}^{-1} = \sum_s \int D^3p \left(\mathscr{P}d_{\boldsymbol{p},s}\mathscr{P}^{-1}u_{\boldsymbol{p},s}e^{-ip\cdot x} \right.$$
$$\left. + \mathscr{P}\widehat{d}^\dagger_{\boldsymbol{p},s}\mathscr{P}^{-1}v_{\boldsymbol{p},s}e^{+ip\cdot x} \right). \tag{6.79}$$

The expression on the left hand side is related to $\psi(\widetilde{x})$ through Eq. (6.23). In order to simplify the right hand side of Eq. (6.23), we need to use the following properties of the spinors:

$$\gamma_0 u_{\boldsymbol{p},s} = u_{-\boldsymbol{p},s}\,, \tag{6.80a}$$
$$\gamma_0 v_{\boldsymbol{p},s} = -v_{-\boldsymbol{p},s}\,, \tag{6.80b}$$

which have been proved in the Appendix, in §F.2.4.

□ **Exercise 6.14** *Derive the identities given in Eq. (6.80) by first showing that they hold with a particular choice of the Dirac matrices. Then, use the fact that any two representations are unitarily related to prove that the relation is independent of the representation.*

We now put in the plane wave expansion of the fermion field, appearing in Eq. (4.65, p 72), to obtain

$$\gamma_0 \psi(\widetilde{x}) = \sum_s \int D^3 p \left(d_{\boldsymbol{p},s} u_{-\boldsymbol{p},s} e^{-ip\cdot\widetilde{x}} - \widehat{d}^\dagger_{\boldsymbol{p},s} v_{-\boldsymbol{p},s} e^{+ip\cdot\widetilde{x}} \right). \qquad (6.81)$$

Using Eq. (6.73) and changing the integration variable from $\boldsymbol{p}$ to its negative, we obtain

$$\gamma_0 \psi(\widetilde{x}) = \sum_s \int D^3 p \left(d_{-\boldsymbol{p},s} u_{\boldsymbol{p},s} e^{-ip\cdot x} - \widehat{d}^\dagger_{-\boldsymbol{p},s} v_{\boldsymbol{p},s} e^{+ip\cdot x} \right). \qquad (6.82)$$

Putting the expressions obtained in Eqs. (6.82) and (6.79) into Eq. (6.23) and using the orthogonality of different Fourier components, we obtain the relations

$$\mathscr{P} d_{\boldsymbol{p},s} \mathscr{P}^{-1} = \eta_P d_{-\boldsymbol{p},s}, \qquad (6.83a)$$

$$\mathscr{P} \widehat{d}^\dagger_{\boldsymbol{p},s} \mathscr{P}^{-1} = -\eta_P \widehat{d}^\dagger_{-\boldsymbol{p},s}. \qquad (6.83b)$$

Suppose we now apply both sides of Eq. (6.83b) on the vacuum, recalling Eq. (6.72). The creation operator, acting on the vacuum, creates a state with one antiparticle, which we will denote by $\big| \widehat{F}_s(\boldsymbol{p}) \big\rangle$. So we obtain

$$\mathscr{P} \big| \widehat{F}_s(\boldsymbol{p}) \big\rangle = -\eta_P \big| \widehat{F}_s(-\boldsymbol{p}) \big\rangle, \qquad (6.84)$$

which tells us that the intrinsic parity of the antiparticle is $-\eta_P$. To compare it with the same property of the particle, we take the hermitian conjugate of Eq. (6.83a). Remembering that parity is a unitary operator, i.e., $\mathscr{P}^\dagger = \mathscr{P}^{-1}$, we obtain

$$\mathscr{P} d^\dagger_{\boldsymbol{p},s} \mathscr{P}^{-1} = \eta_P d^\dagger_{-\boldsymbol{p},s}. \qquad (6.85)$$

Applying both sides on the vacuum, we get

$$\mathscr{P} \big| F_s(\boldsymbol{p}) \big\rangle = \eta_P \big| F_s(-\boldsymbol{p}) \big\rangle, \qquad (6.86)$$

where $\big| F_s(\boldsymbol{p}) \big\rangle$ is a state containing one particle. So we obtain that the intrinsic parity of a particle state is η_P, i.e., same as that for the field. Thus, a fermion and its antifermion have opposite intrinsic parity.

We can prove the same result in an alternative and instructive way. The intrinsic parity of the fermion state can be defined by the quantity η_P appearing in Eq. (6.23). Now, the antiparticle bears the same relation with $\widehat{\psi}$ that the particle bears with ψ. Thus, the intrinsic parity of the antifermion, denoted by $\widehat{\eta}_P$, can be defined by the relation

$$\mathscr{P} \widehat{\psi}(x) \mathscr{P}^{-1} = \widehat{\eta}_P \gamma_0 \widehat{\psi}(\widetilde{x}). \qquad (6.87)$$

Using the definition $\widehat{\psi}$ from Eq. (6.49) to compare Eqs. (6.23) and (6.87), we can find the relation between $\widehat{\eta}_P$ and η_P. We outline the steps here. First, we take the hermitian conjugate of Eq. (6.23). Remembering that η_P must be real, we obtain

$$\mathscr{P} \psi^\dagger(x) \mathscr{P}^{-1} = \eta_P \left(\gamma_0 \psi(\widetilde{x}) \right)^\dagger. \qquad (6.88)$$

Since the components of $\psi^\dagger$ are the same as those of ψ^*, we can also write this equation as

$$\mathscr{P}\psi^*(x)\mathscr{P}^{-1} = \eta_P\gamma_0^*\psi^*(\tilde{x}) \,. \tag{6.89}$$

Using the definition of Eq. (6.49) now, we obtain

$$\mathscr{P}\widehat{\psi}(x)\mathscr{P}^{-1} = \eta_P\gamma_0 C\gamma_0^*\psi^*(\tilde{x}) = \eta_P\gamma_0 C\gamma_0^*\left(\gamma_0 C\right)^{-1}\widehat{\psi}(\tilde{x}) \,. \tag{6.90}$$

Recall that γ_0 is hermitian, so that $\gamma_0^* = \gamma_0^\mathsf{T}$. Using Eq. (6.57) now, it is easy to see that the previous equation gives

$$\mathscr{P}\widehat{\psi}(x)\mathscr{P}^{-1} = -\eta_P\gamma_0\widehat{\psi}(\tilde{x}) \,. \tag{6.91}$$

Comparing with Eq. (6.87), we obtain

$$\widehat{\eta}_P = -\eta_P \,, \tag{6.92}$$

which shows that the intrinsic parity of the antifermion should be opposite to that of the fermion.

□ **Exercise 6.15** We have not obtained the analog of Eq. (6.83) for the scalar case. Follow the arguments for the fermion case to prove that

$$\mathscr{P}a(p)\mathscr{P}^{-1} = \eta_P a(-p) \tag{6.93}$$

for a scalar field. Use it to obtain Eq. (6.77).

□ **Exercise 6.16** Complete the parallelism between the derivation of intrinsic parity of scalars and fermions by finding Eqs. (6.84) and (6.86) in the way shown for the scalars. In other words, apply both sides of Eq. (6.79) on the vacuum state and prove Eq. (6.84). Follow the similar procedure on the plane wave expansion of $\psi^\dagger(x)$ and obtain Eq. (6.86).

□ **Exercise 6.17** Strictly speaking, we have left out an important step in arriving at Eq. (6.83). Orthogonality of Fourier components gives directly the relations

$$\sum_s \mathscr{P}d_{p,s}\mathscr{P}^{-1}u_{p,s} = \sum_s d_{-p,s}u_{p,s} \,, \tag{6.94a}$$

$$\sum_s \mathscr{P}\widehat{d}_{p,s}^\dagger\mathscr{P}^{-1}v_{p,s} = -\sum_s \widehat{d}_{-p,s}^\dagger v_{p,s} \,. \tag{6.94b}$$

Show that the equalities in Eq. (6.83) follow from these two equations and the normalization conditions for the spinors, Eq. (4.63, p 72).

6.4.3 Orbital parity

For single-particle systems, intrinsic parity is the only contribution to parity. For multi-particle systems, however, there are other contributions. We elaborate these contributions in case of a two-particle system.

The dynamics of two-particle system can be broken up into the evolution of the center of mass and that of the relative co-ordinate r. The interaction between the two particles should be a function of r only. If the interaction is isotropic, it depends only on the magnitude r and not on the angular coordinates θ and ϕ. In this case, the angular part of the wavefunction of a stationary state is an eigenfunction of orbital angular momentum, i.e., is a spherical harmonic $Y_l^m(\theta, \phi)$ for some particular values of l and m. Apart from an overall normalization factor, the spherical harmonics can be written as

$$Y_l^m(\theta, \phi) = P_l^m(\cos\theta)e^{im\phi}, \qquad (6.95)$$

where P_l^m denotes the associated Legendre functions, defined in terms of the Legendre polynomials P_l through the relation

$$P_l^m(\xi) = (-1)^m(1 - \xi^2)^{m/2}\frac{d^m}{d\xi^m}P_l(\xi). \qquad (6.96)$$

The Legendre polynomials have the property

$$P_l(-\xi) = (-1)^l P_l(\xi). \qquad (6.97)$$

The exact forms for the Legendre polynomials are not necessary for the present discussion.

If we make a parity transformation on the state, the position of each particle changes sign, and so does the relative co-ordinate vector. In the spherical co-ordinates, the magnitude r remains unaffected so that the radial part of the wavefunction is not changed under parity. But on the angular co-ordinates, the parity transformation implies the changes

$$\theta \to \pi - \theta, \qquad \phi \to \pi + \phi. \qquad (6.98)$$

With the changed co-ordinates, the exponential factor present in Eq. (6.95) becomes

$$e^{im(\pi+\phi)} = (-1)^m e^{im\phi}. \qquad (6.99)$$

On the other hand, since $\cos\theta$ changes sign, the associated Legendre functions become

$$P_l^m(-\xi) = (-1)^m(1 - \xi^2)^{m/2}\frac{d^m}{d(-\xi)^m}P_l(-\xi)$$
$$= (-1)^l(1 - \xi^2)^{m/2}\frac{d^m}{d\xi^m}P_l(\xi) = (-1)^{l+m}P_l^m(\xi), \qquad (6.100)$$

using Eq. (6.97) on the way. Multiplying the two contributions, we obtain

$$Y_l^m(\pi - \theta, \pi + \phi) = (-1)^l Y_l^m(\theta, \phi). \qquad (6.101)$$

Thus, if two particles have a relative orbital angular momentum l between them, the wavefunction of the two-particle system changes by a factor of $(-1)^l$. This may be called the orbital parity factor.

6.5 Charge conjugation properties of particle states

We now undertake the same exercise for charge conjugation that we took for parity in §6.4, viz., identify the relation between charge conjugation properties of a field and the particles it represents. We start with a scalar field, as always. Eq. (6.40), acting on a vacuum state that is invariant under charge conjugation symmetry, gives

$$\mathscr{C}\phi(x)\,|0\rangle = \eta_C \phi^\dagger(x)\,|0\rangle \ . \tag{6.102}$$

Recalling the plane wave expansion of the scalar field from Eq. (4.12, *p 64*) and using the independence of different Fourier components, we can write

$$\mathscr{C}\widehat{a}^\dagger(\boldsymbol{p})\,|0\rangle = \eta_C a^\dagger(\boldsymbol{p})\,|0\rangle \ , \tag{6.103}$$

or,

$$\mathscr{C}\left|\widehat{B}(\boldsymbol{p})\right\rangle = \eta_C\,|B(\boldsymbol{p})\rangle \ . \tag{6.104}$$

This equation shows that the operator $\mathscr{C}$, acting on an antiparticle state, turns it into a particle state of the same momentum, and with an overall phase that is the same as that which occurs for the transformation of the field under charge conjugation. Acting on a particle state, however, the phase is opposite, since

$$\mathscr{C}\,|B(\boldsymbol{p})\rangle = \eta_C^*\left|\widehat{B}(\boldsymbol{p})\right\rangle \ , \tag{6.105}$$

which follows by operating both sides of Eq. (6.104) by $\mathscr{C}$ and recalling that $\mathscr{C}$ is a toggle operator, i.e., $\mathscr{C}^2 = 1$.

For a Dirac particle, we combine the definitions in Eqs. (6.48) and (6.49) to write

$$\mathscr{C}\psi(x)\mathscr{C}^{-1} = \eta_C \gamma_0 \mathbb{C}\psi^*(x)\,. \tag{6.106}$$

Taking the plane wave expansion of the Dirac field from Eq. (4.65, *p 72*), we can write

$$\gamma_0\mathbb{C}\psi^*(x) = \sum_s \int D^3p\left(d_s^\dagger(\boldsymbol{p})\gamma_0\mathbb{C}u_s^*(\boldsymbol{p})e^{+ip\cdot x} + \widehat{d}_s(\boldsymbol{p})\gamma_0\mathbb{C}v_s^*(\boldsymbol{p})e^{-ip\cdot x}\right). \tag{6.107}$$

In order to make a connection with the left hand side of Eq. (6.106), we need to use the conjugation properties of Dirac spinors. Each spinor can be defined with an overall arbitrary phase, and this phase affects its complex conjugation property. We can also extract an overall phase from the matrix

C. In Appendix F, we show that these phases can be chosen in a way such that the following relations hold:

$$\gamma_0 \mathbb{C} u_s^*(\boldsymbol{p}) = v_s(\boldsymbol{p}), \tag{6.108a}$$

$$\gamma_0 \mathbb{C} v_s^*(\boldsymbol{p}) = u_s(\boldsymbol{p}). \tag{6.108b}$$

This allows us to write

$$\gamma_0 \mathbb{C} \psi^*(x) = \sum_s \int D^3 p \left(d_s^\dagger(\boldsymbol{p}) v_s(\boldsymbol{p}) e^{+ip \cdot x} + \widehat{d}_s(\boldsymbol{p}) u_s(\boldsymbol{p}) e^{-ip \cdot x} \right), \tag{6.109}$$

so that, comparing with the left hand side of Eq. (6.106), we get

$$\mathscr{C} d_s(\boldsymbol{p}) \mathscr{C}^{-1} = \eta_C \widehat{d}_s(\boldsymbol{p}), \tag{6.110a}$$

$$\mathscr{C} \widehat{d}_s^\dagger(\boldsymbol{p}) \mathscr{C}^{-1} = \eta_C d_s^\dagger(\boldsymbol{p}). \tag{6.110b}$$

The conclusion is the same as that for scalar fields:

$$\mathscr{C} \left| F_s(\boldsymbol{p}) \right\rangle = \eta_C^* \left| \widehat{F}_s(\boldsymbol{p}) \right\rangle, \tag{6.111a}$$

$$\mathscr{C} \left| \widehat{F}_s(\boldsymbol{p}) \right\rangle = \eta_C \left| F_s(\boldsymbol{p}) \right\rangle. \tag{6.111b}$$

☐ **Exercise 6.18** Show that the two relations given in Eq. (6.110) are equivalent.

☐ **Exercise 6.19** Perform the same exercise for a vector field (not necessarily self-conjugate like the photon field) and show that the conclusions are the same as in Eqs. (6.104) and (6.105).

6.6   Multi-photon states

Multi-photon states were discussed in §5.6. Here we discuss properties of these states under the operations of parity and charge conjugation.

Charge conjugation properties are easy. Since photons are odd under charge conjugation, states with an even number of photons are even, and states with an odd number of photons are odd.

As for parity, we need to remember that 3-momentum is odd under parity. The polarization vectors are also odd, as noted in Table 6.1 *(p 157)*. If we now look at the state given in Eq. (5.102, *p 137*), we clearly see that it is even under parity transformation, since the magnitude of momentum is invariant under parity. On the other hand, the state shown in Eq. (5.103, *p 137*) is parity odd. Among the three photon states, the one shown in Eq. (5.108, *p 138*) is parity odd, whereas the one shown in Eq. (5.109, *p 138*) is parity even.

6.7   Positronium

Positronium is a bound state of a positron and an electron. It should be recalled that 'positron' is the name for the antiparticle of the electron. Thus positronium is a bound state of a fermion and its antiparticle.

In a sense the positronium resembles the hydrogen atom: it is a bound state of a positively charged particle with the electron. For one who is familiar with the energy states of the hydrogen atom, it is trivial to find out the same for positronium. For the hydrogen atom, the reduced mass is almost the same as the electron mass: more precisely, it is $m_e m_p/(m_e + m_p)$. For positronium, we should replace the proton mass by the positron mass. And since the mass of the positron is the same as that of the electron, the reduced mass is $\frac{1}{2}m_e$. Once this adjustment has been made, energy states and eigenvalues of positronium can be derived from those of the hydrogen atom. Like a hydrogen atom, the positronium ground state will be a $^1S_0$ state. If the total spin of the electron and the positron happens to be 1, we can obtain the excited state $^3S_1$. If the relative orbital angular momentum is 1, we obtain the P-states. Depending on the spin and depending on how the spin combines with the orbital angular momentum, we can obtain the excited states $^1P_1$, $^3P_0$, $^3P_1$ and $^3P_2$. Then there can be other excited states with larger values of the orbital angular momentum L.

Like hydrogen, excited states can decay to a lower state by emitting a photon. Since this is electromagnetic interaction, parity is conserved in the process. The same is true for decays of excited states of the hydrogen atom. But positronium decays are more exciting for two reasons.

First, note that under the operation of charge conjugation, the electron changes to positron, and vice versa. Therefore, any positronium state is an eigenstate of the charge conjugation operator as well, with eigenvalue either $+1$ or -1. Since electromagnetic interactions respect charge conjugation symmetry, any positronium state must decay only in a way that the final products, all taken together, have the same charge conjugation properties. This restricts the possibilities.

Second, a fermion-antifermion pair can undergo pair annihilation and turn to photons. Such final states are not possible for the hydrogen atom. For positronium, depending on the charge conjugation properties of the initial state, the final state can contain two or three photons. In principle it can contain more photons, but those final states will be suppressed because the emission of each photon costs a factor of e in the amplitude.

With this in mind, let us now discuss the parity and charge conjugation properties of a positronium state. If the electron and the positron are in a state with relative orbital angular momentum L, the state should have orbital parity $(-1)^L$, as discussed in §6.4.3 earlier. Thus the parity of the positronium state should be given by

$$P_{e^-e^+} = \eta_{e^-}\eta_{e^+}(-1)^L, \tag{6.112}$$

where η_{e^-} and η_{e^+} represent the intrinsic parities of the electron and the positron. However, we found in Eq. (6.92) that the two quantities differ by a sign, and by general considerations each of these quantities must be ± 1. Thus the product of the two intrinsic parities is -1, and we obtain

$$\mathscr{P} \left| \Psi \right\rangle = (-1)^{L+1} \left| \Psi \right\rangle , \qquad (6.113)$$

where $\left| \Psi \right\rangle$ represents a positronium state. This means that the states are eigenstates of parity, and the eigenvalue is determined by the orbital angular momentum quantum number.

Next we look at the charge conjugation property of a positronium state. For this purpose, we write a general positronium state in the following form in its center-of-mass frame:

$$\left| \Psi \right\rangle = \sum_{\boldsymbol{k},s_1,s_2} f(\boldsymbol{k},s_1,s_2) \widehat{d}^{\dagger}_{-\boldsymbol{k},s_2} d^{\dagger}_{\boldsymbol{k},s_1} \left| 0 \right\rangle . \qquad (6.114)$$

The action of the two creation operators on the vacuum creates a state consisting of an electron and a positron with opposite momenta $\boldsymbol{k}$ and $-\boldsymbol{k}$ respectively, and with spin components s_1 and s_2 along any particular direction. Any positronium state, in its center-of-mass frame, will be a superposition of such states, and the exact form of the superposition is determined by the function $f(\boldsymbol{k},s_1,s_2)$. This function also determines how the spins of the electron and the positron are combined in the state $\left| \Psi \right\rangle$. If the momentum values are continuous, the sum over momenta should be interpreted as an integration.

Charge conjugation operation will change the electron to a positron, and vice versa. This means that the action of the operator $\mathscr{C}$ on the state $\left| \Psi \right\rangle$ is given by

$$\begin{aligned}
\mathscr{C} \left| \Psi \right\rangle &= \sum_{\boldsymbol{k},s_1,s_2} f(\boldsymbol{k},s_1,s_2) d^{\dagger}_{-\boldsymbol{k},s_2} \widehat{d}^{\dagger}_{\boldsymbol{k},s_1} \left| 0 \right\rangle \\
&= \sum_{\boldsymbol{k},s_1,s_2} f(-\boldsymbol{k},s_1,s_2) d^{\dagger}_{\boldsymbol{k},s_2} \widehat{d}^{\dagger}_{-\boldsymbol{k},s_1} \left| 0 \right\rangle ,
\end{aligned} \qquad (6.115)$$

where use has been made of Eq. (6.110). In the last step, we have redefined the dummy momentum. Doing the same to the dummy spin components and using the anticommutation property of fermion creation operators, we can write

$$\mathscr{C} \left| \Psi \right\rangle = - \sum_{\boldsymbol{k},s_1,s_2} f(-\boldsymbol{k},s_2,s_1) \widehat{d}^{\dagger}_{-\boldsymbol{k},s_2} d^{\dagger}_{\boldsymbol{k},s_1} \left| 0 \right\rangle . \qquad (6.116)$$

Comparing Eqs. (6.114) and (6.116), we find that the effect of the charge conjugation operator on the energy eigenstate can be summarized as

$$f(\boldsymbol{k},s_1,s_2) \overset{\mathscr{C}}{\longrightarrow} -f(-\boldsymbol{k},s_2,s_1) . \qquad (6.117)$$

This result can be expressed in a more convenient manner. The total spin of the electron and the positron must be either 0 or 1. A spin-0 combination of two spin-$\frac{1}{2}$ particles is antisymmetric, whereas a spin-1 combination is symmetric. Thus, interchanging the spins would imply multiplying the original wavefunction by -1 in case of spin-0, and by $+1$ in case of spin-1. In short, we can say that for a system with two spin-$\frac{1}{2}$ particles, interchange of spins multiplies the wavefunction by $(-1)^{S+1}$, where S is the total spin. On the other hand, changing the sign of $\boldsymbol{k}$ in the Fourier space means changing the sign of the position vectors in the co-ordinate space, i.e., an orbital parity transformation. This gives a factor of $(-1)^L$, as we have already discussed. Summarizing, we can say that

$$f(-\boldsymbol{k}, s_2, s_1) = (-1)^{L+S+1} f(\boldsymbol{k}, s_1, s_2) \qquad (6.118)$$

for a state with total orbital angular momentum L and spin S. Putting this back into Eq. (6.116), we obtain

$$\mathscr{C} |\Psi\rangle = (-1)^{L+S} |\Psi\rangle , \qquad (6.119)$$

i.e., the charge conjugation eigenvalue of the energy eigenstate is $(-1)^{L+S}$.

□ **Exercise 6.20** Consider a state consisting of a spinless boson and its antiparticle. Show that the state is a charge conjugation eigenstate with eigenvalue $(-1)^L$.

So let us look at the ground state of positronium, or the $^1S_0$ state. The notation means that the state has $L = 0$ and $S = 0$. From Eqs. (6.113) and (6.119), we find that the parity and charge conjugation eigenvalues of the $^1S_0$ state are -1 and $+1$ respectively. The $\mathscr{C}$ eigenvalue tells us that this state can decay into two photons:

$$^1S_0 \to \gamma + \gamma, \qquad (6.120)$$

but not into three photons. On the other hand, the $^3S_1$ has a negative $\mathscr{C}$ eigenvalue, so it can decay into three photons but not into two. Parity conservation does not stand in the way because we know from §6.6 there are parity-even as well as parity-odd states of two photons. Parity conservation dictates that the $^1S_0$ state should decay to the parity-odd state shown in Eq. (5.103, p 137).

If we now want to discuss decays of a positronium state into another positronium state of lower energy by the emission of a photon, we can use the selection rules developed in the context of atomic physics. These rules are based upon the fact that the wavefunctions of the states are localized in a small region whose radius is of the order of the Bohr radius, $a_0 = (m\alpha)^{-1}$, whereas the energy eigenvalues, and therefore their differences, are of the order of $\alpha/a_0 = m\alpha^2$. Thus, the wavelengths of the radiation emitted in the transitions will be much larger than the size of the system. In evaluating the matrix element of transition, it is therefore a good approximation to expand the factor of $e^{i\boldsymbol{k}\cdot\boldsymbol{r}}$ appearing in the expression for the photon field in powers of $\boldsymbol{k}$ and keeping up to the first order term only. The transition amplitude will then involve matrix elements of $\boldsymbol{k} \cdot \boldsymbol{r}$ between the initial and the final states, i.e., $\boldsymbol{k} \cdot \boldsymbol{d}_E$, where $\boldsymbol{d}_E$ is the electric dipole moment operator. Since the operator has angular dependence of $\cos\theta$ with respect to the direction of $\boldsymbol{k}$, and $\cos\theta$ can be written as $P_1(\cos\theta)$ where P_l's denote Legendre polynomials, the matrix element vanishes unless the initial and final states differ by 1 in their orbital angular momentum quantum number L, i.e., unless

$$|\Delta L| = 1 \,. \tag{6.121}$$

Similar arguments show that the quantum number m_l can change by at most 1, something that we won't need too much. Finally, since total angular momentum must be conserved and photon has angular momentum equal to 1, the total angular momenta of the initial and final states, $\boldsymbol{J}_i$ and $\boldsymbol{J}_f$, must be such that $\boldsymbol{J}_f$ and the photon angular momentum can add up to $\boldsymbol{J}_i$. According to the rules of angular momentum addition, this can happen if

$$|\Delta J| = 0 \text{ or } 1 \,, \tag{6.122}$$

with the rider that both $\boldsymbol{J}_i$ and $\boldsymbol{J}_f$ cannot be zero, i.e.,

$$J = 0 \rightarrow J = 0 \quad \text{transition not allowed.} \tag{6.123}$$

It should be noted that when we say "allowed" or "forbidden" transitions in this context, we just make a comment about whether the transitions are possible through the first order term in $\boldsymbol{k}$. A transition branded "forbidden" is not really impossible: can still take place through higher order terms in the expansion of $e^{i\boldsymbol{k}\cdot\boldsymbol{r}}$. However, such transitions will be much suppressed compared to the transitions that we brand "allowed".

With this machinery, we can easily conclude that

$$^3S_1 \nrightarrow {}^1S_0 + \gamma \,, \tag{6.124}$$

where the cross mark implies that the process is forbidden. This process violates the selection rule of Eq. (6.121). Another example of a decay forbidden by angular momentum is

$$^3P_0 \nrightarrow {}^1S_0 + \gamma \,, \tag{6.125}$$

where Eq. (6.123) is violated.

The case of the $^3S_1$ state is interesting. We just saw that it cannot radiatively decay into any lower-lying positronium state. It can only decay into three photons. The ground state, $^1S_0$, on the other hand, decays into two photons. The amplitude for the three photon decay must have an extra power of e, and therefore the rate will be down roughly by a factor of the fine-structure constant α. Thus, the rate for $^3S_1$ decay will be much smaller than that for $^1S_0$ decay. In other words, the $^3S_1$ will be much more stable than the ground state. Indeed, experimental searches reveal that

$$\tau(^3S_1) = (142.05 \pm 0.02) \times 10^{-9}\,\text{s}\,,$$
$$\tau(^1S_0) = (125.14 \pm 0.02) \times 10^{-12}\,\text{s}\,. \tag{6.126}$$

□ **Exercise 6.21** Consider the transitions from each of the following states of positronium

$$^3P_0, \quad ^3P_1, \quad ^3P_2, \quad ^1P_1 \qquad (6.127)$$

to any of the states

$$^1S_0 + \gamma, \quad ^3S_1 + \gamma, \quad \gamma + \gamma, \quad \gamma + \gamma + \gamma. \qquad (6.128)$$

For each of the 16 transitions, tell whether it is allowed or forbidden by

a) Angular momentum conservation,

b) P conservation,

c) C conservation.

6.8 Parity assignment of different particles

We now examine the intrinsic parity of some particles. Already, in Eq. (6.36), we have shown that the electromagnetic interactions dictate that the intrinsic parity of the photon must be negative. This conclusion is obtained irrespective of the intrinsic parity of any fermion. The reason is that the interaction Lagrangian of any fermion with the photon contains a pair of fermion fields, and their effects cancel out no matter whether we assign positive or negative intrinsic property for the fermion. The result also extends to fundamental scalar fields, and we can say in general that electromagnetic interactions cannot fix the intrinsic parity of any elementary particle except the photon.

Let us then turn to particles which have strong interactions, or hadrons. As mentioned earlier, hadrons can be fermions or bosons. There is something special about fermions. If the number of fermions in the initial state of a process is odd, the total angular momentum is half-integral, and so angular momentum conservation law tells us that the number of fermions in the final state must also be odd. If the number is even in the initial state, it is even in the final state as well. Therefore, if we change the intrinsic parities of *all* fermions by a sign, the effect of this change would cancel out between the initial and the final states. The equality of initial and final parities in a parity conserving process will not be affected by this exercise.

This observation tells us that there must be two different ways of assigning intrinsic parity to all fermions: if anyone makes a consistent choice, we can make another consistent choice by changing the intrinsic parities of all fermions by a sign. We can use this observation to fix the intrinsic parity of any one fermion arbitrarily, and we choose the proton to have positive parity:

$$\eta_P^{(\text{proton})} = +1. \qquad (6.129)$$

Protons and neutrons have identical strong interactions, a phenomenon that will be discussed in detail in Ch. 8. It means that these two particles should have the same properties in everything that the strong interaction respects.

Strong interactions respect parity symmetry, so, in particular, parity properties of the proton and the neutron should be the same:

$$\eta_P^{(\text{neutron})} = +1 \, . \tag{6.130}$$

Let us now look at the intrinsic parity of pions. Pions are of three types: π^-, π^0 and π^+, where the superscripts indicate the electric charges of the particles. The neutral pion decays to a pair of photons:

$$\pi^0 \longrightarrow \gamma\gamma \, . \tag{6.131}$$

The two photons must be in a state with total angular momentum zero, since the pion has no spin. As we have seen in §5.6, there are two such states of two photons. It was further discussed in §6.6 that one of these two states is odd under parity whereas the other is parity even. In the parity even state, the two polarization vectors appear in the combination $\epsilon_1 \cdot \epsilon_2$, so that the two polarization vectors are most likely to be parallel. In the other state, $\epsilon_1 \times \epsilon_2$ is involved, so that the two polarization vectors are most likely to be perpendicular to each other. Experiments show that the second alternative is what really happens in nature, which means that the two photons are in a parity odd state. We then conclude that

$$\eta_P^{(\pi^0)} = -1 \, . \tag{6.132}$$

The intrinsic parity of the negative pion can be determined through the reaction

$$\pi^- + d \longrightarrow n + n \, . \tag{6.133}$$

Here d denotes *deuteron*, a bound state of a proton and a neutron. In the ground state of deuteron, the proton and the neutron are in a relative $L = 0$ state, so their orbital parity is $+1$. Putting in the intrinsic parities of the proton and the neutron, we find that the deuteron itself has a positive intrinsic parity. The deuteron absorbs the π^- in a relative $L = 0$ state, so that the orbital parity in the initial state is $+1$. Thus the parity of the initial state has the sign of the intrinsic parity of the negative pion.

Let us now look at the parity of the final state. In the initial state, the pion is spinless, and the spin of the deuteron is 1. Since they are combined in an $L = 0$ state, the total angular momentum of the initial state is 1. The final state angular momentum must then be 1 as well. With two spin-$\frac{1}{2}$ particles, the total spin must be 0 or 1, so the possible $J = 1$ states are $^3S_1$, $^1P_1$ and $^3P_1$. Since the final state contains two identical fermions, the state must be antisymmetric under the interchange of these fermions. The properties of these states, under the interchange of position and spin, are summarized below:

State	L	S	Symmetry property under interchange		
			Position	Spin	Total
$^3S_1$	0	1	symm	symm	symm
$^1P_1$	1	0	antisymm	antisymm	symm
$^3P_1$	1	1	antisymm	symm	antisymm

From this table, it is clear that only the state $^3P_1$ is admissible for two neutrons. The parity of the two neutrons in this state is -1 coming from the orbital contribution, whereas the intrinsic parities of the two neutrons is $+1$. Thus, the final state has a parity eigenvalue -1, which means that in the initial state, we must have

$$\eta_P^{(\pi^-)} = -1 \,. \tag{6.134}$$

The intrinsic parity of the positively charged pion, π^+, is the same, as can be argued from the reaction

$$\pi^+ + d \longrightarrow p + p \,. \tag{6.135}$$

Let us consider the question of the intrinsic parity of the electron. As explained earlier, this cannot be determined from electromagnetic interactions. Electrons do not have strong interactions. Thus, we have only one option left: weak interactions. However, we will discuss later in the book that weak interactions do not conserve parity. Therefore, we cannot assign an intrinsic parity to the electron through weak interactions either, and we need to make another arbitrary choice for the intrinsic parity of the electron like we made for the proton. The same comment applies to any lepton like the muon or the tau, which do not have any strong interaction.

6.9 Signature of parity violation

6.9.1 Correlations in experiments

Earlier in this chapter, we have given some examples of interactions which would not conserve parity. This was concluded by examining parity transformation properties of various parts of the interaction Lagrangian.

In an experiment, we do not see the Lagrangian. So how would we know that parity is not conserved in certain processes? We see the particles taking part in the process. We can measure the momenta, the spins and other such properties of particles. If we have to know about parity violation, we will have to know it through such measurements.

The most straightforward way of approaching the problem would be to measure the rate of some process and its parity transformed process. By the parity transformed process, we mean a process involving the same particles, but with reversed 3-momenta (and other possible kinematical variables which are odd under parity). Table 6.2 shows the behavior of common kinematical variables under parity.

Table 6.2: Behavior of common kinematical variables under parity. Behavior under time reversal has also been included for later chapters.

Quantity	Behavior under		Justification
	Parity	Time reversal	
Momentum	$-$	$-$	Involves velocity, which is $d\boldsymbol{r}/dt$
Energy	$+$	$+$	Kinetic energy, e.g., involves square of velocity
Angular momentum	$+$	$-$	Orbital angular momentum is $\boldsymbol{r} \times \boldsymbol{p}$

In practice, it is not necessary to perform two sets of measurements, one on the direct process and the other on the parity transformed process. Performing measurements on a single process, it is possible to tell whether the corresponding parity transformed process would give a different result by noting some parity odd correlations between kinematical variables. We explain this statement now.

As an example, consider a scattering process in which the final state has two particles. We perform the experiment in the CM frame, in which the 3-momentum of one the initial particles is $\boldsymbol{p}$, and that of one of the final particles is $\boldsymbol{p}'$. Of course, for a fixed $\boldsymbol{p}$, infinitely many directions are possible for $\boldsymbol{p}'$, and we measure the differential cross-section. Suppose we find that the differential cross-section depends on the angle between $\boldsymbol{p}$ and $\boldsymbol{p}'$. Because of rotational symmetry, the dependence must come through a combination like $\boldsymbol{p} \cdot \boldsymbol{p}'$. Symbolically, we denote this situation by writing

$$\frac{d\sigma}{d\Omega} = A_0 + A_1 \boldsymbol{p} \cdot \boldsymbol{p}', \qquad (6.136)$$

where A_0 is an angle-independent part. We can now ask: will such a dependence signal parity violation? The answer is 'no', because the combination $\boldsymbol{p} \cdot \boldsymbol{p}'$ is even under parity.

The situation changes if one of the initial particles is spin-polarized, i.e., its beam has a net non-zero spin. We denote the spin by $\boldsymbol{s}$. Making observations at different directions for $\boldsymbol{p}'$, we will find a differential cross-section which will in general have the form

$$\frac{d\sigma}{d\Omega} = A_0 + A_1 \boldsymbol{p} \cdot \boldsymbol{p}' + A_2 \boldsymbol{s} \cdot \boldsymbol{p}' + A_3 \boldsymbol{s} \cdot (\boldsymbol{p} \times \boldsymbol{p}'). \qquad (6.137)$$

The A_2 term is different from the others because the combination $\boldsymbol{s} \cdot \boldsymbol{p}'$ is odd under parity, as can be confirmed by looking at Table 6.2. Thus, if we observe such a correlation in the differential cross-section, it would signal parity violation.

This kind of correlation constituted the first ever detection of parity violation in 1957 by Wu and her collaborators. The experiment involved beta decay of radioactive ^{60}Co. The ^{60}Co nucleus has spin $s = 5$. In the experiment, these spins were aligned by applying a magnetic field. Of course thermal agitations would destroy such alignment unless the entire system is kept at a very low temperature. In Wu's experiment, temperatures of the order of $0.01\,\text{K}$ were required.

The beta decay reaction for ^{60}Co is

$$^{60}\text{Co} \rightarrow {}^{60}\text{Ni} + e + \widehat{\nu}_e. \tag{6.138}$$

The final electron was detected at various angles, and the differential decay rate I was obtained with respect to the direction of its momentum. The results showed a variation of the form

$$I(\theta) = a(1 + b\cos\theta), \tag{6.139}$$

where θ is the angle between the spin of the ^{60}Co nucleus and the final electron. If b were zero, it would have meant that parity was conserved. But the experiment found

$$b \approx -0.4, \tag{6.140}$$

signalling parity violation. Since then, signatures of parity violation have been observed in many experiments involving polarized source.

A variant of the idea would be to look for correlation between spin of a particle and its own momentum. In the same issue of the same journal, right after Wu's paper, appeared a paper by Garwin, Lederman and Weinrich, who used this idea. They considered the correlation of spin and momentum of antimuons (μ^+) coming from the decay of positively charged pion:

$$\pi^+ \rightarrow \mu^+ + \nu_\mu. \tag{6.141}$$

Suppose an antimuon is produced with 3-momentum along the z-direction with spin-component $+\frac{1}{2}$ along the z-direction. Since momentum is odd under parity and angular momentum even, as shown in Table 6.2, the parity transformed situation will have the antimuon with S_z still along the z-direction but the 3-momentum in the opposite direction. The quantity called *helicity*, which is the projection of spin along the direction of momentum, would be positive in the first situation and negative in the parity reversed situation. If parity is conserved, the two situations should occur with equal probabilities and therefore the net helicity of the antimuon should be zero, i.e., there should not be any correlation between spin and momentum. The experiment found a correlation, proving that parity is violated. Moreover, the helicity was found to be consistent with $+1$, i.e., the spin projection seemed to be always in the direction of the momentum, suggesting that parity is violated maximally.

There could be other ways of detecting parity violation, but one needs more than two particles in the final state for such methods. For example, consider a decay in which the final state contains four particles. Because of momentum conservation, three of the final state momenta would be independent. Let us denote them by p_1, p_2 and p_3. And now suppose the differential decay rate shows a variation like

$$I = a(1 + b\, p_1 \cdot p_2 \times p_3)\,. \qquad (6.142)$$

This would also signal parity violation, since the momenta are odd with respect to parity.

6.9.2 Parity violating transitions

Generally speaking, parity violation can be inferred when we see a state with given parity properties evolves into another state with different parity properties. We can look for such phenomena in decays of particles. For example, suppose we have a spin-0 particle ϕ, and through some means, we know that its intrinsic parity is negative, i.e., the particle is a pseudoscalar. Now suppose we find that it decays into two identical spin-0 particles, i.e., a decay of the form

$$\phi \to \phi'\phi'\,. \qquad (6.143)$$

No matter what the intrinsic parity of ϕ' is, the intrinsic parity of the two particles on the right hand side is $+1$. Angular momentum conservation tells us that the two particles in the final state must be in an $L = 0$ combination, so that the orbital parity is $+1$ as well. Thus the parity of the final state is $+1$. Since we started from a pseudoscalar particle in the initial state, such a decay would imply parity violation.

Needless to say, the conclusion would be unchanged even if the final state contains two different spin-0 particles, as long as they have the same intrinsic parity. In fact, it is not even necessary that the final state is an eigenstate of parity. Even if a parity eigenstate evolves into a state which is not an eigenstate of parity, we know that there must be a parity violating part in the Hamiltonian that governs the evolution.

Historically, the existence of parity violation was hinted from a situation like this. In the early 1950s, scientists discovered many new particles. One of them was a charged particle, dubbed θ^+, which was seen to decay into $\pi^+\pi^0$. By the argument above, it should have an intrinsic parity of $+1$. Another particle, discovered the same way, was called τ^+. It was seen to decay into three pions, and therefore it was concluded that it should have an intrinsic parity of -1. However, it was found that the θ^+ and the τ^+ have equal masses and lifetimes, within the limits of error of the experiments, and this suggested that there are not really two particles, but only one. This created a situation that was called the τ-θ puzzle. In 1956, Lee and Yang finally broke the impasse by suggesting that parity conservation had not been tested in weak interactions before, and that the two kinds of decays can occur for the same particle if parity is violated. Within a year, the test of parity violation came through the experiments that we have already discussed. The names θ and τ were abandoned: the particle is now called the (charged) kaon and denoted by K^+. The symbol τ, thus freed from the puzzle, is used now to denote a lepton like the electron or the muon.

The initial state need not be a decaying particle. For example, one can take the energy eigenstates of an electron in an atom. In absence of parity violation, these states have definite parity. The selection rules discussed in §6.7 then apply. However, if there is parity violating interaction between the nucleus and the electron, the selection rules will be violated. Such violations have been observed, establishing the existence of parity violation.

While checking for parity invariance, we seem to be multiplying the different contributions coming from intrinsic as well as orbital parity in determining the parity property of a state containing more than one particles. One may ask why we are doing this. After all, while checking momentum or angular momentum conservation, we add, rather than multiply, different contributions. So, is something different being done here?

The answer is 'no'. We are doing the same thing, really. Group elements should always be multiplied. For discrete symmetries, this is what we are doing directly. For continuous symmetries like translation or rotation, remember that momentum and angular momentum are the generators, which appear in the exponent in the expression for the group element. So, multiplying group elements means adding exponents. This is what we do for momentum. For angular momentum, since the different components do not commute, the exponents cannot be simply added: they should be subjected to the Baker–Campbell–Hausdorff formula, Eq. (3.18, p 44).

6.9.3 Parity violating coupling with external fields

In §5.7, we showed that interactions of fermions with the electromagnetic field can be expressed in terms of four form factors. This result was obtained by assuming Lorentz invariance and gauge invariance only, without paying any attention to discrete symmetry. Let us now see what is to be expected of these form factors in a theory where parity is a conserved symmetry.

The effective electromagnetic vertex of a fermion has been given in Eq. (5.121, p 140). In the co-ordinate space, this vertex can be interpreted as an effective interaction term

$$\mathscr{L}_{\text{eff}}(x) = \overline{\psi}(x)\mathscr{O}_\mu \psi(x) A^\mu(x)\,, \tag{6.144}$$

where $\mathscr{O}_\mu$ contains derivatives and matrices, but no field operator. The parity transformation of this effective interaction would be

$$\mathscr{P}\mathscr{L}_{\text{eff}}(x)\mathscr{P}^{-1} = \overline{\psi}(\widetilde{x})\gamma_0 \mathscr{O}_\mu \gamma_0 \psi(\widetilde{x})\ \mathscr{P}A^\mu(x)\mathscr{P}^{-1}\,, \tag{6.145}$$

using Eq. (6.23, p 155). Note that the intrinsic parity of the fermion field cancels out. Parity invariance would require that this expression equals $\mathscr{L}_{\text{eff}}(\widetilde{x})$.

Using the transformation of the photon field under parity that was given in
Eqs. (6.14) and (6.36), this would imply that parity invariance requires

$$\gamma_0 \mathcal{O}_0 \gamma_0 = \mathcal{O}_0, \qquad \gamma_0 \mathcal{O}_i \gamma_0 = -\mathcal{O}_i. \qquad (6.146)$$

Coming back to the vertex function, it is easily seen that the conditions of
parity invariance reduce down to

$$\gamma_0 \Gamma_0 \gamma_0 = \Gamma_0, \qquad \gamma_0 \Gamma_i \gamma_0 = -\Gamma_i. \qquad (6.147)$$

This is obviously satisfied by the combination that goes with the form factor
F_1 because of the basic definitions of the Dirac matrices given in Eq. (4.49,
p 70). It can be easily seen that the same is true if the 4×4 matrix sandwiched
between the spinors is $\sigma_{\mu\nu}$. But the combinations involving γ_5 do not satisfy
Eq. (6.147). Hence, presence of the electric dipole or anapole moment form
factor of a fermion would signal parity violation.

☐ **Exercise 6.22** Verify that the combinations involving the matrix γ_5
in the fermion vertex function do not satisfy Eq. (6.147).

6.9.4 Connection with field theory

Earlier, we talked how parity violation shows in a Lagrangian. Now we discussed how parity violation shows up in an experiment. The question that
arises is this: what is the connection between the two kinds of specification
of parity violation?

The answer can be guessed easily. The parity violating correlations can
occur in differential decay rates only if the underlying Lagrangian is parity
violating, as Ex. 6.23 will demonstrate.

☐ **Exercise 6.23** Consider an interaction connecting five scalar fields:

$$\mathscr{L}_{\text{int}} = h\phi\phi_1\phi_2\phi_3\phi_4 + h'\phi\varepsilon_{\mu\nu\lambda\rho}(\partial^\mu \phi_1)(\partial^\nu \phi_2)(\partial^\lambda \phi_3)(\partial^\rho \phi_4). \qquad (6.148)$$

a) Find the dimensions of the coupling constants h and h'. [**Note :**
   You will find that the theory is not renormalizable. But never mind. We
   do not need to calculate any loop diagram, so we can obtain well-defined
   answers.]

b) Show that the interactions shown above must violate parity.

c) The interactions can induce, at the tree-level, the decay

$$B(p) \rightarrow B_1(p_1) + B_2(p_2) + B_3(p_3) + B_4(p_4), \qquad (6.149)$$

   where B etc. denote the particles corresponding to the fields,
   and the notations in parentheses are the 4-momenta. Write
   down the amplitude for the decay.

d) Evaluate $|\mathscr{M}|^2$ in the rest frame of the decaying particle. Show
   that it contains a triple product of the momenta of final particles, like the one shown in Eq. (6.142).

If we want to see an example that produces spin-momentum correlations which violate parity, we first need to learn how to perform calculations involving spin-polarized particles. Note that in all calculations in Ch. 5, we have used unpolarized fermions and summed over the spins. While calculating the absolute square of the Feynman amplitude, this allowed us to express the result in terms of traces, which are easy to calculate.

Indeed, the machinery is so comfortable that we don't want to do away with it. So, when we have polarized states, we keep summing over all spin states, but insert a projection matrix in the Feynman amplitude so that only one spin state contributes to the sum.

In order to see explicitly how the idea works, let us consider an illustrative example. Suppose there are two fermion fields and a scalar field, with an interaction

$$\mathscr{L}_{\text{int}} = \overline{\psi}_2(a + b\gamma_5)\psi_1\phi + \text{h.c.}, \tag{6.150}$$

where we assume the constants a and b to be real. Suppose the masses of the particles are such that it is possible for the first fermion (denoted by ψ_1) to decay into the other fermion and the scalar. For unpolarized states, we would write the Feynman amplitude as

$$\mathscr{M} = \overline{u}_2(p_2)\left(a + b\gamma_5\right)u_1(p_1), \tag{6.151}$$

where u_1 and u_2 denote spinor solutions for the fields ψ_1 and ψ_2 respectively, and the momenta of the particles are given in parentheses. If we calculate the differential cross-section from this Feynman amplitude by summing over all spin states, we will get nothing that will signal parity violation.

However, suppose now that the initial fermion is spin-polarized along a direction denoted by the unit 3-vector $\widehat{s}$. Instead of using only the spinor solution for that direction, we will write the matrix element as

$$\mathscr{M} = \overline{u}_2(p_2)\left(a + b\gamma_5\right)P_{\widehat{s}}u_1(p_1), \tag{6.152}$$

where $P_{\widehat{s}}$ is a projection matrix defined in such a way that if it acts on the spinor solution whose spin eigenvalue is $+\frac{1}{2}$ along the direction $\widehat{s}$, the spinor will be unaffected; but if it acts on the spinor solution with the spin eigenvalue $-\frac{1}{2}$ along the same direction, the result would vanish. In Appendix F, we present arguments leading to Eq. (F.118, *p 751*) to show that this operator is given by

$$P_{\widehat{s}} = \frac{1}{2}(1 + \gamma_5\slashed{s}), \tag{6.153}$$

where s^μ is a 4-vector whose components are given by

$$s^\mu = (0, \widehat{s}) \tag{6.154}$$

in the rest frame of the particle. While squaring the Feynman amplitude, this $\widehat{s}$ will appear in the expression, dotted with some momentum. That will constitute the signal for parity violation.

☐ **Exercise 6.24** *Show that $P_{\hat{s}}$, defined in Eq. (6.153), is indeed a pro-jection matrix, i.e., $(P_{\hat{s}})^2 = P_{\hat{s}}$.*

☐ **Exercise 6.25** *Consider the amplitude given in Eq. (6.152), for de-caying fermions polarized in the direction of $\hat{s}$.*

    a) Show that

$$\frac{1}{2}\sum_{\text{spin}}\left|\mathcal{M}\right|^2 = (a^2 + b^2)p_1 \cdot p_2 + (a^2 - b^2)m_1m_2 + 2abm_{1\hat{s}} \cdot p_2\,.$$

(6.155)

    b) Note that the parity violating term is non-zero only if both a and b are non-zero. Using the discussion of §6.2, argue why it must be so.

    c) Show that the total decay rate is the same as that obtained for unpolarized initial particles.

6.10 Consequences of charge conjugation symmetry

The operation of charge conjugation takes particles to antiparticles, and vice versa. Therefore, charge conjugation symmetry would relate properties of a particle to those of its antiparticle. Most naively, one might expect that this symmetry would imply that the masses of the particle and the antiparticle would be equal, and so will be the decay rates if the particle happens to be unstable. As it turns out, these properties follow from a much weaker assumption. Even if the charge conjugation symmetry is violated, mass and decay rate of a particle are equal to the same properties of its antiparticle as long as the product CPT is a good symmetry. We discuss these issues in §7.5.

What would constitute a violation of charge conjugation symmetry alone, without an accompanying violation of either parity or time reversal? A few examples appear in §6.7, involving decays of various states of positronium. For example, the $^1S_0$ state cannot decay into three photons, although the decay would have been allowed by parity and angular momentum conservation. Decay of the neutral pion provides another such example. The neutral pion is seen to decay into two photons:

$$\pi^0 \to \gamma\gamma\,.$$

(6.156)

However, the decay to three photons has never been observed:

$$\mathcal{B}(\pi^0 \to 3\gamma) < 3.1 \times 10^{-8}\,.$$

(6.157)

This has a very simple explanation in terms of charge conjugation symmetry. Photons are odd under charge conjugation, as mentioned in Eq. (6.69). A

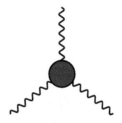

Figure 6.1: The blob denotes an effective vertex, i.e., any number of lines of any kind might be present in the blob. The external lines are three photon lines. Furry's theorem says that such a vertex is impossible.

state of two photons is therefore even under charge conjugation. Since the neutral pion decays to two photons, we conclude that

$$\eta_C^{(\pi^0)} = +1\,, \tag{6.158}$$

if charge conjugation symmetry is not violated in the decay. And in fact charge conjugation symmetry should not be violated, because the only relevant interactions here are the strong and the electromagnetic ones, the first for the binding of the pion and the second for producing the photons. We have shown that the basic QED interactions conserve the charge conjugation symmetry. The strong interactions also respect this symmetry. A state of three photons is odd under charge conjugation. Therefore, π^0 cannot decay into three photons. In fact, considerations like this one led to the realization of the existence of the charge conjugation symmetry.

The $\mathscr{C}$-odd property of the photon has other important consequences. For example, consider an interaction term involving three photons. Of course such a term does not occur in the Lagrangians of QED, or even scalar QED. The question is, can such an effective interaction arise out of other interactions, in a manner symbolized in Fig. 6.1? The answer is 'no', unless charge conjugation symmetry is violated. In fact, charge conjugation symmetry implies the absence of any effective interaction that contains only an odd number of photons. This statement is called *Furry's theorem*.

6.11 CP symmetry

The CP operation is nothing but a combination of charge conjugation and parity. If the parity operator operates on a particle state with momentum $\boldsymbol{p}$, we obtain a particle state with momentum $-\boldsymbol{p}$, as shown in Eqs. (6.77) and (6.86), for example. On the other hand, if we apply the charge conjugation operator on the same state, we would obtain an antiparticle state with momentum $\boldsymbol{p}$. The operation CP, acting on a particle state of momentum $\boldsymbol{p}$ will then give an antiparticle state with momentum $-\boldsymbol{p}$.

The effect of the CP operator on different fields can be written down by combining the effects of C and P separately on these operators. For example, from Eqs. (6.4) and (6.40), we obtain that for a scalar field,

$$(\mathscr{C}\mathscr{P})\phi(x)(\mathscr{C}\mathscr{P})^{-1} = \mathscr{C}(\mathscr{P}\phi(x)\mathscr{P}^{-1})\mathscr{C}^{-1}$$
$$= \eta_P \mathscr{C}\phi(\widetilde{x})\mathscr{C}^{-1} = \eta_{CP}\phi^\dagger(\widetilde{x}), \qquad (6.159)$$

where

$$\eta_{CP} = \eta_P \eta_C \qquad (6.160)$$

for the field, which can be called the intrinsic CP-phase for the field. Similarly, for a vector field $V_\mu(x)$, Eqs. (6.14) and (6.45) imply that

$$(\mathscr{C}\mathscr{P})V_0(x)(\mathscr{C}\mathscr{P})^{-1} = -\eta_{CP}V_0^\dagger(\widetilde{x}),$$
$$(\mathscr{C}\mathscr{P})\boldsymbol{V}(x)(\mathscr{C}\mathscr{P})^{-1} = +\eta_{CP}\boldsymbol{V}^\dagger(\widetilde{x}), \qquad (6.161)$$

where again the intrinsic CP-phase of the field is given in terms of its intrinsic phases under parity and charge conjugation through Eq. (6.160). Finally, for a Dirac field, the corresponding equation is

$$(\mathscr{C}\mathscr{P})\psi(x)(\mathscr{C}\mathscr{P})^{-1} = \eta_{CP}\mathbb{C}\psi^*(\widetilde{x}), \qquad (6.162)$$

which can be read from Eqs. (6.23), (6.48) and (6.49).

As indicated in earlier sections of this chapter, it is helpful to know the transformation properties of fermion bilinears under a certain symmetry. With that in mind, let us define, for a constant matrix F, the matrix F_{CP} by the relation

$$(\mathscr{C}\mathscr{P})\overline{\psi}_1(x)F\psi_2(x)(\mathscr{C}\mathscr{P})^{-1} = \eta_{12}\overline{\psi}_2(\widetilde{x})F_{CP}\psi_1(\widetilde{x}), \qquad (6.163)$$

where

$$\eta_{12} = \eta_{CP}^{(1)*}\eta_{CP}^{(2)}. \qquad (6.164)$$

Then, consulting Eqs. (6.26) and (6.68), we can easily write

$$\begin{array}{c|c|c|c|c|c|c|c|c}
F & \mathbb{1} & \gamma_0 & \gamma_i & \sigma_{0i} & \sigma_{ij} & \gamma_0\gamma_5 & \gamma_i\gamma_5 & \gamma_5 \\
\hline
F_{CP} & \mathbb{1} & -\gamma_0 & \gamma_i & \sigma_{0i} & -\sigma_{ij} & -\gamma_0\gamma_5 & \gamma_i\gamma_5 & -\gamma_5
\end{array}. \qquad (6.165)$$

□ **Exercise 6.26** *Show that*

$$F_{CP} = \gamma_0 \mathbb{C} F^\top \mathbb{C}^{-1}\gamma_0. \qquad (6.166)$$

Let us find out how CP might be conserved or violated in an interaction between fermions and vector bosons. We take the interaction in the general form, involving a polar and an axial vector bilinear:

$$\mathscr{L}_{\text{int}} = \overline{\psi}_1 \gamma^\mu (a + b\gamma_5)\psi_2 W_\mu + \overline{\psi}_2 \gamma^\mu (a^* + b^*\gamma_5)\psi_1 W_\mu^\dagger, \qquad (6.167)$$

where the second term in the hermitian conjugate of the first. Let us assume, for the sake of convenience, that the two fermions have the same intrinsic CP-phase. Then, the CP transform of the first term is

$$(\mathscr{C}\mathscr{P})\overline{\psi}_1 \gamma^\mu (a + b\gamma_5)\psi_2 W_\mu (\mathscr{C}\mathscr{P})^{-1} = \overline{\psi}_2 \gamma^\mu (a + b\gamma_5)\psi_1 \eta_{CP}^{(W)} W_\mu^\dagger,$$
$$(6.168)$$

obtained by using Eqs. (6.161) and (6.163), and using the bilinear transformation rules from Eq. (6.165). Note that CP is a unitary operator, so it does not affect the numerical constants a and b.

Looking at Eq. (6.168), we find that the CP conjugate of the first term of Eq. (6.167) is equal to the second term provided the intrinsic CP-phase of the vector boson W is $+1$, and the constants a and b are real. Thus, if these conditions are satisfied, the two interaction terms of Eq. (6.168) would transform into each other under the action of CP, and consequently the two terms taken together would be CP invariant.

This is an interesting conclusion. Earlier, we noted that a combination of polar and axial vector couplings cannot be invariant under C or P. However, we now see that if such combination of couplings occur, the violations of C and P may just compensate each other so that CP can remain invariant. In Ch. 21, we will see that this has a profound impact on models of particle interactions.

□ **Exercise 6.27** ♪ Consider an interaction of fermions with a complex scalar field ϕ of the form

$$\mathscr{L}_{\text{int}} = \overline{\psi}_1 (a + b\gamma_5)\psi_2 \phi + \text{h.c.} \qquad (6.169)$$

Assume the two fermions to have the same intrinsic CP-phase. Find the conditions that must be satisfied in order that the interaction is CP conserving.

Chapter 7

Time-reversal and CPT symmetries

7.1 Anti-unitary operators

7.1.1 Definition

We mentioned some "intricacies" regarding the time reversal symmetry in §6.1. Let us now explain the statement.

What do we mean by a symmetry? Suppose we perform an experiment that measures, in some way, the matrix element between a certain initial state $|\phi_1\rangle$ and a certain final state $|\phi_2\rangle$. It is customary to denote this matrix element by the notation $\langle\phi_2|\phi_1\rangle$. Any operator $\mathscr{U}$ would change the initial and the final states, and we fix the notation for the changed states as follows:

$$\mathscr{U}|\phi_1\rangle = |\mathscr{U}\phi_1\rangle\,, \qquad \mathscr{U}|\phi_2\rangle = |\mathscr{U}\phi_2\rangle\,. \tag{7.1}$$

For arbitrary operators $\mathscr{U}$, the matrix element of these changed states will not be equal to the matrix element between the original states. However, if we find some operator $\mathscr{U}$ such that it satisfies the relation

$$\langle\mathscr{U}\phi_2|\,\mathscr{U}\phi_1\rangle = \langle\phi_2|\,\phi_1\rangle\,, \tag{7.2}$$

for arbitrary $|\phi_1\rangle$ and $|\phi_2\rangle$, then we say that $\mathscr{U}$ is a symmetry operator. In fact, since Eq. (7.1) implies

$$\langle\mathscr{U}\phi_2| \equiv \left(|\mathscr{U}\phi_2\rangle\right)^\dagger \equiv \left(\mathscr{U}|\phi_2\rangle\right)^\dagger = \langle\phi_2|\,\mathscr{U}^\dagger\,, \tag{7.3}$$

it is easily seen that Eq. (7.2) can be written as

$$\langle\phi_2|\mathscr{U}^\dagger\mathscr{U}|\phi_1\rangle = \langle\phi_2|\,\phi_1\rangle\,. \tag{7.4}$$

And if this relation has to be obeyed for arbitrary states $|\phi_1\rangle$ and $|\phi_2\rangle$, the conclusion is that

$$\mathscr{U}^\dagger\mathscr{U} = 1\,, \tag{7.5}$$

which defines a *unitary operator*.

This definition of a symmetry operator would do for most symmetries, such as the ones discussed in Ch. 6. For time reversal, however, this definition is inadequate. The problem appears from a tacit assumption that goes into writing Eq. (7.2). In the original situation, $|\phi_1\rangle$ was the initial state and $|\phi_2\rangle$ was the final state. In the transformed situation, we are assuming that $|\mathscr{U}\phi_1\rangle$ still remains the initial state, and $|\mathscr{U}\phi_2\rangle$ the final state. Certainly this cannot be true when we are talking of time reversal! If the direction of time is reversed, the erstwhile initial state would become the final state and vice versa.

In order to accommodate such symmetries, we can broaden our definition of symmetry operations. The requirement for a symmetry operation should be this: the matrix element between arbitrary states should remain the same whether or not we perform a symmetry operation on the states. If the symmetry operation does not change initial states into final states and vice versa, the definition of Eq. (7.2) is quite adequate. However, if a symmetry operation $\widetilde{\mathscr{U}}$ interchanges initial and final states, the statement about the equality of the matrix element should have to be written as

$$\left\langle \widetilde{\mathscr{U}}\phi_1 \,\middle|\, \widetilde{\mathscr{U}}\phi_2 \right\rangle = \langle \phi_2 \,|\, \phi_1 \rangle . \tag{7.6}$$

An operator satisfying this equation is called an *anti-unitary operator*. Time reversal is obviously an operator of this kind. The same comment applies to any symmetry operation involving time reversal, such as its product with either charge conjugation or parity, or both. It should be noted that for unitary operators, the basic definition of Eq. (7.2) could be shorn of the states, and Eq. (7.5) could be written, because the ordering of the states was the same on both sides of the equation. It is impossible to write down an analog of Eq. (7.5) for anti-unitary operators since the ordering of states is different on the right and left sides of Eq. (7.6).

7.1.2 Rules for working with operators

We are so accustomed to working with unitary operators that we often forget which relations involving operators require the unitarity property explicitly, and cannot be used for operators in general. A perfect example of this sort of relations is the condition

$$\mathscr{O}\mathscr{H}\mathscr{O}^\dagger = \mathscr{H} , \tag{7.7}$$

which is often used as the definition of a symmetry operator $\mathscr{O}$ for a system with a Hamiltonian $\mathscr{H}$. It is easy to see that this condition is not quite general. Symmetry operators commute with the Hamiltonian of a system, which means a symmetry operation would always satisfy the relation

$$\mathscr{O}\mathscr{H} = \mathscr{H}\mathscr{O} . \tag{7.8}$$

It should be noted that $\mathscr{O}^{-1}$ must exist. In fact, symmetry operations always have inverses, and the inverse of a symmetry operator is also a symmetry operator by itself. Thus, multiplying both sides by $\mathscr{O}^{-1}$ from the right, we obtain the relation

$$\mathscr{O}\mathscr{H}\mathscr{O}^{-1} = \mathscr{H} . \tag{7.9}$$

Obviously the statements in Eqs. (7.7) and (7.9) are equivalent for a unitary operator by dint of Eq. (7.5). But in general, only Eq. (7.9) applies, and this is the form that must be used while working with anti-unitary symmetry operators.

Notations about states should be independent of the nature of operators acting on them, and they indeed are. We can certainly use the relation

$$\left(|\phi\rangle \right)^{\dagger} = \langle \phi | . \tag{7.10}$$

In fact, this is the definition of the bra vector. Relations like

$$\langle \phi_1 | \phi_2 \rangle = \langle \phi_2 | \phi_1 \rangle^{*} \tag{7.11}$$

follow from Eq. (7.10), and are applicable in general. Thus, the definition of anti-unitary operators in Eq. (7.6) could equivalently be written as

$$\left\langle \widetilde{\mathscr{U}}\phi_1 \,\middle|\, \widetilde{\mathscr{U}}\phi_2 \right\rangle = \langle \phi_1 | \phi_2 \rangle^{*} . \tag{7.12}$$

Let us now consider some relations between states where operators are also involved. The relations in Eq. (7.3) are universal. In fact, together with the definition with the bra vectors in Eq. (7.10), this equation constitutes the definition of the hermitian conjugate of an operator. Similarly, Eq. (7.1) is just a notation and can be used irrespective of the nature of the operator and the states.

When a unitary operator acts on a superposition of states, we use the relation

$$\mathscr{U}\left(\alpha_1 |\Psi_1\rangle + \alpha_2 |\Psi_2\rangle \right) = \alpha_1 \mathscr{U} |\Psi_1\rangle + \alpha_2 \mathscr{U} |\Psi_2\rangle , \tag{7.13}$$

where α_1 and α_2 are arbitrary complex numbers and Ψ_1 and Ψ_2 are arbitrary states. This relation defines a *linear operator*, and is not applicable to anti-unitary operators. There is a very important theorem by Wigner which states that if a transformation on the states in a Hilbert space leaves all probabilities unchanged, the transformation can be one of the following two types:

a) Unitary and linear

b) Anti-unitary and anti-linear

Anti-unitary operators like time reversal should therefore be anti-linear, which means that they should satisfy the following relation while acting on super-positions of states:

$$\widetilde{\mathscr{U}}\left(\alpha_1 |\Psi_1\rangle + \alpha_2 |\Psi_2\rangle \right) = \alpha_1^{*}\widetilde{\mathscr{U}} |\Psi_1\rangle + \alpha_2^{*}\widetilde{\mathscr{U}} |\Psi_2\rangle . \tag{7.14}$$

It is important to keep this distinction in mind while working with anti-unitary operators. Of course, the generalization of Eqs. (7.13) and (7.14) to superpositions of three or more states is obvious. It is also useful to remember that these equations also tell us about the result obtained when an operator acts on a single state multiplied by a numerical co-efficient. These results are obtained by taking one of the co-efficients to be zero in Eqs. (7.13) and (7.14). As we see, for unitary operators, it implies

$$\mathscr{U}\left(\alpha\left|\Psi\right\rangle\right) = \alpha\mathscr{U}\left|\Psi\right\rangle, \tag{7.15}$$

i.e., unitary operators leave the numerical co-efficients undisturbed. On the other hand, for an anti-linear operator, the result is:

$$\widetilde{\mathscr{U}}\left(\alpha\left|\Psi\right\rangle\right) = \alpha^{*}\widetilde{\mathscr{U}}\left|\Psi\right\rangle, \tag{7.16}$$

which means that numerical co-efficients are complex conjugated. Because time reversal is an anti-unitary operation, it must also be anti-linear, and we need to be careful about this aspect while dealing with time reversal.

7.2 Time reversal transformation on fields

In the notation introduced in Eq. (6.2, *p 151*), time reversal operation means the transformation

$$x^{\mu} \rightarrow -\widetilde{x}^{\mu}. \tag{7.17}$$

A Lagrangian will be invariant under time reversal, $\mathscr{T}$, if

$$\mathscr{T}\mathscr{L}(x)\mathscr{T}^{-1} = \mathscr{L}(-\widetilde{x}). \tag{7.18}$$

Let us see how it acts on fields.

7.2.1 Free fields

Acting on a scalar field, time reversal transformation should yield

$$\phi_{T}(x) \equiv \mathscr{T}\phi(x)\mathscr{T}^{-1} = \eta_{T}\phi(-\widetilde{x}), \tag{7.19}$$

where η_{T} is a phase factor, which can be called the intrinsic time reversal phase for the field. It can be easily seen, by more or less following the steps of §6.2.2, that the free scalar Lagrangian is invariant under time reversal for any choice of η_{T}.

☐ **Exercise 7.1** *Time reversal is a toggle operator. Show that this only says that $|\eta_{T}|^{2} = 1$, i.e., η_{T} can be an arbitrary phase factor.* [**Note :** *While applying time reversal the second time, remember that $\mathscr{T}$ is an anti-linear operator.*]

We next turn to fermions. In analogy with Eq. (6.23, *p 155*), we write

$$\mathscr{T}\psi(x)\mathscr{T}^{-1} = \mathbb{T}\psi(-\tilde{x}), \tag{7.20}$$

where $\mathbb{T}$ is a matrix. Our task is to find out if there is any such matrix that will keep the free Dirac Lagrangian invariant.

To this end, first note that

$$\mathscr{L}(-\tilde{x}) = \overline{\psi}(-\tilde{x})\Big[i\gamma_0\partial_{-t} + i\boldsymbol{\gamma}\cdot\boldsymbol{\nabla} - m\Big]\psi(-\tilde{x}). \tag{7.21}$$

Next, we need to evaluate the left hand side of Eq. (7.18). For this, we need to use Eq. (7.20). But we also need to be careful about the anti-unitary property of time reversal. The Dirac matrices that appear in the Lagrangian are all numerical co-efficients, and hence they have to be complex conjugated. For example,

$$\begin{aligned}
\mathscr{T}\overline{\psi}(x)\mathscr{T}^{-1} &= \mathscr{T}\Big(\psi^\dagger(x)\gamma_0\Big)\mathscr{T}^{-1} \\
&= \mathscr{T}\Big(\psi^\dagger(x)\Big)\mathscr{T}^{-1}\gamma_0^* = \Big(\mathbb{T}\psi(-\tilde{x})\Big)^\dagger\gamma_0^*.
\end{aligned} \tag{7.22}$$

Thus,

$$\mathscr{T}\mathscr{L}(x)\mathscr{T}^{-1} = \psi^\dagger(-\tilde{x})\mathbb{T}^\dagger\gamma_0^*\Big[-i\gamma_0^*\partial_t - i\boldsymbol{\gamma}^*\cdot\boldsymbol{\nabla} - m\Big]\mathbb{T}\psi(-\tilde{x}). \tag{7.23}$$

Comparing the time derivative terms in Eqs. (7.21) and (7.23), we find that

$$\mathbb{T}^\dagger\mathbb{T} = 1, \tag{7.24}$$

i.e., the matrix $\mathbb{T}$ must be unitary. In order to extract information from the other terms, we use the hermiticity property of Dirac matrices from Eq. (4.49, *p 70*) as well as their transposition properties from Eq. (6.57, *p 161*) to write

$$\gamma_0^* = \gamma_0^\mathsf{T} = -\mathbb{C}^{-1}\gamma_0\mathbb{C}, \tag{7.25a}$$
$$\gamma_i^* = -\gamma_i^\mathsf{T} = \mathbb{C}^{-1}\gamma_i\mathbb{C}. \tag{7.25b}$$

Using these, we obtain the relations

$$\Big[\mathbb{C}\mathbb{T}, \gamma_\mu\Big]_+ = 0. \tag{7.26}$$

There is only one matrix, apart from an overall numerical factor, which anti-commutes with all four Dirac matrices: it is γ_5. Thus, we can write

$$\mathbb{T} = \eta_T\mathbb{C}^{-1}\gamma_5, \tag{7.27}$$

where η_T is the numerical factor. The only constraint on η_T comes from the toggle nature of time reversal operation, which says that $|\eta_T| = 1$, i.e., η_T has to be a phase factor.

On a spin-1 field such as the photon field, time reversal operation works as follows:

$$\mathscr{T} A_0(x) \mathscr{T}^{-1} = -\eta_T A_0(-\tilde{x}),$$
$$\mathscr{T} \boldsymbol{A}(x) \mathscr{T}^{-1} = +\eta_T \boldsymbol{A}(-\tilde{x}). \tag{7.28}$$

It is easy to see that the Lagrangian for the free photon field is invariant under this transformation.

7.2.2 Interactions

Let us start with fermion bilinears, which appear in all interaction terms involving fermions. Using Eq. (7.22), we can write

$$\mathscr{T} \left(\overline{\psi}_1(x) F \psi_2(x) \right) \mathscr{T}^{-1} = \psi_1^\dagger(-\tilde{x}) \mathbb{T}^\dagger \gamma_0^* F^* \mathbb{T} \psi_2(-\tilde{x}), \tag{7.29}$$

recalling that F, the matrix of numerical co-efficients, has to be complex conjugated because of anti-linearity. Using Eq. (7.27) now and allowing for different intrinsic phases for the two fields, we obtain

$$\mathscr{T} \left(\overline{\psi}_1(x) F \psi_2(x) \right) \mathscr{T}^{-1} = \eta_T^{(1)*} \eta_T^{(2)} \, \overline{\psi}_1(-\tilde{x}) F_T \psi_2(-\tilde{x}), \tag{7.30}$$

where

$$F_T = \gamma_5 \mathbb{C} F^* \mathbb{C}^{-1} \gamma_5. \tag{7.31}$$

More explicitly, we arrange the transformation rules for various independent possibilities for F:

$$\begin{array}{c|c|c|c|c|c|c|c|c}
F & \mathbb{1} & \gamma_0 & \gamma_i & \sigma_{0i} & \sigma_{ij} & \gamma_0\gamma_5 & \gamma_i\gamma_5 & \gamma_5 \\
\hline
F_T & \mathbb{1} & \gamma_0 & -\gamma_i & \sigma_{0i} & -\sigma_{ij} & \gamma_0\gamma_5 & -\gamma_i\gamma_5 & \gamma_5
\end{array}. \tag{7.32}$$

This list immediately tells us that the QED interaction term is invariant under time reversal provided we take

$$\eta_T^{(A)} = -1. \tag{7.33}$$

Another interesting observation is that the polar and axial vector bilinears transform the same way under time reversal. Thus, interactions shown in Eq. (6.168, *p 187*) would be invariant under time reversal. It implies that the intrinsic time reversal phase of the vector boson should be +1. Of course, one also needs the coupling constants to be real so that they are not affected by anti-unitarity of the time reversal operation.

□ **Exercise 7.2** *The interactions given in Ex. 6.23 (p 182) produced triple momentum correlations. As discussed in §6.9, such correlations indicate not only parity violation but also time reversal violation. Show that Eq. (6.148, p 182) violates time reversal symmetry.*

7.3   CPT transformation on fields

CPT transformation is, as the name implies, a combination of the transformations C, P and T. Since P changes the spatial co-ordinates by a sign and T does the same thing to time,

$$x^\mu \to -x^\mu \tag{7.34}$$

under CPT. Because the combination includes time reversal, it is an anti-unitary operation.

Transformation of various fields under this transformation can be deduced easily by combining the transformation rules under CP given in §6.11 and those under time reversal given in §7.2. Writing

$$\Theta \equiv \mathscr{C}\mathscr{P}\mathscr{T} \tag{7.35}$$

for the sake of brevity, we obtain, using Eqs. (6.159) and (7.19),

$$\Theta\phi(x)\Theta^{-1} = \mathscr{T}\left((\mathscr{C}\mathscr{P})\phi(x)(\mathscr{C}\mathscr{P})^{-1}\right)\mathscr{T}^{-1}$$
$$= \mathscr{T}\left(\eta_{CP}\phi^\dagger(\widetilde{x})\right)\mathscr{T}^{-1} = \eta_\Theta\phi^\dagger(-x), \tag{7.36}$$

where η_Θ is a combination of intrinsic phases under each of the three discrete operations. Similarly, from Eqs. (6.162), (7.20) and (7.27), we obtain

$$\Theta\psi(x)\Theta^{-1} = \eta_\Theta\gamma_5^{\mathsf{T}}\psi^*(-x). \tag{7.37}$$

☐ **Exercise 7.3** Prove Eq. (7.37). [**Note :** *Remember the anti-unitary nature of CPT.*]

☐ **Exercise 7.4** *Show that a fermion bilinear transforms in the following way under CPT:*

$$\Theta\overline{\psi}_1(x)F\psi_2(x)\Theta^{-1} = \eta_\Theta^{(1)*}\eta_\Theta^{(2)}\,\overline{\psi}_2(-x)F_\Theta\psi_1(-x), \tag{7.38}$$

where

$$F_\Theta = \gamma_5 F^\ddagger\gamma_5, \tag{7.39}$$

with $F^\ddagger = \gamma_0 F^\dagger\gamma_0$, a notation introduced in Eq. (5.64, p 127).

Finally, for any vector boson V^μ that behaves like the photon under P, C and T, we have

$$\Theta V_\mu(x)\Theta^{-1} = -V_\mu^\dagger(-x), \tag{7.40}$$

which can be obtained by combining the behavior of the photon under each of the individual symmetries, shown in Eqs. (6.36), (6.69) and (7.33).

7.4 CPT theorem

These various CPT transformation rules suggest something very interesting. Suppose we take the intrinsic CPT phases of all scalar and fermion fields to be equal to 1. A scalar field then just transforms to its hermitian conjugate. A vector field, as seen in Eq. (7.40), also does the same, but with a negative sign.

The corresponding statement is not so simple if we consider only a fermion field, but fermion fields must appear in bilinears. We can ask: what happens to fermion bilinears? The hermitian conjugate of a fermion bilinear is given by

$$\left(\overline{\psi}_1 F \psi_2\right)^\dagger = \overline{\psi}_2 F^\ddagger \psi_1 \,. \tag{7.41}$$

Compare this with the CPT transform of the bilinear, given in Eqs. (7.38) and (7.39). The CPT transform contains $\gamma_5 F^\ddagger \gamma_5$ in the place where the hermitian conjugate contains just $F^\ddagger$. What is the difference between these two combinations? If F, and therefore $F^\ddagger$, contains odd number of γ^μ's, we get $\gamma_5 F^\ddagger \gamma_5 = -F^\ddagger$. On the other hand, if the number of γ^μ's is even in F, we get $\gamma_5 F^\ddagger \gamma_5 = +F^\ddagger$.

Odd numbers of Dirac matrices mean odd numbers of Lorentz indices in the bilinear. Noting this fact, we can now summarize the results obtained for fermion bilinears with those of scalars and vector bosons, and say that whenever we take the CPT conjugate of a quantity which contains an odd number of Lorentz indices, the result contains a negative sign on the hermitian conjugate. For CPT conjugates of quantities which contain an even number of Lorentz indices, there is no negative sign. More succinctly, we can say that the sign is $(-1)^n$ if the number of Lorentz indices is n.

Any term in a Lagrangian must be Lorentz invariant. Hence it cannot have any unmatched Lorentz index. For any factor in a term carrying a Lorentz index, there must be another factor which should have a matching index to contract it. The total number of Lorentz indices in any term must therefore be even. Hence, the CPT transform of any term in a Lagrangian must be just the hermitian conjugate of the term itself. But the Lagrangian must be hermitian, so the hermitian conjugate of any term must also be part of the Lagrangian. Taking any particular term and its hermitian conjugate, we see that under CPT, they transform into each other. Thus, the Lagrangian must be CPT invariant. This is called the *CPT theorem*.

The way we stated it, it might seem that Lorentz invariance implies CPT invariance. That is not quite true. We have tacitly made other assumptions. The most important one is that we have assumed that fermion fields, which have half-integral spin, have anticommuting creation and annihilation operators, whereas for bosonic fields such operators are commuting.

7.5   Consequences of CPT symmetry

From a purely empirical point of view, we should not accept CPT theorem blindly. The assumptions that go into it can certainly be questioned, like any other assumption in any physical theory. However, the assumptions entering into the proof of the CPT theorem are so minimal that, like Lorentz invariance, we will take them to be granted for the discussions in this book and will therefore assume that CPT is a good symmetry for all interactions.

To see the general form of the consequence of this symmetry, consider an operator $\mathcal{O}$ that is CPT invariant, i.e.,

$$\mathcal{O}\Theta = \Theta\mathcal{O}. \tag{7.42}$$

The question is, what is the relation of the matrix element of the operator $\mathcal{O}$ between the states $|a\rangle$ and $|b\rangle$, and between the states $|\Theta a\rangle$ and $|\Theta b\rangle$? As before, we are using the shorthand Θ for CPT and $|\Theta a\rangle$ means $\Theta|a\rangle$. Now,

$$|\mathcal{O}\Theta b\rangle \equiv \mathcal{O}\,|\Theta b\rangle = \mathcal{O}\Theta\,|b\rangle = \Theta\mathcal{O}\,|b\rangle = |\Theta\mathcal{O}b\rangle\,, \tag{7.43}$$

using Eq. (7.42) on the way. The anti-unitary property of the operation Θ will imply

$$\langle\Theta a\,|\,\Theta\mathcal{O}b\rangle = \langle a\,|\,\mathcal{O}b\rangle^{*}\,, \tag{7.44}$$

which means

$$\langle\Theta a\,|\mathcal{O}|\,\Theta b\rangle = \langle a\,|\mathcal{O}|\,b\rangle^{*} = \langle b\,|\mathcal{O}^{\dagger}|\,a\rangle\,. \tag{7.45}$$

This general result can have striking consequences for various choices of $\mathcal{O}$ and the states, as we will see now.

As a first example, consider the mass of a particle and its antiparticle. Let the particle be called a, and its antiparticle as $\hat{a}$. By definition, the mass of a is its energy when it is free and at rest. We can write this in the form of an equation:

$$m_a \equiv \langle a(\mathbf{0})\,|\mathcal{H}|\,a(\mathbf{0})\rangle\,, \tag{7.46}$$

where $\mathcal{H}$ is the Hamiltonian, including all interactions of all kinds. In what follows, we will omit the momentum label on the states.

Now consider an antiparticle state of the same momentum and denote it by $|\hat{a}\rangle$. This should be the CPT conjugate of $|a\rangle$, except a possible phase factor:

$$|\Theta a\rangle = \Theta\,|a\rangle = e^{i\theta}\,|\hat{a}\rangle\,. \tag{7.47}$$

Eq. (7.45) then gives, with the Hamiltonian as the operator $\mathcal{O}$, the result

$$\langle\hat{a}\,|\mathcal{H}|\,\hat{a}\rangle = \langle a\,|\mathcal{H}^{\dagger}|\,a\rangle\,. \tag{7.48}$$

The left hand side of this expression, in analogy with Eq. (7.46), is the mass of the antiparticle. The right hand side, because of hermiticity of the Hamiltonian, is the mass of the particle. So we have proved that

$$m_a = m_{\hat{a}} \,. \tag{7.49}$$

The proof in fact entails some more general results. For example, by considering the matrix element of the Hamiltonian between states of non-zero 3-momentum and following the same states, it can be easily proved that the energy of a free particle with a given momentum is equal to the energy of its antiparticle with the same momentum. More generally, even if the state $|a\rangle$ is not an eigenstate of the Hamiltonian, the same analysis shows that the expectation values of the Hamiltonian in the states $|a\rangle$ and $\Theta|a\rangle$ are equal.

☐ **Exercise 7.5** Argue along the same lines that the decay rates of a particle and its antiparticle must be equal.

A similar proof can be constructed to show that the electromagnetic properties like the electric charge and the magnetic moment of a particle and its antiparticle must be opposite. This is because all electromagnetic properties are derived from the interaction with the photon, which can be written as

$$\mathscr{L}_{\text{int}} = -j^\lambda A_\lambda \,, \tag{7.50}$$

by combining the results of Eq. (5.12, *p 115*) and Eq. (5.17, *p 116*). The CPT transformed form of this Lagrangian term is

$$\Theta \mathscr{L}_{\text{int}} \Theta^{-1} = -\Theta j^\lambda \Theta^{-1} \,\Theta A_\lambda \Theta^{-1} \,. \tag{7.51}$$

The Lagrangian should be invariant under CPT, meaning that

$$\Theta \mathscr{L}_{\text{int}}(x) \Theta^{-1} = \mathscr{L}_{\text{int}}(-x) \,. \tag{7.52}$$

We know that the photon field transforms as

$$\Theta A_\lambda(x) \Theta^{-1} = -A_\lambda(-x) \,, \tag{7.53}$$

as shown earlier, in a more general notation, in Eq. (7.40). Combining these pieces of information, we obtain

$$\Theta j^\lambda(x) \Theta^{-1} = -j^\lambda(-x) \,. \tag{7.54}$$

Electromagnetic properties of particles occur as matrix elements of the current operator j^λ between suitably defined states. For the properties of a particle a, we should consider the matrix element $\langle a|j^\lambda|a\rangle$, whereas for an antiparticle we should consider $\langle \hat{a}|j^\lambda|\hat{a}\rangle$. We can now follow the same steps as we did for the Hamiltonian, the only difference coming from the minus sign in Eq. (7.54), and the final answer would be

$$\langle \hat{a}|j^\lambda|\hat{a}\rangle = -\langle a|j^\lambda|a\rangle \,, \tag{7.55}$$

which would imply that the electromagnetic properties like the charge and the magnetic moment should be opposite for a particle and its antiparticle.

It should be noted that the crucial element in the deduction is the minus sign coming in the transformation property of the photon field in Eq. (7.53). The photon field is negative under the charge conjugation symmetry itself, and so the same conclusion about particles and antiparticles can be reached with charge conjugation symmetry alone. The difference is that the charge conjugation symmetry is violated by weak interactions, and so it will not be clear from such a deduction that the results would not be disturbed when weak corrections are taken into account. CPT, however, is a symmetry respected by weak interactions as well, so the deduction through CPT is guaranteed to be maintained even after weak corrections are considered.

□ **Exercise 7.6** *The differential decay rate for the $\mu^\pm$ can be written in the form*

$$R^\pm = R_0^\pm \left[1 + A^\pm \mathbf{S}_e \cdot \mathbf{p} + B^\pm \mathbf{S}_\mu \cdot \mathbf{p} + C^\pm \mathbf{S}_e \cdot \mathbf{S}_\mu + D^\pm \mathbf{p} \cdot \mathbf{S}_e \times \mathbf{S}_\mu \right],$$

(7.56)

where boldface letters denote 3-vectors, with $\mathbf{p}$ the momentum and $\mathbf{S}_e$ the spin of the electron or positron emitted in the decay, and $\mathbf{S}_\mu$ the spin polarization of decaying particle. Find conditions on or relations between the quantities $R_0^\pm$, $A^\pm$, $B^\pm$, $C^\pm$ and $D^\pm$ assuming

  a) *♩CP invariance only*
  b) *T invariance only*
  c) *CPT invariance only*

Verify that, if CPT invariance holds, the consequences of CP and T invariances are identical.

7.6   Time reversal transformation on states

As proposed before, we will take CPT invariance for granted in this book. But time reversal symmetry might be violated. In order to obtain information about that from particle interactions, we have to know how time reversal symmetry transforms different particle states. We give the details for fermion states here, leaving scalar and vector states as exercises.

Combining Eqs. (7.20) and (7.27), we obtain

$$\mathscr{T}\psi(x)\mathscr{T}^{-1} = \eta_T \mathbb{C}^{-1} \gamma_5 \psi(-\tilde{x}) \tag{7.57}$$

for a fermion field. Using the plane wave expansion of fermion fields from Eq. (4.65, *p 72*) and remembering the anti-linear nature of time reversal, the left hand side of Eq. (7.57) can be written as

$$\mathscr{T}\psi(x)\mathscr{T}^{-1} = \sum_s \int D^3p \left(\mathscr{T}d_s(\mathbf{p})\mathscr{T}^{-1} u_s^*(\mathbf{p})e^{+ip\cdot x} \right.$$

$$\left. + \mathscr{T}\widehat{d_s^\dagger}(\mathbf{p})\mathscr{T}^{-1} v_s^*(\mathbf{p})e^{-ip\cdot x} \right). \tag{7.58}$$

As for the right side, we first note the spinor identities which have been proved in Appendix F, using certain conventions about the phases of different spinors which have been elaborated in the said appendix:

$$\mathbb{C}^{-1}\gamma_5 u_s(\boldsymbol{p}) = -s u^*_{-s}(-\boldsymbol{p}), \tag{7.59a}$$

$$\mathbb{C}^{-1}\gamma_5 v_s(\boldsymbol{p}) = -s v^*_{-s}(-\boldsymbol{p}), \tag{7.59b}$$

where $s = \pm$ in our notation. Thus,

$$\mathbb{C}^{-1}\gamma_5\psi(-\widetilde{x}) = \sum_s \int D^3p \left(- s d_s(\boldsymbol{p}) u^*_{-s}(-\boldsymbol{p}) e^{iE_p t + i\boldsymbol{p}\cdot\boldsymbol{x}} \right.$$

$$\left. - s \widehat{d}^\dagger_s(\boldsymbol{p}) v^*_{-s}(-\boldsymbol{p}) e^{-iE_p t - i\boldsymbol{p}\cdot\boldsymbol{x}} \right)$$

$$= \sum_s \int D^3p \left(s d_{-s}(-\boldsymbol{p}) u^*_s(\boldsymbol{p}) e^{+i\boldsymbol{p}\cdot x} \right.$$

$$\left. + s \widehat{d}^\dagger_{-s}(-\boldsymbol{p}) v^*_s(\boldsymbol{p}) e^{-i\boldsymbol{p}\cdot x} \right). \tag{7.60}$$

In writing the last step, we have changed the dummy index s and the dummy momentum $\boldsymbol{p}$ to their negatives. Comparing this form with Eq. (7.58) and using Eq. (7.57), we obtain

$$\mathscr{T} d_s(\boldsymbol{p}) \mathscr{T}^{-1} = s\eta_T d_{-s}(-\boldsymbol{p}), \tag{7.61a}$$

$$\mathscr{T} \widehat{d}^\dagger_s(\boldsymbol{p}) \mathscr{T}^{-1} = s\eta_T \widehat{d}^\dagger_{-s}(-\boldsymbol{p}). \tag{7.61b}$$

Clearly then, by the action of time reversal operation on a one-particle state, the 3-momentum $\boldsymbol{p}$ and spin s of the state change sign. This is exactly what we expect from classical mechanics as well, and was presented in Table 6.2 *(p 178)*.

□ **Exercise 7.7** *Starting from Eq. (7.19) and using the plane wave expansion for a scalar field, show that the creation and annihilation operators satisfy the relations*

$$\mathscr{T} a(\boldsymbol{p}) \mathscr{T}^{-1} = \eta_T a(-\boldsymbol{p}), \tag{7.62a}$$

$$\mathscr{T} a^\dagger(\boldsymbol{p}) \mathscr{T}^{-1} = \eta_T a^\dagger(-\boldsymbol{p}). \tag{7.62b}$$

7.7   Signature of time reversal violation

In §6.9, we discussed some correlations which indicate parity violation, It can be noted that some of these, like correlations of three different momenta, is not invariant under time reversal symmetry as well. However, such correlations indicate only the violation of what can be called the naïve time reversal symmetry. A true test of time reversal would constitute the reversal of initial and final states as well, and therefore cannot be performed using a decay process.

It is possible to find tests relating a scattering process to its reverse. However, the tests are not very clean because final-state interactions can also mimic time reversal violation.

Earlier, we showed that the most general vertex function between a fermion and the photon contains four form factors. Two of them respect parity invariance. It can be seen that these two respect time reversal invariance as well. Among the other two which violate parity invariance, it can be shown easily that the anapole moment term in fact respects time reversal invariance. However, the electric dipole moment violates time reversal invariance as well. Measurement of the electric dipole moment (EDM) of a fermion would constitute a clean test for violation of time reversal invariance. So far, there are only upper limits on such measurements. For the electron, the limit is

$$|d_{(e)}| < 10.5 \times 10^{-28} \, e \, \text{cm} \,. \tag{7.63}$$

There are also limits on the EDM of the muon and the tau, but they are much weaker. In the hadronic sector, the best limit is on the neutron EDM. It will be discussed in Ch. 21.

☐ **Exercise 7.8** Consider the electric dipole moment interaction. If the form factor is a constant, the effective interaction can be written as

$$\overline{\psi}\sigma_{\mu\nu}\gamma_5\psi F^{\mu\nu} \tag{7.64}$$

times a constant. Show that this interaction violates time reversal invariance.

☐ **Exercise 7.9** Use the non-relativistic reduction of the anapole moment term given in Eq. (5.147, p 145) to verify that this interaction violates parity, but not time reversal.

Once CPT invariance is taken for granted, there can be other types of tests. If CPT is conserved, T violation is equivalent to CP violation. Thus, we can look for tests of CP violation, which will be discussed at length in Ch. 21. A class of CP violation tests have also inspired some novel ways of testing T violation directly. Such tests will also be discussed in Ch. 21.

Chapter 8

Isospin

As mentioned earlier, quantum field theories are based on Poincaré symmetry of the Lagrangian. This includes translational symmetries and the proper Lorentz group. In earlier chapters, we found that electromagnetic interactions are invariant under some discrete symmetries as well. All of these fall under the general class of spacetime symmetries.

In Ch. 5, we also found that QED is governed by a gauge symmetry, or a local symmetry, which is internal. It turns out that strong interactions obey a large number of internal symmetries. We begin the discussion of these symmetries in this chapter with the isospin symmetry.

8.1 Nuclear energy levels

Heisenberg noticed, in the early 1930s, that nuclei with equal mass numbers show remarkable similarity regarding their energy levels. Check, for example, some of the energy levels of ^{7}Li and ^{7}Be, shown in Fig. 8.1. The first one has three protons and four neutrons, whereas the second one has four protons and three neutrons. Such pairs of nuclei are called *mirror nuclei*, i.e., the number of protons in one nucleus is the same as the number of neutrons in the other.

The similarity is striking, to say the least. There is a difference between the ground state energies of the two nuclei, which has been indicated by drawing the two ground states at different heights in the diagram. But, apart from that, the states appear at more or less equal spacings for the two nuclei. Even the spin and parity of the states are perfectly matched between the two nuclei.

More examples of this sort can be given, but the main point remains the same. There is overwhelming evidence to suggest that the nuclear energy levels depend only on the mass number, and not on the number of protons and neutrons separately. Since the dominant force between protons and neutrons comes from strong interactions, it implies that the strong interactions cannot distinguish between a proton and a neutron. In other words, if we replace a neutron by a proton in a nucleus, or vice versa, it makes no difference insofar as the strong interaction is concerned. This idea took ground in the mid-

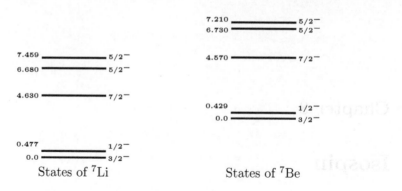

Figure 8.1: Low-lying states of the nuclei ^{7}Li and ^{7}Be. On the left side of the levels, we have given the energy of the level in MeV, measured from the ground state of the corresponding nucleus. On the right side, we give the spin and parity of the state.

1930s, and has proved extremely successful. This is the basis of the isospin symmetry, which we will discuss in §8.2.

We can ask, if the interaction makes no difference between the proton and the neutron, why aren't the energy levels of nuclei of equal mass number exactly the same? The answer to this question is that the strong interaction is not the only interaction in a collection of protons and neutrons. Between two protons, there is also Coulomb repulsion. Since there are three protons in ^{7}Li as opposed to four in ^{7}Be, the Coulomb repulsion is higher in the latter. As a result, energy levels of ^{7}Be are higher than the corresponding energy levels of ^{7}Li, as seen in Fig. 8.1.

We can extend our discussion to mass number equal to 1. Here, we have just the two particles, the neutron and the proton. We can compare their masses. Experimental data show that

$$\frac{m_n - m_p}{\frac{1}{2}(m_n + m_p)} = 1.38 \times 10^{-3}. \tag{8.1}$$

The two particles are remarkably degenerate. Of course, it is also true that they are not exactly degenerate, and this raises the question: what causes the difference between their masses? This question will be examined in §8.9.

8.2 Isospin symmetry

8.2.1 Group structure of isospin transformations

So far, we have talked about replacing a neutron by a proton, or vice versa, and commented that the change is imperceptible to strong interactions. But if the change is really imperceptible, it does not really matter whether we replace an entire neutron by an entire proton. For example, if we replace a neutron by a state which is a superposition of neutron and proton, this replacement should also go unnoticed by strong interactions. If we write the proton and the neutron in the form of a doublet,

$$N = \begin{pmatrix} p \\ n \end{pmatrix},$$ (8.2)

then the most general such replacement can be represented by a transformation of the form

$$N \longrightarrow N' = UN,$$ (8.3)

where U is a 2×2 matrix. Since we do not want the normalization of the states to be affected, the matrix must be unitary. In Ch. 3, we discussed that such matrices form a group which can be called U(2). We also showed there that

$$\mathrm{U}(2) = \mathrm{SU}(2) \times \mathrm{U}(1),$$ (8.4)

which means that the group U(2) of the general transformations has an SU(2) part and a U(1) part, which are mutually commuting. The SU(2) part is called isospin, in the sense described below. Strong interaction has this symmetry and therefore conserves isospin. The U(1) part will be discussed later in the chapter, in §8.10.

It may be worthwhile to make a few comments about the notation introduced in Eq. (8.2), since such notation will be used often in the rest of the book. The object N has been shown as a doublet. It contains two fields, viz., those of the proton and the neutron. Each of these is a fermion field, which can be expressed in the form of a 4-component column. Thus, if we want to write down N in its full glory, it should have eight components, the first four corresponding to the proton field and the last four to the neutron field. If we make a Lorentz transformation, only the first four components mix among themselves and the last four among themselves. Here, we are talking of a different kind of symmetry whose transformations mix the upper block and the lower block, without any intermixing within the blocks. This is the reason we have suppressed the component structure of p and n while writing Eq. (8.2).

When both kinds of transformations are involved, we will continue to use the economic notation of Eq. (8.2), assuming that the reader will understand from the context and from the notation what kind of transformations we are talking about. For example, we might write a bilinear of the nucleon field N in the form

$$\overline{N}\gamma_\mu \tau^a N$$ (8.5)

where the τ^a denotes the Pauli matrices. It would mean that the matrix τ^a should act on the two blocks of four as a whole, whereas each of the two blocks has to be multiplied by γ_μ. For example, with $a = 3$, the bilinear shown in Eq. (8.5) implies the combination $(\overline{p}\gamma_\mu p - \overline{n}\gamma_\mu n)$.

8.2.2 Isospin representations

We find ourselves in a fortuitous situation because the isospin symmetry group is SU(2). We learn about this group in any course of quantum mechanics in connection with rotational symmetry. The group of proper rotations in

3-dimensional space is called SO(3), where the "O" in the middle stands for "orthogonal", reminding us that any rotation can be represented by an orthogonal 3×3 matrix, which tells us how the new co-ordinates of any point, with respect to the rotated axes, are related to the old ones. But SO(3) is a group that is not simply connected, so it is much better to use SU(2), which is identical to SO(3) so long as infinitesimal transformations are concerned, but has a different global structure. And that is what we do while discussing representations of the rotation group.

The necessity and advantage of using SU(2) is most easily seen with spin-$\frac{1}{2}$ particles. Suppose we rotate the x and y axes of our co-ordinate system by an angle θ, keeping the z-axis unchanged. The wavefunction of a spin-$\frac{1}{2}$ particle in the new system of axes, ψ', should be related to the wavefunction ψ with respect to the original axes by the equation

$$\psi' = \exp(i\theta\sigma_z/2)\psi = \left(\cos\frac{\theta}{2} + i\sigma_z \sin\frac{\theta}{2}\right)\psi. \tag{8.6}$$

This shows that even when $\theta = 2\pi$, i.e., even when one has performed a complete rotation, ψ' does not become equal to ψ. This is exactly a manifestation of the fact that SO(3) is not simply connected. We are not getting into the explanation further because in the case of isospin, we need not think about SO(3) at all. The physics of the situation automatically leads us to think about an SU(2) symmetry.

We can apply all our knowledge of rotation group to obtain information about the isospin group and its representations. In fact, the mathematics is exactly the same, although it is now being applied in a different physical context. The group has three generators. We can take the generators to be hermitian, in which case they satisfy the following commutation relations:

$$\left[I_i, I_j\right] = i\varepsilon_{ijk}I_k, \tag{8.7}$$

where ε_{ijk} is totally antisymmetric in the indices, as defined in Eq. (3.22, p 44).

The irreducible representations of an SU(2) group are given by the eigenvalue of the operator

$$I^2 = I_1^2 + I_2^2 + I_3^2. \tag{8.8}$$

As we know from similar exercises with angular momentum multiplets, the eigenvalues of I^2 are in the form of $I(I+1)$, where the latter I is either an integer or a half-integer. We denote the irreducible representation by this number. Thus, just like spin-0, spin-$\frac{1}{2}$ or spin-1 particles, we can talk of representations with isospin equal to 0, $\frac{1}{2}$, 1, etc. With isospin I, there are $2I+1$ states in the representation, like the angular momentum projection states. An isospin-0 representation is thus just one state, and is often called an *isosinglet*. Similarly, an *isodoublet* is a representation with $I = \frac{1}{2}$, which has two states. $I = 1$ gives an isotriplet. Alternatively, representations with $I = 0, \frac{1}{2}, 1$ are called *isoscalar, isospinor* and *isovector* respectively, because their transformation under isospin is similar to the transformations of scalars, spinors and vectors under spatial rotation.

Since each of the generators I_1, I_2 and I_3 commutes with I^2, we can diagonalize any one of them, along with I^2 itself, to denote the states. Conventionally, one takes I_3, without any loss of generality. Within an irreducible representation, the eigenvalues of I_3 range from $-I$ to $+I$ with intervals of unity, and the eigenstates are non-degenerate. All isospin states can thus be written in terms of the basis states $|I, I_3\rangle$, which are eigenstates of both I^2 and I_3. The statement made in Eq. (8.2) means that the proton and the neutron are eigenstates of I^2 and I_3, with eigenvalues given by the following table:

$$
\begin{array}{c|cc}
 & p & n \\
\hline
I & \frac{1}{2} & \frac{1}{2} \\
I_3 & \frac{1}{2} & -\frac{1}{2}
\end{array}
\tag{8.9}
$$

The proton and the neutron can thus be thought of the $I_3 = +\frac{1}{2}$ and $I_3 = -\frac{1}{2}$ manifestations of a single particle, which is called the *nucleon*.

Usually, in quantum mechanics textbooks, the angular momentum operator is denoted by $\boldsymbol{J}$. The eigenvalues of the operator J^2 are written as $j(j + 1)$, using the lower case letter. The eigenvalues of the operator J_z are denoted by a different letter, m. For isospin, we are using the same letter I for writing the operator I^2 and the eigenvalues $I(I + 1)$. Also, we are using I_3 to denote both the operator as well as its eigenvalues. This usage is customary, and we hope that the meaning intended in any given formula will be obvious from the context.

The states $|I, I_3\rangle$ cannot be eigenstates of I_1 or I_2, because according to Eq. (8.7) both of these generators have non-zero commutation relations with I_3. The ladder operators

$$
I_\pm = I_1 \pm i I_2 \,,
\tag{8.10}
$$

acting on the $|I, I_3\rangle$ states, can change the I_3 value by one unit:

$$
I_\pm |I, I_3\rangle = \sqrt{I(I + 1) - I_3(I_3 \pm 1)} \; |I, I_3 \pm 1\rangle \,.
\tag{8.11}
$$

Of course the right hand side can have an arbitrary phase factor. We have chosen it to be unity.

□ **Exercise 8.1** Show that

$$
\left| I_\pm |I, I_3\rangle \right|^2 = I(I + 1) - I_3(I_3 \pm 1) \,,
\tag{8.12}
$$

assuming that the ket appearing on the left hand side is normalized to unity.

8.3 Pions

Pions were conjectured theoretically by Yukawa in order to explain the interaction between nucleons. He took the analogy of electromagnetic interactions, where a photon is exchanged between charged fermions such as electrons.

Figure 8.2: Basic vertices in the Yukawa theory.

Similarly, he thought, strong interaction is mediated by the exchange of some bosons between the nucleons. To keep things simple, he assumed that these bosons are spinless. These are the pions. According to the classification of strongly interacting particles, or *hadrons*, that was introduced in §1.3, pions therefore fall into the class called *mesons*, which are nothing but hadrons with integer spin. The proton and the neutron are, on the other hand, *baryons*, i.e., hadrons of half-integer spin.

Yukawa's theory did not turn out to be a description of strong interactions at a fundamental level. However, pions turned out to be real: they were discovered a few years after Yukawa gave his theory. For this reason, and also because Yukawa's theory laid the prototype of fermion interactions with possible fundamental scalars, we will describe the theory qualitatively here, and more seriously in §8.8.

The main interaction vertices of Yukawa theory have been shown in Fig. 8.2. It is easy to see that such vertices can give rise to nucleon-nucleon interaction where the pion propagates as an internal line, much like the electron–electron interaction diagrams given in Ch. 5 which use the photon as the mediator. These vertices also show that there should be three kinds of pions: one with a positive electric charge, another with negative, and a third one which is electrically neutral. The pions thus form an isotriplet, i.e., an isospin multiplet with $I = 1$. The I_3 values for the three pions are given below:

$$
\begin{array}{c|ccc}
 & \pi^+ & \pi^0 & \pi^- \\
\hline
I & 1 & 1 & 1 \\
I_3 & 1 & 0 & -1
\end{array}
\qquad (8.13)
$$

One might ask, what determines the I_3 values for the pions? Or one might go back one step further and ask why we took the proton, rather than the neutron, to have $I_3 = +\frac{1}{2}$? Could we not have done otherwise?

Indeed, we could. As far as the nucleons are concerned, assignment of I_3 is merely a convention. But once that is fixed, the assignment for the pions is fixed as well. This can be seen as follows. Consider any process where there are only nucleons and pions in the initial and the final states. Total electric charge must be conserved in the process, i.e.,

$$
\Delta Q \equiv \Delta Q_N + \Delta Q_\pi = 0 , \qquad (8.14)
$$

where ΔQ_N is defined to be the difference of charges of the nucleons in the initial and the final states, and ΔQ_π the same difference for pions. Isospin symmetry of strong interactions implies a similar relation for I_3:

$$\Delta I_3 \equiv \Delta I_{3N} + \Delta I_{3\pi} = 0. \tag{8.15}$$

With proton being assigned $I_3 = +\frac{1}{2}$, in the nucleon sector we find

$$\Delta Q_N = \Delta I_{3N}. \tag{8.16}$$

Therefore, the same relation must hold for pions:

$$\Delta Q_\pi = \Delta I_{3\pi}. \tag{8.17}$$

This tells us that the I_3-value of π^+ should exceed the I_3-value of π^0 by one unit, and the latter in turn should be 1 higher than the I_3-value of π^-. Hence the I_3 assignments for pions shown in Eq. (8.13) are forced upon us.

If isospin were conserved in all kinds of interactions, components of any isospin multiplet would have been degenerate. To be specific, the three pions would have had the same mass. In reality, the π^+ and π^- have the same mass, but this fact does not owe its origin to isospin invariance. The two charged pions are antiparticles of each other, and CPT invariance ensures the equality of their masses, as shown in §7.5. The neutral pion has a slightly lower mass.

$$\frac{m_{\pi^\pm} - m_{\pi^0}}{\frac{1}{2}(m_{\pi^\pm} + m_{\pi^0})} = 3.35 \times 10^{-2}. \tag{8.18}$$

The smallness of this splitting, as well as the splitting for the nucleon given in Eq. (8.1), is reassuring. It is consistent with the idea that isospin is indeed conserved in strong interactions. The small splittings among the members of an isomultiplet will be discussed in §8.9, as promised earlier.

8.4 Isospin relations

8.4.1 Forbidden processes

Most simply, isospin symmetry of strong interactions tells us that some processes are allowed in strong interactions while some others are forbidden. Take for example the reaction

$$d + d \longrightarrow {}^4\text{He} + \pi^0, \tag{8.19}$$

which is allowed by charge conservation, where d denotes the deuteron nucleus, ^{2}H. But the deuteron is an isospin singlet, so the state on the left hand side is an overall isosinglet. On the right hand side, we have the ^{4}He nucleus which is again an isosinglet, whereas the π^0 is part of an isotriplet. A triplet and a singlet cannot combine into a singlet state, so this process would violate isospin, and hence will not be mediated by strong interactions. It can be mediated through electromagnetic interaction though, which does not obey isospin invariance.

8.4.2   Relative strengths

Isospin symmetry can also help us find the relative strengths of two or more processes, each of which conserves isospin. As an example, consider the processes

$$\pi + d \to N + N \,, \tag{8.20}$$

where π now stands generically for pions and N for nucleons. Since deuteron has $I = 0$ as already mentioned, the state on the left hand side has $I = 1$. On the right hand side, two nucleons, each having $I = \frac{1}{2}$, can combine into an $I = 1$ state, so isospin is conserved in such processes. Owing to the conservation of electric charge, the following possibilities can only arise that fall under the general notation of Eq. (8.20):

$$\pi^+ + d \to p + p \,, \tag{8.21a}$$
$$\pi^0 + d \to p + n \,, \tag{8.21b}$$
$$\pi^- + d \to n + n \,. \tag{8.21c}$$

The states on the left hand sides of this set have the following values of I and I_3:

$$|\pi^+ d\rangle = |1, +1\rangle \,, \tag{8.22a}$$
$$|\pi^0 d\rangle = |1, 0\rangle \,, \tag{8.22b}$$
$$|\pi^- d\rangle = |1, -1\rangle \,. \tag{8.22c}$$

On the right hand side of Eq. (8.20), we have two nucleons which can combine into an $I = 0$ or an $I = 1$ state. A state with two protons, for example, has $I_3 = 1$ and therefore cannot be an $I = 0$ state. Thus,

$$|pp\rangle = |1, +1\rangle \,, \tag{8.23a}$$
$$|nn\rangle = |1, -1\rangle \,. \tag{8.23b}$$

The transition matrix elements to the pp and the nn channels thus have essentially the following isospin structure:

$$\langle pp \,|\, \mathscr{H}_{\rm st} \,|\, \pi^+ d\rangle = \langle 1, +1 \,|\, \mathscr{H}_{\rm st} \,|\, 1, +1\rangle \,, \tag{8.24a}$$
$$\langle nn \,|\, \mathscr{H}_{\rm st} \,|\, \pi^- d\rangle = \langle 1, -1 \,|\, \mathscr{H}_{\rm st} \,|\, 1, -1\rangle \,, \tag{8.24b}$$

where $\mathscr{H}_{\rm st}$ is the strong interaction Hamiltonian. Conservation of isospin implies that the amplitudes of the processes do not depend on the I_3 values. The amplitudes for the processes $\pi^+ d \to pp$ and $\pi^- d \to nn$ are thus equal, which implies that, barring corrections coming from electromagnetic effects as well as from the small mass difference between the neutron and the proton, the cross-sections are also equal:

$$\sigma(\pi^+ d \to pp) = \sigma(\pi^- d \to nn) \,. \tag{8.25}$$

On the other hand, the combination of two nucleons with $I_3 = 0$ can be of two types:

$$|1,0\rangle = \frac{1}{\sqrt{2}}(|pn\rangle + |np\rangle),$$

$$|0,0\rangle = \frac{1}{\sqrt{2}}(|pn\rangle - |np\rangle). \tag{8.26}$$

Thus

$$|pn\rangle = \frac{1}{\sqrt{2}}(|1,0\rangle + |0,0\rangle). \tag{8.27}$$

The $I = 0$ part does not contribute because of isospin conservation, only the $I = 1$ part does. So we obtain

$$\langle pn |\mathcal{H}_{\mathrm{st}}| \pi^0 d\rangle = \frac{1}{\sqrt{2}} \langle 1,0 |\mathcal{H}_{\mathrm{st}}| 1,0\rangle. \tag{8.28}$$

It should be noted that when π^0 interacts with deuteron, the final state need not be $|pn\rangle$. By $|pn\rangle$, we denote a state in which the first particle is a proton and the second one a neutron, in some way of numbering them. When we wrote the $\pi^0 d$ reaction in Eq. (8.21), we did not really care which of the final state particles is the proton and which one neutron, so $|np\rangle$ would fit the bill just as well. Note that

$$|np\rangle = \frac{1}{\sqrt{2}}(|1,0\rangle - |0,0\rangle). \tag{8.29}$$

In this language, the final state of the reaction $\pi^0 + d \to p + n$ can be either $|pn\rangle$ or $|np\rangle$. The total amplitude of transition into either of these two states is

$$\langle pn |\mathcal{H}_{\mathrm{st}}| \pi^0 d\rangle + \langle np |\mathcal{H}_{\mathrm{st}}| \pi^0 d\rangle = \sqrt{2} \langle 1,0 |\mathcal{H}_{\mathrm{st}}| 1,0\rangle. \tag{8.30}$$

Thus

$$\frac{\mathfrak{a}(\pi^0 d \to pn)}{\mathfrak{a}(\pi^+ d \to pp)} = \sqrt{2} \frac{\langle 1,0 |\mathcal{H}_{\mathrm{st}}| 1,0\rangle}{\langle 1,1 |\mathcal{H}_{\mathrm{st}}| 1,1\rangle}, \tag{8.31}$$

where $\mathfrak{a}(\cdots)$ stands for the amplitude of the process inside the parentheses, without considering the order of the particles in the states. Remembering that isospin conservation implies that the amplitudes do not depend on the I_3 values, the ratio on the right hand side is simply $\sqrt{2}$. If we neglect electromagnetic effects, as well as the mass difference between the neutron and the proton and between the neutral and the charged pions, we obtain

$$\frac{\sigma(\pi^0 d \to pn)}{\sigma(\pi^+ d \to pp)} = 2. \tag{8.32}$$

This is a non-trivial example where the ratio is neither zero (if one of the processes is not allowed) nor unity, and comes basically through the Clebsch–Gordan co-efficients which occur in Eq. (8.26).

□ **Exercise 8.2** The mirror nuclei ^{3}H and ^{3}He form an isodoublet. Show that

$$\frac{a(p+d \to {}^3\text{H} + \pi^+)}{a(p+d \to {}^3\text{He} + \pi^0)} = \sqrt{2}.\qquad(8.33)$$

□ **Exercise 8.3** Consider the following reactions, proceeding through strong interactions:

$$\pi^+ + p \longrightarrow \pi^+ + p,\qquad(8.34a)$$
$$\pi^- + p \longrightarrow \pi^- + p,\qquad(8.34b)$$
$$\pi^- + p \longrightarrow \pi^0 + n.\qquad(8.34c)$$

A pion and a nucleon can be either in a $I = \frac{1}{2}$ state or in a $I = \frac{3}{2}$ state. Show that, if the matrix element for the $I = \frac{1}{2}$ state dominates over the other, the cross-sections of these processes should be in the ratio 0:2:1. In the reverse case, i.e., if the matrix element for the $I = \frac{3}{2}$ state dominates, show that the ratio should be 9:1:2.

8.4.3   Smushkevich's method

This is a method for calculating relations among branching ratios and cross-sections. In many cases, it is easier to apply than the Clebsch–Gordan co-efficients, but it does not contain all the information that the Clebsch–Gordan co-efficients carry. To introduce the method, we first need some definitions.

Definition 1:   A collection of particles will be called *uniform* if it contains equal number of each member of any isospin multiplet. For example, a collection of 75 pions will be uniform if the numbers of π^+, π^0 and π^- are 25 each.

Definition 2:   The *charge symmetry transformation* (C_S) on a particle is defined by

$$C_S \, |I, I_3\rangle = |I, -I_3\rangle.\qquad(8.35)$$

Thus, for example, C_S acting on the proton will give the neutron, and vice versa.

Isospin conservation is equivalent to finding the unique way of satisfying the two following rules:

Rule 1:   Uniformity of a collection of particles is preserved in isospin conserving processes.

Rule 2:   Cross-sections and branching ratios are the same for a process and its charge symmetric process.

To see examples of application of this rule, consider first the decay of an uncharged isoscalar meson f_0 which can decay to two pions. Conservation of electric charge tells us that there might be two possible decay modes:

$$f_0 \to \begin{cases} \pi^+ \pi^-, \\ 2\pi^0. \end{cases}\qquad(8.36)$$

Since f_0 is an isosinglet, a collection of any number of these mesons is a uniform collection. The final population of pions will then have to be uniform as well, i.e., the number of π^+, π^- and π^0 should be equal. This implies that

$$\frac{\Gamma(f_0 \to 2\pi^0)}{\Gamma(f_0 \to \pi^+\pi^-)} = \frac{1}{2} . \tag{8.37}$$

Now suppose there is another isoscalar uncharged meson ω which cannot decay to two pions for some reason, but can decay to three pions. Possible channels, allowed by conservation of electric charge, are:

$$\omega \to \begin{cases} \pi^+\pi^-\pi^0 , \\ 3\pi^0 . \end{cases} \tag{8.38}$$

A collection of any number of ω mesons is uniform. The final population of pions can be uniform only if the decay occurs exclusively to the $\pi^+\pi^-\pi^0$ channel. Thus we obtain the result that $\omega \to 3\pi^0$ is forbidden by isospin invariance.

☐ **Exercise 8.4** *Using Clebsch–Gordan co-efficients, verify the following statements.*

    a)   *Eq. (8.37) gives the correct branching ratios;*

    b)   *The meson ω does not decay to $3\pi^0$.*

8.5   *G*-parity

Since strong interactions are invariant under the operations of charge conjugation and isospin rotation, any combination of these symmetries is also respected by the strong interactions. One such combination is very useful. It is called the *G*-parity, and defined as

$$G = \mathscr{C} e^{i\pi I_2} . \tag{8.39}$$

Let us see what it means. The symbol $\mathscr{C}$, of course, is the charge conjugation operation. It is multiplied by an isospin rotation: a rotation by 180 degrees about the I_2 axis. Such a rotation will change the directions of the other two axes of isospin. In particular, I_3 changes to $-I_3$. So, this part is similar to the operation C_S defined earlier.

    To see why this definition of *G*-parity is useful, consider what this operation does to the pions. The charge conjugation operation $\mathscr{C}$, acting on the π^+, changes it to π^- which has $I_3 = -1$. But the specified isospin rotation inverts the value of I_3, returning it back to an $I_3 = +1$ state, i.e., a π^+ back again. Thus π^+, though not an eigenstate of either $\mathscr{C}$ or C_S, is an eigenstate of the *G* operation. So is π^-. The neutral pion is an eigenstate of both $\mathscr{C}$ and C_S, and is therefore an eigenstate of *G* as well. Thus the entire isomultiplet of pions is an eigenstate of *G*-parity. Similar conclusions can be reached for other isomultiplets of mesons. This is very convenient for analyzing processes where the initial or the final state contains only mesons, as we are about to see.

To obtain the eigenvalue of G for the neutral pion state, we first note that the $\mathscr{C}$ eigenvalue of π^0 is $+1$, because it can decay to two photons, as noted earlier in Eq. (6.158, *p 185*). Since π^0 is a state with $I = 1$ and $I_3 = 0$, its behavior under an isospin rotation will be exactly similar to that of an orbital angular momentum state with $L = 1$ and $L_z = 0$ under spatial rotations.

We know that the eigenfunctions of orbital angular momentum are given by the spherical harmonics Y_l^m, given in Eq. (6.95, *p 168*). The rotation operator analogous to $\exp(i\pi I_2)$ is $\exp(i\pi L_y)$, which changes the sign of the x and z co-ordinates but keeps the y co-ordinate unchanged. In the spherical co-ordinate system, it implies the changes

$$\theta \to \pi - \theta, \qquad \phi \to \pi - \phi. \tag{8.40}$$

Thus, under this rotation,

$$P_l^m(\cos\theta)e^{im\phi} \to P_l^m(-\cos\theta)e^{im(\pi-\phi)}. \tag{8.41}$$

This shows that Y_l^m is not an eigenfunction of $\exp(i\pi L_y)$ for arbitrary values of m. But it also shows that the functions with $m = 0$ are eigenfunctions, since the exponential factor is unity in this case, and the associated Legendre functions reduce to the Legendre polynomials $P_l(x)$, which have the property that $P_l(-x) = (-1)^l P_l(x)$. Thus, under the co-ordinate changes given in Eq. (8.40),

$$Y_l^0(\theta, \phi) \to (-1)^l Y_l^0(\theta, \phi). \tag{8.42}$$

Carrying the analogy to isospin rotations, we conclude that

$$e^{i\pi I_2} |I, I_3 = 0\rangle = (-1)^I |I, I_3 = 0\rangle. \tag{8.43}$$

Since π^0 has $I = 1$, it follows that its eigenvalue is negative for the operator $\exp(i\pi I_2)$. Combining this with the $\mathscr{C}$ eigenvalue of π^0 given in Eq. (6.158, *p 185*), we conclude that

$$G\left|\pi^0\right\rangle = -\left|\pi^0\right\rangle. \tag{8.44}$$

Let us now look at the charged pions. Since the π^+ is an $|1, 1\rangle$ state in the $|I, I_3\rangle$ notation, the application of the operator $\exp(i\pi I_2)$ will turn it into a $|1, -1\rangle$ state, i.e., into a π^-. This doesn't, however say that $\exp(i\pi I_2)|\pi^+\rangle = |\pi^-\rangle$, because there might be a phase in the equation. Let us fix the phases by the convention of Eq. (8.11). Note that the definition of the ladder operators in Eq. (8.10) implies that

$$I_2 = \frac{1}{2i}(I_+ - I_-). \tag{8.45}$$

Then,

$$\begin{aligned} I_2\left|\pi^\pm\right\rangle &= \mp\frac{1}{\sqrt{2}i}\left|\pi^0\right\rangle, \\ I_2\left|\pi^0\right\rangle &= \frac{1}{\sqrt{2}i}\left(\left|\pi^+\right\rangle - \left|\pi^-\right\rangle\right). \end{aligned} \tag{8.46}$$

Consequently,

$$(I_2)^2 \left|\pi^\pm\right\rangle = \pm\frac{1}{2}\Big(\left|\pi^+\right\rangle - \left|\pi^-\right\rangle\Big),$$
$$(I_2)^2 \left|\pi^0\right\rangle = \left|\pi^0\right\rangle. \tag{8.47}$$

It can be easily shown that the actions of all higher odd powers yield the same result as that given by I_2 itself, and all higher even powers yield the same result as that given by $(I_2)^2$. We can then write

$$\begin{aligned}
\exp(i\pi I_2)\left|\pi^\pm\right\rangle &= \left(1 + i\pi I_2 - \frac{\pi^2}{2!}(I_2)^2 + \cdots\right)\left|\pi^\pm\right\rangle \\
&= \left|\pi^\pm\right\rangle \pm (\cos\pi - 1)\frac{\left|\pi^+\right\rangle - \left|\pi^-\right\rangle}{2} \mp \sin\pi\frac{\left|\pi^0\right\rangle}{\sqrt{2}} \\
&= \left|\pi^\mp\right\rangle.
\end{aligned} \tag{8.48}$$

As expected, the action of $\exp(i\pi I_2)$ changes a charged pion to a pion of opposite charge. The phase has been fixed by the convention taken in Eq. (8.11).

Of course when we operate a charged pion state by the charge conjugation operator $\mathscr{C}$, that also changes the charge. If we fix the phases here by the convention

$$\mathscr{C}\left|\pi^\pm\right\rangle = -\left|\pi^\mp\right\rangle, \tag{8.49}$$

then we find that

$$G\left|\pi^\pm\right\rangle = -\left|\pi^\pm\right\rangle. \tag{8.50}$$

Comparing with Eq. (8.44), we find that all three kinds of pions are eigenstates of G-parity, and the eigenvalues are -1.

□ **Exercise 8.5** *An alternative definition of G-parity is*

$$G = \mathscr{C}e^{i\pi I_1}. \tag{8.51}$$

Show that with this definition also one can reach the G-parity assignments of Eq. (8.50) provided one defines the charge conjugation phases by

$$\mathscr{C}\left|\pi^\pm\right\rangle = +\left|\pi^\mp\right\rangle. \tag{8.52}$$

The G-parity is very helpful in any discussion about final states containing pions only. For example, we discussed that the ω particle can decay into three pions. This means that the ω is negative under G-parity. Since ω is an isosinglet which is unaffected by $\exp(i\pi I_2)$, it implies that ω is odd under charge conjugation. Its negative G-parity tells us that it cannot decay into two pions, for example.

Of course that is true if G-parity is conserved in the interactions. Strong interactions are invariant under both isospin and charge conjugation, and therefore under G-parity as well. Electromagnetic interactions are invariant under charge conjugation but not under isospin, as will be discussed in some detail in §8.9. Thus, electromagnetic interactions do not respect G-parity. The ω particle can decay into $\pi^+\pi^-$ through electromagnetic interactions.

This is an interesting observation. If the ω (or any other particle, for that matter) can decay into a two pion state as well as into a three pion state, the latter is expected to be suppressed from phase space considerations. However, experimental data shows that

$$\mathscr{B}(\omega \to \pi^+\pi^-\pi^0) = (89.1 \pm 0.7)\%,$$
$$\mathscr{B}(\omega \to \pi^+\pi^-) = (1.70 \pm 0.27)\%. \tag{8.53}$$

Clearly, the conclusion is that the amplitude for the two-pion decay mode is much smaller. We have seen the reason why: the two-pion decay can occur only through electromagnetic interactions, whereas the three-pion decay can occur through strong interactions, as can be seen through G-parity considerations.

8.6   Generalized Pauli principle

The Pauli exclusion principle, in its simplest form, says that in a multiparticle system, it is not possible to put two or more identical fermions in the same one-particle state. A more sophisticated and more general statement is this: if one interchanges two identical fermions in the wavefunction of a multiparticle system, the wavefunction changes by a sign. If we do the same thing on two bosons, the exercise will not change the wave function at all. This can be called a more generalized form of the Pauli principle in terms of the exchange of two particles, and it includes fermions as well as bosons.

Now, what exactly is meant by the interchange of two particles? It certainly includes the interchange of their co-ordinates, which can be called the spatial interchange. But that is not all. One needs to interchange other characteristics of the particles as well. For example, if the particles have spin, the spins need to be interchanged as well. Thus, we can summarize the Pauli principle by the following statement:

$$\begin{pmatrix} \text{Spatial} \\ \text{interchange} \end{pmatrix} \times \begin{pmatrix} \text{Spin} \\ \text{interchange} \end{pmatrix} \longrightarrow \begin{cases} -1 & \text{for fermions,} \\ +1 & \text{for bosons.} \end{cases} \tag{8.54}$$

Under isospin, the proton and the neutron are considered to be two states of the same "particle" called the *nucleon*: one with $I_3 = +\frac{1}{2}$ and the other with $I_3 = -\frac{1}{2}$, as announced right after Eq. (8.9). Therefore, in a system of nucleons, we can extend the Pauli principle to include isospin interchange as well. And of course, the exercise need not stop at nucleons. For any system

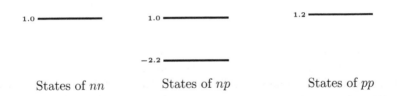

Figure 8.3: States of the dineutron, the deuteron and the diproton. The energy values given to the left of the lines are in MeV, and their signs indicate that the deuteron has a bound state, whereas all other states shown in the diagram are not bound.

of particles, this generalized form of the Pauli principle should mean that

$$\left(\begin{array}{c} \text{Spatial} \\ \text{interchange} \end{array} \right) \times \left(\begin{array}{c} \text{Spin} \\ \text{interchange} \end{array} \right) \times \left(\begin{array}{c} \text{Isospin} \\ \text{interchange} \end{array} \right)$$

$$\longrightarrow \begin{cases} -1 & \text{for fermions,} \\ +1 & \text{for bosons.} \end{cases} \quad (8.55)$$

Below, we explore some simple consequences of this principle.

a) Example 1

Consider two-pion states at relative orbital angular momentum zero. Pions are bosons, so the states must be symmetric under the generalized Pauli principle. Pions do not have spin, so there is no spin part of the wavefunction of a two-pion state. The orbital part is symmetric under the interchange of the two pions since $L = 0$ as stated above. Thus the isospin part of the wavefunction must also be symmetric under the interchange of the two pions. A pion is an isotriplet, so the combination of two pions can be in the isospin states $I = 0, 1, 2$. Of these, the $I = 1$ state is antisymmetric under the interchange, so it is not allowed. Two pions with $L = 0$ can only be in $I = 0$ or $I = 2$ states.

b) Example 2

Now consider states of two nucleons. One well-known bound state of this sort is the deuteron, which contains one neutron and one proton. The nucleons are fermions, so the generalized Pauli principle dictates that the state should be antisymmetric under their interchange. The bound state has $L = 0$, so the spatial part is symmetric under the interchange. The spin of deuteron is 1, which means that it is the symmetric combination of the spins of two nucleons, each of which has spin-$\frac{1}{2}$. Thus, the isospin part must be antisymmetric, which means that the two nucleons must combine into an $I = 0$ state. The deuteron, thus, is an isosinglet: it is a stand-alone state.

If we consider spin-singlet states, the isospin part must be symmetric. In this case, we will obtain an isotriplet, as shown in Fig. 8.3. For nn and np systems, this state has the same energy. For the diproton system, however, the energy is slightly higher because of Coulomb repulsion between the protons. Of course, none of these three states is bound, as is seen from their positive energy eigenvalues. However, unlike usual scattering states, the wavefunctions of these states are normalizable. The existence of these states is inferred from measurements of scattering phase shifts. The details are of not much importance for the subjects that we want to discuss, so we are not getting into it.

c)   Example 3

As a final example of the application of the generalized Pauli principle, consider three-nucleon states. Since nucleons have $I = \frac{1}{2}$, the combination of three nucleons can have either $I = \frac{3}{2}$ or $I = \frac{1}{2}$. To be specific, we talk of the $I = \frac{1}{2}$ states. Obviously, such states will contain two protons and one neutron if the total I_3 value is $+\frac{1}{2}$, in which case it will be the nucleus of ^{3}He. Alternatively, if total I_3 is $-\frac{1}{2}$, there will be one proton and two neutrons. This is the nucleus of ^{3}H, also called the tritium nucleus. Both of them have spin $S = \frac{1}{2}$. For both, the ground state has no orbital angular momentum.

Let us try to write the spin-isospin part of the ground state wavefunction of ^{3}He in the state $S_z = +\frac{1}{2}$. We first look at the spin part of the wavefunction. Here we have the task of starting with three particles of spin-$\frac{1}{2}$, and constructing the state containing those three particles which has $S = \frac{1}{2}$ and $S_z = \frac{1}{2}$, a state which we will denote by $|\frac{1}{2}, \frac{1}{2}\rangle$. We start with two particles, which can be combined into a spin-1 or a spin-0 combination. In either case, we can add the third particle and obtain the total spin to be equal to $\frac{1}{2}$.

If the first two particles are in a spin-1 combination, the overall combination after the addition of the third particle is given by

$$\left|\tfrac{1}{2}, \tfrac{1}{2}\right\rangle_{123} = \sqrt{\tfrac{1}{3}}\,|1,0\rangle_{12}\left|\tfrac{1}{2}, \tfrac{1}{2}\right\rangle_3 - \sqrt{\tfrac{2}{3}}\,|1,1\rangle_{12}\left|\tfrac{1}{2}, -\tfrac{1}{2}\right\rangle_3 \equiv \psi_S . \quad (8.56)$$

The subscripts on the states denote the particles contained in them. Thus, the state on the left hand side of this equation is the combined state of all three particles, whereas the spin-1 states on the right hand side contains only the first two particles. We have used the standard Clebsch–Gordan co-efficients to combine the angular momenta. The state is called ψ_S, where the subscript indicates that the state is symmetric under the interchange of the first two particles. This can be seen by the explicit forms of the combinations of these two particles:

$$|1,0\rangle_{12} = \frac{1}{\sqrt{2}}\left(\left|\tfrac{1}{2},\tfrac{1}{2}\right\rangle_1\left|\tfrac{1}{2},-\tfrac{1}{2}\right\rangle_2 + \left|\tfrac{1}{2},-\tfrac{1}{2}\right\rangle_1\left|\tfrac{1}{2},\tfrac{1}{2}\right\rangle_2\right),$$
$$|1,1\rangle_{12} = \left|\tfrac{1}{2},\tfrac{1}{2}\right\rangle_1\left|\tfrac{1}{2},\tfrac{1}{2}\right\rangle_2. \tag{8.57}$$

Using the notations of up-arrow and down-arrow for the states $\left|\tfrac{1}{2},\tfrac{1}{2}\right\rangle$ and $\left|\tfrac{1}{2},-\tfrac{1}{2}\right\rangle$ of a single particle, we can thus write

$$\psi_S = \frac{1}{\sqrt{6}}\Big((\blacktriangle\blacktriangledown + \blacktriangledown\blacktriangle)\blacktriangle - 2\blacktriangle\blacktriangle\blacktriangledown\Big), \tag{8.58}$$

where we have not put in the subscripts for the number of particles, because each term has been arranged with particle number 1 first, number 2 second and number 3 third.

The first two particles might also be in the antisymmetric spin-0 combination. In this case, the desired combination of the three particles is obtained as

$$\psi_A \equiv \left|\tfrac{1}{2},\tfrac{1}{2}\right\rangle_{123} = |0,0\rangle_{12}\left|\tfrac{1}{2},\tfrac{1}{2}\right\rangle_3 = \frac{1}{\sqrt{2}}(\blacktriangle\blacktriangledown - \blacktriangledown\blacktriangle)\blacktriangle. \tag{8.59}$$

Remember that ψ_S and ψ_A are symmetric and antisymmetric respectively with respect to the interchange of the first two particles only. Neither shows any particular symmetry when any of the first two particles is interchanged with the third. Thus the spin parts of the wavefunctions do not satisfy the Pauli principle by themselves.

But that is not necessary either. We have the isospin parts as well, and only the combination needs to satisfy the Pauli principle. The mathematical steps for the construction of the isospin parts of the wavefunction are exactly the same, since we are aiming at the wavefunction of ^{3}He, which is a state with $I = \tfrac{1}{2}$, $I_3 = \tfrac{1}{2}$. We can thus identify the following two combinations:

$$\chi_S = \frac{1}{\sqrt{6}}\Big((pn + np)p - 2ppn\Big),$$
$$\chi_A = \frac{1}{\sqrt{2}}(pn - np)p. \tag{8.60}$$

We can now write the wavefunction of ^{3}He in the spin-up state as

$$|^3\text{He}\blacktriangle\rangle = a_1\psi_S\chi_A + a_2\psi_A\chi_S, \tag{8.61}$$

where the co-efficients a_1 and a_2 need to be determined. Note that we have not included the combinations $\psi_S\chi_S$ and $\psi_A\chi_A$. Those are symmetric in the interchange of the first two particles and are clearly unacceptable.

The relative magnitude of the co-efficients a_1 and a_2 can now be determined by imposing the Pauli principle. For this, we note that

$$\psi_S\chi_A = \frac{1}{2\sqrt{3}}\Big(\blacktriangle\blacktriangledown\blacktriangle + \blacktriangledown\blacktriangle\blacktriangle - 2\blacktriangle\blacktriangle\blacktriangledown\Big)(pnp - npp)$$

$$= \frac{1}{2\sqrt{3}} \Big(p\blacktriangle n\blacktriangledown p\blacktriangle + p\blacktriangledown n\blacktriangle p\blacktriangle - 2p\blacktriangle n\blacktriangle p\blacktriangledown$$

$$-n\blacktriangle p\blacktriangledown p\blacktriangle - n\blacktriangledown p\blacktriangle p\blacktriangle + 2n\blacktriangle p\blacktriangle p\blacktriangledown \Big). \tag{8.62}$$

In the last step, we have written the terms in a different notation: by putting the isospin and the spin characteristics of any one particle together. In a similar notation,

$$\psi_{A\chi S} = \frac{1}{2\sqrt{3}} \Big(p\blacktriangle n\blacktriangledown p\blacktriangle - p\blacktriangledown n\blacktriangle p\blacktriangle + n\blacktriangle p\blacktriangledown p\blacktriangle$$

$$-n\blacktriangledown p\blacktriangle p\blacktriangle - 2p\blacktriangle p\blacktriangledown n\blacktriangle + 2p\blacktriangledown p\blacktriangle n\blacktriangle \Big). \tag{8.63}$$

Putting these expressions back into Eq. (8.61) we obtain

$$|^3\text{He}\blacktriangle\rangle \propto (a_1 + a_2)\Big(p\blacktriangle n\blacktriangledown p\blacktriangle - n\blacktriangledown p\blacktriangle p\blacktriangle + n\blacktriangle p\blacktriangle p\blacktriangledown$$

$$- p\blacktriangle n\blacktriangle p\blacktriangledown + p\blacktriangledown p\blacktriangle n\blacktriangle - p\blacktriangle p\blacktriangledown n\blacktriangle \Big)$$

$$+(a_1 - a_2)\Big(p\blacktriangledown n\blacktriangle p\blacktriangle - n\blacktriangle p\blacktriangledown p\blacktriangle + n\blacktriangle p\blacktriangle p\blacktriangledown$$

$$- p\blacktriangle n\blacktriangle p\blacktriangledown - p\blacktriangledown p\blacktriangle n\blacktriangle + p\blacktriangle p\blacktriangledown n\blacktriangle \Big). \tag{8.64}$$

Clearly, the multiplier of $a_1 + a_2$ is not antisymmetric under the $1 \leftrightarrow 3$ exchange, so we need to take

$$a_1 = -a_2. \tag{8.65}$$

Then, normalizing, we can write down the wavefunction as

$$|^3\text{He}\blacktriangle\rangle = \frac{1}{\sqrt{6}} \Big(p\blacktriangledown n\blacktriangle p\blacktriangle - n\blacktriangle p\blacktriangledown p\blacktriangle + n\blacktriangle p\blacktriangle p\blacktriangledown$$

$$- p\blacktriangle n\blacktriangle p\blacktriangledown - p\blacktriangledown p\blacktriangle n\blacktriangle + p\blacktriangle p\blacktriangledown n\blacktriangle \Big). \tag{8.66}$$

Once this is obtained, it is trivial to find the wavefunction of ^{3}He in the spin-down state: we merely have to interchange the up-arrows and down-arrows in the expression in Eq. (8.66). If, on the other hand, we need the wavefunction for ^{3}H, we only need to interchange neutrons with protons in the ^{3}He wavefunction.

□ **Exercise 8.6** Find the spin-isospin part of the wavefunction of ^{3}He in the state with $S_z = \frac{1}{2}$, assuming that this part of the wavefunction is completely *symmetric* under the interchange of any two nucleons. [**Note :** The result is not relevant for ^{3}He or ^{3}H, but will be important in a different context later.]

□ **Exercise 8.7** From the discussion on deuterons given above, write down the spin-isospin part of the deuteron wave function in the states $S_z = +1$ and $S_z = 0$.

8.7   Isospin and quarks

Properties of nucleons and pions can be easily understood if it is assumed that
they are composite particles built of just two different kinds of fundamental
fermions which are called quarks. These two quarks, like the proton and the
neutron, transform like a doublet of isospin:

$$q = \begin{pmatrix} u \\ d \end{pmatrix}. \tag{8.67}$$

The upper component, or the $I_3 = +\frac{1}{2}$ component, is called the u-quark or
the up-quark. The $I_3 = -\frac{1}{2}$ component is called the down-quark, or d-quark
in short. Instead of saying that the up and the down are two different "kinds"
of quark, one often says that they are two different *flavor*s of quark.

8.7.1   Nucleons

It is easy to see how the isospin properties of nucleons can be understood in
terms of these quarks. Since the nucleons are fermions, they must contain an
odd number of quarks. They cannot possibly contain just one quark because
then the nucleons would have to be the quarks themselves, and it would be
useless talking about quarks. So the smallest possible number of quarks in
a nucleon is 3, and we will stick to this simplest choice. The proton has
$I_3 = +\frac{1}{2}$. The only combination of three u and d quarks that can give this
value is uud. For the neutron, we must have the combination udd since it
gives $I_3 = -\frac{1}{2}$. The charges of these quarks can now be obtained by matching
with the known charges of the nucleons:

$$2Q_u + Q_d = 1\,,$$
$$Q_u + 2Q_d = 0\,. \tag{8.68}$$

The solution is

$$Q_u = \frac{2}{3}\,, \qquad Q_d = -\frac{1}{3}\,. \tag{8.69}$$

The quarks seem to have fractional electric charge in the unit of the proton charge. Historically
speaking, this was a shocking aspect of the quark hypothesis. Whenever anyone has performed
any measurement of the charge of anything, the result has always turned out to be an integral
multiple of the proton charge, positive or negative. This led to the belief that electric charge
has quantized values, and the unit is the proton charge.

   The structure of the U(1) group, which is the gauge group for electromagnetism, does not
give any hint about why such quantization should be there. When we wrote down the phase
factor $e^{-ieQ\theta}$ in Eq. (5.2, *p 112*), there was no restriction on the number Q. All measured
charges are indeed multiples of the proton charge, and some other theoretical ideas were needed
to explain this phenomenon. The most notable was Dirac's idea that if magnetic monopoles
exist, consistency of electromagnetic theory demands that the product of magnetic and electric
charges must be quantized. Much later, grand unified theories have justified charge quantization.

   But none of these theoretical ideas really contradicts the presence of fractionally charged
particles. Even if charge is really quantized, there is nothing that tells us that it must be
quantized in units of the proton charge. After all, if we take the charge of the d-quark as the
unit, all charges turn out to be integral multiples of it.

8.7.2   Pions

Let us now talk about pions. These are spinless bosons. So they must contain an even number of constituent fermions. Consideration of baryon number, to be discussed in §8.10, tells us that pions cannot contain only quarks. They must contain quarks as well as antiquarks, and the simplest option is to have just one of each.

The necessity of including antiquarks can be understood even without invoking the argument about baryon number. Suppose the neutral pion can be composed of n_u up-quarks and n_d down quarks, without any antiquarks. Then the electric charge of the π^0 implies the relation

$$\frac{2}{3}n_u - \frac{1}{3}n_d = 0, \tag{8.70}$$

whereas the I_3 value of π^0 tells us that

$$\frac{1}{2}n_u - \frac{1}{2}n_d = 0. \tag{8.71}$$

The solution of these two equations is $n_u = n_d = 0$. This is meaningless, and we are back to the conclusion that pions cannot contain only quarks: they must contain antiquarks as well.

Once this is decided, it is easy to find the content of the charged pions. In fact, there is only one way to make a unit of positive charge from a quark and an antiquark, viz., $u\widehat{d}$. This must therefore be the structure of π^+. The negative pion, π^-, will contain $d\widehat{u}$.

What about the neutral pion? The combinations $u\widehat{u}$ and $d\widehat{d}$ both are chargeless, and can both be candidates. The wavefunction of π^0 will in fact have to be a superposition of these two combinations, constructed in such a way that the resulting state has $I = 1$ and $I_3 = 0$. The task is now clear cut: we have an isospin doublet of quarks given in Eq. (8.67). The antiquarks must also form a doublet of isospin. Given the two doublets, we can combine them into a symmetric combination to obtain the states in an isotriplet.

Before proceeding, it will be worthwhile to clarify a comment just made. We said that the antiquarks $\widehat{u}$ and $\widehat{d}$ form a doublet of isospin just because the quarks do. Let us see why this is so, and what is the exact analog of Eq. (8.67) for antiquarks.

Let us take a more general notation and suppose that the object

$$\phi \equiv \begin{pmatrix} \phi_1 \\ \phi_2 \end{pmatrix} \tag{8.72}$$

transforms like a doublet under some SU(2) symmetry. This means that, under an SU(2) transformation denoted by the parameters θ_a ($a=1,2,3$), we can write the transformation rule for the doublet as

$$\phi \to \phi' = \exp\left(-i\frac{\tau_a}{2}\theta_a\right)\phi, \tag{8.73}$$

where the τ_a's are the Pauli matrices. This tells us that, under the same SU(2) transformation,

$$\phi^* \to \phi'^* = \exp\left(i\frac{\tau_a^*}{2}\theta_a\right)\phi^*. \tag{8.74}$$

We can ask ourselves: is this the transformation law of a doublet? The answer is 'no', because all Pauli matrices do not satisfy the condition $\tau_a^* = -\tau_a$.

We therefore change the question and ask ourselves whether it is possible to find a matrix ε such that $\varepsilon\phi^*$ transforms like a doublet. The matrix ε will have to be a constant matrix, in the sense that it should be invariant under SU(2) transformations. Eq. (8.74) implies that, under an SU(2) transformation,

$$\varepsilon\phi^* \to \varepsilon\exp\left(i\frac{T_a^*}{2}\theta_a\right)\phi^* . \tag{8.75}$$

On the other hand, we want $\varepsilon\phi^*$ to transform exactly as in Eq. (8.73), i.e.,

$$\varepsilon\phi^* \to \exp\left(-i\frac{T_a}{2}\theta_a\right)\varepsilon\phi^* . \tag{8.76}$$

These two equations would be consistent with each other if the matrix ε satisfies the condition

$$\varepsilon T_a^* = -T_a\varepsilon \tag{8.77}$$

for each Pauli matrix. It is easy to see that this condition can in fact be satisfied if we choose

$$\varepsilon = i\tau_2 = \begin{pmatrix} 0 & 1 \\ -1 & 0 \end{pmatrix} . \tag{8.78}$$

Clearly, the factor i in this choice is arbitrary. We have made this choice only to make the elements of ε real. We have thus proved that if ϕ transforms like a doublet, so does

$$\tilde{\phi} \equiv \varepsilon\phi^* = \begin{pmatrix} 0 & 1 \\ -1 & 0 \end{pmatrix} \begin{pmatrix} \phi_1^* \\ \phi_2^* \end{pmatrix} = \begin{pmatrix} \phi_2^* \\ -\phi_1^* \end{pmatrix} . \tag{8.79}$$

We now get back to the quarks. For any quark field, the antiquark field will be defined through Eq. (6.49, *p 160*). Apart from the matrix $\gamma_0 C$ which shuffles the different components of a quark field, it contains a complex conjugation of the field. Our exercise above then tells us that, since the quark doublet is given by Eq. (8.67), the antiquark doublet will be given by

$$\hat{q} = \begin{pmatrix} \hat{d} \\ -\hat{u} \end{pmatrix} . \tag{8.80}$$

Combining two doublets into a triplet is easy. We can use the tables of Clebsch–Gordan co-efficients and find that

$$\begin{aligned}
|1,+1\rangle &= \left|\tfrac{1}{2},\tfrac{1}{2}\right\rangle\left|\tfrac{1}{2},\tfrac{1}{2}\right\rangle , \\
|1,0\rangle &= \frac{1}{\sqrt{2}}\left(\left|\tfrac{1}{2},\tfrac{1}{2}\right\rangle\left|\tfrac{1}{2},-\tfrac{1}{2}\right\rangle + \left|\tfrac{1}{2},-\tfrac{1}{2}\right\rangle\left|\tfrac{1}{2},\tfrac{1}{2}\right\rangle\right) , \\
|1,-1\rangle &= \left|\tfrac{1}{2},-\tfrac{1}{2}\right\rangle\left|\tfrac{1}{2},-\tfrac{1}{2}\right\rangle ,
\end{aligned} \tag{8.81}$$

where the states of the combined system are on the left hand sides of these equations. From the two doublets given in Eqs. (8.67) and (8.80), we find that the combination $u\hat{d}$ will transform like $|1,1\rangle$. Thus,

$$\left|\pi^+\right\rangle = \hat{d}u , \tag{8.82a}$$

as already mentioned from considerations of electric charge only. Similarly, the π^- is the state that transforms like $|1,-1\rangle$ of isospin, and so

$$\left|\pi^-\right\rangle = \hat{u}d , \tag{8.82b}$$

apart from an overall sign. The neutral pion is given by

$$\left|\pi^0\right\rangle = \frac{1}{\sqrt{2}}\left(\hat{u}u - \hat{d}d\right) . \tag{8.83}$$

Of course there might be an overall phase in the definitions of each of these bound states.

One property of the pions is very easily explained in this model. The pions are bound states of a quark and an antiquark, and in this respect are like the positronium states discussed in §6.7. We saw, in Eq. (6.113, *p 172*), that the parity of such states is given by $(-1)^{L+1}$, where L is the relative orbital angular momentum between the fermion and the antifermion. Pions are the lightest mesons known, so the quark and the antiquark are expected to be in an $L = 0$ state in pions. This tells us that the intrinsic parity of the pions should be -1, consistent with the value deduced from experiments mentioned in §6.8.

8.7.3   More hadrons

We have discussed the nucleons and the pions, but many other hadrons can be built with the two flavors of quarks and their antiparticles. We can have baryons, i.e., hadrons with half-integral spin, from three quarks. The other kind of hadrons, which have integral spin and are called the mesons, can be obtained by combining a quark and an antiquark. Let us get acquainted with some of them.

In constructing the π^0 state, we combined the quark and the antiquark into an isospin triplet. We can also combine them in an isosinglet. The only difference would be in the relative sign between the two terms appearing in the expression. Thus, the isosinglet state can be written as

$$|\eta_0\rangle = \frac{1}{\sqrt{2}}\left(\widehat{u}u + \widehat{d}d\right). \qquad (8.84)$$

In Ch. 10, we will see that this is not really the wavefunction of a physical particle. Nevertheless, this combination appears in the wavefunction of two mesons called η and η'.

As already remarked, the quark and the antiquark in pions are in a relative $L = 0$ state. The combined spin must be zero, since the spin of the pion has to be zero. Thus, the space-spin part of the pion wavefunction is $^1S_0$. One can easily contemplate other possibilities. For example, we can have mesons whose isospin part of the wavefunctions will be the same as that for the pions, but whose space-spin part has a $^3S_1$ wavefunction. Spin is 1, orbital angular momentum is 0, so the total angular momentum has to be 1, as indicated in the notation. When we look at the whole combination as a single unit, these details are irrelevant: we see a total angular momentum of the constituents of the composite unit, which should be treated as the intrinsic angular momentum of the meson. These will thus be spin-1 mesons, or vector mesons. These are called the ρ mesons. Obviously this is an isotriplet, with masses around 770 MeV. One can have higher values of L in the space-spin configuration as well, which will give heavier mesons.

There are baryons heavier than the nucleons. In nucleons, the spins of the three quarks combine to make a spin-$\frac{1}{2}$ particle. Obviously, the spins can be combined into a total spin of $\frac{3}{2}$ as well. It turns out that the isospin of the resulting baryons is also $\frac{3}{2}$, and consequences of this fact will be discussed in detail in §10.11.1. Thus, these baryons form a isoquartet, i.e., an isospin multiplet with four members. Collectively, they are denoted by Δ, with masses around 1232 MeV. If all the quarks are u-quarks, the hadron will be doubly positively charged and will be denoted by Δ^{++}. If there are two u-quarks and one d-quark, the hadron is Δ^{+}. With two d-quarks and one u-quark, we obtain Δ^{0}, and finally, with three d-quarks, we obtain Δ^{-}.

A word of caution on the names of the baryons. Unlike the pions where π^{-} is the antiparticle of π^{+}, the Δ^{-} is *not* the antiparticle of the Δ^{+}. The superscript on Δ denotes the charge of the baryon. Different charges occur due to different combinations of up and down quarks. The antiparticles of baryons should contain antiquarks, and will be adorned with hats according to our general scheme of denoting antiparticles. Thus, for example, the antiparticle of Δ^{+} will be denoted by $\widehat{\Delta}^{-}$, and so on.

□ **Exercise 8.8** The Δ particles decay mainly to $N\pi$, i.e., to a nucleon and a pion. Using the Smushkevich method or the Clebsch–Gordan co-efficients, show that

$$\Gamma(\Delta^{++} \to p\pi^{+}) : \Gamma(\Delta^{+} \to p\pi^{0}) : \Gamma(\Delta^{+} \to n\pi^{+}) = 3 : 2 : 1. \quad (8.85)$$

[**Hint :** *To apply Smushkevich's method, start with equal numbers of Δ^{++}, Δ^{+}, Δ^{0} and Δ^{-}, which will make it a uniform collection.*]

8.8 Pion–nucleon interaction

Let us go back to Yukawa's theory where the nucleons are supposed to interact via exchange of pions. We are talking of strong interactions, so the theory should conserve isospin. We can ask, what would be the exact form of this interaction?

Even before that, let us settle some other issues. The interaction must be Lorentz invariant. Since the pions are spinless, the nucleon fields must also appear in a spinless combination. There are two such combinations, as listed in Eq. (4.93, *p 79*), called the scalar and the pseudoscalar. Among these, the pseudoscalar is odd under parity, as noted in Eq. (6.26, *p 156*). Since the pions are odd under parity, they should couple to the pseudoscalar current in order that the interaction is parity invariant. This means that the interaction should involve terms like

$$\overline{p}\gamma_{5}n\pi^{+}, \quad \overline{n}\gamma_{5}p\pi^{-}, \quad \overline{p}\gamma_{5}p\pi^{0}, \quad \overline{n}\gamma_{5}n\pi^{0}. \quad (8.86)$$

Isospin invariance would imply certain relations between the strengths of the couplings of these different terms.

The interaction itself must be isospin invariant, i.e., an isosinglet. The pion is an isotriplet. It can give an isosinglet only if it is coupled to another isotriplet. Since the nucleon is an isodoublet, two nucleon fields must appear in an isospin-symmetric combination in the interaction.

What is a symmetric combination? Consider two doublets N and N', written as column vectors. Obviously, the symmetric combination is of the form $M_{ij}N'_i N_j$, where $M_{ij} = M_{ji}$. In matrix notation, we can say that a combination of the form $N'^\top M N$ will be symmetric when M is a symmetric matrix. The matrix will have to be hermitian so that the interaction is hermitian. Thus, the only choices are the unit matrix, and two of the three Pauli matrices. Writing the Pauli matrices by τ_a (with $a = 1, 2, 3$), we can write these choices for M as

$$M = \varepsilon \tau_a , \tag{8.87}$$

with ε defined in Eq. (8.78). We need a combination with the nucleon doublet N introduced in Eq. (8.2), and its complex conjugate. As shown in the argument leading from Eq. (8.73) to Eq. (8.79), the object involving N^* that transforms like a doublet is the combination εN^*, which we identify with N'. Thus, the combination will be of the form $(\varepsilon N^*)^\top M N$, or $N^\dagger \varepsilon^\top M N$, which will be SU(2) invariant and symmetric, where M is given by Eq. (8.87). However, in order to make it Lorentz invariant as well, we need to use $\overline{N}$, not $N^\dagger$. Thus, finally, we can write the nucleon-pion interaction term of the Yukawa theory as

$$\mathscr{L}_{\text{int}} = g \sum_a \overline{N} \tau_a \gamma_5 N \pi_a , \tag{8.88}$$

where the quantity g is a coupling constant. This is a compact form of writing the interaction. If we expand it by explicitly writing down both components of N as well as the explicit forms of the Pauli matrices, we get interaction terms like those shown in Eq. (8.86).

Although Eq. (8.88) is consistent with the known symmetries of strong interactions and should describe the dynamics of pion–nucleon interactions, it is difficult to test it against experiments. The reason is that strong interactions are, well, strong, which means that the coupling constant g appearing in the interaction Lagrangian is not small. Thus it is meaningless to perform calculations using perturbation theory as we have done for electromagnetic interactions earlier in Ch. 5. Theoretical predictions are therefore hard to obtain using Eq. (8.88).

□ **Exercise 8.9** The objects π_a appearing in Eq. (8.88) are not necessarily the pions with specific electric charges that appear in Eq. (8.86). Identify the charged and neutral pions in terms of π_1, π_2 and π_3.

8.9   Isospin breaking

Even though isospin symmetry is respected by strong interactions, it is quite obvious that it cannot be a good symmetry for all types of particle interactions. Electromagnetic interactions, e.g., are not isospin invariant. Making an isospin rotation, we can turn a proton into a neutron, but a neutron has a different electric charge and therefore does not have the same electromagnetic interactions as the proton.

A little thought confirms that this cannot be the only source of isospin violation in nature. Of course weak interactions also do not respect isospin, but the violations coming from it would be tiny. If electromagnetic interactions were the only dominant sources of isospin violation, the proton would have been heavier than the neutron, since Coulomb repulsion among the quarks in a proton is higher than the same quantity in a neutron. In reality, however, the neutron is heavier than the proton by about 1.3 MeV.

One possibility is that the d-quark is somewhat heavier than the u-quark. Since the quark content of the neutron is udd as opposed to the proton's uud, a heavier d-quark would certainly make the neutron heavier than the proton.

The credibility to this hypothesis is strengthened by looking at the Δ baryons. The average mass of the Δ's, as mentioned earlier, is about 1232 MeV. But there are minute mass differences between the four particles collectively known as Δ, signifying isospin violation. The Δ^{++}, which contains three u quarks, is about an MeV lighter than the Δ^+, where one of the quarks is d. Once again, this cannot be explained by electromagnetic interactions, and can be explained by a heavier mass of the d-quark compared to the u-quark.

This should not be taken to imply that the mass difference between the d and the u quarks is the mass difference between the neutron and the proton, or, for that matter, that between Δ^{++} and Δ^+. As already stated, electromagnetic interactions between the quarks also introduce a mass difference between the hadrons. Thus, we can write

$$m_n - m_p = (m_d - m_u) + (\text{electromagnetic contribution}). \qquad (8.89)$$

Since the electromagnetic contributions are negative, as discussed above, $m_d - m_u$ is expected to be somewhat larger than $m_n - m_p$.

8.10 Baryon number

So far, we have talked about the isospin SU(2) symmetry of strong interactions. We have not talked about the U(1) part that appeared in Eq. (8.4). Let us now discuss that.

As we said earlier, the U(1) part commutes with the SU(2) part. It means that the U(1) quantum number must be the same for all components of an SU(2) multiplet. In particular, we can consider the nucleon, which is an isodoublet introduced first in Eq. (8.2). The proton and the neutron should have identical charge under this U(1). This charge is called *baryon number*, which we will denote by B. Since the scale of any U(1) charge is arbitrary, we can set this charge to be $+1$ for both proton and neutron.

We know the electric charges of both proton and neutron. We know their I_3 values. Now we assign their baryon numbers. Obviously, this new number can be written as a combination of Q and I_3. The relation can be easily obtained, and it reads

$$Q = I_3 + \frac{B}{2}. \tag{8.90}$$

If we apply this formula to pions as well, we find that pions have $B = 0$. Anything with $B = 0$ is not a baryon. Indeed, we said earlier that pions are classified as mesons. Looking at the quark level structure, it is easy to see why it must be so: the mesons have quarks and antiquarks, so the baryon number cancels between them.

If we apply Eq. (8.90) for the up and down quarks as well, using their isospin assignment from Eq. (8.67) and their charges from Eq. (8.69), we find that

$$B = \frac{1}{3} \tag{8.91}$$

for both quarks. This makes sense, since nucleons have the quantum numbers of three quarks, and baryon numbers for quarks add up to give the baryon number equal to $+1$ for nucleons. Under the same token, pions, which have one quark and one antiquark, should have $B = 0$, because the baryon number of the quark must cancel with that of the antiquark.

More generally, consider a hadron made up of the u and d quarks and their antiquarks. Now define the following quantities:

$$N_u = (\text{number of } u\text{-quarks}) - (\text{number of anti-}u) \tag{8.92}$$

in the hadron, and similarly the quantity N_d. Then, obviously the total charge of the hadron is given by

$$Q = \frac{2}{3} N_u - \frac{1}{3} N_d, \tag{8.93}$$

and the I_3 of the hadron is given by

$$I_3 = \frac{1}{2} N_u - \frac{1}{2} N_d. \tag{8.94}$$

And of course the baryon number of the hadron is given by

$$B = \frac{1}{3} N_u + \frac{1}{3} N_d. \tag{8.95}$$

Eliminating N_u and N_d from these equations, we obtain Eq. (8.90). This means that Eq. (8.90) is true not only for nucleons and pions, but for any hadron made up of u and d quarks and their antiquarks.

□ **Exercise 8.10** Verify that Eq. (8.90) holds for the Δ baryons.

In Eq. (8.4), we talked about the invariance of strong interactions under the group U(2). Since baryon number is a part of it, it implies that baryon number is conserved. This principle of *conservation of baryon number* in fact holds for all interactions, as we will see.

One striking consequence of baryon number conservation is the absolute stability of the proton. The reason is that the proton is the lightest baryon. It cannot pass on its baryon number to a lighter particle. If anyone observes an instance of proton decay, that would signal baryon number violation. No one has ever seen anything to this effect.

We would like to make a comment about the expressions appearing in Eqs. (8.93), (8.94) and (8.95). As mentioned clearly, they were intended only for systems made from the two flavors of quarks and their antiquarks, and nothing else. We know quite well that the total electric charge of any system cannot be given by Eq. (8.93): there will be contributions from other fundamental particles like the electron. Similarly, baryon number also receives contribution to other flavors of quarks which will be discussed in subsequent chapters. The expression for I_3 is however the most general one, i.e., the only fundamental particles that carry isospin are the u and the d quarks. Thus, from the conservation of isospin in strong interactions, we can conclude that strong interactions conserve the combination $N_u - N_d$. In fact, strong interactions conserve more. The individual quark numbers, N_u and N_d, defined in the sense of Eq. (8.92), are conserved in strong interactions. This will be discussed later in §12.3 when we write the fundamental Lagrangian of strong interactions in terms of quark fields.

Notice that electromagnetic interactions cannot change N_u or N_d either. Fundamental electromagnetic interaction, whose theory was developed in Ch. 5, always contains a $\overline{\psi}$ for every fermion field ψ, or a $\phi^\dagger$ for every scalar field ϕ. Therefore, electromagnetic interactions cannot change the number of any fundamental particle. The case is the same with strong interactions: the number of each kind of fundamental particle (minus the corresponding antiparticle) is conserved in strong interactions.

Chapter 9

Discovering particles

In Ch. 1, we mentioned names of various leptons and hadrons. Some of the leptons appeared in our discussion of Ch. 5. In Ch. 8, some hadrons started making their importance felt. At this point, let us make an experimental interlude and ask ourselves how these particles were discovered. Of course we cannot perform an experiment on the pages of a book. We will do something that is next best: explain the principles that go behind the discovery of particles. As can be guessed, not all particles are discovered the same way. Here, we will outline some basic techniques only. Specifics for individual particles may be discussed in the later chapters as and when they are needed.

9.1 Discoveries of electron, proton and neutron

This section summarizes the prehistory of particle physics. Among the particles which are today recognized as elementary, the earliest one to be identified was the electron. In the first half of the nineteenth century, Faraday noticed a luminescence when an electric current was passed through two electrodes in a tube of rarefied air. The luminescence was termed *cathode rays*, and its origin was debated. Some people thought that the rays came from some kind of ions, while some others thought that the source was some kind of wave phenomenon.

In 1896, Thomson thought of identifying the nature of cathode rays by making a hole in the anode. Some of the particles coming from the cathode would escape through this hole. Thomson passed the escaping cathode rays through a crossed electric and magnetic field. On a screen placed at the far end, he could see that the ray was deflected by this process, implying that the ray consisted of charged particles. He found that the charge was negative. Moreover, from the amount of deflection, he could determine the ratio of charge and mass of the particle and found that it was three orders of magnitude larger than that of any known ion. He concluded that the cathode rays consisted of very light particles with a negative electric charge, and called them *electrons*.

□ **Exercise 9.1** Consider Thomson's experiment, with the electrons traveling in $+x$ direction. We find where it hits a screen placed on its path. Call this point P. If we apply an electric field $\boldsymbol{E} = E\hat{y}$, the electrons will hit the screen at a point displaced from P along the y-direction. Now we apply a magnetic field $\boldsymbol{B} = B\hat{z}$ and adjust its magnitude such that an electron hits the point P on the screen again. Show that the velocity of the electrons is given by

$$v = E/B. \tag{9.1}$$

Now we turn off the electric field so that the electron moves in a circular path. If the radius of this path is R, show that the charge to mass ratio of the electron is given by

$$\frac{|q|}{m} = \frac{v}{BR}. \tag{9.2}$$

[**Note :** *This is how Thomson found the ratio q/m for the electron.*]

No person's name is attached to the discovery of the proton. Once the existence of the electron was confirmed, it was obvious that an atom must also contain some positively charged particle, since an atom is electrically neutral. Geiger and Marsden published the results of their α-scattering experiments in 1909, which showed scattering through large angles. In a paper published in 1911, Rutherford demonstrated that the results of that experiment implied the existence of an atomic nucleus that is positively charged. It was then naturally concluded that the positively charged building blocks of atoms reside in the nucleus.

The presence of some electrically neutral particles in the nucleus was anticipated by the existence of nuclear isotopes. In 1932, Irene Curie and Frederic Joliot showed that some neutral radiation coming out of nuclei in certain processes cannot be gamma rays, i.e., electromagnetic radiation. Because the radiation was neutral, one could not determine the mass of the particle from its charge and its q/m ratio, as was done for the electron. In the same year, James Chadwick determined the mass of the particles emitted by these processes and found that the mass was very close to that of the proton. With this, the existence of the neutron was established.

Although this happened in 1932, the idea of the neutron was around for more than a decade by that time. Scientists started to believe that the electron, proton and the neutron are the only fundamental fermions in the universe. The proton and the neutron make the atomic nucleus, and the electrons move around the nuclei. Electromagnetic interactions are mediated by the photon. That is all there is to it! But this worldview was shattered almost as soon as it was born, with the advent of experiments with high energy particles.

9.2 New particles in cosmic rays

In §1.5, we explained why an exploration into smaller and smaller objects required the control of higher and higher energies. In the early part of the twentieth century, the kind of energy that could be produced in a laboratory was not enough to produce any of the unstable particles in a collision process.

In the beginning of twentieth century, it was observed that various detectors for charged particles, like the gold-leaf electroscope, produced a small but non-vanishing signal even when not near any known source of charged particles. At first it was thought that it was caused by radioactive substances in the earth's crust. However, Hess found that the flux increased in balloon-borne experiments sent high up into the atmosphere. It was then concluded that the signals came from processes taking place outside the earth, and were therefore named *cosmic rays*. In various astronomical environments, processes involving very high energies take place and produce particles. These particles, on hitting our atmosphere, produce secondary particles.

Cosmic rays gave scientists an opportunity for studying high energy interactions between particles to see what particles are involved. All that was needed were detectors. Various kinds of detectors were developed in the first half of the twentieth century. For example, Wilson developed the *cloud chamber* in the 1910s. It consists of a chamber where supersaturated vapor of water or alcohol is kept. The vapor cannot condense into liquid because there is nothing like a dust particle that will provide a seed for condensation. If a charged particle passes through the vapor, it ionizes the atoms along the path, and these ions can act as seeds. Condensation of the vapor occurs along the path, and the path has a foggy appearance. This can be photographed and the paths analyzed later. The bubble chamber, developed in 1952 by Glaser, utilized roughly the same kind of technique, except that a superheated liquid took the place of the supersaturated vapor. With the need for detectors with faster response and better resolution, detection techniques based on other properties of matter were also invented, like the scintillation counter, photographic emulsions and so on.

The first new particle discovered in cosmic rays was the *positron*. In 1932, Carl Anderson analyzed some cloud chamber photographs and found tracks of a positively charged particle whose charge-over-mass ratio had exactly the same magnitude as that for the electron. The absolute value of the charge-to-mass ratio was determined by measuring the radius of curvature in an applied magnetic field, which caused the charged particle tracks to be curved. In order to find the sign of the charge, Anderson put a lead plate in the chamber, as seen in Fig. 9.1. Any particle, after passing through the lead plate, would slow down, so that its radius of curvature would decrease. Seeing the tracks, he could tell which direction a particle's path bent in the magnetic field, thus determining the sign of its charge.

After Yukawa's theory of nucleon-nucleon interaction was proposed in 1935, a search for pions in cosmic rays was launched. Indeed, within a cou-

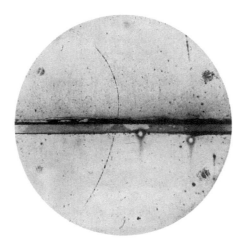

Figure 9.1: Discovery of the positron. The lead plate shows in the middle. The curvature of the track clearly shows that the particle went from the lower side to the upper side of the plate. [Reprinted with permission from: C. D. Anderson, Phys. Rev. **43** (1933) 491; © (1933) by the American Physical Society.]

ple of years of Yukawa's proposal, Anderson and Neddermeyer found some cloud chamber tracks which corresponded to particles which were much heavier compared to the electron but lighter than the proton. They thought that they had found the pion. However, on closer scrutiny, it was revealed that the properties of this new particle were rather similar to that of the electron. In particular, the particle did not have strong interactions and had the same charge as the electron. Later, in 1947, this particle was seen in the decay track of yet another new particle, somewhat heavier. It was then concluded that the heavier of these two new particles was the charged pion, and its decay gave rise to the other particle, which is now called the muon. The existence of the muon was completely unexpected from any theory at that time. When the muon's existence was finally confirmed, it still felt so much out of the scheme of everything else that was known at that time that I. I. Rabi asked his famous question about the muon: "Who ordered that?"

Discoveries continued. After the end of the second world war, the endeavor got boosts from accelerators in which controlled experiments could be performed. Let us introduce the accelerators before getting into a discussion of the newer and newer particles which were discovered.

9.3   Accelerators

There are limitations with cosmic rays: they cannot be used to search for new particles beyond a few hundred MeV. Not that there are no processes with energies higher than that range in cosmic rays. There are, but their numbers

Figure 9.2: The basic idea of a cyclotron.

fall with energy, and become small enough beyond a few hundred MeVs so that it is difficult to accumulate enough data to conclude the existence of new particles.

It was therefore necessary, for the advancement of the field, that particles be accelerated to high energies in a manner that their energy and flux can be controlled within a range. The simplest way of accelerating charged particles would be to let them travel through an electric field. Basically, this is what was done in devices called *generators*, like the Cockcroft–Walton generator or the van de Graaf generator. A huge potential difference was created and maintained in such machines. Particles injected into one part of the machine pass through the potential difference and come out at a higher energy. If the charge of the particle is q and it is passed through a potential difference V, the gain of energy would be qV.

Experiments were performed by accelerating particles in such machines, but the limitation of these machines was quickly realized. In order to produce particles with higher and higher energies, one needed higher and higher potential differences. It is difficult to maintain huge voltages. Even the best insulators break down beyond the megavolt range: sparks and breakdowns occur. It was desirable to accelerate particles to high energies without actually giving them a big push with a high voltage. Rather, if the same thing could be done by giving numerous pushes of smaller magnitude, one could reach high energies without having to deal with the perils of high voltage.

Lawrence found a way of doing this around 1930. The basic idea is simple. If a charged particle moves in a magnetic field, the force on it is $q\boldsymbol{v} \times \boldsymbol{B}$, which is perpendicular to the velocity. If the magnetic field $\boldsymbol{B}$ is homogeneous, the magnitude of the force is constant. If the initial velocity of the particle is perpendicular to $\boldsymbol{B}$, the particle moves in a circular path whose radius is given by Eq. (9.2). The time taken for the particle to go around half the

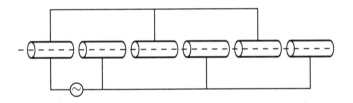

Figure 9.3: Schematic diagram of a linear accelerator.

circle would be

$$\frac{\pi R}{v} = \frac{\pi m}{|q|\,B}, \tag{9.3}$$

which is independent of v. Lawrence now made the breakthrough by contemplating a mechanism to provide a periodic kick to the particle to increase its speed. The result was his brainchild, the *cyclotron*. The first cyclotron was built in 1931.

The cyclotron consists of two D-shaped cavities, with a magnetic field perpendicular to the plane of the D's. There is a little gap between the two D's. The two D's are kept at different potentials, so that there is an electric field in the gap. Whenever a charged particle finishes a half circle in a D, it comes near the gap and gets a kick from the electric field there. The potential on the D's is alternating, with a frequency adjusted so that every time the particle comes near a gap, the electric field is in the right direction to accelerate the particle. As the energy of the particle increases, it travels in bigger and bigger circles. Thus, repeated application of the electric field increases the energy of the particle.

Lawrence's argument about the fixed time taken for every half rotation breaks down when relativistic effects become important. For this reason, and for others which will be clear shortly, cyclotrons are rarely used in present day high-energy experiments. They have been superseded by two different kinds of accelerating machines.

The first kind of machines are called *linear accelerator*s. In these machines, a charged particle passes through cylindrical cavities, with a small gap between successive cavities. Alternate cylinders are connected to be on the same potential, and all such cylinders are connected to an alternating source. The frequency of the source is so adjusted that each time a particle comes to the end of a cylinder, the potential difference between this cylinder and the next one is of the right sign in order that the electric field accelerates the particle.

The second kind of machines are called *synchrotron*s. Here the basic geometry is circular, like the cyclotron. The difference is that, whereas for a cyclotron the magnetic field is kept constant and the radius of the particle's path increases with energy, exactly the opposite occurs in a synchrotron. Particles orbit in a fixed circular ring, and the magnetic field is adjusted so that the particles stay on the same path. There are also magnets which focus the paths of the particles. The particles are accelerated by an electric field which is applied at certain segments of the path.

There are two big problems in obtaining high energies in synchrotrons. The first is the magnetic field available in the bending magnets. The second is that the charged particles in a circular path are always accelerating because the direction of their velocity is changing continuously. Any accelerating charged particle emits electromagnetic radiation, known as synchrotron radiation. For a particle of mass m moving in a circular path of radius R, the amount of power radiated in the form of radiation is proportional to $E^4/(m^4 R^2)$ when the particle's energy is E. Thus, both problems are ameliorated if one goes to bigger and bigger values of R. This is the reason bigger and bigger machines are built to explore higher and higher energies. Starting from hundreds of MeVs of energy in the 1950s, GeV scale energies were available already the 1960s. In the 1980s, protons could be accelerated to TeV (i.e., 10^{12} eV or tera-eV) energies.

There was another aspect of accelerator technology that helped us explore high energy effects. In the early days, experiments were carried out by accelerating a beam of particles and dumping it on a fixed target. In other words, the lab frame was the fixed target (FT) frame. In this case, only a fraction of the energy of the incident beam can be utilized to produce heavy particles in the final state. This is because the particles in the initial state have some net 3-momentum. Therefore, the particles in the final states will also carry the momentum, and will have some kinetic energy in addition to their mass energy.

□ **Exercise 9.2** Consider a scattering experiment $A + B \rightarrow C + D$, performed in the laboratory where the A particles are at rest. Suppose $m_C + m_D > m_A + m_B$. Show that the minimum energy E that the B particle must possess in order for this scattering to happen must satisfy the relation

$$E > \frac{(m_C^2 + m_D^2) - (m_A^2 + m_B^2)}{2m_A}. \tag{9.4}$$

To get a feeling for the magnitudes, let us look at Eq. (9.4). To make the exercise simple, let us consider that m_B and m_D are negligible compared to the other masses in this equation. Then, it shows that, in order to produce a C-particle through this reaction that is r times heavier than A, (i.e., $m_C = rm_A$), the incident B particle should have energies in excess of $\frac{1}{2}(r^2 - 1)m_A$. The seriousness of this threshold can be understood if we consider, say, $r = 10$. It shows that, in order to produce a particle 10 times heavier than A, we need energies about 50 times the mass of A. Most of the energy is wasted: it is not utilized in producing heavy particles.

This loss can be avoided if, instead of taking one of the initial particles to be stationary, we can make two particles collide head to head. This means that the lab frame should essentially be the CM frame, in which there is no net 3-momentum in the initial state. The final state particles can then be produced even with vanishing 3-momentum, which means that we won't waste energy by giving kinetic energy to the final particles.

In order to achieve this in an experiment, we need to produce not one but two oppositely moving beams. Two beams can be accelerated in opposite directions with the same set-up if the charges of the two types of particles are opposite and the masses equal, i.e., the two beams are composed of a particle and its antiparticle. Otherwise, different mechanisms for acceleration have to be installed for the two beams. Finally, the two beams are made to collide. The idea of such machines developed in the 1960s. At first, they were called *storage rings*, but the word *collider* has become more common gradually. Because of these machines, the task of producing high mass particles has become easier and more efficient. We will discuss some colliders in §9.8.

9.4 Detectors

Accelerators only accelerate particles and prepare them for a high energy reaction. One still needs detectors to find out what are the products of the reactions. Some forms of detectors like the cloud chamber or the bubble chamber have been discussed earlier. Some of these are still used, especially the bubble chamber. In addition, newer technologies have developed for detecting particles.

The particular type of detector, or detectors, used in an experiment depends not only on the energy range for which an experiment is intended, but also on the type of particles that one wants to detect. For detecting photons, photomultiplier tubes are widely used. For detecting charged particles, a variety of methods are employed. The principle of operation of a bubble chamber has been described already. In scintillation detectors, one uses organic or plastic material which generates optical photons when hit by an ionizing particle. Čerenkov radiation, described in §5.2, is also utilized for the identification of fast moving charged particles in a medium as well as for calorimetric measurements. Compressed gas is used as the medium usually, but water has been used as well. Since the refractive index of water is about 1.33, a particle will emit Čerenkov radiation if its speed exceeds three-quarters of the speed of light in vacuum.

There are also methods for detecting electrically neutral particles. Such detectors are, in a broad sense of the term, calorimetric, meaning that the method of detection hinges on the energy deposited in a detector when a fast particle passes through it. Details vary about how the deposited energy is measured. Of course, such methods can be used for charged particles as well. The materials comprising the detectors have to be chosen according to

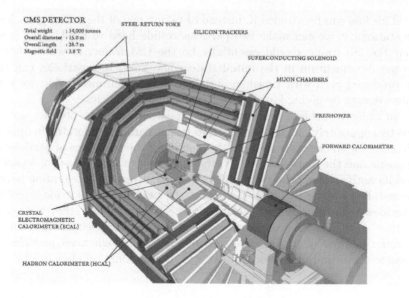

Figure 9.4: Section of the CMS detector at the large hadron collider. The detector is cylindrical, with small openings along the axis through which the two colliding beams enter. The interaction point is at the center of the cylinder. The overall size of the detector is indicated by drawing a person in the same scale. [Image: Tai Sakuma © CERN, for the benefit of the CMS Collaboration. Reprinted with permission.]

the needs, i.e., one has to make sure that the particles that one intends to detect lose their energy fast enough to be registered. For detecting electrons or photons, a light crystal, usually of sodium iodide, may be used if the energy is not very large. For highly energetic electrons or photons, one has to use metals of high atomic number, like lead or tungsten. For detecting hadrons, a material like iron or copper is used.

In one aspect, detectors used in high energy collider experiments are very different from those used in any other experiment. Think, e.g., of an experiment where one is trying to study the gas discharge spectrum from a certain element. One would take the gas in a discharge tube, would pass electricity through it, and would set up a detector some place away from the discharge tube to study the spectrum. This way, the experiment area subtends a small solid angle in the detector, and a lot of information from the experiment never reaches the detector. One can be nonchalant about this loss if a lot of data can be collected despite such loss, and the experiments can be repeated easily to obtain more. High energy experiments involve huge accelerators which are difficult to set up and are thus not many in number. Besides, in such experiments one also often studies very rare processes, for which the event count is very low. Some processes are not seen at all, and the experiment puts upper bounds on the rates of these processes. For these reasons, high

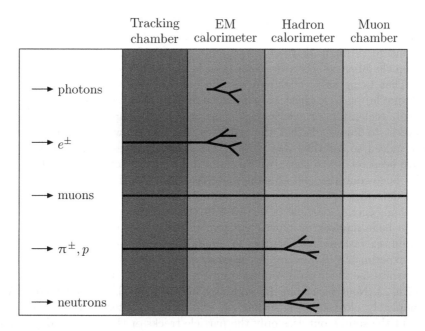

Figure 9.5: Parts of a modern detector, showing the functions of the different parts as explained in the text. Regions have not been drawn to scale.

energy experimentalists do not want to lose data. They, therefore, do not build small detectors which cover a small solid angle around the system under study. Instead, the detectors in high energy experiments *surround* the collision region of the experiment. For collider experiments, the detector is set up all around the interaction region, covering almost the entire 4π solid angle around the interaction point. It cannot possibly be the full 4π because one has to leave some room for the beam to enter the interaction point. But efforts are made to minimize the opening due to the beam path. In Fig. 9.4, we show the schematic diagram of one modern detector, surrounding the beam path completely, except for the small opening through which the beams come in.

In a modern detector, detection of different particles takes place in different coaxial cylinders carrying different kinds of detector materials. In Fig. 9.5, we have schematically shown the functions of different zones of the detector. The left end of the figure is the interaction region where the beams collide and different particles come out in the process. First they have to go through the tracking chamber, which records the paths of charged particles. After crossing this zone, the particles enter the electromagnetic calorimeter, where the electrons and the positrons are absorbed, and one can measure the energy deposited by them. Photons are also absorbed in this zone, but they are distinguished from the electrons and the positrons by the fact that they leave no track in the previous zone. Hadrons go through and enter the next zone, where they lose their energy by interaction with the detector material. The muons go through all these zones and are identified by their tracks in the muon chamber, which is the outermost layer of a detector.

That being said, it should also be mentioned that not *all* experiments relevant for particle physics are based on particle beams coming from a collider or an accelerator, or even nuclear reactors. For example, it will be of great interest to know whether the particles that we know as absolutely stable are indeed so, or they decay with very large lifetimes. Many experiments have been set up to check whether protons decay. Such experiments consist of a huge container full of some material which invariably contains protons, and detectors around the sample to see whether any of the protons decay. No high energy particle is involved in the experiment. In addition, there are a huge number of experiments where one tries to detect particles from various extra-terrestrial sources in somewhat modern-day versions of cosmic ray experiments. For example, there have been elaborate efforts to detect solar neutrinos, i.e., neutrinos coming from the sun, a topic that will be discussed in Ch. 22. Neutrinos from supernova explosions have also been detected. And completely old-fashioned cosmic ray experiments, where one does not care about the source but sees only the particle tracks or their energy deposition, are still performed. In such experiments, the detector is obviously not near the region of production of the particles.

9.5 Hadronic zoo

We said earlier that when the second world war ended in 1945, only a handful of particles were known. Apart from the electron, proton and the neutron, the muon was seen but not understood, and the positron was discovered in cosmic rays. Neutrinos were suggested in order to explain the continuous spectrum of electrons in nuclear beta decay and pions were suggested as carriers of strong interactions, but there was no experimental confirmation of their existence.

The situation started to change in the late 1940s, and by the end of the 1950s, a great many new hadrons were found. We have already described the discovery of pions. The antiproton was discovered in 1955. The pions were expected on the basis of Yukawa's theory, and the antiprotons on the basis of Dirac's theory of antiparticles, and the experiments confirmed the theoretical expectations. Neutrinos were detected in 1956, some details of which will be given in §9.7.

The first big taste of the unexpected came in 1947, the same year that charged pions were discovered, in the form of some particles which appeared always in pairs. It was conjectured that they carried a quantum number that the earlier-known particles did not carry. This quantum number was dubbed *strangeness*. The point was that, since the nucleons and pions did not have any strangeness, a strong interaction involving them cannot produce

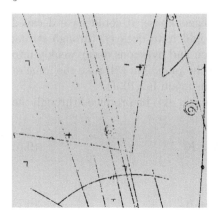 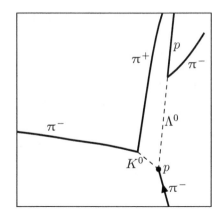

Figure 9.6: The picture on the left shows bubble chamber tracks originated by an incoming π^- hitting a proton in the target. [Reprinted with permission from the Lawrence Berkeley National Laboratory.] Some of these tracks have been shown in the diagram on the right in solid lines, with dashed lines to show the inferred neutral particles which have not left any tracks. The reaction at the lower right vertex is the reaction shown in Eq. (9.6).

just one particle carrying this quantum number. There must be two, carrying opposite values of strangeness, so that their contributions cancel. The first strange particles found were the *kaons* (earlier called K-mesons) which are spinless, and *hyperons* or Λ-particles which are spin-$\frac{1}{2}$ fermions, produced in reactions like

$$\pi^+ + n \to \Lambda + K^+ . \tag{9.5}$$

The charged particles were detected by their tracks, and the tracks of the neutral particles were reconstituted from their production point and decay point, where the decay particles left tracks. An example is shown in Fig. 9.6.

Here comes another important point. The tracks of these particles are long enough to be seen or reconstituted. In §9.6, we argue that this means that these particles do not decay via strong or electromagnetic interactions. Nevertheless, they decay, which means in addition that the property strangeness can be violated in weak interactions.

It is clear why the hyperons or kaons cannot decay through strong or electromagnetic interactions. The kaons have masses around 495 MeV. They are the lightest particles that carry any strangeness. Obviously, they cannot decay unless strangeness is violated. The case for hyperons is a little more complicated. They are certainly heavier than the kaons, so strangeness alone cannot prevent their decay. As mentioned earlier, hyperons are fermions. So, angular momentum conservation would demand that the decay product should contain some fermions. The hyperon mass is 1115 MeV, and nucleons are lighter. However, the mass is not heavy enough to produce a nucleon

and a kaon in the decay. With a much bigger mass, it could have decayed
into a nucleon and a kaon, and then it could have decayed through strong
interactions. However, that does not happen and hyperons decay weakly into
non-strange particles like the nucleon and the pion. We will take the issue of
decays of hyperons and many other hadrons again in §10.9.

The K^+ is not alone. A neutral particle can also be produced through the
reaction

$$\pi^- + p \to \Lambda + K^0 . \tag{9.6}$$

In fact, the reaction

$$\pi^0 + n \to \Lambda + K^0 \tag{9.7}$$

can also happen, although it would be difficult to observe because π^0 has a
very short lifetime. This K^0 forms an isodoublet along with the K^+, and
both of them have the same value of strangeness. The Λ, on the other hand,
proved to be an isosinglet. A convention was set such that the strangeness
(S) of K^+ is +1, so that Λ has $S = -1$.

More baryons with $S = -1$ were found soon. An isotriplet of baryons was
found, with masses around 1190 MeV. They were collectively denoted by the
symbol Σ. The three particles in the isomultiplet have electric charges equal
to +1, 0 and −1, so that their individual names were decided to be Σ^+, Σ^0
and Σ^- respectively. Like for the Δ baryons mentioned in §8.7.3, it should
be noted that the Σ^+ and the Σ^- are *not* antiparticles of each other. Both of
them carry one unit of baryon number.

Baryons with $S = -2$ followed. They were called the cascades, and denoted by Ξ (the Greek capital letter 'xi'). They form an isodoublet, consisting
of Ξ^0 and Ξ^-, with masses around 1320 MeV. The story of the discovery of
an $S = -3$ baryon will be discussed in Ch. 10.

In addition, there were discoveries of many hadronic states which did not
involve any novel quantum number like strangeness. The Δ particles were
discovered in the early 1950s. As mentioned in Ch. 8, they do not carry any
strangeness, and seem to be composed of the u and the d quarks, just as
the nucleons are. At least the Δ's differ from nucleons in their spin and
isospin (both of which are $\frac{3}{2}$); some of the hadrons looked tantalizingly similar to other hadronic states of lower mass. For example, isodoublet baryons
carrying no strangeness have been discovered around the mass of 1440 MeV,
around 1520 MeV, and 1535 MeV and so on. Similarly, non-strange $I = \frac{3}{2}$
baryons have been found at 1600 MeV, 1620 MeV, and so on. The list of
mesons show similar features as well. We have mentioned the ρ mesons in
Ch. 8. We also commented that such a variety of particles is expected in the
quark model because the quarks can have many different space-spin configurations within a hadron. There are many such strange hadrons as well, like
an isodoublet of mesons around 892 MeV which carry the same strangeness
as the kaons, although their spin is 1. The two particles in this isodoublet

are collectively called K^*. As for baryons, there are the isotriplet Σ^* particles and the isodoublet Ξ^*, having the same values of strangeness as the Σ and Ξ particles respectively.

The purpose of the section is not to give a complete catalog of each new hadron that has been detected so far. In fact, it is not even possible to do so, because there are literally hundreds of hadronic states which have been discovered. We have presented only a few here to give examples of new hadrons, and new quantum numbers that came with them. In Ch. 10, we will explain that the quantum number strangeness is related to the existence of a third flavor of quark, called the *strange quark*, apart from the up and down introduced in Ch. 8. Beginning in the 1970s, scientists discovered hadrons with three other quarks beyond these three. These quarks have been named the *charm*, the *bottom* and the *top*, as was mentioned in Ch. 1. Physics involving these three quarks will be discussed in Ch. 20.

9.6 Detecting short-lived particles

If a stable charged particle is created in a collision, one can see its track through the detector. Even for unstable particles, one can see the tracks if the tracks are long enough to be seen. For weakly decaying particles like the charged pion or the muon, the tracks are indeed quite long. Charged pions, for example, have a lifetime of the order of 10^{-8} s. If they travel at a speed close to the speed of light, they would travel a few metres and will leave a track. Even with a lower speed, the track will be appreciable. Muons have even larger lifetimes: their tracks are long.

This is not the case if the lifetime is much shorter. Hadrons have strong interactions. If a hadron can decay through strong interactions, its lifetime will be very short. To obtain an idea of the shortness, we can make a very rough order-of-magnitude estimate. In natural units, the decay rate (inverse of lifetime) has the dimensions of mass. Without giving much thought about anything else, we can write down a formula like

$$\Gamma \sim g^2 m \,, \tag{9.8}$$

where m is the mass of the decaying particle, and g is the strong coupling constant without which the particle would not have decayed. We ignored the masses of the decay products in order to obtain this naïve estimate.

The strong interactions are strong because the strong coupling is not small. If we use $g \sim 1$, the lifetime of a particle of mass of a few GeV turns out to be of order 10^{-23} s. Needless to say, within such a small time the particle cannot travel any distance for which a track can be seen in a detector. Of course a particle can be produced with an energy much larger than its mass so that its lifetime is longer in the lab frame because of time dilation. But even in the most energetic machines, the effect is not large enough to be helpful in this regard. For example, if a GeV-scale particle is produced with an energy of the order of $10\,\mathrm{TeV}$, the time dilation factor is about 10^4, and the particle can travel a distance of about $10^{-9}\,\mathrm{cm}$ in its entire lifetime.

Even if a hadron decays dominantly through electromagnetic interactions, the problem is the same. For electromagnetic interactions, the naive decay rate formula given in Eq. (9.8) should be modified: the factor g^2 should be replaced by e^2, or by the fine-structure constant α. Even after considering the fact that $\alpha \approx 1/137$, the lifetimes are too short to leave any track of any sort. It is therefore clear that hadrons which decay through strong or electromagnetic interactions cannot be detected through their tracks. Some indirect methods have to be applied, which we discuss now.

9.6.1   Resonances

Scattering cross-sections sometimes show a big increase in some energy range. Such phenomena are called *resonances*.

To understand why resonance happens and what it indicates, let us take a specific example of π^+ collision with protons. Starting from low momenta, suppose we keep increasing the incoming energy. In order to make the discussion simple, we consider the collision in the CM frame. The total 3-momentum in this frame is zero. As the energy increases, the cross-section changes in some manner that depends on the interaction dynamics. In the course of increasing the energy, we reach a point when the total energy, including the mass energies of the pion and the proton, is equal to the mass of the Δ^{++}. Certainly, at this energy, a Δ^{++} particle can be created at rest.

Once created, the Δ^{++} particle lives for a very short time, decaying to $p\pi^+$. Thus, the final result of the whole process is a proton and a pion, the same as the initial particles in the scattering. The process, with the intermediate production of the Δ^{++} and its subsequent decay, can therefore be thought of as elastic scattering between a proton and a pion. Our goal would be to understand the effect of the intermediate production of the Δ^{++} particle on the elastic scattering cross-section.

We can think of the proton-pion interaction to be mediated by an intermediate Δ^{++}. If we now try to write down the amplitude of the scattering process using the normal prescriptions that were outlined in Ch. 4, the amplitude will involve the propagator of the Δ^{++}. Our experiences so far show that the propagator of any particle of mass m contains a denominator of $(p^2 - m^2)$. When the incoming energy of the proton and the pion is equal to the mass of the Δ^{++} in the CM frame, this denominator vanishes since the total 3-momentum is zero. It thus seems that the amplitude blows up at this energy.

Of course, an amplitude cannot become infinite. We seem to be staring at infinity because we have not considered the propagator properly. When we wrote down expressions for propagators of different fields in Ch. 4, we tacitly assumed that the particles corresponding to those fields are stable particles. The Δ^{++} that we are encountering here is certainly not stable. If it were, i.e., if it had not decayed after being produced, we would not have obtained an elastic scattering between the proton and the pion. In that case, the Δ^{++} would have been the final state particles, and its propagator would not have been necessary.

To see the behavior of propagators of unstable particles, consider a scalar field $\phi(x)$ for the sake of simplicity. The Klein–Gordon equation will not describe such a particle, because that would give plane wave solutions, for which $|\phi(x)|^2$ is constant. When the particle decays with a lifetime Γ in its rest frame, the probability should go like

$$|\phi(x)|^2 \sim \exp\left(-\frac{m}{E}\Gamma t\right) \tag{9.9}$$

in a frame where it has an energy E. The factor E/m present in this expression is the usual time dilation factor for the lifetime, which is $1/\Gamma$ in the rest frame. This suggests that we should have solutions of the form

$$\phi(x) \sim \exp\left(-iEt + i\boldsymbol{p}\cdot\boldsymbol{x} - \frac{m}{2E}\Gamma t\right) . \tag{9.10}$$

It is straightforward to check that this implies a differential equation

$$(\Box + m^2)\phi(x) = im\Gamma\phi(x), \tag{9.11}$$

ignoring a term that contains Γ^2, the cause for which will be given shortly. Eq. (9.11) can be obtained from a Lagrangian

$$\frac{1}{2}(\partial_\mu\phi)(\partial^\mu\phi) - \frac{1}{2}(m^2 - im\Gamma)\phi^2 . \tag{9.12}$$

Using the method outlined in §4.10, we then find that the propagator of the unstable particle is given by

$$\frac{i}{p^2 - m^2 + im\Gamma} . \tag{9.13}$$

Before proceeding, let us discuss why the Γ^2 term was neglected in writing Eq. (9.11). The term would have been $(m\Gamma/2E)^2\phi$ in the equation, and would be largest in the rest frame where $E = m$. Denoting by T_1 and T_2 these two terms which are respectively linear and quadratic in Γ, we see $T_2/T_1 = \frac{1}{4}\Gamma/m$. Now, if a measurement of the energy of the particle has an uncertainty ΔE, according to the time-energy uncertainty relation the time Δt required for making the measurement must satisfy the relation $\Delta t \Delta E \gg 1$. For an unstable particle we need $\Delta t < 1/\Gamma$ because otherwise the particle will disappear, and $\Delta E < m$ in the rest frame where the energy is m. This gives $m/\Gamma \gg 1$, which shows that $T_2/T_1 \ll 1$.

The exact expression for the propagator is different for particles with different values of spin, but all of them contain a factor of the expression given in Eq. (9.13). Going back to the example of proton-pion scattering, we can now say that such a factor is present in the amplitude, where m is the mass and Γ is the decay rate of the Δ^{++}. In the CM frame of the proton and the

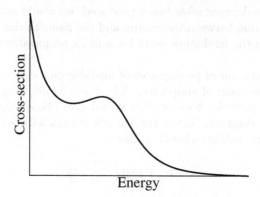

Figure 9.7: Schematic nature of variation of cross-section with energy near the mass of an unstable intermediate particle.

pion that we have been considering for the sake of simplicity, the amplitude therefore has the form

$$\frac{a(E)}{E^2 - m^2 + im\Gamma} + b(E). \tag{9.14}$$

Here, $a(E)$ represents all other factors that appear in the amplitude coming from the intermediate Δ^{++} contribution, and $b(E)$ represents amplitudes coming from all other kinds of intermediate states.

The cross-section contains the absolute square of the amplitude, and so

$$\sigma \propto \frac{a^2 + 2ab(E^2 - m^2)}{(E^2 - m^2)^2 + (m\Gamma)^2} + b^2, \tag{9.15}$$

if a and b are real. When E is not close to m, the b^2 term dominates. As E approaches m, the contribution from the a term increases, and becomes maximum at $E = m$. The schematic nature of the variation of cross-section with energy is shown in Fig. 9.7 with some arbitrary choice of the parameters and functions that appear in the expression for cross-section. As we see, because of the finite value of the lifetime, i.e., non-zero value of Γ, the cross-section does not really blow up to infinity, but rather shows a hump. Such humps are called *resonances*.

Now let us look at the whole problem from an experimental point of view. Suppose we measure the cross-section for proton-pion scattering and see its variation with the CM energy. We see that near 1232 MeV, the cross-section shows a hump. We would then conclude that there is an unstable particle with mass 1232 MeV that can decay to proton and pion. In fact, this is exactly the way that the Δ particles and numerous other hadrons were discovered.

At first people were reluctant to call these humps 'particles'. The name 'resonance' seemed appropriate. However, gradually it was realized that there is no good reason for not calling them particles. A particle has a definite mass:

so does the peak of the resonance, which can be taken as its mass. Certainly the resonance has a definite value of the electric charge, equal to the sum of the electric charges of the particles whose scattering produces the resonance. A hadron is expected to have a definite value of isospin. Comparison of the decay rates of the resonances, like those shown in Eq. (8.85, *p 223*), confirmed that the resonances indeed have a specific value of isospin. So there is really no difference in the properties of a particle and a resonance: only that a resonance can be a very short-lived particle. The width of the resonance indicates the lifetime of this particle. One can study the properties of such particles by studying the scattering phenomenon which shows the resonance.

One might wonder why the hump is bell-shaped rather than like a spike. Naively, it might seem that the resonance particle will be produced only for a particular value of the total energy of the initial particles. For example, if we consider the process in the CM frame of the colliding particles, this particular value should be equal to the mass of the resonance, since the resonance will be produced with zero 3-momentum. True. But the point is that the time-energy uncertainty relation tells us that the mass cannot be defined with an accuracy better than the inverse lifetime, in natural units. This is the reason for the width of the hump, and this is also the reason that the width of the hump gives the lifetime of the resonance. The width of the hump, in energy units, is often called the *decay width* for the same reason.

9.6.2   Reconstruction of events

Consider the following situation. In a scattering experiment, we see events where four particles (say, A, B, C and D) are produced. To make things easy, let us assume that each product particle is charged and lives long enough so that we can see its track and determine its energy and 3-momentum. We also notice the following feature of the particles produced. We take the particles C and D, and calculate what the momenta and energies of these two particles would be in a moving frame in which the total 3-momentum of these two particles is zero, i.e., in the center-of-mass frame of these two particles. In this frame, we calculate the total energy that these two particles have, and find that its value is the same in all scattering events of the type described.

If the initial states directly produced the four particles A, B, C and D, such a coincidence could never have occurred. Consider, however, the following chain of events:

$$\text{(initial state particles)} \longrightarrow A + B + X$$
$$\downarrow$$
$$\overbrace{C + D}. \qquad (9.16)$$

In other words, the scattering process really produced the A and B particles and another particle X. This X is very short-lived, so we cannot see its track and can only see its decay products, which are C and D. In this case, the total

energy of the C and D particles in their CM frame should be equal to the mass
of X. Thus, by measuring the energies and momenta of the decay products,
we can infer whether a short-lived particle was produced in the intermediate
state even though we don't see the particle directly, and can also deduce the
mass of the particle.

> □ **Exercise 9.3** *Suppose in an experiment we find two particles whose
> energies are E_1 and E_2 in the lab frame, and the 3-momenta are p_1
> and p_2 (not necessarily along the same line). What would be the total
> energy of these two particles in a frame in which they have equal and
> opposite 3-momenta?*

In practice things are often not so simple. For example, the final state
might contain not three but more particles. All of them may not be charged,
so we may not see the tracks of all decay particles. Also, if some of the particles
are identical, that would cause problems. For example, in the example above,
suppose C and A denote the same particle. This means that we got an A
particle from the scattering experiment directly, and another A particle from
the decay of X. As we look at the final tracks, we have no way of knowing
which A track came from the decay and which one from the earlier stage. In
this case, we will have to perform the analysis involving each AD pair. There
will now be a distribution of energies (by 'energy', we mean the total energy
of the AD pair in a frame in which their total 3-momentum is zero). However,
if we plot the number of pairs obtained with any given energy, there would
be a peak in the plot where energy equals X mass.

9.7   Discovering leptons

As mentioned in Ch. 1, there are three known charged leptons, and three
associated neutrinos. Among the charged leptons, the story of the discovery
of the electron and the muon have already been described in this chapter.
The story of the tau will be taken up later in this section, after we describe
the neutrinos.

Leptons do not have strong interactions. Neutrinos, among the leptons,
are electrically neutral and therefore do not have any electromagnetic interaction either. Thus the only interaction they experience is the weak interaction.
Earlier in §9.6, we described problems in detecting strongly interacting particles: some of them which decay strongly decay very fast. Weakly interacting
particles cannot possible have this problem, but they pose problems of a different kind. Their scattering cross-section is so small with any kind of detector
material that they escape almost unscathed through any detector. To get a
feel for the magnitudes involved, consider neutrino interactions with nucleons.
We will derive formulas for the cross-section in Ch. 14. If 10 MeV neutrinos
fall upon nucleons at rest, the cross-section is of order $10^{-43}\,\mathrm{cm}^2$, about 19
orders of magnitude smaller than the Thomson cross-section, introduced in
Ex. 5.9 *(p 129)*, which gives a benchmark value for electromagnetic scattering

cross-section. With the cross-sections that they have, the mean free path of neutrinos through matter of normal densities is of the order of 10^{18} to 10^{19} cm, about a million times larger than the earth-to-sun distance.

This is the main reason why neutrinos remained undetected for about a quarter of a century after being proposed theoretically by Pauli. Finally, with the advent of nuclear reactors, one could attempt to detect them in the 1950s. Since reactors produce enormous numbers of antineutrinos, it was hoped that, despite the feeble probability of interaction with the detection material, a few neutrinos would in fact interact with their detector material and therefore be detected. And this turned out to be right — from about 10^{20} antineutrinos that are produced in a reactor per second, Reines and Cowan succeeded in detecting about 2.88 per hour at an average.

The signal by which they performed their detection was indirect, but ingenious. According to Pauli's hypothesis, a nuclear beta decay process involves three particles in the final state. Essentially, a neutron in a nucleus decays:

$$n \to p + e + \widehat{\nu}_e, \tag{9.17}$$

the decay products being a proton, an electron and a particle that is now called the electron-type antineutrino. Because of crossing symmetry, one can also contemplate the scattering process

$$p + \widehat{\nu}_e \to n + e^+, \tag{9.18}$$

which is often called the *inverse beta process*. Thus, if antineutrinos are allowed to pass through a tank of water, a hydrogen nucleus (which is nothing but the proton) will occasionally interact with the antineutrino to produce a neutron and a positron. The task of detection is complete if one can detect the neutron and the positron.

A positron, once it is produced, quickly undergoes pair annihilation with an electron within the detecting material, releasing energy in the form of gamma rays. To detect the neutron, Reines and Cowan kept cadmium in their detector. This metal absorbs neutrons very efficiently by going to an excited state, and relaxes by emitting a photon:

$$n + {}^{108}\text{Cd} \to {}^{109}\text{Cd}^* \to {}^{109}\text{Cd} + \gamma. \tag{9.19}$$

Thus, once the inverse beta reaction occurs, there will be two spurts of energy, once from the positron annihilation and once from the neutron absorption. From the rates of these processes, one can estimate the time difference between the two spurts. If one sees two spurts of energy separated by that length of time, that would constitute a signature for the antineutrino. That is what Reines and Cowan succeeded in doing.

By the time this detection experiment was performed, muons and pions had been identified. The decay of π^+ to positron and neutrino is very suppressed, for reasons that will be elaborated in §17.5.1. The dominant decay channel turned out to be μ^+ and a neutral fermion, implying that the muon is also accompanied with a neutrino. It was not known whether this neutrino was different from the neutrino that accompanies the electron. In 1962, a group led by Lederman, Schwartz and Steinberger showed that neutrinos produced from decays of charged pions in flight cannot induce inverse beta reactions to produce electrons. This proved conclusively that the muon and the electron are not associated with the same neutrino. The symbols ν_e and ν_μ have been employed to denote the respective neutrinos ever since then.

Thus the list of elementary particles came to contain two charged leptons and two associated neutrinos. The count changed in 1975, when Perl and his team observed that in e^+e^- collisions with CM energy exceeding 3.5 GeV, there were final states containing muon without any antimuon, or the opposite. In other words, starting from the initial e^+e^- beams, they found final states where μ^+e^- or μ^-e^+ were the only charged particles, and these two options seemed to have equal branching ratios. The explanation of this phenomenon was that in the collisions a third kind of charged lepton was produced along with its antilepton. This new lepton, named τ^-, decayed with a lifetime short enough so that its track could not be seen. The τ^- decay can occur through electrons or muons, and some neutrinos. The τ^+ decay would occur to e^+ or μ^+. If the τ is much heavier than the muon, the phase spaces going to e or μ final states would be roughly equal. In that case, the decay of τ^- would produce e^- or μ^- with equal probability, and similarly the decay of τ^+ would produce e^+ or μ^+ with equal probability. Overall then, charged leptons and antileptons coming from the decay of a $\tau^+\tau^-$ pair can be e^+e^-, μ^+e^-, μ^-e^+ or $\mu^+\mu^-$, with roughly equal probability. Of course $\mu^+\mu^-$ can be produced directly from e^+e^-, and there are e^+e^- pairs can be produced in elastic scattering of the original beam. But μ^+e^- or μ^-e^+ pairs cannot be produced otherwise, so observation of such final states with equal probability indicated the existence of a third charged lepton, the τ.

With the knowledge that neutrinos associated with the muon and the electron are different, it was natural to assume that the τ-lepton has a separate neutral particle, the ν_τ, associated with it. There was indirect evidence for this belief that came from multiple sources. First, such a neutrino was required to cancel gauge anomalies, a subject that will be discussed in Ch. 18. The Z boson decay width came out to be consistent with three neutrino species that the boson can decay to, an issue that will be discussed in some detail in Ch. 19. There were also the neutrino oscillation experiments, to be described in Ch. 22. Such experiments, with neutrinos from different kinds of sources, show that there are two different non-zero mass differences among the neutrinos. Certainly, this needs at least three different neutrinos. The direct experimental proof of the existence of ν_τ came about a quarter of a century after the discovery of the τ-lepton, in a paper by the DONUT collaboration published in 2001.

There is some reason to believe that there isn't any other neutral fermion similar to these three varieties of neutrinos. The biggest reason comes in the form of the decay rate of the Z bosons, one of the bosons that mediate weak interactions. One can measure its total decay width as well as its decay rates to all charged fermion-antifermion pairs. There is a mismatch, to be described in more detail in Ch. 19, which can be accounted for if there are three different neutral channels to which the Z boson can decay. Two such channels are provided by the ν_e and the ν_μ, and the third one should be ν_τ. If there are any more neutral leptons, they must either be heavier than $\frac{1}{2}M_Z$, or have very little coupling with the Z boson, so that they do not contribute to the decay width of the Z.

9.8   Overview of particle physics experiments

In this section, we describe different kinds of experiments that are currently in operation, or had been in operation in the recent past, which aim at understanding basic properties and interactions of fundamental particles. As might be suspected, this will involve some overlap with the material presented in earlier sections of this chapter.

Broadly speaking, all experiments related to particle physics can be divided into two categories:

1. Non-accelerator experiments

2. Accelerator experiments

We discuss them separately.

9.8.1   Non-accelerator experiments

This category of experiments can be divided into two subclasses: those that do not involve any high energy particle at all, and those that involve some high energy particle in the initial state. The phrase "high energy particle" in this context means that the kinetic energy of the particle is greater than, or at least comparable to, the particle's mass energy.

Let us first describe some of the experiments which do not involve any high energy particle at all. These are usually experiments to test various conservation laws. For example, we have talked about the conservation of baryon number in §8.10, and said that it implies that the proton is absolutely stable. To check whether this number is really conserved, we can test whether the proton decays. This type of experiment is carried out by taking a huge container or tank of matter and putting detectors all around, trying to detect whether any signal of possible decay channels are seen. Many such experiments were set up beginning in late 1970s, when the grand unified theories made their appearance and predicted finite proton lifetimes, a topic that will be discussed in Ch. 23. In another example of this sort, lepton number violation is searched by trying to observe whether neutrinoless double beta decay occurs in certain nuclei. The importance of this process will be described in some detail in Ch. 22.

The second sub-class of non-accelerator experiments are done with cosmic rays. This is still a vigorous area of research, although there has been a shift in the emphasis since the early days of cosmic ray research. For energies which are accessible to accelerators, cosmic rays are not very interesting. However, sources outside the earth can provide particles with energies beyond the reach of present day accelerators. Very high energy photons, or high energy neutrinos, can tell us about fundamental interactions at very high energies. They can also be used to investigate possible sources of such high energy processes in the universe.

Sometimes, the table is turned around in such investigations. For example, since the late 1960s many experiments were set up to detect neutrinos coming from the sun. The basic aim was to test models of the sun. Discrepancies were observed between solar model predictions and the number of detected *solar neutrinos*. Finally, it was found out that the key to the discrepancy lay in some properties of neutrinos. The story will be described in Ch. 22.

9.8.2   Accelerator experiments

Accelerator experiments are of two types, as has been indicated earlier. In one type, there is a fixed target, and it is hit by some kind of fast-moving particle. In fact, this was the only kind of accelerator experiment until storage rings, or colliders, were invented. In a collider, two opposite moving beams of particles are made to collide.

The colliding beams must be beams of charged particles so that they can be accelerated to high energies. The particles must be stable, at least in the time scale needed to accelerate them, because otherwise the beams would disappear through decay. An easy choice of stable charged particle is the electron. Indeed, electron–positron colliders have been used for many experiments, and many particles have been discovered through such experiments. Another popular choice is the proton, for obvious reasons.

The machine at CERN which discovered the W and the Z bosons, carriers of weak interactions, was called the *SpS*, or the *super proton synchrotron*. It was a proton-antiproton collider, about 7 km long. After it discovered the W and Z bosons, a more energetic machine called the *LEP*, or the *large electron–positron* collider, was built, which had a circumference of about 27 km. It made accurate measurements on the masses and decay widths of the bosons. The mark of TeV as the CM energy was crossed by the machine called *Tevatron* at Fermilab. This was also a proton-antiproton collider, and the top quark was discovered at this machine. The most energetic machine built so far is the *LHC* or the *Large Hadronic Collider*, which uses the LEP tunnel that is 27 km long. As the name suggests, it is a hadronic collider where two proton beams are collided. There are plans for building an *International Linear Collider* or *ILC*, which will be a electron–positron machine. It is proposed to be between 30 to 50 km long, more than ten times longer than the longest linear accelerator that exists now, which is a 3.2 km long machine at the Stanford Linear Accelerator Center (SLAC).

Note that the LHC machine has two proton beams, unlike the Tevatron which is a proton-antiproton collider. The reason for this will be explained in Ch. 13.

It seems that, while making plans for building bigger and bigger machines, one alternates between electron–positron and hadronic projectiles. There is a reason behind it. Hadronic colliders have an advantage that protons are roughly two thousand times heavier than the electrons. Thus, in order to obtain colliding beams where the CM energy of each particle should have a certain value E, the Lorentz factor E/m needed for the protons will be much less than that needed for an electron. Put another way, using the same resources, a proton-antiproton pair can be raised to much higher CM energy compared to an electron–positron pair.

This advantage is counterbalanced by the fact that, as mentioned already in Ch. 8, protons are not fundamental particles, while the electrons are. Protons are composed of quarks. From a fundamental point of view, when a proton collides with an antiproton, the collision really occurs between a quark in the proton and an antiquark in the antiproton. In fact, protons have a substantial share of other fundamental particles in their constituents, like the gluon, as will be shown in Ch. 13, and the collision can also take place between such constituents. We do not know the momenta of individual quarks and gluons inside the proton. The bottom line is that, when a proton-antiproton collision occurs, we do not really know the momenta of the fundamental constituents which are actually experiencing the collision. On the one hand, it can be seen as an advantage: collisions occur for a broad range of momenta of the constituent particles. But on the other hand, this fact introduces a lot of uncertainty in interpreting the final state particles. Conversely, there is no such problem in an electron–positron collision. We know precisely the energy of the electron and of the positron which are undergoing collision.

Because of the advantage of hadronic colliders mentioned above, it is easier to break energy barriers with them. Thus, exploration into higher energy always starts with hadronic machines. Once some new particle is found with the hadronic machine, an electron–positron machine is used to make accurate measurements on it. This is why the two types of machines are called *discovery machines* and *precision machines* respectively, and this is why the two types of machines are alternately built.

There are also machines that use nuclei, rather than single electrons or protons, for acceleration. A dedicated machine of this sort is the *RHIC*, or *relativistic heavy ion collider*, at the Brookhaven National Laboratory. The LHC can also use heavy ions. Collisions between heavy ions involve many particles at a time, and one can obtain glimpses of collective behavior of particles in such experiments.

Earlier, we mentioned that if a hadron decays through strong or electromagnetic interactions, its path will be too short to be observed. If a hadron decays via weak interactions, this is not in general the case. For example, the charged pions have a lifetime of the order of 10^{-8} s. The Λ baryon has a lifetime that is a few times 10^{-10} s, so that even it can travel a few centimetres before decaying. However, as we go to study heavier and heavier hadrons, this advantage quickly fades away. For example, B^0 mesons, which will be discussed in detail in Ch. 20, have a mass of 5.28 GeV. Partly because of this mass which is much larger than that of the Λ, the lifetime of this particle is much shorter, about 1.5×10^{-12} s, even though the decay occurs through weak interactions. Obviously, if the B^0 particles are produced almost at rest, the amount of path they travel in the entire lifetime would be too small to be seen.

This was a problem in studying the properties of B mesons. In electron–positron colliders, one produces *bottomonium* states, which are $b\widehat{b}$ states denoted by Υ. The ground state energy of the quark-antiquark pair is not enough to produce the pair of mesons B^0 and $\widehat{B}^0$. However, there is an excited state, called $\Upsilon(4S)$, whose mass is just above the mass of the B^0-$\widehat{B}^0$ pair. Thus, if $\Upsilon(4S)$ is produced at rest, or nearly at rest, the B^0-$\widehat{B}^0$ pair obtained from its decay would also be nearly at rest, and hence the problem with the short lifetime. It was therefore proposed that an *asymmetric collider* be built, where the electron and the positron beam would have very different energies, and therefore different magnitudes of momentum. Two such facilities started operating around the year 2000, one in SLAC and one in the KEK laboratories in Japan. In the SLAC facility, the electrons are accelerated to an energy of 9 GeV whereas the positrons in the beam have an energy of 3 GeV each. Thus the momentum imbalance is 6 GeV whereas the total energy is 12 GeV, and therefore the $\Upsilon(4S)$ meson is produced with a velocity of 6/12 or half the speed of light. The resulting B^0 and $\widehat{B}^0$ mesons also have roughly the same speeds. In the lab frame, their lifetimes look longer because of time dilation, and hence the track can also be longer and observable. The two detectors mentioned have unearthed a lot of data, to be discussed later in the book.

We should mention that smaller machines are not rendered useless with the advent of higher energy machines. The smaller machines act as injectors, or pre-accelerators, to higher energy machines. For example, the SpS collider at CERN has not been made obsolete with the installation of LHC. Rather, the particles are pre-accelerated through SpS and then injected into the main LHC tunnel. This chain, or hierarchy, goes on much further. At the beginning of this chapter, we mentioned van de Graaf generators and Cockcroft–Walton generators, which cannot be used these days to obtain the highest of energies at which experiments are performed. However, they are still used in the first stage of acceleration in almost all experiments. Once the projectiles are accelerated through these generators, they are injected into other parts of the machinery where further acceleration is obtained through more sophisticated means.

The effectiveness of an accelerator depends not only on the energy that can be obtained with it, but also on the number of particles that can be stuffed in the beams. The relevant physical parameter in this regard is called the *luminosity*, which gives the number of particles in each beam that crosses a unit area in unit time. Sometimes it is called *instantaneous luminosity* in order to distinguish it from integrated luminosity, which will be introduced presently. Every machine is built with the plan of achieving a certain luminosity, which is called the *design luminosity*. The machine first starts operating with a lower luminosity, and then the luminosity is gradually increased toward the design luminosity. For example, the design luminosity of the LHC is $10^{34}\,\mathrm{cm}^{-2}\,\mathrm{s}^{-1}$, meaning that each beam contains a flow of 10^{34} protons per square centimetre per second.

Certainly, the longer the beam is kept on, the more data is obtained. Thus, the amount of data collected in a machine is given by its *integrated luminosity*, which is nothing but the time integral of the luminosity at which the machine operates. Consider, e.g., that the LHC operates at its design luminosity for $10^6\,\mathrm{s}$. The integrated luminosity is then $10^{40}\,\mathrm{cm}^{-2}$. Experimentalists often use the unit *barn* for denoting cross-sections, the definition being

$$1\ \mathrm{barn} = 10^{-24}\,\mathrm{cm}^2\,. \tag{9.20}$$

The unit is motivated by the Thomson cross-section given in Eq. (5.77, *p 129*), whose magnitude is of this order. Thus,

$$1\ \mathrm{fb} = 10^{-39}\,\mathrm{cm}^2\,, \tag{9.21}$$

where 'fb' denotes 'femtobarn', the prefix 'femto' implying 10^{-15}. An integrated luminosity of $10^{40}\,\mathrm{cm}^{-2}$ will therefore be usually referred to as $10\,\mathrm{fb}^{-1}$. As said above, the integrated luminosity gives a measure of the amount of data accumulated in an experiment. For example, if the data accumulated is $1\,\mathrm{fb}^{-1}$, it means that if some process has the cross-section of 1 fb, there will be at an average one such event in the accumulated data.

Chapter 10

SU(3) quark model

We introduced two quarks in Ch. 8, and discussed hadrons which are made up of these two quarks. We also saw how an SU(2) symmetry, viz., the isospin symmetry, helps us understand various properties of these hadrons. At the time when the idea of quarks was introduced, there already existed some hadrons which could not be understood with only the u and the d quarks. So the quark model was born with three types of quarks. We begin this chapter with the reasons for assuming the third quark, and discuss the physics of different hadrons made from the three types of quarks.

10.1 Strange quark

In Ch. 9, we talked about the discovery of a property called strangeness that some hadrons possess. This property is conserved in strong and electromagnetic interactions. Electromagnetic interactions are obtained by replacing the derivatives occurring in the free Lagrangian of a field by $ieQA_\mu$, and hence such interactions cannot contain two different fields other than the photon field. As a result electromagnetic interactions cannot change one particle into another. Strong interaction cannot change flavors of quarks as well, as has been hinted earlier and will be explicitly seen when we write down the basic Lagrangian of strong interactions in Ch. 12. So, probably the easiest way to understand the conservation of strangeness quantum number is to imagine that this quantum number is carried by a new particle. Since the strangeness quantum number occurs in hadrons, it is natural to imagine that the new particle needed to understand this conservation law is a quark, like the u and the d quarks that constitute the nucleons and pions. This quark is called the *strange quark*, and is denoted by s. This is then a third type of quark. Using a word that was introduced in §8.7, we can say that it is a third *flavor* of quark.

The K^+ and the K^0 constitute an isodoublet, as mentioned in Ch. 9. It is natural to assume that the isospin property comes from the u and the d quarks introduced in §8.7. Mesons, as seen earlier, have the composition of a quark and an antiquark. Thus we can write

$$\left|K^+\right\rangle = u\widehat{s}, \qquad \left|K^0\right\rangle = d\widehat{s}. \tag{10.1}$$

Their antiparticles will contain an s-quark:

$$\left|K^-\right\rangle = \widehat{u}s, \qquad \left|\widehat{K}^0\right\rangle = \widehat{d}s. \tag{10.2}$$

From the charges of the mesons, it is clear that the electric charge of the s-quark should be $-\frac{1}{3}e$, i.e., exactly equal to that of the d-quark.

The quark compositions of the kaons and pions are quite similar. The similarity does not end there. Both kinds of particles are spinless. Moreover, measurement of parity of the particles shows that the kaons have odd parity, just like the pions. In a succinct notation, one says that both pions and kaons have $J^P = 0^-$, where J stands for the spin and the subscripted sign gives the intrinsic parity of the particle.

10.2 Hypercharge

10.2.1 Hypercharge and charge of hadrons

Clearly, the charge formula of Eq. (8.93, $p\,226$) does not hold for kaons. The kaons are quark-antiquark states, so that their baryon number is zero. But K^+, for example, has $I_3 = +\frac{1}{2}$.

However, it is also easy to see that a little modification of the formula will make it applicable to kaons as well. Consider a hadron made up of nothing but u, d and s quarks, and their antiquarks. Now define the following quantities:

$$N_u = \text{number of } u\text{-quarks} - \text{number of anti-}u \tag{10.3}$$

in the hadron, and similarly the quantities N_d and N_s. Then, obviously the total charge of the hadron is given by

$$Q = \frac{2}{3}N_u - \frac{1}{3}N_d - \frac{1}{3}N_s. \tag{10.4}$$

As for the I_3 value, notice that the strange quark and its antiparticle do not carry any isospin, so that I_3 of the hadron is given by

$$I_3 = \frac{1}{2}N_u - \frac{1}{2}N_d. \tag{10.5}$$

And of course the baryon number of the hadron is given by

$$B = \frac{1}{3}N_u + \frac{1}{3}N_d + \frac{1}{3}N_s, \tag{10.6}$$

since all quarks carry a baryon number of $\frac{1}{3}$ and all antiquarks $-\frac{1}{3}$.

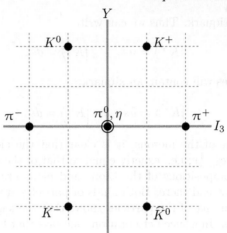

Figure 10.1: Isospin and hypercharge of $J^P = 0^-$ mesons. The grid marks are at distances of half units on both axes. Two circles (one filled and another around it) at the center indicate that there are two particles there.

☐ **Exercise 10.1** For non-strange hadrons, i.e., for hadrons with $N_s = 0$, verify that Eq. (8.93, *p 226*) follows from the above expressions for Q, I_3 and B.

It is not possible to eliminate N_u, N_d and N_s from these three equations to obtain a relation between Q, I_3 and B. However, there is a fourth quantity, the strangeness S, which we have introduced in §9.5. We mentioned there that according to the accepted convention, the K^+ has $S = +1$. Since K^+ contains the combination $u\widehat{s}$, as discussed in §10.1, we identify

$$S = -N_s.\qquad(10.7)$$

Taking help of Eq. (10.7), we can now eliminate N_u, N_d and N_s to obtain a relation of the form

$$Q = I_3 + \frac{Y}{2},\qquad(10.8)$$

where the quantity Y is defined by

$$Y = B + S,\qquad(10.9)$$

and is called the *hypercharge*. The relation of Eq. (10.8) is called the *Gell-Mann–Nishijima relation*.

10.2.2   Hypercharge and isospin

It is interesting to see the hypercharge and isospins of different particles that we have already encountered. To be specific, we concentrate on the pseudoscalar mesons, i.e., mesons with $J^P = 0^-$. These include the isotriplet of

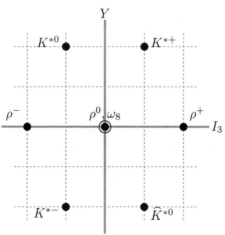

Figure 10.2: Isospin and hypercharge of $J^P = 1^-$ mesons.

pions, the two isodoublets of kaons and antikaons, and the isosinglet state encountered in Eq. (8.84, *p 222*). In Fig. 10.1, we show these particles on a plot of I_3 versus Y.

We see a pretty pattern, and this is not a coincidence. The same pattern occurs for other mesons as well. The mesons appearing in Fig. 10.1 all have $J^P = 0^-$ because the quark and the antiquark in any such meson are in a relative $^1S_0$ state. These are the states of lowest energy, as is usual in any two-body bound state. But there are other possibilities. For example, the quark and the antiquark might be in a relative $^3S_1$ state. From the formulas derived for positronium states in §6.7, it is obvious that the intrinsic parities of such states would also be negative, so that these will be mesons with $J^P = 1^-$. We show the I_3 and Y values of such mesons in Fig. 10.2. It has the same pattern as was seen in Fig. 10.1. Similar patterns can also be seen for baryons, and we will discuss them shortly.

10.3   SU(3)

Gell-Mann and Neeman realized, in 1961, that the structure shown in Fig. 10.1 represents an irreducible representation of the group SU(3). Historically, it is this realization which ushered the idea of quarks, which we have already introduced, anachronistically, in our discussion. To see the connection, it would be necessary to have a discussion on the mathematical structure of this group and its representations.

10.3.1   Some general properties of SU(N) groups

By definition, SU(3) is a group of 3×3 unitary matrices of unit determinants. This means that, to each element of the abstract group, we can associate a 3×3 unitary matrix. More generally, the group SU(N) is defined by the abstract group with which $N \times N$ unitary matrices can be associated. In §3.4.1, it was shown that the number of generators of the group SU(N) is $N^2 - 1$. This means that the group elements of SU(N) are written in terms of $N^2 - 1$ parameters θ_a ($a = 1, \cdots, N^2 - 1$) as

$$U = \exp\left(-iT_a\theta_a\right), \tag{10.10}$$

where the T_a's are called *generators* of the group, and a sum over the index a is implied. The determinant of U can be written as

$$\det U = \det \exp\left(-iT_a\theta_a\right) = \exp\left(-i\operatorname{tr} T_a\theta_a\right). \tag{10.11}$$

Since the left hand side is to equal 1 for arbitrary values of θ_a, we obtain

$$\operatorname{tr} T_a = 0. \tag{10.12}$$

The generators of SU(N) groups are therefore always traceless matrices.

If we take all the parameters θ_a to be real, it easily follows that the generators will be hermitian. The generators of SU(N) can thus be represented by traceless hermitian matrices. As is obvious from Eq. (10.10), the number of generators is equal to the number of parameters, i.e., $N^2 - 1$ for the group SU(N).

It should be emphasized that it is not necessary to take the parameters θ_a to be real, or equivalently, the generators to be hermitian. For example, for the rotation group, we can use the hermitian generators J_x, J_y and J_z, but for some problems it is convenient to use the ladder operators $J_\pm = J_x \pm iJ_y$, which are not hermitian. Of course, if we want to define a general element of the rotation group in terms of the generators $J_\pm$ and J_z, the parameters multiplying J_+ and J_- will be complex, and in fact they should be complex conjugates of each other.

□ **Exercise 10.2** Prove that the structure constants of SU(N) are real when the parameters are chosen to be real.

10.3.2   Fundamental representation

From the general discussion on SU(N) groups, we learned that one requires eight parameters to represent SU(3) group elements and consequently eight generators. The generators are traceless, and can be chosen to be hermitian.

By the very definition of the group, the elements of SU(3) can be represented by 3×3 matrices. This representation, which follows from the definition of the group, is called the *fundamental representation* of the group. To obtain the generators in this representation, we can try to find 3×3 traceless hermitian matrices. Here is a set:

$$\lambda_1 = \begin{pmatrix} 0 & 1 & 0 \\ 1 & 0 & 0 \\ 0 & 0 & 0 \end{pmatrix}, \quad \lambda_2 = \begin{pmatrix} 0 & -i & 0 \\ i & 0 & 0 \\ 0 & 0 & 0 \end{pmatrix},$$

$$\lambda_3 = \begin{pmatrix} 1 & 0 & 0 \\ 0 & -1 & 0 \\ 0 & 0 & 0 \end{pmatrix},$$

$$\lambda_4 = \begin{pmatrix} 0 & 0 & 0 \\ 0 & 0 & 1 \\ 0 & 1 & 0 \end{pmatrix}, \quad \lambda_5 = \begin{pmatrix} 0 & 0 & 0 \\ 0 & 0 & -i \\ 0 & i & 0 \end{pmatrix},$$

$$\lambda_6 = \begin{pmatrix} 0 & 0 & 1 \\ 0 & 0 & 0 \\ 1 & 0 & 0 \end{pmatrix}, \quad \lambda_7 = \begin{pmatrix} 0 & 0 & -i \\ 0 & 0 & 0 \\ i & 0 & 0 \end{pmatrix},$$

$$\lambda_8 = \frac{1}{\sqrt{3}} \begin{pmatrix} 1 & 0 & 0 \\ 0 & 1 & 0 \\ 0 & 0 & -2 \end{pmatrix}. \tag{10.13}$$

We have normalized all of them such that they satisfy the relation

$$\operatorname{tr}(\lambda_a \lambda_b) = 2\delta_{ab}. \tag{10.14}$$

It is customary to choose the generators in the fundamental representation in such a way that they satisfy the relation

$$\operatorname{tr}(T_a T_b) = \frac{1}{2}\delta_{ab}. \tag{10.15}$$

For example, for the group SU(2), this normalization tells us that the generators in the 2-dimensional representation are given by $\tau_a/2$, where the τ_a's are the Pauli matrices. Similarly, for the group SU(3), if we denote the representation of the generators by t_a, we can write

$$t_a^{(f)} = \lambda_a/2, \tag{10.16}$$

where the superscript is the first letter of the word 'fundamental', which is the representation described by this relation. This means that the 3×3 unitary matrices of unit determinant can be written in the form

$$\exp\left(-i\frac{\lambda_a}{2}\theta_a\right) \tag{10.17}$$

using the matrices defined in Eq. (10.13).

□ **Exercise 10.3** The λ-matrices shown in Eq. (10.13), taken together with the unit matrix, can be used as a basis for writing any 3×3 matrix. Writing an arbitrary matrix A as a linear combination of these nine basis matrices, show that

$$\lambda_{\alpha\beta}^a \lambda_{\alpha'\beta'}^a + \frac{2}{3}\delta_{\alpha\beta}\delta_{\alpha'\beta'} = 2\delta_{\alpha\beta'}\delta_{\beta\alpha'}. \tag{10.18}$$

One general property of the group becomes obvious during the exercise of constructing the representation of the generators. Notice that in Eq. (10.13), there are two diagonal generators, which have been given the names λ_3 and λ_8. The situation is different from SU(2), where in any basis we can have at most one diagonal generator. In SU(3), there are two generators which commute with each other.

It is easy to see that we cannot have more than two diagonal generators by changing the basis of the representation. A diagonal 3×3 matrix has three elements, only two of which are independent because of the constraint of tracelessness. Thus, there can be only two linearly independent 3×3 matrices which are diagonal, hermitian and traceless at the same time. Equivalently, we can say that in SU(3), we can choose two generators to commute with each other. Introducing a terminology, we say that SU(3) group has *rank* 2. SU(2) has rank 1, and SU(N) has rank $N - 1$.

10.3.3   Conjugate representation

There is, in fact, another inequivalent 3-dimensional representation of SU(3). This can be seen as follows.

Consider two arbitrary elements G_1 and G_2 of the abstract group. Their product is also an element of the group, and let us call it $G_1 G_2$. A representation is a mapping from the group elements to any collection of matrices that preserves the group multiplication property. In other words, if the elements G_1 and G_2 are represented by the matrices R_{G_1} and R_{G_2} and the product by $R_{G_1 G_2}$, matrix multiplication should obey the result

$$R_{G_1} R_{G_2} = R_{G_1 G_2} . \tag{10.19}$$

Clearly, Eq. (10.19) implies the relation

$$R^*_{G_1} R^*_{G_2} = R^*_{G_1 G_2} . \tag{10.20}$$

Thus, if we associate a different set of matrices $\overline{R}$ to the abstract group elements by the rule

$$\overline{R}_G = R^*_G \tag{10.21}$$

for all elements G, the matrices $\overline{R}$ would also constitute a representation of the group. This representation would have the same dimension as R, and can be called the conjugate representation of R for obvious reasons. We will denote the fundamental representation of SU(3) by 3, and the conjugate of the fundamental by 3^*.

In the 3-representation, any group element G is represented by a matrix of the form given in Eq. (10.17), i.e.,

$$R_G = \exp\left(-i\frac{\lambda_a}{2}\theta^G_a\right) \tag{10.22}$$

for a set of real parameters θ_a^G. Thus,

$$R_G^* = \exp\left(i\frac{\lambda_a^*}{2}\theta_a^G\right). \tag{10.23}$$

Comparing these two equations, we find that the generators in the 3^*-representation are given by $-\lambda_a^*/2$.

These comments about conjugate representations apply to any representation of any group. It is also obvious, from the discussion above, that if the generators are represented by the matrices t_a in the representation R, they will be represented by the matrices $-t_a^*$ in the representation $\overline{R}$.

In some cases, however, the distinction between the representations R and $\overline{R}$ vanishes. Suppose, for a certain representation, we find that that there is a constant unitary matrix ϵ such that

$$-t_a^* = \epsilon t_a \epsilon^{-1} \tag{10.24}$$

for *all* generators. In other words, we find that the matrices $-t_a^*$ are related to the matrices t_a by a similarity transformation. This would mean that they are essentially the same set of operators written in two different bases, and they can be called equivalent. In this case, the representation can be called self-conjugate, or simply real. Obviously this happens if all generators are represented by matrices which are purely imaginary, in which case Eq. (10.24) is satisfied by taking ϵ as the unit matrix. But this is not a necessity. For example, in Eq. (8.78, *p 221*), we found that a relation such as Eq. (10.24) can be satisfied for the 2-dimensional representation of the group SU(2), whose generators are not all imaginary. In fact, all representations of SU(2) obey such a relation, so that all its representations are real. For this reason, SU(2) is sometimes called a *real group*.

For SU(3), all representations are not real. In fact, for the matrices λ_a given in Eq. (10.13), it is impossible to find a matrix ϵ that would satisfy a relation of the form of Eq. (10.24), so that the fundamental representation is indeed complex, i.e., 3 and 3^* are inequivalent. However, all representations are not complex either. We will see that there are some real representations of SU(3), and as a matter of fact for any group.

10.3.4   Examples of other representations

All groups have a 1-dimensional representation, or a singlet representation. If we associate each element of the group with the number 1, or the 1×1 matrix whose only element is 1, group multiplication property of Eq. (10.19) will be automatically satisfied. The generators in this representation are given by

$$t_a^{(s)} = 0 \qquad \forall a. \tag{10.25}$$

Obviously the generators do not have any real part and therefore this representation is real.

All continuous groups must have at least another real representation. This is defined through the relation

$$\left(t_a^{(\text{ad})}\right)_{bc} = -if_{abc}. \tag{10.26}$$

On the left hand side, we have written a particular matrix element of the matrix $t_a^{(\text{ad})}$. On the right hand side, we have the structure constants of the group which were defined in Eq. (3.19, *p44*). The equation defines what is called the *adjoint representation*, and the superscript on the left hand side bears a shortened version of the name. Clearly, the dimension of the representation is the number of values that the indices b and c can take, which is equal to the number of generators of the group.

□ **Exercise 10.4** *Follow the steps indicated below to prove that Eq. (10.26) defines a representation.*

　a)　*In order to qualify as a representation, the matrix forms of the generators must satisfy the algebra, Eq. (3.19, p44). By evaluating an arbitrary matrix element of both sides, show that this equation is equivalent to the following relation among the structure constants:*

$$f_{abp}f_{cpq} + f_{bcp}f_{apq} + f_{cap}f_{bpq} = 0. \tag{10.27}$$

　b)　*Because of the associative property of group composition, the generators must satisfy the identity*

$$\left[t_a, \left[t_b, t_c\right]\right] + \left[t_b, \left[t_c, t_a\right]\right] + \left[t_c, \left[t_a, t_b\right]\right] = 0. \tag{10.28}$$

Show that this implies that the structure constants must satisfy Eq. (10.27), which completes the proof that we were after.
*[**Note :** Eq. (10.28) is an example of Jacobi identity, which is any relation that depends on associativity properties only, irrespective of the commutative properties of the quantities involved.]*

There are, of course, many other representations: infinitely many. To get a taste of some of these, let us start by considering what the fundamental representation does. The matrices in this representation act on states to create new states. The states in this representation should then be column matrices with three entries. We can denote them by the symbol Ψ^i. The action of the group element G on such a state creates a state Ψ', whose elements are given by

$$\Psi'^i = \left(R_G\right)^i_{\ j} \Psi^j, \tag{10.29}$$

where R_G is the matrix associated with the element G in the fundamental representation. If we think of Ψ^j as components of a "vector", we can follow the discussion of §3.5.3 to convince us that any tensorial combination of state vectors with a well-defined symmetry property must also have an associated set of matrices that will form a representation of the group elements. Thus, for example, we will have a representation that inflicts group transformations on symmetric rank-2 tensors, another representation that does the same on antisymmetric rank-2 tensors, and so on. Sometimes, in order not to make the sentences too complicated, we will use the states interchangeably with the matrices while talking about a representation and would use phrases like 'the representation of rank-2 tensors'.

There is one tricky point. Unlike the rotation group discussed in §3.5.3, groups like SU(3) have conjugate representations which are inequivalent. The easiest way to accommodate this fact in the tensorial notation is to treat states in the complex conjugate representation by a lower index. For example, while Ψ^i can represent states in the $\mathbf{3}$ representation of SU(3), Ψ_i can represent the states in $\mathbf{3}^*$.

Obviously this opens up new possibilities for representations. Earlier, we talked about a representation of rank-2 symmetric tensors. Clearly, it is not enough to say this. There can be rank-2 symmetric tensors with upper indices, and likewise with lower indices. Same for antisymmetric tensors. Moreover, there can be rank-2 tensors with one upper and one lower index. Of course, no question of symmetrization or antisymmetrization arises for such tensors since the two indices are of different types.

10.3.5   Young tableaux

There is a neat graphical way of talking of representations of any SU(N) group, which is quite helpful in performing some mathematical operations with them. These are called *Young tableaux*. The word "tableaux" (pronounced tāblō, with 'ā' as in bār and 'ô' as in rôle) is the plural of "table" in French. The graphics involve tabular structures, as we will see.

a)   Tableaux for tensors

The basic idea consists of using a box to stand for each index in the states. For example, the Young tableaux corresponding to the fundamental representation will consist of just one box, since this representation acts on states with one index. For any other representation, there will be more boxes. If any two of the indices are symmetric, we put the corresponding boxes in the same row. If the indices are antisymmetric, the boxes are in the same column. For example, the tableaux corresponding to the rank-2 symmetric and antisymmetric tensor representations would be

$$\square\square \quad \text{and} \quad \begin{array}{c}\square\\\square\end{array} . \tag{10.30}$$

For higher rank tensors, we will have more boxes. We will always arrange them such that the number of boxes in any given row is not smaller than the number in any of the lower rows.

So far, what we have said about Young tableaux applies for all SU(N). For SU(3), the antisymmetric rank-2 tensor has a special significance. Whenever we are dealing with a system where the indices can take N possible values, we can define a tensor with N indices that is completely antisymmetric in the indices. This tensor is called the Levi-Civita tensor. For SU(3), this is a rank-3 tensor which we will write as ϵ_{ijk}. For any rank-2 tensor with upper indices, we can define the combinations

$$t_i \equiv \epsilon_{ijk} T^{jk} . \tag{10.31}$$

For symmetric rank-2 tensors, this contraction would vanish and would therefore be useless. But the contraction says something useful about an antisymmetric rank-2 tensor, viz., that it is equivalent to a vector with lower indices. Since the conjugate of the fundamental representation acts on vectors with lower indices, it implies that this representation is equivalent to that of the rank-2 antisymmetric tensors in SU(3). Thus, one lower index translates, for SU(3) Young tableaux, into two boxes in a column. Said otherwise, the tableaux with two boxes in a column corresponds to the representation whose states carry one lower index, i.e., the $\mathbf{3}^*$.

Indeed, what would happen if we had three boxes in a column? It would signify a rank-3 antisymmetric tensor. And for SU(3), there is only one rank-3 antisymmetric tensor apart from an overall numerical factor, and it is none other than the Levi-Civita tensor. And this tensor is also invariant, as has been hinted in Ex. 10.5. Thus, a column with three boxes represents an invariant. Effectively, it is therefore like a scalar, which is an object with no indices at all. So three boxes in a column is equivalent to no box at all, which implies that any Young tableaux for SU(3) can have at most two rows of boxes. For $SU(N)$, the generalization is obvious: there can be $N-1$ rows at most.

For SU(3) then, a Young tableaux is a collection of boxes in two rows, where the number of boxes in the upper row is larger or equal to that in the lower row. For SU(N), the generalization is obvious.

□ **Exercise 10.5** *A rank-3 tensor T_{ijk} transform as*

$$T'_{lmn} = U_{li} U_{mj} U_{nk} T_{ijk} \tag{10.32}$$

*under the action of the group elements U_{ij}. Show that, if we consider $T_{ijk} = \varepsilon_{ijk}$, then $T'_{lmn} = \varepsilon_{lmn}$, signifying that the Levi-Civita tensor is invariant under group transformations. [**Hint :** A similar result, for the tensors of the Lorentz group, has been proved in Eq. (D.17, p 730). The difference for the present case is that the determinant of the group elements is equal to 1.]*

b)   Dimension from tableaux

Given a tensor of a certain rank with given symmetry properties, we can find its dimensionality, i.e., the number of independent components, by using combinatorial mathematics. One can also obtain the same answer from the associated Young tableaux. The formula is of the form

$$d = \frac{\prod_{j,k} A_{j,k}}{\prod_{j,k} B_{j,k}} , \tag{10.33}$$

where the products run over all boxes: each box is characterized by the number j of the row it appears in, and its place k in that row. For an SU(N) tableaux, the factor in the numerator will be given by

$$A_{j,k} = N + k - j. \qquad (10.34)$$

On the other hand, the factor contributed to the denominator is given by

$$B_{j,k} = n_j + n'_k - j - k + 1, \qquad (10.35)$$

Figure 10.3: Pictorial meaning of Eq. (10.35). In the example shown, the value of $B_{j,k}$ is 5.

where the total number of boxes in row j is n_j, and that in column k is n'_k. This number for any given box can be obtained pictorially for any box by drawing an L-shaped line that starts vertically below the box, comes up to the center of the box, and then continues straight to the right, out of the tableaux. The number $B_{j,k}$ is equal to the number of boxes that this L-shaped line goes through, as illustrated in Fig. 10.3.

☐ **Exercise 10.6** Try applying this formula to the Young tableaux given in Eq. (10.30) and check your results by using combinatorial techniques on the choice of tensor indices.

☐ **Exercise 10.7** For a tableaux of three vertical boxes, apply the formula and show that in SU(3), the dimensionality of the representation is indeed 1.

☐ **Exercise 10.8** How does the Young tableaux of SU(2) look? Describe the tableaux for the $2j + 1$ dimensional representation of SU(2).

c) Kronecker products from tableaux

The biggest advantage of Young tableaux is in evaluating Kronecker products of any two representations. The general recipe for finding such products in SU(N) is as follows.

1. Draw the two tableaux corresponding to the two representations whose Kronecker product is being sought. It is convenient if the one with more boxes is drawn to the left.

2. Mark all boxes in the first row of the right tableaux by the letter a, all in the second row by the letter b etc. (With some experience, this step may seem unnecessary.)

3. Take one box at a time from the right tableaux and attach it to the tableaux on the left. While attaching any box, remember two things:

   a) After attaching, the collection of boxes should look like a Young tableaux. In other words, no row should be longer than a row above it.

   b) Two boxes carrying the same letter cannot sit in the same vertical
      line.

4. If the resulting tableaux has a column of N boxes, disregard it and treat
   the rest as the tableaux.

5. In general, one will be able to form multiple tableaux by following the
   rules above. Each such solution will represent an irreducible representation in the Kronecker product.

6. Use Eq. (10.33) to obtain the dimensionalities of each irreducible representation present in the product.

Let us try some examples of SU(3) Kronecker products. Consider the
product 3×3. Each 3 is a single box. The box from the right member can
attach to the right or to the bottom of the box from the left member. Thus
we obtain

$$\square \times \square = \square\square + \begin{array}{c}\square\\\square\end{array}. \tag{10.36}$$

Evaluation of the dimensions yield the following rule for SU(3):

$$3 \times 3 = 6 + 3^*. \tag{10.37}$$

Take another example. This time, we take the Kronecker product of the
fundamental representation of SU(3) and its conjugate. Following the prescription outlined above, we obtain

$$\begin{array}{c}\square\\\square\end{array} \times \square = \begin{array}{c}\square\square\\\square\end{array} + \begin{array}{c}\square\\\square\end{array} = \begin{array}{c}\square\square\\\square\end{array} + \bullet. \tag{10.38}$$

Notice that in the process, we obtained one tableaux with three boxes in a
column. For SU(3), such combinations can be crossed out. Since nothing
remained after that, we represent the remaining "tableaux" with zero boxes
as a dot. This dot then represents a singlet, and we obtain the product

$$3^* \times 3 = 8 + 1. \tag{10.39}$$

It might be useful to verify this Kronecker product rule from tensorial manipulations. States in the representation 3^* carry one lower index, whereas those
in 3 carry one upper index. Let us denote them by ψ_i and ϕ^i respectively.
The product states would carry one upper and one lower indices, and there
will be nine such states. Of them, we can single out the combination $\psi_i\phi^i$,
which is invariant under SU(3) transformations. The rest transform as an 8,
or octet.

From this discussion, it is clear that whenever we will take the Kronecker
product of a certain representation and its conjugate in any SU(N), we will
have a singlet in which all indices will be contracted. This means that, given
the Young tableaux of a representation, the tableaux for the conjugate representation must be such that all columns of the one can fit with the columns

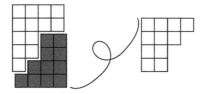

Figure 10.4: How to find the Young tableaux of the conjugate of a representation. The example is in SU(6).

of the other to produce columns of N boxes, which reduce to no boxes or the singlet representation. In other words, if the Young tableaux corresponding to a certain representation contains c columns of lengths $n_1, \cdots, n_c$, the conjugate representation should have a Young tableaux whose columns will have the lengths $N - n_c, \cdots, N - n_1$. This is illustrated in Fig. 10.4.

□ **Exercise 10.9** Find the dimension of the representations of SU(3) which have the following Young tableaux:

$$\square\square\square, \qquad \text{□□}, \qquad \text{□□□.} \tag{10.40}$$

In each case, draw the Young tableaux of the conjugate representation and verify that it has the same dimension as the original one.

□ **Exercise 10.10** Verify the following product rules for SU(3) representation by using Young tableaux:

$$6 \times 3 = 10 + 8, \tag{10.41}$$
$$8 \times 8 = 27 + 10 + 10^* + 8 + 8 + 1. \tag{10.42}$$

□ **Exercise 10.11** Find the dimensions of the original tableaux and its conjugate shown in Fig. 10.4 and show that they are the same, as it should be for the complex conjugate representation.

10.3.6 Decompositions in a subgroup

An SU(3) must have an SU(2) subgroup. This can be seen very easily.

SU(3), by definition, is the group of unitary 3×3 matrices of determinant 1. The group elements include matrices of the following type:

$$\mathcal{U}_2 = \left(\begin{array}{cc|c} & U_2 & 0 \\ & & 0 \\ \hline 0 & 0 & 1 \end{array} \right) \tag{10.43}$$

where U_2 is a 2×2 matrix with determinant unity. It is easy to see that such elements form a group by themselves. They therefore constitute a subgroup of SU(3). We now take a shorthand notation and denote the matrix of Eq.

(10.43) by U_2 itself, getting rid of the zeros on the side and the 1 in the lowest diagonal element which are common for all elements in this subgroup. Formally, we can say that we map the matrices of Eq. (10.43) to the 2×2 matrices U_2, which will be a one-to-one mapping. It shows that the subgroup is SU(2). The generators for this SU(2) can be taken, in terms of the matrices given in Eq. (10.13), as $\lambda_1/2$, $\lambda_2/2$ and $\lambda_3/2$.

But this is not a *maximal subgroup* of SU(3). The statement means that there is a subgroup of SU(3) which is smaller than SU(3) itself but bigger than SU(2), in the sense that it contains all elements of SU(2) and more, but does not contain all elements of SU(3). To see this, consider 3×3 unitary matrices of the form

$$U_2 X , \qquad (10.44)$$

where

$$X = \exp(i\theta\lambda_8/2) \qquad (10.45)$$

for arbitrary real numbers θ. The set of matrices denoted by X form a group by themselves, which depends only on one parameter θ and is a U(1) group. Moreover, for any value of θ, the matrix X will commute with the matrix U_2, because Eq. (10.13) shows that the upper 2×2 block of the diagonal matrix λ_8 is in fact a multiple of the unit matrix. Using this information, it is easy to convince oneself that matrices of the form $U_2 X$ do indeed form a group. Certainly SU(2) is a subgroup of it, obtained by taking only the elements with $\theta = 0$ or $X = 1$. Since X spans a U(1) group, the group of elements of the form given in Eq. (10.44) is denoted by

$$\mathrm{SU(3)} \supset \mathrm{SU(2)} \times \mathrm{U(1)} . \qquad (10.46)$$

This is a maximal continuous subgroup of SU(3).

Subgroups are important because sometimes physical interpretation of a certain result is more easily understood by considering a subgroup rather than the entire group. The reason is that, as shown in §3.5.4, the states forming an irreducible representation of a group $\mathcal{G}$ break up, in general, into smaller irreducible representations of its subgroup $\mathcal{G}'$, and it is easier to interpret these smaller units.

In the present case, the decomposition of SU(3) representations under its SU(2) × U(1) subgroup is easily obtained from the embedding of the subgroup shown in Eq. (10.43). The 3×3 matrices form the fundamental, or **3**, representation of SU(3). We can think of the states as three-element column vectors on which the group elements act. The subgroup SU(2) can be identified with the isospin group, and can be assumed to consist of matrices of the form given in Eq. (10.43). Thus, elements of this SU(2) mix only the first two entries of the column vector, leaving the third one undisturbed. This means that under this subgroup, the **3** representation of SU(3) breaks into a doublet, represented by the two upper elements of the column vector, and a singlet, which is the lowest element. The U(1) quantum number of each entry is proportional to the corresponding element of the diagonal matrix λ_8. Looking at the explicit form of λ_8 in Eq. (10.13), we find that the U(1) quantum number is equal for the two states in the SU(2) doublet. It should be, because U(1) commutes with SU(2) and so the states in any SU(2) representation should have the same U(1) property. So finally we can write

$$\mathbf{3} \rightarrow (\mathbf{2}, \frac{1}{3}) + (\mathbf{1}, -\frac{2}{3}) . \tag{10.47}$$

Note that the U(1) quantum numbers taken in this equation are proportional, but not equal, to the diagonal elements of λ_8. It needs to be said that this is just a matter of convention. In fact, all U(1) quantum numbers are defined only up to some arbitrary multiplicative factor, as noted in §5.1.1. We have chosen the factor in a way that the U(1) quantum number agrees with the value of hypercharge defined in Eq. (10.9), as will be clear from our upcoming discussion in §10.4.

Once the decomposition of the fundamental representation is obtained, it is easy to write down how the $\mathbf{3}^*$-representation decomposes. It is the complex conjugate of the $\mathbf{3}$-representation, so we just need to take the complex conjugate of Eq. (10.47). As we already know, all SU(2) representations are essentially real so that their complex conjugates are the same as the representations themselves. As for the U(1) part, complex conjugation yields something with the opposite charge. Thus

$$\mathbf{3}^* \rightarrow (\mathbf{2}, -\frac{1}{3}) + (\mathbf{1}, \frac{2}{3}) . \tag{10.48}$$

For another representation, the decomposition is trivial to obtain. This is the singlet representation. A singlet is an invariant of the group, i.e., if it is acted upon by any element of the group, it remains unchanged. It then follows that it must be invariant under the action of any of its subgroup. Thus the SU(3) invariant should be a singlet of SU(2) and should have the U(1) charge equal to zero:

$$\mathbf{1} \rightarrow (\mathbf{1}, 0) . \tag{10.49}$$

Since other representations can be obtained from the fundamental one, their decompositions can also be obtained the same way. For example, imagine taking the product $\mathbf{3} \times \mathbf{3}^*$. Already from Eq. (10.39), we know the irreducible representations of SU(3) that appear in the product, viz., an octet and a singlet. But how do these irreducible representations of SU(3) transform under the subgroup SU(2) × U(1)? This can be worked out easily using the decompositions of the $\mathbf{3}$ and $\mathbf{3}^*$ representations given in Eqs. (10.47) and (10.48). From the viewpoint of the subgroup, the product $\mathbf{3} \times \mathbf{3}^*$ looks like the following:

$$\left[(\mathbf{2}, \frac{1}{3}) + (\mathbf{1}, -\frac{2}{3}) \right] \times \left[(\mathbf{2}, -\frac{1}{3}) + (\mathbf{1}, \frac{2}{3}) \right] . \tag{10.50}$$

This contains four products, and each can be evaluated easily. For the product of the SU(2) parts of the representation, we use the familiar angular momentum addition rules, since angular momentum and isospin are identical in their mathematical structures. And for the U(1) part, the quantum numbers should just add. Thus,

$$\left(2, \frac{1}{3}\right) \times \left(2, -\frac{1}{3}\right) = (3,0) + (1,0), \tag{10.51}$$

since two isodoublets can make either an isotriplet or an isosinglet. Proceeding similarly, we find the following irreducible representations of SU(2) × U(1) in the product of $3 \times 3^*$ of SU(3):

$$3 \times 3^* \Rightarrow (3,0) + (1,0) + (2,1) + (2,-1) + (1,0). \tag{10.52}$$

Comparing it with Eq. (10.39) and using Eq. (10.49), we obtain

$$8 \Rightarrow (3,0) + (1,0) + (2,1) + (2,-1). \tag{10.53}$$

□ **Exercise 10.12** *Applying similar techniques, obtain the following decompositions of SU(3) irreducible representations under the SU(2) × U(1) subgroup:*

$$6 \Rightarrow \left(3, \frac{2}{3}\right) + \left(2, -\frac{1}{3}\right) + \left(1, -\frac{4}{3}\right) \tag{10.54}$$
$$10 \Rightarrow (4,1) + (3,0) + (2,-1) + (1,-2). \tag{10.55}$$

10.4   Mesons from three flavors of quarks

Let us now see what we got in Eq. (10.53). We find that an octet of SU(3) contains a triplet, a singlet and two doublets of the SU(2) subgroup. The quantum numbers of these states in the direction orthogonal to the SU(2) are also given in Eq. (10.53).

Now imagine that this SU(2) subgroup is isospin. We have mentioned before that all particles in an isospin multiplet have the same value of hypercharge. This means that the hypercharge generator commutes with all isospin generators. Once we take the SU(2) subgroup as isospin, the U(1) generator which commutes with it must be taken as the hypercharge. Thus, e.g., Eq. (10.53) would imply that in the SU(3) octet, we have an isotriplet and an isosinglet, both of which have zero hypercharge, and two isodoublets with hypercharges +1 and −1.

But this is exactly what we see in Fig. 10.1 *(p 256)*, for example. The isotriplet pion has zero hypercharge. The kaons appear in two isodoublets, with hypercharges +1 and −1. And the η-particle is an isosinglet with hypercharge zero. In summary, we can say that the $J^P = 0^-$ mesons shown in Fig. 10.1 *(p 256)* form an octet of SU(3). The same applies to the $J^P = 1^-$ mesons shown in Fig. 10.2 *(p 257)*. Nucleons are members of an octet of baryons, as we will describe shortly.

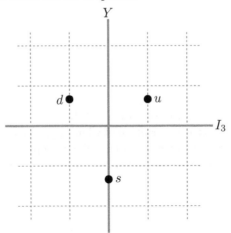

Figure 10.5: Isospin and hypercharge of quarks.

We mentioned in §10.3 that all representations of a group can be constructed from the fundamental representation by taking Kronecker products. The octet representation, mathematically, is not fundamental. So, when Gell-Mann and Neeman saw these octet representations of SU(3), it was almost inevitable to conjecture that these hadrons must be made up of something more fundamental, something which belongs to the fundamental representation of SU(3). The fundamental representation is of course 3-dimensional, and so it was hypothesized that there are three objects which transform like a triplet of SU(3). This was proposed by Gell-Mann and by Zweig independently, in 1964. These objects were called *quarks* by Gell-Mann and *aces* by Zweig, and the first name stuck.

The three states in the fundamental representations are three different types, or *flavors*, of quarks that we have introduced earlier, viz.,

$$\Psi \equiv \begin{pmatrix} u \\ d \\ s \end{pmatrix} \sim \mathbf{3} \,. \tag{10.56}$$

The representation is written at the right end of this equation. This is the representation in the flavor SU(3) group. In Fig. 10.5, we show these three quarks in an I_3 versus Y plot that was used earlier to represent mesons in Fig. 10.1 *(p 256)* and Fig. 10.2 *(p 257)*.

Obviously, the antiquarks transform as an antitriplet, i.e., as the representation $\mathbf{3}^*$:

$$\widehat{\Psi} \equiv \begin{pmatrix} \widehat{u} \\ \widehat{d} \\ \widehat{s} \end{pmatrix} \sim \mathbf{3}^* \,. \tag{10.57}$$

Consider now what happens if we have a quark and an antiquark in a system. The quark transforms like **3**, and the antiquark like **3*** of the flavor SU(3). There can be nine states like this, and they would transform like **3** × **3*** of the flavor SU(3). These nine states would not form an irreducible representation. From Eq. (10.39), notice that the product **3** × **3*** transforms like an octet plus a singlet. Therefore, of the nine states obtained, eight will transform like an octet, and the other will be an SU(3) singlet.

This, of course, is the description of what happens in the flavor sector. The physical characteristics of these states will also depend on the space and spin configurations of the quark and the antiquark. For example, if the space-spin configuration is a $^1S_0$ state, i.e., the relative angular momentum L and the combined spin S are both zero, we obtain mesons with spin $J = 0$. The flavor octet with these characteristics is the one encountered in Fig. 10.1 *(p 256)*. The flavor contents of the particles at the boundaries of this octet have been given earlier, in Eqs. (8.82), (10.1) and (10.2). At the center, we have the particles π^0, the flavor part of whose wavefunction was given in Eq. (8.83, *p 221*). The other state at the center, η, is given by

$$|\eta\rangle = \frac{1}{\sqrt{6}}\left(\widehat{u}u + \widehat{d}d - 2\widehat{s}s\right). \tag{10.58}$$

One might wonder why we have put the factor of 2 in front of the strange quark component in the wavefunction of η. This can be understood in two ways. First, note that the flavor part of the wavefunction of all mesons in the octet can be written in the form $\widehat{\Psi}^\top \lambda_i \Psi$, where Ψ and $\widehat{\Psi}$ have been given in Eqs. (10.56) and (10.57), and λ_i is some linear combination of the matrices shown in Eq. (10.13). For example, π^+ can be written, apart from a normalization constant, as $\widehat{\Psi}^\top(\lambda_1 + i\lambda_2)\Psi$. Similarly, π^0 can be written as $\widehat{\Psi}^\top \lambda_3 \Psi$, and η as $\widehat{\Psi}^\top \lambda_8 \Psi$. This explains the factor of 2 in Eq. (10.58).

In the second way of understanding the factor of 2, we recall that, in addition to this octet, we should also have an SU(3) singlet meson. This is called η'. Obviously, the flavor part of the wavefunction of this particle would be

$$|\eta'\rangle = \frac{1}{\sqrt{3}}\left(\widehat{u}u + \widehat{d}d + \widehat{s}s\right), \tag{10.59}$$

since this is the combination that transforms as an SU(3) singlet. The state η must be orthogonal to this state, as well as to the state π^0 given in Eq. (8.83, *p 221*). This justifies the form given in Eq. (10.58).

As discussed in §8.7.3, one can contemplate other possibilities for the space-spin parts of the wavefunctions. If the space-spin part has a $^3S_1$ configuration, we get the octet of mesons shown in Fig. 10.2 *(p 257)* that includes the ρ mesons. The isosinglet particle in this octet has been denoted by ω_8. The flavor parts of the wavefunctions of the ρ's and the ω are exactly the same as those for the pions and the η. The flavor wavefunctions of the K^*'s are also exactly the same as those for the K-mesons. In addition to the octet, there will also be an SU(3)-singlet state, which can be denoted by ω_1, whose flavor wavefunction will be that given in Eq. (10.59). There are thus two isosinglet states: one is an SU(3) singlet, and the other belongs to the SU(3) octet. We will see later in §10.7 that the physical isosinglet particles are linear superpositions of these two states.

One aspect of the properties of mesons is easily explained with the quark model. Mesons are quark-antiquark systems. From our discussion in §6.7, we know that any fermion-antifermion system has an overall parity of $(-1)^{L+1}$, where L is the orbital angular momentum between the fermion and the antifermion. We have discussed the space-spin configurations of the quark-antiquark pair in both octets of mesons, and found that in both cases $L = 0$. Thus these states have negative parity. The spin-0 mesons of Fig. 10.1 *(p 256)* have negative parity and are thus pseudoscalar, whereas the spin-1 mesons of Fig. 10.2 *(p 257)* also have negative parity are thus pseudovectors.

10.5   Baryons from three flavors of quarks

With two quarks, one can form a sextet and an antitriplet of SU(3), as given in Eq. (10.37). If we now bring in a third quark, the irreducible representations will be contained in 6×3 and $3^* \times 3$. Using the results of Eqs. (10.41) and (10.39), we can thus write

$$3 \times 3 \times 3 = 10 + 8 + 8 + 1. \tag{10.60}$$

Among these, the 10 is completely symmetric in the three indices. The singlet, or 1, must be a column of three boxes, which means that the indices are completely antisymmetric in this representation. The octets have mixed symmetries. One can take the two octets in a way that one of them is symmetric in the first two indices, and the other antisymmetric in the same two indices. The situation is reminiscent of the isospin part of the wavefunctions of ^{3}H and ^{3}He that appeared in Eq. (8.60, *p 217*).

Of course, this is only the flavor part. The wavefunction must be completely antisymmetric in the three quarks when we take the symmetry properties from other characteristics into account, something that will be done in §10.11. For the purpose of this section, it is enough to keep an eye on the symmetry properties of the flavor wavefunctions with respect to the interchange of the first two particles. As we said, we have a 10 and an 8 which are symmetric under this interchange, and an 8 and a 1 which are antisymmetric.

The amazing thing is that we observe baryons that correspond only to the 10 and the 8, representations which are symmetric under the interchange of the first two quarks. The octet contains the nucleons, spin-$\frac{1}{2}$ and positive parity particles. Other $J^P = \frac{1}{2}^+$ particles that complete the octet are the isotriplet Σ (Sigma), the isosinglet Λ (Lambda), and the isodoublet consisting of Ξ^0 and Ξ^-, which are usually called *cascade* particles. All these particles have been shown in Fig. 10.6. The isotriplet and the isosinglet have

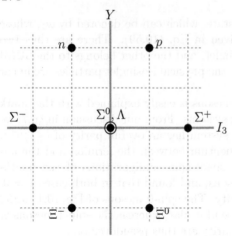

p	uud	938
n	udd	939
Λ	$(ud - du)s$	1116
Σ^+	uus	1189
Σ^0	$(ud + du)s$	1193
Σ^-	dds	1197
Ξ^0	uss	1315
Ξ^-	dss	1321

Figure 10.6: The $J^P = \frac{1}{2}^+$ baryon octet. The numbers in the box are the approximate masses in MeV units.

strangeness $S = -1$. The nucleons are not strange, i.e., they have $S = 0$. The cascades have $S = -2$. In the language of the quark model, nucleons do not contain any strange quark: in fact, their quark structure has already been given in §8.7.1 before we introduced the s-quark in this book. The Σ's and the Λ contain one s-quark, whereas the cascades contain two s-quarks. These facts have been summarized in Fig. 10.6.

Let us now look at the **10** representation, or the decuplet, of SU(3). From Eq. (10.55), we understand that it should contain an isoquartet, an isotriplet, an isodoublet and an isosinglet. Using the hypercharge values of these isomultiplets, we can also find out the electric charge of each component by using Eq. (10.8). In the isoquartet, we find the charges to be $+2$, $+1$, 0 and -1. These four particles are collectively called Δ, which we earlier encountered in §8.7.3. The isotriplet is similar to the isotriplet of Σ encountered in the baryon octet, and is called Σ^*. By the same token, the isodoublet is called Ξ^*. At the time the quark model was proposed, all these particles were known from experiments, and they fitted nicely into the decuplet. All these particles have been shown in Fig. 10.7.

There was one apparent disagreement with experiments though. The decuplet can be completed with the isosinglet Ω^-, but no such particle was experimentally known at that time. So, based on the quark model, a prediction of this particle was made. Considering the mass differences between different isomultiplets in the decuplet, the mass of this particle was also roughly predicted. And very soon, the particle was discovered, vindicating the quark model and the underlying SU(3) symmetry.

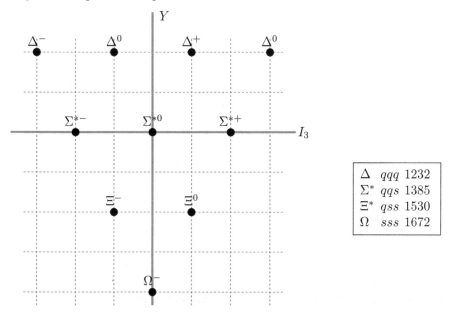

Figure 10.7: The $J^P = \frac{3}{2}^+$ baryon decuplet. In the box to the right, q denotes either u or d quark, and the numbers are the approximate masses in MeV units.

10.6 *U*-spin and *V*-spin

It is true that SU(3) contains an SU(2) × U(1) subgroup, as has been explained in §10.3.6. In the discussion so far, we have always identified the SU(2) with the group of isospin transformations, and U(1) as the group that is generated by hypercharge.

But this is not a unique choice for the embedding. This can be easily understood by visualizing a similar problem. The group of 3-dimensional rotations has a subgroup which is the group of 2-dimensional rotations. However, there is no unique way of identifying this subgroup. If we denote an orthogonal set of axes in the 3-dimensional space by x, y and z, the subgroup of 2-dimensional rotations might correspond to the rotation in the x-y plane, or in the y-z plane, or in the z-x plane, or, indeed, in any plane contained in the 3-dimensional space.

In SU(3) also, there can be different embeddings of the SU(2) × U(1) subgroup. The embedding of isospin and hypercharge is obtained by taking the SU(2) generators to be $\lambda_1/2$, $\lambda_2/2$ and $\lambda_3/2$ from the list of generators of SU(3) presented in Eq. (10.13). With this choice, the diagonal generator of SU(2) gives an eigenvalue of $+\frac{1}{2}$ for the u-quark, $-\frac{1}{2}$ for the d-quark and zero for the s-quark, which are the correct eigenvalues for the third component of isospin for these quarks.

We can find a different embedding by choosing the non-diagonal SU(2) generators to be $\lambda_4/2$ and $\lambda_5/2$ in the fundamental representation. The SU(2) subgroup corresponding to this choice is called U-spin, in analogy with I-spin or isospin. The third or the diagonal generator of this U-spin group can be obtained by using the usual commutation relations of SU(2). Denoting this generator by $\lambda_3^U/2$ in the fundamental representation, we can write

$$\left[\lambda_4/2, \lambda_5/2\right] = i\lambda_3^U/2. \tag{10.61}$$

The left hand side of this equation can be easily evaluated by using the matrices given in Eq. (10.13), and we obtain

$$\lambda_3^U = \begin{pmatrix} 0 & 0 & 0 \\ 0 & 1 & 0 \\ 0 & 0 & -1 \end{pmatrix}. \tag{10.62}$$

Certainly this was not one of the eight matrices given in Eq. (10.13), but it is not outside the representation either. It can easily be expressed as a linear combination of the two diagonal generators present in Eq. (10.13):

$$\lambda_3^U = \frac{\sqrt{3}}{2}\lambda_8 - \frac{1}{2}\lambda_3. \tag{10.63}$$

The generator λ_8 commutes with all generators of the isospin SU(2). Similarly, there should be a generator that commutes with all generators of U-spin. It is easily seen that this generator is represented by the matrix

$$\lambda_8^U = \frac{1}{\sqrt{3}} \begin{pmatrix} -2 & 0 & 0 \\ 0 & 1 & 0 \\ 0 & 0 & 1 \end{pmatrix}. \tag{10.64}$$

We have put an overall factor so that the normalization condition of Eq. (10.14) is obeyed by it. Also, it is obvious that this is no independent matrix, but is rather a different linear combination of the diagonal matrices λ_3 and λ_8:

$$\lambda_8^U = -\frac{1}{2}\lambda_8 - \frac{\sqrt{3}}{2}\lambda_3, \tag{10.65}$$

which can be easily checked from the matrices given earlier.

Since the lower 2×2 block of the matrix in Eq. (10.64) is just a multiple of the unit matrix, and all U-spin generators have non-zero elements only within this block, it is obvious that this matrix commutes with all generators of U-spin. Thus, this generator bears the same relation to U-spin as hypercharge does to I-spin.

The importance of isospin or I-spin lies in the fact that the members of the same isomultiplet are nearly degenerate. This is not true for the members of the same U-spin multiplet, as will be discussed in detail in §10.7. However, there is a different reason why U-spin is physically important. Looking at the diagonal generator λ_8^U given in Eq. (10.64), we see that it is equal to the charge of the quarks apart from an overall factor:

$$Q = -\frac{1}{\sqrt{3}} \lambda_8^U . \qquad (10.66)$$

In other words, the charge operator commutes with all three generators of U-spin. All particles in the same U-spin multiplet should therefore have the same charge. This is clearly seen in the fundamental representation of SU(3), where the choice of λ_3^U in Eq. (10.62) tells us that the d and the s quarks transform like a U-spin doublet, and they indeed have the same charge. The u-quark has a different charge, but it belongs to a different U-spin representation, viz., a singlet.

We can look for U-spin multiplets in the octet as well. For example, let us look at the mesons in Fig. 10.1 *(p 256)*. The π^+ and the K^+ belong to a U-spin doublet. So do the π^- and the K^-. There are four neutral particles, and they are distributed in a triplet and a singlet representation of U-spin. Obviously, the states on the outer edge, K^0 and $\widehat{K}^0$, belong to the triplet. The two states at the center have $U_3 = 0$. One of them belongs to the $U = 1$ multiplet and the other has $U = 0$.

But it has to be realized that neither π^0 nor η is an eigenstate of U-spin. The neutral pion or π^0 belongs to an isotriplet and therefore behaves like the generator λ_3. On the other hand, η, an isosinglet, behaves like λ_8. A superposition of these two states will have $U = 1, U_3 = 0$. Let us call this state π_U^0. Its relation to π^0 and η is exactly similar to the relation between the generators in Eq. (10.63), i.e.,

$$\left| \pi_U^0 \right\rangle = \frac{\sqrt{3}}{2} \left| \eta \right\rangle - \frac{1}{2} \left| \pi^0 \right\rangle . \qquad (10.67)$$

The U-spin singlet state η_U, on the other hand, is defined through a relation similar to Eq. (10.65):

$$\left| \eta_U \right\rangle = -\frac{1}{2} \left| \eta \right\rangle - \frac{\sqrt{3}}{2} \left| \pi^0 \right\rangle . \qquad (10.68)$$

Similar relations would hold for the pseudovector meson or the baryon octet.

We can also similarly discuss another SU(2) embedding of SU(3) for which the multiplets will be along the lines tilted at 60 degrees with the horizontal axes in the figures for octets and decuplets. This could be called the V-spin, but it is not very useful.

10.7 SU(3) breaking and mass relations

In Ch. 8, we remarked that isospin is not an exact symmetry of nature. Although it is a symmetry of strong interactions, it is not respected by electromagnetic and weak interactions. For this reason, members of an isomultiplet are not exactly degenerate. However, the fractional differences between the masses of the members of an isomultiplet are very small, indicating that isospin is respected to a very good accuracy.

We can ask the same question about the SU(3) symmetry: how good is it? If it were exact, all hadrons within the same SU(3) multiplet would have been degenerate in mass. In reality, this is far from being true. We have already presented the masses of different hadrons, and there are huge differences in the masses in any SU(3) multiplet, for mesons as well as for baryons.

As in the case of isospin, we can ask what causes these differences. Of course the effects of electromagnetic and weak interactions do not respect the SU(3) symmetry, so they can induce differences of the said sort. But these effects would be small, nothing compared to what we see in an SU(3) multiplet.

We commented in §8.9 that, even for isospin breaking, the dominant contributions do not come from electromagnetic interactions. Rather, they come from the mass difference between the u-quark and the d-quark. Let us take this cue and investigate whether a heavy s-quark can be responsible for the SU(3) breaking effects.

The mass terms for the three flavors of quarks in the Lagrangian are as follows:

$$-\mathscr{L}_m = m_u \bar{u}u + m_d \bar{d}d + m_s \bar{s}s \,, \tag{10.69}$$

where u, d and s stand for the field operators of the three types of quarks. In order to keep the discussion simple, we will ignore isospin breaking effects by taking

$$m_u = m_d \equiv m_0 \,. \tag{10.70}$$

Then the expression of Eq. (10.69) can be rewritten as

$$-\mathscr{L}_m = \frac{2m_0 + m_s}{3}\left(\bar{u}u + \bar{d}d + \bar{s}s\right) - \frac{m_s - m_0}{3}\left(\bar{u}u + \bar{d}d - 2\bar{s}s\right). \tag{10.71}$$

The masses that appear in this equation need not be the masses present in the fundamental Lagrangian that describes quarks and their interactions. Because of interactions, a quark can develop an effective mass in a hadron, which is the relevant quantity here. Such effective masses are usually called *constituent masses* in the context of quarks, and in this sense the Lagrangian shown in Eq. (10.69) is an effective Lagrangian. The actual mass terms that appear in the fundamental Lagrangian will be discussed in Ch. 18.

Using the notation Ψ introduced in Eq. (10.56), we can write it as

$$-\mathscr{L}_m = \frac{2m_0 + m_s}{3}\overline{\Psi}\Psi - \frac{m_s - m_0}{\sqrt{3}}\overline{\Psi}\lambda_8\Psi\,. \tag{10.72}$$

Between the $\overline{\Psi}$ and the Ψ, we have the unit matrix in one term, and the matrix λ_8 in the other. Both these matrices commute with all generators of the isospin SU(2) group. Thus, these mass terms cannot produce any difference for particles in the same isomultiplet, a fact that we have taken as our starting point by taking the $m_u = m_d$.

In order to obtain some non-trivial relations, consider the relation

$$\lambda_8 = -\frac{1}{2}\lambda_8^U + \frac{\sqrt{3}}{2}\lambda_3^U , \qquad (10.73)$$

which can be obtained by inverting Eqs. (10.63) and (10.65). Using this, we can write

$$-\mathcal{L}_m = \frac{2m_0 + m_s}{3}\,\overline{\Psi}\Psi + \frac{m_s - m_0}{2}\,\overline{\Psi}\Big(\frac{1}{\sqrt{3}}\lambda_8^U - \lambda_3^U\Big)\Psi . \qquad (10.74)$$

The term $\overline{\Psi}\Psi$ is an SU(3) singlet, and therefore cannot be responsible for any SU(3) breaking effect. The next term contains λ_8^U. This commutes with all generators of U-spin and therefore cannot inflict any difference between members of the same U-spin multiplet. The last term contains λ_3^U, which is a representation of U_3. This is the only term that can induce differences among different members of a U-spin multiplet. Thus, if we consider the action of the mass terms on the members of a U-spin multiplet, we can write the mass terms symbolically as $a - bU_3$, where a and b are constants for a given multiplet. Now note that

$$\langle U, U_3 | a - bU_3 | U, U_3 \rangle = a - bU_3 , \qquad (10.75)$$

where on the right hand side we have the eigenvalue of the operator U_3. It shows that if we take the expectation value of the quark mass terms among different members of an U-spin multiplet, the results will be equispaced for increasing values of U_3.

Let us apply this on some particles in the baryon octet. This octet contains a U-spin triplet whose extreme members are the neutron ($U_3 = +1$) and the Ξ^0 ($U_3 = -1$), whereas the middle member is Σ_U^0, which is a combination of Σ^0 and Λ that resembles the combination π_U^0 given in Eq. (10.67) for the pseudoscalar octet, i.e.,

$$|\Sigma_U^0\rangle = \frac{\sqrt{3}}{2}|\Lambda\rangle - \frac{1}{2}|\Sigma^0\rangle . \qquad (10.76)$$

The first thing to notice is that the Ξ^0 is heavier than the neutron, as is expected from Eq. (10.75) if $b > 0$, i.e.,

$$m_s > m_0 . \qquad (10.77)$$

Second, the equispacing rule implies

$$m(n) + m(\Xi^0) = 2m(\Sigma_U^0) . \qquad (10.78)$$

On the left hand side, we have the masses of the neutron and the Ξ^0. But what is meant by $m(\Sigma_U^0)$ in this formula? The mass of any physical particle is the expectation value of the Hamiltonian in an eigenstate of vanishing 3-momentum. This definition of mass can be extended to states which are not eigenstates of the Hamiltonian. Thus

$$m(\Sigma_U^0) = \langle \Sigma_U^0 \,|\, \mathcal{H} \,|\, \Sigma_U^0 \rangle$$
$$= \left\langle \frac{\sqrt{3}}{2}\Lambda - \frac{1}{2}\Sigma^0 \,\middle|\, \mathcal{H} \,\middle|\, \frac{\sqrt{3}}{2}\Lambda - \frac{1}{2}\Sigma^0 \right\rangle, \qquad (10.79)$$

where the states are implied to have vanishing 3-momentum. The off-diagonal matrix elements of the Hamiltonian would vanish in the eigenbasis, and the diagonal elements would give the masses of the eigenstates. Putting the result into Eq. (10.78), we obtain

$$m(n) + m(\Xi^0) = \frac{3}{2}m(\Lambda) + \frac{1}{2}m(\Sigma^0). \qquad (10.80)$$

This is called the Gell-Mann–Okubo mass formula for the baryons. If we use the experimentally measured mass values of the relevant hadrons, the left hand side of this equation comes out to be 2270 MeV, whereas the right hand side becomes 2254 MeV. Thus, the formula is good to better than 1%.

Let us now consider the similar relation in the meson octet containing the pion. The places occupied by n, Ξ^0, Σ and Λ in Fig. 10.6 *(p 274)* are occupied respectively by K^0, $\widehat{K}^0$, π^0 and η in Fig. 10.1 *(p 256)*. The masses of K^0 and $\widehat{K}^0$ must be equal to each other by CPT invariance, as discussed in §7.5. So the resulting mass relation should involve only three different masses. However, a straightforward replacement of the baryon masses by the corresponding meson masses in Eq. (10.80) does not produce an acceptable relation. It was subsequently proposed that for mesons, the squares of masses should be used rather than mass. In other words, the analog of Eq. (10.80) for mesons should be

$$2m^2(K^0) = \frac{3}{2}m^2(\eta) + \frac{1}{2}m^2(\pi^0), \qquad (10.81)$$

or

$$m^2(\eta) = \frac{4}{3}m^2(K^0) - \frac{1}{3}m^2(\pi^0). \qquad (10.82)$$

This is called the Gell-Mann–Okubo mass formula for the pseudoscalar mesons. If we put in the experimentally known masses for the K^0 and π^0 and calculate the mass of the η from this relation, we obtain 569 MeV, whereas the measured mass is 548 MeV for the η.

This raises two questions: first, why squared masses had to be used in the relation, and second, although Eq. (10.82) is not dismally disobeyed by experimental mass values, why is the agreement not as good as it was for the baryons?

The first question can be answered only heuristically. For a fermion field, the only parameter that appears in the free Lagrangian is the mass itself. In the Lagrangian of a free bosonic field, the parameter that appears is, in fact, the square of the particle's mass. This leads us to suspect that for baryons (which are fermions), it should be the mass that would appear in the mass formula, whereas for mesons (which are bosons), the square of mass would appear.

As far as the second question is concerned, it is to be noted that there is a big difference in the spectrum of the baryons and the mesons. There is no baryon which is an SU(3) singlet. But for the mesons, the SU(3) singlet state very much exists, as was mentioned in Eq. (10.59). The mass of the η' is 940 MeV, about twice the mass of the η. This should not surprise us, because the η' belongs to a different SU(3) multiplet, and its mass would be unrelated to the masses in the octet even if flavor SU(3) were an exact symmetry. It might be possible that the physical η and η' states are not the ones shown in Eqs. (10.58) and (10.59). Rather, the η particle is predominantly the octet state of Eq. (10.58) but has a slight admixture of the singlet state shown in Eq. (10.59), whereas the physical η' is predominantly the singlet with a slight admixture of the octet. That would make the η-mass a little different from the prediction of the Gell-Mann–Okubo mass formula, which is based on the octet character of the η.

This point becomes quite important for the pseudovector octet of mesons that was shown in Fig. 10.2 (p 257). The isosinglet state present in the octet was denoted by ω_8 there. The analog of Eq. (10.82) for this SU(3) multiplet would be

$$m^2(\omega_8) = \frac{4}{3}m^2(K^{*0}) - \frac{1}{3}m^2(\rho^0). \tag{10.83}$$

Putting in the experimental values $m(K^*) = 892\,\text{MeV}$ and $m(\rho) = 770\,\text{MeV}$, we obtain

$$m(\omega_8) = 929\,\text{MeV} \tag{10.84}$$

from the Gell-Mann–Okubo formula given above. But there is no isosinglet meson with $J^P = 1^-$ with such mass. The nearest ones are a particle called ω (omega) with mass 783 MeV, and a particle called ϕ with mass 1018 MeV. Let us then assume that these two particles are superpositions of the octet state ω_8 described above, and an SU(3) singlet state ω_1. The expectation value of the Hamiltonian in the state ω_8 is the value obtained from the Gell-Mann–Okubo formula and quoted in Eq. (10.84). Thus, in the ω_1-ω_8 basis, the Hamiltonian will have the form

$$\begin{pmatrix} a & b \\ b & 929 \end{pmatrix} \tag{10.85}$$

in units of MeV. The values of a and b can be determined by imposing the condition that the eigenvalues of this matrix should correspond to the masses of the particles ω and ϕ. The solution is

$$a = 872, \qquad b = 114. \tag{10.86}$$

One can also find the matrix that diagonalizes the Hamiltonian of Eq. (10.85) and obtain that the eigenstates ω and ϕ are given by

$$\omega = \omega_1 \cos\theta + \omega_8 \sin\theta\,,$$
$$\phi = -\omega_1 \sin\theta + \omega_8 \cos\theta\,, \tag{10.87}$$

with

$$\theta = \frac{1}{2}\tan^{-1}\frac{2b}{929-a} = 38°\,. \tag{10.88}$$

There is thus a huge mixing between the SU(3) octet and singlet states. Because of this, one often contemplates this singlet along with the entire octet and call the collection of nine particles a *nonet*.

It turns out that the value of the mixing encountered in Eq. (10.88) is tantalizingly close to a very important value. Since the octet and the singlet combinations are given by the right hand sides of Eqs. (10.58) and (10.59), the combination ϕ defined in Eq. (10.87) would be given by

$$\phi = \left(-\frac{\sin\theta}{\sqrt{3}} + \frac{\cos\theta}{\sqrt{6}}\right)\left(\widehat{u}u + \widehat{d}d\right) - \left(\frac{\sin\theta}{\sqrt{3}} + \frac{2\cos\theta}{\sqrt{6}}\right)\widehat{s}s\,. \tag{10.89}$$

For $\sin\theta = \sqrt{1/3}$ or $\theta \approx 35°$, we find that the meson ϕ is given by

$$\phi = \widehat{s}s\,, \tag{10.90}$$

whereas the meson ω does not contain any $\widehat{s}s$:

$$\omega = \frac{1}{\sqrt{2}}\left(\widehat{u}u + \widehat{d}d\right)\,. \tag{10.91}$$

This set of affairs is termed *ideal mixing*. As we said earlier, the real octet-singlet mixing is very close to ideal mixing. There is no convincing theory of why it is so.

10.8 Electromagnetic properties in SU(3)

10.8.1 Fundamental interaction

The SU(3) flavor symmetry implies interesting relations between electromagnetic properties of different members of a multiplet, which we now discuss.

All electromagnetic properties arise through the interaction involving photons in the fundamental Lagrangian. Quarks are fermions, and for any fermion field ψ the interaction term with the photon is of the form $-eQ\overline{\psi}\gamma^\mu\psi A_\mu$, as discussed in Ch. 5. Thus, for the three flavors of quarks, the interaction term with the photon in the fundamental Lagrangian has the form

$$\mathscr{L}_{\text{em}-\text{int}} = -e\left(\frac{2}{3}\overline{u}\gamma^\lambda u - \frac{1}{3}\overline{d}\gamma^\lambda d - \frac{1}{3}\overline{s}\gamma^\lambda s\right)A_\lambda\,. \tag{10.92}$$

Using the notation for the SU(3) triplet introduced in Eq. (10.56), we can write it as

$$\mathcal{L}_{\text{em-int}} = \frac{1}{\sqrt{3}} e \overline{\Psi} \gamma^\lambda \lambda_8^U \Psi A_\lambda , \qquad (10.93)$$

where the matrix λ_8^U was introduced in Eq. (10.64). It is the generator of SU(3) that commutes with all U-spin generators.

Clearly, the electromagnetic interaction is a U-spin singlet, and therefore, in the limit of unbroken SU(3), all members of a U-spin multiplet should have the same electromagnetic properties. This is obviously true for electric charges: all members of a U-spin multiplet have the same charge. It should be true for other electromagnetic properties as well, and we show some consequences of this property below.

10.8.2   Baryon magnetic moments

a)   SU(3) relations

Magnetic moment is, of course, an electromagnetic property. The pseudoscalar mesons cannot have magnetic moment because they do not have spin, but the baryons can. By the argument given earlier, magnetic moment should be the same for all states in a given U-spin multiplet. Denoting the magnetic moments by the customary symbol μ, we can then write down the following relations in the baryon octet:

$$\begin{aligned}
\mu_p &= \mu_{\Sigma^+} , \\
\mu_n &= \mu_{\Xi^0} = \mu_{\Sigma_U^0} , \\
\mu_{\Sigma^-} &= \mu_{\Xi^-} .
\end{aligned} \qquad (10.94)$$

Notice that in the middle relation for the uncharged baryons, the quantity $\mu_{\Sigma_U^0}$ appears, which should interpreted as $\langle \Sigma_U^0 | O | \Sigma_U^0 \rangle$, where O is the magnetic moment operator and Σ_U^0 is of course the combination shown in Eq. (10.76), the uncharged baryon state which transforms like the neutral member of a U-spin triplet.

Moreover, since the electric charge is part of the SU(3) group generators, it should be traceless in any representation. For the triplet representation, this is clearly true for the charges of the three flavors of quarks. A similar relation should hold for the magnetic moments of all particles in the baryon octet. Thus, in the limit of exact SU(3), we get

$$\mu_p + \mu_n + \mu_{\Sigma^+} + \mu_{\Sigma^0} + \mu_{\Sigma^-} + \mu_\Lambda + \mu_{\Xi^0} + \mu_{\Xi^-} = 0. \qquad (10.95)$$

A further relation between the magnetic moments can be obtained by taking the form of λ_8^U given in Eq. (10.65) and substituting it into Eq. (10.93). This enables us to write the electromagnetic current into two parts. The part that contains λ_8 is an isosinglet, and produces the same effect on all members of an isomultiplet. The other part contains λ_3 and should therefore produce a contribution to the electromagnetic properties which should be equally spaced among different members of an isomultiplet. For the isotriplet Σ, we thus obtain

$$2\mu_{\Sigma^0} = \mu_{\Sigma^+} + \mu_{\Sigma^-}. \tag{10.96}$$

Then Eqs. (10.94), (10.95) and (10.96) give six relations among the eight magnetic moments. The quantity $\mu_{\Sigma^0_U}$ appearing in Eq. (10.94) can be interpreted in terms of the magnetic moments of physical particles by noting that Σ^0 and Λ have the SU(3) transformation properties of λ_3 and λ_8. On the other hand, Σ^0_U, defined in Eq. (10.76), transforms like λ_3^U. We can define an orthogonal combination Λ_U which transforms like λ_8^U and is therefore a U-spin singlet. The states in these two different basis are related by the equations

$$\Sigma^0 = -\frac{1}{2}\Sigma^0_U - \frac{\sqrt{3}}{2}\Lambda_U,$$

$$\Lambda = \frac{\sqrt{3}}{2}\Sigma^0_U - \frac{1}{2}\Lambda_U. \tag{10.97}$$

Consider now taking the matrix element of the magnetic moment operator between Σ^0 states on both sides. Note that

$$\left\langle \Sigma^0_U \left| O \right| \Lambda_U \right\rangle = 0, \tag{10.98}$$

since the U-spin scalar operator O cannot connect between two states belonging to different U-spin multiplets. Thus we obtain

$$\mu_{\Sigma^0} \equiv \left\langle \Sigma^0 \left| O \right| \Sigma^0 \right\rangle = \frac{1}{4}\mu_{\Sigma^0_U} + \frac{3}{4}\mu_{\Lambda_U},$$

$$\mu_\Lambda \equiv \left\langle \Lambda \left| O \right| \Lambda \right\rangle = \frac{3}{4}\mu_{\Sigma^0_U} + \frac{1}{4}\mu_{\Lambda_U}. \tag{10.99}$$

Solution of these two equations shows that we should use

$$\mu_{\Sigma^0_U} = \frac{1}{2}\left(3\mu_\Lambda - \mu_{\Sigma^0}\right) \tag{10.100}$$

in Eq. (10.94).

We can now take the magnetic moments of the proton and the neutron to be independent quantities and express all other magnetic moments in terms of these two. Experimentally measured values of these two quantities are

$$\mu_p = 2.79\mu_N, \qquad \mu_n = -1.91\mu_N, \tag{10.101}$$

where μ_N is the *nuclear magneton* which is related to the nucleon mass m_N by the standard formula

$$\mu_N = \frac{e}{2m_N}. \tag{10.102}$$

Table 10.1: Magnetic moments of baryons in the octet. All numerical values are in the unit of nuclear magneton, $e/2m_N$, where m_N is the nucleon mass.

Quantity	Measured value	SU(3) value		Mass corrected
		Relation	Numerical	
μ_p	2.79	(input)		
μ_n	-1.91	(input)		
μ_{Σ^+}	2.46	μ_p	2.79	2.20
μ_{Σ^0}	not known	$-\frac{1}{2}\mu_n$	0.95	0.75
μ_{Σ^-}	-1.16	$-(\mu_p+\mu_n)$	-0.88	-0.69
μ_Λ	-0.61	$\frac{1}{2}\mu_n$	-0.95	-0.80
μ_{Ξ^0}	-1.25	μ_n	-1.91	-1.36
μ_{Ξ^-}	-0.65	$-(\mu_p+\mu_n)$	-0.88	-0.63

Taking these inputs, we calculate the SU(3) prediction of all other magnetic moments and show them in Table 10.1 under the column heading of "SU(3) relation".

Comparing with the measured values, we find that the SU(3) predictions are not very good really. In fact, they are not supposed to be. Such relations for magnetic moments are true in the limit of exact SU(3) symmetry where the masses of all particles are equal because they belong to the same SU(3) multiplet. In reality, the masses of the baryons are not very close, as has been discussed in §10.7. It is therefore expected that these relations should not be valid to a high degree of accuracy. The magnetic moment of a particle is inversely proportional to its mass, so the SU(3) relations are expected to overestimate the absolute values of magnetic moments of heavier particles. If we use the real mass instead of a SU(3) invariant mass, the relations get modified, as also shown in Table 10.1.

There is another relevant quantity which can be evaluated in this context. In analogy with the quantities appearing in Eq. (10.99), we can evaluate the *transition magnetic moment* between the Σ^0 and the Λ. Using Eq. (10.98) and the expressions for the Σ^0 and the Λ from Eq. (10.97), we find

$$\langle \Lambda \,|O|\, \Sigma^0 \rangle = \frac{\sqrt{3}}{4}\left(\mu_{\Lambda_U} - \mu_{\Sigma_U^0}\right). \qquad (10.103)$$

Using Eq. (10.99), this can also be written as

$$\langle \Lambda \,|O|\, \Sigma^0 \rangle = \frac{\sqrt{3}}{2}\left(\mu_{\Sigma^0} - \mu_\Lambda\right) = -\frac{\sqrt{3}}{2}\mu_n, \qquad (10.104)$$

where in the last step we have used the SU(3) expression of the magnetic moments of the Σ^0 and the Λ which have been summarized in Table 10.1.

b) Quark model predictions

The relations given so far in this section depend only on the fact that the baryons form an octet of SU(3): they do not depend on the inherent quark structure. With the quark structure, we can do a bit more.

The baryon wavefunction can be obtained in terms of the quarks. The exercise is similar to what was done earlier in §8.6 for ^{3}He and ^{3}H nuclei. Those nuclei contain three nucleons. The baryons contain three quarks. Thus, apart from trivial changes in notation, the wavefunction of the proton and the neutron can be written in terms of quarks by using the formulas of §8.6. If we assume that the spatial part of the wavefunction is symmetric and the spin-flavor part is antisymmetric, the wavefunction for the proton in spin-up state can be written as

$$|p\blacktriangle\rangle = \frac{1}{\sqrt{6}}\Big(u\blacktriangledown d\blacktriangle u\blacktriangle - d\blacktriangle u\blacktriangledown u\blacktriangle + d\blacktriangle u\blacktriangle u\blacktriangledown$$

$$- u\blacktriangle d\blacktriangle u\blacktriangledown - u\blacktriangledown u\blacktriangle d\blacktriangle + u\blacktriangle u\blacktriangledown d\blacktriangle \Big), \qquad (10.105)$$

which is basically Eq. (8.66, $p\,218$) with appropriate changes in notation. The magnetic moment of the proton in this state would then be given by

$$\langle p\blacktriangle\,|O|\,p\blacktriangle\rangle = \frac{1}{6}\Big[\langle u\blacktriangledown d\blacktriangle u\blacktriangle\,|O|\,u\blacktriangledown d\blacktriangle u\blacktriangle\rangle + \langle d\blacktriangle u\blacktriangledown u\blacktriangle\,|O|\,d\blacktriangle u\blacktriangledown u\blacktriangle\rangle$$

$$+ \langle d\blacktriangle u\blacktriangle u\blacktriangledown\,|O|\,d\blacktriangle u\blacktriangle u\blacktriangledown\rangle + \langle u\blacktriangle d\blacktriangle u\blacktriangledown\,|O|\,u\blacktriangle d\blacktriangle u\blacktriangledown\rangle$$

$$+ \langle u\blacktriangledown u\blacktriangle d\blacktriangle\,|O|\,u\blacktriangledown u\blacktriangle d\blacktriangle\rangle + \langle u\blacktriangle u\blacktriangledown d\blacktriangle\,|O|\,u\blacktriangle u\blacktriangledown d\blacktriangle\rangle\Big].$$

$$(10.106)$$

The cross terms will vanish. Now, in each term, we see that the spins of the two up quarks are opposite, so their contributions cancel, and we are left only with the contribution from the down quark. Since the down quark spin is up in each case, we obtain

$$\langle p\blacktriangle\,|O|\,p\blacktriangle\rangle = \mu_d\,, \qquad (10.107)$$

where μ_d is the magnetic moment of the down quark. Since the magnetic moment of the proton is defined to be the expectation value of the magnetic moment operator in the spin-up state, we can then write

$$\mu_p = \mu_d\,. \qquad (10.108)$$

A similar exercise will give

$$\mu_n = \mu_u\,. \qquad (10.109)$$

If we assume that the masses of the u and the d quarks are roughly equal so that their magnetic moments are proportional to their charges:

$$\mu_u = \frac{2}{3}\mu_0, \qquad \mu_d = -\frac{1}{3}\mu_0, \qquad (10.110)$$

then we obtain the relation

$$\frac{\mu_p}{\mu_n} = -\frac{1}{2}. \tag{10.111}$$

This is a disastrous relation when we compare it with the experimentally measured values of these quantities, presented in Eq. (10.101).

At this point, it would be interesting to check what result might be obtained if we assume that the spin-flavor parts of the nucleon wavefunctions are not antisymmetric as given in Eq. (10.105), but are symmetric instead. For overall symmetry, the spin-flavor wavefunction of the proton in spin-up state is given by

$$|p\!\uparrow\rangle = \frac{1}{3\sqrt{2}}\Big(2u\!\uparrow u\!\uparrow d\!\downarrow + 2u\!\uparrow d\!\downarrow u\!\uparrow + 2d\!\downarrow u\!\uparrow u\!\uparrow$$
$$- u\!\uparrow u\!\downarrow d\!\uparrow - u\!\uparrow d\!\uparrow u\!\downarrow - d\!\uparrow u\!\uparrow u\!\downarrow$$
$$- u\!\downarrow u\!\uparrow d\!\uparrow - u\!\downarrow d\!\uparrow u\!\uparrow - d\!\uparrow u\!\uparrow u\!\downarrow\Big). \tag{10.112}$$

This gives

$$\langle p\!\uparrow |O| p\!\uparrow\rangle = \frac{1}{18}\Big(12(2\mu_u - \mu_d) + 6\mu_d\Big), \tag{10.113}$$

or

$$\mu_p = \frac{4}{3}\mu_u - \frac{1}{3}\mu_d. \tag{10.114}$$

In the corresponding formula for the neutron, the roles of the u and the d quarks should be interchanged:

$$\mu_n = \frac{4}{3}\mu_d - \frac{1}{3}\mu_u. \tag{10.115}$$

If we now use Eq. (10.110), we obtain

$$\frac{\mu_p}{\mu_n} = -\frac{3}{2}, \tag{10.116}$$

which is in very good agreement with the experimental values presented in Eq. (10.101).

It therefore seems, from these considerations, that the wavefunctions of the members of the baryon octet are symmetric in spin-flavor. Indeed, this has to be trivially true for the Δ's of the decuplet as well, which have spin-$\frac{3}{2}$ and isospin-$\frac{3}{2}$, and therefore must be symmetric individually in the spin part and the isospin part of the wavefunction. This issue will be discussed in §10.11.

□ **Exercise 10.13**   a) *Using SU(3) alone, express the magnetic moments of the baryon decuplet members in terms of that of the Δ^-.*

b)   Use the quark model to find μ_{Δ^-} in terms of the magnetic mo-
     ments of the u and d quarks. Hence express μ_{Δ^-} in terms of
     the magnetic moments of the proton and the neutron.

c)   For which member of the decuplet is the magnetic moment eas-
     iest to measure?

□ **Exercise 10.14**   a)   Using the quark model, find the magnetic dipole
     transition moment connecting Δ^+ to the proton.

b)   Using $\langle\Lambda|O|\Sigma^0\rangle$ from Eq. (10.104), find the ratio of the mag-
     netic dipole transition rates $\Gamma(\Delta^+ \to p\gamma)/\Gamma(\Sigma^0 \to \Lambda\gamma)$. Take the
     phase space differences into account in the calculation.

c)   Compare the result with experimental data.

10.8.3   Electromagnetic mass differences

Particles in the same isospin multiplet have nearly the same mass. The small
mass differences within a multiplet can be ascribed to electromagnetic interactions, which break isospin symmetry. Once the electromagnetic effects
between the quarks are included, the neutron and the proton, for example,
are expected to have different masses. However, in an SU(3) multiplet, the
corrections to all particles with the same charge are expected to be equal,
since electromagnetic interactions depend on electric charge.

Let us consider the consequence of this expectation on the baryon octet.
The octet contains hadrons with three different charges: +1, 0 and −1. If
the electromagnetic contributions to the masses of these differently charged
baryons are denoted by δ_+, δ_0 and δ_- respectively, we can write

$$m_n = m_N + \delta_0, \qquad m_p = m_N + \delta_+,$$
$$m_{\Sigma^-} = m_\Sigma + \delta_-, \qquad m_{\Sigma^+} = m_\Sigma + \delta_+,$$
$$m_{\Xi^-} = m_\Xi + \delta_-, \qquad m_{\Xi^0} = m_\Xi + \delta_0, \tag{10.117}$$

where m_N, m_Σ and m_Ξ are the isospin-obeying contributions to the masses
that come from strong interactions between quarks. We can eliminate the
electromagnetic contributions from these equations to write

$$m_n - m_p + m_{\Sigma^+} - m_{\Sigma^-} + m_{\Xi^-} - m_{\Xi^0} = 0. \tag{10.118}$$

The central values of the experimentally measured masses show that

$$m_n - m_p = 1.3\,\text{MeV},$$
$$m_{\Sigma^+} - m_{\Sigma^-} = -8.0\,\text{MeV},$$
$$m_{\Xi^-} - m_{\Xi^0} = 6.6\,\text{MeV}, \tag{10.119}$$

so that the combination on the left hand side of Eq. (10.118) comes out to
be −0.1 MeV, in excellent agreement where the individual masses are around

1000 MeV. If we consider the error bars on the experimental values, experimental data is certainly consistent with Eq. (10.118).

It should be noted that, while writing Eq. (10.117), we have not put down the expressions for the masses of the Σ^0 and the Λ. They could not have been included in a relation like that in Eq. (10.118). Moreover, the electromagnetic mass corrections are not linear in the electromagnetic interactions, so there cannot be any linearity relation like Eq. (10.96) that we had used for relating magnetic moments of different baryons.

10.9 Decays of hadrons

We now discuss decays of the mesons appearing in the $J^P = 0^-$ and $J^P = 1^-$ octets, as well as baryons in the octet and decuplet.

10.9.1 Decays of members of pseudoscalar meson octet

As far as strong interactions are concerned, the mesons in the $J^P = 0^-$ octet cannot decay. Strong decays would only yield hadronic final states. The neutral pion, π^0, is the lightest hadron, and therefore cannot decay into any other hadrons. We have discussed before in §6.8 that π^0 decays into two photons:

$$\pi^0 \longrightarrow \gamma\gamma. \tag{10.120}$$

The photons in the final state indicate that this decay must be electromagnetic. In fact, the π^0 contains quark-antiquark pairs of the same flavor. They are antiparticles of each other, and they undergo pair annihilation to give two photons. The lifetime is of order 10^{-16} s.

The charged pion, π^+, is the lightest hadron carrying electric charge. Since electric charge is conserved, it cannot possibly decay into π^0 or to photons. It can only decay to leptons, and the decay has to be governed by weak interactions. The dominant decay mode is

$$\pi^+ \longrightarrow \mu^+ \nu_\mu. \tag{10.121}$$

It is also possible to have $e^+\nu_e$ in the final state. Although energetically favourable, this mode is very suppressed, for reasons which will be discussed in §17.5.

The kaons cannot decay strongly because they are the lightest strange particles. As mentioned in §10.1, strong as well as electromagnetic interactions conserve strangeness. Thus kaons can decay only via weak interactions. The decay products may be purely leptonic as for the case of charged pions, or semileptonic, i.e., containing both leptons and hadrons, or purely hadronic. For example, K^+ can have the leptonic decay mode of $\mu^+\nu_\mu$, or the semileptonic mode of $\pi^0\mu^+\nu_\mu$, or the hadronic mode of $\pi^+\pi^0$.

This leaves us with the isoscalar η. Like π^0, it contains quarks and anti-quarks of the same flavor, so pair annihilation into two photons is possible. In fact, this is one of the major decay modes of the η, and

$$\mathscr{B}(\eta \to \gamma\gamma) \equiv \frac{\Gamma(\eta \to \gamma\gamma)}{\Gamma(\eta \to \text{anything})} = 39.31\% . \qquad (10.122)$$

The symbol $\Gamma(\cdots)$ stands for the rate of process given in the parentheses, and $\mathscr{B}$, as defined here, is called the *branching ratio* of a given channel.

What about decays of η into pions? Since η has a mass of 548 MeV which is more than three but less than four pion masses, at most three pions are allowed by the phase space. Decay into two pions breaks parity, and is therefore forbidden through strong as well as electromagnetic interactions. Moreover, η cannot decay into three pions through strong interactions. This can best be seen through G-parity. When we introduced G-parity in §8.5, we discussed that pions have negative G-parity. On the other hand, η has a C-eigenvalue $+1$ since it can decay into two photons. It is an isosinglet, so the isospin rotation appearing in the definition of G-parity leaves it unchanged. The conclusion is that the G-parity of η is $+1$, and therefore it cannot decay strongly into three pions. Electromagnetic interactions, however, break isospin and consequently G-parity, so decays of this kind can occur through electromagnetic interactions. The branching ratios are somewhat smaller than the two-photon mode because of reduced phase space:

$$\mathscr{B}(\eta \to 3\pi^0) = 32.57\% , \qquad (10.123\text{a})$$
$$\mathscr{B}(\eta \to \pi^+\pi^-\pi^0) = 22.74\% . \qquad (10.123\text{b})$$

Note that the branching ratio of the two-photon mode is roughly comparable to that of any of the three-pion modes. The reason is that both amplitudes are second order in electromagnetic interactions. The two-photon mode clearly needs two electromagnetic vertices with quarks. The three-pion modes also need a virtual photon exchange since, as argued, strong interactions alone cannot induce these processes. A virtual photon will have to connect to quarks at both ends, and so the amplitudes for these processes will also be second order in e.

☐ **Exercise 10.15** ♪ Show that the decay $\eta \to 2\pi$ must violate parity. Moreover, show that the decay conserves charge conjugation symmetry, so that it is also CP violating.

☐ **Exercise 10.16** Use only isospin argument to show that $\eta \to 3\pi$ is forbidden in strong interactions. [**Hint :** *The isosinglet combination of three isotriplets is completely antisymmetric.*]

☐ **Exercise 10.17** The decay $\eta \to \pi^0 + \gamma$ is forbidden. Why?

☐ **Exercise 10.18** Consider a meson ψ which is an SU(3) singlet, so that $U = V = 0$, and that is is odd under $\mathscr{C}$ with $J^P = 1^-$.

  a) Show that the decay $\psi \to \pi^0\rho^0$ is allowed by $\mathscr{C}$ and also by G-parity and so occurs strongly.

b) Use charge conjugation symmetry as well as I-spin, U-spin and V-spin symmetries to show that the amplitudes of the decays of ψ into $\pi^+\rho^-$, $\pi^0\rho^0$, $\pi^-\rho^+$, $K^+K^{\star-}$, $K^-K^{\star+}$, $K^0\widehat{K}^{\star0}$ and $\widehat{K}^0K^{\star0}$ should all be equal in magnitude.

c) Verify the validity of the above result from data on the decay of the meson $\psi = \widehat{c}c$ which has a mass of $3100\,\mathrm{MeV}$. Comment on the phase spaces in the different channels.

10.9.2 Decays of members of baryon octet

The proton is the lightest baryon, and therefore is stable because baryon number is absolutely conserved. Decay of the neutron must involve a proton in the final state because of baryon number conservation, but the mass difference of the neutron and the proton is only about $1.3\,\mathrm{MeV}$, which is not enough to accommodate even one pion. Thus, hadronic decay modes of the neutron are not available. The neutron can of course decay through weak interactions:

$$n \to p + e + \widehat{\nu}_e. \qquad (10.124)$$

This is the basic process underlying all nuclear beta decay.

The isosinglet Λ is the lightest strange baryon. It cannot decay strongly because of a combination of baryon number conservation and strangeness: there is simply no combination of hadrons which has baryon number 1 and strangeness -1, and is lighter than the Λ. The same argument applies for the Σ^+ and the Σ^-. The decay of these particles involves a change of strangeness, and therefore cannot be mediated by strong or electromagnetic interactions. They decay through weak interactions, almost exclusively to $N\pi$, i.e., a nucleon and a pion. The cascades cannot decay strongly or electromagnetically for exactly the same reason. They decay almost exclusively into $\Lambda\pi$, mediated be weak interactions.

The Σ^0 is somewhat exceptional. It has the same flavor content as the Λ, and is heavier than the latter. Therefore it can decay into Λ with the emission of a photon:

$$\mathscr{B}(\Sigma^0 \to \Lambda + \gamma) = 100\%. \qquad (10.125)$$

Electromagnetic decays obviously occur much faster than weak decays. The lifetime of Σ^0 is of order $10^{-20}\,\mathrm{s}$, whereas Σ^+, Σ^-, the cascades and the Λ, which decay by weak interactions, have lifetimes of order $10^{-10}\,\mathrm{s}$. For details, see Table B.3 *(p 720)*.

The neutron is a special case. Although it decays weakly, its lifetime is nowhere near that of, say, Λ. A free neutron has a very long lifetime of $887\,\mathrm{s}$. Its decay rate must therefore be much smaller than that of other weakly decaying baryons in the octet. But this is not caused by any symmetry of any sort. The point is that the neutron is only very slightly heavier than the combined mass of its decay products. Thus it gets very little phase space. For other baryons like the Λ, the phase space factor is enormous compared to that of the neutron. This is what suppresses neutron decay.

☐ **Exercise 10.19** To obtain an idea of the phase space suppression for neutron decay, suppose that the neutron can decay to the proton and a negatively charged hypothetical meson whose mass is the same as that of the electron. Assuming that the matrix element for this decay is the same as that for the decay $\Lambda \to p\pi^-$, calculate the ratio of the decay rates by using the masses of the particles involved.

10.9.3   Decays of members of baryon decuplet

The members of the baryon decuplet come in four different isospin multiplets. The largest multiplet has $I = \frac{3}{2}$, and the particles in this multiplet are called Δ. All these particles decay strongly:

$$\Delta \to N\pi, \tag{10.126}$$

where N denotes the nucleon. Thus, for example, Δ^{++} decays to $p\pi^+$, whereas Δ^- decays to $n\pi^-$. For the other two Δ particles, two different decay channels are possible: Δ^+ decays to $p\pi^0$ and $n\pi^+$, and Δ^0 decays to $p\pi^-$ and $n\pi^0$. The branching ratios of the two channels are determined by the isospin Clebsch–Gordan co-efficients, barring small corrections coming from neutron-proton mass difference and π^+-π^0 mass difference, which are isospin-breaking effects.

☐ **Exercise 10.20** Using the Clebsch–Gordan co-efficients given in Appendix E, find the ratio of the amplitudes of the decays $\Delta^+ \to p\pi^0$ and $\Delta^+ \to n\pi^+$ and check your result against the statement of Ex. 8.8 (p 223).

Members of the $I = 1$ multiplet, Σ^*, decay to $\Lambda\pi$ and $\Sigma\pi$. These are strong decays as well. Decays of Ξ^* are also strong decays, almost exclusively to $\Xi\pi$.

For all these particles, the lifetimes are very small, of the order of 10^{-20} s, as expected for strong decays of particles with masses around 1 GeV. However, the remaining particle in the decuplet, the Ω^-, is much longer lived, with a lifetime of 0.821×10^{-10} s. The reason is that this particle carries strangeness $S = -3$, and there is no combination of lighter hadrons which has the same strangeness and the same electric charge. Therefore, it decays through weak interactions, to ΛK^- and $\Xi\pi$, violating strangeness by 1 unit.

10.9.4   Decays of members of vector meson nonet

All these particles can decay through strong interactions. The ρ-mesons decay almost exclusively to two pions. The K^*-mesons carry strangeness, and therefore cannot decay to pions only; rather, they decay to $K\pi$. The only remaining octet member is ω_8. As mentioned earlier in §10.7, this state mixes with the SU(3) singlet state ω_1 and the resulting eigenstates are ω with mass 783 MeV and ϕ with mass 1020 MeV. Both these states have negative G-parity and therefore cannot decay into two pions. The dominant decay mode of ω is into $\pi^+\pi^-\pi^0$. As illustrated in §8.4.3, the decay mode to three neutral pions is forbidden by isospin symmetry. Decay to two pions is possible, but only electromagnetically, because it breaks isospin symmetry as well. Electromagnetic decay to $\pi^0\gamma$ is also possible.

The ϕ-meson can also decay to states containing pions and rho particles, but the more dominant channels involve two kaons. This is a peculiar property of mesonic decays. If the decaying particle has a quark-antiquark pair of the same flavor, the dominant decay products would contain the quark and the antiquark in two different particles. In other words, processes where a certain flavored quark and its own antiquark of the initial state undergo pair annihilation are suppressed. This is called the Okubo–Zweig–Iizuka (OZI) rule. After we develop the theory of strong interactions further, a qualitative explanation of the rule will be given in Ch. 20 in connection with decays of heavier mesons. Right now, it is only important to recognize that we are seeing an example of the application of the rule. The ϕ-meson, as mentioned in §10.7, contains almost purely the combination $s\hat{s}$. If the decay products contain no particle carrying strangeness, the $s\hat{s}$ in ϕ has to pair annihilate into virtual gluons or photons, and new quarks and antiquarks should be created from them to produce the final state particles. This is what is suppressed by the OZI rule. The dominant final states are K^+K^-, or a pair of neutral kaons. Among the non-strange modes which are OZI suppressed, $\rho\pi$ or three-pion modes are preferred.

☐ **Exercise 10.21** The spin of the ρ-meson is 1. The ρ^0 is composed of $u\hat{u}$ and $d\hat{d}$, like the π^0. Using the formulas derived in connection with positronium states, find the charge conjugation eigenvalue of ρ^0. Hence show that the ρ-meson cannot decay into three neutral pions via strong interactions.

☐ **Exercise 10.22** Find the G-parity of the ρ-mesons and hence (or otherwise) show that the none of the members of the isomultiplet of ρ-mesons can decay into any three-pion state via strong interactions.

☐ **Exercise 10.23** Use G-parity arguments to show that the ϕ cannot decay to two pions via strong interaction.

10.10 Summary of conservation laws

In §5.1.3, we said that each kind of particle can come with only one number conservation law. Thus, with three quarks, we can at most three different 'charges', and we can take them to be N_u, N_d and N_s, which were used in §10.2. In practice, we do not use the conservation laws in this form. Rather, we make linear combinations of these three quantities and give them names. These are the names that appear in Eqs. (10.4), (10.5) and (10.6).

To be more precise, these three equations give only the contributions of the three flavors of quarks to the electric charge, I_3 and baryon number. When we include contributions from all elementary particles, Eq. (10.5) still remains the same, i.e., the only elementary particles that carry the property isospin are the u and the d quarks. Baryon number comes only from quarks. There are other flavors of quarks apart from the u, the d and the s, and they contribute to the baryon number equally. As regards electric charge, there are contributions from particles which are not quarks, like the electrons. Once we consider all particles, electric charge is indeed conserved. Its conservation is related to a U(1) gauge symmetry, as we saw in §5.1. Baryon number is also absolutely conserved, so far as direct evidences are concerned. However, it is not connected to any gauge symmetry so far as we know.

Strong and electromagnetic interactions obey all these symmetries that contain, in their currents, terms involving only one kind of particle at a time.

10.11 Color

In §10.8.2, while calculating magnetic moments of the proton and the neutron, we noticed that their wavefunctions need to be symmetric under spin and flavor interchanges in order that the quark model calculations produce the correct ratio between the magnetic moments. Let us analyze this situation a bit further, including the baryons in the decuplet.

10.11.1 Symmetry of baryon states

Notice that the proton wavefunction given in Eq. (10.112) does not have any particular symmetry property if we interchange only the spin orientations of two of the quarks. They also do not maintain their forms if only the flavors of two quarks are interchanged. The symmetry property is manifest only when *both* spin and flavor tags of two quarks are changed simultaneously. We constructed such wavefunctions in §8.6 and §10.8.2 by taking combinations which have definite symmetry properties under the interchange of particles 1 and 2, and then making appropriate linear combinations of them to obtain the results like in Eqs. (8.61) and (10.112). This is a rather round-about way, and can be avoided if one can treat spin and flavor together.

This can be done by considering all spin and flavor variations together in a multiplet of some group. Unlike Eq. (10.56) which contains the three flavors of quarks in a multiplet, we will now take a multiplet whose members contain all these three flavors in both spin-up and spin-down states, i.e., something like

$$\Psi \equiv \begin{pmatrix} u\blacktriangle \\ d\blacktriangle \\ s\blacktriangle \\ u\blacktriangledown \\ d\blacktriangledown \\ s\blacktriangledown \end{pmatrix} \sim 6 \,. \tag{10.127}$$

Since this is a 6-dimensional representation, it can be the fundamental representation of an SU(6) group. This is called the *spin-flavor SU(6)* group.

Each element of the fundamental representation belongs to a triplet of the flavor SU(3) and to a doublet of the spin SU(2). This fact is summarized by saying that

$$6 \longrightarrow (3,2) \tag{10.128}$$

under the embedding

$$SU(6) \supset SU(3) \otimes SU(2). \tag{10.129}$$

Earlier, in Eq. (10.46) for example, we used the symbol '×' in the notation for subgroups. Here, in Eq. (10.129), we are using a different notation, an encircled cross. This difference in notations indicates an important difference in the nature of embeddings of the subgroups. In the embedding discussed in Eq. (10.46), only two of the elements of the fundamental representation of SU(3) transform as a doublet under the subgroup SU(2). The other element is neutral under SU(2). This is not the case in Eq. (10.129). Here, every element of the fundamental representation of SU(6) transforms like a triplet, and there are two such triplets. Similarly, every element of Ψ of Eq. (10.127) transforms like a doublet of SU(2), and there are three such doublets. None of the elements is a singlet either under SU(3) or under SU(2).

Quarks thus transform like a 6 under the spin-flavor $SU(6)$ group. It will be easy to see how a baryon might transform under the same group. A baryon contains three quarks, so it must transform like a $6 \times 6 \times 6$. This Kronecker product can be performed by using Young tableaux. Since 6 is just one box in $SU(6)$, we have the product of three boxes. By calculating the product, we obtain

$$\square \times \square \times \square = \square\square\square + {\square\square \atop \square} + {\square\square \atop \square} + {\square \atop {\square \atop \square}}, \tag{10.130}$$

or

$$6 \times 6 \times 6 = 56 + 70 + 70 + 20. \tag{10.131}$$

The 56-dimensional representation is the one whose Young tableaux appears first on the right hand side of Eq. (10.130). It shows that it is a representation of rank-3 tensors which are completely symmetric in its indices. In other words, states appearing in this representation would be completely symmetric under the interchange of $SU(6)$ indices. Since $SU(6)$ indices contain flavor and spin, these states are spin-flavor symmetric.

How do these states transform under the subgroup $SU(3) \otimes SU(2)$? We can use the methods described in §10.3.6 to obtain

$$56 \longrightarrow (10,4) + (8,2). \tag{10.132}$$

The first one is a 4-dimensional representation of $SU(2)$, i.e., represents spin-$\frac{3}{2}$ particles. And they come in a decuplet of $SU(3)$. These are exactly the particles shown in Fig. 10.7 *(p 275)*. The other representation contains an $SU(3)$ octet of spin-$\frac{1}{2}$ particles, i.e., the particles shown in Fig. 10.6 *(p 274)*. Thus the 56 of $SU(6)$ contains all the baryons in the octet and the decuplet. And remember that this representation is completely symmetric with respect to interchanges.

□ **Exercise 10.24** Verify the Kronecker product rule of Eq. (10.131) in SU(6), and find the dimensions of various representations given in Eq. (10.130).

□ **Exercise 10.25** Consider the particles Δ^{++}, Δ^- and Ω^-, each of which contains three quarks of the same flavor. Without getting into SU(6), argue that the spin-flavor wavefunction must be symmetric for these particles.

10.11.2   Generalized Pauli principle

Obviously, this creates a problem with the generalized Pauli principle. In the lowest lying baryons, the quarks should not have any relative orbital angular momentum. Thus, they should be symmetric with respect to spatial interchange. We just saw that they are symmetric in spin-flavor as well. Taken together, we can conclude that the space, spin and flavor parts of their wavefunctions are symmetric with respect to the interchange of the quarks. But baryons are fermions, and the generalized Pauli principle demands that their wavefunction should be antisymmetric under the interchange of quarks!

To avoid the impasse, it was proposed that quarks carry another kind of characteristics as well, apart from the ones already mentioned. This new property came to be known as *color*. In a sense, the name is unfortunate, because this new property or characteristic that we are talking about has nothing to do with the physiological response to electromagnetic radiation of different wavelengths in the optical range — a sensation that is termed 'color' in usual parlance. However, the name was suggested in this case by analogy with some property of optical color, as we will comment on presently.

Since all other parts of the wavefunctions of the baryons in the octet and the decuplet seem to be symmetric under quark exchanges, the generalized Pauli principle can be salvaged if the color parts of their wavefunctions are antisymmetric. If three quarks will have to be antisymmetric in color, it is simplest to think that there are three different colors. Let us denote these colors by α, β etc. Then, for example, the wavefunction of Δ^- in the $S_z = \frac{3}{2}$ state can be written in the form

$$\varepsilon_{\alpha\beta\gamma} d^\alpha_\blacktriangle d^\beta_\blacktriangle d^\gamma_\blacktriangle \,, \tag{10.133}$$

where ε denotes the completely antisymmetric tensor, or the Levi-Civita tensor, in the color indices. For other baryons, the flavor content or the spin content might be different, but as far as the color indices are concerned, it will be the same story: three different color indices contracted by the Levi-Civita tensor of color.

In the end, then, there is no free color index in Eq. (10.133). Said another way, the baryon is not a colored object. This is where the optical analogy comes in. Light of three primary colors — red, green and blue — when mixed in equal amounts, produces the sensation of white color on our eyes.

In analogy, the different quark colors are also called red, green and blue, and their superposition in a baryon produces a colorless object.

Mesons contain a quark and an antiquark. They are also colorless. The color of the quark is exactly counterbalanced by that of the antiquark. Carrying the analogy with optical color, we can say that the antiquarks carry complementary colors. A red quark, along with a blue-green (or cyan) antiquark, can form a colorless object. Similarly, the complementary colors of green and blue are magenta and yellow respectively.

Thus the idea of color saves Pauli exclusion principle for quarks. But this should not be taken to mean that color is only a trick designed to solve that problem. In fact, the entire mystery of strong interactions lies in the color properties of objects. Just like the electric charge plays a fundamental role in electromagnetic interactions, color plays a fundamental role in strong interactions. This led to the theory of strong interactions called *quantum chromodynamics*. After some infrastructural work of Ch. 11, this theory will be discussed in Ch. 12.

Chapter 11

Non-abelian gauge theories

The idea of gauge invariance was introduced in Ch. 5, where it was shown that the interactions of QED, or quantum electrodynamics, arise out of the requirement of the symmetry of the Lagrangian under a local U(1) transformation. One can ask whether the same idea can be extended to other groups. Yang and Mills showed that it can be done, and we discuss the general structure of the resulting theories in this chapter. In some of the later chapters, we will see that this idea is crucial in understanding the dynamics of strong and weak interactions.

11.1   Local SU(N) invariance

As an example, let us consider SU(N) transformations on a set of objects. The objects, taken all together, must form some representation R of SU(N), otherwise we are going to get new objects outside the set while performing the SU(N) transformations. Let us denote these objects by

$$
\Psi = \begin{pmatrix} \psi_1 \\ \psi_2 \\ \vdots \\ \psi_{d_R} \end{pmatrix},
\tag{11.1}
$$

where d_R is the dimensionality of the representation R. Let us further suppose that each element of Ψ is a Dirac field, and that each of them has the same mass m. Then the free Lagrangian containing these N fields is given by

$$
\mathscr{L}_0 = \sum_k \left(\overline{\psi}_k i \gamma^\mu \partial_\mu \psi_k - m \overline{\psi}_k \psi_k \right).
\tag{11.2}
$$

With the help of the notation introduced in Eq. (11.1), we can write this Lagrangian in a compact form as follows:

$$
\mathscr{L}_0 = \overline{\Psi} i \gamma^\mu \partial_\mu \Psi - m \overline{\Psi} \Psi.
\tag{11.3}
$$

Since each Dirac field consists of four components, the object Ψ has really $4d_R$ number of components. But for most of the arguments contained in this chapter, it is not really necessary to remember the different components of each ψ_k. Only when something like a Dirac matrix appears in an expression, as it does in Eq. (11.3), we need to realize that the matrix appears with each of the d_R fields written down in Eq. (11.1).

Under an $SU(N)$ transformation, Ψ will change as

$$\Psi \longrightarrow \Psi' = U\Psi \,, \tag{11.4}$$

where U is a matrix that represents a group element in the representation R. Let us see how the Lagrangian of Eq. (11.3) behaves under this transformation. The transformed Lagrangian would be

$$\begin{aligned}
\mathscr{L}_0' &= \overline{\Psi}' i\gamma^\mu \partial_\mu \Psi' - m\overline{\Psi}' \Psi' \\
&= (U\Psi)^\dagger \gamma_0 i\gamma^\mu \partial_\mu U\Psi - m(U\Psi)^\dagger \gamma_0 U\Psi \,..
\end{aligned} \tag{11.5}$$

As we said, matrices like γ_0 can be attached to each element of Ψ, so we can write

$$(U\Psi)^\dagger \gamma_0 = \Psi^\dagger \gamma_0 U^\dagger = \overline{\Psi} U^\dagger \,. \tag{11.6}$$

The mass term of $\mathscr{L}_0'$ can then be written as

$$-m\overline{\Psi} U^\dagger U\Psi \,, \tag{11.7}$$

which is the same as the mass term in the original Lagrangian, since

$$U^\dagger U = 1 \,. \tag{11.8}$$

For the derivative term, with the same argument, we obtain

$$\overline{\Psi} i\gamma^\mu U^\dagger \partial_\mu U\Psi \,. \tag{11.9}$$

The derivative acts on everything to its right. If U is independent of the position, it can be taken outside the derivative, and we can apply Eq. (11.8) once again to conclude that this term is also the same as the derivative term in Eq. (11.3). This means that the Lagrangian of Eq. (11.3) has a global $SU(N)$ invariance.

However, the previous discussion also shows that the Lagrangian of Eq. (11.3) does not have a local $SU(N)$ invariance. If the elements of the matrix U are functions of spacetime, then we would obtain

$$\mathscr{L}_0' - \mathscr{L}_0 = \overline{\Psi} i\gamma^\mu U^\dagger (\partial_\mu U)\Psi \,, \tag{11.10}$$

where the parentheses limit the range of applicability of the derivative. Earlier in §3.4.1, we showed that an $SU(N)$ group contains $N^2 - 1$ independent parameters, so the derivative of U would contain the derivatives of all these parameters.

If we really want a local invariance, we will have to change the Lagrangian. We mimic what we did in §5.1.2 and write the new Lagrangian in terms of a covariant derivative as

$$\mathscr{L} = i\overline{\Psi}\gamma^\mu D_\mu \Psi - m\overline{\Psi}\Psi, \qquad (11.11)$$

where

$$D_\mu = \partial_\mu + igT_a A_\mu^a. \qquad (11.12)$$

Here, g is a constant, like the constant e that appears in Eq. (5.8, $p\,114$). The T_a's denote generators of the group SU(N) in the representation R to which the components of Ψ belong. These are the only items on the right hand side of Eq. (11.12) that carry the properties of Ψ, and are analogous to the quantity Q of Eq. (5.8, $p\,114$). And finally, A_μ^a's are a new set of fields, analogous the photon field that appeared in the corresponding equation for electrodynamics. The number of such fields is equal to the number of generators, i.e., equal to the number of independent parameters necessary to denote the group elements. These fields, for a general *gauge group*, are called *gauge fields*. Alternatively, the quanta of the gauge fields are called *gauge bosons*, since obviously these particles have the same spin as the photon, and are therefore bosons. For specific gauge groups, specific names are used, e.g., we used the name *photon* for the gauge field of electromagnetic U(1) gauge symmetry.

For the sake of convenience, let us define the object

$$\mathbb{A}_\mu \equiv T_a A_\mu^a, \qquad (11.13)$$

which is a collection of four matrices corresponding to different possible values for the Lorentz index, each matrix having the same size as any of the T_a's. Thus,

$$D_\mu = \partial_\mu + ig\mathbb{A}_\mu. \qquad (11.14)$$

We now ask the question, would the Lagrangian of Eq. (11.11) be invariant under the transformation of Eq. (11.4), even if the matrix U is a function of spacetime? Obviously, the mass term would be invariant just as it was with constant U. The other term would be invariant only if, under the transformation,

$$(D_\mu\Psi)' = UD_\mu\Psi. \qquad (11.15)$$

This means that

$$\partial_\mu\Psi' + ig\mathbb{A}_\mu'\Psi' = U(\partial_\mu\Psi + ig\mathbb{A}_\mu\Psi). \qquad (11.16)$$

Using the transformation property of Ψ from Eq. (11.4), this can be rewritten as

$$(\partial_\mu U)\Psi + ig\mathbb{A}'_\mu U\Psi = igU\mathbb{A}_\mu\Psi. \tag{11.17}$$

Since this equality has to be satisfied for arbitrary Ψ, we obtain the condition

$$\mathbb{A}'_\mu = U\mathbb{A}_\mu U^{-1} + \frac{i}{g}(\partial_\mu U)U^{-1}, \tag{11.18}$$

or equivalently,

$$\mathbb{A}'_\mu = U\mathbb{A}_\mu U^{-1} - \frac{i}{g}U(\partial_\mu U^{-1}), \tag{11.19}$$

which shows how $\mathbb{A}_\mu$ has to transform, along with the transformation of Ψ defined in Eq. (11.4), in order that the Lagrangian of Eq. (11.11) remains invariant under the combined transformation of Ψ and $\mathbb{A}_\mu$.

☐ **Exercise 11.1** Show the equivalence of Eqs. (11.18) and (11.19) by verifying that, for a matrix U whose elements depend on x^μ,

$$\partial_\mu U^{-1} = -U^{-1}(\partial_\mu U)U^{-1}. \tag{11.20}$$

☐ **Exercise 11.2** Use $U = \exp(-ieQ\theta)$ as would be appropriate for a $U(1)$ symmetry and verify that the transformation rule for A_μ in this case is the same as that obtained in §5.1.

11.2   Gauge fields

11.2.1   Transformation properties of gauge fields

We started with an arbitrary representation Ψ of $SU(N)$. The transformation on Ψ was given in Eq. (11.4), where U is a matrix in the representation R. Then we used the Lagrangian of Eq. (11.11) which contains the gauge fields. If we claim that this Lagrangian has to be gauge invariant, that should fix how the gauge fields should behave under the gauge transformation.

In a sense, this transformation has been given in Eq. (11.18). But looking at it, one feels a little uneasy. Transformation of the gauge fields should not depend on the representation of Ψ that we started with. In fact, the same gauge fields should be able to work with all representations, otherwise the program of localizing the symmetry becomes meaningless. But the transformation rule in Eq. (11.18) contains the matrices U which belong to a particular representation.

This is true, but it should be noted that Eq. (11.18) expresses the transformation rule for the matrix $\mathbb{A}_\mu$, which was defined in Eq. (11.13). These matrices obviously depend on the representation of the generators used in the definition. So, there is no wonder in the fact that the transformation rule for these matrices depends on the representation.

Let us try to strip the transformation rule, Eq. (11.18), of the representation matrices and find out how the gauge fields themselves behave under a gauge transformation. For this, we use Eq. (11.13) into Eq. (11.18) and write

$$T_a A_\mu^{a\prime} = U T_a U^{-1} A_\mu^a + \frac{i}{g}(\partial_\mu U)U^{-1} . \tag{11.21}$$

Let us now recall that the matrix U can be written in terms of the generators in the form

$$U = \exp(-ig T_b \theta^b) , \tag{11.22}$$

where θ^b's are the parameters of the transformation. Thus,

$$\partial_\mu U = (-ig T_b \partial_\mu \theta^b)U , \tag{11.23}$$

so that the last term on the right hand side of Eq. (11.21) is $T_a \partial_\mu \theta^a$, which resembles the gauge transformation term for the U(1) gauge field that was shown in Eq. (5.10, *p 114*). The first term on the right hand side of Eq. (11.21) has no analogy for the U(1) case, for reasons that will be clear soon. For small θ^a's, we can expand U to first order in these parameters. Then

$$\begin{aligned} U T_a U^{-1} &= (1 - ig T_b \theta^b)T_a(1 + ig T_c \theta^c) + \mathcal{O}\left(\theta^2\right) \\ &= T_a + ig[T_a, T_b]\theta^b + \mathcal{O}\left(\theta^2\right) \\ &= T_a - g f_{abc} T_c \theta^b + \mathcal{O}\left(\theta^2\right) , \end{aligned} \tag{11.24}$$

where the quantities denoted by f_{abc} are defined by

$$[T_a, T_b] = i f_{abc} T_c , \tag{11.25}$$

i.e., they are the structure constants of the group. Putting the results of Eqs. (11.23) and (11.24) into Eq. (11.21) and changing the dummy variables at places, we obtain

$$T_a A_\mu^{a\prime} = T_a A_\mu^a + g f_{bca} T_a \theta^b A_\mu^c + T_a \partial_\mu \theta^a , \tag{11.26}$$

by using the antisymmetry of the structure constants in their first two indices, and neglecting higher order terms in the parameters θ^a. This means that the gauge transformation rule for the gauge fields is given by

$$A_\mu^{a\prime} = A_\mu^a + g f_{bca} \theta^b A_\mu^c + \partial_\mu \theta^a . \tag{11.27}$$

The structure constants appear in the algebra of the group, Eq. (11.25), and are properties of the group itself. Thus the transformation law of Eq. (11.27) is free from any reference of the representation of the field Ψ that we started with.

It is clear that we are *not* doing something completely different from the U(1) case. It would be more reasonable to say that we are performing an extension to that theory. If the gauge group were abelian, all structure constants would vanish, and Eq. (11.27) would look like Eq. (5.10, *p 114*) which showed how the photon field behaves under gauge transformations. The new features appear through the structure constants, and hence are relevant only for non-abelian gauge groups. Because of this reason, the field theories we are describing here are called *non-abelian gauge theories*. Alternatively, they are called *Yang–Mills theories*, after the names of the two persons who first described such theories.

Turning back to Eq. (11.27), we find that when a gauge transformation is performed, the gauge fields mix with one another through the term containing the structure constants. The nature of this transformation can be understood by considering global transformations, for which the term $\partial_\mu \theta^a$ vanishes. Using the definition of the adjoint representation given in Eq. (10.26, *p 262*), we can write

$$A_\mu^{a\,\prime} = A_\mu^a - ig\left(t_b^{(\mathrm{ad})}\right)_{ac} \theta^b A_\mu^c \,. \qquad (11.28)$$

Compare it now with the gauge transformation of any other multiplet, for example the one given in Eq. (11.4). Using infinitesimal parameters and putting in the indices explicitly, the rule can be written as

$$\Psi_i' = \Psi_i - ig(T_b)_{ij}\theta^b \Psi_j \,. \qquad (11.29)$$

This equation is valid for any multiplet Ψ transforming according to any representation. When some multiplet transforms according to the adjoint representation, the matrix indices i, j appearing in this equation become the same as the indices for the group parameters, and then we get exactly Eq. (11.28). This proves that the gauge fields transform according to the adjoint representation of the gauge group.

While talking about U(1) gauge groups, we commented that there is a multiplicative arbitrariness in the definition of the gauge coupling constant. Here, g is the gauge coupling constant, and it appears everywhere multiplied by some generator of the gauge group. So it seems that there is a multiplicative arbitrariness in the definition of the gauge coupling constant and the gauge group generators as well. Strictly speaking, this is correct. But it is also true that in practice, this arbitrariness is defunct for a non-abelian group, because the generators have a 'natural' normalization. This comes about from the so-called *ladder operators* that can be formed as combinations of group generators, which can change the eigenvalues of a commuting set of operators for a state. As a familiar example, we can go back to the rotation group, where the states can be taken as eigenstates of J^2 and J_z. The ladder operator $J_x + iJ_y$, acting on such an eigenstate, changes it to a state with an increased eigenvalue of J_z. The amount of change in the eigenvalue is the same no matter which state $J_x + iJ_y$ operates on, and this sets up a natural scale. It would be silly not to take this change as some sort of a unit, and everyone does that. Even when the group is not the rotation group but some sort of internal group, the same comments apply. This sets up the normalization of the generators, and effectively it fixes the normalization of the gauge coupling constant as well.

11.2.2 Field-strength tensor

The gauge fields, we recall, were introduced to establish local invariance to the free Lagrangian of Ψ. Once they are introduced, we should allow also other terms in the Lagrangian which may not contain Ψ but involve the gauge fields. Here again, we can take a clue from the U(1) case, where there was a term which contained only the photon field and nothing else. These involved the rank-2 tensor $F_{\mu\nu}$, called the *field-strength tensor*. For gauge fields as well, there might be such *pure gauge terms*, and in order to construct them, we should first find the field strength tensor for the non-abelian case.

In Eq. (4.16, *p 65*), we defined the field-strength tensor for the abelian case. This equation cannot be used directly, because it involves partial derivatives. We need to find a definition in terms of the covariant derivative. For this, we note that the covariant derivative of the abelian case satisfies the equation

$$[D_\mu, D_\nu]f = ieQF_{\mu\nu}f \tag{11.30}$$

for any function f.

□ **Exercise 11.3** *Check and verify Eq. (11.30), with D_μ defined in Eq. (5.8, p 114).*

Let us try to extend this to the case of non-abelian theories and write

$$[D_\mu, D_\nu]f = igT^a F^a_{\mu\nu} f \,. \tag{11.31}$$

Note that

$$D_\mu D_\nu f = (\partial_\mu + igT_b A^b_\mu)(\partial_\nu + igT_c A^c_\nu)f$$
$$= \partial_\mu \partial_\nu f + igT_b A^b_\mu \partial_\nu f + igT_c \partial_\mu(A^c_\nu f) - g^2 T_b T_c A^b_\mu A^c_\nu f \,. \tag{11.32}$$

Subtracting the expression for $D_\nu D_\mu f$ which can be obtained merely by interchanging the Lorentz indices in the formula above, we obtain

$$igT^a F^a_{\mu\nu} f = igT_a(\partial_\mu A^a_\nu)f - igT_a(\partial_\nu A^a_\mu)f - g^2[T_b, T_c]A^b_\mu A^c_\nu f \,, \tag{11.33}$$

which shows that

$$F^a_{\mu\nu} = \partial_\mu A^a_\nu - \partial_\nu A^a_\mu - g f_{bca} A^b_\mu A^c_\nu \,. \tag{11.34}$$

Obviously, there is a new feature appearing in this definition in the form of a quadratic term in the gauge fields, which was not there for abelian gauge theories because of the vanishing of the structure constants.

11.2.3   Pure gauge Lagrangian

Let us now check how the field-strength tensor behaves under a gauge transformation. It is easy to see it through the combination

$$\mathbb{F}_{\mu\nu} = T^a F^a_{\mu\nu} \,, \tag{11.35}$$

a matrix defined in analogy with the matrix associated with the gauge fields that was defined in Eq. (11.13). We can write it as

$$\mathbb{F}_{\mu\nu} = \partial_\mu \mathbb{A}_\nu - \partial_\nu \mathbb{A}_\mu + ig[\mathbb{A}_\mu, \mathbb{A}_\nu] \,. \tag{11.36}$$

If we perform a gauge transformation on it, the result will be denoted by $\mathbb{F}'_{\mu\nu}$, which can be obtained by replacing all $\mathbb{A}$ on the right hand side by $\mathbb{A}'$, which is the gauge transform of $\mathbb{A}$. Using Eq. (11.18) for $\mathbb{A}$, it is straightforward to obtain the result

$$\mathbb{F}'_{\mu\nu} = U\mathbb{F}_{\mu\nu}U^{-1}. \tag{11.37}$$

Thus, the field-strength tensor transforms homogeneously, unlike the gauge fields which have an inhomogeneous term in their transformation rule, as shown by the last term of Eq. (11.18).

□ **Exercise 11.4** Verify Eq. (11.37).

The transformation law of Eq. (11.37) also shows that the quantity

$$\text{Tr}\left(\mathbb{F}^{\dagger}_{\mu\nu}\mathbb{F}^{\mu\nu}\right) \equiv \text{Tr}(T_a^{\dagger}T_b)\, F_{\mu\nu}^{a\dagger}F_b^{\mu\nu} \tag{11.38}$$

is gauge invariant, and is therefore a good candidate for the Lagrangian of the gauge fields. The trace of the product of two generators depends on the representation of the generators. However, in §11.2.4 we show that in any representation the generators can be made to satisfy a relation of the sort

$$\text{Tr}(T_a^{\dagger}T_b) = C\delta_{ab} \tag{11.39}$$

where C is a non-negative number which determines the normalization of the generators in the representation. Putting this in, we can write the Lagrangian for the gauge fields as

$$\mathscr{L}_{\text{gauge}} = -\frac{1}{4}F_{\mu\nu}^{a\dagger}F_a^{\mu\nu}, \tag{11.40}$$

where the numerical co-efficient has been adjusted by invoking the analogy with an abelian gauge field theory. This is called the *pure gauge Lagrangian* because it contains no other field except the gauge fields.

It should be noted that the pure gauge Lagrangian cannot contain any non-derivative term except the non-derivative terms in $F_{\mu\nu}^a$. Using the gauge fields A_μ^a directly, we cannot produce any term that is gauge invariant, because of the inhomogeneous term in the transformation law of the gauge fields. In particular, a term like $A_\mu^a A_a^\mu$, though Lorentz invariant, is not gauge invariant. Therefore, in a gauge invariant theory, the non-abelian gauge fields must be massless, just like their abelian counterparts.

There is another convention of defining the gauge field and the field strength tensor that is used in many texts. The gauge fields defined in this convention is the combination gA_μ of our convention. In order to distinguish it from our convention, let us use a different typeface and write $\mathsf{A}_\mu = gA_\mu$. One then defines the field-strength tensor as

$$\mathsf{F}_{\mu\nu}^a = \partial_\mu \mathsf{A}_\nu^a - \partial_\nu \mathsf{A}_\mu^a - f_{bca}\mathsf{A}_\mu^b \mathsf{A}_\nu^c, \tag{11.41}$$

which turns out to be g times the object that we defined in Eq. (11.34). In this convention, the pure gauge Lagrangian is therefore given by

$$\mathscr{L}_{\text{gauge}} = -\frac{1}{4g^2}\mathsf{F}_{\mu\nu}^{a\dagger}\mathsf{F}_a^{\mu\nu}. \tag{11.42}$$

We will *not* use this convention. It is mentioned here only to make the reader conscious of the fact that some texts do indeed use it, and it should be remembered while comparing formulas derived here with those derived in this other convention.

☐ **Exercise 11.5** Add the pure gauge Lagrangian of Eq. (11.40) to the terms involving fermion fields given in Eq. (11.11). Using the infinitesimal changes in the fermion and gauge fields given earlier in the text, show that the Noether currents for gauge transformations are given by

$$J_a^\mu = \overline{\Psi}\gamma^\mu T_a \Psi + f_{abc} F_b^{\mu\nu} A_\nu^{c\dagger} . \qquad (11.43)$$

[**Note :** *The last term in the current shows that the gauge fields themselves carry the gauge charge.*]

☐ **Exercise 11.6** Show that the Euler–Lagrange equation for the gauge fields is given by

$$\partial_\mu F_a^{\mu\nu} = J_a^\nu . \qquad (11.44)$$

Using the form of the Noether current from Eq. (11.43), show that this equation can also be written in the form

$$D_\mu F_a^{\mu\nu} = j_a^\nu , \qquad (11.45)$$

where j_a^ν is the part of J_a^ν that involves all fields other than the gauge fields, and D_μ is the gauge-covariant derivative.

11.2.4   Mathematical interlude

In §10.3.1, we argued that the generators of any SU(N) group can be taken to be traceless hermitian matrices. With hermitian generators, the structure constants would be real numbers, and they will also be completely antisymmetric in the three indices, as we will show a little later. Thus, we can be somewhat nonchalant about the particular order of the gauge indices in the definition of the field strength tensor, e.g., we can write f_{abc} in place of f_{bca} in Eq. (11.34). Similarly, one could do away with the dagger signs in Eqs. (11.39) and (11.40) if one uses hermitian generators, which also implies that A_μ^a are hermitian fields.

Despite these notational simplifications, we will adhere to a more general notation in what follows. The primary reason for this is that, in some cases, it is convenient to use non-hermitian generators, as will be seen in Ch. 12, Ch. 16 and Ch. 19.

It is worthwhile to note that the generators can always be chosen so that they satisfy Eq. (11.39). As a first step, note that in any representation,

$$\mathrm{Tr}(T_a^\dagger T_a) = \sum_{i,j} \left|(T_a)_{ij}\right|^2 \qquad \text{(no sum on } a). \qquad (11.46)$$

This is the sum of the absolute square of all elements of the matrix T_a, and therefore can be zero if and only if each element of T_a is zero. All T_a vanish only in the singlet representation of the group, and in this representation Eq.

(11.39) will be trivially satisfied with $C = 0$. In any other representation, suppose we start with a set of generators called T'_a which satisfy the relation

$$\text{Tr}(T'^\dagger_a T'_b) = \alpha_{ab} \tag{11.47}$$

for some set of numbers α_{ab}. Obviously, α_{ab} will be real for $a = b$. We can now make linear combinations

$$T_1 = \sqrt{\frac{C}{\alpha_{11}}}\, T'_1,$$

$$T_2 = \sqrt{\frac{C}{\alpha_{11}(\alpha_{11}\alpha_{22} - |\alpha_{12}|^2)}}\left(\alpha_{12}T'_1 - \alpha_{11}T'_2\right), \tag{11.48}$$

and so on, so that the unprimed generators satisfy Eq. (11.39). This is the Gram-Schmidt orthogonalization algorithm that can be applied to any vector space. The resulting objects are orthogonal, which means that the inner product of two different objects is always zero. In this case where we are dealing with matrices which form the representation of generators of a group, the inner product is defined by the left hand side of Eq. (11.39).

□ **Exercise 11.7** *Show that, with the definition given in Eq. (11.47), $\alpha_{11} \geq 0$, $\alpha_{11}\alpha_{22} \geq |\alpha_{12}|^2$ etc., so that the numerical co-efficients appearing on the right-hand sides of Eq. (11.48) are all real.*

□ **Exercise 11.8** *The inner product of two members A and B of a vector space is a number (complex in general), which can be denoted by $\langle A, B \rangle$. The definition must satisfy the following conditions:*

a) $\langle A, B \rangle = \langle B, A \rangle^*$. This implies that $\langle A, A \rangle$ is real.

b) $\langle A, A \rangle = 0$ if and only if A is a null vector, i.e., if $A + C = C$ for any vector C.

c) If α and β are two numbers, $\langle C, \alpha A + \beta B \rangle = \alpha \langle C, A \rangle + \beta \langle C, B \rangle$.

d) The triangle inequality is satisfied, i.e., $|\langle A, B \rangle| \leq |\langle A, C \rangle| + |\langle C, B \rangle|$.

Show that all these conditions are satisfied if we define the left hand side of Eq. (11.39) as the inner product of T_a and T_b. (Property 'b' has already been proved in the text.)

In order to discuss group invariance properties with possible non-hermitian generators, we will introduce the notation

$$T_{\bar{a}} \equiv T^\dagger_a. \tag{11.49}$$

The summation convention on repeated indices would work irrespective of the presence or absence of a bar on the index. Thus, for example, $T_{\bar{a}} T_a$ will be equivalent to $T^\dagger_a T_a$, with an implied sum over the generators.

Taking the hermitian conjugates of both sides of Eq. (11.25), we obtain

$$f_{\bar{a}\bar{b}\bar{c}} = f^*_{abc}. \tag{11.50}$$

As for the symmetry properties of the structure constants, we have already mentioned the antisymmetry with respect to the first two indices. For interchanges involving the third index, we can multiply both sides of Eq. (11.25) by $T_d^\dagger$, i.e., $T_{\bar{d}}$, from the left and take traces of both sides. Using Eq. (11.39), we obtain

$$iC f_{abd} = \text{Tr}\left(T_{\bar{d}} T_a T_b\right) - \text{Tr}\left(T_{\bar{d}} T_b T_a\right). \tag{11.51}$$

Replacing b and $\bar{d}$, and therefore d by $\bar{b}$, in this equation, we obtain

$$iC f_{a\bar{d}\bar{b}} = \text{Tr}\left(T_b T_a T_{\bar{d}}\right) - \text{Tr}\left(T_b T_{\bar{d}} T_a\right). \tag{11.52}$$

Using the cyclic property of the traces, we can now conclude that $f_{abd} = -f_{a\bar{d}\bar{b}}$. The properties of the structure constants under interchange of indices are therefore as follows:

$$f_{abc} = -f_{bac} = f_{b\bar{c}\bar{a}} = -f_{\bar{c}\bar{b}\bar{a}}. \tag{11.53}$$

Evidently, if we choose hermitian generators for which the indices a and $\bar{a}$ are the same, these relations tell us that the structure constants will be completely antisymmetric in the indices.

We want to point out here that the orthogonalization process, shown in Eq. (11.48), does not really determine the normalization constant C. It only assumes that C is positive, which can be easily inferred from Eq. (11.46). But the value of C can be adjusted by scaling the generators. We will normalize the generators such that, for any SU(N),

$$C^{(f)} = \frac{1}{2} \tag{11.54}$$

in the fundamental representation, which has been indicated by the parenthesized superscript on the left hand side. This agrees with the choice of the fundamental representation of SU(3) generators given in Eq. (10.16, *p 259*), and of the usual representation of SU(2) generators in terms of the Pauli matrices.

Once this is fixed, the normalizations for all other representations are fixed because all other representations can be built up from the fundamental, as explained in §10.3. For the adjoint representation of SU(N), this normalization implies

$$C^{(\text{ad})} = N, \tag{11.55}$$

which will be proved presently.

☐ **Exercise 11.9** Verify Eq. (11.55) for SU(2), where the structure constants constitute the Levi-Civita symbol.

☐ **Exercise 11.10** With the help of the interchange properties summarized in Eq. (11.53), show that the Jacobi identity, shown in Eq. (10.28, *p 262*), can be written as

$$f_{abe} f_{cde} + f_{ace} f_{dbe} + f_{ade} f_{bce} = 0. \tag{11.56}$$

Another important quantity, related to the normalization of the generators, is the so-called *Casimir invariant* of rank-2, denoted by $C_2^{(R)}$ for a representation R. To motivate it, note that the combination $T_a^\dagger T_a$ commutes with any of the generators. The reason is that

$$[T_a^\dagger T_a, T_b] = [T_{\bar a} T_a, T_b] = T_{\bar a}[T_a, T_b] + [T_{\bar a}, T_b]T_a$$
$$= i f_{abc} T_{\bar a} T_c + i f_{\bar a bc} T_c T_a, \qquad (11.57)$$

Since the index a is summed over, we can change the dummy and write the first term as $i f_{\bar a bc} T_a T_c$ as well, obtaining $i f_{\bar a bc}(T_a T_c + T_c T_a)$. Since the combination of the generators in the parentheses is symmetric under the exchange of the indices a and c, whereas according to Eq. (11.53), $f_{\bar a bc}$ is antisymmetric under the same exchange, the result vanishes.

If T_a denotes generators in an irreducible representation, this implies, through Schur's lemma, that the combination $T_a^\dagger T_a$ must be a multiple of the unit matrix:

$$T_a^\dagger T_a = C_2 \mathbb{1}, \qquad (11.58)$$

which defines the Casimir invariant C_2 for the irreducible representation.

It is easy to see that the value of C_2 will depend on the normalization of the generators. This can be established by contracting both sides of Eq. (11.39) by δ_{ab}. On the right hand side, the combination $\delta_{ab}\delta_{ab}$ will be equal to the number of generators of the group, which is the dimensionality of the adjoint representation, $d_{(\mathrm{ad})}$. On the left hand side, we will obtain the Casimir invariant multiplied by the trace of the unit matrix, and this trace is the dimensionality of the representation of the generators. We can write this as

$$C_2^{(R)} d_{(R)} = C^{(R)} d_{(\mathrm{ad})}. \qquad (11.59)$$

For the fundamental representation, the value of C_2 can be obtained through a generalization of Eq. (10.18, *p 259*). For the SU(N) group with the normalization of the generators fixed through Eq. (11.54), the generalization is given by

$$\delta_{il}\delta_{jk} = \frac{1}{N}\delta_{ij}\delta_{kl} + 2\left(t_a^\dagger\right)_{ij}\left(t_a\right)_{kl}, \qquad (11.60)$$

where t_a's stand for generators in the fundamental representation, and the indices i, j, k, l run from 1 to N. Contracting this identity by δ_{jk}, one obtains

$$N\delta_{il} = \frac{1}{N}\delta_{il} + 2\left(t_a^\dagger t_a\right)_{il}, \qquad (11.61)$$

which shows that

$$C_2^{(f)} = \frac{N^2 - 1}{2N}. \qquad (11.62)$$

With the help of this equation, we can now find the normalization and the Casimir invariant for any other representation through Eq. (11.59). For example, putting $d_{(\text{ad})} = N^2 - 1$, we find

$$C_2^{(\text{ad})} = N. \tag{11.63}$$

Also, Eq. (11.59) shows that C should equal C_2 for the adjoint representation, which verifies Eq. (11.55). Once again, we emphasize that the values of the normalization constant and Casimir invariant of any representation depends on the the normalization constant $C^{(f)}$ of the fundamental representation generators, and are fixed once $C^{(f)}$ is chosen.

☐ **Exercise 11.11** ♪ *Show that, in any representation*

$$T_a T_b T_{\bar{a}} = \left(C_2 - \frac{1}{2} C_2^{(\text{ad})} \right) T_b. \tag{11.64}$$

☐ **Exercise 11.12** *Prove Eq. (11.60). How will this equation be modified if we do not commit to the convention of Eq. (11.54)?*

11.2.5   Free gauge Lagrangian

The pure gauge Lagrangian has been shown in Eq. (11.40). It contains $F_{\mu\nu}^a$ and its conjugate. The form of $F_{\mu\nu}^a$ was given in Eq. (11.34). From it, it follows that

$$F_{\mu\nu}^{a\dagger} = \partial_\mu A_\nu^{a\dagger} - \partial_\nu A_\mu^{a\dagger} - g f_{bca}^* A_\mu^{b\dagger} A_\nu^{c\dagger}. \tag{11.65}$$

Note that if we use non-hermitian generators, the gauge fields are also non-hermitian, and the structure constants can be complex.

The free Lagrangian of the non-abelian gauge field is the collection of terms which are quadratic terms in these fields. These are:

$$\mathscr{L}_{\text{free}} = -\frac{1}{4} (\partial_\mu A_\nu^a - \partial_\nu A_\mu^a)^\dagger (\partial^\mu A_a^\nu - \partial^\nu A_a^\mu). \tag{11.66}$$

For each value of the gauge index 'a', the terms present here are exactly similar to the terms in the Lagrangian for the free photon field. Therefore, for any gauge field, the Feynman rules coming from the quadratic terms are the same as those for the photon field. For example, we need to introduce a gauge-fixing term in order to obtain the propagator, and we choose it to be

$$\mathscr{L}_{\text{fix}} = -\frac{1}{2\xi} (\partial_\mu A_a^\mu)^\dagger (\partial_\nu A_a^\nu). \tag{11.67}$$

Once this is done, it is trivial to see that the propagator for any of the gauge bosons is of the form given in Eq. (4.148, *p 92*), and we can use the 't Hooft–Feynman gauge to write it in the form given in Eq. (4.149, *p 92*). As for physical or on-shell gauge bosons, we will be able to define a polarization vector which will satisfy the properties shown for the photon in Eq. (4.24, *p 67*) through Eq. (4.33, *p 68*). The Feynman rules for external gauge boson lines will be exactly the same as those given in Table 4.1 *(p 91)*.

11.3   Self-interaction of gauge bosons

The difference between abelian and non-abelian gauge fields lies in the term containing the structure constant in the expression for the field strength tensor. Note that this term is quadratic in the gauge fields. Thus, in the Lagrangian which is quadratic in the field strength tensor, there will be cubic as well as quartic terms in the gauge fields. These constitute interaction vertices involving gauge bosons only, a feature that was absent in abelian gauge theories. Let us discuss these vertices now.

As a prelude to this discussion, note that the last term of Eq. (11.65), the one involving the structure constant, can be written in a slightly different manner which will be helpful for us. Using the notation advocated in Eq. (11.49) and the relation in Eq. (11.50), we can write

$$f^*_{bca} A^{b\dagger}_\mu A^{c\dagger}_\nu = f_{\bar b \bar c \bar a} A^{\bar b}_\mu A^{\bar c}_\nu = f_{bc\bar a} A^b_\mu A^c_\nu \,, \tag{11.68}$$

the last step following by renaming the dummy indices. Thus

$$F^{a\dagger}_{\mu\nu} = \partial_\mu A^{\bar a}_\nu - \partial_\nu A^{\bar a}_\mu - g f_{bc\bar a} A^b_\mu A^c_\nu \,. \tag{11.69}$$

11.3.1   Quartic vertices of gauge bosons

We first discuss vertices involving four gauge bosons, because they do not contain any derivatives and are consequently somewhat simpler than the cubic vertices. From Eqs. (11.34) and (11.69), we can easily identify the quartic interaction present in Eq. (11.40). With slight changes in the names of the dummy indices, we write it as

$$\mathscr{L}_{\text{quartic}} = -\frac{1}{4} g^2 f_{a'b'\bar e} f_{c'd'e} \, g_{\alpha\gamma} g_{\beta\delta} \, A^\alpha_{a'} A^\beta_{b'} A^\gamma_{c'} A^\delta_{d'} \,. \tag{11.70}$$

The Feynman rule obtained from this Lagrangian term has been given in Fig. 11.1, where we have used the shorthand notation

$$E_{\alpha\beta\gamma\delta} \equiv g_{\alpha\gamma} g_{\beta\delta} - g_{\alpha\delta} g_{\beta\gamma} \,. \tag{11.71}$$

Remember that in our notation, the gauge fields can be complex, so it is important to distinguish between incoming and outgoing gauge bosons, which is the same as distinguishing between a particular gauge boson and its antiparticle. If hermitian generators and real gauge fields are used, this distinction need not be maintained.

Let us discuss how the single term of Eq. (11.70) gives rise to so many different terms in the Feynman rule for the vertex. It all depends on the different possibilities of assigning the particles in the vertex to the field operators in the Lagrangian. First, suppose that the gauge bosons carrying the gauge indices (a, b, c, d) in Fig. 11.1 have been created by the field operators in Eq. (11.70)

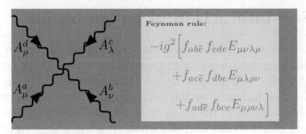

Figure 11.1: Feynman rule for quartic interactions of gauge bosons. The tensor E with four indices has been defined in Eq. (11.71).

which carry the gauge indices (a', b', c', d') respectively. This possibility gives an amplitude

$$-\frac{1}{4}g^2 f_{ab\bar{e}} f_{cde} g_{\mu\lambda} g_{\nu\rho}. \tag{11.72}$$

It is easy to see that if (a, b, c, d) correspond to (b', a', d', c'), the result would be the same since the structure constants are antisymmetric in the first two indices. Also, if the first pair is exchanged with the last pair, the result remains unchanged. So there are four contributions like that shown in Eq. (11.72), and they add up to give a contribution four times as large as the expression written in Eq. (11.72). Other correspondences between the set (a, b, c, d) and the gauge indices appearing in Eq. (11.70) provide other terms in the Feynman rule given in Fig. 11.1.

☐ **Exercise 11.13** Check the other terms of Fig. 11.1.

11.3.2   Cubic vertices of gauge bosons

We now look at the cubic terms present from Eq. (11.40). We first pick up such terms from Eqs. (11.34) and (11.69):

$$\mathscr{L}_{\text{cubic}} = \frac{1}{4}g \left[f_{bc\bar{a}} \left(\partial_\mu A_\nu^a - \partial_\nu A_\mu^a \right) + f_{bca} (\partial_\mu A_\nu^{\bar{a}} - \partial_\nu A_\mu^{\bar{a}}) \right] A_b^\mu A_c^\nu. \tag{11.73}$$

Changing some dummy indices and using the antisymmetry of the structure constants in the first two indices, we can write it as

$$\mathscr{L}_{\text{cubic}} = g f_{b'c'\bar{a}'} (\partial_\alpha A_\beta^{a'}) A_{b'}^\alpha A_{c'}^\beta. \tag{11.74}$$

As in the case of the quartic coupling, the Feynman rule for the cubic coupling will contain various terms, depending on the six possible ways that the field operators present in the cubic term can be associated with the three gauge bosons present at the vertex. There is an additional complication here, viz., that the interaction contains a derivative. In §4.10.3, we have discussed that one has to specify the momentum in order to define the Feynman rule for

Figure 11.2: Feynman rule for cubic interactions of gauge bosons. The momentum associated with each leg is given in the parentheses. The tensor E with four indices has been defined in Eq. (11.71).

such interactions. For the sake of definiteness, we take all momenta to be incoming, as indicated by the arrows in Fig. 11.2. In that case, as explained in §4.10.3, the derivative will give a factor of $-ip$ if p is the momentum of the associated particle.

Let us now go through the different ways that the three gauge bosons can be annihilated via the interaction of Eq. (11.74). For example, we have

$$
\begin{aligned}
(a,b,c) \leftarrow (a',b',c') &\quad:\quad gf_{bc\bar{a}}(-ip_\nu)g_{\mu\lambda}\,, \\
(a,b,c) \leftarrow (c',b',a') &\quad:\quad gf_{ba\bar{c}}(-ir_\nu)g_{\mu\lambda}\,, \\
(a,b,c) \leftarrow (b',a',c') &\quad:\quad gf_{ac\bar{b}}(-iq_\mu)g_{\nu\lambda}\,,
\end{aligned}
\tag{11.75}
$$

and so on. Combining all six terms like these ones, we obtain the Feynman rule for the cubic vertex shown in Fig. 11.2 after using the symmetry properties of the structure constants given in Eq. (11.53). More explicitly, the Feynman rule for the cubic interaction can be written as

$$
gf_{ab\bar{c}}\Big[(p_\nu - r_\nu)g_{\mu\lambda} + (r_\mu - q_\mu)g_{\nu\lambda} + (q_\lambda - p_\lambda)g_{\mu\nu}\Big]\,.
\tag{11.76}
$$

In §6.10, we talked about Furry's theorem, which basically says that the 3-point function involving three photon fields should vanish. For gauge bosons in a non-abelian gauge theory, we find cubic couplings in the Lagrangian, signifying the existence of 3-point functions right at the tree level. There is no conflict between these two statements, and no need for concluding that the charge conjugation invariance, pivotal in the proof of Furry's theorem, does not apply to non-abelian gauge theories. The important thing to notice is that the structure constants of the group appear in the coupling of Eq. (11.76). So, if the three legs of the vertex pertain to the same gauge boson, the vertex will still vanish, as was the case for the 3-point function of the photon.

11.4   Fadeev–Popov ghosts

The gauge-fixing term introduced in Eq. (11.67) is not enough to obtain consistent quantization of a non-abelian gauge theory. For reasons that we do not elaborate here, we have to add some unphysical fields which cannot be created or annihilated in any experiment, but are essential for the consistency

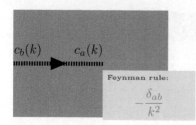

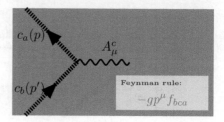

Figure 11.3: Feynman rules involving ghosts. Momenta on different legs are shown in parentheses.

of the quantum theory. These are called *Fadeev–Popov ghost fields*, which do not appear as the outer leg of any Feynman diagram: they can only appear as internal lines. They contribute a term to the Lagrangian which is of the form

$$\mathscr{L}_{\text{ghost}} = -ic_a^\dagger M_{ab} c_b \,, \tag{11.77}$$

where the c_a's are called *ghosts* and the $c_a^\dagger$'s *antighosts*, and the index a goes over all the generators. In other words, the ghosts transform like the adjoint representation of the gauge group. They are scalar fields which obey the Pauli exclusion principle: so in this sense they are weird objects. But, as we said, they are not real particles that can be produced or annihilated in a physical process.

We have not explained what M_{ab} is in Eq. (11.77). Its form depends on the gauge-fixing condition. The general prescription can be given by taking the gauge-fixing condition to be of the form

$$f^a(A_\mu) = 0 \,, \tag{11.78}$$

where the functional form can involve all gauge fields A_μ^a. Now, suppose we make an infinitesimal gauge transformation that changes A_μ^a to $A_\mu'^a$, as given in Eq. (11.27). This will change the content of the gauge-fixing functions, and let us write

$$\delta f^a(A_\mu) = M_{ab}\theta_b \,, \tag{11.79}$$

where the θ_b's are the parameters in the gauge transformation.

For the covariant gauge condition of Eq. (11.67), we have

$$f^a(A_\mu) = \partial_\mu A_a^\mu \,. \tag{11.80}$$

Using Eq. (11.27), we then obtain

$$\delta f^a(A_\mu) = \partial_\mu(\delta A_a^\mu) = \partial^\mu \Big(g f_{bca} \theta^b A_\mu^c + \partial_\mu \theta^a \Big) \,. \tag{11.81}$$

Thus the Fadeev–Popov ghost Lagrangian becomes

$$\mathscr{L}_{\text{ghost}} = -ic_a^\dagger \left(g f_{bca} \partial^\mu A_\mu^c + \partial^\mu \partial_\mu \delta_{ab} \right) c_b$$
$$= -ic_a^\dagger \partial^\mu \partial_\mu \delta_{ab} c_b + ig f_{bca} (\partial^\mu c_a^\dagger) A_\mu^c c_b \,, \tag{11.82}$$

where in the last step, we have added a total derivative term which should not affect the action. From this Lagrangian, we can find out the propagator of the ghost fields, as well as their interaction vertex with the gauge fields. The Feynman rules have been shown in Fig. 11.3.

In passing, we want to make two comments about ghosts. First, the ghost number is conserved, which means that a ghost line cannot begin or end within a diagram. In conjunction with the comment made earlier that ghosts appear only as internal lines, this statement means that ghosts can only make closed loops within a diagram.

The second comment is about ghosts in abelian theories. In Ch. 5, we have always worked with the gauge condition given in Eq. (4.21, *p 66*), which means that the function set to zero at the classical level is the same as that shown in Eq. (11.80). As seen from Fig. 11.3, the only interaction vertex involving ghosts is proportional to the structure constants of the group, which means that they vanish in an abelian theory. Thus, for the gauge-fixing term used throughout Ch. 5, ghosts have no interaction with anything in an abelian theory. This is the reason we have never discussed them while talking about the abelian gauge theory of QED. In principle, one can take other gauge-fixing terms as well. If we take a gauge-fixing term that contains higher powers of the abelian gauge field, the ghosts will have interaction vertices with the gauge fields, and their effects will have to be taken into account while performing loop calculations.

☐ **Exercise 11.14** *Consider an abelian gauge theory with a gauge-fixing term*

$$f(A_\mu) = \partial_\mu A^\mu + A_\mu A^\mu \,. \tag{11.83}$$

Find the ghost Lagrangian for this choice.

11.5   Interaction of gauge bosons with other particles

In a gauge theory, the gauge bosons fields must be there. In addition, there might be other fields. Such fields need not be spin-1 fields like the gauge fields. They can be spin-0 or spin-$\frac{1}{2}$. They will have to appear in multiplets that would transform like a representation of the gauge group. The gauge interactions of these fields would be determined by their representation under the gauge group.

In fact, in §11.1, we started with a multiplet like this. It was a multiplet of spin-$\frac{1}{2}$ fields. The gauge fields were introduced in Eq. (11.11) through the gauge covariant derivative defined in Eq. (11.12). The covariant derivative has

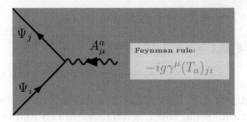

Figure 11.4: Gauge boson vertex with fermions.

two terms. One of them is the ordinary derivative, and this term contributes to the free Lagrangian of the fermion fields. The other term, containing the gauge fields, give the interaction of gauge fields with fermions. This term is

$$\mathscr{L}_{\text{int}} = -g\overline{\Psi}\gamma^{\mu}T_a\Psi A^a_{\mu}\,. \tag{11.84}$$

More explicitly, if we announce the components of the multiplet Ψ, we should write

$$\mathscr{L}_{\text{int}} = -g(\overline{\Psi})_j\gamma^{\mu}(T_a)_{ji}(\Psi)_i A^a_{\mu}\,. \tag{11.85}$$

From this, we can write the Feynman rule for the interaction of fermions with gauge bosons, as has been done in Fig. 11.4.

For gauge interactions of scalars, we will follow the same algorithm. In the free Lagrangian of a multiplet Φ of scalar fields, if we replace the ordinary derivatives by covariant derivatives, we obtain

$$\mathscr{L} = (D_\mu\Phi)^{\dagger}(D^\mu\Phi) - M^2\Phi^{\dagger}\Phi\,. \tag{11.86}$$

Once again, using the expression for D_μ from Eq. (11.12), we can extract the interaction terms that appear in this Lagrangian:

$$\mathscr{L}_{\text{int}} = ig(\partial_\mu\Phi)^{\dagger}(T_aA^\mu_a\Phi) - ig(T_aA^a_\mu\Phi)^{\dagger}(\partial^\mu\Phi)$$
$$+g^2(T_aA^a_\mu\Phi)^{\dagger}(T_bA^\mu_b\Phi)\,. \tag{11.87}$$

This includes vertices with one as well as two gauge bosons. Putting in the indices for the components of the multiplet explicitly, we can write

$$\mathscr{L}_{\text{int}} = ig(\partial_\mu\Phi^{\dagger}_j)(T_a)_{ji}A^\mu_a\Phi_i - ig\Phi^{\dagger}_j(T_a)_{ji}A^a_\mu(\partial^\mu\Phi_i)$$
$$+g^2g_{\lambda\rho}\Phi^{\dagger}_j(T_{a'}T_{b'})_{ji}\Phi_i A^\lambda_{a'}A^\rho_{b'}\,. \tag{11.88}$$

The Feynman rules for these vertices have been given in Fig. 11.5. Note that two terms appear in the quartic vertex, coming from the ways that the gauge indices a' and b' present in the interaction Lagrangian can be equal to the gauge indices a and b in the figure.

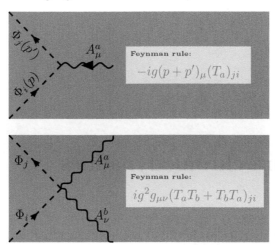

Figure 11.5: Gauge boson vertices with scalars. Momenta on different legs are shown in parentheses.

It should be noted that gauge bosons can connect particles within the same irreducible multiplet only. In other words, the two fermions that appear in Fig. 11.4 or the two scalars that appear in Fig. 11.5 must belong to the same multiplet of the gauge group. Of course, even if the two non-gauge particles belong to the same multiplet, their vertex may not exist with a particular gauge boson if the relevant matrix element of the generator(s) vanish. But if the two particles do *not* belong to the same multiplet, the gauge interaction vertex vanishes with all gauge bosons.

Chapter 12

Quantum chromodynamics

We introduced quark colors in Ch. 10 and mentioned that the idea is essential for keeping the baryon wavefunctions completely antisymmetric in the quarks. At the time this idea was introduced, probably no one thought that this idea would become crucial in the understanding of the nature of strong interactions. The modern theory of strong interactions is in fact based on the color property of quarks, and the theory is called *quantum chromodynamics* in analogy with quantum electrodynamics. And, following the same analogy, the acronym *QCD* is used for this theory.

12.1 SU(3) of color

The basic idea of QCD is that there is a local symmetry involving quark colors. In Ch. 11, we discussed how local symmetries give rise to interactions mediated by gauge bosons. When the local symmetry involves color transformations, the corresponding interactions mediated by the gauge bosons are strong interactions.

To present the idea in more detail, let us recall that any quark comes in three colors. With the optical analogy presented in §10.11.2, let us call these three colors red, green and blue. For a given flavor of quark q, let us consider these three colors to constitute the triplet representation of some group:

$$q \equiv \begin{pmatrix} q_r \\ q_g \\ q_b \end{pmatrix} \sim \mathbf{3} \,. \tag{12.1}$$

Obviously, the gauge group must admit an irreducible triplet representation. Among the unitary groups, only two groups have this property: SU(2) and SU(3). For SU(2), the triplet is the adjoint representation, which is real. If we identify the quarks in this representation, the quarks and the antiquarks would behave the same way under color. In reality, we know that quarks and antiquarks are not the same, but rather opposite so far as their electric charges or their baryon numbers are concerned. So it makes no sense to assume that their color properties would be the same. If we need a complex representation, we are forced to take the gauge group to be SU(3) in which the three colors would then constitute the fundamental representation. This is the gauge group of QCD.

In Ch. 10, we discussed an SU(3) symmetry involving quarks. Here we are discussing one. Mathematically the two structures are the same, but it should be clearly understood that the physical contexts are very different. In the SU(3) symmetry discussed in Ch. 10, the fundamental multiplet consists of the up, the down and the strange quarks, which are three different quark flavors. In the SU(3) that we introduced in this section, the fundamental multiplet consists of the three possible colors of a single flavor of quark. For this reason, the former symmetry is called the *flavor SU(3)* and denoted by $SU(3)_F$, whereas the new one is called the *color SU(3)* and denoted by $SU(3)_c$. We will often omit the subscript, hoping that it should be understandable from the context anyway.

Although both groups have the same structure, their status in the Lagrangian are not the same. The flavor symmetry, $SU(3)_F$, is an approximate symmetry which is realized only if we ignore the mass differences between the u, the d and the s quarks. Even in the limit when we ignore these details, the flavor symmetry is a global symmetry. On the other hand, the symmetry $SU(3)_c$ is an exact symmetry of the Lagrangian. And it is a gauge symmetry, which means that it governs the dynamics of the interactions.

From the discussion of §11.1, it is quite obvious how gauge symmetry can be implemented in this case. Consider first an SU(3) transformation on the quark colors. Recalling the generators of SU(3) in the fundamental representation given in Eq. (10.16, *p 259*), we can write such a transformation as

$$q(x) \rightarrow q'(x) = \exp\left(-i\frac{\lambda_a}{2}\theta_a\right) q(x), \qquad (12.2)$$

where the index a runs from 1 to 8. As described in Ch. 11, such a transformation will be a symmetry of the free Dirac Lagrangian for quarks if the transformation parameters θ_a do not depend on x. But if the symmetry is made local, i.e., the θ_a's depend on x, the same statement cannot be made. In that case, we must introduce eight vector bosons, in the manner shown in §11.1, in order to implement the symmetry. For the color SU(3) symmetry, these eight gauge bosons are called *gluons*.

The interaction of these gluons with the quarks can easily be deduced from the general formula given in §11.5. Note that there are two diagonal generators in SU(3), e.g., λ_3 and λ_8 shown in Eq. (10.13, *p 259*). When the corresponding gauge bosons interact with quarks, the color of the quark is not changed. However, different colors are treated differently because the diagonal elements are different. When any of the other six gauge bosons interact with the quarks, the quark color changes. For example, consider λ_1 and λ_2 of Eq. (10.13, *p 259*). Each of these two generators has two non-zero elements, in the 12 and 21 positions. This means that the gauge boson associated with these generators can change a red quark into a green quark and vice versa. Similarly, the gauge boson associated with λ_4 and λ_5 can interchange red and blue colors, whereas those associated with λ_6 and λ_7 can interchange green and blue colors of quarks.

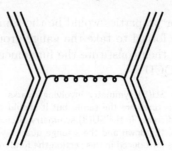

Figure 12.1: Gluon exchange mediates strong interactions. The three solid lines on each side are supposed to represent three quarks in a baryon like the nucleons or the Δ's. The exchanged line represents a gluon. We do not attempt to show how strong interactions are responsible in binding three quarks into a baryon. Only the interaction *between* baryons has been shown.

The physical interpretation of the gluons becomes much more clear if, instead of dealing with the hermitian generators that have been presented in Eq. (10.13, *p 259*), we take the six off-diagonal generators as

$$\lambda_{rg} = \frac{1}{\sqrt{2}}(\lambda_1 + i\lambda_2),$$

$$\lambda_{gb} = \frac{1}{\sqrt{2}}(\lambda_4 + i\lambda_5),$$

$$\lambda_{rb} = \frac{1}{\sqrt{2}}(\lambda_6 + i\lambda_7), \tag{12.3}$$

and their hermitian conjugates which will be called λ_{gr}, λ_{bg} and λ_{br} respectively. Thus, e.g., we will have

$$\lambda_{rg} = \begin{pmatrix} 0 & \sqrt{2} & 0 \\ 0 & 0 & 0 \\ 0 & 0 & 0 \end{pmatrix}. \tag{12.4}$$

If we associate a gauge boson to this generator, this gluon will be able to do only one thing: turn a red quark into a green quark. The hermitian conjugate gauge boson will be able to do just the opposite: turn a green quark into a red quark. Similarly, there will be other two pairs of gluons which will perform other color changes. The gluons associated with the diagonal generators do not change color, as we commented earlier.

Gluon exchanges between quarks would give rise to strong interactions, in the same manner that photon exchanges between two charged particles give rise to electromagnetic interactions. For example, consider the lowest order electron–electron interaction diagrams that were shown in Fig. 5.3 (*p 120*). Strong interaction between two hadrons would be mediated by similar-looking diagrams. The only big difference is that hadrons are not elementary particles

Figure 12.2: How gluon exchange between two nucleons can be seen as a meson exchange.

like the electron. A baryon contains three quarks. A possible diagram which induces interaction between two baryons has been shown in Fig. 12.1.

Of course, this is not a lone diagram. It is an example of what are called *spectator diagrams*, where one considers interaction between two hadrons by a gluon exchange between only one quark (or antiquark) in one hadron and one in the other, while the other quarks and antiquarks do not really participate in the process: they behave like spectators. Obviously, there can be more than one spectator diagrams, because the virtual gluon can be emitted from any of the quarks in one baryon and absorbed by any of the quarks in the other baryon. Different gluons can induce different color changes for the quarks. In addition, there will be infinitely many more non-spectator diagrams. These will include multiple gluon exchanges involving different quarks in either baryon. In fact, there must be gluon exchanges between the three quark lines that constitute each baryon, because the baryon is a bound state formed because of strong interaction between quarks.

Earlier, in §8.8, we described the interactions between nucleons by the exchange of mesons. Now we seem to preach that the interaction should be described by exchange of gluons. There is no conflict between these two descriptions. To see this, let us look at the diagram in Fig. 12.2. Needless to say, topologically this diagram is equivalent to that of Fig. 12.1: we have only added the arrows on one of the quark lines for the sake of clarity. As it appears now, the exchange part of the diagram consists of a quark line going one way and a quark line going the other. Alternatively, we can also say that we have a quark-antiquark pair going one way. Since such a pair represents a meson, it appears that a meson has been interchanged between the two nucleons. The gluon exchange between the nucleon and this meson can be absorbed in a definition of the nucleon-meson vertex. In the simplest diagram with just a single gluon exchanged, the meson must have spin 1 because it must carry the angular momentum of the gluon. However, there can be more diagrams with more gluons exchanged, and then spinless mesons would also be allowed. When the exchanged momentum is small, such processes will be dominated by the meson with smallest mass, i.e., the pion. As the exchanged momentum becomes larger, one needs to include mesons of higher and higher mass in order to continue with such effective pictures.

In the more general case where the exchanged momentum is large, the effective meson exchange picture does not work, and we must describe strong interactions in terms of gluon exchanges. As already indicated, the Lagrangian for gluon interactions is known through the general formalism of Yang-Mills theories. We might be tempted to believe that the rest is trivial: just as we had calculated the rates of different processes in QED, we should be able to calculate strong interaction effects. But it is not so simple. The reason is an obvious one: strong interactions are strong. In other words, the gauge coupling constant, the quantity that was denoted by g in Ch. 11, is not small. In the case of QED, the coupling constant e was small, and going to a higher order in perturbation theory automatically guaranteed a smaller result. So, in order to calculate any process with a given accuracy, diagrams up to a certain order were necessary. But in QCD, that is not the case. So it makes no sense to apply perturbation theory. This is the big problem.

12.2 Running parameters

But there is a solace: the coupling "constant" is not really a constant: it changes with the momentum scale of the interaction. This phenomenon is described, rather picturesquely, by the statement that the "coupling parameters run". It happens not just for the QCD coupling constant, but for other parameters in a Lagrangian. We will demonstrate the phenomenon with the example of the QED coupling constant.

12.2.1 Why parameters run

The essential reason for the running of parameters was encountered in §5.7, while discussing higher order effects in QED. To be precise, we found that the form factors that appear in the QED vertex function are functions of the momentum transfer.

There is one aspect of this momentum dependence that was intentionally not brought up in the earlier discussion. Let us look back at the expression for the charge form factor, $F_1(q^2)$, that was given in Eq. (5.158, *p 147*). Consider the region of momentum integration in which the mass and the momentum transfer can be neglected. In this region, the momentum integration that appears in this formula is

$$\int \frac{d^4k}{(k^2)^2} \, , \tag{12.5}$$

apart from constant factors. The indefinite integral is proportional to $\ln(k^2)$. This form is valid for very large values of k^2, ranging all the way to infinity.

When we put the limits, the result clearly diverges at the upper end. This divergence cannot be canceled from the contribution of the integral coming from smaller values of the integration variable. This is called *ultraviolet (or UV) divergence*. The one-loop contribution to the charge form factor is therefore infinite.

Certainly, this cannot be the final result in a physical theory, since the charge of a particle can be measured. To rectify this problem, we say that once we add the quantum corrections in the form of loop diagrams, we need to take into account corrections to the classical Lagrangian as well. In other words, the complete vertex function is given by an expression of the form

$$\Gamma_\lambda(q^2) = Q\gamma_\lambda + \Gamma_\lambda^{(\text{loop})}(q^2) + \Gamma_\lambda^{(\text{CT})} , \qquad (12.6)$$

where the contribution $\Gamma_\lambda^{(\text{CT})}$ comes from the extra terms in the Lagrangian, which are called *counterterms*. The counterterms are infinite, and cancel the infinity coming from the loop contributions.

It should be noted that the divergence appears only in the F_1 form factor. So we can take

$$\Gamma_\lambda^{(\text{CT})} = -F_1^{(\text{loop})}(\mu^2)\gamma_\lambda , \qquad (12.7)$$

where μ in the parentheses indicates a mass scale. This counterterm contribution can come from a counterterm Lagrangian of the form

$$\mathscr{L}_{\text{CT}}^{(1)} = -(Z_1 - 1)eQ\overline{\psi}\gamma_\mu\psi A^\mu , \qquad (12.8)$$

where Z_1 is related to $F_1(\mu^2)$. Adding this with the original interaction term, which was of the same form but without the factor $Z_1 - 1$ in front, we find the total interaction term in the Lagrangian as

$$\mathscr{L}_{\text{int}} = -Z_1 eQ\overline{\psi}\gamma_\mu\psi A^\mu . \qquad (12.9)$$

The free terms are also not free from infinities. These can be evaluated by considering diagrams with two external lines, called *2-point functions* or *self-energy functions*. In QED, there are two kinds of 2-point functions: one kind where the external lines correspond to the charged fermion, and another kind where the external lines correspond to photons. We will need counterterms for both kinds. In fact, for the fermion self-energy, we will need two counterterms: one for the derivative term and one for the mass term. After adding all these counterterms, the QED Lagrangian, presented in Eq. (5.11, *p 114*), is modified to

$$\mathscr{L} = Z_2\Big(i\overline{\psi}\gamma^\mu\partial_\mu\psi - (m + \delta m)\overline{\psi}\psi\Big) - Z_1 eQ\overline{\psi}\gamma^\mu\psi A_\mu - \frac{1}{4}Z_3 F_{\mu\nu}F^{\mu\nu} , \qquad (12.10)$$

apart from the gauge-fixing term which is not important for the present discussion. This expression can also be written as

$$\mathscr{L} = \Big(i\overline{\psi}_B\gamma^\mu\partial_\mu\psi_B - m_B\overline{\psi}_B\psi_B\Big) - e_B Q\overline{\psi}_B\gamma^\mu\psi_B A_{B\mu} - \frac{1}{4}F_{B\mu\nu}F_B^{\mu\nu} , \qquad (12.11)$$

where the subscript 'B' has been added to denote the so-called *bare fields*

$$\psi_B = \sqrt{Z_2}\,\psi\,, \qquad A_B^\mu = \sqrt{Z_3}\,A^\mu\,, \tag{12.12}$$

and the *bare parameters*

$$m_B = m + \delta m\,, \qquad e_B = \frac{Z_1}{Z_2\sqrt{Z_3}}\,e\,. \tag{12.13}$$

Thus the Lagrangian with the counterterms looks exactly like the classical Lagrangian which did not have the counterterms, but the definitions of the fields and couplings have to be changed by a normalizing factor. This is why the process is called *renormalization*.

This is a loaded statement. Usually when we change the normalization of something, it is only for the sake of convenience of some sort. This is not exactly the case here. It should be noted that the quantities called Z_1, Z_2 and Z_3 that appear in the bare Lagrangian of Eq. (12.10) are all infinite. Thus, the bare fields and bare parameters are infinite. However, this is not a matter of practical concern since the bare parameters are never seen. Interactions are taking place, and anything that we observe is a result of interactions. The infinities of the bare parameters are canceled by the infinities that appear in loop corrections, and as a result we see finite results in all experiments.

Once the redefinitions are made and renormalization performed, one might wonder whether we have changed the classical theory as well. The answer is negative, but is obscured by our choice of units with $\hbar = 1$. To clarify the point, let us, for a moment, use some more conventional units where $\hbar$ is not used to set up the units. We can keep using $c = 1$, and take the basic units to be L (for length) and $\hbar$ (in lieu of a mass unit). The action $\mathscr{A}$ has the dimensions of $\hbar$, and the dynamics is governed by the dimensionless ratio $\mathscr{A}_{\rm cl}/\hbar$, where $\mathscr{A}_{\rm cl}$ is the classical action.

Let us now count how many powers of $\hbar$ will enter the expression of the amplitude of a diagram. From each vertex, there will be a factor of $1/\hbar$. Since propagators come from the inverses of the quadratic part of the Lagrangian, each propagator will contribute a factor $\hbar$. So, if there are v vertices and n internal lines in a diagram, the power of $\hbar$ in the amplitude will be $n - v$. If the diagram is planar, i.e., can be drawn on a plane without any crossing of the lines where there is not interaction vertex, Euler's topological formula reads

$$n - v = \ell - 1\,, \tag{12.14}$$

where ℓ denotes the number of loops. This is the power of $\hbar$ that goes into $\mathscr{A}_{\rm eff}/\hbar$, which means that the power in $\mathscr{A}_{\rm eff}$ is simply equal to the power of loops in the diagram. This means that loop corrections always come multiplied by some power of $\hbar$ compared to the classical action. In the classical limit $\hbar \to 0$, these corrections therefore vanish. For non-planar diagrams, $n - v > \ell - 1$, so they vanish even faster in the classical limit.

In Ch. 5, we mentioned that the classical Lagrangian has a gauge symmetry. In fact, we built the theory by banking on this symmetry. The symmetry depends on the relative strengths of the interaction term and the free term containing the derivative of the fermion field. This symmetry will be lost in the modified Lagrangian unless

$$Z_1 = Z_2\,. \tag{12.15}$$

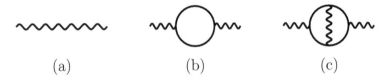

(a)   (b)   (c)

Figure 12.3: Examples of diagrams that contribute to the 2-point function of the photon in QED.

This relation must then be a consequence of the gauge symmetry, which can be proved by using the *Ward–Takahashi identity*, which is a relation between the vertex function and the fermion self-energy function. We do not derive the identity here. Rather, we use Eq. (12.15) and find that it implies

$$e_B = \frac{e}{\sqrt{Z_3}} \qquad (12.16)$$

through Eq. (12.13). This is a very interesting result, and we are now going to explore its consequences.

Recall how we defined the propagator of a free particle in §4.10: it was the inverse of the *quantity* that appears, apart from the fields themselves, in the quadratic part of the field Lagrangian written in momentum space. The 'quantity' mentioned in the previous sentence can be called the *2-point function*. In fact, if we think of the simple line of Fig. 12.3a as a diagram with two external legs, one coming in and one going out, and want to write its amplitude following the same prescription that we adopt for vertices, this is exactly what we are going to get. The propagator would be the inverse of this 2-point function.

The first of these diagrams, the unadorned photon line, represents the 2-point function in absence of interactions. Referring back to Eq. (4.146, *p 92*) and disregarding the gauge-fixing term which is not important in the present context, we find that this contribution to the 2-point function is given by

$$\Pi_{\lambda\rho}^{(0)}(q) = -(q^2 g_{\lambda\rho} - q_\lambda q_\rho), \qquad (12.17)$$

where the parenthesized superscript '0' indicates that this is the contribution from zero loops, i.e., tree diagrams.

Suppose now we want to add to it the contribution to the photon 2-point function coming from the loop diagrams like those shown in Fig. 12.3b,c or innumerable other diagrams which have not been shown. Loop diagrams will give divergent results of the form

$$\Pi_{\lambda\rho}^{(\text{loop})}(q) = (q^2 g_{\lambda\rho} - q_\lambda q_\rho) \times \Pi^{(\text{loop})}(q^2), \qquad (12.18)$$

where $\Pi^{(\text{loop})}(q^2)$ is a Lorentz invariant function of the 4-momentum. The sum of the contributions in Eqs. (12.17) and (12.18) is infinite, and so, in order to cancel the infinity, we need to introduce a counterterm. Suppose we introduce

$$\Pi^{(CT)}_{\lambda\rho}(q) = -(q^2 g_{\lambda\rho} - q_\lambda q_\rho) \times \Pi^{(\text{loop})}(-\mu^2),\qquad(12.19)$$

then the total result, after adding this term, would be finite.

The reason for taking the argument of $\Pi^{(\text{loop})}$ with a negative sign in Eq. (12.19) is the following. For an on-shell charged particle emitting or absorbing a photon, the value of q^2 must always be negative. Thus, if we call the argument $-\mu^2$, we can take μ to be a positive real number. We can then define

$$\ell \equiv \ln\mu,\qquad(12.20)$$

and rewrite functions of $-\mu^2$ as functions of ℓ, which we will use soon.

□ **Exercise 12.1** Verify the statement that if q_μ is the momentum carried by a gauge boson at a QED vertex with an incoming particle and an outgoing particle both being on-shell, then $q^2 < 0$.

Once we add the counterterm, the co-efficient of $-\frac{1}{4}F^{\mu\nu}F_{\mu\nu}$ in the total Lagrangian changes. Originally, the co-efficient was 1. After adding the counterterm, it becomes

$$Z_3 = 1 + \Pi^{(\text{loop})}(-\mu^2).\qquad(12.21)$$

We see that Z_3 depends on μ. We now look back at Eq. (12.16). Since e_B is a parameter that appears in the full Lagrangian, it cannot depend on energy transfer, or anything else for that matter. So the inevitable conclusion is that e must also depend on μ.

To understand the physical significance of this statement, consider what happens if a photon carries a momentum with $q^2 = -\mu^2$. Contributions given in Eqs. (12.18) and (12.19) cancel in this case. Thus, if we try to write the Feynman amplitude of the interaction vertex, the loop-induced contribution cancels with the counterterm, and only the tree-level contribution remains, which is just $e\gamma_\mu$. In other words, the parameter e in the Lagrangian is the coupling of the fermion to the photon if the photon momentum satisfies $q^2 = -\mu^2$. This e depends on μ, which means the coupling depends on μ, which characterizes the momentum transfer. In short, the coupling runs.

It must be noted that the running depends only on Z_3, which represents corrections to the 2-point function for the photon, i.e., to terms in the Lagrangian which are quadratic in the photon field. So the energy dependence of e depends only on the photon line, i.e., is independent of which particle the photon attaches to. In fact, this is the reason why the dependence can be seen as the dependence of e, and not of Q. The universal coupling constant e changes in a universal manner through the photon 2-point function, leaving Q unchanged for any particle.

12.2.2   Running of QED coupling

Let us continue with our argument for QED and quantify the running of the QED coupling constant. The most dominant contribution to $\Pi^{(\text{loop})}$ comes from the one-loop graph of Fig. 12.3b. Let us denote the contribution of this diagram by $\alpha \Pi^{(1)}$. Because of the two vertices, this contribution must have a factor e^2 in it. Then, squaring Eq. (12.16) and expanding in powers of α, we can write

$$\alpha_B = \alpha(t)\left[1 - \Pi^{(1)}(t)\right] + \mathcal{O}\left(\alpha^3\right) . \tag{12.22}$$

If we use a different counterterm to cancel the loop contribution at $q^2 = -\mu'^2$, the left hand side should remain the same, so that neglecting terms of $\mathcal{O}\left(\alpha^3\right)$ we can write

$$\alpha(t)\left[1 - \Pi^{(1)}(t)\right] = \alpha(t')\left[1 - \Pi^{(1)}(t')\right], \tag{12.23}$$

or

$$\alpha(t) - \alpha(t') = \alpha\left[\Pi^{(1)}(t) - \Pi^{(1)}(t')\right]. \tag{12.24}$$

Taking t' very close to t, we obtain

$$\frac{d\alpha}{dt} = \alpha\frac{d\Pi^{(1)}}{dt} . \tag{12.25}$$

This shows how α changes because of momentum dependence of the photon self-energy function. Note that on the right hand side of Eqs. (12.24) and (12.25), we have not specified any momentum scale for α. The point is that because $\Pi^{(1)}$ itself is $\mathcal{O}\left(\alpha\right)$, the terms on the right hand side are already $\mathcal{O}\left(\alpha^2\right)$. Thus the factor of α present here can be either $\alpha(t)$ or $\alpha(t')$ or some sort of average of the two: the difference between these different choices produces corrections $\mathcal{O}\left(\alpha^3\right)$ which we have neglected in writing the equation.

In order to proceed, we need to calculate $\Pi^{(1)}$. We will assume that the momentum scale is much higher than the mass of the fermion in the loop, and therefore neglect the mass altogether. Using the Feynman rules for QED we obtain, for the diagram in Fig. 12.3b,

$$i\Pi^{(1)}_{\lambda\rho}(q) = -(ie)^2 \int \frac{d^4l}{(2\pi)^4} \text{Tr}\left(\gamma_\lambda \frac{i\slashed{l}}{l^2} \gamma_\rho \frac{i(\slashed{l} + \slashed{q})}{(l + q)^2}\right) , \tag{12.26}$$

where we assume that the fermion in the loop is the electron. Applying Eq. (G.2, *p 756*) to combine the factors in the denominator, we obtain

$$\Pi^{(1)}_{\lambda\rho}(q) = ie^2 \int \frac{d^4l}{(2\pi)^4} \int_0^1 d\zeta \frac{\text{Tr}\left(\gamma_\lambda \slashed{l} \gamma_\rho (\slashed{l} + \slashed{q})\right)}{[l^2 + 2\zeta l \cdot q + \zeta q^2]^2} . \tag{12.27}$$

Redefining $l + \zeta q$ as the new loop integration variable, this expression can be written as

$$\Pi^{(1)}_{\lambda\rho}(q) = ie^2 \int \frac{d^4l}{(2\pi)^4} \int_0^1 d\zeta \, \frac{\text{Tr}\left(\gamma_\lambda \slashed{l} \gamma_\rho \slashed{l}\right) - \zeta(1-\zeta) \, \text{Tr}\left(\gamma_\lambda \slashed{q} \gamma_\rho \slashed{q}\right)}{[l^2 + \zeta(1-\zeta)q^2]^2} \tag{12.28}$$

omitting terms odd in l in the integrand because they vanish on integration. Evaluating the traces through Eq. (F.38, *p 739*) and reducing the terms involving l in the numerator to invariant integrals as shown in Eq. (G.10, *p 758*), we obtain

$$\Pi_{\lambda\rho}(q) = 4\pi i\alpha \int_0^1 d\zeta \left(2g_{\lambda\rho}J_1 + 8\zeta(1-\zeta)(g_{\lambda\rho}q^2 - q_\lambda q_\rho)J_2\right), \tag{12.29}$$

where

$$J_1 = -\int \frac{d^4l}{(2\pi)^4} \frac{l^2 + 2\zeta(1-\zeta)q^2}{[l^2 + \zeta(1-\zeta)q^2]^2},$$

$$J_2 = \int \frac{d^4l}{(2\pi)^4} \frac{1}{[l^2 + \zeta(1-\zeta)q^2]^2}. \tag{12.30}$$

The technique for evaluating loop integrals like these have been described in Appendix G. The integral J_1 can be evaluated by employing *dimensional regularization*, i.e., by turning the 4-dimensional momentum integration into a D-dimensional one and taking the limit $D \to 4$ in the end. Using Eq. (G.34, *p 762*), we obtain that in D dimensions,

$$J_1 = -i \, \frac{\Gamma(1 + D/2)\Gamma(1 - D/2) + 2\Gamma(D/2)\Gamma(2 - D/2)}{(4\pi)^{D/2} \, \Gamma(D/2)\Gamma(2)\left(-\zeta(1-\zeta)q^2\right)^{1-D/2}}. \tag{12.31}$$

Using the property of the gamma function, $z\Gamma(z) = \Gamma(z+1)$, we can write $\Gamma(2 - D/2) = (1 - D/2)\Gamma(1 - D/2)$. Then, taking the limit $D \to 4$, we find that the integral J_1 vanishes.

This is good news, because if it hadn't, we would have to put in a counterterm proportional to $A^\mu A^\nu g_{\lambda\rho}$ to cancel its infinities, and we would have required a mass term for the photon to absorb this infinity. Gauge invariance protects us from having such a term. The remaining term has precisely the form advocated in Eq. (12.18), with

$$\Pi^{(1)}(q^2) = 32\pi i\alpha \int_0^1 d\zeta \, \zeta(1-\zeta)J_2, \tag{12.32}$$

where the q^2-dependence is in J_2. Eq. (12.25) then tells us that

$$\frac{d\alpha}{d\ell} = 32\pi i\alpha^2 \frac{d}{d\ell} \int_0^1 d\zeta \, \zeta(1-\zeta) \int \frac{d^4l}{(2\pi)^4} \frac{1}{[l^2 - \zeta(1-\zeta)\mu^2]^2}, \tag{12.33}$$

Figure 12.4: Schematic view of virtual electron–positron pairs around a bare electron. Note that the diagram is deceptive in the sense that it shows a finite size of the electrons and positrons, which are point-like objects to the best of our knowledge.

where μ and t are related through Eq. (12.20). Using this relation, we can write (d/dt) as $\mu(d/d\mu)$ on the right hand side and obtain

$$
\begin{aligned}
\frac{d\alpha}{dt} &= 32\pi i\alpha^2 \mu \frac{d}{d\mu} \int_0^1 d\zeta \, \zeta(1-\zeta) \int \frac{d^4l}{(2\pi)^4} \frac{1}{[l^2 - \zeta(1-\zeta)\mu^2]^2} \\
&= 128\pi i\alpha^2 \mu^2 \int_0^1 d\zeta \, [\zeta(1-\zeta)]^2 \int \frac{d^4l}{(2\pi)^4} \frac{1}{[l^2 - \zeta(1-\zeta)\mu^2]^3} \, . \text{(12.34)}
\end{aligned}
$$

The momentum integral in this last equation is finite. We can use Eq. (G.35, p 762) to write

$$
\frac{d\alpha}{dt} = \frac{4\alpha^2}{\pi} \int_0^1 d\zeta \, \zeta(1-\zeta) = \frac{2\alpha^2}{3\pi} \, . \tag{12.35}
$$

This is the contribution to the evolution of the fine-structure constant coming from the electron loop. There will be similar contributions coming from other fermions in the loop. The notable feature is that, for any fermion, the contribution is positive, since the self-energy function contains square of the electric charge of the fermion. Thus, the fine-structure constant, or equivalently the QED coupling constant, increases with energy.

Intuitively, it can be understood in the following way. Suppose there is a bare electron at some point in space. There is an electric field around it, which, in the language of quantum field theory, is a collection of virtual photons. These virtual photons produce virtual electron–positron pairs at every instant, which recombine within a small time to produce virtual photons. At any given instant, there are some virtual photons and a cloud of virtual electron–positron pairs. Among the members of any of these pairs, the positron will be attracted by the bare electron, whereas the electron will be repelled. The situation would look schematically like the diagram in Fig. 12.4.

If we now probe this cloud with a projectile having a certain momentum, the projectile will reach up to a smallest distance r given by the uncertainty principle. It will then sense the charge that is contained within a sphere whose radius is r. This sphere will contain more positrons from the cloud than electrons, so the net cloud charge would be positive. We will observe the total charge of the bare electron and the cloud. The more energetic the projectile, the nearer it will go to the bare electron, and we will see more and more of the bare charge. This is equivalent to the statement that the coupling constant rises with energy.

12.2.3   Beta function

All parameters present in the Lagrangian of a quantum field theory can undergo evolution with energy scale, as demonstrated for the case of QED. If we denote the coupling constants generically by g, their variation with respect to the scale μ can be expressed in the form

$$\frac{dg}{dt} = \beta(g), \qquad (12.36)$$

where g is the physical coupling constant at the scale μ, and t is defined through Eq. (12.20).

The right hand side is a function which gives the nature of variation, and is called the *beta function*. Obviously it depends on g, and we have shown the functional dependence in the equation. In general, it can depend on other parameters of the theory as well. For an SU(n) gauge theory, the one-loop beta function for the gauge coupling constant is given by

$$\beta(g) = -\frac{g^3}{48\pi^2}\left(11C_2^{(G)} - 4\sum_F C_2^{(F)} - \sum_S C_2^{(S)}\right) \qquad (12.37)$$

if all couplings other than the gauge couplings are negligibly small.

The expression contains various *Casimir invariants*, which were defined in Eq. (11.58, p 309). $C_2^{(G)}$, for example, denotes the Casimir invariant for the representation that the gauge bosons belong to, i.e., the adjoint representation. The last term contains $C_2^{(S)}$, which is the Casimir invariant for the representation of the gauge group that any complex scalar field might belong to. If we have a real scalar field instead, the corresponding co-efficient should be halved because a real scalar field contains half as many degrees of freedom as a complex scalar field. The other term contains $C_2^{(F)}$, which is the Casimir invariant for the representation of the internal symmetry group that a Dirac fermion belongs to. Obviously, there are sums over all scalars and fermions.

We recall that, in deriving Eq. (12.35), we neglected the mass of the fermion in the loop. In other words, Eq. (12.35) is valid only when the scale

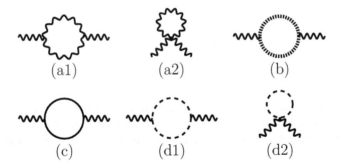

(a1)   (a2)   (b)

(c)   (d1)   (d2)

Figure 12.5: One-loop self-energy diagrams for gauge bosons in a non-abelian gauge theory. In the loop, we have (a) gauge bosons, (b) Fadeev–Popov ghosts, (c) fermions, (d) scalars.

μ is much larger than the mass in the loop. The same restriction applies to Eq. (12.37). At any given scale, one should use only those particles whose masses satisfy the relation $m \ll \mu$. In practice, one includes all masses which satisfy the relation $m < \mu$. If $m > \mu$, the contribution is so small that it can be neglected for all practical purposes.

In writing the co-efficient for the fermion term in Eq. (12.37), we have assumed that the entire fermion field transforms like the same representation of the gauge group. This is not really necessary for an arbitrary gauge theory. The internal symmetries commute with spacetime symmetries, which means that by performing Lorentz transformations, one cannot change the property of any field under an internal symmetry group. Therefore, all components of an irreducible representation of the Lorentz group must transform the same way under an internal symmetry group. A Dirac field, however, is not an irreducible representation. As mentioned in §4.4.2, it contains two irreducible representations of the Lorentz group. In general, these two irreducible parts can transform differently under an internal symmetry. This issue will be discussed again in Ch. 16. Here, it is irrelevant because entire fermion fields indeed transform like the same representation in QCD. Theories which have this property are called *vector-like theories*. In the more general case, if the two irreducible parts of any Dirac field transform differently under an internal symmetry, the theory is called a *chiral theory*. In this case, Eq. (12.37) should be modified to

$$\beta(g) = -\frac{g^3}{48\pi^2} \left(11 C_2^{(G)} - 2 \sum_F C_2^{(F_L)} - 2 \sum_F C_2^{(F_R)} - \sum_S C_2^{(S)} \right), \qquad (12.38)$$

where F_L and F_R refer to the two irreducible parts.

Looking at the definition of the Casimir invariant C_2, it is not difficult to guess why they appear in the evolution formula of Eq. (12.37). As described in §12.2.1, only the self-energy of the gauge boson is responsible for the running of the gauge coupling parameters. Consider now a non-abelian gauge theory. The self-energy of a gauge boson obtains contributions from the diagrams shown in Fig. 12.5. To be specific, let us look at Fig. 12.5c, with fermions in the loop. Each of the two vertices has a factor of the generator, in the representation appropriate for the fermions. Because of the closed loop, there is a trace over these two generators, which is exactly what the Casimir invariant is. The same can be said about Fig. 12.5d1, with scalars in the loop, except that this time the contribution will contain the Casimir invariant of whatever representation the scalars transform like. The amplitude of the diagram of Fig. 12.5d2, on the other hand, is proportional to the metric tensor $g_{\lambda\rho}$ that comes from the quartic coupling. This merely helps cancel some gauge-noninvariant contribution coming from Fig. 12.5d1. The same thing can be said about the gauge boson loop in Fig. 12.5a2 that contains a quartic coupling. It cancels some gauge-noninvariant contribution coming from Fig. 12.5a1 and Fig. 12.5c. Since the gauge bosons and the Fadeev–Popov ghosts running in the loop transform as the adjoint representation of the gauge group, they give the Casimir invariant of the adjoint.

The beta function for a U(1) gauge theory can easily be guessed from Eq. (12.37) and the analogy between U(1) and SU(N) gauge theories given in Ch. 11. There will be no contribution coming from gauge bosons in the loop, because the trilinear gauge boson coupling does not exist for a U(1) theory. There is no coupling to the ghosts as well, as mentioned in §11.4. The beta function would therefore be given by

$$\beta(e) = \frac{e^3}{48\pi^2} \left(4 \sum_F Q_F^2 + \sum_S Q_S^2 \right). \tag{12.39}$$

For fermions, this is exactly the result that we got earlier in Eq. (12.35).

☐ **Exercise 12.2** *The low energy value of the fine-structure constant* α *is* 1/137. *Use the QED beta function to show that at a scale* $M_Z =$ 91 GeV, *the value of* α *is roughly* 1/128, *assuming that it crosses the thresholds of only the following fermions on the way:*

- *Charged leptons:* e, μ, τ
- *Neutrinos (how many? should you care?)*
- *The quarks* u, d, s, c *and* b

[**Note** : *Read the masses of these particles from Table B.2 (p 720) and Table B.5 (p 723). You need not be too fussy about very accurate values of the masses, since the dependence is only logarithmic.*]

☐ **Exercise 12.3** *Take the Lagrangian of scalar QED from* §5.5. *Calculate the contribution of one charged scalar to the evolution of electric charge at the one-loop level. The answer is given in Eq. (12.39).*

But the important point is that, even when we are above the thresholds for all known quarks, the one-loop beta function of QCD is negative. This means that, at higher and higher energies, the gauge coupling constant becomes smaller and smaller. And this will be the case no matter how much we increase the energy scale, unless there exists a huge number of hitherto unknown fermions and scalars which can offset the contribution coming from the gauge bosons in the beta function and make it positive. Thus, at higher and higher energies, quarks will be more and more free, a property termed *asymptotic freedom*.

It can of course be said that, since the QCD coupling constant is large at low energies, one should not trust a perturbative result such as the expression for the beta function in Eq. (12.37). True. But one can take its value at an energy scale where it is already in the perturbative region, and see whether the value goes down at even higher energies. Experimental results show that this is exactly what happens. For example,

$$\alpha_3(m_\tau) = 0.327\,,$$
$$\alpha_3(m_b) = 0.20\,,$$
$$\alpha_3(M_Z) = 0.1184 \pm 0.0007\,, \tag{12.40}$$

where $M_Z = 91\,\text{GeV}$, $m_b \simeq 4.18\,\text{GeV}$, $m_\tau = 1.78\,\text{GeV}$. Note that in Eq. (12.40) we have not given the values of the gauge coupling constant itself, but of the strong interaction analog of the fine-structure constant, defined as

$$\alpha_3 = \frac{g_3^2}{4\pi}\,, \tag{12.41}$$

where g_3 is the QCD coupling constant. The value of α_3 has been quoted at various scales, with the scales, i.e., the values of μ, appearing in the parentheses in Eq. (12.40). The values are determined in various ways, comparing theoretical predictions with experimental measurements in τ-decay, jet production in e^+-e^- collisions, and many other processes.

12.2.4 Running of the QCD coupling

As mentioned in Eq. (12.37), the one-loop beta function is proportional to g^3. Contributions to the beta function coming from higher loops depend on higher powers of g. If g is small, one can neglect the higher order corrections and write

$$\frac{dg}{d(\ln\mu)} = -b_3 g^3\,. \tag{12.42}$$

This equation can be easily integrated. In terms of the analog of fine-structure constant, the result is

$$\frac{1}{\alpha_3(\mu_2)} = \frac{1}{\alpha_3(\mu_1)} - 8\pi b_3 \ln\frac{\mu_1}{\mu_2}\,. \tag{12.43}$$

In this form, the evolution equation for the coupling constant is valid for any group as long as the coupling constant is small enough so that approximation of Eq. (12.42) is valid. For QCD in particular, we can put $C_2^{(F)} = \frac{1}{2}$ in Eq. (12.37) since the quarks are in the fundamental representation of $SU(3)_c$. Calling the number of quark flavors below a certain scale to be N_F, we obtain

$$b_3 = \frac{1}{48\pi^2}\left(33 - 2N_F\right) \tag{12.44}$$

at that scale.

☐ **Exercise 12.4** Use the value of α_3 at the scale M_Z given in Eq. (12.40) and the values of M_Z and m_b given below that equation to calculate the value of α_3 at the scale m_b. Show that the result given in Eq. (12.40) is obtained approximately, confirming that α_3 is small enough in this range that the higher order corrections can be neglected. Remember that for this problem $N_F = 5$, corresponding to u, d, s, c and b quarks.

What is the lowest energy where we can use perturbative methods? The results of Ex. 12.4 indicate that one can use perturbative methods even when the value of μ is as low as the mass of the b quark. When the scale is near the mass of the c quark, the method begins to be suspect, which can be seen by using Eq. (12.43) to calculate $\alpha_3(m_\tau)$ from $\alpha_3(m_b)$, and comparing the result with the experimental value. Exactly where the line is to be drawn depends on the accuracy sought. A benchmark value for this borderline can be set as follows. Suppose we start from the value of α_3 at some high energy, and pretend that the evolution equation is exactly of the form given in Eq. (12.42) at all scales. In other words, we neglect all higher order terms in the beta function, even though they might well be comparable to the $\mathcal{O}\left(g^3\right)$ term when g is not very small. The solution of the evolution equation would then be given by Eq. (12.43). With this solution, if we start with the value of α_3 at some scale μ_1, we can find a value for μ_2 where the coupling constant will become infinite. This value of μ_2 is called the *QCD scale parameter*, and usually denoted by Λ_{QCD}. In other words, putting in $1/\alpha_3(\mu_2) = 0$, the QCD coupling constant at any scale μ is given by

$$\frac{1}{\alpha_3(\mu)} = 8\pi b_3 \ln \frac{\mu}{\Lambda_{\mathrm{QCD}}}, \tag{12.45}$$

which can be taken as a definition of Λ_{QCD}. Equivalently, we can say that the value of α_3 at a scale μ is given by

$$\alpha_3(\mu) = \frac{12\pi}{\left(33 - 2N_F\right) \ln \frac{\mu^2}{\Lambda_{\mathrm{QCD}}^2}}. \tag{12.46}$$

Needless to say, the coupling constant will not really diverge at the scale Λ_{QCD} as Eq. (12.46) might suggest. The reason is that Eq. (12.42) will not be a good equation to describe the real running once the coupling constant g becomes sufficiently large: higher order terms in g will be as important in describing the evolution of the coupling constant. Still, Λ_{QCD} gives a benchmark value, in the sense that for energy scales much larger than it, the coupling constant can be safely assumed to be small.

The value of the QCD scale parameter can now be estimated, using the value of α_3 at the scale m_τ given in Eq. (12.40). Although the charm quark is slightly lighter than the tau lepton, let us assume, for a rough estimate, that there are only three flavors of quarks lighter than the tau, so that we can write

$$\alpha_3(m_\tau) = \frac{12\pi}{27 \ln \dfrac{m_\tau^2}{\Lambda_{\text{QCD}}^2}} \, . \qquad (12.47)$$

Putting in the value of m_τ and the value of $\alpha_3(m_\tau)$ given in Eq. (12.40), one obtains $\Lambda_{\text{QCD}} \approx 210 \, \text{MeV}$. However, considering that the charm quark mass threshold has not been taken into account, and that there is a huge uncertainty in the definition of m_c, one usually takes Λ_{QCD} somewhere in the range of 100 to 200 MeV.

12.3   QCD Lagrangian

Suppose we have only the QCD gauge symmetry and no other internal symmetry in the Lagrangian. What is the most general form of the Lagrangian in this case?

As far as the particle content is concerned, there are of course the gauge bosons, which are absolutely necessary for building up the gauge symmetry. For QCD, these gauge bosons are the gluons. Then there are some quark fields. We will denote different flavors of quark fields by q_A, where A is a flavor index. Each flavor comes in three colors, and the fields corresponding to the three colors form a triplet of the gauge group $\text{SU}(3)_\text{c}$. Then the most general renormalizable Lagrangian consistent with the gauge symmetry is:

$$\mathscr{L}_{\text{QCD}} = -\frac{1}{4} \mathsf{G}_{\mu\nu}^{a\dagger} \mathsf{G}_a^{\mu\nu} + \sum_A \bar{q}_A \Big(i\gamma^\mu D_\mu - m_A \Big) q_A \,, \qquad (12.48)$$

where D_μ is the covariant derivative defined in the way shown in Ch. 11, and $\mathsf{G}_{\mu\nu}^a$ is the notation specifically for the $\text{SU}(3)_\text{c}$ field-strength tensors. The color indices on the quark fields have been kept implicit. Of course we can add free Lagrangians of other fields which are singlets of the color gauge group, e.g., the leptons. But they, being singlets of the gauge group, would not have any gauge interaction.

There are three kinds of interaction vertices in the QCD Lagrangian. There are interactions between gluons, and these interactions can have a cubic or a quartic vertex, as discussed for general gauge theories in §11.3. The gluons also interact with the quarks through their presence in the covariant derivative. This interaction term is

$$\mathscr{L}_{\text{int}} = -g_3 \sum_A \bar{q}_A \gamma_\mu \Big(\frac{1}{2}\lambda_a\Big) q_A \mathsf{G}_a^\mu \,, \qquad (12.49)$$

where G_a^μ denotes the gluon fields, and the matrices $\frac{1}{2}\lambda_a$ (with $a = 1, 2, \cdots 8$) are the generators of $\text{SU}(3)$ in the fundamental representation that the quarks belong to. If we make the color indices explicit and hide the flavor indices for the sake of clarity, we can write the interaction term as

$$\mathscr{L}_{\text{int}} = -g_3 \sum_q \overline{q}_\alpha \gamma_\mu \left(\frac{1}{2}\lambda_a\right)_{\alpha\beta} q_\beta G_a^\mu, \tag{12.50}$$

where α, β are color indices.

The explicit form for the λ-matrices can be found in Eq. (10.13, *p 259*), where they were introduced to represent some different physical transformations, viz., transformations among different flavors of quarks. Here, they do nothing to the flavors, but rather inflict changes in quark colors. But the mathematical structure is the same, which is why the same matrices can be used.

There is one aspect of the Lagrangian that requires some clarification. Since all quark flavors transform the same way under the gauge group, it seems that the mass term could have a more general form like

$$\sum_{A,B} m_{AB}\overline{q}_A q_B. \tag{12.51}$$

On closer scrutiny, it seems unnecessarily cumbersome. Even if we are faced with such 'more general' mass terms, we can always make linear combinations of the fields so that the mass terms are diagonal in the redefined fields. Such linear transformations would not affect the covariant derivative term of the quark fields in the Lagrangian, since those terms have the same co-efficient for all quark fields. In other words, in the flavor space the co-efficient of the covariant derivative term is like a unit matrix, which remains unaffected in a similarity transformation involving the fields.

The diagonal form of the mass terms means that the QCD Lagrangian does not have any term that changes quark flavor. Of course quark color can change through gauge interactions, as explained in §12.1, but flavor cannot change through strong interactions. Flavor numbers like N_u or N_d, discussed in earlier chapters, are conserved so far as only strong interaction effects are concerned. Later in Ch. 17, we will see that they can be violated by weak interactions.

12.4   Perturbative QCD

The discussion of §12.2 makes it clear that at very high energies, i.e., high compared to Λ_{QCD}, we can use perturbative techniques to describe strong interactions. In this section, we discuss QCD calculations in the perturbative region.

No one has seen any free quark or gluon. Physical states, as far as we know, are hadrons, which are color singlets. So, in order to compare with physical observables, one needs to calculate rates of processes involving hadrons in the initial as well final state. But to get there, we first need to analyze the basic processes involving quarks and gluons.

12.4.1   Cross-sections involving quarks and gluons

In Ch. 5, we have calculated cross-sections of various processes mediated by the QED interaction. Discussions of QCD mediated cross-sections are similar in the perturbative domain, and we need not go through the same amount of detail again. We only give a few examples to show how the results of QCD calculations differ from their QED counterparts.

A few comments about the notations used. First of all, since free quarks and gluons are not available, it is useless to talk about the CM frame, or any other frame, of the incoming particles. It is therefore best to present the differential cross-sections in the invariant form given in §4.13.2. We will use the symbol q as a generic quark, and numbered subscripts to distinguish between different flavors of quarks. We will neglect the quark masses, so that the formula for invariant differential cross-section given in Eq. (4.215, *p 108*) reduces to the following:

$$\frac{d\sigma}{dt} = \frac{1}{16\pi s^2} \left| \mathscr{M} \right|^2 . \tag{12.52}$$

Since QCD does not distinguish between different flavors, the formulas derived below would apply for any flavor as long as its mass can be neglected with respect to the center of mass energy. We present the calculations by assuming that the gauge group is $SU(N)$. The results for QCD can be obtained by putting $N = 3$ at the end.

a) $q_1\bar{q}_1 \to q_2\bar{q}_2$ and $q_1q_2 \to q_1q_2$

The QED analog of $q_1\bar{q}_1 \to q_2\bar{q}_2$ is the process $e^-e^+ \to \mu^-\mu^+$, discussed in §5.4.2. Recalling the expression for $\left| \mathscr{M} \right|^2$ from Eq. (5.93, *p 133*) and using Eq. (12.52), we can write

$$\frac{d\sigma}{dt}(e^-e^+ \to \mu^-\mu^+) = \frac{\pi\alpha^2}{s^2}(1 + \cos^2\theta) = \frac{2\pi\alpha^2}{s^2}\frac{t^2 + u^2}{s^2}, \tag{12.53}$$

neglecting the masses of the fermions. The differential cross-section of $e\mu \to e\mu$ would also be governed by the diagram of Fig. 5.6 *(p 132)*, except that the outgoing μ^+ line should now be treated as the incoming μ^- line, whereas the incoming e^+ line should be treated as an outgoing e^- line. With these modifications, the photon propagator will have a denominator of the Mandelstam variable t rather than s, so that the corresponding result for this process can be obtained by interchanging t and s in the part coming from the matrix element:

$$\frac{d\sigma}{dt}(e\mu \to e\mu) = \frac{2\pi\alpha^2}{s^2}\frac{s^2 + u^2}{t^2}. \tag{12.54}$$

When we consider the QCD version of this latter process, i.e., the process $q_1q_2 \to q_1q_2$ mediated by gluons, the diagrams remain the same as those given in Fig. 5.3 *(p 120)*, except that now the fermion lines represent quarks and the gauge boson line represents gluons. In the amplitude, the propagators and external line factors remain the same as in its QED analog. As a result, kinematical dependence on the Mandelstam variables have the same structure. The only differences occur in the vertex factors. First of all, the QCD coupling constant g_3 appears instead of e. This means that in the final result, α_3 would appear in place of its QED analog, the fine-structure constant α. The other difference is that for QCD, each vertex would involve the generators of SU(3) in the fundamental representation, since the quarks belong to this representation. In the amplitude for the process, this factor would be

$$(T_{\bar{a}})_{\beta_1\alpha_1}(T_a)_{\beta_2\alpha_2}\,, \tag{12.55}$$

if the initial and final colors for the quark q_A (for $A = 1, 2$) are α_A and β_A. Note that the repeated indices on the generators imply a summation, which means that any gluon would contribute to the process as long the relevant matrix element of the generator is non-zero.

When we square the amplitude, we need to multiply the quantity appearing in Eq. (12.55) by its complex conjugate. Note that

$$\left[(T_a)_{\beta\alpha}\right]^* = (T_a^\dagger)_{\alpha\beta} = (T_{\bar{a}})_{\alpha\beta}\,, \tag{12.56}$$

using a notation of barred gauge indices that we had introduced in §11.2.4. In the expression for the matrix element squared, we obtain a factor

$$(T_{\bar{a}})_{\beta_1\alpha_1}(T_a)_{\beta_2\alpha_2}(T_b)_{\alpha_1\beta_1}(T_{\bar{b}})_{\alpha_2\beta_2} = \mathrm{Tr}\left(T_{\bar{a}}T_b\right)\mathrm{Tr}\left(T_aT_{\bar{b}}\right)$$

$$= [C^{(f)}]^2\,\delta_{ab}\delta_{ab}\,, \tag{12.57}$$

using the normalization formula introduced in Eq. (11.39, *p 305*). The normalization of the fundamental representation was specified in Eq. (11.54, *p 308*). The combination $\delta_{ab}\delta_{ab}$ equals δ_{aa}, with a sum on the index a. It therefore merely counts the number of generators, which is $N^2 - 1$. And finally, when we put the matrix element in the calculation for the cross-section, we need to average over the initial color states, which gives a factor of $1/N^2$. Combining these factors and borrowing the momentum-dependent factors from Eq. (12.54), we can write the differential cross-section:

$$\frac{d\sigma}{dt}(q_1q_2 \to q_1q_2) = \frac{\pi(N^2-1)\alpha_3^2}{2N^2\mathfrak{s}^2}\frac{\mathfrak{s}^2+\mathfrak{u}^2}{\mathfrak{t}^2}\,. \tag{12.58}$$

The analogous formula for the process $q_1\widehat{q}_1 \to q_2\widehat{q}_2$ can be obtained by interchanging s and t, as discussed earlier. Thus,

$$\frac{d\sigma}{dt}(q_1\widehat{q}_1 \to q_2\widehat{q}_2) = \frac{\pi(N^2-1)\alpha_3^2}{2N^2\mathfrak{s}^2}\frac{\mathfrak{t}^2+\mathfrak{u}^2}{\mathfrak{s}^2}\,. \tag{12.59}$$

b) $qq \to qq$ **and** $q\widehat{q} \to q\widehat{q}$

If the two quarks in the initial state have the same flavor, we should look for the QCD analog of electron–electron scattering, described in §5.3.1. The difference, compared with $q_1 q_2$ elastic scattering, is that now there is an extra diagram, which is obtained by interchanging the two quarks in the initial state. The contribution to the cross-section coming from the square of this extra term can be easily obtained by interchanging the Mandelstam variables t and u in the formula of Eq. (12.58).

But, because of this extra diagram that adds to the amplitude, there will now be an interference term as well. The momentum dependence of this term would be exactly similar to the interference term that appears in Eq. (5.40, *p 122*), i.e., of the form s^2/tu. But the color factor, coming from the generators, needs to be calculated separately. Suppose we label the colors in a way such that the color factor from one of the diagrams is given by Eq. (12.55). In the other diagram, since the initial quarks are interchanged, the corresponding factor would be

$$(T_{\bar{b}})_{\beta_1 \alpha_2} (T_b)_{\beta_2 \alpha_1} . \tag{12.60}$$

Thus the interference term will contain the factor

$$(T_{\bar{a}})_{\beta_1 \alpha_1} (T_a)_{\beta_2 \alpha_2} (T_b)_{\alpha_2 \beta_1} (T_{\bar{b}})_{\alpha_1 \beta_2} + \text{c.c.} = 2 \operatorname{Tr} \left(T_{\bar{a}} T_{\bar{b}} T_a T_b \right) . \tag{12.61}$$

To evaluate the trace, note that

$$T_{\bar{b}} T_a T_b = T_{\bar{b}} T_b T_a - T_{\bar{b}} [T_b, T_a] = C_2 T_a - i f_{bac} T_{\bar{b}} T_c , \tag{12.62}$$

where the Casimir invariant C_2 must be evaluated in the representation that the generators T_a belong to. Now, using the symmetry properties of the structure constants given in Eq. (11.53, *p 308*), we can write

$$f_{bac} T_{\bar{b}} T_c = -f_{\bar{c} a \bar{b}} T_{\bar{b}} T_c = -f_{bac} T_c T_{\bar{b}} , \tag{12.63}$$

where in the last step we have just renamed the dummy indices $c \to \bar{b}, b \to \bar{c}$. Adding the two equal contributions and dividing by 2, we can write

$$f_{bac} T_{\bar{b}} T_c = \frac{1}{2} f_{bac} \left[T_{\bar{b}}, T_c \right] = \frac{i}{2} f_{bac} f_{\bar{b} cd} T_d . \tag{12.64}$$

Further, using the definition of the adjoint representation from Eq. (10.26, *p 262*) and the definition of the Casimir invariant, it is easy to show that

$$f_{bac} f_{\bar{b} cd} = - \left(T_b^{(\mathrm{ad})} \right)_{ca} \left(T_{\bar{b}}^{(\mathrm{ad})} \right)_{dc} = - \left(T_{\bar{b}}^{(\mathrm{ad})} T_b^{(\mathrm{ad})} \right)_{da} = -C_2^{(\mathrm{ad})} \delta_{da} . \tag{12.65}$$

Thus, for any representation R,

$$\left(T_{\bar{b}} T_a T_b \right)^{(R)} = \left(C_2^{(R)} - \frac{1}{2} C_2^{(\mathrm{ad})} \right) T_a^{(R)} , \tag{12.66}$$

and so

$$\text{Tr}\left(T_{\bar{a}}T_{\bar{b}}T_aT_b\right)^{(R)} = \left(C_2^{(R)} - \frac{1}{2}C_2^{(\text{ad})}\right)C_2^{(R)} \text{ Tr } \mathbb{1}, \qquad (12.67)$$

where the factor $\text{Tr }\mathbb{1}$ gives the dimension of the representation, denoted by $d_{(R)}$ earlier in Eq. (11.59, $p\,309$).

Since the quarks are in the fundamental representation, we can borrow the result of Eqs. (11.63) and (11.62) to write

$$\text{Tr}\left(T_{\bar{a}}T_{\bar{b}}T_aT_b\right)^{(f)} = \left(\frac{N^2-1}{2N} - \frac{N}{2}\right)\frac{N^2-1}{2} = -\frac{N^2-1}{4N}. \qquad (12.68)$$

So, in the end, we obtain the differential cross-section in the form

$$\frac{d\sigma}{dt}(qq \to qq) = \frac{\pi(N^2-1)\alpha_3^2}{2N^2\mathsf{s}^2}\left(\frac{\mathsf{s}^2+\mathsf{u}^2}{\mathsf{t}^2} + \frac{\mathsf{s}^2+\mathsf{t}^2}{\mathsf{u}^2} - \frac{2}{N}\frac{\mathsf{s}^2}{\mathsf{tu}}\right). \qquad (12.69)$$

For the process $q\widehat{q} \to q\widehat{q}$, the differential cross-section can be written easily, using crossing symmetry arguments:

$$\frac{d\sigma}{d\Omega}(q\widehat{q} \to q\widehat{q}) = \frac{\pi(N^2-1)\alpha_3^2}{2N^2\mathsf{s}^2}\left(\frac{\mathsf{s}^2+\mathsf{u}^2}{\mathsf{t}^2} + \frac{\mathsf{u}^2+\mathsf{t}^2}{\mathsf{s}^2} - \frac{2}{N}\frac{\mathsf{u}^2}{\mathsf{st}}\right). \qquad (12.70)$$

c) $q\mathsf{g} \to q\mathsf{g}$, $q\widehat{q} \to \mathsf{gg}$ and $\mathsf{gg} \to q\widehat{q}$

Consider $q\mathsf{g} \to q\mathsf{g}$ first. This can be called quark-gluon scattering, and is the QCD analog of Compton scattering, which was discussed in §5.3.3. Obviously, contributions to this process come from the diagrams of Fig. 5.5 $(p\,127)$, where the fermion lines are now to be thought of as quark lines, and the gauge boson lines as gluon lines. If the initial and final colors of the quarks are denoted by α and β and the two gluons correspond to the generators T_a and T_b, we see that the generators appearing in the amplitude of any one of these diagrams should have the form

$$(T_{\bar{b}})_{\beta\gamma}(T_a)_{\gamma\alpha}, \qquad (12.71)$$

where the index γ represents the color of the intermediate quark line. The square of the amplitude will then contain the factor

$$(T_aT_{\bar{a}})_{\gamma\delta}(T_bT_{\bar{b}})_{\delta\gamma} = [C_2^{(f)}]^2 \text{ Tr } \mathbb{1} = \frac{(N^2-1)^2}{4N}. \qquad (12.72)$$

We should also remember that, for the QCD process at hand, we should average over N possible quark colors and N^2-1 possible gluons. If we redress the expression obtained in Eq. (5.75, $p\,129$) with these modifications, we arrive at the formula

$$\frac{d\sigma}{dt} = -\frac{\pi(N^2-1)\alpha_3^2}{2N^2\mathsf{s}^2}\frac{\mathsf{s}^2+\mathsf{u}^2}{\mathsf{su}}. \qquad (12.73)$$

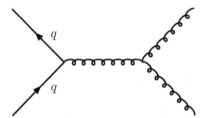

Figure 12.6: Quark-gluon scattering at the second order in perturbation theory. There are two other diagrams, which are exactly like those presented in Fig. 5.5 *(p 127)*. This diagram, involving cubic couplings of gauge bosons, has no analog in QED.

But this is not the correct answer. The reason is that, unlike the processes discussed so far in this section, there is a qualitative difference for this one between QED and QCD. Because of the presence of a trilinear gauge boson coupling, there is an extra diagram at this order, which has been shown in Fig. 12.6. Clearly, the denominator coming from the gluon propagator of this diagram would be t, and the coupling factors would be

$$(T_c)_{\beta\alpha} f_{abc}, \tag{12.74}$$

where T_c represents a generator in the fundamental representation. The factor coming from it in the amplitude square can be determined in the way shown for earlier examples, and turns out to be $\frac{1}{2}N(N^2 - 1)$. After putting in the momentum factors, one obtains the differential cross-section as

$$\frac{d\sigma}{dt}(qg \to qg) = \frac{\pi\alpha_3^2}{s^2}\left[-\frac{(N^2-1)}{2N^2}\frac{s^2+u^2}{su} + \frac{u^2+s^2}{t^2}\right]. \tag{12.75}$$

Here, the term with su in the denominator is the same as that in Eq. (12.73). The other term is the square of the amplitude coming from Fig. 12.6. Note that there is no interference term. As commented in §5.3.3, the interference between the two diagrams of Fig. 5.5 *(p 127)* vanishes in the limit that the fermion mass is neglected. The same is true about the interference terms involving the new diagram, Fig. 12.6.

□ **Exercise 12.5** *Write the amplitude from Fig. 12.6 and verify that its square gives the $1/t^2$ term in the differential cross-section.*

□ **Exercise 12.6** *Show that, for the QCD Compton process, the two diagrams of Fig. 5.5 (p 127) are not gauge invariant in the sense of Eq. (5.67, p 128). When the diagram of Fig. 12.6 is added, show that the amplitude is indeed gauge invariant.*

Once this is done, it is easy to write the expressions for differential cross-section for some other processes using crossing symmetry. Take first $q\bar{q} \to gg$. This can be obtained by interchanging s and t in the matrix element squared,

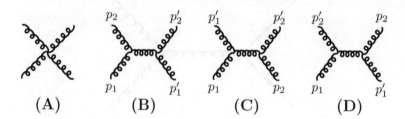

Figure 12.7: Tree-level diagrams for gluon–gluon elastic scattering. The momenta of the external legs are indicated. The primed momenta should be considered outgoing, and the unprimed ones incoming.

and adding an overall minus sign since one fermion has changed sides. In addition, the initial state averaging factor is different: it is $1/N^2$ for this case, instead of $1/N(N^2 - 1)$ that we used for $qg \to qg$. So we can write

$$\frac{d\sigma}{dt}(q\widehat{q} \to gg) = \frac{\pi(N^2 - 1)\alpha_3^2}{N\mathsf{s}^2} \left[\frac{(N^2 - 1)}{2N^2} \frac{t^2 + u^2}{tu} - \frac{u^2 + t^2}{\mathsf{s}^2} \right]. \quad (12.76)$$

Crossing symmetry can be used, either on Eq. (12.75) or on Eq. (12.76), to obtain

$$\frac{d\sigma}{dt}(gg \to q\widehat{q}) = \frac{\pi N\alpha_3^2}{(N^2 - 1)\mathsf{s}^2} \left[\frac{(N^2 - 1)}{2N^2} \frac{t^2 + u^2}{tu} - \frac{u^2 + t^2}{\mathsf{s}^2} \right]. \quad (12.77)$$

d)   Gluon–gluon elastic scattering

In QED, photons do not have any self-interaction. So photon-photon scattering can occur only at loop level, mediated by charged particles in the loop. In contrast, non-abelian theories have cubic and quartic interactions among gauge bosons. So gluons can scatter off gluons at the tree level. The diagrams are shown in Fig. 12.7.

□ **Exercise 12.7** *Draw a one-loop diagram for photon-photon scattering in QED, and estimate the powers of α that will appear in the cross-section. How many powers of α_3 will appear in the expression for the gluon–gluon scattering cross-section?*

We denote the process as

$$g(p_1) + g(p_2) \to g(p_1') + g(p_2'). \quad (12.78)$$

The amplitude can be written in the form

$$\mathscr{M} = \epsilon_1^\mu \epsilon_2^\nu \epsilon_1'^\lambda \epsilon_2'^\rho \mathscr{M}_{\mu\nu\lambda\rho}, \quad (12.79)$$

where ϵ_1, for example, denotes the polarization vector of the gluon with momentum p_1. Each diagram gives a contribution to $\mathscr{M}^{\mu\nu\mu'\nu'}$. For example the contribution of Fig. 12.7A comes directly from the quartic coupling given in Fig. 11.1 *(p 312)*, and is given by

$$i\mathcal{M}^{(A)}_{\mu\nu\lambda\rho} = -ig_3^2 \Big[f_{abc}f_{a'b'\overline{c}} E_{\mu\nu\lambda\rho} + f_{aa'c}f_{bb'\overline{c}} E_{\mu\lambda\nu\rho} + f_{ab'c}f_{ba'\overline{c}} E_{\mu\rho\nu\lambda} \Big],$$
(12.80)

where $E_{\alpha\beta\gamma\delta}$ was defined in Eq. (11.71, *p 311*), and a,b are the gauge indices of the initial state gluons which will be averaged over, whereas a',b' are the gauge indices of the final state gluons which will be summed over.

For the other diagrams, we need to use the cubic couplings and the propagators for gluons. For the latter, we use the 't Hooft–Feynman gauge. This gives

$$\mathcal{M}^{(B)}_{\mu\nu\lambda\rho} = \frac{2g_3^2}{\mathfrak{s}} f_{abc}f_{a'b'\overline{c}} \Big[2p_{2\mu}p'_{2\lambda}g_{\nu\rho} - 2p_{2\mu}p'_{1\rho}g_{\nu\lambda} + p_{2\mu}(p'_1 - p'_2)_\nu g_{\lambda\rho}$$
$$-2p_{1\nu}p'_{2\lambda}g_{\mu\rho} + 2p_{1\nu}p'_{1\rho}g_{\mu\lambda} - p_{1\nu}(p'_1 - p'_2)_\mu g_{\lambda\rho}$$
$$+(p_1 - p_2)_\rho p'_{2\lambda}g_{\mu\nu} - (p_1 - p_2)_\lambda p'_{1\rho}g_{\mu\nu} + \frac{1}{2}(\mathfrak{u} - \mathfrak{t})g_{\mu\nu}g_{\lambda\rho} \Big].$$
(12.81)

The amplitudes $\mathcal{M}^{(C)}_{\mu\nu\lambda\rho}$ and $\mathcal{M}^{(D)}_{\mu\nu\lambda\rho}$ can be obtained by making suitable interchanges. For example, to obtain $\mathcal{M}^{(C)}_{\mu\nu\lambda\rho}$, change $p_2 \leftrightarrow -p'_1$ and $\nu \leftrightarrow \lambda$ in the expression for $\mathcal{M}^{(B)}_{\mu\nu\lambda\rho}$.

☐ **Exercise 12.8** Write $\mathcal{M}^{(C)}_{\mu\nu\lambda\rho}$ and $\mathcal{M}^{(D)}_{\mu\nu\lambda\rho}$. Hence, show that the total amplitude satisfies the gauge invariance conditions

$$p_1^\mu \mathcal{M}_{\mu\nu\lambda\rho} = 0,$$
(12.82)

and similar ones involving contractions with other momenta. [**Note :** *The combinations of pairs of structure constants of the group that appear in the amplitude are not independent: they are related through the Jacobi identity, Eq.* (11.56, p 308).]

☐ **Exercise 12.9** *Had we used a general R_ξ-gauge instead of the 't Hooft–Feynman gauge, the propagator would have contained extra terms. Show that these terms do not contribute to the amplitude.*

In the square of the matrix element, there will be squares of each of these terms, and there will be the interference terms. As an example, we present some details of the square of $\mathcal{M}^{(A)}$, summed over final polarization states and averaged over the initial ones. Using the polarization sum given in Eq. (4.36, *p 68*), we can write

$$\sum_{\text{pol}} \overline{|\mathcal{M}_A|^2} = \frac{g_3^4}{4(N^2 - 1)^2}$$
$$\times \Big[f_{abc}f_{a'b'\overline{c}} E_{\mu\nu\lambda\rho} + f_{aa'c}f_{bb'\overline{c}} E_{\mu\lambda\nu\rho} + f_{ab'c}f_{ba'\overline{c}} E_{\mu\rho\nu\lambda} \Big]$$
$$\times \Big[f_{\overline{a}\overline{b}\overline{e}} f_{\overline{a}'\overline{b}'e} E^{\mu\nu\lambda\rho} + f_{\overline{a}\overline{a}'\overline{e}} f_{\overline{b}\overline{b}'e} E^{\mu\lambda\nu\rho} + f_{\overline{a}\overline{b}'\overline{e}} f_{\overline{b}\overline{a}'e} E^{\mu\rho\nu\lambda} \Big],$$
(12.83)

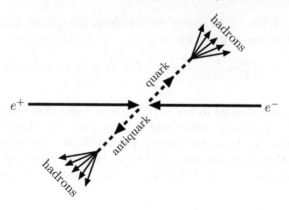

Figure 12.8: Schematic diagram for two-jet production in e^+e^- collision.

where the overall factor of $\frac{1}{4}$ comes from averaging over the physical polarizations of the initial state, and $1/(N^2-1)^2$ comes from averaging of color states of initial gluons. Notice that we have used the formula for complex conjugate of the structure constants given in Eq. (11.50, *p 307*). From the definition of the quantity $E_{\mu\nu\lambda\rho}$, it is easy to obtain their contractions:

$$E_{\mu\nu\lambda\rho}E^{\mu\nu\lambda\rho} = 24\,,$$
$$E_{\mu\nu\lambda\rho}E^{\mu\lambda\nu\rho} = 12\,, \tag{12.84}$$

and so on. Finally, the factors of structure constants can be arranged in the form of some Casimir invariants, e.g.,

$$f_{abc}f_{a'b'\bar{c}}\,f_{\bar{a}\bar{a}'\bar{e}}\,f_{\bar{b}\bar{b}'e} = f_{abc}f_{b'c\bar{a}'}\,f_{\bar{a}\bar{a}'\bar{e}}\,f_{\bar{b}'\bar{e}b}$$
$$= \left(t_a^{(ad)}\right)_{bc}\left(t_{b'}^{(ad)}\right)_{c\bar{a}'}\left(t_{\bar{a}}^{(ad)}\right)_{\bar{a}'\bar{e}}\left(t_{\bar{b}'}^{(ad)}\right)_{\bar{e}b}$$
$$= \mathrm{Tr}\left(t_a^{(ad)}t_{b'}^{(ad)}t_{\bar{a}}^{(ad)}t_{\bar{b}'}^{(ad)}\right)\,. \tag{12.85}$$

As proposed, this is a Casimir invariant, whose value was calculated in Eq. (12.66). Similarly the other terms can be calculated, and the final result is:

$$\frac{d\sigma}{dt}(gg\to gg) = \frac{4\pi N^2\alpha_3^2}{(N^2-1)\mathfrak{s}^2}\left(3 - \frac{ut}{\mathfrak{s}^2} - \frac{\mathfrak{s}u}{t^2} - \frac{\mathfrak{s}t}{u^2}\right)\,. \tag{12.86}$$

One remarkable feature of this calculation is that the interference terms all vanish, which can be seen easily from the final expression, which does not have any term with denominators like $\mathfrak{s}t$.

12.4.2 Jet production

Suppose we consider hadron production in e^+e^- scattering. At very high energies, a large number of hadrons can be produced. In general, if a final state

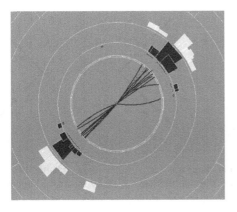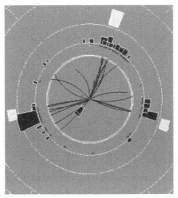

Figure 12.9: Pictures of two-jet and three-jet events. The tracks are shown in the middle part of the pictures. The histogram-like drawings on the outer circles show the amount of energy deposited in the corresponding part of the electromagnetic and hadronic calorimeters. [From the OPAL collaboration website at http://www.hep.phy. cam.ac.uk/opal].

consists of many particles, the particles are likely to be distributed over varied directions. However, if quarks are the fundamental particles that constitute hadrons, at the basic level the interaction that takes place is of the form $e^+e^- \to q\widehat{q}$ for some quark flavor. In the CM frame, the quark-antiquark pair will be produced back to back. Subsequently, hadrons will be formed out of these quarks, a process that is often called *hadronization*. This phenomenon of formation of hadrons out of quarks (and also secondary gluons) is governed by strong interaction only. If the original e^+e^- beams have a high CM energy, hadronization will not be very prompt because the QCD coupling constant is not very large at those energies. By the time hadronization starts efficiently, the quark and the antiquark produced directly from e^+e^- both move some distance, defining the direction of the final momenta. Hadronization then takes place around these two directions. So the hadrons will appear in two narrow cones whose axes will be back to back, as shown schematically in Fig. 12.8. A collection of hadrons within such a small solid angle is collectively called a *jet*. And so two such back-to-back jets will be seen. Production of two-jet events, instead of hadrons spread over large solid angles, is a signature of the quark substructure of hadrons.

Sometimes, either from the quark or from the antiquark produced in the basic level process, a gluon can be emitted. In such cases, hadrons will be formed along and around the gluon line as well, and one would observe three-jet events. Such processes would be rarer than two-jet events, because certainly the basic level process involves an extra power of the QCD coupling constant, which is not large at high energies. Fig. 12.9 shows pictures of real two-jet and three-jet events.

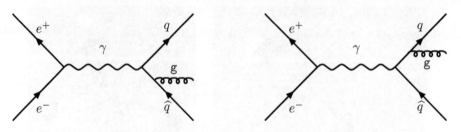

Figure 12.10: Lowest order Feynman diagrams for $e^+e^- \to q\widehat{q}g$. Remember our arrow convention: the outgoing arrow on the line marked e^+ really means an incoming positron. The final state particles are responsible for three-jet events.

A quantitative test of these statements would involve measurement of the angle that the jet axes make with the line of the original e^+e^- beams. For two-jet events, a quark-antiquark pair is produced at the basic level. It is an electromagnetic process mediated by a virtual photon, and is similar to the muon pair production that we discussed in §5.4.2. The angular distribution of the quark should have exactly the same nature as that of the muon shown in Eq. (5.94, $p\ 133$), i.e., neglecting quark masses, we should have

$$\frac{d\sigma}{d\Omega} \propto (1 + \cos^2 \theta). \tag{12.87}$$

Experimental data agree quite well with this expectation.

Three-jet events occur from the basic process $e^+e^- \to q\widehat{q}g$. Lowest order Feynman diagrams for this process have been shown in Fig. 12.10. The part of the diagrams that involves the quark, the antiquark and the gluon can be written as $\gamma^* \to q\widehat{q}g$, where γ^* is the virtual photon. From the result of Ex. 5.10 $(p\ 130)$, it is straightforward to write the matrix element squared of the process $\gamma^* q \to qg$. Then, using crossing symmetry, we can obtain the matrix element squared of the processes shown in Fig. 12.10. The result is

$$\overline{|\mathscr{M}|^2} = K \left(\frac{\mathfrak{s}^2 + \mathfrak{t}^2 + 2\mathfrak{u}\mathfrak{Q}^2}{\mathfrak{s}\mathfrak{t}} \right), \tag{12.88}$$

where the factor K contains contribution from the leptonic spinors, the vertices and the photon propagator. This part is of no interest to us in the present context. In Eq. (12.88),

$$\mathfrak{s} = (p_\gamma - p_q)^2, \qquad \mathfrak{t} = (p_\gamma - p_{\widehat{q}})^2, \qquad \mathfrak{u} = (p_\gamma - p_g)^2, \tag{12.89}$$

where the subscripts denote the particle whose momentum is represented by the notation, and $\mathfrak{Q}^2 = p_\gamma^2$. Neglecting the masses of the quarks and introducing the dimensionless variables

$$x_q = \frac{2E_q}{\mathfrak{Q}}, \qquad x_{\widehat{q}} = \frac{2E_{\widehat{q}}}{\mathfrak{Q}}, \qquad x_g = \frac{2E_g}{\mathfrak{Q}}, \tag{12.90}$$

we can rewrite Eq. (12.88) in the form

$$|\mathcal{M}|^2 = K \, \frac{x_q^2 + x_{\hat{q}}^2}{(1 - x_q)(1 - x_{\hat{q}})} \, . \tag{12.91}$$

If we insert it in the expression for deriving the cross-section, we would obtain

$$\frac{d\sigma}{dx_q dx_{\hat{q}}} = K' \, \frac{x_q^2 + x_{\hat{q}}^2}{(1 - x_q)(1 - x_{\hat{q}})} \, , \tag{12.92}$$

where K' is another uninteresting factor, obtained by integrating K over all other kinematical variables except x_q and $x_{\hat{q}}$.

Because the gluon is emitted, the quark and the antiquark will not be back to back. The interesting parameter therefore is the transverse momentum of the antiquark with respect to the direction of motion of the quark. Taking the quark direction of motion to be the z-axis, we can write the 4-momenta of the particles as follows:

$$p_q^\mu = \frac{1}{2}\mathcal{Q} \cdot (x_q, 0, 0, x_q) \, ,$$

$$p_{\hat{q}}^\mu = \frac{1}{2}\mathcal{Q} \cdot (x_{\hat{q}}, x_T, 0, -x_L) \, ,$$

$$p_g^\mu = \frac{1}{2}\mathcal{Q} \cdot (x_g, -x_T, 0, x_L - x_q) \, . \tag{12.93}$$

The masslessness of the antiquark and the gluon imposes extra constraints:

$$x_{\hat{q}}^2 - x_T^2 - x_L^2 = 0 \, , \tag{12.94a}$$

$$x_g^2 - x_T^2 - (x_L - x_q)^2 = 0 \, . \tag{12.94b}$$

In addition, the definitions given in Eq. (12.90) imply the condition

$$x_q + x_{\hat{q}} + x_g = 2 \, . \tag{12.95}$$

Using Eq. (12.94), we can eliminate x_L and obtain

$$x_T^2 = \frac{4}{x_q^2} \, (1 - x_q)(1 - x_{\hat{q}})(1 - x_g) \, . \tag{12.96}$$

□ **Exercise 12.10** Verify Eq. (12.96).

If the gluon hadn't been emitted at all, we would have obtained $x_g = 0$, and $x_q = x_{\hat{q}} = 1$. Because of the gluon emission, both x_q and $x_{\hat{q}}$ would differ from the value unity. Let us consider the case when the gluon is very soft, so that we can still use

$$x_q \approx x_{\hat{q}} \approx 1 \, , \qquad x_g \approx 0 \, , \tag{12.97}$$

except in factors where the differences from these approximate values are important. With this approximation, Eq. (12.96) gives

$$\left|\frac{\partial x_T^2}{\partial x_q}\right| \approx 4(1 - x_{\hat{q}}).$$ (12.98)

Then Eq. (12.92) can be rewritten as

$$\frac{d\sigma}{dx_T^2 dx_{\hat{q}}} = K' \frac{x_q^2 + x_{\hat{q}}^2}{4(1 - x_q)(1 - x_{\hat{q}})^2} \approx \frac{2K'}{x_T^2(1 - x_{\hat{q}})},$$ (12.99)

using Eq. (12.97) in writing the last step.

In order to obtain the differential cross-section with respect to the variable x_T^2 only, we need to integrate over $x_{\hat{q}}$. The limits of this integration depend on x_T. We note that $x_{\hat{q}}$ cannot reach the value unity because the gluon has been emitted. The softer the gluon, the higher the value of $x_{\hat{q}}$. Eq. (12.94b) tells us that, for a fixed x_T, the gluon is softest if $x_q = x_L$, and the smallest value is $x_g = x_T$. Putting this value of x_L into Eq. (12.94a) and using Eq. (12.95), we obtain

$$\left(x_{\hat{q}}\right)_{\text{max}} \approx 1 - \frac{1}{2}x_T,$$ (12.100)

neglecting higher order terms in x_T since we are considering small transverse momenta. Integration of Eq. (12.99) would then give

$$\frac{d\sigma}{dx_T^2} = \frac{2K'}{x_T^2} \ln(x_T^2) + \cdots,$$ (12.101)

where the dots represent the sub-leading terms for small x_T, which include terms coming from the lower end of the $x_{\hat{q}}$ integration. Since the transverse momentum is given by $p_T = \frac{1}{2}\Omega x_T$, we conclude that

$$\frac{d\sigma}{dp_T^2} \propto \frac{1}{p_T^2} \ln\left(\frac{\Omega^2}{4p_T^2}\right) + \cdots.$$ (12.102)

This prediction of transverse momentum distribution has been tested for various values of Ω^2, and the results are in excellent agreement.

12.5   The $1/N$ expansion

The problem with QCD is that, at low energies, the coupling constant is not small enough, and therefore perturbation methods cannot be used. It would therefore be of much help if we can find a small parameter with which perturbation expansion can be performed.

One strategy is to treat QCD as an $SU(N)$ gauge theory and treat this N, the number of colors, as a free parameter. To see how this might help, consider

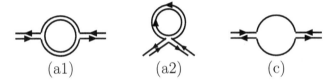

(a1)   (a2)   (c)

Figure 12.11: Examples of one-loop self-energy diagrams for gauge bosons in the double-line notation. The labels on the diagrams are the same as those used in Fig. 12.5 *(p 331)*.

the gluon self-energy diagrams shown in Fig. 12.5 *(p 331)*. In particular, let us first discuss diagram 'a', which has a gluon loop. Suppose this self-energy diagram contains the gluons g_a and g_b as external lines, where a and b are gauge indices as usual. The couplings will involve the generators in the adjoint representation. This is more easily seen if we write the field-strength tensor in the form

$$F^a_{\mu\nu} = \partial_\mu A^a_\nu - \partial_\nu A^a_\mu + ig\left(t^{(\mathrm{ad})}_b\right)_{ac} A^b_\mu A^c_\nu \qquad (12.103)$$

instead of what was written in Eq. (11.34, *p 304*). The two expressions are equivalent because of the definition of the adjoint representation given in Eq. (10.26, *p 262*). The trilinear coupling of the gluons arises from the quadratic term in the expression for the field-strength tensor, and hence the generator. The diagram will contain two such generators from two vertices, and there will be a trace on them because the internal gluon lines are in a loop. Looking at Eq. (11.39, *p 305*) now, we find that the amplitude contains the normalization factor C in the adjoint representation, which is N, according to Eq. (11.55, *p 308*).

 If we consider very large values of N, this will be a large factor, and will become infinite in the limit $N \to \infty$. To compensate for this factor and to obtain a smooth limit for the self-energy diagram for $N \to \infty$, we can suppose that the coupling constant is of the form $g/\sqrt{N}$. Since there are two vertices in the diagram, the factor of $1/N$ coming from the vertices will cancel the factor of N from the loop amplitude will be independent of N. However, for large N, $g/\sqrt{N}$ is small, so now we have a small coupling constant and we can use perturbation theory. This is the basic strategy: to perform calculations with large N and coupling constant $g/\sqrt{N}$, and take the limit $N \to \infty$ to see what are the results that survive in this limit.

 What will happen to the diagram in Fig. 12.5c *(p 331)* that contains a quark loop? In this case, one will obtain a trace over the generators in the fundamental representation, and that is independent of N. Therefore, because of the factor of $1/N$ coming from the two vertices, this diagram will vanish in the limit of infinite N.

There is an easy way to find out which diagrams would survive in the limit of large N without getting into the details of the Feynman rules and associated traces. For that, we have to introduce the so-called *double-line notation* for gluons. The idea is to represent each color index, rather than each particle, by a line in the diagram. This means that a quark is still represented by a single line. But a gluon, which belongs to the adjoint representation of the gauge group which can be obtained by the multiplication $N \times N^*$ in $SU(N)$, is represented by two lines in opposite directions. So, in effect, each color index corresponds to a line in any such diagram, and a name such as *color-flow diagram* would be appropriate for it. Various diagrams of Fig. 12.5 *(p 331)* have been redrawn in Fig. 12.11 using this new notation.

Looking at the diagrams, we see that each of the diagrams marked 'a1' and 'a2' contains one closed line that is not at all connected with the external lines. Any such closed line would contribute a factor N in the amplitude because N different colors can run through this loop irrespective of the external colors. The factors of N coming from vertices is easy to determine, just by looking at the number of vertices. Thus, for example, diagrams a1 and a2 contain the factor $1/N$ from vertices and N from the closed color loop, and therefore these amplitudes are independent of N. The diagram in Fig. 12.11c, on the other hand, has no closed color loop. So, because of the vertex factors, this diagram will be of order $1/N$, and would be irrelevant in the limit $N \to \infty$.

There is one class of diagrams which are even more suppressed. One such diagram has been shown in Fig. 12.12. The two gluon lines within the loop do not meet at any point. In other words, there is no quartic vertex in the middle of the diagram. One gluon line, say the one going from top left to bottom right, can be thought of as going above the plane of the paper, whereas the other is going below the plane, somewhat like a fly-over crossing a road below. This

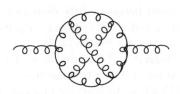

Figure 12.12: A non-planar diagram for gluon self-energy. The discontinuity on one of the internal lines implies that the two lines do not meet in a vertex there: they cross, one over the other.

feature of the diagram cannot be faithfully drawn on a plane, and therefore such diagrams are called *non-planar diagrams*. Any non-planar diagram will be suppressed by at least $1/N^2$.

    ☐ **Exercise 12.11** Redraw the diagram in Fig. 12.12 using the double-line notation and hence show that its amplitude is suppressed by $1/N^2$ for large N.

    ☐ **Exercise 12.12** Show that, if we add more gluon lines within the first two one-loop diagrams of Fig. 12.11 such that the resulting diagram is still planar, the self-energy function is still independent of N.

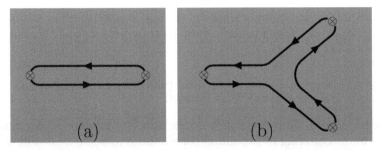

Figure 12.13: Color lines for (a) meson 2-point function; (b) diagram for the decay of one meson into two mesons. The crosses imply a meson wavefunction.

☐ **Exercise 12.13** Use the double-line notation to draw the self-energy diagram for a quark and argue that the amplitude will have a finite limit for $N \to \infty$.

Now that we have set up the notation, let us discuss some features of strong interactions that can be easily understood using large-N QCD. As a first example, we consider the decay widths of mesons. Mesons, as said earlier, contain a quark and an antiquark. Thus, a meson state can be written as

$$|M\rangle = \frac{1}{\sqrt{N}} \, q^\dagger_{1\alpha} \hat{q}^\dagger_{2\alpha} |0\rangle \,, \tag{12.104}$$

where $q^\dagger_1$ and $\hat{q}^\dagger_2$ represent creation operators of a quark of flavor 1 and an antiquark of flavor 2 (the two flavors may be the same for neutral mesons), and α is a color index. Since there are N colors, there are N different terms in the implied sum over the color index, and therefore there is a normalization factor $1/\sqrt{N}$. There may be extra normalization factors present for unflavored mesons in which q_1 and q_2 represent the same flavor, such as the factor of $1/\sqrt{2}$ in the definition of the π^0 state in Eq. (8.83, *p 221*), but these are irrelevant for our present discussion.

Consider now the mass of a meson. The mass appears in the 2-point function, which has been shown in Fig. 12.13a in a color-flow diagram. The flavors of the lines are not shown, and are not relevant either, so the discussion applies for any meson. There is one color loop, which gives a factor of N. Introducing the meson wave function introduces a factor of $1/\sqrt{N}$, as seen in Eq. (12.104). Therefore, the overall power of N, coming from the loop and the two wavefunctions, is N^0.

Consider now the decay width for a meson decaying into two mesons, for which the color-flow diagram has been shown in Fig. 12.13b. Here also there is one color loop, and hence a factor of N. However, there are three meson wavefunctions, which give $N^{-3/2}$. So the overall N-dependence of the amplitude is $N^{-1/2}$. In the limit $N \to \infty$, this implies that the decay amplitudes vanish, i.e., mesons are stable.

Of course mesons are not stable in the real world, but then N is also not infinity. What this extreme limit shows us is that the ratio Γ/M, where Γ is the decay rate of a meson of mass M, is small. And indeed that is true for any meson.

> ☐ **Exercise 12.14** *Evaluate the ratio Γ/M for $\pi^{\pm}$, π^{0} and $K^{\pm}$ to appreciate the fact that the ratio is small.*

Meson-meson scattering will be even more suppressed. Color-flow diagrams for a scattering of two initial-state mesons into two final-state mesons would look similar to that in Fig. 12.13b, except with four lobes instead of three. The meson wavefunctions now contribute a factor N^{-2}, whereas the single color loop contributes N. Thus, the overall amplitude behaves as N^{-1}, which shows that the mesons are also non-interacting in the large-N limit.

The argument above also shows why exotic meson states are not found in nature. By the term *exotic meson states*, we imply states with two or more quarks and an equal number of antiquarks. Since normal mesons, with one quark and one antiquark, are non-interacting, they cannot possibly bind into more complicated structures.

Baryons are much more complicated structures in the large-N theory. As seen in Ch. 8 and Ch. 10, quarks in a baryon are completely antisymmetric in their color indices. Thus, in an $SU(N)$ color theory, a baryon will contain N quarks. They can be described as topologically non-trivial states. We do not get into their detail in this book.

12.6   Lattice gauge theory

It is true that we cannot perform analytical calculations in the low energy regime of QCD, where the coupling constant is large. But can we at least get numerical solutions to some problems in that regime? Attempts toward this direction improved dramatically with the improvement of computers and algorithms, and have flowered into a subfield called *lattice gauge theory*. In this section, we will try to give a brief introduction to the methods and the problems of this subfield. Although we focus our attention on QCD, it should be understood that the any quantum field theory can be treated on a lattice. Such treatment provides a natural way of eliminating all infinities that arise from momentum integration in a continuum.

12.6.1   Scope and basics

The QCD Lagrangian was given in Eq. (12.48). It contains gluon fields and quark fields. Among the six known quarks, the top quark is too heavy to form any bound state, as will be discussed in some detail in Ch. 20. The masses of the other five quarks, as well as the gauge coupling constant for QCD, form the set of parameters present in the Lagrangian. With some input values of these parameters, one can try to find various properties of hadrons and match them with experimental results available. This way, one can find the values of the parameters, and then use them to predict other hadronic properties which may not be well-known or may even be unobserved.

There are two kinds of basic calculations that one can perform. The first kind consists of calculations of static properties of various hadrons like their masses and lifetimes. For example, one can try to calculate the lowest lying meson state that appears from the Lagrangian. Once this is found, it can be identified with the mass of the pion. Similarly, heavier hadron masses can be calculated with some specific values for the input parameters. The hadronic spectrum need not contain only bound states of quarks and/or antiquarks which were discussed in Ch. 10. There might also be states consisting of gluons only. Such states are called *glueballs*.

The second kind of calculations involve strong corrections to weak and electromagnetic processes. Consider, e.g., the decay of the tau lepton. This lepton is heavy enough for the final state to contain hadrons. The basic decay, at the quark level, would produce a tau-neutrino, along with a quark and an antiquark. The decay is governed by weak interactions, and we will discuss such decays in Ch. 16. However, once a quark and an antiquark are produced in the final state, they will experience strong interaction. The weak interaction mediated amplitude will be modified by strong interaction corrections. Evaluation of these corrections is an important ingredient in checking the validity of weak interaction theory when hadrons are involved. Such corrections are also estimated through lattice calculations.

The *lattice* mentioned in the name of the method corresponds to spacetime. In this formulation, spacetime is supposed not to be a continuum, but rather a collection of lattice points. We will confine our discussion to *hypercubic* lattices (which are just 4-dimensional analogs of 3-dimensional *cubic* lattices), which are used in most calculations anyway. The lattice spacing, a, is a parameter introduced into the theory. The physical limit, or the continuum limit, can be taken by performing the calculations for various values of a and finally extrapolating the result to the limit $a \to 0$.

There is a big advantage of performing calculations on a lattice with a non-zero lattice spacing. Earlier we mentioned how we can regularize a theory by introducing an extra parameter in it. In particular, we mentioned that, by introducing the number of spacetime dimensions as a parameter, loop integrals encountered in perturbation theory can be evaluated. Similarly, the lattice spacing a can act as a regularization parameter, i.e., it can save the amplitudes from becoming infinite because the momentum integrations run only over a finite range that is inversely proportional to a, as will be shown in Eq. (12.172). In the continuum theory, since the loop corrections are infinite, the bare parameters are also infinite, as discussed in §12.2. On a lattice, the loop amplitudes will be finite, and therefore the bare parameters defined from them would also be finite. Thus it would be possible to take the bare parameters — the coupling constant and the quark masses — as the input parameters of the calculations.

12.6.2   Path integral

In performing calculations on the lattice, one uses the *path integral approach* to quantum field theories developed by Feynman. The approach does not depend on a lattice structure of spacetime, so we outline the basics of this approach in the continuum language before going over to the lattice picture.

The main idea is to write something like the partition function in statistical mechanics, from which all physical properties of the system can be derived. For a quantum field theory, this similar object is called the path integral, and is defined as

$$Z = \int \mathscr{D}\Phi_1 \int \mathscr{D}\Phi_2 \cdots \int \mathscr{D}\Phi_n \ \exp(i\mathscr{A}), \qquad (12.105)$$

where $\mathscr{A}$ is the classical action of the fields Φ_1 to Φ_n. The integration measure $\mathscr{D}\Phi$ for any field Φ is defined as

$$\mathscr{D}\Phi = \lim_{n\to\infty} \prod_{i=1}^{n} d\Phi(x_i), \qquad (12.106)$$

where $\Phi(x_i)$ denotes the value of Φ at the point x_i, and we consider n such points in the entire spacetime. Consider what we mean when we write an integration measure dx over a real variable x: we mean that x can take any value (maybe within a range) and we must sum the integrand over all such values. Similarly, here we consider the sum of the integrand over all possible values of $\Phi(x_i)$ at n different points. In the limit of n going to infinity, we obtain the sum over all possible configurations of the field Φ. The action itself is the integral of the Lagrangian over the entire spacetime. From the path integral, one can obtain the vacuum expectation value of any operator through the prescription

$$\langle \mathscr{O} \rangle = \frac{1}{Z} \int \mathscr{D}\Phi_1 \int \mathscr{D}\Phi_2 \cdots \int \mathscr{D}\Phi_n \ \mathscr{O} \exp(i\mathscr{A}). \qquad (12.107)$$

For the QCD Lagrangian, the relevant fields are the quark and gluon fields, and we can write

$$Z = \int \mathscr{D}\Psi \int \mathscr{D}\overline{\Psi} \int \mathscr{D}\mathsf{G}_a^\mu \ \exp\left(i \int d^4x \ \mathscr{L}_{\mathrm{QCD}} \right), \qquad (12.108)$$

where Ψ represents all quark fields taken together as a huge column vector. In other words, integration over all configurations of Ψ implies integration over the configurations of all quark fields. The basic task in lattice calculations is therefore to estimate the path integral and quantities like that given in Eq. (12.107) for various operators.

For the gauge fields, the procedure produces some complication. The point is that two different configurations of gauge fields might in fact be related through a gauge transformation and be therefore equivalent physically. In other words, one really shouldn't integrate over *all* configurations of the gluon fields: just the ones that are physically inequivalent. In order to filter out the inequivalent configurations of gauge fields, one has to introduce a gauge-fixing term in the form of a delta function in the integrand of Eq. (12.105) or Eq. (12.108). It can be shown that this delta function can be absorbed in the action by introducing some extra unphysical fields. These are the Fadeev–Popov ghost fields mentioned in Ch. 11.

We will not elaborate on this procedure because, as we shall see, in the lattice formulation one does not use the gauge fields directly. Rather, one makes use of the *path-ordered integrals*. In a continuum theory, the path-ordered integral of a gauge field between two spacetime points x_1 and x_2 is defined as:

$$\mathscr{U}(x_1, x_2) \equiv \mathscr{P} \exp\left(ig \int_{x_1}^{x_2} dx^\mu \, \mathbb{A}_\mu\right), \tag{12.109}$$

where $\mathbb{A}_\mu$ is a 3×3 matrix $\frac{1}{2}\lambda^a A_\mu^a$, as defined in Eq. (11.13, *p 300*). The symbol $\mathscr{P}$ indicates that it is a path-ordered object, i.e., if we divide the path into small segments by inserting some intermediate points $y_1, \cdots, y_{N-1}$ and identify x_1 as y_0 and x_2 as y_N, then

$$\mathscr{U}(x_1, x_2) = \lim_{N \to \infty} \prod_{r=1}^{N} \mathscr{U}(y_{r-1}, y_r). \tag{12.110}$$

To see how such quantities transform under a gauge transformation, we first consider one of the factors of infinitesimal path length appearing in Eq. (12.110), going from x^μ to $x^\mu + \epsilon^\mu$. Keeping only first order term in ϵ, we can write

$$\mathscr{U}(x, x + \epsilon) = 1 + ig\epsilon^\mu \mathbb{A}_\mu(x). \tag{12.111}$$

Let us now recall the gauge transformation properties of the gauge fields from Eq. (11.19, *p 301*). Clearly, for a gauge transformation that changes a fermion field Ψ to

$$\Psi'(x) = U\Psi(x), \tag{12.112}$$

the quantity $\mathscr{U}(x, x + \epsilon)$ will be changed to

$$\mathscr{U}'(x, x + \epsilon) = 1 + ig\epsilon^\mu U(x)\left(\mathbb{A}_\mu(x)U^{-1}(x) - \frac{i}{g}\partial_\mu U^{-1}(x)\right). \tag{12.113}$$

Note that we can write

$$\epsilon^\mu \partial_\mu U^{-1}(x) = U^{-1}(x + \epsilon) - U^{-1}(x). \tag{12.114}$$

Using this, we obtain

$$\mathscr{U}'(x, x + \epsilon) = ig\epsilon^\mu U(x)\mathbb{A}_\mu(x)U^{-1}(x) + U(x)U^{-1}(x + \epsilon). \tag{12.115}$$

Since the term containing $\mathbb{A}_\mu$ is already first power in smallness, it does not matter to this order if we change $U^{-1}(x)$ to $U^{-1}(x+\epsilon)$ in this term and write

$$\mathscr{U}'(x, x+\epsilon) = U(x)\Big(1 + ig\epsilon^\mu \mathbb{A}_\mu(x)\Big)U^{-1}(x+\epsilon)$$

$$= U(x)\mathscr{U}(x, x+\epsilon)U^{-1}(x+\epsilon). \tag{12.116}$$

From Eq. (12.110), it is now obvious that for a finite path, we will obtain

$$\mathscr{U}'(x_1, x_2) = U(x_1)\mathscr{U}(x_1, x_2)U^{-1}(x_2). \tag{12.117}$$

From Eqs. (12.112) and (12.117), it is clear that there can be two kinds of gauge invariants involving the gauge fields. They are of the following generic forms:

$$\text{Type 1} \ : \overline{\psi}(x_1)\mathscr{U}(x_1, x_2)\psi(x_2),$$

$$\text{Type 2} \ : \text{Tr}\Big(\mathscr{U}(x_1, x_2)\mathscr{U}(x_2, x_3)\mathscr{U}(x_3, x_4)\mathscr{U}(x_4, x_1)\Big). \tag{12.118}$$

We will see later that the QCD action on the lattice can be built up by using only these two kinds of invariants.

12.6.3 Quantum field theory on Euclidean spacetime

As we remarked earlier, lattice calculations are done by taking the spacetime points to be points on a hypercubic lattice. A hypercubic lattice is defined in a Euclidean space, so first of all we need to define the theory on a Euclidean spacetime. From the Minkowskian spacetime, we therefore go over to a Euclidean space, whose spatial co-ordinates are the same as the spatial co-ordinates of the Minkowski spacetime, and whose temporal co-ordinate, to be denoted by $\underline{t}$, is defined by

$$\underline{t} = it. \tag{12.119}$$

The co-ordinate 4-vector in the Euclidean space is defined as

$$\underline{x}^\mu \equiv \{\underline{t}, x, y, z\} = \{it, x, y, z\}. \tag{12.120}$$

Superscripted and subscripted vector indices will be taken to denote the same thing, which means that the Euclidean space metric is taken to be

$$\underline{g}_{\mu\nu} = \text{diag}(1, 1, 1, 1). \tag{12.121}$$

Thus,

$$x^\mu x_\mu = -\underline{x}_\mu \underline{x}_\mu = -\underline{x}^2. \tag{12.122}$$

The Lagrangian depends on fields which are functions of spacetime points, and therefore varies from one spacetime point to another. We can denote this dependence by writing the Lagrangian of Minkowski space as $\mathscr{L}(t, \boldsymbol{x})$. We now define the Lagrangian in the Euclidean space by the relation

$$\underline{\mathscr{L}}(\underline{t}, \boldsymbol{x}) = -\mathscr{L}(t, \boldsymbol{x}), \tag{12.123}$$

where $\underline{t}$ is defined through Eq. (12.119). The Euclidean action, $\underline{\mathscr{A}}$, is defined by the obvious formula

$$\underline{\mathscr{A}} = \int d^4\underline{x}\, \underline{\mathscr{L}}(\underline{t}, \boldsymbol{x}). \tag{12.124}$$

Since Eq. (12.119) tells us that the Euclidean element of spacetime volume is given by

$$d^4\underline{x} = id^4x, \tag{12.125}$$

we can combine it with Eq. (12.123) to write

$$i\underline{\mathscr{A}} = -\mathscr{A}. \tag{12.126}$$

Therefore the path integral can be written in this Euclidean space in the form

$$Z = \int \mathscr{D}\Psi \int \mathscr{D}\overline{\Psi} \int \mathscr{D}G_a^\mu \, e^{-\underline{\mathscr{A}}}, \tag{12.127}$$

and the vacuum expectation value of an operator $\mathscr{O}$ as

$$\langle \mathscr{O} \rangle = \frac{1}{Z} \int \mathscr{D}\Psi \int \mathscr{D}\overline{\Psi} \int \mathscr{D}G_a^\mu \, \mathscr{O}e^{-\underline{\mathscr{A}}}. \tag{12.128}$$

One might wonder why we put a minus sign on the right hand side of the definition in Eq. (12.123). We want to emphasize that it is only a matter of convention. If we had omitted the minus sign there, Eq. (12.126) would not have the minus sign as well. In this case, we would have written the path integral with a factor of $e^{\underline{\mathscr{A}}}$ rather than $e^{-\underline{\mathscr{A}}}$. But, because of the change of notation in Eq. (12.123), the new $\underline{\mathscr{A}}$ would have been negative of the action that appears in Eq. (12.127). So the path integral would have contained the same exponent, whether we call it $-\underline{\mathscr{A}}$ and define the Euclidean Lagrangian through Eq. (12.123), or call it $\underline{\mathscr{A}}$ and define $\underline{\mathscr{L}}$ with the opposite sign on the right hand side of Eq. (12.123).

Let us now discuss how the Euclidean Lagrangian looks. We start with the Lagrangian of a free Dirac field in Minkowski space and use Eq. (12.119):

$$\begin{aligned}
\mathscr{L}_{\text{Dirac}} &= \overline{\psi}\left(i\gamma^0 \frac{\partial}{\partial t} + i\gamma^i \frac{\partial}{\partial x^i} - m \right)\psi \\
&= \overline{\psi}\left(-\gamma^0 \frac{\partial}{\partial \underline{t}} + i\gamma^i \frac{\partial}{\partial x^i} - m \right)\psi.
\end{aligned} \tag{12.129}$$

We can now define the Dirac matrices in the Euclidean space as

$$\underline{\gamma}^0 = \gamma^0, \qquad \underline{\gamma}^i = -i\gamma^i, \tag{12.130}$$

so that they satisfy the anticommutation relation

$$\left[\gamma^\mu,\gamma^\nu\right]_+ = 2\delta^{\mu\nu}\,. \tag{12.131}$$

The Euclidean Dirac matrices are therefore all hermitian. Since lower and upper vector indices are equivalent in the Euclidean space, γ_μ and γ^μ are numerically the same matrices.

Then we can write

$$\mathscr{L}_{\text{Dirac}} = \overline{\psi}\left(-\gamma^0\frac{\partial}{\partial t} - \gamma^i\frac{\partial}{\partial x^i} - m\right)\psi$$
$$= -\overline{\psi}\left(\gamma_\mu\partial_\mu + m\right)\psi\,. \tag{12.132}$$

Eq. (12.123) now tells us that the free Dirac Lagrangian in the Euclidean space is given by

$$\mathscr{L}_{\text{Dirac}} = \overline{\psi}\left(\gamma_\mu\partial_\mu + m\right)\psi\,. \tag{12.133}$$

In passing, it should be noted that $\overline{\psi}$ is not defined as $\psi^\dagger\gamma_0$ in the Euclidean space. Since all γ_μ's have the same hermiticity property, such a definition would not make the right hand side of Eq. (12.133) invariant under rotations in the Euclidean space. Instead, $\overline{\psi}$ in independently defined Euclidean space such that $\overline{\psi}\psi$ is invariant.

□ **Exercise 12.15** *Taking all Euclidean gamma matrices to be hermitian is just a matter of convention. We could have also taken all γ_μ's to be anti-hermitian. Make such a choice and find the Euclidean Lagrangian for a free Dirac field in this convention.*

We now perform the same exercise on the interaction term between the fermions and the gauge bosons. In the Minkowski space, we have

$$\mathscr{L}_{\text{int}} = -g\overline{\psi}\left(\gamma^0\mathbb{G}^0 - \gamma^i\mathbb{G}^i\right)\psi\,, \tag{12.134}$$

where $\mathbb{G}_\mu$ are 3×3 matrices defined with the gluon fields,

$$\mathbb{G}_\mu = \frac{1}{2}\lambda^a\mathbb{G}_\mu^a\,, \tag{12.135}$$

defined in accordance with the more general notation introduced in Eq. (11.13, *p 300*). Since the gluon field is a vector field, we should define its components in the Euclidean space in such a way that they bear the same relation with the Minkowski space components as the components of the co-ordinate vector does, i.e.,

$$\mathbb{G}^0 = i\mathbb{G}^0\,, \qquad \mathbb{G}^i = \mathbb{G}^i\,. \tag{12.136}$$

The interaction Lagrangian can be written as

$$\mathscr{L}_{\text{int}} = ig\overline{\psi}\left(\gamma^0\mathbb{G}^0 + \gamma^i\mathbb{G}^i\right)\psi$$
$$= ig\overline{\psi}\gamma_\mu\mathbb{G}_\mu\psi\,. \tag{12.137}$$

Combining with the Euclidean Lagrangian of the free Dirac field given in Eq. (12.133), we can now write

$$\mathscr{L}_{\text{Dirac}} + \mathscr{L}_{\text{int}} = \overline{\psi} \left(\gamma_\mu \underline{D}_\mu + m \right) \psi, \tag{12.138}$$

where

$$\underline{D}_\mu = \partial_\mu - ig\underline{\mathbb{G}}_\mu. \tag{12.139}$$

Because of the Euclidean space definitions given in Eqs. (12.119) and (12.136), it is easy to see that the Euclidean components of the field-strength tensor are given by

$$\underline{\mathsf{G}}_{0i}^a = i\mathsf{G}_{0i}^a, \qquad \underline{\mathsf{G}}_{ij}^a = \mathsf{G}_{ij}^a. \tag{12.140}$$

Therefore, the pure gauge Lagrangian can be written as

$$
\begin{aligned}
\mathscr{L}_{\text{gauge}} &= -\frac{1}{4} \mathsf{G}_{\mu\nu}^{a\dagger} \mathsf{G}_a^{\mu\nu} \\
&= -\frac{1}{4} \left(-2\mathsf{G}_{0i}^{a\dagger} \mathsf{G}_{0i}^a + \mathsf{G}_{ij}^{a\dagger} \mathsf{G}_{ij}^a \right) \\
&= -\frac{1}{4} \left(2\underline{\mathsf{G}}_{0i}^{a\dagger} \underline{\mathsf{G}}_{0i}^a + \underline{\mathsf{G}}_{ij}^{a\dagger} \underline{\mathsf{G}}_{ij}^a \right) \\
&= -\frac{1}{4} \underline{\mathsf{G}}_{\mu\nu}^{a\dagger} \underline{\mathsf{G}}_{\mu\nu}^a.
\end{aligned}
\tag{12.141}
$$

This implies that the Euclidean action of QCD is given by

$$\underline{\mathscr{A}} = \int d^4\underline{x} \left(\frac{1}{4} \underline{\mathsf{G}}_{\mu\nu}^{a\dagger} \underline{\mathsf{G}}_{\mu\nu}^a + \overline{\Psi} \left(\gamma_\mu \underline{D}_\mu + M \right) \Psi \right). \tag{12.142}$$

In writing this equation, we have gone back to the notation where Ψ stands for all quark fields, and M is a huge square matrix, with appropriate entries for the masses of the different quark fields.

☐ **Exercise 12.16** Prove Eq. (12.140). [**Hint** : *Beware of the fact that Eq. (12.136) implies*

$$\underline{\mathsf{G}}_0 = i\mathsf{G}_0, \qquad \underline{\mathsf{G}}_i = -\mathsf{G}_i, \tag{12.143}$$

where the minus sign appears because the spatial components of a Euclidean vector has the same value whether the index is upper or lower, whereas in Minkowski space there is a relative negative sign.]

☐ **Exercise 12.17** From the definition of co-ordinates and gauge fields in the Euclidean space, show that the analog of the gauge transformation rule of Eq. (11.19, *p 301*) is given by

$$\underline{\mathbb{A}}_\mu' = U\underline{\mathbb{A}}_\mu U^{-1} + \frac{i}{g} U(\partial_\mu U^{-1}) \tag{12.144}$$

in the Euclidean space. [**Note** : *The sign of the A-independent term is different.*]

12.6.4   Quarks and gluons on a lattice

So far, we have worked with a spacetime continuum. We have been developing the tools which would make it easier for us to formulate the path integral on a spacetime lattice. We are now ready to do so. Our first task is to know how to write the lattice action with quarks and gluons.

The action will have to be gauge invariant. Already in Eq. (12.118), we have identified two kinds gauge invariant combinations involving fermion fields and gauge fields. The fermion fields appeared directly in the type-1 invariants, and the gauge fields appeared in both types of invariants through the path-ordered integrals.

On the lattice, then, we can take a direct approach for the quark fields and represent them by anticommuting variables on each lattice site. They belong to the fundamental representation of the QCD gauge group SU(3).

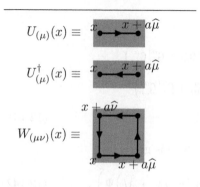

$$U_{(\mu)}(x) \equiv$$

$$U^{\dagger}_{(\mu)}(x) \equiv$$

$$W_{(\mu\nu)}(x) \equiv$$

Figure 12.14: Pictorial representations of basic link and plaquette variables defined in Eqs. (12.147), (12.148) and (12.149).

In order to introduce the gauge fields, i.e., the gluons, we need something that resembles the path-ordered integrals discussed earlier. There is however no continuous path now from one point to another, because the lattice represents spacetime, and it is discrete. Thus, the best we can do is to assign a function of two different lattice points that will work the same way as a path-ordered integral would do in the continuum, i.e., will have the gauge transformation property like that shown in Eq. (12.117). The simplest such thing would be a function of two nearest neighbors sites on the lattice. We call this function a *link variable* and denote it by the notation $U_{(\mu)}(x)$. This would stand for a link variable for the two points on the lattice, one at x, and the other its nearest neighbors in the μth direction. Note that we put the index μ within parentheses because it should not be thought that the quantity transforms like a spacetime vector. In fact, there is no concept of a spacetime vector now, because we are on the lattice where the Lorentz symmetry is not present.

The connection between the link variables and the gauge fields can be established through the analogy with the continuum theory. Remember that the path-ordered integral in the Minkowski space was defined through Eq. (12.109). In the Euclidean space, the corresponding expression should read

$$\mathscr{U}(\underline{x}_1, \underline{x}_2) \equiv \mathscr{P} \exp\left(-ig \int_{\underline{x}_1}^{\underline{x}_2} d\underline{x}_\mu \, \mathbb{A}_\mu\right). \tag{12.145}$$

The reason for the appearance of the minus sign was explained in Eq. (12.122). Between the points x and $x + a\widehat{\mu}$, we can neglect the variation of the gauge field and obtain

$$\mathscr{U}(x, x + a\widehat{\mu}) = \exp\left(-iag\mathbb{A}_{\mu}(x)\right). \qquad (12.146)$$

where $\widehat{\mu}$ is the unit vector on the spacetime lattice in the μth direction, so that $x + a\widehat{\mu}$ is the nearest neighbor of the point x in the μth direction. It is this quantity that we identify as our link variable:

$$U_{(\mu)}(x) \equiv \mathscr{U}(x, x + a\widehat{\mu}) = \exp\left(-iag\mathbb{A}_{\mu}(x)\right). \qquad (12.147)$$

Clearly,

$$U_{(\mu)}^{\dagger}(x) = \mathscr{U}(x + a\widehat{\mu}, x) = \exp\left(iag\mathbb{A}_{\mu}(x)\right). \qquad (12.148)$$

Pictorial representations of the objects $U_{(\mu)}(x)$ and $U_{(\mu)}^{\dagger}(x)$ have been shown in Fig. 12.14.

In writing Eqs. (12.147) and (12.148) and all subsequent equations of this section, we omit the special symbol for denoting Euclidean space variables. Our transition from Minkowski space to Euclidean space is now complete. We will now talk exclusively about field theory on a lattice, which is defined in the Euclidean space, and therefore all variables are supposed to be in the Euclidean space from now on, without any special notational indication. The special notation will come back later where necessary.

While these link variables are the basic units of type-1 gauge invariants on the lattice, the type-2 invariants described in Eq. (12.118) are formed from the *plaquette variables*, which are defined for a unit planar face of a lattice cell. For a lattice cell in the μ-ν plane, the plaquette variable is defined as

$$W_{(\mu\nu)}(x) \equiv U_{(\mu)}(x)U_{(\nu)}(x + a\widehat{\mu})U_{(\mu)}^{\dagger}(x + a\widehat{\nu})U_{(\nu)}^{\dagger}(x), \qquad (12.149)$$

where x is the vertex with the minimum values of μ as well as ν co-ordinates. Fig. 12.14 contains a pictorial representation of such variables as well.

It should be clear that neither the link variables nor the plaquette variables are gauge invariant. For example, comparing Eq. (12.118) with Eq. (12.149), it is clear that the $W_{(\mu\nu)}$'s are not gauge invariant, but rather their traces are. The transformation properties of the link variables are obvious from Eq. (12.117). We now show how the link variables and the plaquette variables help us build the gauge invariant lattice action.

12.6.5 Lattice action

The pure gauge action can be written in terms of the plaquettes. To show this, we take the easy case of a U(1) gauge theory. Writing e for the coupling constant and the gauge fields as A_μ, we can use Eq. (12.147) to rewrite Eq. (12.149) in the form

$$W_{(\mu\nu)}(x) = \exp\left(iaeA_\mu(x)\right)\exp\left(iaeA_\nu(x+a\hat\mu)\right)$$
$$\times \exp\left(-iaeA_\mu(x+a\hat\nu)\right)\exp\left(-iaeA_\nu(x)\right). \quad (12.150)$$

The exponents all commute for a U(1) theory, so we can add the powers. Expanding the total power now about the point x, we find

$$W_{(\mu\nu)}(x) = \exp\left(iae\left[A_\mu + \left(A_\mu + a\partial_\mu A_\nu\right) - \left(A_\mu + a\partial_\nu A_\mu\right) - A_\nu\right] + \ldots\right)$$
$$= \exp\left(ia^2 eF_{(\mu\nu)} + \ldots\right), \quad (12.151)$$

where all field values now refer to the point x. The dots refer to higher order terms in the lattice spacing a.

☐ **Exercise 12.18** *Because of the non-abelian nature of the QCD gauge group, the exponents cannot be simply added as has been done in Eq. (12.151). One must use the Baker–Campbell–Hausdorff formula in order to combine the indices. Show that the commutator terms in the Baker–Campbell–Hausdorff formula give rise to the non-linear terms of the field-strength tensor, so that one still obtains the last step of Eq. (12.151).*

For QCD, we will obtain a similar expression:

$$W_{(\mu\nu)} = \exp\left(ia^2 g\mathbb{G}_{(\mu\nu)} + \ldots\right). \quad (12.152)$$

Here and henceforth, we use the notation for the matrix representation of the gluon fields, $\mathbb{G}$, instead of the more general notation $\mathbb{A}$ used for general gauge fields in earlier sections. Expanding the exponential, we obtain

$$W_{(\mu\nu)} = 1 + ia^2 g\mathbb{G}_{(\mu\nu)} - \frac{1}{2}a^4 g^2 \mathbb{G}_{(\mu\nu)}\mathbb{G}_{(\mu\nu)} + \cdots. \quad (12.153)$$

Notice the parentheses around the directional indices. They imply that these are not vector indices, and no summation should be implied on them. Note that the notation defined in Eq. (12.135) implies that

$$\mathrm{Tr}\left(\mathbb{G}_{(\mu\nu)}\mathbb{G}_{(\mu\nu)}\right) = \frac{1}{2}\mathbb{G}^a_{(\mu\nu)}\mathbb{G}^{a\dagger}_{(\mu\nu)}, \quad (12.154)$$

using the normalization of the fundamental representation decided upon in Eq. (11.54, *p 308*). Therefore,

$$\frac{1}{g^2}\,\mathrm{Tr}\,\mathbb{Re}\left(1 - W_{(\mu\nu)}\right) = \frac{1}{4}a^4 \mathbb{G}^a_{(\mu\nu)}\mathbb{G}^{a\dagger}_{(\mu\nu)} + \ldots, \quad (12.155)$$

where $\mathbb{Re}$ denotes the real part of the quantity that follows.

☐ **Exercise 12.19** Find the next higher order term in Eq. (12.155).

Let us now see what is the relation of this quantity with the discretized version of the continuum action. In the continuum Euclidean space, the pure gauge action is given in the fermion-independent term of Eq. (12.142). In the lattice, we can approximate the spacetime integral of a function by the prescription

$$\int d^4x \rightarrow a^4 \sum_x \, , \tag{12.156}$$

where the sum on the right hand side runs over each cell in the lattice, and a^4 is the volume of a cell. Thus, the discretized limit of the pure gauge action should be

$$\frac{1}{4} a^4 \sum_x \sum_{\mu<\nu} 2G^a_{(\mu\nu)} G^{a\dagger}_{(\mu\nu)} \, , \tag{12.157}$$

where the factor of 2 is coming from the fact that the continuum action has unrestricted sum over the Lorentz indices, whereas in the discretized version we sum only over $\mu < \nu$. Comparing with Eq. (12.155), we find that the pure gauge part of the lattice action can be written as

$$\mathscr{A}_{\text{gauge}} = \frac{2}{g^2} \sum_x \sum_{\mu<\nu} \text{Tr} \, \text{Re} \left(1 - W_{(\mu\nu)} \right) . \tag{12.158}$$

This expression is often written in the form

$$\mathscr{A}_{\text{gauge}} = \frac{2}{g^2} \sum_P \text{Tr} \left(1 - \frac{1}{2}(W_P + W_P^\dagger) \right) , \tag{12.159}$$

where the subscript P denotes a plaquette, and the sum is over all plaquettes.

Let us now look at the terms involving fermions. The mass terms are straightforward. The derivative term of a fermion field can be written, in terms of quantities on the lattice, by the prescription

$$\overline{\psi}\gamma_\mu \partial_\mu \psi \rightarrow \frac{1}{2a} \sum_\mu \overline{\psi}(x)\gamma_\mu \left(\psi(x + a\widehat{\mu}) - \psi(x - a\widehat{\mu}) \right) . \tag{12.160}$$

Adding to it the interaction term between fermions and gauge fields:

$$-ig\overline{\psi}(x)\gamma_\mu \mathbb{G}_\mu(x)\psi(x) \, , \tag{12.161}$$

we obtain

$$\overline{\psi}\gamma_\mu D_\mu \psi \rightarrow \frac{1}{2a} \sum_\mu \overline{\psi}(x)\gamma_\mu \left(\psi(x + a\widehat{\mu}) - 2iag\mathbb{G}_\mu(x)\psi(x) - \psi(x - a\widehat{\mu}) \right) . \tag{12.162}$$

Note that, according to Eq. (12.147),

$$U_{(\mu)}(x) = 1 - iag\mathbb{G}_\mu(x) + \mathcal{O}\left(a^2\right) . \tag{12.163}$$

Multiplying both sides to the right by $\psi(x + a\hat{\mu})$, we obtain

$$U_{(\mu)}(x)\psi(x + a\hat{\mu}) = \psi(x + a\hat{\mu}) - iag\mathbb{G}_\mu(x)\psi(x) + \mathcal{O}\left(a^2\right) . \tag{12.164}$$

Similarly,

$$U^\dagger_{(\mu)}(x)\psi(x - a\hat{\mu}) = \psi(x - a\hat{\mu}) + iag\mathbb{G}_\mu(x)\psi(x) + \mathcal{O}\left(a^2\right) . \tag{12.165}$$

Therefore the part of the action that involves the fermions can be written in the form

$$\begin{aligned}
\mathscr{A}_{\text{fermion}} &= -\int d^4\underline{x}\left(\overline{\psi}\underline{\gamma_\mu D_\mu}\psi + m\overline{\psi}\psi\right) \\
&\rightarrow -a^4\Bigg[\frac{1}{2a}\sum_x\sum_\mu\left(\overline{\psi}(x)\gamma_\mu U_{(\mu)}(x)\psi(x + a\hat{\mu})\right. \\
&\qquad\qquad\left. - \overline{\psi}(x)\gamma_\mu U^\dagger_{(\mu)}(x)\psi(x - a\hat{\mu}) + \ldots\right) \\
&\qquad + m\sum_x\overline{\psi}(x)\psi(x)\Bigg] ,
\end{aligned} \tag{12.166}$$

where the dots represent terms with higher powers of a. On the lattice, the Lagrangian therefore can be taken to be only the terms written explicitly in the equation above, discarding the terms represented by the dots. The total action will be a sum of the contributions given in Eqs. (12.158) and (12.166). This lattice action would reduce to the continuum action in the limit $a \rightarrow 0$.

It should be noted that the lattice action contains only the gauge invariant combinations shown in Eq. (12.118). For the pure gauge action, this is obvious since we used only the traces of the plaquette variables, which are type-2 gauge invariants. In Eq. (12.166), we might note that the combination used in the first term is nothing but $\overline{\psi}(x)\gamma_\mu \mathscr{U}(x, x + a\hat{\mu})\psi(x + a\hat{\mu})$. The Euclidean Dirac matrices γ_μ are irrelevant while considering gauge transformations. Other than that, this is exactly like a type-1 invariant. Similarly, the combination in the second term is $\overline{\psi}(x)\gamma_\mu \mathscr{U}(x + a\hat{\mu}, x)\psi(x - a\hat{\mu})$, which can be written as $\overline{\psi}(x)\gamma_\mu \mathscr{U}(x, x - a\hat{\mu})\psi(x - a\hat{\mu})$ by neglecting higher order terms in a, and this is also a type-1 invariant.

12.6.6   Fermion doubling problem

However, the simple procedure described above has a serious problem. To see this, let us use Eq. (12.160) to write the free part of the lattice action, implied in Eq. (12.138), in the form

$$\mathscr{A}_0 = a^4\sum_x\sum_y\overline{\psi}(x)K(x, y)\psi(y) , \tag{12.167}$$

where

$$K(x,y) = \frac{1}{2a} \sum_\mu \gamma_\mu \left(\delta_{y,x+a\hat\mu} - \delta_{y,x-a\hat\mu} \right) + m\delta_{y,x} . \tag{12.168}$$

Although we have written the left hand side as a function of the two co-ordinates x and y, the right hand side is really a function of $y-x$. So, we can define the Fourier transform as

$$\begin{aligned} K(p) &= \sum_{y-x} \exp\left(i \sum_\mu p_\mu (y_\mu - x_\mu) \right) K(x,y) \\ &= \frac{1}{2a} \sum_\mu \left[\exp\left(i\gamma_\mu p_\mu a \right) - \exp\left(-i\gamma_\mu p_\mu a \right) \right] + m \\ &= i \sum_\mu \gamma_\mu \tilde p_\mu + m , \end{aligned} \tag{12.169}$$

where

$$\tilde p_\mu = \frac{1}{a} \sin\left(p_\mu a \right) . \tag{12.170}$$

The inverse of this expression will be the propagator, as discussed in §4.10.2. It is easy to see that the propagator will be given by

$$K^{-1}(p) = \frac{-i \sum_\mu \gamma_\mu \tilde p_\mu + m}{\sum_\mu \tilde p_\mu^2 + m^2} . \tag{12.171}$$

The propagator in the co-ordinate space will be obtained by taking the Fourier transform.

It is there that the problem lies. To appreciate this, first note that since the co-ordinates are now the lattice points, which are points with a well-defined periodicity, each p_μ is bounded in the region

$$-\frac{\pi}{a} < p_\mu \le \frac{\pi}{a} . \tag{12.172}$$

Eq. (12.172) is reminiscent of the fact that if we have a function $f(x)$ which is periodic with period 2π, then it can be decomposed into a Fourier series of the form

$$f(x) = \frac{1}{2}C_0 + \sum_{n=1}^\infty \left(C_n \cos nx + S_n \sin nx \right), \tag{12.173}$$

with

$$C_n = \frac{1}{\pi} \int_{-\pi}^\pi dx\, f(x) \cos nx ,$$

$$S_n = \frac{1}{\pi} \int_{-\pi}^\pi dx\, f(x) \sin nx . \tag{12.174}$$

We can state this in reverse by saying that if we have a function defined only for integers (like C_n and S_n), it is the Fourier transform of a function that has the property $f(x-\pi) = f(x+\pi)$, i.e., is restricted to the region $-\pi < x \le \pi$. On a lattice, since $y_\mu - x_\mu$ can only be of the form na where n is an integer, it follows that ap_μ is restricted in the region $-\pi$ to $+\pi$, which is what Eq. (12.172) shows.

Next, consider what will happen if we try to find the propagator in the co-ordinate space. It will be given by an integral:

$$K^{-1}(x-y) = \int \frac{d^4p}{(2\pi)^4} \, \exp\left(-i\sum_\mu \underline{p}_\mu(\underline{x}_\mu - \underline{y}_\mu)\right) K^{-1}(p). \quad (12.175)$$

For each component of $\underline{p}_\mu$, the integral extends to the region given in Eq. (12.172). Dominant contributions to this integral will come from values of $\underline{p}_\mu$ for which the denominator of $K^{-1}(p)$ is minimum, i.e., when $\widetilde{p}_\mu = 0$ for all μ. For each μ, this can happen when $\underline{p}_\mu = 0$ or when $\underline{p}_\mu = \pi/a$. The first of these two solutions is expected, since even in the continuum limit, i.e., $a \to 0$, the contribution from the region near $\underline{p}_\mu = 0$ will dominate the integral that defines the propagator in co-ordinate space. But the other solution is a shocker. In the limit $a \to 0$, this means that infinite momentum contributions would dominate the said integral. This is unphysical. Occurrence of this extra solution is called the *fermion doubling problem*. For 4 dimensions, there are therefore 2^4 or 16 different solutions for $\widetilde{p}_\mu = 0$ for all μ: only one of them is physically acceptable in the continuum limit, the other 15 are not.

☐ **Exercise 12.20** *Consider the action of a complex scalar field ϕ on a lattice. Show that, apart from terms which are integrals of total derivatives in the continuum limit, the action can be written in the form given in Eq. (12.167), with the obvious replacements of ψ by ϕ and $\overline{\psi}$ by $\phi^\dagger$, and*

$$K(x,y) = -\frac{1}{a^2}\sum_\mu\left(\delta_{y,x+a\widehat{\mu}} + \delta_{y,x-a\widehat{\mu}} - 2\delta_{y,x}\right) + m^2\delta_{y,x}. \quad (12.176)$$

Hence show that the propagator is given by

$$K_S^{-1}(p) = \frac{1}{\dfrac{4}{a^2}\sum_\mu \sin^2(ap_\mu/2) + m^2}, \quad (12.177)$$

so that the doubling problem does not occur for a scalar field.

To avoid the problem of doubling, one usually takes the fermion terms in the lattice action as follows:

$$\begin{aligned}
\mathscr{A}_{\text{fermion}} = a^4 \Bigg[&m\sum_x \overline{\psi}(x)\psi(x)\\
&+\frac{1}{2a}\sum_{x,\mu}\overline{\psi}(x)\underline{\gamma}_\mu\left(U_{(\mu)}(x)\psi(x+a\widehat{\mu})\right.\\
&\hspace{4cm}\left. -U_{(\mu)}^\dagger(x-a\widehat{\mu})\psi(x-a\widehat{\mu})\right)\\
&-\frac{r}{2a}\sum_{x,\mu}\overline{\psi}(x)\left(U_{(\mu)}(x)\psi(x+a\widehat{\mu}) - 2\psi(x)\right.\\
&\hspace{4cm}\left. +U_{(\mu)}^\dagger(x-a\widehat{\mu})\psi(x-a\widehat{\mu})\right)\Bigg], \quad (12.178)
\end{aligned}$$

where r is called the *Wilson parameter*. The extra term should not affect the continuum limit because it vanishes in that limit. There are other prescriptions for avoiding the same problem, but they will not be discussed here.

The expression in Eq. (12.178) has been written only for one quark flavor. In the total action, we need to sum over such terms for all quark flavors. The resulting expression, along with the pure gauge contribution given in Eq. (12.158), constitutes the full QCD action on a lattice.

☐ **Exercise 12.21** *Argue that the term involving the Wilson parameter should not affect the continuum limit by showing that this term is of higher order in the lattice spacing a compared to the other term.*

12.6.7   Path integral on a lattice

We have already defined the action on a lattice. In order to define the path integral, we now have to know how to integrate $e^{-\mathscr{A}}$ over the field configurations. For the fermion field, the answer is straightforward. For the gauge field, we mentioned that the continuum formulation has some complication because of gauge invariance.

For the lattice, we notice that the gauge fields do not directly appear anywhere in the lattice action. They appear only through the link variables. The link variables, as defined in Eq. (12.147), are matrices. For QCD, these are in fact SU(3) matrices in which the gauge fields take the role of the parameters that define an SU(3) transformation. Thus, all we need to do is to integrate over all possible SU(3) matrices. This integration can be defined irrespective of the gauge choice, in a gauge invariant manner.

As shown earlier, a general element of SU(3) has eight parameters. In other words, any SU(3) matrix can be written as a function of eight parameters, which we called θ_a's. The allowed values of these θ_a's span an 8-dimensional space. We need to integrate over this space. The value of any θ_a is bounded. For example, the U(1) group has only one parameter which is bounded in the region $0 \le \theta < 2\pi$. Thus, the integration over all SU(3) matrices would mean an integration over a finite region of the 8-dimensional parameter space of SU(3). However, we cannot simply write the integration as $\int d\theta_1 \cdots \int d\theta_8$. In general there exists a non-trivial factor in the measure on this parameter space.

An analogy might help. Consider spherical polar co-ordinates in 3-dimensional space, with the parameters r, θ and ϕ. We know that the integral of a function $f(r, \theta, \phi)$ over the space cannot be written as $\int dr \int d\theta \int d\phi \, f(r, \theta, \phi)$. Rather, we write $\int dr \int d\theta \int d\phi \, r^2 \sin\theta f(r, \theta, \phi)$. The factor $r^2 \sin\theta$ has to be put in because it makes this measure independent of co-ordinate transformation. For a general system of co-ordinates, this factor is given by $\sqrt{|\det g|}$, where g is the metric, i.e., the distance between two neighboring points q_i and $q_i + dq_i$ is given by ds, where

$$ds^2 = \mathsf{g}_{ij} dq_i dq_j \,. \tag{12.179}$$

Similarly, if we can define the notion of a *distance* between two unitary matrices, we can find an invariant measure on the parameter space of SU(3), a measure that will remain invariant under gauge transformations.

Consider an arbitrary matrix $\mathbb{G}$ that belongs to the group SU(3). It is realized by some particular values of the parameters θ_a. Consider now another matrix where the parameters have been increased by infinitesimal amounts, and call it $\mathbb{G} + d\mathbb{G}$. The *distance* of two matrices $\mathbb{G}$ and $\mathbb{G} + d\mathbb{G}$ in the parameter space can be defined by the relation

$$ds^2 = \text{tr}\left(d\mathbb{G}^\dagger d\mathbb{G}\right). \tag{12.180}$$

Obviously, it is gauge invariant, in the sense that it does not change if both $\mathbb{G}$ and $\mathbb{G} + d\mathbb{G}$ are multiplied by the same matrix $\mathbb{G}'$, either from the left or from the right. It is also non-negative, and vanishes only when $d\mathbb{G}$ vanishes.

If we put in the functional form of $\mathbb{G}$ in terms of the θ_a's, this will give us an equation of the form

$$ds^2 = g_{ab} d\theta_a d\theta_b, \tag{12.181}$$

where g_{ab} will depend on the details of the functional dependence of $\mathbb{G}$ on the θ_a's. This is the metric in the parameter space. As outlined before,

$$\int d\theta_1 \cdots \int d\theta_8 \sqrt{\left|\det(g_{ab})\right|} \tag{12.182}$$

will then constitute a gauge invariant measure. This is called the *Haar measure*. We will denote this whole expression by dU, and then will define $\mathscr{D}U$ according to the prescription given in Eq. (12.106). In finding the path integral, we therefore need not integrate over all gauge field configurations: we need to use the measure $\mathscr{D}U$ for integrating over the gauge fields.

□ **Exercise 12.22** *Consider the group* SU(2). *The most general element of this group can be written in the form*

$$\begin{pmatrix} A & B \\ -B^* & A^* \end{pmatrix}, \tag{12.183}$$

with

$$A = \cos\frac{\theta}{2} + i\cos\alpha\sin\frac{\theta}{2}, \qquad B = \sin\frac{\theta}{2}\sin\alpha\, e^{i\beta}. \tag{12.184}$$

Find the invariant integration measure on the SU(2) *group space.*

12.6.8 Correlation functions and physical observables

We have described how the lattice action as well as the path integral is set up. We now outline how some calculations are performed to obtain physical quantities. We already mentioned in Eq. (12.107) how the expectation value of any operator is defined through the path integral. Suppose now we consider the correlation function of $\mathscr{O}(\mathbf{x}, t)\mathscr{O}(\mathbf{0}, 0)$ for $t > 0$, and where the operator $\mathscr{O}$ is chosen to be $\overline{\psi}\gamma_0\gamma_5\psi$. Introducing a complete set of energy eigenstates $|n\rangle$ between the two factors of the operator $\mathscr{O}$, we can write

$$\langle 0|\mathscr{O}(\mathbf{x}, t)\mathscr{O}(\mathbf{0}, 0)|0\rangle = \sum_n \frac{\langle 0|\mathscr{O}(\mathbf{x}, 0)|n\rangle\langle n|\mathscr{O}(\mathbf{0}, 0)|0\rangle}{2E_n} e^{-E_n t}, \tag{12.185}$$

where the energy eigenstates have been normalized by the relation

$$\langle n \mid n \rangle = 2E_n \, . \tag{12.186}$$

The exponential factor appearing in Eq. (12.185) is just the time evolution factor $e^{-iE_n t}$, written in terms of the Euclidean time defined in Eq. (12.119). To avoid the dependence on the spatial co-ordinate, we can sum over $\boldsymbol{x}$, which will give us only the zero-momentum modes:

$$\left\langle 0 \left| \sum_{\boldsymbol{x}} \mathscr{O}(\boldsymbol{x},\underline{t}) \mathscr{O}(\boldsymbol{0},0) \right| 0 \right\rangle = \sum_n \frac{\langle 0|\mathscr{O}|n\rangle \langle n|\mathscr{O}|0\rangle}{2m_n} e^{-m_n \underline{t}} \, . \tag{12.187}$$

The right hand side has non-zero contribution from all states which have the transformation properties of $\overline{\psi}\gamma_0\gamma_5\psi$. The pion is the lightest among such states. Thus, if we can evaluate the left hand side for large values of $\underline{t}$, the contribution from other states will be negligible, and the exponential decay of the result with respect to $\underline{t}$ will give the value of the pion mass.

The procedure described above would work if we knew all the basic parameters in the QCD Lagrangian. In practice, this is not the case. So one takes the values a few experimentally observed quantities (like the pion mass) as inputs, and use them to obtain values of the basic parameters. These basic parameters are then used to predict other observables.

12.6.9 Continuum limit

Of course, all lattice calculations are done with an implied non-zero lattice spacing a, and the result of any physical quantity depends on this spacing. In order to extract the physical value of the quantity, one needs to extrapolate the results to the limit of vanishing a.

There is one important issue that we did not point out when we wrote the lattice action in Eq. (12.178). In any theory, the action must be dimensionless in the natural units. For a field theory in the continuum, the Lagrangian contains fields and their derivatives and has dimensions of $[M^4]$, and this dimension is compensated by the integration measure over the 4-dimensional spacetime when we calculate the action. For a lattice, there is no integration over spacetime points: only a sum. The dimension of the Lagrangian is compensated by an explicit factor of the elementary volume a^4 of the lattice. This brings in a new feature: there is a fundamental unit of length on a lattice. We can use it to define dimensionless fields on the lattice, e.g.,

$$\check{\psi} = a^{3/2} \psi \, . \tag{12.188}$$

We can also define a dimensionless mass parameter through the relation

$$\check{m} = am \, . \tag{12.189}$$

We can say that $\check{\psi}$ and $\check{m}$ denote the fermion field and its mass in *lattice units*. The interesting point is that we can write Eq. (12.178) by using $\check{\psi}$ and

$\breve{m}$ only, without having to write the lattice spacing a explicitly anywhere. It is to be noted that the link variables contain the combination

$$\breve{A}_\mu = aA_\mu, \tag{12.190}$$

which can also be thought of as the gauge field in the lattice units. The pure gauge action also can be written in terms of $\breve{A}_\mu$ only, without having to use a explicitly. Thus, all parameters and fields can be considered as dimensionless in a lattice action. Equivalently, we can say that the lattice spacing does not appear explicitly in the lattice action. Even where a appears as the argument of the field variables, as e.g. $\psi(x+a\hat{\mu})$ or $\psi(x-a\hat{\mu})$ in Eq. (12.178), the lattice only knows them as the field at the *next* lattice point along the positive or negative $\hat{\mu}$ direction, without realizing anything about the physical distance between the two points.

This is the reason that going to the continuum limit $a \to 0$ is tricky. It is like killing a person who does not exist!

However, things are not as bad as the analogy suggests. Note that Eq. (12.189) implies going to the limit $a \to 0$ is the same as going to the limit where $\breve{m}$, the fermion mass on the lattice, vanishes. The fermion propagator in the co-ordinate space, $K(r)$, represents correlation function, and the correlation length ξ is defined by the relation

$$\lim_{r\to\infty} K(r) \sim \frac{1}{r} e^{-r/\xi}. \tag{12.191}$$

As indicated in Ex. 12.23 below, this implies that

$$\xi = \frac{1}{\breve{m}} = \frac{1}{am}. \tag{12.192}$$

On the lattice therefore, as we go toward the continuum limit, the correlation length diverges. This is indicative of a critical point, in terms of the language of the physics of *phase transitions*. To summarize, the continuum limit is realized for the values of the parameters for which the correlation length becomes infinite. One therefore needs to tune the bare coupling constant to a value that shows critical behavior. In numerical computations of lattice QCD one can never reach the critical point where the correlation length is really infinite. Rather, one reaches a region of parameter space where the correlation length is sufficiently large so that critical behavior emerges, and the system loses the memory of the discrete lattice.

□ **Exercise 12.23** *Take the propagator of a scalar field, given in Eq. (4.143, p 91). Take the Fourier transform in the static limit and show that in the co-ordinate space, the propagator has the form given in Eq. (12.191), with ξ identified by Eq. (12.192).*

It should be noted that the correlation length given in Eq. (12.192) is the dimensionless correlation length, or in other words, the correlation length in lattice units. As we go to larger and larger values of ξ and therefore smaller and smaller values of the lattice spacing a, we require larger and larger

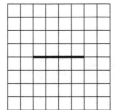 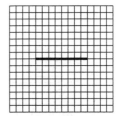

Figure 12.15: The figure on the left shows a coarse lattice, the figure on the right shows a finer lattice. Obviously, in order to cover physics of a fixed physical length indicated by the thick line, one needs larger number of lattice points.

numbers of points on the lattice to understand the physics of a definite physical length scale. The idea is indicated in Fig. 12.15. A minimum lattice size is therefore essential for understanding a particular phenomenon. For example, if a typical hadron size is roughly 0.5 fm (*fm* means *femtometer*, i.e., 10^{-15} m) and we want to study the correlation function of two hadrons, we will have to take a lattice which at least extends beyond a couple of femtometers. If we are working with a value of ξ that amounts to $a = 0.05$ fm, we need a lattice which has at least a few dozens of points along any direction. In principle, it is best to take lattices with a very large number of points. However, it is not possible to take lattices with thousands, or even hundreds, of points in any given direction, because then the calculations would require prohibitively large computer time. So we must be satisfied so far with a maximum of a few dozen points along each direction.

12.7 Confinement

We have talked about asymptotic freedom, i.e., the fact that at higher and higher energies, quarks behave more and more like free particles. High energy implies that we are probing small distance, as explained in Ch. 1. What happens at the other extreme? Behavior of quarks is certainly very different at large distances. We know this because, although there is ample evidence that the hadrons are made of quarks and gluons, no experiment has ever detected a quark or a gluon in isolation. It seems that the more we try to separate the constituents of a hadron, the more they resist separation, remaining inseparable. This property is called *confinement*, i.e., the fact that quarks want to be confined in bound states.

It seems counterintuitive that the interaction is small at small distances and large at large distances. But it is also true that our intuition is largely based on electromagnetic and gravitational interactions, whose macroscopic effects we see all around, and these effects fall off at large distances. Considering only electromagnetic interactions since we decided not to talk about

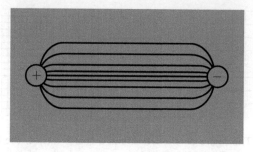

Figure 12.16: Schematic lines of force between two magnetic monopoles placed in a superconductor.

gravitational interactions in this book, we notice that there is a big difference between electromagnetic and strong interactions, viz., the dynamics of the former is governed by an abelian gauge theory whereas that of the latter by a non-abelian gauge theory. As we noted earlier, it is the crucial difference which ensures that the sign of the beta function is negative for strong interactions. This, in turn, implies asymptotic freedom at high energies and large coupling at low energies.

Nevertheless, there is an analogy from electromagnetism that might help us understand the phenomenon of confinement. It has to do with magnetic poles rather than electric charges. A magnet has two opposite poles, which we call the north pole and the south pole usually. We cannot separate the poles. In other words, if we pull them apart, the two poles do not become isolated monopoles. In the extreme case that the magnet breaks into two pieces, each piece becomes a dipole consisting of a north pole and a south pole. The same is true, for example, with a meson made from a quark and an antiquark. If we put in a lot of energy to separate the pair, a new quark-antiquark pair can be produced, and each of the two quarks — the original one and the one newly produced in the pair — would pair with an antiquark to form two mesons.

To extend the analogy further, we can consider what would have happened if magnetic monopoles existed and could be placed in a superconductor. Superconductors show Meissner effect, i.e., they expel magnetic flux. If there is a magnetic monopole placed within the superconductor, the flux lines have to come out and therefore there must be flux lines within the superconductor. However, because of the abhorrence of superconductors toward magnetic field lines, the lines should not be spread uniformly within the superconducting medium, but would rather exit through the nearest boundary of the superconductor. If we now consider two monopoles with opposite magnetic charges placed in a superconducting medium, the lines coming out of one monopole will be squeezed and end up on the other, occupying minimum possible area within the superconductor. The state of affairs is shown schematically in Fig. 12.16.

The same picture can be a representation of the color field lines between a quark and an antiquark, provided the vacuum behaves like a superconducting medium so far as the color forces are concerned. However, there is no analytic derivation from the QCD Lagrangian that confirms that the vacuum indeed behaves this way. This should be taken only as an analogy to understand confinement.

Although there is no analytic proof of confinement for QCD theory in the spacetime continuum, lattice formulation of QCD has a simple argument in favor of it. The argument involves the calculation of the static potential energy $V(R)$ between a quark and an antiquark, separated by a spatial distance R. Without loss of generality, we take the quarks to be separated along one of the co-ordinate axes, say the x-axis. Consider now a Wilson loop of length R along the x-axis and length T along the Euclidean time axis. It will correspond to a quark-antiquark pair created at a certain time, separated instantaneously by the distance R, allowed to stay like that for time T, and then annihilate. Let us denote the value of this Wilson loop by $W(R,T)$. If we average over all possible configuration of gauge fields, we obtain

$$\langle W(R,T) \rangle = \exp\left(-\int dt\, H_{\text{int}}\right) = e^{-V(R)T}, \qquad (12.193)$$

a statement that we do not prove here.

Now we try to find the value of $\langle W(R,T)\rangle$ on the lattice. For this, we first need to write the basic equations, Eqs. (12.127) and (12.128), in a form that is appropriate for a lattice formulation.

We can, for the present purpose, forget about the integration over fermionic variables. As for the integration over gauge fields, we argued earlier that the proper measure is the Haar measure, which we denoted by $\mathscr{D}U$. Thus, path integral on the lattice can be written as

$$Z_{\text{lat}} = \int \mathscr{D}U\, e^{-\mathscr{A}_{\text{gauge}}}, \qquad (12.194)$$

where the pure gauge action is given in Eq. (12.158). The expression for expectation value of $W(R,T)$ would then be

$$\langle W(R,T)\rangle = \frac{1}{Z_{\text{lat}}} \int \mathscr{D}U\, W(R,T) e^{-\mathscr{A}_{\text{gauge}}}. \qquad (12.195)$$

The factor Z_{lat} in the denominator is a constant, and is not important for our discussion. In the numerator, we use the pure gauge action given in Eq. (12.159). This action has an overall factor of $1/g^2$. For large g, we can make an expansion of the exponential in powers of $1/g^2$ and keep only the first non-zero term.

The question is: what is the first non-zero term? Let us start with the zeroth order term, which is proportional to the integral

$$\int \mathscr{D}U\, W(R,T). \qquad (12.196)$$

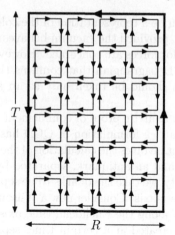

Figure 12.17: Schematic representation of the first non-zero term that appears in the strong-coupling expansion for the expectation value of $W(R,T)$.

This integral vanishes. The reason is that $W(R,T)$ contains various link variables. Any link variable, when integrated over all possible values of U, gives zero.

This is easy to see. Start from the 2×2 orthogonal matrix in its most general form, whose elements are of the form $\cos\theta$ and $\sin\theta$. If we integrate over all possible matrices, i.e., over all possible values of θ, we obtain zero for each element. The conclusion remains unchanged even if there are phases with different elements so that the matrix is not orthogonal, but still unitary. With very little effort, one can see that the argument holds for any $N \times N$ unitary matrix, i.e., for any any element U_{ij},

$$\int dU \, U_{ij} = 0 \,. \tag{12.197}$$

Already, this equation shows why the integral of $W(R,T)$ will vanish. A non-zero integral can be obtained if the integrand contains element of U as well as $U^\dagger$:

$$\int dU \, U_{ij}(U^\dagger)_{kl} = \frac{1}{N}\delta_{il}\delta_{jk} \,. \tag{12.198}$$

We will have to use this result presently, in order to obtain a non-vanishing term that contributes to $\langle W(R,T)\rangle$.

It follows from Eq. (12.198) that the integral will have non-zero contribution when, corresponding to any link present in the $R \times T$ loop, there is a neutralizing contribution from one of the plaquette factors present in the exponential. This will mean that we need to consider the term with at least all plaquettes bordering the boundary of the loop. If we include only these plaquettes, the inner links of these plaquettes will not have a matching contribution from anywhere else, and the integration over these links will give zero. Continuing the argument, it is clear that the first non-zero contribution comes from the term in the exponential that contains all plaquettes enclosed by the loop, and only the single power of each of them. This point has been schematically represented in Fig. 12.17.

The number of plaquettes enclosed by the loop is RT/a^2. Each plaquette will bring one power of $1/g^2$ with it. Thus, the calculation must produce a result of the form

$$\langle W(R,T) \rangle \propto \left(\frac{1}{g^2} \right)^{RT/a^2}. \tag{12.199}$$

Comparing with Eq. (12.193), we find that the quark-antiquark static potential is given by

$$V(R) = \frac{R}{a^2} \ln g^2 \tag{12.200}$$

in the strong-coupling limit, i.e., for large g.

This is the important result. It shows that the potential grows linearly with distance. Thus the potential is larger and larger at greater distances. It is therefore impossible to pull apart a quark and an antiquark from a bound state of the two, which implies that quarks are confined. The only problem is that this result cannot be continued to the continuum limit because of the factor a^2 in the denominator of Eq. (12.200), so that for the continuum theory, we really do not have a proof, only an indication.

As an aside, we can note that the proof does not depend at all on the non-abelian nature of the interaction. In fact, the same result would follow for an abelian gauge theory, if the coupling constant is large enough so that the strong-coupling expansion can be made. QED, however, has a small coupling constant, which is why electrons and positrons are not confined.

12.8 Asymptotic properties of color gauge fields

12.8.1 Dual of field-strength tensor

In Ch. 6 and Ch. 7, we introduced various discrete symmetries and showed that the QED Lagrangian is invariant under parity, time reversal and charge conjugation. Using exactly the same kind of arguments, we can show that the QCD Lagrangian, as given in Eq. (12.48), respects all the discrete symmetries mentioned in the previous sentence. In other words, discrete symmetries P, T and C are consequences of the gauge invariant Lagrangian of Eq. (12.48).

At this point, we can stop and ask ourselves how the Lagrangian was obtained in the first place. We introduced the field-strength tensor in §11.2.2 and constructed the Lagrangian of pure gauge fields in §11.2.3. In performing this exercise, we depended heavily on the analogy of QED. The question is: have we written the most general Lagrangian that is allowed by gauge invariance?

Of course we have seen that mass terms for gauge fields are not gauge invariant. We also don't want to write non-renormalizable terms. However, other than what we have already written in Eq. (11.40, *p 305*), there is at least one possible term that is both renormalizable and gauge invariant. To show this, let us first introduce the *dual* to the field-strength tensor:

$$\widetilde{F}^a_{\mu\nu} = \frac{1}{2}\varepsilon_{\mu\nu\lambda\rho}F_a^{\lambda\rho} \,. \tag{12.201}$$

Basically, this interchanges the space-space components with the time-space components. If we do this for the electromagnetic field tensor, for which the F^{0i}'s are the electric field components whereas the F^{ij}'s are the magnetic field components, we see that for $\widetilde{F}$, the roles are reversed.

Once we identify the dual, we can ask whether we can construct terms in the Lagrangian involving the dual. Like the field-strength tensor, the dual contains linear as well as quadratic terms in the gauge fields. Therefore, we can contemplate a term of the form

$$(\text{constant}) \times \widetilde{F}^a_{\mu\nu}\widetilde{F}_a^{\mu\nu} \tag{12.202}$$

in the Lagrangian. Certainly this term will be gauge invariant, which can easily be seen from the gauge transformation property of the field strength tensor given earlier. It will also be a renormalizable term. However, there is no need to worry about the reason for not writing such a term in the Lagrangian, because this term is not independent from what we have already written, as elaborated in Ex. 12.24.

☐ **Exercise 12.24** Using the properties of the antisymmetric tensor, in particular Eq. (D.11, *p 729*), show that

$$\widetilde{F}^a_{\mu\nu}\widetilde{F}_a^{\mu\nu} = F^a_{\mu\nu}F_a^{\mu\nu} \,. \tag{12.203}$$

However, it is also possible to consider a term of the following form in the Lagrangian:

$$\mathscr{L}_{\text{new}} = (\text{constant}) \times F^a_{\mu\nu}\widetilde{F}_a^{\mu\nu} \,. \tag{12.204}$$

Again, this term will be gauge invariant and renormalizable. It is not related to any other term that we have considered so far in the Lagrangian. Then why did we not write such a term in the Lagrangian for pure gauge fields?

Let us go back to abelian gauge theories. Could the Lagrangian of QED contain a term like that shown in Eq. (12.204)? Here we note that, using the complete antisymmetry of the Levi-Civita tensor, we can write

$$F_{\mu\nu}\widetilde{F}^{\mu\nu} = \frac{1}{2}\varepsilon^{\mu\nu\lambda\rho}F_{\mu\nu}F_{\lambda\rho} = 2\varepsilon^{\mu\nu\lambda\rho}(\partial_\mu A_\nu)(\partial_\lambda A_\rho)$$
$$= \partial_\mu\left(2\varepsilon^{\mu\nu\lambda\rho}A_\nu\partial_\lambda A_\rho\right) \,. \tag{12.205}$$

This is a total divergence term. And we said in Ch. 4 that such terms are irrelevant.

We now consider the same exercise for non-abelian fields. Using the anti-symmetry of the Levi-Civita tensor and the expressions for the field-strength tensor given in Eqs. (11.34) and (11.69), we can write

$$F_{\mu\nu}^{\bar{a}}\widetilde{F}_a^{\mu\nu} = \frac{1}{2}\varepsilon^{\mu\nu\lambda\rho}\left(2\partial_\mu A_\nu^{\bar{a}} - gf_{bc\bar{a}}A_\mu^b A_\nu^c\right)\left(2\partial_\lambda A_\rho^a - gf_{dea}A_\lambda^d A_\rho^e\right).$$
(12.206)

In this expression, first it should be noticed that the term quartic in the gauge fields is in fact zero. To see this, we apply the Jacobi identity, Eq. (11.56, *p 308*), and then rename the dummy gauge indices to write this term in the form

$$\text{Quartic term} \propto \varepsilon^{\mu\nu\lambda\rho} f_{bc\bar{a}} f_{dea} A_\mu^b A_\nu^c A_\lambda^d A_\rho^e$$
$$= -\varepsilon^{\mu\nu\lambda\rho}\left(f_{be\bar{a}} f_{cda} + f_{bd\bar{a}} f_{eca}\right) A_\mu^b A_\nu^c A_\lambda^d A_\rho^e$$
$$= -\varepsilon^{\mu\nu\lambda\rho} f_{bc\bar{a}} f_{eda} A_\mu^b \left(A_\nu^e A_\lambda^d A_\rho^c + A_\nu^d A_\lambda^c A_\rho^e\right). \quad (12.207)$$

Finally, using $f_{eda} = -f_{dea}$ as well as the antisymmetry of the Levi-Civita symbol on both terms, we obtain that the expression is -2 times the original expression, which means that it vanishes. The remaining terms of Eq. (12.206) can be written in the form

$$F_{\mu\nu}^{\bar{a}}\widetilde{F}_a^{\mu\nu} = 2\varepsilon^{\mu\nu\lambda\rho}\left((\partial_\mu A_\nu^{\bar{a}})(\partial_\lambda A_\rho^a) - gf_{bc\bar{a}}(\partial_\mu A_\nu^a)A_\lambda^b A_\rho^c\right). \quad (12.208)$$

This shows that, even for the non-abelian case, we can write the $\widetilde{F}F$ term as a total derivative. In fact, if we define

$$K^\mu = 2\varepsilon^{\mu\nu\lambda\rho}\left(A_\nu^a \partial_\lambda A_\rho^{\bar{a}} - \frac{1}{3}gf_{bc\bar{a}}A_\nu^a A_\lambda^b A_\rho^c\right), \quad (12.209)$$

or alternatively

$$K^\mu = \varepsilon^{\mu\nu\lambda\rho} A_\nu^a \left(F_{\lambda\rho}^{\bar{a}} + \frac{1}{3}gf_{bc\bar{a}} A_\lambda^b A_\rho^c\right), \quad (12.210)$$

it can be shown that

$$\partial_\mu K^\mu = F_{\mu\nu}^{\bar{a}}\widetilde{F}_a^{\mu\nu}. \quad (12.211)$$

□ **Exercise 12.25** *Show that the expressions in Eqs. (12.209) and (12.210) are indeed equivalent.*

□ **Exercise 12.26** *Starting from Eq. (12.209), verify Eq. (12.211), using the complete antisymmetry of the Levi-Civita symbol, as well as the symmetry properties of the structure constants given in Eq. (11.53, p 308).*

□ **Exercise 12.27** *With the matrix fields defined in Eq. (11.13, p 300), show that Eq. (12.209) can be written in the form*

$$K^\mu = 2C\varepsilon^{\mu\nu\lambda\rho}\,\text{Tr}\left(\mathbb{A}_\nu\partial_\lambda\mathbb{A}_\rho^\dagger - \frac{2}{3}g\mathbb{A}_\nu\mathbb{A}_\lambda\mathbb{A}_\rho\right), \quad (12.212)$$

where C is the normalization constant for the generators defined in Eq. (11.39, p 305).

This seems like the end of the story: we have shown that the term is a total derivative, and therefore should be neglected, according to what we said in Ch. 4. However, let us ask ourselves why any total derivative term present in the Lagrangian has to be irrelevant? An answer was provided in Ch. 4, viz., that all fields vanish at the spacetime surface at infinity. But that just brings in a new question: why should fields vanish at infinity?

The simplest answer to this question is that, unless fields vanish at infinity, the total energy of the field becomes infinite, which would not correspond to a physical configuration of the field. This is indeed the right answer for scalar fields and fermion fields. For gauge fields, however, the situation is somewhat different. The basic fields are the A_μ^a's, whereas the pure gauge Lagrangian involves the field-strength tensor $F_{\mu\nu}^a$. Finiteness of energy would dictate that the field-strength tensor should vanish at infinity, not necessarily the fields themselves.

In the abelian case, this requirement is enough to render the $F\widetilde{F}$ term irrelevant. Using the complete antisymmetry of the Levi-Civita tensors, we can write Eq. (12.205) in the form

$$F_{\mu\nu}\widetilde{F}^{\mu\nu} = \partial_\mu\left(\varepsilon^{\mu\nu\lambda\rho}A_\nu F_{\lambda\rho}\right).\tag{12.213}$$

Thus the action coming from this term can be transformed into a surface integral of $A_\nu F_{\lambda\rho}$, which vanishes since $F_{\lambda\rho}$ vanishes at infinity.

This is not so for non-abelian gauge fields. From the gauge transformation property of the fields A_μ^a given in Eq. (11.21, p 302), it is clear that if we have a gauge field for which

$$g\mathbb{A}_\mu = i(\partial_\mu U)U^{-1}\tag{12.214}$$

at some spacetime point, then by making a suitable gauge transformation we can make the field vanish at that point. That would mean that the field-strength tensor would vanish as well, and so will the energy density of the field. Alternatively, we can say that if at each spacetime point at infinity the field has the form given in Eq. (12.214) for some U, the energy density would vanish for all asymptotically far points, and therefore the total energy of the field cannot be infinite. However, using Eq. (12.211), we can write

$$\int d^4x\, F_{\mu\nu}^{a\dagger}\widetilde{F}_a^{\mu\nu} = \int dS_\mu\, K^\mu,\tag{12.215}$$

where dS_μ denotes an element of the boundary of spacetime. Even though the field-strength tensor vanishes at the boundary, K^μ does not necessarily do so because of the extra term in Eq. (12.210) which does not contain the field-strength tensor. So one obtains

$$\int d^4x\, F_{\mu\nu}^{a\dagger}\widetilde{F}_a^{\mu\nu} = \frac{1}{3}g\varepsilon^{\mu\nu\lambda\rho}f_{bc\bar{a}}\int dS_\mu\, A_\nu^a A_\lambda^b A_\rho^c.\tag{12.216}$$

Obviously, there is no guarantee that the surface integral of K^μ vanishes at infinity, which means that surface effects cannot be neglected for such a term. Said another way, even if the $F\widetilde{F}$ term is a total divergence, it cannot be ignored in the Lagrangian for a non-abelian gauge theory. It can have real physical effects.

The nature of some of these effects can be easily guessed if we notice that the term contains a Levi-Civita symbol in the definition of the dual of the field-strength tensor. Because of this, such a term would inflict violation of parity and time reversal symmetries. If we take CPT invariance for granted, violation of time reversal symmetry is equivalent to CP violation. For this reason, we will discuss the implication of an $F\widetilde{F}$ term in Ch. 21, where CP violation will be discussed. Before that, in the rest of this section, we want to do some groundwork by discussing some properties of such a term.

12.8.2 Topological invariants

We now discuss an interesting property of the surface integral mentioned in Eq. (12.216). First, we note that taking surface integrals is not an easy task in the Minkowski space, where the *far* points cannot be identified through the invariant distance since the metric is not positive definite. Even a point with very large values of spacetime co-ordinates might be at a very small invariant distance, or even zero distance, from the origin of co-ordinates. In contrast, in a Euclidean space, the surface of spacetime at infinity is indeed a surface that is at infinite distance from the origin. So we use the Euclidean space, introduced in §12.6, for our work. In Eq. (12.142) we have shown the Euclidean form of the pure gauge action. Similarly, if we consider the left hand side of Eq. (12.215) as a term in the Minkowski space action, the corresponding term in the Euclidean action will be

$$I \equiv - \int d^4\underline{x}\, F^{a\dagger}_{\underline{\mu\nu}} \widetilde{F}^{\mu\nu}_a \,, \tag{12.217}$$

where the dual of the field-strength tensor is related to the field-strength tensor itself in exactly the same way as it is in the Minkowski space, i.e., through Eq. (12.201), and the Levi-Civita symbol in the Euclidean space is defined by $\underline{\varepsilon}_{0123} = \underline{\varepsilon}^{0123} = +1$. Making the same arguments that we have made for the Minkowski space, we can write

$$I = -\frac{1}{3} g \underline{\varepsilon}^{\mu\nu\lambda\rho} f_{bc\bar{a}} \int d\underline{S}_\mu\, \underline{A}^a_\nu \underline{A}^b_\lambda \underline{A}^c_\rho \,. \tag{12.218}$$

The next thing is to notice that pure gauge fields are easily identified in the matrix notation, as in Eq. (12.214). In order to use the matrix notation, we need to rewrite the integrand. Note that

$$\mathrm{Tr}\left(T_a T_b T_c\right) = \frac{1}{2}\,\mathrm{Tr}\left(T_a[T_b,T_c]\right) + \frac{1}{2}\,\mathrm{Tr}\left(T_a[T_b,T_c]_+\right)$$
$$= \frac{1}{2} i f_{bcd}\,\mathrm{Tr}\left(T_a T_{\bar{d}}\right) + (\text{symmetric in } b \leftrightarrow c)\,. \tag{12.219}$$

Using the normalization of Eq. (11.39, *p 305*) and specializing to the fundamental representation for which $C = \frac{1}{2}$, we obtain

$$\mathrm{Tr}\left(T_a T_b T_c\right) = \frac{1}{4} i f_{bc\bar{a}} + (\text{symmetric in } b \leftrightarrow c). \tag{12.220}$$

If we now put this back into Eq. (12.216), the term symmetric in the interchange of the indices b and c does not contribute, and we can write

$$I = \frac{4}{3} i g \varepsilon^{\mu\nu\lambda\rho} \int d\underline{S}_\mu \ \mathrm{Tr}\left(\underline{A}_\nu \underline{A}_\lambda \underline{A}_\rho\right)$$

$$= \frac{4}{3g^2} \varepsilon^{\mu\nu\lambda\rho} \int d\underline{S}_\mu \ \mathrm{Tr}\left((\partial_\nu U)U^{-1}(\partial_\lambda U)U^{-1}(\partial_\rho U)U^{-1}\right), \tag{12.221}$$

using the form of the pure gauge fields from Eq. (12.144). Taking the Lorentz
index μ along the timelike direction, this integral can be written as

$$I = \frac{4}{3g^2} \varepsilon^{ijk} \int d^3x \ \mathrm{Tr}\left((\partial_i U)U^{-1}(\partial_j U)U^{-1}(\partial_k U)U^{-1}\right)\Bigg|_{\underline{t}=-\infty}^{\underline{t}=+\infty}. \tag{12.222}$$

It is the value of this integral that concerns us.

In order to make the argument without getting into unnecessary details,
we consider the gauge group to be SU(2). Clearly, the value of the integral
I can be zero if U is a constant at the boundary of spacetime. To see that
other values are possible as well, consider gauge fields of the form

$$\mathbb{A}_i = \frac{\underline{t}^2 + r^2}{\underline{t}^2 + r^2 + a^2} \times \frac{i}{g} (\partial_i U)U^{-1}, \tag{12.223}$$

where a is a positive constant, r denotes the spatial distance from the origin
of co-ordinates, and

$$U = \frac{t + i\boldsymbol{x} \cdot \boldsymbol{\sigma}}{\sqrt{\underline{t}^2 + r^2}}, \tag{12.224}$$

where $\boldsymbol{\sigma}$ denotes the Pauli matrices. Note that, at large distances from the
Euclidean origin, the gauge fields are of the form mentioned in Eq. (12.214).

Evaluation of Eq. (12.222) is now straightforward. First, we note that

$$(\partial_i U)U^{-1} = \frac{i}{\underline{t}^2 + r^2} (\delta_{im}\underline{t} - \varepsilon_{imm'}x_{m'})\sigma_m. \tag{12.225}$$

There are three such terms in the integrand. The trace is over the Pauli
matrices:

$$\mathrm{Tr}(\sigma_m \sigma_n \sigma_p) = 2i\varepsilon_{mnp}. \tag{12.226}$$

Putting this into the integrand of Eq. (12.222) and noting the fact that the
integral involving an odd number of spatial co-ordinates would vanish, we

obtain

$$I = \frac{8}{3g^2} \varepsilon^{ijk} \int \frac{d^3x}{(\underline{t}^2 + r^2)^3} \left[\underline{t}^3 \varepsilon_{ijk} - 3\underline{t}\varepsilon_{inp}\varepsilon_{jnn'}\varepsilon_{kpp'} x_{n'} x_{p'} \right] \Bigg|_{t=-\infty}^{\left| t=+\infty \right.} .$$

(12.227)

Since the integrand is odd in $\underline{t}$, we can write this result as

$$I = \frac{16}{3g^2} \lim_{\underline{t}\to\infty} \varepsilon^{ijk} \int \frac{d^3x}{(\underline{t}^2 + r^2)^3} \left[\underline{t}^3 \varepsilon_{ijk} - \underline{t}\varepsilon_{inp}\varepsilon_{jnn'}\varepsilon_{kpn'} r^2 \right] \quad (12.228)$$

using also the result that $\int d^3x \, f(r) x_i x_j = \frac{1}{3}\delta_{ij} \int d^3x \, f(r) r^2$. The rest of the integration can be performed easily and one obtains

$$I = \frac{32\pi^2}{g^2} . \qquad (12.229)$$

If we do not restrict ourselves to the choice of U given in Eq. (12.224), we obtain that the integral given in Eq. (12.216) is always of the form

$$I = \frac{32\pi^2 n}{g^2} \qquad (12.230)$$

for some integer n.

□ **Exercise 12.28** Supply the missing steps for obtaining the result given in Eq. (12.229). [**Note :** *You may use the result*

$$\int_0^\infty dz \, \frac{z^{n-1}}{(z+1)^m} = B(n, m-n) \qquad (12.231)$$

where B denotes the beta function. The result has been discussed in some detail in §G.5 of Appendix G.]

□ **Exercise 12.29** Redo the integrals for the choice

$$U = \frac{t - i\boldsymbol{x}\cdot\boldsymbol{\sigma}}{\sqrt{t^2 + r^2}} , \qquad (12.232)$$

which is in fact the inverse of the choice taken in Eq. (12.224). Show that Eq. (12.230) is obtained with $n = -1$.

12.8.3 QCD vacuum

The integer n encountered in Eq. (12.230) is called the *winding number* or the *Pontryagin index* of a function. We can gain an intuitive feeling for this number by considering the 1-dimensional analogy of functions on a circle. We will denote the co-ordinate on the circle by ϕ. Obviously, $0 \le \phi < 2\pi$, and the point $\phi = 2\pi$ is identical to the point $\phi = 0$. So, obviously the mappings from the circle will have to have the property $f(\phi) = f(\phi + 2\pi)$. Consider then

$$f(\phi) = e^{in\phi} \qquad (12.233)$$

where n is an integer. Obviously,

$$\int_0^{2\pi} d\phi \, \frac{f'(\phi)}{f(\phi)} = 2\pi i n \,, \tag{12.234}$$

where $f'(\phi)$ is the derivative of $f(\phi)$. This equation is the 1-dimensional analog of Eq. (12.230).

The integer n appearing in Eq. (12.233) can be interpreted as the number of points on the circle that gives the same functional value. For example, with $n = 2$, both $\phi = \pi/2$ and $\phi = 3\pi/2$ will have $f(\phi) = -1$. With $n = 3$, the same functional value is obtained for the points where ϕ equals $\pi/3$, π, or $5\pi/3$. In this sense, the value of the function winds around n times on the set of complex numbers of unit modulus: hence the name *winding number*. This kind of visualization is not possible for 4-dimensional spacetime, but the idea is still the same. Here we consider mapping from the boundary of the 4-dimensional spacetime, which is a 3-sphere, denoted in the mathematics literature by S^3. The functional values are SU(2) group elements (remember we have been talking about the SU(2) gauge group for a while), which also encompass an S^3. Thus, the mappings are from S^3 to S^3, and are in perfect analogy with the mappings from S^1 to S^1 (i.e., from a circle to the elements of the U(1) gauge group) which were used in the 1-dimensional example. As we have seen a little while ago, these mappings are characterized by an integer n, and different values of n belong to different topological sectors of the theory.

We can digress a little bit here to explain the statement that the elements of an SU(2) group encompass an S^3. In the fundamental representation, the most general element of an SU(2) is of the form given in Eq. (12.183), with $|A|^2 + |B|^2 = 1$. Writing $A = a_1 + ia_2$ and $B = b_1 + ib_2$, the condition can be expressed in the form $a_1^2 + a_2^2 + b_1^2 + b_2^2 = 1$. This defines a 3-sphere in a 4-dimensional Euclidean space, just as the constraint $x^2 + y^2 + z^2 = 1$ defines the surface of a sphere (or 2-sphere, if one insists on a more general nomenclature) in a 3-dimensional space.

The group of QCD is SU(3), not SU(2) that we have been discussing so far in the context of topological properties of gauge field configurations. Fortunately, there is no need to get into a similar discussion about SU(3). There is a theorem by Bott that says that the topological structure of gauge fields of any SU(N) group (and many other groups) in 4-dimensional spacetime is the same. In other words, different topological sectors can be characterized by an integer. Gauge field configurations with $n > 0$ are called *instantons*, and those with $n < 0$ are called anti-instantons.

The vacuum state can therefore have gauge fields in any of the topological sectors. Let us call the state with gauge field configurations with winding number n by $|n\rangle$. Clearly, such a state is not an eigenstate of the Hamiltonian: instantons can change the winding number. The vacuum state is in fact given by

$$|\theta\rangle = \sum_n e^{-in\theta} |n\rangle \,. \tag{12.235}$$

The sum goes over all integers, from $-\infty$ to $+\infty$.

There is a very simple and well-known analogy from non-relativistic quantum mechanics that tells us that the vacuum state must be of the form given in Eq. (12.235). Consider the non-relativistic Schrödinger equation for the energy eigenvalues:

$$\frac{d^2}{dx^2}\psi + V(x)\psi = E\psi\,,\tag{12.236}$$

which we have written for a 1-dimensional system in order not to deal with unnecessary notational complication, and also absorbed factors of $2m/\hbar^2$ into the definitions of $V(x)$ and E for the same reason. Now suppose we are dealing with a periodic potential, i.e., $V(x) = V(x+a)$ for all x. This means that the Hamiltonian commutes with the translation operator T_a which is defined by $T_a f(x) = f(x+a)$ for any function $f(x)$. The solutions for the energy eigenstates will then also be eigenstates of the operator T_a. It can then be shown that these solutions are of the form

$$\psi(x) = e^{ikx} F_k(x)\,,\tag{12.237}$$

where F_k is some function which has the same periodicity as the potential, i.e., $F_k(x) = F_k(x+a)$. Obviously, this implies

$$\psi(x+a) = e^{ika}\psi(x)\,,\tag{12.238}$$

so that the probability is the same at x and $x+a$ for any value of x. These are usually called *Bloch solutions* in the parlance of condensed matter physics, and correspond to wavefunctions of electrons in a lattice.

In Eq. (12.235), we found that the vacuum state is characterized by a parameter θ. This is truly an eigenstate of the Hamiltonian because a $|\theta\rangle$ state cannot evolve into another state which has a different value of the parameter. To see this, consider the matrix element $\langle\theta'|e^{-iHt}|\theta\rangle$.

$$\begin{aligned}\langle\theta'|e^{-iHt}|\theta\rangle &= \sum_{n,n'} e^{i(n'\theta'-n\theta)}\langle n'|e^{-iHt}|n\rangle\\ &= \sum_{n,k} e^{in(\theta'-\theta)}e^{ik\theta'}\langle n+k|e^{-iHt}|n\rangle\,,\end{aligned}\tag{12.239}$$

replacing n' by $n+k$ in the last step. The matrix element between the two states would depend only on the difference between the winding numbers of the states, so we can write

$$\begin{aligned}\langle\theta'|e^{-iHt}|\theta\rangle &= \sum_{n,k} e^{in(\theta'-\theta)}e^{ik\theta'}\langle k|e^{-iHt}|0\rangle\\ &= \delta(\theta'-\theta)\sum_k e^{ik\theta}\langle k|e^{-iHt}|0\rangle\,,\end{aligned}\tag{12.240}$$

performing the sum over n. This shows that the parameter θ is not changed by time evolution, proving that the state $|\theta\rangle$ is an eigenstate of the Hamiltonian for any value of θ.

The explicit factor of $e^{ik\theta}$ appearing in the expression above can be absorbed into the Lagrangian. For this, we write the matrix element of the time-evolution operator,

$$\langle k|e^{-iHt}|0\rangle = \int\mathscr{D}\Psi\int\mathscr{D}\overline{\Psi}\int\mathscr{D}G_a^\mu\,\exp\left(i\int d^4x\,\mathscr{L}_{\text{QCD}}\right)\,,\tag{12.241}$$

where the integral runs only over gauge configurations that have winding numbers equal to zero at $t = -\infty$ and equal to k at $t = +\infty$. Using the expression for the winding number in terms of the gauge fields,

$$k = \frac{g^2}{32\pi^2} \int d^4x \, F_{\mu\nu}^{a\dagger} \tilde{F}_a^{\mu\nu} \qquad (12.242)$$

that follows from Eq. (12.230), we can therefore write

$$e^{ik\theta} \left\langle k \left| e^{-iHt} \right| 0 \right\rangle = \int \mathscr{D}\Psi \int \mathscr{D}\bar{\Psi} \int \mathscr{D}G_a^{\mu}$$

$$\exp \left(i \int d^4x \left(\mathscr{L}_{\text{QCD}} + \frac{\theta g^2}{32\pi^2} F_{\mu\nu}^{a\dagger} \tilde{F}_a^{\mu\nu} \right) \right). \quad (12.243)$$

Thus θ is an additional parameter in the Lagrangian. The value of this parameter can be determined by experiments. So far there has been no measurement of this parameter. There are only bounds, which imply that θ must be very small. This bound will be discussed in Ch. 21.

Chapter 13

Structure of hadrons

In Ch. 10, our entire discussion revolved around the so-called static properties of hadrons. We argued that such properties of low-lying hadrons can be explained by assuming that the hadrons have quarks as their constituents. In order to make the argument more convincing, we need to show that the dynamical properties of hadrons can also be explained with the help of quarks. This is what we want to do in this chapter.

Dynamical properties, as discussed in Ch. 1, are the characteristics shown in various interactions. Thus, in this chapter, we analyze the phenomena of various scattering processes involving hadrons. Remember that this does not mean processes where the initial and the final states contain *only* hadrons. Hadrons will be there, but there might be leptons as well. In fact, it is easier to analyze a scattering of a lepton and a hadron, because in this case we need to worry about the structure of only one of the initial state particles. That is certainly easier than having two hadrons in the initial state, and having to analyze two unknown structures at the same time.

13.1   Electron–proton elastic scattering

The most easily available stable lepton is the electron, and the most abundant hadron is the proton. Thus, it is natural that we begin with the discussion of their scattering. Since the electron does not have any strong interaction, the dominant contribution to this process will come from a photon exchange between the electron and the proton. The relevant diagram has been shown in Fig. 13.1.

The diagram is the same as the diagram for muon–antimuon production, Fig. 5.6 *(p 132)*, or, more precisely, to the diagram of electron-muon elastic scattering. The only difference is that the proton, unlike the muon, is not a fundamental particle. Hence we cannot use the simple Dirac vertex for the proton. To remind us of this, we have drawn the proton lines with a thicker pen in the figure.

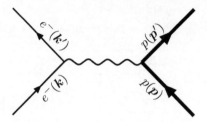

Figure 13.1: Diagram for electron–proton elastic scattering at the lowest order in perturbation theory. We used thick lines for the proton to remind us that it is not an elementary particle.

The amplitude of the process can be written in the form

$$i\mathcal{M} = \left[\bar{u}(k')ie\gamma^\mu u(k)\right]\frac{-ig_{\mu\nu}}{q^2}\left[\bar{u}(p')(-ie\Gamma^\nu)u(p)\right], \tag{13.1}$$

where $e\Gamma^\nu$ is the electromagnetic vertex function of the proton, and q^μ is the 4-momentum carried by the photon. Note that we have used the same symbol for the spinors pertaining to the proton and to the electron field. They can be distinguished only by the momenta associated with them:

$$\mathcal{M} = -\frac{e^2}{q^2}\left[\bar{u}(k')\gamma^\mu u(k)\right]\left[\bar{u}(p')\Gamma_\mu u(p)\right]. \tag{13.2}$$

This gives

$$\overline{|\mathcal{M}|^2} = \frac{e^4}{q^4}\ell^{\mu\nu}h_{\mu\nu}, \tag{13.3}$$

where $\ell^{\mu\nu}$ and $h_{\mu\nu}$ come respectively from the bilinears involving the electron and the proton spinors, after averaging over initial spins and summing over final spins. Evaluation of $\ell^{\mu\nu}$ is straightforward, and the result is

$$\ell^{\mu\nu} = \frac{1}{2}\operatorname{Tr}\left((\slashed{k}+m)\gamma^\nu(\slashed{k}'+m)\gamma^\mu\right) = 2\left(k^\mu k'^\nu + k^\nu k'^\mu + \frac{1}{2}q^2 g^{\mu\nu}\right) \tag{13.4}$$

where m is the electron mass, and

$$q = k - k' \tag{13.5}$$

is the momentum carried by the internal photon line, a relation which ensures that $q^2 = 2(m^2 - k \cdot k')$ that has been used in the evaluation of $\ell^{\mu\nu}$.

For the part involving the proton spinors, we obtain

$$h_{\mu\nu} = \frac{1}{2}\operatorname{Tr}\left((\slashed{p}+M)\Gamma^\ddagger_\nu(\slashed{p}'+M)\Gamma_\mu\right), \tag{13.6}$$

where M is the mass of the proton, and the notation of double dagger was introduced in Eq. (5.64, *p 127*). In order to proceed, we need to know what Γ_μ is. For this, we take a hint from the discussion of form factors in §5.7.1. The electromagnetic current is, after all, conserved. We showed that the most general form of the matrix element subject to this condition contains four form factors. Later, in §6.9.3, we showed that two of these form factors violate parity. With pure QED interactions, parity will not be violated, so we can write

$$\Gamma_\mu = F_1(q^2)\gamma_\mu - F_2(q^2)i\sigma_{\mu\nu}q^\nu . \tag{13.7}$$

As explained in §5.7.1, $F_1(q^2)$ and $F_2(q^2)$ denote the charge form factor and the anomalous magnetic moment form factor of the proton. [Note the apparent sign difference of the F_2 term between Eq. (5.121, *p 140*) and this equation, which has been caused by the fact that in this case, $p - p' = -q$ because of our choice of notation in Eq. (13.5).] Since the expression appears sandwiched between spinors, we can use Gordon identity to rewrite it as

$$\Gamma_\mu = \left(F_1(q^2) + 2MF_2(q^2)\right)\gamma_\mu - (p+p')_\mu F_2(q^2). \tag{13.8}$$

Putting this in, we obtain

$$h_{\mu\nu} = (-g_{\mu\nu} + \frac{q_\mu q_\nu}{q^2})H_1 + (p_\mu - \frac{\nu}{q^2}q_\mu)(p_\nu - \frac{\nu}{q^2}q_\nu)\frac{H_2}{M^2}, \tag{13.9}$$

where

$$\nu = p \cdot q, \tag{13.10}$$

and both H_1 and H_2 are functions of q^2:

$$H_1 = -q^2(F_1 + 2MF_2)^2, \tag{13.11a}$$
$$H_2 = 4M^2(F_1^2 - q^2F_2^2). \tag{13.11b}$$

These can be called the form factors that appear in electron–proton elastic scattering.

It is to be noted that, since $p' = p+q$, by squaring both sides we obtain $2p \cdot q = -q^2$, which means that the ratio ν/q^2 appearing in Eq. (13.9) could have been written simply as $-\frac{1}{2}$. But we keep ν with a view to future applications, and never use the value of ν/q^2 in what follows. Moreover, keeping ν explicitly in the expression for $h_{\mu\nu}$, it is easy to see that

$$q^\mu h_{\mu\nu} = 0, \qquad q^\nu h_{\mu\nu} = 0, \tag{13.12}$$

a property that is a consequence of the fact that $q^\mu \Gamma_\mu$ vanishes between the spinors, as was argued in §5.7.

Using the expressions in Eqs. (13.4) and (13.9), we obtain

$$\ell^{\mu\nu}h_{\mu\nu} = (4k^\mu k^\nu + q^2 g^{\mu\nu})h_{\mu\nu}$$

$$= -(4m^2 + 2q^2)H_1 + \left\{(2k\cdot p - \nu)^2 + (q^2 M^2 - \nu^2)\right\}\frac{H_2}{M^2}, \quad (13.13)$$

where we have used kinematical relations like

$$q^2 = 2k\cdot q = -2k'\cdot q = 2m^2 - 2k\cdot k', \quad (13.14)$$

which follow from Eq. (13.5).

So far in the discussion, we did not assume any particular frame. From now on, let us specialize to the fixed target frame of the proton, i.e., in the frame in which the initial proton is at rest. In this frame,

$$\nu = p\cdot(k - k') = M(E - E'), \quad (13.15)$$

where E and E' are the energies of the incoming and the outgoing electron. Thus, in terms of the parameters defined in this frame, we obtain

$$\ell^{\mu\nu}h_{\mu\nu} = -\left(4m^2 + 2q^2\right)H_1 + \left(4EE' + q^2\right)H_2. \quad (13.16)$$

We now further assume that the energies involved are much higher than the electron mass, so that the electron mass can be ignored. In that case, if θ is the angle between k and k', we get

$$q^2 = -2k\cdot k' = -2EE'(1 - \cos\theta), \quad (13.17)$$

so that

$$\ell^{\mu\nu}h_{\mu\nu} = 4EE'\left[2H_1\sin^2\frac{\theta}{2} + H_2\cos^2\frac{\theta}{2}\right], \quad (13.18)$$

and consequently

$$\overline{|\mathscr{M}|^2} = \frac{4\pi^2\alpha^2}{EE'\sin^4\frac{\theta}{2}}\left[2H_1\sin^2\frac{\theta}{2} + H_2\cos^2\frac{\theta}{2}\right]. \quad (13.19)$$

Using the value of E' from the relation in Eq. (4.226, *p 109*), the angular distribution of the cross-section can now be written down directly by using Eq. (4.225, *p 109*). In the notation used here, the expression would be

$$\frac{d\sigma}{d\Omega} = \frac{\alpha^2}{16ME^2(M + E - E\cos\theta)}\left[\frac{2H_1\sin^2\frac{\theta}{2} + H_2\cos^2\frac{\theta}{2}}{\sin^4\frac{\theta}{2}}\right]. \quad (13.20)$$

As an aside, we can also discuss the cross-section of the process $e^-e^+ \to p\bar{p}$. The problem is similar to that of the muon pair production discussed in §5.4.2, except that for the electromagnetic vertex of the proton, we should use the vertex function given in Eq. (13.7). The Feynman amplitude will therefore be given by

$$\mathcal{M} = \frac{e^2}{\mathfrak{s}} \left[\overline{v}_{\boldsymbol{p}_2} \gamma^\mu u_{\boldsymbol{p}_1} \right] \left[\overline{u}_{\boldsymbol{p}_1'} \left(F_1 \gamma_\mu - i F_2 \sigma_{\mu\nu} q^\nu \right) v_{\boldsymbol{p}_2'} \right], \qquad (13.21)$$

instead of the simpler expression given in Eq. (5.87, *p 132*). Performing the calculations in a straightforward manner, one obtains

$$\sigma = \frac{4\pi\alpha^2}{3\mathfrak{s}} \sqrt{1 - \frac{4M^2}{\mathfrak{s}}} \left(G_M^2(\mathfrak{s}) + \frac{2M^2}{\mathfrak{s}} G_E^2(\mathfrak{s}) \right), \qquad (13.22)$$

where M is the proton mass, and

$$G_E(q^2) = F_1(q^2) + q^2 F_2(q^2), \\ G_M(q^2) = F_1(q^2) + 2M F_2(q^2). \qquad (13.23)$$

☐ **Exercise 13.1** Verify Eq. (13.22).

13.2 Deep inelastic scattering

When energies are high, electron–proton scattering will not remain elastic. In the final state, one can obtain multiple hadrons. Let us call this reaction figuratively as

$$e + p \rightarrow e + X, \qquad (13.24)$$

where X can mean 'anything that can be consistent with all conservation laws'. This is schematically shown in Fig. 13.2. Such processes are often called *inclusive*, because the final state includes everything ponderable. As opposed to it, a scattering is called *exclusive* when we take only one or a few specific outgoing channels into consideration.

The expression for the cross-section of an inclusive process would be given in the form

$$d\sigma = \frac{1}{4\sqrt{(k \cdot p)^2 - m^2 M^2}} \frac{d^3 k'}{(2\pi)^3 2E'} F, \qquad (13.25)$$

where

$$F = \sum_f \left(\prod_a \frac{d^3 p_a'}{(2\pi)^3 2E_a'} \right) (2\pi)^4 \delta^4 \left(k + p - k' - \sum_a p_a' \right) \overline{|\mathcal{M}_{fi}|^2}. \quad (13.26)$$

This is basically the same as Eq. (4.178, *p 99*). The first factor on the right hand side of Eq. (13.25) is the initial state factor. In the final state factors, we have singled out one of the particles in the final state, viz., the electron, and written its phase space factor explicitly in Eq. (13.25). The phase space

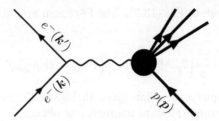

Figure 13.2: Pictorial representation of electron–proton inelastic scattering. The blob on the right vertex implies that it is not a fundamental vertex coming from a Lagrangian. The number of outgoing lines at the proton vertex is symbolic: there can be any number of particles there.

factors of all other particles, along with the 4-momentum conserving delta function and the amplitude squared of the process, have been dumped into something called F, as seen in Eq. (13.26). Note that this expression contains a summation over all possibilities regarding other particles. The quantity denoted by $\overline{|\mathscr{M}_{fi}|^2}$ denotes the amplitude squared for a particular final state f, summed over all possible final spins and polarizations, and averaged over the initial ones.

If we want to analyze experimental data in the fixed target (FT) frame of the proton, the initial state factor becomes $1/(4M\mathrm{k})$. The phase space factor for the electron can be written as

$$d^3 k' = d\Omega' d\mathrm{k}' \, \mathrm{k}'^2 = d\Omega' dE' \, E'\mathrm{k}' , \qquad (13.27)$$

using $\mathrm{k}' \, d\mathrm{k}' = E' dE'$, as shown in Eq. (4.162, *p 96*). Putting these into Eq. (13.25), we can write the differential cross-section with respect to the electron energy as well as the solid angle of scattering as

$$\frac{d\sigma}{d\Omega' dE'} = \frac{1}{(4\pi)^3 M} \frac{\mathrm{k}'}{\mathrm{k}} F . \qquad (13.28)$$

Since inelastic scattering takes places at energies much higher than the proton mass, we will neglect the electron mass and write E and E' in places of k and k' in what follows.

Usually, when we write any derivative of order two or more, we write the order of the derivative with the d that appears in the numerator, e.g., $d^2 y/dx^2$. Here, we have deviated from this custom. Or, more precisely, we are defying it. Partly the reason is that the notation is superfluous: we can determine the order of derivatives by counting the independent variables which appear downstairs. Also, for cross-sections, the usual custom poses a huge problem. For a multi-particle final state, there are too many variables on the right hand side, and it is cumbersome to keep track of the order of the derivatives with respect to a number of them. For example, even in a simple formula like the definition of cross-section in Eq. (13.25), if we want to attach the order of smallness to the cross-section on the left hand side, it would be a different number for different final states. So, whatever is the order of the derivative, we will write the numerator as $d\sigma$. In fact, that's what we have done in many places earlier in the book where we denoted the angular distribution of the cross-section by $d\sigma/d\Omega$, knowing fully well that the differential of the solid angle contains two independent angles.

How much can we say about F without being bothered about the products in the final state? From the discussion of §4.12.1, we know that F is a Lorentz invariant quantity. It must contain a factor of e from the electron vertex, and at least another factor of e from the vertex where the other end of the photon line attaches. The photon propagator has a denominator of q^2. These two together provide a factor of $e^2/q^2 = 4\pi\alpha/q^2$ to the amplitude. The contribution to the amplitude coming from the electron spinors and the electron-photon vertex is the same as obtained for the elastic scattering, so in the amplitude squared, we should have the tensor $\ell^{\mu\nu}$, as defined in Eq. (13.4). To contract its indices, we need another tensor $W_{\mu\nu}$. Let us normalize this tensor $W_{\mu\nu}$ such that

$$F = 4\pi M \left(\frac{4\pi\alpha}{q^2}\right)^2 \ell^{\mu\nu} W_{\mu\nu}\,, \qquad (13.29)$$

so that

$$\frac{d\sigma}{d\Omega' dE'} = \frac{\alpha^2}{q^4}\frac{E'}{E}\ell^{\mu\nu}W_{\mu\nu}\,. \qquad (13.30)$$

Now the question is, what more can we say about $W_{\mu\nu}$? Well, we know that it should be composed of the 4-vectors p and q, since all vectors for the outgoing particles have been integrated over. Further, since the electromagnetic current is conserved, it should obey the relations

$$q^\mu W_{\mu\nu} = 0\,, \qquad q^\nu W_{\mu\nu} = 0\,. \qquad (13.31)$$

The most general tensor that satisfies these conditions is of the form

$$W_{\mu\nu} = \left(-g_{\mu\nu} + \frac{q_\mu q_\nu}{q^2}\right)W_1 + \left(p_\mu - \frac{p\cdot q}{q^2}q_\mu\right)\left(p_\nu - \frac{p\cdot q}{q^2}q_\nu\right)\frac{W_2}{M^2}$$
$$+ i\varepsilon_{\mu\nu\lambda\rho}p^\lambda q^\rho\frac{W_3}{M^2}\,. \qquad (13.32)$$

Here, W_1, W_2 and W_3 are Lorentz invariant form factors. Or, in this case, it is better to call them *structure functions* because they tell us something about the structure of the proton, as we will see presently. We have put in inverse powers of the proton mass with the last two terms so that all three structure functions have the same dimensions.

□ **Exercise 13.2** *What are the mass dimensions of the structure functions H_1, H_2, and of the functions W_1, W_2?*

Since parity is conserved in electromagnetic interactions, the form factor involving the Levi-Civita symbol must be zero. In any case, it does not contribute to the cross-section since the leptonic tensor is symmetric in the two vector indices. The other two structure functions are both non-zero in general. And notice that, with only these two structure functions, the form of the hadronic tensor is exactly the same as that encountered in the electron–proton elastic scattering in §13.1. In performing the contraction $\ell^{\mu\nu}W_{\mu\nu}$, we can use our experience from the elastic scattering case. The only difference is that in the present case, $p \cdot q$ and q^2 are unrelated, whereas in the case of elastic scattering they obeyed the relation $p \cdot q = -\frac{1}{2}q^2$. But we have never used this special relation for performing the contraction in the case of elastic scattering. So we can use the analogy with Eq. (13.18) to write

$$\ell^{\mu\nu}W_{\mu\nu} = 4EE'\left[2W_1 \sin^2 \frac{\theta}{2} + W_2 \cos^2 \frac{\theta}{2}\right],\qquad(13.33)$$

Putting this result into Eq. (13.30), we obtain

$$\frac{d\sigma}{d\Omega' dE'} = \frac{\alpha^2}{4E^2 \sin^4 \frac{\theta}{2}}\left[2W_1 \sin^2 \frac{\theta}{2} + W_2 \cos^2 \frac{\theta}{2}\right].\qquad(13.34)$$

At this point, it is worthwhile to make a comparison between the structure functions W_1, W_2 and the form factors H_1, H_2 of elastic scattering used in §13.1. As said earlier, W_1 and W_2 are Lorentz invariants which depend on the 4-vectors p and q. Lorentz invariant combinations from these two 4-vectors are p^2, q^2 and $p \cdot q$. Of these, $p^2 = M^2$, so it is not a variable. Thus, W_1 and W_2 are in general functions of both q^2 and $p \cdot q$. Of the two, q^2 is necessarily negative, so it is convenient to define the positive variable

$$\mathcal{Q}^2 \equiv -q^2 = -q^\mu q_\mu \,.\qquad(13.35)$$

☐ **Exercise 13.3** *Argue that $q^2 < 0$ for the process under consideration.*

The functional dependence of the structure functions can then be summarized by writing

$$W_{1,2} = W_{1,2}(\mathcal{Q}^2, \nu)\,,\qquad(13.36)$$

where $\nu = p \cdot q$ as defined previously. On the other hand, the form factors H_1, H_2 are functions of $\mathcal{Q}^2$ only, since $p \cdot q = \frac{1}{2}\mathcal{Q}^2$ for elastic scattering. This fact can be made explicit if we try to construct the quantity F, defined in Eq. (13.25), for the case of elastic scattering. From the expression in Eq. (13.26), it is clear that for elastic scattering, F takes the value

$$F_{\rm el} = \int \frac{d^3 p'}{(2\pi)^3 2p'_0}(2\pi)^4\delta^4(k+p-k'-p')\overline{|\mathcal{M}|^2}$$

$$= \int \frac{d^4 p'}{(2\pi^3)}\delta(p'^2 - M^2)\Theta(p'_0)(2\pi)^4\delta^4(q+p-p')\overline{|\mathcal{M}|^2},\qquad(13.37)$$

using Eq. (4.158, *p 96*). Performing the integration over the 4-momentum p' now, we obtain

$$F_{\rm el} = 2\pi\delta\Big((p+q)^2 - M^2\Big)\Theta(p_0+q_0) \times \frac{e^4}{q^4}\ell^{\mu\nu}h_{\mu\nu}\,,\qquad(13.38)$$

using the square of the amplitude from Eq. (13.3). Comparing this with the general expression for F in Eq. (13.29), we find that for elastic scattering, $W_{\mu\nu}$ is related to $h_{\mu\nu}$:

$$W_{\mu\nu}^{(\mathrm{el})} = \frac{1}{4M}\delta(\nu - \frac{1}{2}\mathcal{Q}^2)h_{\mu\nu}. \tag{13.39}$$

The delta function present here explicitly shows that the form factors appearing in $h_{\mu\nu}$ can be taken as functions of either $\mathcal{Q}^2$ or ν. Note that we have omitted the step function in this expression. The reason is that it is really not necessary. Once we have taken all other particles on-shell and imposed 4-momentum conservation, the proton can only gain in energy, and therefore the energy of the scattered proton has to be positive.

In terms of the form factors F_1 and F_2 defined in Eq. (13.7), we can write Eq. (13.39) as:

$$W_1^{(\mathrm{el})} = \frac{\nu}{2M}(F_1 + 2MF_2)^2\delta(\nu - \frac{1}{2}\mathcal{Q}^2),$$
$$W_2^{(\mathrm{el})} = M(F_1^2 + \mathcal{Q}^2 F_2^2)\delta(\nu - \frac{1}{2}\mathcal{Q}^2). \tag{13.40}$$

Further, consider what would happen if the electron were scattering off some pointlike particle. Then F_2 would be negligible, being proportional to the fine-structure constant α, as shown in §5.7.4. Barring $\mathcal{O}(\alpha)$ corrections, F_1 is the electric charge of the particle, which would be 1 for the proton. So we could write

$$W_1^{(\mathrm{pl})} = \frac{1}{2M}\delta(1 - \frac{\mathcal{Q}^2}{2\nu}),$$
$$\nu W_2^{(\mathrm{pl})} = M\delta(1 - \frac{\mathcal{Q}^2}{2\nu}), \tag{13.41}$$

where *pl* in the superscript stands for *pointlike*. Note that on the right hand sides of these two equations, $\mathcal{Q}^2$ and ν appear exclusively in the combination $\mathcal{Q}^2/2\nu$. This fact will be very important for our subsequent discussion.

13.3 Structure functions and charge distribution

How does a structure function carry information about an object? To obtain a general idea, let us consider an experiment where an electron is scattered off an extended object with a static charge distribution given by $e\rho(\boldsymbol{x})$ in a certain frame. The electrostatic potential $\varphi(\boldsymbol{x})$ will be related to the charge distribution by the Poisson equation:

$$\nabla^2\varphi(\boldsymbol{x}) = -e\rho(\boldsymbol{x}). \tag{13.42}$$

The electron, which is used as the probe in this experiment, will interact with this charge distribution through the interaction Lagrangian

$$\mathcal{L}_{\text{int}} = e\overline{\psi}\gamma_\mu\psi A^\mu = e\overline{\psi}\gamma_0\psi\varphi, \tag{13.43}$$

since for the static charge distribution, $\boldsymbol{A} = 0$. The Feynman amplitude for the transition coming from this interaction would be

$$\mathcal{M} = e\Big[\overline{u}(\boldsymbol{k}')\gamma_0 u(\boldsymbol{k})\Big]\varphi(\boldsymbol{q}), \tag{13.44}$$

where $\boldsymbol{q} = \boldsymbol{k} - \boldsymbol{k}'$, as defined in Eq. (13.5). Here, $\varphi(\boldsymbol{q})$ is the Fourier transform of $\varphi(\boldsymbol{x})$. Using the Poisson equation, Eq. (13.42), we can write

$$\varphi(\boldsymbol{q}) = \frac{e}{\boldsymbol{q}^2}F(\boldsymbol{q}), \tag{13.45}$$

where $F(\boldsymbol{q})$ is the Fourier transform of the charge distribution:

$$F(\boldsymbol{q}) = \int d^3x \, e^{i\boldsymbol{q}\cdot\boldsymbol{x}}\rho(\boldsymbol{x}). \tag{13.46}$$

Putting Eq. (13.45) into Eq. (13.44), we obtain

$$\mathcal{M} = \frac{e^2}{\boldsymbol{q}^2}\Big[\overline{u}(\boldsymbol{k}')\gamma_0 u(\boldsymbol{k})\Big]F(\boldsymbol{q}). \tag{13.47}$$

In the expression for cross-section, we will have to used the absolute square of this amplitude, which will contain $|F(\boldsymbol{q})|^2$. If we were dealing with a point charge, equal to the charge of the proton, located at a point $\boldsymbol{x} = \boldsymbol{a}$, we would have used $\rho(\boldsymbol{x}) = \delta^3(\boldsymbol{x} - \boldsymbol{a})$, which would have given $|F(\boldsymbol{q})|^2 = 1$. So we obtain the result

$$d\sigma = |F(\boldsymbol{q})|^2\Big(d\sigma\Big)_{\text{pl}}, \tag{13.48}$$

where the subscripted letters *pl* stand for *pointlike*. The differential of the cross-section, denoted by $d\sigma$, may or may not contain the differentials of all possible kinematical variables. As long as we do not integrate over the kinematical variables that occur in $\boldsymbol{q}$, the result is valid.

Let us summarize the lesson. If we measure the differential cross-section of scattering between a point particle (e.g., the electron) hitting a target, the proportional deviation from the result expected for a pointlike target gives us the Fourier transform of the charge distribution in the target.

13.4   Scaling

It is now clear that the Lorentz invariant structure functions W_1 and W_2 contain information about the structure of the proton. In order to extract this body of information, it is useful to use a different set of independent kinematic parameters. We define a few such variables, paying no attention at this stage whether they are independent:

$$x = \frac{\mathcal{Q}^2}{2p \cdot q} = \frac{\mathcal{Q}^2}{2\nu} \,,$$
$$y = \frac{p \cdot q}{p \cdot k} \,. \tag{13.49}$$

The first one is often called *Bjorken-x*. Note that for the case of elastic scattering, its value would be 1. The second one, y, has an easy interpretation in the FT frame: it is $(E-E')/E$, i.e., the fractional energy loss of the electron. Obviously, the kinematical range of this parameter is given by

$$0 \le y \le 1 \,. \tag{13.50}$$

As for the variable x, we first note that it must be positive since both its numerator and its denominator are positive. Moreover, note that

$$(p+q)^2 = M^2 + 2p \cdot q - \mathcal{Q}^2 \,. \tag{13.51}$$

The quantity on the left hand side, the invariant mass squared of the hadronic part of the final state, cannot be smaller than M^2, which is the value obtained for elastic scattering. Thus we find that the variable x also has the same range for a general scattering:

$$0 \le x \le 1 \,. \tag{13.52}$$

An interpretation of this variable will be given in §13.5. As we will see, this holds the key to the goal that we are after, viz., understanding the structure of hadrons.

□ **Exercise 13.4** Show that, in the deep inelastic region where the masses of the incoming particles and the outgoing electron can be neglected, the invariant variables x and y can be written in terms of the Mandelstam variables as

$$x = -\frac{t}{s+u} \,, \qquad y = \frac{s+u}{s} \,. \tag{13.53}$$

[**Note :** *Although this is not a 2-to-2 scattering, the Mandelstam variables can be defined through the incoming electron and proton and the outgoing electron.*]

Earlier in Eq. (13.34), we expressed the differential cross-section for the inelastic scattering with respect to the energy and direction of the final electron in the FT frame. Let us now transform the result in terms of variations with respect to the Lorentz invariant variables x and y. As a first step, we integrate Eq. (13.34) over the azimuthal angle and write

$$\frac{d\sigma}{d(\cos\theta)dE'} = \frac{\pi\alpha^2}{2E^2 \sin^4\frac{\theta}{2}} \left[2W_1 \sin^2\frac{\theta}{2} + W_2 \cos^2\frac{\theta}{2} \right] , \tag{13.54}$$

a relation that uses variables of the FT frame. The relation between the variables (E', θ) and (x, y) can be obtained by using Eqs. (13.10) and (13.17) in the equations defining x and y. Using these, we obtain

$$\frac{d\sigma}{dx\,dy} = \frac{2\pi\alpha^2}{xyME}\left(y\widetilde{W}_1 + \left[\frac{1-y}{xy} - \frac{M}{2E}\right]\widetilde{W}_2\right),\qquad(13.55)$$

where we have defined the new combinations

$$\widetilde{W}_1 = MW_1,$$
$$\widetilde{W}_2 = \frac{\nu}{M}W_2,\qquad(13.56)$$

which are both dimensionless structure functions. Both of these can be seen as functions of the kinematical variables $\mathcal{Q}^2$ and ν, or equivalently of x and y.

Eq. (13.55) seems to be an expression of differential cross-section written entirely in terms of Lorentz invariant quantities, except for the presence of E, which is the energy of the incoming electron in the FT frame. However, this shortcoming can easily be fixed. Eq. (13.15) shows that ν is equal to $M(E - E')$ in the FT frame. Thus, the maximum possible value of ν is ME, and we can use this to replace E by $\nu_{\max}/M$ in Eq. (13.55) if we wish so.

□ **Exercise 13.5** Show that

$$dE'\,d(\cos\theta) = \frac{My}{1-y}\,dx\,dy.\qquad(13.57)$$

Use this to derive Eq. (13.55) from Eq. (13.54).

A function of two variables is said to exhibit the property of *scaling* if it depends only on one combination of the two variables. Experimentally, this is exactly what is seen of the structure functions $\widetilde{W}_1$ and $\widetilde{W}_2$ when $\mathcal{Q}^2$ is very large. In particular, it is seen that in this kinematical region, the structure functions $\widetilde{W}_1$ and $\widetilde{W}_2$ depend not on $\mathcal{Q}^2$ and ν independently, but only on their ratio, i.e., on the Bjorken-x variable. In other words, if one analyzes scattering data at different sets of values of $\mathcal{Q}^2$ and ν such that their ratio is a constant, the deduced values for the structure functions would show no variation at all. In mathematical notation, we can write

$$\widetilde{W}_1(\mathcal{Q}^2, \nu) \xrightarrow{\mathcal{Q}^2 \gg M^2} \mathscr{F}_1(x),$$
$$\widetilde{W}_2(\mathcal{Q}^2, \nu) \xrightarrow{\mathcal{Q}^2 \gg M^2} \mathscr{F}_2(x).\qquad(13.58)$$

This scaling property of the structure functions is crucial in understanding the structure of hadrons, as we will presently see.

13.5 Partons

At this point, let us recall Eq. (13.41) which says that if the electrons scatter off pointlike particles, the structure functions $\widetilde{W}_1$ and $\widetilde{W}_2$ should depend only on the combination $\mathcal{Q}^2/2\nu$, or the Bjorken-x. Scattering data then seem to be telling us that, for large $\mathcal{Q}^2$, the electrons "see" pointlike particles in the proton. The constituent parts of the proton, or of a hadron in general, were given the name *partons* at the beginning, because there was no a priori reason to believe that these were the same as the quarks introduced by Gell-Mann and Zweig to explain the static properties of hadrons.

Let us assume that the partons are quarks, i.e., they are fermions with some charge which is not necessarily equal to the proton charge. A proton has more than one quark. Let us assume that a particular quark is carrying a momentum fraction x' of the whole proton, which means that its 4-momentum is $x'p^\mu$. Since the quark is assumed to be pointlike at the energies in consideration, we can easily write the cross-section of its elastic scattering with an electron. The absolute square of the amplitude in this case would still be given by an expression of the form given in Eq. (13.3), with $\ell^{\mu\nu}$ given by Eq. (13.4). The difference occurs in other parts. First, since the quark is also pointlike, the tensor $h_{\mu\nu}$ should now have a form similar to that of $\ell^{\mu\nu}$, except that it should involve the initial and the final quark momenta. Second, the electric charge of the quark is *not* the same as that of the proton. If the charge is eQ, we should have an extra factor of Q^2 in the expression. In summary, we would obtain

$$\overline{|\mathcal{M}|^2} = \frac{Q^2 e^4}{q^4} \ell^{\mu\nu} \times 2\left(x'p_\mu p'_\nu + x'p_\nu p'_\mu + \frac{1}{2}q^2 g_{\mu\nu}\right), \qquad (13.59)$$

where p' is the 4-momentum of the final quark and q is defined, as before, by Eq. (13.5). In the leptonic part, the electron mass can be neglected, as announced before. The quark mass has been neglected as well, assuming the quark to be sufficiently light.

Suppose we now put this expression into the formula for the scattering cross-section. There will be an integration over the final quark momentum p'. The integration measure can be written as

$$\int \frac{d^3p'}{(2\pi)^3 2p'_0}\cdots = \int \frac{d^4p'}{(2\pi)^3}\,\delta(p'^2)\cdots, \qquad (13.60)$$

where the dots denote the rest of the integrand. So, the differential cross-section with respect to the final electron momentum will be given by

$$\begin{aligned}
d\sigma &= \frac{1}{4x'k\cdot p}\frac{d^3k'}{(2\pi)^3 2E'}\int \frac{d^4p'}{(2\pi)^3}\,\delta(p'^2)(2\pi)^4\delta^4(xp+q-p')\overline{|\mathcal{M}|^2}\\
&= \frac{1}{4x'k\cdot p}\frac{d^3k'}{(2\pi)^2 2E'}\,\delta\left((x'p+q)^2\right)\overline{|\mathcal{M}|^2}. \qquad (13.61)
\end{aligned}$$

Let us now look at the remaining delta function. Ignoring the term $x'^2 p^2$ compared to the magnitude of q^2, we can write this part of the expression as

$$\delta\left(2x'p\cdot q - \mathcal{Q}^2\right) = \frac{1}{2p\cdot q}\delta(x'-x) = \frac{x}{\mathcal{Q}^2}\delta(x'-x), \qquad (13.62)$$

using the definition of x from Eq. (13.49). Note that this factor forces x' to be equal to x. And this, finally, is the physical interpretation of the variable x that we had defined earlier: it is the momentum fraction carried by a quark from which the electron scatters. In view of the delta function found above, we will replace x' by x everywhere else in the expression.

In a general frame, the phrase *momentum fraction* would be meaningless. A given quark in a hadron can have momentum in any direction, not necessarily along the direction of motion of the hadron. In particular, even if the 3-momentum of the hadron is zero, the quarks inside the hadron can be moving. The statement about *momentum fraction* makes sense only in the so-called *infinite momentum frame*, in which a hadron moves with a very large momentum. Therefore, magnitudes of transverse momenta are negligible, and the quarks inside the hadron can also be considered to be moving only in the direction of motion of the hadron. In this frame, x is the fraction of hadronic momentum carried by a quark.

Putting things back into Eq. (13.61), we obtain

$$\frac{d\sigma}{dE'd\Omega'} = \frac{2Q^2\alpha^2}{q^4}\frac{1}{M\Omega^2}\frac{E'}{E}\,\delta(x'-x)\ell^{\mu\nu}\left(xp_\mu p'_\nu + xp_\nu p'_\mu + \frac{1}{2}q^2g_{\mu\nu}\right).$$
(13.63)

Comparing this with Eq. (13.30), we find that for electron-quark scattering,

$$W_{\mu\nu} = \frac{Q^2}{M\Omega^2}\,\delta(x'-x)\left(xp_\mu p'_\nu + xp_\nu p'_\mu + \frac{1}{2}q^2g_{\mu\nu}\right).$$
(13.64)

In order to extract the structure functions from this, we use $p' = xp + q$ to note that

$$x(p_\mu p'_\nu + p_\nu p'_\mu) + \frac{1}{2}q^2g_{\mu\nu} = 2x^2(p_\mu + \frac{1}{2x}\,q_\mu)(p_\nu + \frac{1}{2x}\,q_\nu)$$
$$+\frac{1}{2}q^2\left(g_{\mu\nu} - \frac{q_\mu q_\nu}{q^2}\right).$$
(13.65)

Looking back at the definition of the structure functions from Eq. (13.32), we then identify

$$W_1 = \frac{Q^2}{2M}\delta(x-x'),$$
(13.66a)
$$\frac{W_2}{M^2} = \frac{2Q^2x^2}{M\Omega^2}\delta(x-x') = \frac{Q^2x}{M\nu}\delta(x-x').$$
(13.66b)

Thus, the structure functions defined in Eq. (13.56) are given by

$$\widetilde{W}_1 = \frac{1}{2}Q^2\delta(x-x'),$$
(13.67a)
$$\widetilde{W}_2 = Q^2x\delta(x-x').$$
(13.67b)

We thus obtain an important relation connecting the two structure functions:

$$\widetilde{W}_2(x,\Omega^2) = 2x\widetilde{W}_1(x,\Omega^2).$$
(13.68)

This is called the *Callan–Gross sum rule*, and is obeyed very well by data in the deep inelastic region, characterized by the relation $\mathcal{Q}^2 \gg M^2$. Such agreement shows that the electrons are indeed scattering off quarks inside the proton, and these quark degrees of freedom show up in the deep inelastic region. And since it is the scaling region where the structure functions have the form given in Eq. (13.58), we can also write

$$\mathscr{F}_2(x) = 2x\mathscr{F}_1(x) \,. \tag{13.69}$$

☐ **Exercise 13.6** The Callan–Gross sum rule is a consequence of spin-½ nature of quarks. Show that if the quarks were spinless, one would have obtained $W_1 = 0$ instead of the expression shown in Eq. (13.66a).

13.6   Parton distribution functions

Eq. (13.63) gives the contribution of a single quark of charge eQ with a specific value of the momentum fraction x to the differential cross-section of deep inelastic electron–proton scattering. The proton does not contain just a single quark. So, in order to obtain the structure function for the proton, we need to sum over all kinds of partons that are present in the proton. Apart from that, we also need to integrate over the momentum distribution of the quarks inside the proton. If the probability density of finding the ith kind of quark with a momentum fraction x' within the proton is $f_i(x')$, then we should have

$$\widetilde{W}_2 = \sum_i Q_i^2 \int dx' \; x\delta(x - x')f_i(x') = \sum_i Q_i^2 x f_i(x) \,, \tag{13.70}$$

whereas the structure function $\widetilde{W}_1$ will be given by Eq. (13.68).

There is a summation over all different kinds of partons, and we can now ask what different kinds should be taken into account. While discussing isospin and flavor SU(3) symmetry, we mentioned that the proton contains two up-quarks and one down-quark. So, to obtain the structure functions for the proton, we need to add the contributions these two types of quarks. But that is not all. All we have shown in Ch. 8 and Ch. 10 is that many of the static properties of the proton can be explained by treating the proton as a *uud* bound state. But it may be true that apart from these three quarks, there are other particles in the proton which do not affect the properties we have discussed before.

Indeed, such is the case with atoms. Their chemical properties come from the outer shell electrons, or valence electrons, only. The inner shell electrons, as well as the nucleus, are unexposed in chemical reactions. In the same manner, we can think of the possibility that the two up quarks and single down quark that determine the static properties of the proton are the *valence quarks*. In addition, there can be any number of quarks, antiquarks and even gluons, all of which play a part in the structure of the proton. These other ones, analogues of the inner shell electrons of an atom, are called, somewhat figuratively, *sea quarks* or *ocean quarks*. The first name is more commonly used, but we will use the second one here, because, while speaking, the first one can easily be confused with c-quark, the short name for the charm quark.

In the proton or the neutron, we do not expect any charm quark, or the even heavier quarks. The reason is that these quarks are heavier than the nucleon. But the up, down, and strange quarks can exist in the proton, along with their antiparticles. So, Eq. (13.70) can be written in the more explicit form as

$$\frac{1}{x}\,\mathscr{F}_2^{(ep)}(x) = \frac{4}{9}\Big(u_p(x) + \widehat{u}_p(x)\Big)$$
$$+ \frac{1}{9}\Big(d_p(x) + \widehat{d}_p(x)\Big) + \frac{1}{9}\Big(s_p(x) + \widehat{s}_p(x)\Big). \qquad (13.71)$$

We have made some changes in the notations in writing this equation. First, on the left hand side, we have put parenthesized superscripts to identify the initial states of the scattering process considered, in order to distinguish it from data from other initial states that we will use very soon. Second, we have discontinued the use of the notation f_i that appears in Eq. (13.70). Instead, we are using the first letter in the name of a quark to denote its distribution function. And third, we have put a subscript p on all such distribution functions to remind us that these are the distributions obtained in the proton.

There are a lot of functions in Eq. (13.71), and naively it seems that it would be impossible to deduce the values of all of them. But we will show now that under some reasonable assumptions, the functions can all be derived from the data.

Data need not come only from electron–proton scattering. We can also perform electron scattering off neutrons. There will be a similar set of form factors for this scattering, and in the deep inelastic region, we will be able to write

$$\frac{1}{x}\,\mathscr{F}_2^{(en)}(x) = \frac{4}{9}\Big(u_n(x) + \widehat{u}_n(x)\Big)$$
$$+ \frac{1}{9}\Big(d_n(x) + \widehat{d}_n(x)\Big) + \frac{1}{9}\Big(s_n(x) + \widehat{s}_n(x)\Big). \qquad (13.72)$$

The number of functions seems to have proliferated further, but there is a simplifying feature if we neglect all effects of isospin violation, which are very tiny anyway. If isospin symmetry is considered to be exact, then the u-quark distribution function in the proton should be the same as the d-quark distribution function in the neutron, and vice versa. Also, the strange quark and antiquark distribution functions should be the same in the proton and the neutron. Thus we can write

$$u_p(x) = d_n(x) \equiv u(x)\,,$$
$$d_p(x) = u_n(x) \equiv d(x)\,,$$
$$s_p(x) = s_n(x) \equiv s(x)\,. \qquad (13.73)$$

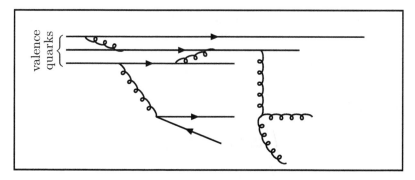

Figure 13.3: Schematic diagram to show how ocean quarks and gluons arise in a baryon.

The functions $u(x)$ and $d(x)$ have some contribution coming from the valence quarks and the rest from the ocean quarks. Let us separate these two contributions and write

$$u(x) = u_v(x) + u_o(x), \tag{13.74}$$

and similarly for $d(x)$. For the antiquarks of these two flavors, as well as for the strange quarks and antiquarks, the entire contribution is from ocean quarks.

Where do the ocean quarks come from? When thinking about the constituents of a nucleon, we can start with the three valence quarks and nothing else. However, we cannot ignore the fact that these quarks are interacting in order that they can form a bound state. These interactions are QCD interactions, whereby gluons are emitted from quarks, and such gluons make quark-antiquark pairs in turn, and those pairs fuse to turn to gluons, and so on. Strong interactions are flavor blind, so all flavors of quarks should be equally produced in such processes unless they are kinematically disfavored. We have disregarded the presence of charm and heavier quarks precisely for such kinematical reasons, as explained earlier. But the up, down and strange quarks are light, and if we disregard their masses all of them should have equal proportions in the ocean. So we write

$$u_o(x) = d_o(x) = \widehat{u}(x) = \widehat{d}(x) = s(x) = \widehat{s}(x) \equiv S(x). \tag{13.75}$$

Eqs. (13.71) and (13.72) acquire much simpler-looking forms under these simplifying assumptions, viz.:

$$\frac{1}{x}\, \mathscr{F}_2^{(ep)}(x) = \frac{1}{9}\Big(4u_v(x) + d_v(x)\Big) + \frac{4}{3}S(x),$$
$$\frac{1}{x}\, \mathscr{F}_2^{(en)}(x) = \frac{1}{9}\Big(u_v(x) + 4d_v(x)\Big) + \frac{4}{3}S(x). \tag{13.76}$$

There are extra constraints on the distribution functions appearing here. The number of valence quarks imply the relations

$$\int_0^1 dx\, u_v(x) = 2\,, \qquad \int_0^1 dx\, d_v(x) = 1\,, \qquad (13.77)$$

or equivalently,

$$\int_0^1 dx \left(u(x) - \widehat{u}(x) \right) = 2\,,$$

$$\int_0^1 dx \left(d(x) - \widehat{d}(x) \right) = 1\,. \qquad (13.78)$$

☐ **Exercise 13.7** Verify, with the help of Eq. (13.75), that Eqs. (13.77) and (13.78) are equivalent.

With the help of these constraints, the parton distribution functions can be found from the analysis of electron–proton and electron–neutron scatterings. The momentum fraction of the quarks and antiquarks in the proton, say, can then be calculated as

$$\epsilon_u \equiv \int_0^1 dx\, x \left(u(x) + \widehat{u}(x) \right) \qquad (13.79)$$

and a similar expression for other flavors of quarks. The findings are surprising. They show that,

$$\epsilon_u = 0.36\,, \qquad \epsilon_d = 0.18\,. \qquad (13.80)$$

In other words, only 54% of a nucleon's momentum comes from the u and the d quarks and their antiparticles. Certainly, it is not possible that the s quark carries the rest of the momentum. In fact, the s quark contribution should be very small. So the inescapable conclusion is that the rest of the momentum is carried by gluons, which are uncharged and therefore do not contribute to electromagnetic scattering. The gluons, therefore, are the dominant constituents of the proton and the neutron. If we use the word *parton* to indicate the fundamental particles which are constituents of hadrons, then it should mean quarks, antiquarks and gluons.

13.7   Parton distribution and cross-section

We now discuss the relation between parton distributions and cross-sections in a different, although related, way. The electron–proton scattering, at the basic level, must be the scattering of an electron with a parton inside the proton. The parton can be a quark or an antiquark. Gluons do not scatter off electrons at the tree-level, hence their effects are higher order. Analysis of electron-quark scattering would be the same as that of electron-muon scattering, except that the quark charge would be different from the muon charge. For a quark or antiquark of charge eQ_f, we can read the differential cross-section from Eq. (12.54, $p\,337$):

$$\frac{d\sigma}{dt} = \frac{2\pi\alpha^2 Q_f^2}{\breve{s}^2} \frac{\breve{s}^2 + \breve{u}^2}{t^2}.$$ (13.81)

Note that we are writing the differential cross-section in the Lorentz invariant manner, something that was discussed in §4.13. Also note that we have added a crescent sign ($\breve{}$) on top of $\breve{s}$ and $\breve{u}$, to indicate that these Mandelstam variables pertain to the quark-level scattering process, and are different from the corresponding Mandelstam variables for the electron–proton scattering which will be denoted without the crescent signs. The variable t is however the same whether we consider the quark-level scattering or the hadron-level scattering, since it involves only the 4-momenta of the initial and final electron. Now, the above formula for differential cross-section can be written as

$$\frac{d\sigma}{dt} = \frac{2\pi\alpha^2 Q_f^2}{t^2}\left(1 + \frac{\breve{u}^2}{\breve{s}}\right) = \frac{2\pi\alpha^2 Q_f^2}{t^2}\left[1 + (1-y)^2\right],$$ (13.82)

using the expression for the variable y given in Eq. (13.53) which can also be written as

$$y = \frac{\breve{s} + \breve{u}}{\breve{s}}$$ (13.83)

in the deep inelastic region where the momentum of the initial quark is taken to be xp^μ where p^μ is the proton momentum.

Eq. (13.82) gives the differential cross-section of electron scattering against a single quark or antiquark with momentum xp^μ. To obtain the differential cross-section for the electron–proton scattering, we need to multiply by the probability of finding that kind of fermion with momentum xp^μ inside the proton, and then integrating over x and summing over all such fermions, i.e., on all types of quarks and antiquarks. Thus,

$$\frac{d\sigma}{dt}\left(e(k)p(p) \to e(k')X\right) = \sum_f \int dx\, f_f(x)\frac{d\sigma}{dt}\left(e(k)f(xp) \to e(k')f\right)$$

$$= \sum_f \int dx\, f_f(x)\frac{2\pi\alpha^2 Q_f^2}{t^2}\left[1 + (1-y)^2\right].$$ (13.84)

Therefore,

$$\frac{d\sigma}{dx\,dt} = \left[\sum_f f_f(x)Q_f^2\right]\frac{2\pi\alpha^2}{t^2}\left[1 + (1-y)^2\right].$$ (13.85)

It is more convenient to exchange t for the dimensionless quantity y on the left hand side of this expression. From the definitions given in Eq. (13.49), we obtain

$$t = q^2 = -2p \cdot qx = -2p \cdot kxy = -sxy.$$ (13.86)

Evaluating the Jacobian for going from the variables $(x, \mathbf{t})$ to (x, y), we obtain

$$dx\, d\mathbf{t} = \mathbf{s}x\, dx\, dy\,. \tag{13.87}$$

So the differential cross-section can be written as

$$\frac{d\sigma}{dx\, dy} = \left[\sum_f x f_f(x) Q_f^2\right] \frac{2\pi\alpha^2 \mathbf{s}}{\mathbf{t}^2}\left[1 + (1-y)^2\right]\,. \tag{13.88}$$

We have used the scaling arguments to assume that the parton distribution function f_f depends only on x and not on q^2. Note that the y dependence of the differential cross-section comes entirely from the scattering of individual partons.

13.8   Fragmentation

In §12.4, we derived cross-sections of various processes involving quarks and gluons in the initial and final states. In a real experiment, we cannot deal with quarks and gluons directly because they are confined. We perform experiments involving hadrons. However, we are beginning to see why the parton level cross-sections are relevant. We can say that the description of any process involving hadrons can be divided into three parts. First, we describe how the hadrons are made of partons. This part is taken care of through the parton distribution functions. The second part consists of the basic scattering process occurring at the parton level. In this process, the final state also contains partons. Once the partons are created, they start losing energy by emitting gluons. When the energies of individual partons become small enough, close to the scale $\Lambda_{\rm QCD}$, the strong coupling constant becomes large enough that these partons can hadronize. This is the third part of the process, which again cannot be described through perturbative QCD. Like the first part, here also we need to use some functions that will bridge the partonic description and the hadronic description. This bridging is done through what are called the *fragmentation functions*.

In order to discuss the fragmentation functions, we can shift our attention from deep inelastic scattering to e^+e^- collisions where there is no hadron in the initial state and therefore no parton distribution function to worry about. As described in §12.4.2, at the basic level $e^+e^- \to q\bar{q}$ is the lowest order scattering process that involves strongly interacting particles in the final state. The cross-section of this process would equal the inclusive cross-section of $e^+e^- \to$ hadrons because the quark and the antiquark produced must hadronize eventually. If we are not interested in the particular hadrons produced, we need not get into the details of the hadronization process.

Things change if we are interested about some exclusive cross-section. For example, suppose we are trying to estimate the e^+e^- cross-section into a final state containing a particular hadron h. We can denote the process by $e^+e^- \rightarrow hX$, where X stands for anything that is consistent with all conservation laws, just as in the case of deep inelastic scattering. The cross-section for this scattering can be written as

$$\sigma(e^+e^- \rightarrow hX) = \sum_q \int dz \, \sigma(e^+e^- \rightarrow q\hat{q}) \times \left[D_q^h(z) + D_{\hat{q}}^h(z) \right] ,(13.89)$$

where $D_q^h(z)$, for example, is the probability that the quark q and the debris resulting from it end up producing the hadron h which carries a fraction z of the momentum of the quark produced at the parton level. These are the *fragmentation functions*. Momentum conservation would imply the constraint

$$\sum_h \int_0^1 dz \, z D_q^h(z) = 1 , \qquad (13.90)$$

implying that all momentum carried by the quark would finally be divided into hadrons. There is another constraint, coming from the fact that each quark-antiquark pair produced must finally end up in some hadron. This constraint can be expressed as

$$\sum_q \int_{z_h}^1 dz \left[D_q^h(z) + D_{\hat{q}}^h(z) \right] = N_h , \qquad (13.91)$$

where N_h is the average number of hadrons h produced in the e^+e^- collision. The lower limit of the integral, z_h, corresponds to the threshold energy needed to produce the hadron h, which is equal to $2m_h/\mathfrak{Q}$, where m_h is the mass of h and $\mathfrak{Q}$ is the total e^+e^- energy in the CM frame, which means that the quark and the antiquark are produced with energies $\frac{1}{2}\mathfrak{Q}$ each.

☐ **Exercise 13.8** The fragmentation functions are usually parametrized in the form

$$D_q^h(z) = A(1 - z)^n/z , \qquad (13.92)$$

where A and n are phenomenological parameters. Show that the leading growth of N_h depends logarithmically on $\mathfrak{Q}$:

$$N_h \sim \ln(\mathfrak{Q}/2m_h) . \qquad (13.93)$$

13.9   Scale dependence of parton distribution

From their definitions, the form factors and consequently the parton distribution functions depend on two kinematical variables. As pointed out in §13.4, these variables can be taken to be $\mathfrak{Q}^2$ and ν, or equivalently of x and y. We can also take x and $\mathfrak{Q}^2$ as the two independent parameters. The advantage of this choice is that x is the scaling parameter, whereas $\mathfrak{Q}^2$ has a simple physical interpretation, viz., negative of the 4-momentum squared exchanged in the process.

Although the distribution functions discussed earlier were considered to be functions of the Bjorken-x only, this is only an approximation. In reality, the distribution functions depend on Ω^2, i.e., on the 4-momentum of the photon that is used to probe them. In this section, we discuss the cause and nature of this dependence.

Let us first try to understand the cause qualitatively. There are quarks in a hadron, and there are gluons as well. If they had not interacted at all, their momentum would not have changed, and the distribution functions would not have changed either. But interactions are there, of course, and they lead to a departure from this idyllic situation. For example, consider two quarks interacting within a hadron. If the interaction is elastic, the total momenta of the quarks do not change. However, it is possible that one of the quarks emits a gluon in the interaction, i.e., the process is effectively $qq \to qqg$. In this case, some momentum and energy are transferred from the quarks to the gluons. There can be other similar processes, resulting in momentum transfer between different kinds of partons in a hadron. Since these processes are energy dependent, the distribution functions depend on the energy scale.

Lately, we had been denoting the distribution functions by the letters denoting the flavors. We can continue doing so, i.e., denote the quark distribution functions by q and so on. The equations governing the scale dependence of the distribution functions of quarks, antiquarks and gluons would then have the following general forms:

$$\frac{dq(x, \Omega^2)}{d \ln \Omega^2} = \frac{\alpha_3}{2\pi} \int_x^1 \frac{dx'}{x'} \left[P_{q \to qg}(\frac{x}{x'})q(x', \Omega^2) + P_{g \to q\widehat{q}}(\frac{x}{x'})g(x', \Omega^2) \right],$$

$$(13.94a)$$

$$\frac{d\widehat{q}(x, \Omega^2)}{d \ln \Omega^2} = \frac{\alpha_3}{2\pi} \int_x^1 \frac{dx'}{x'} \left[P_{q \to qg}(\frac{x}{x'})\widehat{q}(x', \Omega^2) + P_{g \to q\widehat{q}}(\frac{x}{x'})g(x', \Omega^2) \right],$$

$$(13.94b)$$

$$\frac{dg(x, \Omega^2)}{d \ln \Omega^2} = \frac{\alpha_3}{2\pi} \int_x^1 \frac{dx'}{x'} \left[P_{q \to gq}(\frac{x}{x'}) \sum_q \left(q(x', \Omega^2) + \widehat{q}(x', \Omega^2) \right) \right.$$

$$\left. + P_{g \to gg}(\frac{x}{x'})\, g(x', \Omega^2) \right].$$

$$(13.94c)$$

These equations are called the *DGLAP equations* because they were proposed and developed in three different papers: one by Dokshitzer, one by Gribov and Lipatov, and another by Altarelli and Parisi. We now discuss the meanings of different terms in this set of equations.

First of all, let us talk about the factor of α_3 outside the integral sign in each equation. This has to be there, because if α_3 vanished, there would have been no interactions and therefore no evolution of the distribution functions. The numerical factor $1/(2\pi)$ appearing with α_3 does not require an explanation. It is just a convention: it could have been absorbed into the definition of the factors P which appear within the integrals.

The factors P are called *splitting functions*. They represent the probability of the kind of emission indicated in the subscripts. Thus, for example, consider the first term on the right hand side of Eq. (13.94a). The factor $q(x', \mathbf{Q}^2)$ represents quark distribution function with momentum fraction x'. If a quark with this momentum fraction emits a gluon, the momentum of the quark would be reduced. This process would then contribute to the quark distribution $q(x, \mathbf{Q}^2)$ for some $x < x'$. Thus, only values of $x' > x$ can contribute, a fact that is reflected in the limits of the integration. The splitting function, along with the factor $\alpha_3/(2\pi)$, represents the probability that a quark would indeed emit a gluon, i.e., the ratio of the cross-sections of the process with a gluon emission and the process without. Similarly, a gluon splitting into a quark-antiquark pair contributes to the distribution functions of both the quark and the antiquark, represented by the splitting functions $P_{g \to q\widehat{q}}$ in the first two equations. The contribution represented by the first term on the right hand side of Eq. (13.94b) comes really from an antiquark splitting out a gluon, and hence should contain $P_{\widehat{q} \to \widehat{q}g}$. However, because of the invariance of strong interactions under charge conjugation, we must have

$$P_{\widehat{q} \to \widehat{q}g} = P_{q \to qg}, \tag{13.95}$$

which is why we do not use an extra bit of notation for this splitting function. The same splitting function contributes to the scale dependence of the gluon distribution function as well. The last term in Eq. (13.94c) represents the contribution coming from one gluon splitting into two gluons through the cubic vertex that occurs in Yang-Mills theories.

The scale dependence of the parton distribution functions has been determined by various groups. In Fig. 13.4, we show the typical results for two different scales. One clearly sees that the higher the scale, the ocean quarks and gluons become more and more important.

From this comment, we now revisit one question that was left unanswered in Ch. 9. The LHC is a machine with two colliding proton-proton beams, as opposed to other modern collider machines where particle-antiparticle beams are used. The reason for this is the very high energy of the beams. At such high energies, the gluon component of the proton is quite high, so they play a dominant role. This component is the same in a proton and an antiproton. Therefore, there is not much motivation of creating an antiproton beam to collide with a proton beam: two proton beams serve the same purpose, roughly speaking.

13.10 Quark masses

We talked about quark masses in Ch. 10. The precise values of the masses were not necessary in order to derive the mass relations between various hadrons given in §10.7. But the magnetic moment formulas given in §10.8 imply some values of the quark masses. For example, consider Eqs. (10.114) and (10.115),

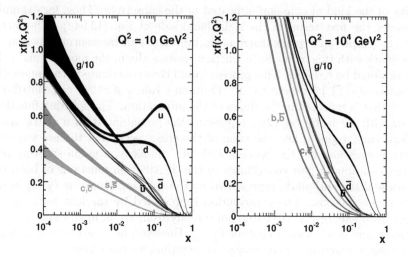

Figure 13.4: Parton distribution functions for the proton at two different scales. [From A. D. Martin, W. J. Stirling, R. S. Thorne, G. Watt: Eur. Phys. J. **C63** (2009) 189; with kind permission from the authors and from Springer Science & Business Media B.V.]

which give magnetic moments of the proton and the neutron in terms of the quark magnetic moments. Inverting these two equations and using the measured values of μ_p and μ_n, we obtain

$$\mu_u = \frac{4}{5}\mu_p + \frac{1}{5}\mu_n = 1.85\mu_N\,,$$
$$\mu_d = \frac{1}{5}\mu_p + \frac{4}{5}\mu_n = -0.97\mu_N\,, \tag{13.96}$$

where μ_N is the nuclear magneton. If the magnetic moment of quarks is given by the standard formula for the magnetic moment of a fermion involving its charge and mass, we obtain the following estimates for the quark masses:

$$m_d = 322\,\text{MeV}\,, \qquad m_u = 338\,\text{MeV}\,. \tag{13.97}$$

If these were really the masses of the up- and down- quarks, we could not have neglected the quark masses in the analysis of deep inelastic scattering of §13.5. Fortunately, there is no reason to believe that these are mass parameters that the quark fields should have in the basic Lagrangian. The reason is that the quarks in a hadron are strongly bound, and the valence quarks are not the only constituents of a hadron. There are ocean quarks, as well as a sizable contingent of gluons. The valence quarks live in this ocean. Whenever a particle is in a non-trivial background of other particles, coherent interactions with the background induce an inertia on the particle, so that its effective mass becomes different from its actual mass. For example, when light passes through a transparent medium, interactions with the particles comprising the medium assign an effective mass to the photon, and as a result light does not travel with the speed that it has in the vacuum. Similarly, the masses quoted in Eq. (13.97) should be seen as the effective masses of the valence quarks while they are inside a nucleon. These values of the masses are called *constituent quark masses*.

In contrast, the mass parameters appearing in the Lagrangian are called *current quark masses*. These masses are lower than the constituent quark masses. For up and down quarks which are the main constituents of the nucleons, agreement of experimental data with the analysis carried out assuming massless quarks indicates that these two quark masses must be negligible.

There is however no way to *measure* these masses directly, because the quarks are always confined in hadrons. One has to take recourse to indirect means of estimating the masses. In Ch. 18, we will describe some techniques that achieve this goal, and the result is

$$m_u = 1 \text{ to } 3 \text{ MeV}, \qquad m_d = 3 \text{ to } 7 \text{ MeV}. \qquad (13.98)$$

Lattice calculations support these estimates.

In §10.7, we argued that the strange quark must be heavier than the up- and the down- quarks, which we summarized in Eq. (10.77, *p 279*). Naively the difference seems to be about 100 MeV, since the mass differences between two adjacent rows in a flavor SU(3) diagram is roughly of that amount. Analysis shows that

$$m_s = 95 \pm 5 \text{ MeV}. \qquad (13.99)$$

These three quarks are therefore much lighter than the nucleon, and in fact lighter than $\Lambda_{\rm QCD}$. The other quarks, starting with the charm, are heavier. We will discuss these heavy quarks in Ch. 20.

13.11   Glueballs

If gluons are present in hadrons, as indicated from the discussion of §13.6, is it possible that there are hadrons which are composed of gluons only? After all, gluons have cubic and quartic self couplings, so they interact with one another. Can they not create bound states through such interactions? There is no reason why they would not be able to bind, and a hadron formed with gluons, without any valence quarks, would be called a *glueball*.

Lattice calculations indicate that such bound states should exist, and the lowest glueball state should have a mass of less than 2 GeV. However, No glueball has been experimentally observed yet. They would be hard to observe because they would mix with meson states.

Chapter 14

Fermi theory of weak interactions

Weak interactions, as the name implies, are very weak compared to the strong and electromagnetic interactions. As a rough estimate of the strengths of different interactions, let us compare the strong, electromagnetic and weak forces between two protons separated by a distance of $1\,\mathrm{fm}$, or $10^{-13}\,\mathrm{cm}$. If the strong force between them defines our unit of force, the electromagnetic force will be of order 10^{-2}, and the weak force will be about 10^{-14} in this unit.

And yet, weak interactions are very important. The reason is that, although they are weak, they can give rise to many kinds of physical phenomena which strong and electromagnetic interaction cannot. Take, for example, parity violation. We have discussed before that both strong and electromagnetic interactions conserve parity. But parity is violated in particle interactions, and weak interactions must be responsible for it. There are many other phenomena which are possible only because of weak interactions. We will encounter many examples of this kind as we discuss weak interactions in this chapter and a few subsequent ones.

14.1 Four-fermion interaction

Fermi was motivated by nuclear beta decays, which are processes in which a nucleus with atomic number Z transforms into another nucleus with atomic number $Z + 1$, emitting an electron and an antineutrino. At the nucleon level, this implies the decay of a neutron into a proton, an electron and an antineutrino:

$$n \rightarrow p + e^- + \widehat{\nu}. \tag{14.1}$$

Obviously, it involves four fermions. Fermi wanted to explain the process by an interaction where all four fermion fields interact at a point. In electrodynamics, the current between fermions interacts with the photon field. Here, Fermi assumed that the current between two fermions interacts with the current between the other two fermions: a current-current interaction. So he wrote the interaction as

$$\mathscr{L}_{\text{int}} = -G\left[\overline{\psi}_{(p)}\gamma^{\mu}\psi_{(n)}\right]\left[\overline{\psi}_{(e)}\gamma_{\mu}\psi_{(\nu)}\right], \tag{14.2}$$

where $\psi_{(n)}$ denotes the neutron field, $\psi_{(p)}$ the proton field, and so on. The neutrino field operator creates an antineutrino, the conjugated electron field operator creates an electron, and thus the process of Eq. (14.1) becomes possible in the first order perturbation using this interaction. Fermi had a Dirac matrix sandwiched between each pair of fermion fields because the result was a vector bilinear, and it was already known that such bilinears play a big role in electromagnetic interactions. The object G appearing in Eq. (14.2) is a constant.

□ **Exercise 14.1** ♪ *What is the mass dimension of the coupling constant appearing in Eq. (14.2)?*

□ **Exercise 14.2** *The interaction written in Eq. (14.2) is of course not hermitian. We said in Ch. 4 that Lagrangians must be hermitian. It means that we need to add the hermitian conjugate of the interaction shown. Construct the conjugate and name one process which can be described by the conjugate term.*

It was soon realized that there is a host of other phenomena for which the same kind of interaction might be responsible. Of course there are processes like

$$n + \nu \to p + e, \tag{14.3}$$

which can occur from the same terms that are responsible for beta decay. But there can be more. For example, the muon decays into an electron, a muon-type neutrino and an electron-type antineutrino: again a process in which four fermions participate. Neutrinos can scatter off electrons, a process which has the neutrino and the electron in both initial and final states, again making a total of four. Thus, the belief grew that Fermi-type interactions, or *four-fermion interactions*, will be able to explain all phenomena involving weak interactions: one only needs different fields assembled in an interaction like that in Eq. (14.2) to describe different processes.

For the nuclear beta decay itself though, Fermi's original idea proved insufficient. It was clearly seen that in some beta decay processes, the spin of the nucleus changes. Fermi took the vector current between two nuclei, in analogy with vector currents that occur in QED. Since the nuclei are quite non-relativistic in beta decay processes, we can consider the non-relativistic limit of the vector bilinear. Such NR limits have been worked out in Appendix F, and from the result given in Eqs. (F.142) and (F.143), we see that such interactions cannot change the spin of a nucleus. On the other hand, if one had used axial vector bilinears in writing the interaction, it could have induced spin change, since in the NR limit such bilinears reduce to spin, as shown in Eq. (F.144, p 755).

So, beta decay transitions were divided into two classes. In one, the emitted electron and antineutrino, taken together, is in a spin-0 combination. Such transitions are called Fermi transitions. The other kind, where the electron and the antineutrino are in a spin-1 combination, is termed Gamow–Teller transition. In the so-called *allowed approximation* where the physical dimensions of the nucleus is completely neglected with respect to the wavelength of the emitted leptons, the leptons do not have any orbital angular momentum. So, in a Fermi transition, the spin of the nucleus does not change. On the other hand, for a Gamow–Teller transition, the nuclear spin can change: $\Delta J_{nuc} = 0, \pm 1$. So, if nuclear spin changes, it has to be a Gamow–Teller transition. On the other hand, if both initial and final nuclear spins are zero, it has to be a Fermi transition. Other cases, where initial and final nuclear spins are equal but non-zero, can be mixed transitions.

> ☐ **Exercise 14.3** In a typical beta decay process, the total energy carried
>   by the emitted leptons is of the order of a few MeV. Estimate the
>   wavelength of the leptons and show that the allowed approximation
>   is a very good one.

Faced with the task of accommodating spin-changing nuclear transitions, it was realized that Fermi over-specified the rules of the game while writing an interaction term like that in Eq. (14.2). All he needed was two bilinears of the same type so that they can give a Lorentz scalar upon contraction. So, instead of the product of two vector currents which we can succinctly denote by [VV], one could also use [SS], [TT], [AA] or [PP] interactions, using the notation for different kind of bilinears that was introduced in Eq. (4.93, *p 79*). Among these, the axial vector interaction between two nucleons would produce a spin-flip in the non-relativistic limit, as is suggested by the non-relativistic reduction of different fermion bilinears given in §F.3.3 of Appendix F. Thus, one should also include [AA] interactions in order to accommodate Gamow–Teller transitions.

About a quarter of a century after Fermi's 1933 paper, when parity violation was discovered, it was realized that Fermi's original interaction cannot give parity violation. Neither can an [AA] kind of interaction. One must have polar and axial vector currents together in a bilinear in order to have parity violation, as was indicated in the discussion of §6.2.5. Therefore, to accommodate parity violating effects, Fermi's interaction involving four fermionic fields was modified to include [VA] and [AV] interactions as well:

$$\mathscr{L}_{int} = -G\Big[\overline{\psi}_1\gamma^\mu(a + a'\gamma_5)\psi_2\Big]\Big[\overline{\psi}_3\gamma_\mu(b + b'\gamma_5)\psi_4\Big], \qquad (14.4)$$

where we have denoted the fermion fields by ψ_1 through ψ_4 in order to make the argument more general than just beta decay. In a sense this is comforting, because it contains both Fermi type and Gamow–Teller type interactions. If the nucleon current is represented by the first bilinear, Fermi-type interaction can be obtained with $a' = 0$ and Gamow–Teller-type interaction can be obtained with $a = 0$.

But it also makes matters uglier, because instead of just one constant that appears in Eq. (14.2), we now seem to need a large number of constants to describe an interaction. And the question is how to know these constants a, a', b, b' in general. We can always set one of them to have any value we want, say 1, by adjusting the definition of the overall coupling G. But that might require a different coupling G for each process that we want to explain through four-fermion interactions. Alternatively, we can stick to one universal G for all processes, and adjust the quantities a, a', b, b' to account for different processes. Either way, we seem to have four constants in Eq. (14.4) for any process that we want to describe.

If any of the fields is a neutrino field, there is a way of telling the relative magnitudes of the polar and axial vector currents in that bilinear. This comes from an experimental input which we discuss in §14.2. If we discuss a process where no neutrinos are involved, we have no such guideline.

We face another burning question when we look at an interaction like that in Eq. (14.4). Suppose from experiments we know that we need an interaction that will involve the creation operators for the particles which are quanta of the fields ψ_1 and ψ_3 and annihilation operators for the particles of ψ_2 and ψ_4. But even then, who told us that ψ_1 and ψ_2 will be combined into one bilinear and ψ_3, ψ_4 in the other? Why don't we have a bilinear like $[\overline{\psi}_1 \cdots \psi_4]$, multiplying another of the form $[\overline{\psi}_2 \cdots \psi_3]$? The answer to this question will be discussed in §14.3.

14.2 Helicity and chirality

14.2.1 Helicity

The Hamiltonian for a free Dirac particle was given in Eq. (4.42, *p 69*). Using the gamma matrices, we can write it as

$$H = \gamma^0 (\gamma^i p^i + m).$$ (14.5)

From general quantum mechanical argument, we know that operators that commute with the Hamiltonian of a system represent conserved quantities for the system. Let us ask what are the kinematical operators that commute with this Hamiltonian.

The momentum operator is an obvious candidate. Any component of the momentum operator commutes with the Hamiltonian, and therefore the momentum of a free particle is conserved.

Let us explore angular momentum operators as well. A free particle cannot have orbital angular momentum. The spin angular momentum comes from the matrix representation of the Lorentz group, as described in §3.5. For Dirac fields, Eq. (4.73, *p 76*) tells us that the matrices $\frac{1}{2}\sigma_{\mu\nu}$ constitute the representation of the generators. A matrix for which both indices of $\sigma_{\mu\nu}$ are spatial would represent an angular momentum generator, as mentioned explicitly in Eq. (3.58a, *p 54*). Thus, we can identify the generators for spin angular momentum for a Dirac particle as $\frac{1}{2}\Sigma^i$ (for $i = 1, 2, 3$), where

$$\Sigma^i = \frac{1}{2}\epsilon^{ijk}\sigma_{jk}.$$ (14.6)

From this, it is clear that all components of spin angular momentum cannot commute with the Hamiltonian of Eq. (14.5), because the sigma matrices do not commute with the gamma matrices, as shown in Eq. (4.96, p 79).

But this is not the end of the story. First of all, notice that the mass term in the Hamiltonian contains the matrix β, or γ_0, which commutes with all sigma matrices with spatial indices. The first term in the Hamiltonian contains the combination $\gamma_0\gamma_l p_l$, and note that

$$\left[\gamma_0\gamma_l p_l \,,\, p_i\sigma_{jk}\right] = \gamma_0 p_l p_i\left[\gamma_l, \sigma_{jk}\right] = -2i\gamma_0 p_l p_i\left(\delta_{jl}\gamma_k - \delta_{kl}\gamma_j\right). \quad (14.7)$$

The combination on the right vanishes when contracted with ϵ^{ijk}, so that we find that the combination $\mathbf{\Sigma} \cdot \mathbf{p}$ commutes with the first term as well, and therefore with the entire Hamiltonian. Thus, although spin is not conserved for a free Dirac particle, its component along the 3-momentum is. We can thus define the quantity

$$h \equiv \frac{\mathbf{\Sigma} \cdot \mathbf{p}}{\mathrm{p}}, \quad (14.8)$$

which is called *helicity*, and which is twice the spin component along the direction of momentum.

An important property of the helicity operator is that its square is the identity operator:

$$h^2 = 1. \quad (14.9)$$

To prove this, it is better to write the spin 3-vector in terms of the covariant components in the form

$$\Sigma_i = \gamma_0\gamma_i\gamma_5, \quad (14.10)$$

which can be shown to be an equivalent definition by taking each of the matrices γ_i and using the definition of γ_5 from Eq. (4.88, p 78). A more formal proof is given in Appendix F, using Eq. (F.48, p 741). With this form, it is easy to see that

$$\left[\Sigma_i, \Sigma_j\right]_+ = 2\delta_{ij}, \quad (14.11)$$

so that

$$\mathbf{p}^2 h^2 = \Sigma_i\Sigma_j p_i p_j = \frac{1}{2}\left[\Sigma_i, \Sigma_j\right]_+ p_i p_j, \quad (14.12)$$

using the fact that momentum components commute with one another. Now we can use the anticommutator of the Σ matrices and find that the right hand side equals $\mathbf{p}^2$, which proves Eq. (14.9). The consequence of Eq. (14.9) is that helicity eigenvalues can be $+1$ or -1. If we somehow produce a Dirac particle in any one of the eigenstates, it will remain in this state until it is influenced by some interaction.

Neutrinos are produced in many reactions. Experiments were designed to measure neutrino helicity in these reactions. Of course it cannot be measured directly, because neutrinos interact only very feebly with all known kinds of matter. But neutrino helicity can be inferred from the data of momentum and spin of other particles. For example, consider the decay of a charged pion:

$$\pi^+ \rightarrow \mu^+ + \nu_\mu. \tag{14.13}$$

Measurements on the antimuon produced in this decay show that its helicity is always -1, within the limits of experimental error. To understand the implication of this result on the neutrino, let us look at Fig. 14.1, where we consider the decay of a particle into two particles. In the rest frame of the decaying particle, the decay products go back to back. If we now measure the spin of the particle going to the right and find that it is opposite to the momentum, the direction of this spin would be as shown in the figure. If the decaying particle is spinless, the total spin of the two decay products should add up to zero. This determines the direction of the spin vector of the other particle as well, and its spin would also be antiparallel to the momentum, as shown in Fig. 14.1. Thus, the helicity of the neutrino, emitted in pion decay, is also -1. And, although different particles show different helicities for different processes, for neutrinos the result is always the same, and for antineutrinos it is just the opposite, i.e., $+1$.

We have two different u-spinors, or positive energy spinors, as the solution of the Dirac equation. These are degenerate, and we can make two linear combinations of them such that one corresponds to helicity -1 and the other to helicity $+1$. For neutrinos, we see that we obtain only one of these two helicity eigenstates. It therefore suggests that in the interaction term, the neutrino field appears with the projection matrix of this helicity.

Figure 14.1: Helicity of two fermions produced in the decay of a spinless particle. Each long arrow denotes the direction of momentum of a decay product, and the short one above it denotes the direction of the component of spin along the momentum.

The projection matrix is easy to obtain, as outlined in Ex. 14.4. But the problem is that the helicity operator, or the associated projection operators, are not Lorentz invariants. In other words, if a fermion has a positive helicity in one frame, in another frame its helicity can be negative. This is easily seen by considering how a particle, which is moving in one frame with a certain velocity, looks from another frame which is moving at a greater velocity in the same direction. Obviously from this second frame the direction of momentum will seem reversed, and so will helicity. So, helicity projection operators cannot be put into Lagrangians, since any Lagrangian must be Lorentz invariant.

□ **Exercise 14.4** If you have a matrix Q with the property that $Q^2 = 1$,
show that $Q_\pm \equiv \frac{1}{2}(1 \pm Q)$ are projection matrices, and that these two
matrices are mutually orthogonal. In other words, show that

$$Q_+^2 = Q_+, \qquad Q_+^2 = Q_+, \qquad Q_+Q_- = Q_-Q_+ = 0. \qquad (14.14)$$

14.2.2 Chirality

We now consider a different kind of projection. We note that the definition
of γ_5, given in Eq. (4.88, *p 78*), implies that

$$\left(\gamma_5\right)^2 = 1. \qquad (14.15)$$

The projection matrices for negative and positive eigenvalues of γ_5 can be
written down in the manner indicated in Ex. 14.4:

$$\mathbb{L} \equiv \frac{1}{2}(1 - \gamma_5), \qquad \mathbb{R} \equiv \frac{1}{2}(1 + \gamma_5). \qquad (14.16)$$

These are called the *left-chiral* and the *right-chiral* projection matrices respectively. By acting these projections on a fermion field or a spinor, we can
obtain chiral projections of the field or the spinor. For example, $\mathbb{L}\psi \equiv \psi_\mathrm{L}$
would be called the left-chiral or left-handed projection of the field ψ, and
$\mathbb{R}\psi \equiv \psi_\mathrm{R}$ the right-chiral or right-handed projection.

The free Dirac Hamiltonian does not commute with γ_5, because it contains
gamma matrices which do not commute with γ_5. Therefore chirality, unlike
helicity, is not conserved even for a free particle. On the other hand, because
γ_5 commutes with the sigma matrices, the chirality of a field is a Lorentz
invariant concept. To see this explicitly, we take the transformation property
of a fermion field under Lorentz transformations from Eq. (4.73, *p 76*), and
multiply both sides of this equation by the projection matrix $\mathbb{L}$. Using the
commutation property of γ_5 with the sigma matrices, we obtain

$$\psi_\mathrm{L}(x) = \mathbb{L}\psi(x) \longrightarrow \mathbb{L}\psi'(x') = \exp\left(-\frac{i}{4}\omega^{\mu\nu}\sigma_{\mu\nu}\right)\psi_\mathrm{L}(x), \qquad (14.17)$$

which says that under Lorentz transformation, a left-chiral field remains left-
chiral. The same is true for right-chiral fields. Thus, any place in a Lagrangian
where a fermion field operator can be used, we can also use a left-chiral or a
right-chiral field operator in its place: it would not affect the Lorentz invariance of the Lagrangian.

Any expression regarding fermionic fields can be written by using the left-
and right-chiral fields. One merely has to observe that

$$\mathbb{L} + \mathbb{R} = 1, \qquad (14.18)$$

so that any fermion field can be written in the form

$$\psi = (\mathbb{L} + \mathbb{R})\psi = \psi_\mathrm{L} + \psi_\mathrm{R}. \qquad (14.19)$$

For the purpose of making Lorentz invariant combinations, we define the objects

$$\overline{\psi}_L \equiv \left(\psi_L\right)^\dagger \gamma_0\,, \qquad \overline{\psi}_R \equiv \left(\psi_R\right)^\dagger \gamma_0\,, \tag{14.20}$$

and note that

$$\overline{\psi}_L = (L\psi)^\dagger \gamma_0 = \psi^\dagger L \gamma_0 = \psi^\dagger \gamma_0 R = \overline{\psi} R\,, \tag{14.21}$$

and a similar equation obtained by interchanging left with right. To see how one can write fermion bilinears by using chiral projections, we can take a scalar bilinear for example, and use Eq. (14.18) to write

$$\overline{\psi}\psi = \overline{\psi} R \psi + \overline{\psi} L \psi\,. \tag{14.22}$$

Using the projection properties

$$L^2 = L\,, \qquad R^2 = R \tag{14.23}$$

and relations like Eq. (14.21), we can then write

$$\overline{\psi}\psi = \overline{\psi} RR \psi + \overline{\psi} LL \psi = \overline{\psi}_L \psi_R + \overline{\psi}_R \psi_L\,. \tag{14.24}$$

☐ **Exercise 14.5** Show the following relations for fermion bilinears:

$$\overline{\psi}\gamma^\mu \psi = \overline{\psi}_L \gamma^\mu \psi_L + \overline{\psi}_R \gamma^\mu \psi_R\,, \tag{14.25}$$
$$\overline{\psi}\sigma_{\mu\nu}\psi = \overline{\psi}_L \sigma_{\mu\nu} \psi_R + \overline{\psi}_R \sigma_{\mu\nu} \psi_L\,. \tag{14.26}$$

☐ **Exercise 14.6** Show that

$$\overline{\psi}_L \psi_L = \overline{\psi}_R \psi_R = 0\,. \tag{14.27}$$

14.2.3 Connection

We thus have the following dilemma. Chirality is Lorentz invariant but not conserved for a massive particle and cannot be measured easily. On the other hand, helicity is conserved for a free particle, and can be measured, but is not a Lorentz invariant and therefore cannot be called an intrinsic property of the particle.

For massless particles, however, the dilemma does not exist. In fact, we now show that for massless particles, helicity and chirality are the same thing. The massless spinor solutions should have the form

$$u_{\boldsymbol{p}} = \frac{\not{p}}{\sqrt{P}}\xi\,, \qquad v_{\boldsymbol{p}} = -\frac{\not{p}}{\sqrt{P}}\chi\,, \tag{14.28}$$

where ξ and χ are normalized eigenvectors of γ_0 with eigenvalues $+1$ and -1 respectively:

$$\gamma_0 \xi = \xi\,, \qquad \gamma_0 \chi = -\chi\,. \tag{14.29}$$

These solutions are obtained by putting $m = 0$ in Eqs. (4.58) and (4.59), and using the normalization prescribed in Eq. (4.62, p 72).

Take the expression for u_p, for example. Since $\not{p} = \gamma_0 \mathrm{p} - \gamma_i p_i$ for a massless particle, we see that

$$u_p = \sqrt{\mathrm{p}}\left(\gamma_0 - \gamma_i n_i\right)\xi, \tag{14.30}$$

where n is a 3-vector of unit length in the direction of p. Using Eq. (14.29), we can further simplify this expression to the form

$$u_p = \sqrt{\mathrm{p}}\left(1 - \gamma_i n_i\right)\xi. \tag{14.31}$$

Consider the action of the helicity operator, $\Sigma_i n_i$, on this spinor. Using the form of Σ_i given in Eq. (14.10), we obtain

$$\Sigma_i n_i u_p = \sqrt{\mathrm{p}}\gamma_5\gamma_0\gamma_i n_i\left(1 - \gamma_j n_j\right)\xi. \tag{14.32}$$

Since $\gamma_i n_i \gamma_j n_j = \frac{1}{2}[\gamma_i, \gamma_j]_+ n_i n_j = -1$ and γ_0 anticommutes with each γ_i, we obtain

$$\Sigma_i n_i u_p = \sqrt{\mathrm{p}}\gamma_5\gamma_0\left(1 + \gamma_i n_i\right)\xi = \sqrt{\mathrm{p}}\gamma_5\left(1 - \gamma_i n_i\right)\gamma_0\xi$$

$$= \sqrt{\mathrm{p}}\gamma_5\left(1 - \gamma_i n_i\right)\xi = \gamma_5 u_p, \tag{14.33}$$

using Eq. (14.29) on the way. The result shows that the actions of the helicity matrix and the matrix γ_5 produce identical results on a massless u-spinor. The same can be shown for a v-spinor. And since these spinors can be taken as a basis to write any spinor, we have proved the equivalence of γ_5 and $\Sigma_i n_i$ on massless spinors.

□ **Exercise 14.7** *Follow similar steps to show that*

$$\Sigma_i n_i v_p = \gamma_5 v_p \tag{14.34}$$

for massless spinors.

14.2.4   Neutrino helicity

Helicity of neutrinos was measured in various experiments. Of course, neutrinos are extremely evasive, so the measurements were not made directly on neutrinos. If a particle has a two-body decay mode of which one is a neutrino, measurements made on the other decay product bear information about the momentum of the neutrino, and its spin projection along the direction of momentum. Earlier in §14.2.1, we mentioned that such experiments have shown that the neutrino helicity is equal to -1 to a remarkable accuracy. For antineutrinos, the helicity was always found to be positive.

If neutrinos are massless, this would imply that in any process, only the left-chiral neutrinos or right-chiral antineutrinos are produced. This can happen if, in the four-fermi interaction of Eq. (14.4), any neutrino field always appears with the combination of polar and axial vector currents that projects out its left-chiral component. For example, if ψ_4 is a neutrino field, one should have $b' = -b$ so that the combination $(b + b'\gamma_5)$ is proportional to the left-chiral projection matrix given in Eq. (14.16). If both factors have a neutrino field, we can take $a = -a' = 1$ and $b = -b' = 1$, and with this normalization of these parameters, the overall coupling parameter is usually denoted by $G_F/\sqrt{2}$, where G_F is called the *Fermi constant*.

Earlier, in Ex. 6.7 *(p 158)*, we argued that only the polar vector interaction or the axial vector interaction cannot produce any parity violating effect, only a combination of them can. We are now encountering a situation where, for neutrinos, the couplings of the polar and the axial vector interactions have equal magnitude. In this sense, we can say that parity is violated maximally.

It is known now that neutrinos are not really massless. Nevertheless, the masses are so small compared to the energies of the neutrinos with which any experiments can be done that the masses can be ignored as a first approximation. The repercussions of non-zero mass will be discussed in Ch. 22.

14.2.5 Weyl fermions

Fermions that can exist in only one chirality state are called *Weyl fermions*. We can decompose a fermion field operator into chiral projections, as in Eq. (14.19). The Dirac equation, Eq. (4.44, *p 70*), takes the following form in terms of these chiral projections of the field:

$$i\gamma^\mu \partial_\mu \psi_L = m\psi_R \,, \tag{14.35a}$$
$$i\gamma^\mu \partial_\mu \psi_R = m\psi_L \,. \tag{14.35b}$$

This shows that the free evolution of a left-chiral field gives rise to a right-chiral field, and vice versa, if the fermion has a mass. A purely left-chiral or right-chiral field can be obtained only if the fermion is massless. Inspired by the fact that the neutrino helicity is negative, we discuss left-chiral Weyl fields here. To discuss right-chiral Weyl fields, one merely needs to interchange the left- and right-chiral projection operators in what follows in this section.

Left-chiral Weyl fields are defined by the relation

$$L\psi(x) = \psi(x) \,, \tag{14.36}$$

or equivalently by

$$R\psi(x) = 0 \,. \tag{14.37}$$

From the discussion on Lorentz transformation properties of chiral projection of fields given in §14.2.2, it is clear that such conditions are Lorentz invariant. The plane wave expansion of such a field will be given by

$$\psi(x) = \int D^3 p \left(d(\boldsymbol{p}) u_L(\boldsymbol{p}) e^{-ip\cdot x} + \widehat{d}^\dagger(\boldsymbol{p}) v_L(\boldsymbol{p}) e^{+ip\cdot x} \right). \tag{14.38}$$

Table 14.1: Feynman rules for external lines related to a left-chiral Weyl field.

	Feynman rule for	
	incoming	outgoing
fermion	$u_{\rm L}(\boldsymbol{p})$	$\overline{u}_{\rm L}(\boldsymbol{p})$
antifermion	$\overline{v}_{\rm L}(\boldsymbol{p})$	$v_{\rm L}(\boldsymbol{p})$

This looks like the plane wave expansion of a Dirac field given in Eq. (4.65, *p 72*), but there is a difference. For a Dirac field, there is a sum over two different u-spinors and v-spinors. Because of the constraint condition of Eq. (14.36) or Eq. (14.37), we cannot get two different solutions of either the u-spinor or the v-spinor. There can be only one of each type, satisfying the conditions

$$\mathbb{L}u = u, \qquad \mathbb{L}v = v. \tag{14.39}$$

Those solutions are denoted by $u_{\rm L}$ and $v_{\rm L}$, and they appear in Eq. (14.38).

It is interesting to note that although both spinors have the same chirality in Eq. (14.38), the states created by $d^{\dagger}(\boldsymbol{p})$ and $\widehat{d}^{\dagger}(\boldsymbol{p})$ acting on the vacuum have opposite helicities. To prove this statement, let us define the states

$$|F(\boldsymbol{p})\rangle \equiv d^{\dagger}(\boldsymbol{p})\,|0\rangle, \qquad \left|\widehat{F}(\boldsymbol{p})\right\rangle \equiv \widehat{d}^{\dagger}(\boldsymbol{p})\,|0\rangle. \tag{14.40}$$

Denoting CPT by Θ as in Ch. 7, we can write

$$\Theta d(\boldsymbol{p})\Theta^{-1} = \widehat{d}(\boldsymbol{p}). \tag{14.41}$$

Now suppose the particle state is left-handed, i.e.,

$$\boldsymbol{\Sigma}\cdot\boldsymbol{p}\,|F(\boldsymbol{p})\rangle = -\mathrm{p}\,|F(\boldsymbol{p})\rangle. \tag{14.42}$$

Applying CPT transformation on both sides of Eq. (14.42), we obtain

$$\boldsymbol{\Sigma}\cdot\boldsymbol{p}\left|\widehat{F}(\boldsymbol{p})\right\rangle = +\mathrm{p}\left|\widehat{F}(\boldsymbol{p})\right\rangle, \tag{14.43}$$

using the fact that under CPT, the helicity operator is odd, since spin changes sign and momentum remain invariant. Thus the left-handed neutrino and right-handed antineutrino are contained in the same field operator. To the list of Feynman rules for external lines given in Table 4.1 *(p 91)*, we can add more for Weyl fermions, which are given in Table 14.1.

14.3   Fierz transformations

In §14.1, we raised the question of how four fermionic fields should be paired into two bilinears. Within the realm of the four-fermion interaction theories, the answer to this question cannot be decided. However, it is also true that the question is somewhat irrelevant in such theories. No matter how they are grouped, it is possible to transform them to a different grouping. Such regroupings are called *Fierz transformations*.

14.3.1   Scalar combinations

In order not to be bothered by the change of sign obtained by commuting two fermionic field operators, we carry out the derivation using spinors whose components are numbers and therefore commute. Suppose w denotes either a u-spinor or a v-spinor, and we define quadrilinears of the form

$$
\begin{aligned}
e_S(1234) &= \left[\overline{w}_1 w_2\right]\left[\overline{w}_3 w_4\right], \\
e_V(1234) &= \left[\overline{w}_1 \gamma^\mu w_2\right]\left[\overline{w}_3 \gamma_\mu w_4\right], \\
e_T(1234) &= \left[\overline{w}_1 \sigma^{\mu\nu} w_2\right]\left[\overline{w}_3 \sigma_{\mu\nu} w_4\right], \\
e_A(1234) &= \left[\overline{w}_1 \gamma^\mu \gamma_5 w_2\right]\left[\overline{w}_3 \gamma_\mu \gamma_5 w_4\right], \\
e_P(1234) &= \left[\overline{w}_1 \gamma_5 w_2\right]\left[\overline{w}_3 \gamma_5 w_4\right],
\end{aligned}
\tag{14.44}
$$

where the four spinors can correspond to four different momenta, and can even be the solutions of the free Dirac equations of four different particles if one wants to be completely general. We can also consider similar quadrilinears where w_1 and w_4 grouped together, and call them $e_I(1432)$ for $I = S, V, T, A, P$. The Fierz rearrangement theorem then says that it is possible to find numerical co-efficients F_{IJ} which satisfy the equation

$$
e_I(1234) = \sum_J F_{IJ}\, e_J(1432). \tag{14.45}
$$

According to our summation convention explained in §2.1, automatic summation convention does not apply on the indices I or J which stand for the five different groupings that appear in Eq. (14.44). The rest of this exercise would consist of the evaluation of the co-efficients F_{IJ}.

We begin this exercise by rewriting the expressions of Eq. (14.44) in the manner

$$
e_I(1234) = n_I^2 \left[\overline{w}_1 \Gamma_{Ir} w_2\right]\left[\overline{w}_3 \Gamma_I^r w_4\right], \tag{14.46}
$$

where the matrices Γ_{Ir} constitute a complete basis for expressing all 4×4 matrices. They have been written with two indices, one for the categories that appear in Eq. (14.44), and the other for the different matrices that occur in each category. Note that although automatic summation convention is not implied on repeated uppercase indices, as noted earlier, it is implied on repeated lowercase indices, since bilinears with the same value of I and different values of r are related by Lorentz transformations. We take these basis matrices in such a way that they satisfy the orthogonality condition

$$\text{Tr}\left(\Gamma_I^r \Gamma_{Jr'}\right) = 4\delta_{IJ}\delta_{r'}^r. \tag{14.47}$$

This means that we can use the matrices shown in Eq. (4.93, *p 79*), with the exception that for Γ_A^r, we should use put in an extra factor of i, i.e., should use $i\gamma_\mu\gamma_5$. The quantity n_I appearing in Eq. (14.46) are then found to be

$$\begin{array}{c|ccccc}
I & S & V & T & A & P \\
\hline
n_I & 1 & 1 & \sqrt{2} & -i & 1
\end{array}. \tag{14.48}$$

Note that $n_T^2 = 2$ because e_T, as defined in Eq. (14.44), contains unrestricted sums over the indices μ, ν, whereas the basis matrices contain only the sigma matrices with $\mu < \nu$.

Using the property mentioned in Eq. (14.47), any 4×4 matrix M can be written as

$$M = \frac{1}{4}\sum_I \Gamma_{Ir}\ \text{Tr}\left(\Gamma_I^r M\right), \tag{14.49}$$

which implies the relation

$$\frac{1}{4}\sum_I \left(\Gamma_{Ir}\right)_{ab}\left(\Gamma_I^r\right)_{cd} = \delta_{ad}\delta_{bc}. \tag{14.50}$$

Multiplying this by $(\Gamma_J^q)_{c'c}(\Gamma_{Jq})_{a'a}$ and summing over the matrix rows or columns whose indices have been repeated, we obtain

$$\frac{1}{4}\sum_I \left(\Gamma_{Jq}\Gamma_{Ir}\right)_{a'b}\left(\Gamma_J^q\Gamma_I^r\right)_{c'd} = (\Gamma_{Jq})_{a'd}(\Gamma_J^q)_{c'b}. \tag{14.51}$$

Next, one should observe that the basis matrices have been chosen in such a way that the product of any two of them always gives another basis matrix. In other words, any two matrices Γ_I^q and Γ_J^r satisfy a relation of the sort

$$\Gamma_I^q\ \Gamma_J^r = (\text{constant}) \times \Gamma_K^s, \tag{14.52}$$

so that Eq. (14.51) can be rewritten in the form

$$(\Gamma_{Iq})_{ad}(\Gamma_I^q)_{cb} = \sum_K C_{IK}\left(\Gamma_{Ks}\right)_{ab}\left(\Gamma_K^s\right)_{cd}. \tag{14.53}$$

To obtain the co-efficients C_{IJ}, we multiply both sides by $\left(\Gamma_{Jr}\right)_{dc}\left(\Gamma_J^r\right)_{ba}$ and use Eq. (14.47) to obtain

$$\left(\Gamma_{Iq}\right)_{ad}\left(\Gamma_I^q\right)_{cb}\left(\Gamma_{Jr}\right)_{dc}\left(\Gamma_J^r\right)_{ba} = 16\sum_K C_{IK}\delta_{JK}\delta_s^r\delta_r^s. \tag{14.54}$$

Recall that the sum over r is implied in our notation, so the right hand side gives $16N_J C_{IJ}$, where N_J is the number of basis matrices in the J-th category, i.e., $N_S = 1$, $N_V = 4$ etc. Thus we obtain

$$C_{IJ} = \frac{1}{16N_J} \operatorname{Tr} \left(\Gamma_{Iq} \Gamma_{Jr} \Gamma_I^q \Gamma_J^r \right). \tag{14.55}$$

Multiplying Eq. (14.53) by $(\overline{w}_1)_a (w_2)_b (\overline{w}_3)_c (w_4)_d$ and by using the definition of the quadrilinears provided in Eq. (14.46), we can easily identify the coefficients F_{IJ} defined in Eq. (14.45):

$$F_{IJ} = \frac{n_I^2}{n_J^2} C_{IJ}. \tag{14.56}$$

The result can be summarized in a matrix form:

$$\boldsymbol{F} = \frac{1}{4} \begin{pmatrix} 1 & 1 & \frac{1}{2} & -1 & 1 \\ 4 & -2 & 0 & -2 & -4 \\ 12 & 0 & -2 & 0 & 12 \\ -4 & -2 & 0 & -2 & 4 \\ 1 & -1 & \frac{1}{2} & 1 & 1 \end{pmatrix}. \tag{14.57}$$

14.3.2 Pseudoscalar combinations

The scalar combinations introduced in Eq. (14.44) are clearly not enough for us, because interactions such as in Eq. (14.4) contain also combinations where one bilinear has a polar vector current whereas the other has an axial vector current. To discuss such combinations, we introduce the notation

$$e'_I(1234) = n_I n_{\widetilde{I}} \left[\overline{w}_1 \Gamma_{Ir} w_2 \right] \left[\overline{w}_3 \Gamma_I^r \gamma_5 w_4 \right], \tag{14.58}$$

where $\widetilde{I}$ denotes the opposite-parity partner of the index I, i.e., $\widetilde{I} = P, A, T, V, S$ respectively for $I = S, V, T, A, P$. Some obvious relations follow from this definition, such as

$$e'_S(1234) = e'_P(3412), \qquad e'_V(1234) = e'_A(3412), \tag{14.59}$$

as well as

$$e'_T(1234) = e'_T(3412), \tag{14.60}$$

which follows by using the property

$$\sigma^{\lambda\rho} \gamma_5 = -\frac{i}{2} \varepsilon^{\lambda\rho\alpha\beta} \sigma_{\alpha\beta}, \tag{14.61}$$

whose proof has been discussed in Ex. F.6 *(p 741)*. Such relations can be summarized by writing

$$e'_I(1234) = \sum_J X_{IJ} \, e'_J(3412), \tag{14.62}$$

where

$$X = \begin{pmatrix} 0 & 0 & 0 & 0 & 1 \\ 0 & 0 & 0 & 1 & 0 \\ 0 & 0 & 1 & 0 & 0 \\ 0 & 1 & 0 & 0 & 0 \\ 1 & 0 & 0 & 0 & 0 \end{pmatrix}. \tag{14.63}$$

It should also be noted that we can write

$$e'_I(1234) = e_I(1234'), \tag{14.64}$$

where, in writing the quadrilinear on the right hand side, we use

$$w'_4 = \gamma_5 w_4 \tag{14.65}$$

instead of the spinor w_4. With the help of this trick, we can now write

$$e'_I(1234) = \sum_J X_{IJ} e'_I(3412) = \sum_J X_{IJ} e_I(3412')$$

$$= \sum_{J,K} X_{IJ} F_{JK} e_K(32'14). \tag{14.66}$$

But the order of the two bilinears is unimportant for the definition of e_I, i.e.,

$$e_I(1234) = e_I(3412). \tag{14.67}$$

Using this, we obtain

$$e'_I(1234) = \sum_J (XF)_{IJ} e_J(1432') = \sum_J (XF)_{IJ} e'_J(1432). \tag{14.68}$$

The relevant matrix in the transformation for the pseudoscalar quadrilinears is therefore

$$XF = \frac{1}{4} \begin{pmatrix} 1 & -1 & \frac{1}{2} & 1 & 1 \\ -4 & -2 & 0 & -2 & 4 \\ 12 & 0 & -2 & 0 & 12 \\ 4 & -2 & 0 & -2 & -4 \\ 1 & 1 & \frac{1}{2} & -1 & 1 \end{pmatrix}. \tag{14.69}$$

14.3.3   Invariant combinations with fields

If we use field operators instead of spinors, derivations with the matrices remain unaffected. The only difference is that, in the two orderings discussed, there would be two field operators whose position need to be interchanged, and that would produce an extra minus sign. Thus, if we introduce the notations

$$E_S(1234) = \left[\overline{\psi}_1 \psi_2\right]\left[\overline{\psi}_3 \psi_4\right] \tag{14.70}$$

and similar ones for E_V, E_T etc, the Fierz transformation rules would read

$$E_I(1234) = -\sum_J F_{IJ} E_J(1432),$$

$$E'_I(1234) = -\sum_J (XF)_{IJ} E'_J(1432). \tag{14.71}$$

Earlier we mentioned that, strictly within the framework of the four-fermion interaction, it is not possible to tell which pairs of fermionic fields should appear in one bilinear. One might be tempted to think that Fierz re-arrangements hold a key to this mystery. If we can identify any quadrilinear that stays invariant under Fierz transformations, the problem disappears, and we reach a theoretically satisfying position. Let us then try to see whether such combinations exist.

We started with polar and axial vector bilinears in §14.1, so let us first check the possibility of Fierz invariants from them. Using Eq. (14.57), we find

$$E_V(1234) + E_A(1234) = E_V(1432) + E_A(1432). \tag{14.72}$$

On the other hand, from Eq. (14.69), we obtain

$$E'_V(1234) + E'_A(1234) = E'_V(1432) + E'_A(1432). \tag{14.73}$$

Combining, we obtain that

$$(E_V + E_A) \pm (E'_V + E'_A) \tag{14.74}$$

are invariant under Fierz transformation. Such combinations would corre-spond to interactions of the form

$$\left[\overline{\psi}_1 \gamma^\mu (1 \pm \gamma_5)\psi_2\right]\left[\overline{\psi}_3 \gamma^\mu (1 \pm \gamma_5)\psi_4\right]. \tag{14.75}$$

Since each bilinear contains a polar current plus or minus an axial current, such interactions are often called $V + A$ or $V - A$ interactions.

Earlier, we said that measurements of neutrino helicity indicate that we should have $V - A$ currents in the bilinears which involve a neutrino field. We now see that four-fermion interactions consisting of a product of $V - A$ interac-tions are invariant under Fierz transformations. These two statements match beautifully, and we might be tempted to think that four-fermion interactions always contain $V - A$ currents for all fermions.

But such a conclusion need not be very reliable. The reason is that, for reasons that will be discussed in §14.8, Fermi theory cannot be taken as a fundamental theory. It can work only if it is the low-energy limit of some deeper theory. That theory may not have any quadrilinear at all — quadri-linears might appear when the low-energy limit is taken. If that is so, the requirement of invariance under Fierz transformations becomes meaningless. Accordingly, the bilinears might involve unequal combinations of polar and axial currents, the combination being determined by the basic requirements of the deeper theory. We will see these statements vindicated when we discuss the standard electroweak model in Ch. 16.

☐ **Exercise 14.8** Show that the combination $E_S + E_P - \frac{1}{2}E_T$ remains invariant under Fierz transformations.

14.4   Elastic neutrino–electron scattering

Not surprisingly, the processes involving neutrinos which have been most extensively studied experimentally are their scattering with electrons. Let us denote the process as follows, indicating our symbols for the 4-momenta of the different particles involved:

$$\nu(k) + e(p) \rightarrow \nu(k') + e(p') \,. \tag{14.76}$$

We want to keep the discussion general in the sense that we do not assume, at the outset, which kind of neutrinos we are talking about. But we do assume that it is the same kind of neutrino in the initial and the final state in order that the process is elastic.

How do we group the four field operators to write the four-fermi interaction? Should we group an electron field with a neutrino field, or the two electron field operators together in a bilinear? If we choose the latter option, i.e., $[ee][\nu\nu]$ combination, then we do not know the co-efficients of the polar and the axial vector currents in the electron bilinear. On the other hand, if we start with one neutrino field operator for each bilinear, i.e., the $[e\nu][\nu e]$ pairing, then each bilinear should be $V - A$ type according to the neutrino helicity argument. But we have seen that such combinations are Fierz invariant, i.e., Fierz transformations guarantee that

$$\left[\overline{\psi}_{(e)}\gamma^\mu(1-\gamma_5)\psi_{(\nu)}\right]\left[\overline{\psi}_{(\nu)}\gamma_\mu(1-\gamma_5)\psi_{(e)}\right]$$
$$= \left[\overline{\psi}_{(e)}\gamma^\mu(1-\gamma_5)\psi_{(e)}\right]\left[\overline{\psi}_{(\nu)}\gamma_\mu(1-\gamma_5)\psi_{(\nu)}\right]. \tag{14.77}$$

The lesson is that, we can take both possibilities into account by writing the four-fermi interaction in the form

$$\mathscr{L}_{\text{int}} = -\frac{G_F}{\sqrt{2}}\left[\overline{\psi}_{(e)}\gamma^\mu(c_V - c_A\gamma_5)\psi_{(e)}\right]\left[\overline{\psi}_{(\nu)}\gamma_\mu(1-\gamma_5)\psi_{(\nu)}\right]. \tag{14.78}$$

Note that the combination occurring in the neutrino bilinear involves left-chiral projections only. The overall constant is taken to be the Fermi constant, and the co-efficients in the electron bilinear have been left unspecified: the parts involving polar and axial currents have co-efficients c_V and c_A respectively.

The amplitude of the process is easily written from the interaction Lagrangian, and it is

$$\mathscr{M} = -\frac{G_F}{\sqrt{2}}\left[\overline{u}_{p'}\gamma^\mu(c_V - c_A\gamma_5)u_p\right]\left[\overline{u}_{k'}\gamma_\mu(1-\gamma_5)u_k\right]. \tag{14.79}$$

In writing this, we have not indicated explicitly which spinor belongs to which kind of field: the momentum corresponding to the spinor should provide answers to such questions.

We now want to square the amplitude, averaging over initial spins and summing over final ones. For neutrinos, the averaging or the summing is trivial, since we have only left-chiral neutrinos. For electrons, the averaging over initial states would mean summing and dividing by 2. Thus we obtain

$$\overline{|\mathscr{M}|^2} = \frac{1}{2} \sum_{\text{spins}} |\mathscr{M}|^2 = \frac{G_F^2}{4} E^{\mu\nu} N_{\mu\nu}, \qquad (14.80)$$

where

$$E^{\mu\nu} = \text{Tr} \left[(\not{p}' + m)\gamma^\mu (c_V - c_A \gamma_5)(\not{p} + m)\gamma^\nu (c_V - c_A \gamma_5) \right],$$
$$N_{\mu\nu} = \text{Tr} \left[\not{k}' \gamma_\mu (1 - \gamma_5) \not{k} \gamma_\nu (1 - \gamma_5) \right]. \qquad (14.81)$$

Note the part that contains the neutrino momenta. The factor $(1 - \gamma_5)$ commutes with the combination $\not{k}\gamma_\nu$, so that we can write

$$N_{\mu\nu} = \text{Tr} \left[\not{k}' \gamma_\mu \not{k} \gamma_\nu (1 - \gamma_5)^2 \right] = 2\,\text{Tr} \left[\not{k}' \gamma_\mu \not{k} \gamma_\nu (1 - \gamma_5) \right], \qquad (14.82)$$

since $(1 - \gamma_5)^2 = 2(1 - \gamma_5)$, which follows easily from Eq. (14.15). The traces can be evaluated using the formulas appearing in §F.1.4 and we end up with the result

$$N_{\mu\nu} = 8 \left(k'_\mu k_\nu + k_\mu k'_\nu - k \cdot k' g_{\mu\nu} - i\varepsilon_{\alpha\mu\beta\nu} k'^\alpha k^\beta \right). \qquad (14.83)$$

The evaluation of $E^{\mu\nu}$, though a little more tedious, is no more complicated. It produces the result

$$E^{\mu\nu} = 4(c_V^2 + c_A^2)(p^\mu p'^\nu + p^\nu p'^\mu - p \cdot p' g^{\mu\nu}) + 4m^2(c_V^2 - c_A^2)g^{\mu\nu}$$
$$-8ic_A c_V \varepsilon^{\lambda\mu\rho\nu} p'_\lambda p_\rho. \qquad (14.84)$$

When we put these expressions back into Eq. (14.80), we note that in the expressions of both $E^{\mu\nu}$ and $N_{\mu\nu}$, there is one part which is symmetric in the indices and one part which is antisymmetric. While taking contractions, we will have one contribution that would involve the symmetric terms from both factors, and another that would involve the antisymmetric parts from both. Thus,

$$\overline{|\mathscr{M}|^2} = 8G_F^2 \left[\left\{ (c_V^2 + c_A^2)(p^\mu p'^\nu + p^\nu p'^\mu - p \cdot p' g^{\mu\nu}) \right. \right.$$
$$\left. + m^2(c_V^2 - c_A^2)g^{\mu\nu} \right\} (k'_\mu k_\nu + k_\mu k'_\nu - k \cdot k' g_{\mu\nu})$$
$$\left. - 2c_A c_V \varepsilon^{\lambda\mu\rho\nu} \varepsilon_{\alpha\mu\beta\nu} k'^\alpha k^\beta p'^\lambda p^\rho \right]$$

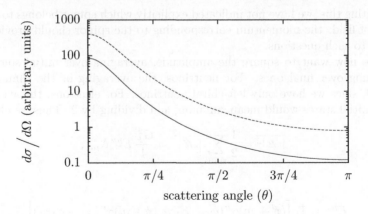

Figure 14.2: Differential cross-section for neutrino–electron scattering as a function of the scattering angle θ for $\omega = 10m$. The overall scale is arbitrary. Two sets of values of the constants c_V and c_A have been taken: one corresponding to Eq. (16.56, *p 476*) and the other to Eq. (16.61, *p 478*). The importance of these values has been described in Ch. 16.

$$= 16G_F^2 \left[(c_V + c_A)^2 (k \cdot p)^2 + (c_V - c_A)^2 (k' \cdot p)^2 \right.$$
$$\left. -2m^2(c_V^2 - c_A^2) k \cdot k' \right], \tag{14.85}$$

using the contraction formula of Eq. (D.11, *p 729*) and the kinematical relations

$$k \cdot p = k' \cdot p', \qquad k \cdot p' = k' \cdot p, \tag{14.86}$$

that follow from 4-momentum conservation.

Neutrino–electron scattering experiments are usually done in fixed target frames where the initial electron is at rest. We can use the result given in Eq. (4.225, *p 109*) to write the differential cross-section for this problem, which is

$$\frac{d\sigma}{d\Omega} = \frac{G_F^2 m^2}{4\pi^2} \frac{1}{(m + \omega - \omega \cos \theta)^2}$$
$$\times \left[(c_V + c_A)^2 \omega^2 + (c_V - c_A)^2 \omega'^2 - (c_V^2 - c_A^2)\omega\omega'(1 - \cos \theta) \right], \tag{14.87}$$

where m is the mass of the electron, and the initial and final energies of the neutrino, ω and ω', are related by

$$\omega' = \frac{m\omega}{m + \omega - \omega \cos \theta}, \tag{14.88}$$

as expounded in Eq. (2.94, p 34).

One characteristic of this differential cross-section is that it is very strongly peaked in the forward direction. In Fig. 14.2, we show a schematic plot of the differential cross-section as a function of the scattering angle θ.

Within the purview of Fermi theory, the constants c_V and c_A that appear in the interaction Lagrangian are completely arbitrary. However, we will see later in Ch. 16 that their values are related to some other properties of the interacting particles.

14.5 Inelastic neutrino–electron scattering

We can also discuss inelastic scattering processes involving neutrinos. For example, consider the process

$$\nu_\mu(k) + e(p) \to \mu(p') + \nu_e(k') . \tag{14.89}$$

This process is very important because it provides an indirect way of detecting muon-neutrinos by detecting the muon produced in the final state. Of course, this is an endergonic reaction, and therefore the muon-neutrino must have sufficiently high energy in order that this reaction takes place in the FT frame where the electrons are kept at rest.

□ **Exercise 14.9** *Find the threshold energy of the muon-neutrinos that is necessary for this reaction to take place in the rest frame of the electron. Use the masses of the electron and the muon from Table B.2 (p 720), and assume the neutrinos to be massless.*

In the four-fermion interaction, the muon will naturally tie up with the muon-neutrino, and the electron with the electron-neutrino. The interaction will therefore have the form

$$\mathscr{L}_{\text{int}} = -\frac{G_F}{\sqrt{2}} \left[\overline{\psi}_{(\mu)} \gamma^\lambda (1 - \gamma_5) \psi_{(\nu_\mu)} \right] \left[\overline{\psi}_{(\nu_e)} \gamma_\lambda (1 - \gamma_5) \psi_{(e)} \right] . \tag{14.90}$$

Since each bilinear contains a neutrino field, we use the combination $(1 - \gamma_5)$, as discussed in §14.2.4. The amplitude, with the notations for the momenta given in Eq. (14.89), is given by

$$\mathscr{M} = -\frac{G_F}{\sqrt{2}} \left[\overline{u}_{p'} \gamma^\lambda (1 - \gamma_5) u_k \right] \left[\overline{u}_{k'} \gamma_\lambda (1 - \gamma_5) u_p \right] . \tag{14.91}$$

This gives

$$\overline{|\mathscr{M}|^2} = \frac{1}{2} \sum_{\text{spins}} \left| \mathscr{M} \right|^2 = \frac{G_F^2}{4} E_{\lambda\rho} M^{\lambda\rho} , \tag{14.92}$$

where $E_{\lambda\rho}$ comes from the bilinear containing the spinors for the electron and the ν_e fields, and $M^{\lambda\rho}$ from the bilinear containing the muon and the ν_μ.

Thus,

$$E_{\lambda\rho} = \left[\bar{u}_{k'}\gamma_\lambda(1-\gamma_5)u_p\right]\left[\bar{u}_{k'}\gamma_\rho(1-\gamma_5)u_p\right]^*$$
$$= \text{Tr}\left[\not{k}'\gamma_\lambda(1-\gamma_5)(\not{p}+m_e)\gamma_\rho(1-\gamma_5)\right]. \qquad (14.93)$$

The term containing the electron mass is traceless because it contains an odd number of Dirac matrices. The rest looks very similar to the expression of $N_{\mu\nu}$ that appears in Eq. (14.81), and we can borrow the result from what we did there:

$$E_{\lambda\rho} = 8\left(k'_\lambda p_\rho + p_\lambda k'_\rho - p\cdot k' g_{\lambda\rho} - i\varepsilon_{\alpha\lambda\beta\rho}k'^\alpha p^\beta\right). \qquad (14.94)$$

The traces involved in $M^{\lambda\rho}$ are similar, and the result is

$$M^{\lambda\rho} = 8\left(p'^\lambda k^\rho + k^\lambda p'^\rho - k\cdot p' g^{\lambda\rho} - i\varepsilon^{\kappa\lambda\tau\rho}p'_\kappa k_\tau\right). \qquad (14.95)$$

Therefore,

$$E_{\lambda\rho}M^{\lambda\rho} = 64\left[\left(k'_\lambda p_\rho + p_\lambda k'_\rho - p\cdot k' g_{\lambda\rho}\right)\left(p'^\lambda k^\rho + k^\lambda p'^\rho - k\cdot p' g^{\lambda\rho}\right)\right.$$
$$\left. -\left(\varepsilon_{\alpha\lambda\beta\rho}k'^\alpha p^\beta\right)\left(\varepsilon^{\kappa\lambda\tau\rho}p'_\kappa k_\tau\right)\right]$$
$$= 64\times 2\left[\left(k\cdot k'\, p\cdot p' + k\cdot p\, k'\cdot p'\right)\right.$$
$$\left. -\left(k\cdot k'\, p\cdot p' - k\cdot p\, k'\cdot p'\right)\right]. \qquad (14.96)$$

This gives

$$\overline{|\mathscr{M}|^2} = 64G_F^2 k\cdot p\, k'\cdot p'. \qquad (14.97)$$

Note that relations of the type given in Eq. (14.86) are not valid here since the initial state particles and the final state particles have different masses. If we analyze the process in the rest frame of the electron, we obtain

$$k\cdot p = m_e\omega, \qquad k'\cdot p' = \left(E' - \sqrt{E'^2 - m_\mu^2}\cos\theta\right)\omega', \qquad (14.98)$$

where ω and ω' are the energies of the initial ν_μ and final ν_e, E' is the energy of the muon, and θ is the angle between k' and p'. The scattering cross-section can be obtained by putting these expressions into Eq. (4.224, *p 109*).

14.6   Muon and tau decay

The products obtained in muon decay have been described in §14.1:

$$\mu^-(p) \rightarrow e^-(k) + \nu_\mu(q_1) + \hat{\nu}_e(q_2), \qquad (14.99)$$

where, as usual, after each particle we have put the notations for its momenta in parentheses. Since the muon goes naturally with its neutrino, and so does the electron, we write the interaction Lagrangian as

$$\mathscr{L}_{\text{int}} = -\frac{G_F}{\sqrt{2}} \left[\overline{\psi}_{(\nu_\mu)} \gamma^\lambda (1 - \gamma_5) \psi_{(\mu)} \right] \left[\overline{\psi}_{(e)} \gamma_\lambda (1 - \gamma_5) \psi_{(\nu_e)} \right]. \quad (14.100)$$

This is just the hermitian conjugate of the interaction term presented in Eq. (14.90). The amplitude of the decay process is then given by

$$\mathscr{M} = -\frac{G_F}{\sqrt{2}} \left[\overline{u}_{q_1} \gamma^\lambda (1 - \gamma_5) u_p \right] \left[\overline{u}_k \gamma_\lambda (1 - \gamma_5) v_{q_2} \right], \quad (14.101)$$

where, as in the previous sections, we have used the momenta to indicate the field corresponding to each spinor. Squaring this, averaging over the initial muon spins and summing over all final spins, we obtain

$$\overline{|\mathscr{M}|^2} = \frac{1}{2} \sum_{\text{spins}} \left| \mathscr{M} \right|^2. \quad (14.102)$$

The evaluation of this quantity is very much similar to what has been done in the previous sections, so we omit the details. The result is

$$\overline{|\mathscr{M}|^2} = 64 G_F^2 \, p \cdot q_1 \, k \cdot q_2. \quad (14.103)$$

Using this expression in the general formula for calculating decay rates, Eq. (4.156, *p 95*), we can write, in the rest frame of the decaying particle,

$$\Gamma = \frac{G_F^2}{\pi^5 m_\mu} \int \frac{d^3k}{2k_0} \int \frac{d^3q_1}{2q_{10}} \int \frac{d^3q_2}{2q_{20}} \, \delta^4(p - k - q_1 - q_2) p \cdot q_1 \, k \cdot q_2. \quad (14.104)$$

The integration is much more complicated than those in earlier examples in this chapter because this process contains three particles in the final state. Let us tackle it first by performing the integrations over q_1 and q_2. Notice that the involvement of these momenta can be summarized in the form

$$I^{\lambda\rho}(q) \equiv \int \frac{d^3q_1}{2q_{10}} \int \frac{d^3q_2}{2q_{20}} \, \delta^4(q - q_1 - q_2) q_1^\lambda q_2^\rho, \quad (14.105)$$

where

$$\Gamma = \frac{G_F^2}{\pi^5 m_\mu} \int \frac{d^3k}{2k_0} I^{\lambda\rho}(p - k) p_\lambda k_\rho. \quad (14.106)$$

As argued earlier in Eq. (4.158, *p 96*), the integration measures in the form $\int d^3p/2p_0$ is Lorentz invariant for any on-shell 4-momentum p. Thus, the expression for $I^{\lambda\rho}$ clearly shows that it is a rank-2 tensor. It can depend only on the 4-vector q that appears in its definition. Thus, we write the most general rank-2 tensor that we can write as the result, which is

$$I^{\lambda\rho}(q) = Aq^2 g^{\lambda\rho} + Bq^\lambda q^\rho , \qquad (14.107)$$

where A and B must be invariant functions of q. To evaluate these functions, we write the results obtained by contracting Eq. (14.107), once by $g_{\lambda\rho}$ and again by $q^\lambda q^\rho$. The results are

$$(4A + B)q^2 = \int \frac{d^3 q_1}{2q_{10}} \int \frac{d^3 q_2}{2q_{20}} \delta^4(q - q_1 - q_2)q_1 \cdot q_2 ,$$

$$(A + B)q^4 = \int \frac{d^3 q_1}{2q_{10}} \int \frac{d^3 q_2}{2q_{20}} \delta^4(q - q_1 - q_2)\left(q_1 \cdot q_2\right)^2 . \qquad (14.108)$$

It should be noted that we have used a crucial property of neutrinos in writing the last factor in the second integral. From the contraction proposed before the equation, the factor comes out to be $q \cdot q_1 \, q \cdot q_2$. But, because of the delta function in the integrand, we can replace q by $q_1 + q_2$. After that, we use $q_1^2 = q_2^2 = 0$, since we take the neutrinos to be massless.

The last set of integrals is Lorentz invariant, and can therefore depend only on q^2. Note that the delta function implies $q^\mu = q_1^\mu + q_2^\mu$, and by squaring it we obtain

$$q^2 = 2q_1 \cdot q_2 , \qquad (14.109)$$

using the masslessness of the neutrinos once again. Thus, we can replace $q_1 \cdot q_2$ in the previous integrals by $\frac{1}{2}q^2$, bring it outside the integration sign and obtain

$$4A + B = \frac{1}{2} \int \frac{d^3 q_1}{2q_{10}} \int \frac{d^3 q_2}{2q_{20}} \delta^4(q - q_1 - q_2) ,$$

$$A + B = \frac{1}{4} \int \frac{d^3 q_1}{2q_{10}} \int \frac{d^3 q_2}{2q_{20}} \delta^4(q - q_1 - q_2) . \qquad (14.110)$$

The integrals on both lines are the same, and they are also identical to the integral that we had encountered in Eq. (4.185, p 101). Using the result from there and solving for A and B, we obtain

$$I^{\lambda\rho}(q) = \frac{\pi}{24}\left(q^2 g^{\lambda\rho} + 2q^\lambda q^\rho\right) . \qquad (14.111)$$

Putting this back into Eq. (14.106), we obtain

$$\Gamma = \frac{G_F^2}{24\pi^4 m_\mu} \int \frac{d^3 k}{2k_0}\left(q^2 p \cdot k + 2q \cdot p \, q \cdot k\right) , \qquad (14.112)$$

where now $q = p - k$. In the rest frame of the muon, denoting the electron energy by E_e, we obtain

$$p \cdot k = m_\mu E_e ,$$
$$q \cdot k = m_\mu E_e - m_e^2 ,$$
$$q \cdot p = m_\mu(m_\mu - E_e) ,$$
$$q^2 = m_\mu^2 + m_e^2 - 2m_\mu E_e . \qquad (14.113)$$

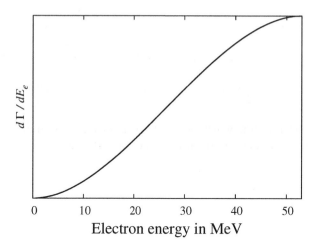

Figure 14.3: Electron energy spectrum in muon decay.

None of these things depends on the direction of $\mathbf{k}$, so we can trivially perform the angular integrations and obtain

$$\Gamma = \frac{G_F^2}{12\pi^3} \int dE_e \; \mathbf{k} \Big[(m_\mu^2 + m_e^2 - 2m_\mu E_e) E_e$$
$$+ 2(m_\mu - E_e)(m_\mu E_e - m_e^2) \Big] . \tag{14.114}$$

This can be written as a differential decay rate with respect to the electron energy E_e:

$$\frac{d\Gamma}{dE_e} = \frac{G_F^2}{12\pi^3} \sqrt{E_e^2 - m_e^2} \Big[(m_\mu^2 + m_e^2 - 2m_\mu E_e) E_e$$
$$+ 2(m_\mu - E_e)(m_\mu E_e - m_e^2) \Big] . \tag{14.115}$$

The function on the right hand side has been plotted in Fig. 14.3.

In order to obtain the total decay rate, we need to integrate over E_e. With the expression shown above, this integration is not easy. However, since the muon, with a mass of 106 MeV, is more than 200 times heavier than the electron, the electron mass can be neglected as a first approximation. This gives

$$\frac{d\Gamma}{dE_e} = \frac{G_F^2 m_\mu}{12\pi^3} E_e^2 \Big(3m_\mu - 4E_e \Big) . \tag{14.116}$$

The lower limit of integration is zero, which corresponds to the situation where the two neutrinos are emitted back to back, and carry away all the energy.

The maximum possible value of E_e corresponds to the situation when the two neutrinos are emitted in the same direction, and the electron balances their momenta by going in the opposite direction. The electron takes half the total available energy in this case. Thus the total decay rate is given by

$$\Gamma = \frac{G_F^2 m_\mu}{12\pi^3} \int_0^{m_\mu/2} dE_e \, E_e^2 \left(3m_\mu - 4E_e \right) = \frac{G_F^2 m_\mu^5}{192\pi^3}. \qquad (14.117)$$

This is the relation which is used to ascertain the value of the Fermi constant. Since the measured value of the muon lifetime is

$$\Gamma^{-1} = 2.2 \times 10^{-6}\,\text{s} \qquad (14.118)$$

and the mass is $106\,\text{MeV}$, one obtains

$$G_F = 1.166 \times 10^{-5}\,\text{GeV}^{-2}. \qquad (14.119)$$

We can similarly discuss the decay of the tau lepton. In this case, there will be two possible decay modes, viz.,

$$\tau^- \to \mu^- + \widehat{\nu}_\mu + \nu_\tau,$$
$$\tau^- \to e^- + \widehat{\nu}_e + \nu_\tau. \qquad (14.120)$$

If we ignore the masses of the muon and the electron compared to that of the tau, the rates to both channels would be the same. In reality, the muon mass, though small, is not completely negligible, so the process involving the muon gets a little less phase space compared to the other one, and is marginally slower. The branching ratios of the two channels are about 17.36% and 17.84% respectively. And notice that the two branching ratios, added together, do not reach anywhere near 100%, because the τ-lepton is heavy enough to have many other decay channels, involving some hadrons in the final states.

14.7 Parity violation

The interactions that we have been using contain both polar and axial vector currents. A mixture of these currents produces parity violating interactions, as noted in the context of Ex. 6.7 *(p 158)*. It would therefore be interesting to find out what sort of signals of parity violation can be obtained from four-fermi interactions.

In §6.9, we noted that signatures of parity violation can be obtained if we take spin-polarized particles in the initial state, and observe whether the directions of the momenta of final particles show any correlation with the direction of that spin. With that in mind, let us revisit the problem of muon decay, but this time with polarized muons.

The interaction Lagrangian is still the same as in Eq. (14.100), and so is the amplitude. The only difference is that now we should not sum over both spin orientations for the muon. If all muon spins are aligned in one direction, we should take the spinor corresponding to that value of spin.

As discussed in §6.9.4, an alternative is to sum over both spins, but introduce a spin projection matrix in the amplitude which will select out only one spin orientation. The spin projection operator was given in Eq. (6.153, *p 183*). Using it, we now write the amplitude as

$$\mathscr{M} = -\frac{G_F}{\sqrt{2}} \left[\overline{u}_{q_1} \gamma^\lambda (1 - \gamma_5) \frac{1}{2} (1 + \gamma_5 \slashed{s}) u_p \right] \left[\overline{u}_k \gamma_\lambda (1 - \gamma_5) v_{q_2} \right] , \quad (14.121)$$

where s^μ is a 4-vector which has the components

$$s^\mu = (0, \widehat{s}) \quad (14.122)$$

in the rest frame of the muon, $\widehat{s}$ being a unit 3-vector along the direction in which the spins of the muons have been aligned.

The amplitude can be written as

$$\mathscr{M} = \frac{1}{2} \left(\mathscr{M}_0 + \mathscr{M}_s \right) , \quad (14.123)$$

where $\mathscr{M}_0$ is the amplitude for unpolarized muons that was written down in Eq. (14.101), and

$$\mathscr{M}_s = +\frac{G_F}{\sqrt{2}} \left[\overline{u}_{q_1} \gamma^\lambda (1 - \gamma_5) \slashed{s} u_p \right] \left[\overline{u}_k \gamma_\lambda (1 - \gamma_5) v_{q_2} \right] , \quad (14.124)$$

using the fact that $(1 - \gamma_5)\gamma_5 = \gamma_5 - 1$. While squaring the amplitude and summing over the spins, we can use the result of Eq. (14.103) for the square of $\mathscr{M}_0$. As for the square of $\mathscr{M}_s$, note that the bilinear containing the electron and the electron-neutrinostill gives the same factors. Writing this part as $E_{\lambda\rho}$, we get

$$\sum_{\text{spins}} |\mathscr{M}_s|^2 = \sum_s \left[\overline{u}_{q_1} \gamma^\lambda (1 - \gamma_5) \slashed{s} u_p \right] \left[\overline{u}_{q_1} \gamma^\rho (1 - \gamma_5) \slashed{s} u_p \right]^* E_{\lambda\rho}$$

$$= \text{Tr} \left[\slashed{q}_1 \gamma^\lambda (1 - \gamma_5) \slashed{s} (\slashed{p} + m_\mu) \slashed{s} \gamma^\rho (1 - \gamma_5) \right] E_{\lambda\rho} . \quad (14.125)$$

The term involving muon mass is traceless because it contains an odd number of Dirac matrices. Also, note that

$$\slashed{s} \slashed{p} \slashed{s} = 2s \cdot p \slashed{s} - \slashed{p} \slashed{s} \slashed{s} . \quad (14.126)$$

From the components of s^μ given in Eq. (14.122), it is clear that $s \cdot p = 0$. Moreover, $\slashed{s}\slashed{s} = s^\mu s_\mu = -1$. Thus, the expression of Eq. (14.125) gives the same value with or without the factor $\slashed{s}$ present in it, which means that $\sum |\mathscr{M}_s|^2 = \sum |\mathscr{M}_0|^2$, the sum being over spins. In the absolute square of the amplitude of Eq. (14.123), the two direct terms taken together would then contribute an amount equal to $\frac{1}{2} \sum |\mathscr{M}_0|^2$, i.e., equal to the amount shown in Eq. (14.103).

Now for the crossed terms.

$$\sum_{\text{spins}} \mathscr{M}_0 \mathscr{M}_s^* = - \text{Tr} \left[\slashed{q}_1 \gamma^\lambda (1 - \gamma_5)(\slashed{p} + m_\mu) \slashed{q} \gamma^\rho (1 - \gamma_5) \right] E_{\lambda\rho} . \quad (14.127)$$

This time, only the term proportional to m_μ has a non-zero trace. And this term is

$$\sum_{\text{spins}} \mathscr{M}_0 \mathscr{M}_s^* = -m_\mu \text{Tr} \left[\slashed{q}_1 \gamma^\lambda (1 - \gamma_5) \slashed{q} \gamma^\rho (1 - \gamma_5) \right] E_{\lambda\rho} . \quad (14.128)$$

Had the $\slashed{q}$ term not been there in Eq. (14.127), we would have obtained $\slashed{p}$ within the trace. Now, we have $-m_\mu \slashed{q}$ in its place. Thus the final evaluation of the traces would contain $-m_\mu$ times the 4-vector s where we obtained the vector p in the direct squared terms. Adding the crossed terms and the direct terms, we then obtain

$$\overline{|\mathscr{M}|^2} = 64 G_F^2 \, (p - m_\mu s) \cdot q_1 \, k \cdot q_2 . \quad (14.129)$$

Compared to Eq. (14.103), there is an overall factor of $\frac{1}{2}$, and p has been replaced by $p - m_\mu s$. The expression for the decay rate for this case can also be written by making the same changes on the result in Eq. (14.112):

$$\Gamma = \frac{G_F^2}{24\pi^4 m_\mu} \int \frac{d^3 k}{2 k_0} \left(q^2 (p - m_\mu s) \cdot k + 2 q \cdot (p - m_\mu s) \, q \cdot k \right) , \quad (14.130)$$

where $q = p - k$ as in §14.6. We will use the various dot products evaluated in Eq. (14.113). In addition, we will need these:

$$s \cdot k = -\mathbf{k} \hat{\mathbf{s}} \cdot \hat{\mathbf{k}} = -\mathbf{k} \cos\theta ,$$

$$s \cdot q = -s \cdot k = \mathbf{k} \cos\theta , \quad (14.131)$$

where θ is the angle between $\hat{\mathbf{s}}$ and $\mathbf{k}$. Working in the approximation where the electron mass can be neglected, we can write Eq. (14.130) as

$$\frac{d\Gamma}{d\Omega} = \frac{G_F^2 m_\mu}{48\pi^4} \int dE_e \, E_e^2 \left(3m_\mu - 4E_e + (m_\mu - 4E_e) \cos\theta \right) . \quad (14.132)$$

The limits of integration for E_e have been discussed in §14.6. Integration over E_e gives

$$\frac{d\Gamma}{d\Omega} = \frac{G_F^2 m_\mu^5}{3 \times 2^8 \pi^4} \left(1 - \frac{1}{3} \cos\theta \right) . \quad (14.133)$$

We see, first of all, that if we integrate overall the solid angle, the $\cos\theta$ term gives no contribution. In fact, the total rate obtained in this case is equal to the total rate for unpolarized muons that was obtained in Eq. (14.117). But the $\cos\theta$ term makes the differential decay rate non-uniform, unlike the situation encountered in §14.6. There is an angle dependence, and the angle in question is the one between the spin of the muon and the momentum of the final electron. In particular, this angle enters the differential decay rate through the function $\cos\theta = \hat{\boldsymbol{s}} \cdot \hat{\boldsymbol{k}}$. This kind of terms signal parity violation, as explained in §6.9. Precisely this kind of correlation was obtained in the earliest experiments of parity violation. Those experiments did not study the decay of the muon, but beta decay of some nucleus, as described in §6.9.1.

☐ **Exercise 14.10** Show that if the muon is not fully polarized but rather has a net polarization P along a certain direction, the angular dependence of electrons is given by the formula

$$\frac{d\Gamma}{d\Omega} = \frac{G_F^2 m_\mu^5}{3 \times 2^8 \pi^4} \left(1 - \frac{1}{3} P \cos\theta \right). \tag{14.134}$$

Study of polarized muon decay played a crucial role in determining the $V - A$ nature of the interaction. To understand how this was done, we can go back to Eq. (14.132) and rewrite it in the form

$$\frac{d\Gamma}{dx\, d\Omega} = \frac{G_F^2 m_\mu^5}{3 \times 2^7 \pi^4} x^2 \left(3 - 2x + (1 - 2x) \cos\theta \right), \tag{14.135}$$

where

$$x = E_e / (E_e)_{\max} = 2 E_e / m_\mu. \tag{14.136}$$

More generally, taking the net muon polarization to be P and without assuming that the interaction is $V - A$ type, one can parametrize the differential decay rate in the form

$$\frac{d\Gamma}{dx\, d\Omega} = \frac{G_F^2 m_\mu^5}{3 \times 2^7 \pi^4} x^2 \left[6(1 - x) - \frac{4}{3}(3 - 4x)\rho \right.$$
$$\left. - \left(2(1 - x) - \frac{4}{3}(3 - 4x)\delta \right) \xi P \cos\theta \right], \tag{14.137}$$

which contains the three parameters ρ, δ and ξ. These are called *Michel parameters* after the name of the scientist who pioneered such analysis. There can be a few more parameters if we include the effects of a non-zero electron mass. Notice that there is no parameter attached to the first term in the square bracket. The reason is that this is the only term that contributes to the total decay rate, and is therefore determined by the lifetime of the muon. The Michel parameters have been experimentally determined by measuring the energy dependence of electrons as a function of the angle made with the direction of muon polarization. The results, to a very good accuracy, give the values

$$\rho = \frac{3}{4}, \qquad \delta = \frac{3}{4}, \qquad \xi = 1, \tag{14.138}$$

which are the values obtained in a $V - A$ theory.

438 *Chapter 14.   Fermi theory of weak interactions*

□ **Exercise 14.11** Consider the most general form for the Feynman amplitude for the muon decay with four-fermion interaction:

$$\mathscr{L}_{\text{int}} = \sum_I \left[\overline{u}_{(e)} O_I u_{(\mu)} \right] \left[\overline{u}_{\nu_\mu} O_I (C_I + C_I' \gamma_5) v_{(\nu_e)} \right], \qquad (14.139)$$

where I can take five different values corresponding to the bilinears shown in Eq. (4.93, p 79). Note that the bilinears can always be put into this form by applying Fierz transformations if necessary. Define the quantities

$$a_I = |C_I|^2 + |C_I'|^2,$$
$$a' = 2\,\text{Re}\left(C_S C_P'^* + C_S' C_P^*\right),$$
$$b' = 2\,\text{Re}\left(C_V C_A'^* + C_V' C_A^*\right),$$
$$c' = 2\,\text{Re}\left(C_T C_T'^*\right). \qquad (14.140)$$

Find the differential muon decay rate, $d\Gamma/dx\,d\Omega$. Show that it is of the form given in Eq. (14.137) with

$$8G_F^2 = a_S + 4a_V + 6a_T + 4a_A + a_P, \qquad (14.141\text{a})$$
$$8G_F^2 \rho = 3a_V + 6a_T + 3a_A, \qquad (14.141\text{b})$$
$$8G_F^2 \xi = -3a' - 4b' + 14c', \qquad (14.141\text{c})$$
$$8G_F^2 \xi\delta = -3b' + 6c'. \qquad (14.141\text{d})$$

[**Note :** *Actually, this is a lot of work. If you are not prepared for it, just take only one of the a_I's to be non-zero and check that your results are consistent with the general results for the Michel parameters given in Eq. (14.141).*]

14.8   Problems with Fermi theory

The Fermi theory is very successful in describing low-energy weak interactions. And yet, it cannot possibly qualify as the fundamental theory of weak interactions. Let us discuss why.

The first reason is that the Fermi constant, G_F, has mass dimension equal to -2. According to the general rules put down in §4.3, this theory is non-renormalizable. It means that, if we take this theory and calculate loop processes with it, we will obtain infinite results for so many things that we will not be able to absorb all these infinities into the redefinitions of the fields and constants of the Lagrangian. Thus, calculation of loop diagrams cannot be performed with the Fermi Lagrangian.

Even if we stick to tree-level diagrams, we encounter difficulties. To appreciate this, let us look at the cross-sections obtained in earlier sections of this chapter. In each case, the total cross-section can be seen to be proportional to $G_F^2 \mathfrak{s}$. So, Fermi theory would predict that the cross-sections should monotonically increase with the Mandelstam variable $\mathfrak{s}$, i.e., with the center of mass energy available for the process.

This is unacceptable from some very basic grounds. In a scattering process, we can write the wavefunction in the form

$$\psi_{\text{tot}} = \psi_{\text{in}} + f(\theta)\,\frac{e^{ikr}}{r}\,, \tag{14.142}$$

where the first term on the right hand side is in the wavefunction of an incident plane wave, and the second term represents spherically outgoing waves centered at the point $r = 0$ which is the point of interaction. The function denoted by $f(\theta)$ is called the *scattering amplitude*, where the angle θ is measured with respect to the direction of the incident wave. For interactions that do not violate isotropy of space, the scattering amplitude cannot depend on the azimuthal angle ϕ. With this general characterization, the differential cross-section comes out to be

$$\frac{d\sigma}{d\Omega} = \left| f(\theta) \right|^2 . \tag{14.143}$$

Further, the scattering amplitude, being a function of θ, can be expanded in terms of the Legendre polynomials $P_l(\cos\theta)$ in the form

$$f(\theta) = \frac{1}{k} \sum_l (2l + 1) a_l P_l(\cos\theta)\,, \tag{14.144}$$

where l can take any integral value including zero. The quantities a_l are dimensionless. From the requirement that the number of scattered particles cannot be more than the number of incident particles, one obtains the constraint

$$|a_l| < 1 \qquad \forall l\,. \tag{14.145}$$

This is called the *partial wave unitarity* condition.

Compare this expression with the angular dependence of cross-section in the CM frame as given in Eq. (4.207, *p 106*). Specializing to elastic scattering for the sake of notational convenience, we find that

$$\left| f(\theta) \right|^2 = \frac{1}{64\pi^2 s} |\mathcal{M}|^2\,, \tag{14.146}$$

where s is the Mandelstam variable whose square root gives the total incident energy in the two particles. Look, e.g., at the Feynman amplitude squared for neutrino–electron elastic scattering, given in Eq. (14.85). The largest power of $\cos\theta$ that appears in this expression is $\cos^2\theta$. Then $f(\theta)$ must have only up to linear terms in $\cos\theta$, which means that only the $l = 0$ and $l = 1$ partial waves contribute. In other words, for this case we can write

$$\frac{d\sigma}{d\Omega} = \frac{1}{k^2} \left| a_0 + 3a_1 \cos\theta \right|^2\,, \tag{14.147}$$

where both a_0 and a_1 must obey the constraint of Eq. (14.145). Obviously, $\left| a_0 + 3a_1 \cos\theta \right| < 4$. Also, we assume that this bound is to be checked for energies which are large so that the masses of interacting particles can be neglected, we can put $k^2 \approx E^2 = \frac{1}{4}\mathfrak{s}$. Then we get

$$\frac{d\sigma}{d\Omega} < \frac{64}{\mathfrak{s}} \,. \tag{14.148}$$

As mentioned before, the cross-sections that come out of Fermi theory are of the form $G_F^2 \mathfrak{s}$ times some factor of order unity. This means that the calculations make sense provided

$$G_F^2 \mathfrak{s} < \text{(some constant)} \times \frac{1}{\mathfrak{s}} \,. \tag{14.149}$$

It implies that at energies much larger than $G_F^{-\frac{1}{2}}$, the cross-sections derived from Fermi theory cannot be valid.

It is therefore clear that the Fermi theory can be used only as a low-energy approximation of some more complete theory. In this low energy regime, cross-sections will of course be proportional to $\mathfrak{s}$, but this state of affairs will not continue to hold for arbitrarily high energies. After some value of $\mathfrak{s}$, the more fundamental theory will produce results which would be significantly different from those of the Fermi theory. And, if this new theory is truly fundamental, the results derived from it should have no problem with renormalizability and unitarity.

14.9   Intermediate vector bosons

There is an easy way to get rid of the Fermi constant which has a negative mass dimension. Instead of thinking of a basic vertex consisting of four fermion fields, we can try to construct a closer analogy of QED. After all, electron–electron or electron–positron elastic scattering also involves four fermions: two in the initial state and two in the final state. But this was obtained from two vertices, each of which contains the basic interaction between two fermion lines and a photon line.

To extend this analogy to the realm of weak interactions, we notice that the Fermi interaction Lagrangian consists of two currents. If, instead of a current coupling to another current, we have a theory where the current couples to a vector boson which in turn couples to the other current, we obtain diagrams which would be analogous to the QED diagrams. The idea, developed in the 1950s, has been illustrated in Fig. 14.4 for a generic process. The vector boson in the intermediate line was first called just the intermediate vector boson. Gradually, the name W boson became popular because of its association with weak processes.

It is clear that the W boson has to be complex. For example, consider beta decay. The basic process involves the conversion of one neutron to a proton, which should be on one side of the intermediate boson line. The vertex therefore involves the neutron, the proton and a W boson. Since the

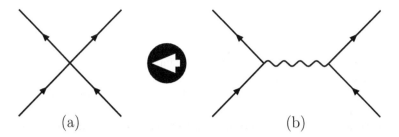

Figure 14.4: On the right panel, we see an interaction between four fermion fields being mediated by an intermediate vector boson. Under certain conditions discussed in the text, this can reduce to the four-fermion interactions shown on the left panel.

neutron is uncharged, the W boson will have to be charged in order that electric charge is conserved. The magnitude of the electric charge must be 1 in units of the proton charge. The W boson with charge $+1$ can be called W^+. Its antiparticle will have charge -1 and can be denoted by W^-. Because they are obviously different, the W boson field must be complex.

To see how the intermediate vector boson can duplicate the successes of the Fermi theory, let us consider the amplitude for the diagram involving the vector boson. Let the vector boson interactions with the fermions be written as

$$\mathscr{L}_{\text{int}} = -g J^\mu W_\mu^\dagger + \text{h.c.}, \tag{14.150}$$

where J^μ is a superposition of fermion field bilinears and g is a coupling constant. The amplitude of Fig. 14.4b is given by

$$i\mathscr{M} = (ig)^2 J^{\mu\dagger} i D_{\mu\nu} J^\nu, \tag{14.151}$$

where in this formula, J^μ means the expression for the current, with the fields substituted by spinors. The symbol $D_{\mu\nu}$ stands for the propagator of the intermediate vector boson. Assume, for the moment, that the vector boson is massive, and its free Lagrangian is given by the Proca Lagrangian given in Eq. (4.150, *p 92*) or its suitable generalization if W is complex. As commented in Ex. 4.20 *(p 92)*, the propagator for such a field can be obtained in a straightforward manner, and the result is

$$D_{\mu\nu}(q) = \frac{1}{q^2 - M_W^2} \left(-g_{\mu\nu} + \frac{q_\mu q_\nu}{M_W^2} \right). \tag{14.152}$$

If we work in a regime where the components of the 4-vector q are much much smaller compared to M_W, we can take the leading term in the propagator only, which is $g_{\mu\nu}/M_W^2$. Putting this into Eq. (14.151), we find that in this regime, the amplitude is of the form

$$\frac{g^2}{M_W^2} J^{\mu\dagger} J_\mu, \tag{14.153}$$

which is exactly how Fermi interactions look. In other words, in the intermediate vector boson theory, the Fermi constant appears as a combination of the
coupling constant and the vector boson mass, and the Fermi theory emerges
as an approximation for momentum transfers small compared to the vector
boson mass. When the momentum becomes high enough, we cannot use the
Fermi theory any longer, but rather should use the full form of the propagator
given in Eq. (14.152).

The coupling constant g is dimensionless in this theory, and so, by the naive
rule given in §4.3, this theory might appear to be renormalizable, just like QED
and QCD are. But the rule given earlier was not complete. There are two
conditions which need to be satisfied in order a theory to be renormalizable:

1. The propagator of any bosonic field falls off like $1/p^2$ for large momenta
   and the propagator of any fermion field falls off like $1/p$.

2. There is no coupling constant with negative mass dimension.

In the intermediate vector boson theory, the second condition is satisfied, but
the first condition is not. The term containing $q_\mu q_\nu$ in the propagator of Eq.
(14.152) does not fall at all for large momenta. Recall that the problem with
UV divergences arises from high values of 4-momenta in the loops. The W
propagator does not decrease when the momentum becomes very high, so its
ultraviolet consequences are much worse. As a result, the infinities cannot be
tamed and the theory cannot be renormalized.

One might wonder at this point about why the $q_\mu q_\nu$ term should matter.
After all, in Eq. (4.148, *p 92*) we obtained such a term in the photon propagator as well, but commented that this term does not contribute to physical
amplitudes. Shouldn't it be the same with massive vector fields as well?

The answer is *no*. The $q_\mu q_\nu$ terms in the photon propagator vanish in the
amplitude because these are arbitrary due to the gauge invariance. This is
clearly seen from the expression for the photon propagator in Eq. (4.148, *p 92*):
the said terms can be altered arbitrarily by changing the gauge parameter ξ.
When we put in a mass term for the gauge boson, the gauge symmetry is lost.
In other words, presence of a gauge symmetry could make the offending terms
of the vector boson propagator irrelevant and make the theory renormalizable,
but the mass term is not gauge invariant. And we definitely need a mass
for the vector boson in order that we obtain Fermi theory as a low energy
approximation. This is the problem that confronted weak interaction theories
in the beginning of the 1960s.

Chapter 15

Spontaneous symmetry breaking

At the end of Ch. 14, we saw that we want some kind of gauge symmetry associated with vector bosons in order that the theory is renormalizable. The problem that comes with it is the mass of the vector boson, which ought to vanish as a consequence of gauge symmetry. In this chapter, we will discuss some scenarios where the consequences of a symmetry are not realized on the physical observables. Such a phenomenon can happen because of a feature of the ground state of the system, and is called *spontaneous symmetry breaking*.

15.1 Examples of spontaneous symmetry breaking

In this section, we present several examples of spontaneous symmetry breaking. First, a general comment. If spontaneous symmetry breaking occurs due to the expectation value of a vector field, the ground state must prefer a certain direction in spacetime, which would mean a breaking of Lorentz invariance as well. The same is true if, instead of a vector field, any non-trivial representation of the Lorentz group is used. We want to deal with Lorentz invariant theories. In this case, we can consider non-trivial ground state configuration for scalar fields only. All examples that we present in this section contain only scalar fields.

15.1.1 Breaking a Z_2 symmetry

Consider the following Lagrangian involving a real scalar field ϕ:

$$\mathscr{L} = \frac{1}{2}(\partial_\mu \phi)(\partial^\mu \phi) - \frac{1}{2}\mu^2 \phi^2 - \frac{\lambda}{4}\phi^4 . \tag{15.1}$$

The Lagrangian is obviously invariant under a transformation

$$\phi \to -\phi . \tag{15.2}$$

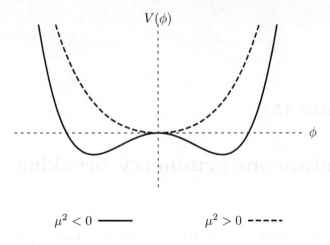

$\mu^2 < 0$ —— $\mu^2 > 0$ - - - - -

Figure 15.1: Shape of the potential of Eq. (15.4).

Along with the identity transformation, this forms a Z_2 symmetry.

Let us look at the parameters in the Lagrangians. There are two of them, which we have called μ^2 and λ. Both need to be real in order that the Lagrangian is hermitian. To obtain more information about these parameters, let us write Eq. (15.1) as

$$\mathcal{L} = \frac{1}{2}(\partial_\mu \phi)(\partial^\mu \phi) - V(\phi), \tag{15.3}$$

where all non-derivative terms have been dumped into a collection

$$V(\phi) = \frac{1}{2}\mu^2 \phi^2 + \frac{\lambda}{4}\phi^4. \tag{15.4}$$

This collection is usually called the *potential* of the theory. To complete the analogy, the derivative terms are sometimes called the *kinetic* terms, in analogy with the fact that the time derivative terms represent the kinetic energy of a non-relativistic particle where the potential energy usually does not contain derivatives of the co-ordinates.

From the exercise leading to Eq. (4.119, *p 84*), we can conclude that the Hamiltonian is given by

$$\mathcal{H} = \frac{1}{2}\left(\frac{\partial \phi}{\partial t}\right)^2 + \frac{1}{2}(\boldsymbol{\nabla}\phi)^2 + V(\phi). \tag{15.5}$$

If λ were negative, the Hamiltonian would become more and more negative with increasing values of ϕ. Classically speaking, the value of the Hamiltonian would go all the way down to the negative infinity. This would be an impossible system to deal with, because it would not have any ground state. Thus, for a physically viable system, we must have $\lambda > 0$.

There is no such constraint on the other parameter in the Lagrangian. We have called it μ^2 in order to indicate that its dimension is that of squared mass. It does not imply that it is the square of a real parameter μ. Therefore μ^2, the parameter in the Lagrangian, can be positive or negative. If μ^2 happens to be positive, it can be interpreted as the square of the mass μ of the particle whose field is ϕ, as we have done before while introducing field theoretical Lagrangians in Ch. 4. What happens if μ^2 is negative?

Let us try to answer this question by first finding the value of ϕ in the ground state. Looking at Eq. (15.5), we see that the derivative terms are non-negative, so they can be minimized when they are zero, i.e., in a configuration where ϕ is independent of the spacetime co-ordinate. The minimum of $V(\phi)$ should occur where

$$\frac{\partial V}{\partial \phi} = 0 \,. \tag{15.6}$$

For the potential given in Eq. (15.4), this condition is

$$\phi(\mu^2 + \lambda\phi^2) = 0 \,. \tag{15.7}$$

The solutions of this equation are

$$\langle \phi \rangle = 0, \quad \pm v \,, \tag{15.8}$$

where

$$v = \sqrt{\frac{-\mu^2}{\lambda}} \,. \tag{15.9}$$

If $\mu^2 > 0$, the only real solution is $\langle \phi \rangle = 0$. When $\mu^2 < 0$, all three are real solutions, as seen from Fig. 15.1, but the solution at vanishing field value is really a local maximum. The minima of the system occur for $\langle \phi \rangle = \pm v$, which means that there are two degenerate minima, and either of them can correspond to the ground state of the system. No matter which one it is, we see that the field ϕ cannot be treated as a quantum field. Quantum fields can be expanded in creation and annihilation operators, as in Eq. (4.12, *p 64*) for example. The vacuum expectation value (VEV) of such a field is always zero, because, as defined in Eq. (4.129, *p 86*), the annihilation operator acting on the vacuum to the right produces zero, and the creation operator does the same by acting on the vacuum state to the left. So the field ϕ that appears in the Lagrangian of Eq. (15.1) cannot be a quantum field.

Suppose we find the system in the vacuum at $\phi = +v$. We now define a field $H(x)$ through the relation

$$\phi(x) = v + H(x) \,. \tag{15.10}$$

Obviously, VEV of the field $H(x)$ vanishes, so that it can be a quantum field. We can say that it represents the fluctuation of the field $\phi(x)$ around its minima denoted by v. In order to use the machinery of quantum field theory with the field $H(x)$, we should rewrite the Lagrangian of Eq. (15.1) in terms of this field. Eliminating the parameter μ^2 with the use of the definition of v, we obtain

$$\mathscr{L} = \frac{1}{2}(\partial_\mu H)(\partial^\mu H) - \lambda v^2 H^2 - \lambda v H^3 - \frac{\lambda}{4} H^4 , \qquad (15.11)$$

apart from a constant proportional to v^4 which is of no importance. Note that the terms linear in H have canceled because of the minimization condition, and the co-efficient of the quadratic term indicates that the quantum of the field H has a mass M_H given by

$$M_H^2 = 2\lambda v^2 . \qquad (15.12)$$

But more importantly, notice that there is a cubic term in the Lagrangian, which means that the Lagrangian is *not* invariant under the transformation

$$H \to -H . \qquad (15.13)$$

It is easy to guess why this happens. As soon as we select one of the two minima and denote the fluctuations around this one as the quantum field, we lose the symmetry. Choice of the vacuum thus results in a lower symmetry in the Lagrangian. This is the essence of the phenomenon called spontaneous symmetry breaking.

One might ask, if the Z_2 symmetry is not there in the Lagrangian of Eq. (15.11) involving the quantum field $H(x)$, what is the point of talking about it? What difference would it make if we started with a Lagrangian of the field $H(x)$ without paying attention to any symmetry at all? The answer lies in the fact that the symmetry of the Lagrangian of Eq. (15.3) contained only two parameters, μ^2 and λ, because of the Z_2 symmetry. The Lagrangian of Eq. (15.11) contains only those two parameters, sometimes in the disguise of the VEV v. Had we tried to write down a Lagrangian of $H(x)$ without paying attention to any symmetry, the co-efficients of the quadratic, cubic and quartic terms would have been independent of each other, and therefore the Lagrangian would have contained three parameters. In the Lagrangian of the model after spontaneous symmetry breaking, the three parameters are related. Thus, even though the original Z_2 symmetry has been broken and therefore absent in Eq. (15.11), it has left its marks in the relation between various parameters of the model.

15.1.2 Breaking a U(1) symmetry

Let us now discuss another example involving a complex scalar field $\phi(x)$, and a Lagrangian

$$\mathscr{L} = (\partial_\mu \phi)^\dagger (\partial^\mu \phi) - V(\phi) , \qquad (15.14)$$

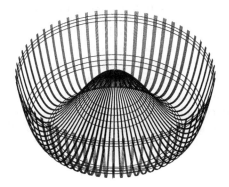

Figure 15.2: Shape of the potential of a complex scalar field.

where

$$V(\phi) = \mu^2 \phi^\dagger \phi + \lambda (\phi^\dagger \phi)^2 \,. \tag{15.15}$$

Clearly, this Lagrangian has a U(1) symmetry, corresponding to phase rotations of the complex scalar field:

$$\phi(x) \to e^{i\theta} \phi(x) \,. \tag{15.16}$$

As for the Lagrangian of §15.1.1, the parameter λ has to be real and positive. The other parameter, μ^2, has to be real but not necessarily positive. If it happens to be negative, the minima of the potential is obtained at

$$\left| \langle \phi \rangle \right| = \frac{v}{\sqrt{2}} \,, \tag{15.17}$$

where v is given by Eq. (15.9). As we see, there are an infinite number of degenerate minima, related by the phase rotation of Eq. (15.16). The shape of the potential has been shown in Fig. 15.2.

Suppose the system chooses the minimum at $\langle \phi \rangle = v/\sqrt{2}$. If we write

$$\phi(x) = \frac{1}{\sqrt{2}} \Big(v + H(x) + i\zeta(x) \Big) \,, \tag{15.18}$$

the fields $H(x)$ and $\zeta(x)$ will be quantum fields with zero vacuum expectation values. We can rewrite the Lagrangian of Eq. (15.14) in terms of these quantum fields. Ignoring constant terms proportional to v^4, we obtain

$$\mathscr{L} = \frac{1}{2} (\partial_\mu H)(\partial^\mu H) + \frac{1}{2} (\partial_\mu \zeta)(\partial^\mu \zeta)$$
$$- \lambda v^2 H^2 - \lambda v H (H^2 + \zeta^2) - \frac{\lambda}{4} (H^2 + \zeta^2)^2 \,. \tag{15.19}$$

Obviously, this Lagrangian does not have the U(1) symmetry because it represents the fluctuations around a minimum along the real axis. Choice of this minimum breaks the symmetry spontaneously.

15.1.3 Breaking a non-abelian symmetry

We now try the same thing with a Lagrangian having a bigger symmetry. The mathematical steps are more or less the same. The difference is that, non-abelian symmetries admit matrix representations of various different dimensions. Depending on which representation of fields develop a VEV, we can expect to see different patterns of symmetry breaking.

As an example, consider three real scalar fields ϕ_1, ϕ_2 and ϕ_3 which transform like a triplet of an SU(2) symmetry. Introducing the notation

$$\Phi = \begin{pmatrix} \phi_1 \\ \phi_2 \\ \phi_3 \end{pmatrix}, \tag{15.20}$$

we can write the Lagrangian as

$$\mathscr{L} = \frac{1}{2}(\partial_\mu \Phi^\top)(\partial^\mu \Phi) - V(\Phi), \tag{15.21}$$

with

$$V(\Phi) = \frac{1}{2}\mu^2 \Phi^\top \Phi + \frac{\lambda}{4}\left(\Phi^\top \Phi\right)^2. \tag{15.22}$$

For $\mu^2 < 0$, the minimum of the potential occurs for

$$\langle \Phi^\top \Phi \rangle = v^2, \tag{15.23}$$

where v is still given by the same expression as in Eq. (15.9). Suppose the system settles down at the minimum

$$\langle \phi_3 \rangle = v, \qquad \langle \phi_1 \rangle = \langle \phi_2 \rangle = 0. \tag{15.24}$$

We can then use ϕ_1 and ϕ_2 as quantum fields, and define

$$\phi_3(x) = v + H(x) \tag{15.25}$$

so that $H(x)$ can be used as a quantum field. In terms of the quantum fields, the Lagrangian is

$$\mathscr{L} = \frac{1}{2}(\partial_\mu \phi_1)(\partial^\mu \phi_1) + \frac{1}{2}(\partial_\mu \phi_2)(\partial^\mu \phi_2) + \frac{1}{2}(\partial_\mu H)(\partial^\mu H)$$
$$+ \frac{1}{2}\lambda v^2 \left(\phi_1^2 + \phi_2^2 + (v + H)^2\right) - \frac{1}{4}\lambda\left(\phi_1^2 + \phi_2^2 + (v + H)^2\right)^2. \tag{15.26}$$

Even without expanding the terms out, we see one feature in what we have written. The Lagrangian is certainly not symmetric in the full SU(2) transformations involving the three quantum fields. However, the fields ϕ_1 and ϕ_2 always appear in the combination $\phi_1^2 + \phi_2^2$, which means that even after spontaneous symmetry breaking, there is a remnant symmetry in the Lagrangian, corresponding to rotations in the ϕ_1-ϕ_2 plane, or equivalently a phase transformation of the form

$$\phi_1 + i\phi_2 \rightarrow e^{i\theta}(\phi_1 + i\phi_2). \tag{15.27}$$

In other words, in this case the symmetry SU(2) breaks to a U(1) symmetry.

☐ **Exercise 15.1** Complete the calculation for the terms in the potential in Eq. (15.26) and show that the fields ϕ_1 and ϕ_2 are massless in the broken theory. Find also the mass of the field H.

☐ **Exercise 15.2** Consider an SU(2) invariant theory involving a doublet φ of complex scalar fields, and

$$V(\varphi) = \mu^2 \varphi^\dagger \varphi + \lambda \left(\varphi^\dagger \varphi\right)^2. \tag{15.28}$$

Show that in this case, there is no remnant symmetry of the Lagrangian, and that three of the four real fields that appear in the two complex components of φ are massless.

☐ **Exercise 15.3** Consider a set of N scalar fields, and a Lagrangian having an O(N) symmetry. Show that a non-trivial vacuum expectation value can break the symmetry to O($N-1$).

15.2 Goldstone theorem

We now turn the attention of the reader to a feature of the Lagrangians obtained after spontaneous symmetry breaking. Start with Eq. (15.19). Note that it has a mass term for the field H, but none for the field ζ: the latter is massless after spontaneous symmetry breaking. Massless particles appeared in other examples as well. As we mentioned, in the model of §15.1.3, the fields ϕ_1 and ϕ_2 turned out to be massless. In Ex. 15.2, we commented that there should be three massless bosons.

All of these are examples of *Goldstone theorem*. The theorem states that, if a Lagrangian is invariant under a continuous symmetry group that has n generators, and if its ground state is symmetric under a continuous group containing n' generators, there should be $n-n'$ massless states in the spectrum of the theory.

☐ **Exercise 15.4** Identify the massless scalars in the problem of Ex. 15.3. Verify that their number is indeed equal to the difference of the numbers of generators of the groups O(N) and O($N-1$). [**Hint :** *The number of generators of an orthogonal group was given in Ex. 3.9 (p 43).*]

Let us give a simple (but incomplete) proof of the theorem. Suppose we have a theory with a number of scalar fields which we denote by ϕ_k. The scalar potential is denoted by $V(\phi)$, where ϕ in parentheses symbolizes all the scalar fields. If the potential is invariant under some continuous group, it means that there exist infinitesimal changes in the fields, $\delta\phi_k$, which keep the potential invariant:

$$\frac{\partial V}{\partial \phi_k} \delta\phi_k = 0. \tag{15.29}$$

In terms of generators, the changes in the fields can be written as

$$\delta\phi_k = -i(T_a)_{kl}\theta_a\phi_l\,,\tag{15.30}$$

where θ_a are the parameters that generate the change, and T_a are the matrix representations of the generators. Since the invariance is guaranteed for arbitrary values of θ_a, Eq. (15.29) implies

$$\frac{\partial V}{\partial\phi_k}\,(T_a)_{kl}\phi_l = 0\,.\tag{15.31}$$

Taking another derivative of this equation with respect to $\phi_{k'}$, we obtain

$$\frac{\partial V}{\partial\phi_k}\,(T_a)_{kk'} + \frac{\partial^2 V}{\partial\phi_k\partial\phi_{k'}}\,(T_a)_{kl}\phi_l = 0\,.\tag{15.32}$$

Let us now ask ourselves what this equation implies at the minimum of the scalar potential, $\langle\phi\rangle$. The first term is zero there. Thus, if

$$\left(T_a\,\langle\phi\rangle\right)_k \neq 0\tag{15.33}$$

for some generator T_a, we must have

$$\left.\frac{\partial^2 V}{\partial\phi_k\partial\phi_{k'}}\right|_{\phi=\langle\phi\rangle} = 0\tag{15.34}$$

for any value of k'. That means that the matrix of second derivatives would have a null row, and accordingly a null eigenvalue. The second derivative of the potential is the mass. So, if we write the quadratic terms involving all scalar fields in the form of a matrix, it has a zero eigenvalue corresponding to any generator for which the inequality of Eq. (15.33) is valid.

Note that if the ground state had the same symmetry as the original Lagrangian, i.e., group elements acting on the ground state would have kept it invariant, we should have had

$$T_a\,\langle\phi\rangle = 0\,.\tag{15.35}$$

This relation is still true for the generators of the part of the original symmetry group which would remain unbroken. For these generators, Eq. (15.34) does not hold. This means that we obtain a zero mass field corresponding to any generator that is not the part of the symmetry group of the ground state.

This is Goldstone theorem, but, as we said, this proof is incomplete. It really does not show the full strength of the theorem. Goldstone theorem predicts that if a continuous symmetry is spontaneously broken, there will be states whose energies go to zero when the 3-momentum goes to zero. If we stick to theories where Lorentz symmetry is not broken, such states must be scalars, as explained earlier. They are called *Goldstone bosons*. But the strength of the theorem lies in the fact that such scalar states would emerge in the spectrum even if the Lagrangian contains no scalar or an insufficient number of scalars. We do not attempt to give a proof for this more general state of affairs.

One should also note that Goldstone theorem says nothing about discrete symmetries. The crucial step in our proof is the use of small changes of the fields characterized by Eq. (15.30), which does not work for discrete transformations. One should not expect any massless modes when discrete symmetries are spontaneously broken. Indeed, in the example of §15.1.1, we did not encounter any massless particle.

> □ **Exercise 15.5** *Consider the examples of symmetry breaking given in §15.1. For each case, explain the number of Goldstone bosons by counting the generators of the original symmetry group and the group that remains intact after spontaneous symmetry breaking.*

15.3   Interaction of Goldstone bosons

We have noticed that, after symmetry breaking, the Lagrangian does not contain any mass term for Goldstone bosons. We now point out a stronger statement about Goldstone bosons. To make the point, we go back to the model of §15.1.2 as an example. We expressed the complex scalar field ϕ in terms of two real fields through Eq. (15.18). Instead, suppose we take the representation

$$\phi(x) = \frac{1}{\sqrt{2}} \left(v + \widetilde{H}(x) \right) \exp\left(i\widetilde{\zeta}(x)/v \right). \tag{15.36}$$

This is the operator analog of writing a complex number in terms of a modulus and a phase, whereas Eq. (15.18) gives the representation of a complex number in terms of its real and imaginary parts. Note that if we expand the exponential of Eq. (15.36) in a power series, we obtain

$$\phi(x) = \frac{1}{\sqrt{2}} \left(v + \widetilde{H}(x) + i\widetilde{\zeta}(x) + \cdots \right), \tag{15.37}$$

i.e., up to first order in the fields, $\widetilde{H} = H$ and $\widetilde{\zeta} = \zeta$. There is a theorem in quantum field theory which says that if we have a set of fields $\Phi^A(x)$ and we define a new set of fields $\widetilde{\Phi}^A(x)$ which are functions of the original fields $\Phi^A(x)$, and the functions are of the form

$$\widetilde{\Phi}^A(x) = \Phi^A(x) + (\text{higher order terms in the fields}), \tag{15.38}$$

then all on-shell matrix elements calculated in the two representations will be equal. We will not prove this *reparametrization theorem* here: we will only use the present model to provide a compelling example for it.

In the present model, the theorem implies that the fields $\widetilde{H}$ and $\widetilde{\zeta}$ are just as good in describing the physical implications as the original fields H and ζ. In view of this, we will omit the tilde sign over the two real fields from now on. If we now try to put the expression of Eq. (15.36) into the Lagrangian given in Eqs. (15.14) and (15.15), we find that the field $\zeta(x)$ disappears in the terms in the potential V. For the kinetic term, we note that

$$\partial_\mu \phi = \frac{1}{\sqrt{2}} \left(\partial_\mu H(x) + \frac{v + H(x)}{v} i\partial_\mu\zeta \right) \exp\left(i\zeta(x)/v \right) . \qquad (15.39)$$

So the Lagrangian can be written as

$$\mathscr{L} = \frac{1}{2}(\partial_\mu H)(\partial^\mu H) + \frac{(v+H)^2}{2v^2}(\partial_\mu\zeta)(\partial^\mu\zeta) - \lambda v^2 H^2 - \lambda v H^3 - \frac{1}{4}\lambda H^4 . \qquad (15.40)$$

Note what has happened! In this form, the field ζ appears in the Lagrangian only through its derivatives.

We could have guessed it earlier. The Lagrangian has a global U(1) symmetry, which means that any constant phase of a field is irrelevant. Our representation of Eq. (15.36) treats the field ζ as a phase. So, if ζ were really constant throughout the spacetime, it should have disappeared altogether because of the U(1) symmetry. Of course ζ is not a constant; it is a field. So its derivatives can contribute to the Lagrangian, but the overall constant part cannot.

This is the stronger statement indicated earlier. The field ζ, which is the Goldstone boson, not only lacks a mass term; it lacks all non-derivative interaction terms as well. There are only terms involving the derivative of ζ, which include the usual kinetic term for the Goldstone boson. In addition, there are the following interactions involving the Goldstone boson:

$$\mathscr{L}_{\text{int}}^{(\zeta)} = \left(\frac{H}{v} + \frac{H^2}{2v^2} \right) (\partial_\mu\zeta)(\partial^\mu\zeta) . \qquad (15.41)$$

No matter which process we consider involving the Goldstone bosons, the derivatives on the fields will deliver momentum factors to the amplitude. In other words, we find the important result that the amplitude of any process involving a Goldstone boson contains factors of momentum and vanishes in the limit where the 4-momentum vanishes.

This is obvious in a representation like that in Eq. (15.36). By the reparametrization theorem, it should also be true in the representation of Eq. (15.18), though not so obvious. To provide an example of how the reparametrization theorem works, consider the decay of the H particle into two Goldstone bosons:

$$H(p) \to \zeta(p_1) + \zeta(p_2) , \qquad (15.42)$$

where in parentheses, we have written the notations for the 4-momenta of the different particles that we are going to use. In the polar representation of the field ϕ, this process occurs at the tree-level through the first interaction

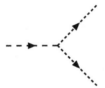

Figure 15.3: Tree-level diagrams contributing to the decay of H into two Goldstone bosons. For the sake of convenience, we have denoted the H line with longer dashes and the ζ lines with shorter dashes.

term appearing in Eq. (15.41). The diagram is shown in Fig. 15.3, and the amplitude is given by

$$i\mathcal{M} = -\frac{2ip_1 \cdot p_2}{v}. \tag{15.43}$$

The momentum factors appear in the Feynman rule of the $H\zeta\zeta$ vertex from the derivatives, as discussed in §4.10. The factor of 2 appears because there are two ζ fields in the interaction term, both of which can create or annihilate the same particle.

Now let us find the same amplitude in the linear representation of the field given in Eq. (15.18). The Lagrangian in terms of the fields H and ζ, given in Eq. (15.19), contains an interaction term $-\lambda v H \zeta^2$, which should be responsible for the decay in question. Noting that there are two factors of the field ζ, the Feynman rule for the vertex can be written as $-2i\lambda v$, i.e., the Feynman amplitude is

$$i\mathcal{M} = -2i\lambda v. \tag{15.44}$$

On the face of it, it does not appear to depend on the 4-momenta of the Goldstone bosons until we recall a few things. First, Eq. (15.19) shows that the H boson has a mass:

$$M_H^2 = 2\lambda v^2. \tag{15.45}$$

Secondly, the on-shell conditions for the particles are given by

$$p^2 = M_H^2, \qquad p_1^2 = p_2^2 = 0. \tag{15.46}$$

Combining these two conditions, we obtain

$$2\lambda v = \frac{1}{v} M_H^2 = \frac{1}{v}(p_1 + p_2)^2 = \frac{2}{v} p_1 \cdot p_2. \tag{15.47}$$

This shows that the expressions for the amplitude given in Eqs. (15.43) and (15.44) are equal, vindicating the reparametrization theorem.

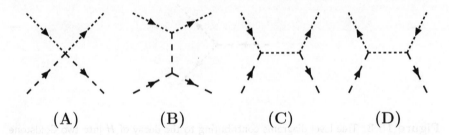

Figure 15.4: Tree-level diagrams contributing to the process of Eq. (15.48). The convention of long and short dashes has been explained in Fig. 15.3. The arrows help us identify which particles are incoming and which are outgoing.

> □ **Exercise 15.6** *Consider the elastic scattering of the Goldstone bosons against the bosons represented by the field H, i.e., the process*
>
> $$H(p) + \zeta(k) \rightarrow H(p') + \zeta(k'). \tag{15.48}$$
>
> *At the tree level, there are four diagrams that contribute to this process, as shown in Fig. 15.4. Calculate the amplitude in both representations and show that they are equal.*

15.4 Higgs mechanism

So far in this chapter, we have talked only about global symmetries. What happens if a theory with a local (or gauge) invariance experiences spontaneous symmetry breaking? Let us try to answer the question by considering a gauge theory with a U(1) symmetry, i.e., a gauged version of the model of §15.1.2.

15.4.1 Gauge boson mass

The scalar potential of this model is the same as that given in Eq. (15.15). The Lagrangian is given by

$$\mathscr{L} = (D_\mu \phi)^\dagger (D^\mu \phi) - V(\phi) - \frac{1}{4} F_{\mu\nu} F^{\mu\nu}, \tag{15.49}$$

where D_μ is the gauge covariant derivative,

$$D_\mu = \partial_\mu + ieA_\mu, \tag{15.50}$$

and $F_{\mu\nu}$ is the field-strength tensor constructed out of the gauge field A_μ.

Let us now say that the scalar potential develops a minimum given by Eq. (15.17), and once again we rewrite everything in terms of the quantum fields shown in Eq. (15.18). The term involving covariant derivatives will now look like

$$(D_\mu \langle \phi \rangle)^\dagger (D^\mu \langle \phi \rangle) + \cdots, \tag{15.51}$$

where the other terms involve the quantum fields. Of course $\partial_\mu \langle \phi \rangle = 0$, since $\langle \phi \rangle$ is defined through the parameters μ^2 and λ and is therefore independent of the spacetime co-ordinates. So this expression can be written as

$$\frac{1}{2} e^2 v^2 A^\mu A_\mu + \cdots . \tag{15.52}$$

This is a non-derivative quadratic term in the gauge field, and should therefore indicate a non-zero mass for the gauge field. In fact, the quadratic terms involving only the gauge field are given by

$$-\frac{1}{4} F_{\mu\nu} F^{\mu\nu} + \frac{1}{2} e^2 v^2 A^\mu A_\mu . \tag{15.53}$$

This is exactly the Proca Lagrangian, introduced in Eq. (4.150, *p 92*), that applies to a vector field of mass

$$M_A = ev . \tag{15.54}$$

We then reach the conclusion that when spontaneous symmetry breaking takes place, the gauge boson obtains a mass. Earlier, in Ch. 14, we mentioned that a direct mass term for a gauge boson is problematic. Here we are faced with a different situation: the gauge invariant Lagrangian of Eq. (15.49) does not have a mass term for the gauge boson, but the ground state after spontaneous symmetry breaking contains such a term. This consequence of spontaneous symmetry breaking of gauge theories was noticed by a number of people around the same time: Englert, Brout, Higgs, Kibble, Guralnik, Hagen. Commonly, from this long list, only the name of Higgs is used to call this phenomenon the *Higgs mechanism*. The physical consequences of this mechanism are different from those of having a mass term in the original Lagrangian, as we will gradually see in this section.

15.4.2   Gauge fixing

So far, we have described only one term of the symmetry-broken Lagrangian, viz., where we used the vacuum expectation value of the scalar field for both occurrences of ϕ in the covariant derivative term. Let us now consider another set of terms, where we take the vacuum expectation value on one side and the derivative terms which act only on the quantum fields on the other. In other words, we are considering the terms

$$\frac{1}{2} \left(D_\mu v \right)^\dagger \left(\partial^\mu (H + i\zeta) \right) + \frac{1}{2} \left(\partial_\mu (H + i\zeta) \right)^\dagger \left(D^\mu v \right) . \tag{15.55}$$

Since $D_\mu v = ie A_\mu v$, as argued above, these terms can be written as

$$-\frac{1}{2} iev A_\mu \left(\partial^\mu (H + i\zeta) \right) + \text{h.c.} = M_A A_\mu \partial^\mu \zeta , \tag{15.56}$$

using Eq. (15.54) in the last step. This is a strange term indeed! In a quantum field theoretic interpretation, this term can annihilate the gauge boson field and create the ζ field. But it is a quadratic term, so it should be part of the free Lagrangian. We seem to have a problem to determine whether it belongs to the free Lagrangian of A_μ or that of ζ. As it stands, it cannot be either.

However, already in Ch. 4 we mentioned that the free Lagrangian of the gauge boson field is problematic, and we must add a gauge-fixing term in order to make it work. Suppose, in this case, we add the gauge-fixing term

$$\mathscr{L}_{\text{gf}} = -\frac{1}{2\xi}(\partial_\mu A^\mu - \xi M_A \zeta)^2 \tag{15.57}$$

to the Lagrangian given in Eq. (15.49). The parameter ξ is similar to the parameter introduced in Eq. (4.106, p 82). It can have arbitrary values, and the physical amplitudes should be free of this parameter. It can be called a *gauge parameter*.

Let us now look at the cross term from Eq. (15.57) along with the pathological quadratic term encountered in Eq. (15.56). Together, they give

$$M_A \partial^\mu (A_\mu \zeta). \tag{15.58}$$

This is a total derivative term, and is therefore of no consequence. This is good news: we have solved the problem of quadratic terms involving different fields.

Let us now look at the quadratic terms involving the gauge field only. These are:

$$\mathscr{L}_0^{(A)} = -\frac{1}{4}F_{\mu\nu}F^{\mu\nu} - \frac{1}{2\xi}(\partial_\mu A^\mu)^2 + \frac{1}{2}M_A^2 A^\mu A_\mu. \tag{15.59}$$

We can follow the method of §4.10.2 to find the propagator of the gauge boson field, and obtain

$$iD_{\mu\lambda}(p) = -\frac{i}{p^2 - M_A^2}\left(g_{\mu\lambda} - \frac{(1-\xi)p_\mu p_\lambda}{p^2 - \xi M_A^2}\right). \tag{15.60}$$

And let us also look at the quadratic terms involving in the scalar field ζ, which was involved with the gauge field in the pathological combination of Eq. (15.56). These terms are

$$\mathscr{L}_0^{(\zeta)} = \frac{1}{2}(\partial_\mu \zeta)(\partial^\mu \zeta) - \frac{1}{2}\xi M_A^2 \zeta^2. \tag{15.61}$$

The first of these terms comes from the derivative term of the scalar field in Eq. (15.49), and the second term comes from the gauge-fixing term of Eq. (15.57). Together, they imply the following propagator for the scalar field:

$$i\Delta^{(\zeta)}(p) = \frac{i}{p^2 - \xi M_A^2}. \tag{15.62}$$

□ **Exercise 15.7** Verify Eqs. (15.60) and (15.62).

15.4.3 Unitary gauge

The gauge parameter ξ introduced in Eq. (15.57) is arbitrary. We can use different values of it to gain insight into the physics of spontaneous symmetry breaking. Here we consider the limit $\xi \to \infty$, which is called the *unitary gauge*. Notice that in this limit, the gauge boson propagator assumes the form of the propagator obtained from the Proca Lagrangian which was given in Eq. (4.151, *p 92*). Also notice that the propagator for the ζ field becomes zero, which means that any Feynman diagram containing the ζ field as an internal line has a vanishing amplitude.

The situation is easily understood if, instead of the representation of the quantum fields given in Eq. (15.18), we consider the *polar* representation

$$\phi(x) = \frac{1}{\sqrt{2}}\left(v + H(x)\right) \exp\left(i\zeta(x)/v\right). \tag{15.63}$$

This is the same as that given in Eq. (15.36), except that we have omitted the tilde signs in view of the reparametrization theorem. Local symmetry means that we change the phase of the field ϕ by any spacetime dependent function without having any effect on the physical consequences of the theory. So suppose we change the phase by $\exp(-i\zeta(x)/v)$. With this choice, we can write

$$\phi(x) = \frac{1}{\sqrt{2}}\left(v + H(x)\right). \tag{15.64}$$

If we substitute this into the Lagrangian, the field ζ would completely disappear from the Lagrangian.

Recall that ζ is the field that would have become the Goldstone boson if we were talking of a global symmetry. For the global symmetry case, we found that the Goldstone bosons can appear in the Lagrangian only through their derivatives. In the case of local symmetry, we find that the restrictions become more severe: the field that could have been the Goldstone boson if the symmetry were global has no physical consequences at all. In the unitary gauge that we have been considering, it does not appear anywhere in the Lagrangian.

There is a connection between the disappearance of one of the scalar fields and the appearance of mass of the gauge boson. A massless gauge boson such as the photon has only two degrees of polarization, as explained in §a). But a vector boson has spin equal to 1, so one should expect three independent components of it. Because of the gauge symmetry, only the components transverse to the direction of the 3-momentum are allowed, the longitudinal component does not exist. Once the gauge symmetry is broken and the gauge boson acquires mass, there is nothing to prevent the longitudinal component. The massive gauge boson has indeed three independent degrees of freedom, or three polarization states. But if the parameter μ^2 appearing in the scalar potential happened to be positive, the gauge boson would have had two degrees of freedom. Imagine that we can somehow change the parameter μ^2 continuously. As it changes from positive to negative, the gauge boson becomes massive and therefore needs three degrees of freedom. Where would it obtain the extra degree of freedom? The answer lies in the disappearance of the would-be Goldstone boson field: it disappears, thereby reducing one degree of freedom in the scalar sector. We can say that one scalar field supplies the longitudinal polarization component of the gauge boson. Figuratively, one says that the gauge boson eats up one scalar field.

15.4.4 Renormalizablity

This figurative speech works only in the unitary gauge, i.e., when the gauge parameter ξ is infinitely large. For any finite value of ξ, the propagator of the ζ field does not vanish, and therefore diagrams with internal ζ lines must be taken into account. As Eq. (15.62) shows, the mass parameter appearing in this propagator is ξM_A^2. Since ξ is an unphysical parameter, this is an unphysical mass. Hence, in any gauge with finite ξ, the field ζ is called the *unphysical Higgs*. The field is *unphysical* because it does not represent any physical particle, and therefore cannot appear as external legs of any Feynman diagrams representing a physical process. Some people also prefer to call this mode the *would-be Goldstone boson*, alluding to the fact that this field would have been the Goldstone boson if the symmetry were global rather than local.

It almost seems that the gauges with finite ξ are more complicated because we need to deal with unphysical internal lines in Feynman diagrams. But there is a pay-off. Notice that for any finite ξ, the gauge boson propagator indeed falls off as $1/p^2$ for large momenta. So does, in fact, the propagator of the unphysical Higgs. Added to the fact that there is no coupling with negative mass dimension in the Lagrangian, we see that this theory satisfies both conditions needed for renormalizability stated on page 442. That is why the gauge condition with an arbitrary finite value of ξ is called the renormalizable ξ-gauge or the R_ξ-*gauge*. For the purpose of performing calculations, it is therefore easier to use these gauges. Unless otherwise stated, we would always use the value $\xi = 1$, which is called the *'t Hooft–Feynman gauge*.

We thus see this dual aspect of these theories. For finite ξ, the theory is renormalizable, although the particle interpretations are somewhat obscure because of the presence of unphysical bosons. On the other hand, for infinite ξ, the particle spectrum is clear, although renormalizability is not obvious. But since the physical amplitudes do not depend on ξ, the theory must be both. This means that we have found a renormalizable theory with massive gauge bosons through spontaneous symmetry breaking.

Chapter 16

Standard electroweak model with leptons

At the end of Ch. 14, we commented that it is necessary to have massive vector particles in order to describe weak interactions. We also commented that without a gauge symmetry, a theory with a massive vector boson is not renormalizable. On the other hand, gauge theory precludes any mass term for gauge bosons. The way out of this apparent impasse was sought through spontaneously broken gauge theories: a gauge symmetry would ensure renormalizability, but it would be broken spontaneously so that the gauge bosons would be massive. The idea was successfully employed by Weinberg and Salam in 1967–68, and 't Hooft proved the renormalizability of such models in 1971. The model that grew out of these ideas is now known as the *standard electroweak model*, because it described both weak and electromagnetic interactions. Along with the theory of strong interactions in the form of quantum chromodynamics or QCD, it provides the *standard model* of all particle interactions except gravity. In this chapter, we will describe the standard electroweak theory in the form that it was originally proposed — involving only the leptons and considering neutrinos to be massless. Quarks will be accommodated in this theory in Ch. 17, and the question of neutrino mass will be discussed in Ch. 22.

16.1   Chiral fermions and internal symmetries

One novel aspect of the standard electroweak theory is spontaneous symmetry breaking, which has been introduced in general terms in Ch. 15 and will be discussed in this particular context in §16.3. The other novel aspect is the subject of this section.

We have introduced chiral projections of fermion fields in §14.2.2. We have shown that any term involving a number of fermion fields can be rewritten in terms of chiral projections by using Eq. (14.19, *p 416*). Example of the mass term was shown in Eq. (14.24, *p 417*).

It was also shown there that a left-chiral or a right-chiral field retains its chirality under Lorentz transformations. Thus, a fermion field with a given chirality forms an irreducible representation of the proper Lorentz group. In fact, these are the irreducible representations $(\frac{1}{2}, 0)$ and $(0, \frac{1}{2})$ mentioned in Ch. 3. If a left-chiral field is said to transform like the $(\frac{1}{2}, 0)$ representation, a right-chiral field would transform like the $(0, \frac{1}{2})$ representation.

An internal symmetry is said to be one which commutes with the Poincaré group. Since the Lorentz group is a subgroup of the Poincaré group, an internal symmetry must commute with the proper Lorentz group. This means that internal symmetry operations should not change the property of any object under Lorentz transformations, and vice versa. For example, if we perform a Lorentz transformation, the electric charge of a particle should not change. By the same token, if we have a multiplet of some non-abelian internal symmetry, each of its components must transform like the same irreducible representation of the proper Lorentz group.

A corollary of this statement is that if we have two different irreducible representations of the proper Lorentz group, there is no reason that they should transform the same way under an internal symmetry. Given a fermion field $\psi(x)$, there is no fundamental requirement that tells us that its left-chiral projection $\psi_{\rm L}(x)$ and the right-chiral projection $\psi_{\rm R}(x)$ should transform the same way under an internal symmetry.

In gauge theories, internal symmetries govern the interactions. Thus, if $\psi_{\rm L}(x)$ and $\psi_{\rm R}(x)$ have different transformations under an internal symmetry, they will have different interactions.

To see why this is of paramount importance, let us recall the parity transformation property of fermion fields described in §6.2.4. Using linearity of the parity operator and Eq. (6.23, *p 155*), we find

$$\mathscr{P}\psi_{\rm L}(x)\mathscr{P}^{-1} = \mathscr{P}{\rm L}\psi(x)\mathscr{P}^{-1} = {\rm L}\mathscr{P}\psi(x)\mathscr{P}^{-1}$$
$$= {\rm L}\eta_P\gamma_0\psi(\widetilde{x}) = \eta_P\gamma_0\psi_{\rm R}(\widetilde{x}) \,. \tag{16.1}$$

This equation means that $\psi_{\rm R}$ is the parity transform of $\psi_{\rm L}$. Thus, if $\psi_{\rm R}$ and $\psi_{\rm L}$ possess different interactions, it signals parity violation. So, if we consider an internal symmetry under which $\psi_{\rm R}$ and $\psi_{\rm L}$ have different transformation properties, the theory would be parity violating. This is the kind of theory we need for weak interactions.

16.2   Leptons and the gauge group

In order to introduce the basic idea without getting into a lot of details, we will consider just one generation of leptons here. The first generation contains the electron and the electron-neutrino. As mentioned in Ch. 14, the measured neutrino helicity is -1, within experimental error bars. Thus we can take the neutrinos to be left-chiral. The electron of course has mass, so we need both chiralities in order to write the mass term in the Lagrangian in the manner shown in Eq. (14.24, $p\,417$). In short, in the first generation of leptons, we have the three chiral fields, ν_L, e_L and e_R, writing the fields by the standard notations for the particles involved. Since we are discussing only one generation of fermions in this section, we are not putting a subscript on the neutrino to indicate that it is the electron-neutrinoor ν_e. That extra subscript is implicit.

In a non-abelian gauge theory, gauge interactions can change one member of a multiplet to another. This is shown, e.g., in the gauge interaction vertex of Fig. 11.4 $(p\,316)$, where the ith member changes to the jth member through the interaction with a gauge boson. Since gauge interactions cannot change chirality, e_R cannot be in the same multiplet with either e_L or ν_L. On the other hand, e_L and ν_L can be in a multiplet, and in fact they should be, because we know that in processes like the beta decay, the electron and the antineutrino are both produced, which means that the interaction involves both the electron and the neutrino.

Having decided this far, let us now wonder what the gauge bosons might be. We have already seen, in Ch. 14, that the W^+ and the W^- bosons can be thought of as mediators of weak interactions. We therefore already have two candidate gauge bosons. The problem is that there is no non-abelian group which has only two generators. If we are willing to go up to three gauge bosons, we can have the gauge group SU(2). But then, what will be the third gauge boson?

Leptons, of course, have electromagnetic interactions as well, which are mediated by photons. So we can think of building a model that will describe both electromagnetic and weak interactions of leptons, with the $W^\pm$ and the photon as the gauge bosons. But this option has a serious problem, and that comes with the leptonic fields. As we said, e_L and ν_L can belong to a multiplet. Suppose the multiplet contains only these two fields, i.e., is a doublet. If we now write the gauge interactions in the standard way mentioned in Ch. 11, we will find that there will be an interaction of the form $\overline{\nu}_L \gamma^\mu \nu_L A_\mu$, i.e., it will imply that the neutrino couples to the photon field directly. We know that the neutrinos have no electric charge, and therefore such interaction terms should not be present. Thus, this possibility is obviously ruled out unless the fermion multiplets are bigger. For example, if the fermions appear in a triplet whose $T_3 = 0$ component is the neutrino, the neutrino does not couple to the photon. However, we then need extra unknown fermions to fill the triplet. This option was viable in the 1960s, but has been ruled out since then by experimental results.

We therefore have to try a gauge group with four gauge bosons. Three of the gauge bosons can be associated with the generators of an SU(2) group, and the fourth one to the generator of a U(1) group, i.e., the gauge group would be SU(2) × U(1). Some people call it U(2). Group theoretically, there is nothing wrong in calling it so, as we have noted in Eq. (3.9, $p\,39$). But it has to be remembered that transformations in the SU(2) part do not affect the U(1) phase, and similarly phase transformations in the U(1) part do not

have any implications on the SU(2) properties of an object. The two parts
of the group commute, and are therefore independent. Among other things,
it means that the two parts can have independent gauge coupling constants,
and all components of an entire SU(2) multiplet must have the same value of
the U(1) charge.

Under the SU(2) part of the gauge group, the left-chiral leptons would
form a doublet representation and the right-chiral one, being alone, will have
to be a singlet. All fields will also have a quantum number that would indicate
how they transform under the U(1) part of the gauge group. Obviously this
quantum number cannot be the electric charge, because it must be the same
for ν_L and e_L, which belong to the same SU(2) multiplet. This quantum
number, in general, is denoted by Y, and is called *weak hypercharge* for reasons
that will be explained later. To make this suggestion explicit, sometimes this
part of the gauge group is called U(1)$_Y$. On the other hand, the SU(2) part
of the gauge group is sometimes called *weak isospin* and denoted by SU(2)$_L$
to indicate the fact that only left-chiral fields transform non-trivially under
it. Taking the two factors together, the gauge group is called SU(2)$_L \times$ U(1)$_Y$
in this more explicit nomenclature.

16.3   Symmetry breaking

16.3.1   Gauge bosons and their masses

The SU(2) part of the gauge group has the $W^{\pm}$ as its gauge bosons. And,
these bosons should be massive, as argued in §14.9. In order to provide masses
to these bosons without making the theory non-renormalizable, we need the
gauge group to be spontaneously broken.

In order to accomplish this, we need some scalar fields in the theory. They
must come in a non-trivial representation of SU(2) in order that their vacuum
expectation values can break SU(2) and give mass to the W boson. We take
the simplest possibility, viz., an SU(2) doublet of scalars, and we denote it as
follows:

$$\phi \equiv \begin{pmatrix} \phi_1 \\ \phi_2 \end{pmatrix} : (2, Y_\phi). \tag{16.2}$$

At the extreme right, we have denoted the group transformation property of
this multiplet: the first number in the parentheses denotes the multiplet of
SU(2) that the fields belong to, and the second number, which has been kept
unspecified for the moment, is the U(1) quantum number.

There will be terms in the Lagrangian which will look like this:

$$\mathscr{L} = (D_\mu \phi)^\dagger (D^\mu \phi) - V(\phi), \tag{16.3}$$

where D_μ, as usual, denotes the gauge covariant derivative and $V(\phi)$ denotes
the scalar potential.

The gauge covariant derivative D_μ will contain the ordinary derivative, of course. In addition, it will have terms involving the gauge bosons. If we denote the SU(2) gauge bosons by W_μ^a, with $a = 1, 2, 3$, and the U(1) gauge boson by B_μ, we can write

$$D_\mu \phi = \left(\partial_\mu + ig \frac{\tau^a}{2} W_\mu^a + ig' Y_\phi B_\mu \right) \phi \,. \tag{16.4}$$

Note that in the term involving the SU(2) gauge bosons, we have put in the Pauli matrices τ^a, because $\frac{1}{2}\tau^a$ are the generators of SU(2) in the doublet representation. The coupling constant for this part of the gauge group is called g. For the U(1) part, the term involving the gauge boson is exactly like the photon term in the gauge covariant derivative of QED given in Eq. (5.8, $p\,114$). The gauge coupling constant for this part of the gauge group has been named g', and Y_ϕ is the weak hypercharge of the multiplet ϕ, as introduced in Eq. (16.2).

Let us now turn to the scalar potential. The most general renormalizable potential with the scalar multiplet ϕ is of the form

$$V(\phi) = \mu^2 \phi^\dagger \phi + \lambda (\phi^\dagger \phi)^2 \,. \tag{16.5}$$

If $\mu^2 < 0$, the minimum of this potential is obtained for

$$\left| \langle \phi_1 \rangle \right|^2 + \left| \langle \phi_2 \rangle \right|^2 = \frac{1}{2} v^2 \,, \tag{16.6}$$

where the angular brackets denote the value at the minimum, and

$$v = \sqrt{\frac{-\mu^2}{\lambda}} \,. \tag{16.7}$$

Clearly, this corresponds to an infinity of degenerate minima. Suppose our system is in the minimum where

$$\langle \phi \rangle = \begin{pmatrix} 0 \\ v/\sqrt{2} \end{pmatrix} \,, \tag{16.8}$$

i.e.,

$$\langle \phi_1 \rangle = 0 \,, \qquad \langle \mathrm{Re}\, \phi_2 \rangle = \frac{v}{\sqrt{2}} \,, \qquad \langle \mathrm{Im}\, \phi_2 \rangle = 0 \,. \tag{16.9}$$

As shown in various examples of Ch. 15, we can now expand around this minimum. The expansion will involve some terms which represent mass terms for various gauge bosons. We can identify these terms easily by following the example of §15.4. For this, we write Eq. (16.4) as

$$D_\mu \phi = \partial_\mu \phi + i G_\mu \phi \,, \tag{16.10}$$

where G_μ is a matrix whose form can be easily found out by inserting the expressions for the Pauli matrices:

$$G_\mu = \begin{pmatrix} \frac{1}{2} g W_\mu^3 + g' Y_\phi B_\mu & \frac{1}{2} g (W_\mu^1 - i W_\mu^2) \\ \frac{1}{2} g (W_\mu^1 + i W_\mu^2) & -\frac{1}{2} g W_\mu^3 + g' Y_\phi B_\mu \end{pmatrix} \,. \tag{16.11}$$

The gauge boson mass terms are then of the form

$$\mathscr{L}_{\text{mass}} = \left(G_\mu \langle \phi \rangle \right)^\dagger \left(G^\mu \langle \phi \rangle \right). \tag{16.12}$$

From Eqs. (16.11) and (16.8), we obtain

$$G_\mu \langle \phi \rangle = \frac{v}{\sqrt{2}} \begin{pmatrix} \frac{1}{2} g (W_\mu^1 - i W_\mu^2) \\ -\frac{1}{2} g W_\mu^3 + g' Y_\phi B_\mu \end{pmatrix}. \tag{16.13}$$

which gives

$$\mathscr{L}_{\text{mass}} = \frac{1}{4} g^2 v^2 W^{+\mu} W_\mu^- + \frac{1}{8} v^2 (-g W_\mu^3 + 2 g' Y_\phi B_\mu)^2, \tag{16.14}$$

where we have defined

$$W_\mu^\pm = \frac{1}{\sqrt{2}} \left(W_\mu^1 \mp i W_\mu^2 \right). \tag{16.15}$$

These are obviously complex fields, appropriate for charged particles.

We see that the charged gauge bosons, $W^\pm$, have acquired a mass:

$$M_W = \frac{1}{2} g v. \tag{16.16}$$

The other term in Eq. (16.14) implies that one combination of the neutral gauge bosons W_μ^3 and B_μ has become massive because of spontaneous symmetry breaking. We will denote this combination by Z_μ. The combination orthogonal to Z_μ remains massless, and can be identified with the photon. Thus, the symmetry breaking process leaves one subgroup of the original gauge group intact. This is the U(1) group of QED, which we can denote by writing U(1)$_{\text{em}}$. The process of symmetry breaking can then be summarized into the statement

$$\text{SU(2)}_{\text{L}} \times \text{U(1)}_{\text{Y}} \rightarrow \text{U(1)}_{\text{em}}. \tag{16.17}$$

This U(1) group, of electromagnetism, remains unbroken. Recalling that the QCD gauge group is also unbroken, the unbroken gauge symmetry of the standard model is SU(3)$_{\text{c}} \times$ U(1)$_{\text{em}}$.

It was inevitable that there will be a residual symmetry group after spontaneous symmetry breaking. We have broken the symmetry by choosing the vacuum denoted in Eq. (16.7), which has a non-zero lower component. This component has the eigenvalue $t_3 = -\frac{1}{2}$ for the neutral generator T_3 of the SU(2) part of the gauge group. Its vacuum expectation value therefore breaks the symmetry associated with the generator T_3. Similarly, it has a non-zero value of the weak hypercharge Y, which is why it breaks U(1)$_{\text{Y}}$. But there must be a linear combination of T_3 and Y for which this lower component of ϕ has the eigenvalue zero. The symmetry associated with this combination cannot be broken by the vacuum expectation value shown in Eq. (16.8). We denote this combination by

$$Q = T_3 + Y. \qquad (16.18)$$

There is no loss of generalization in writing the equation in this manner, without any non-trivial co-efficients in front of the two generators on the right hand side. The reason is that there is a multiplicative arbitrariness in defining U(1) quantum numbers, as discussed in §5.1.1. We can use this freedom on Q and Y, both of which are U(1) generators, such that Eq. (16.18) is valid as it is. Once we have fixed this convention, we see that we must have $Y_\phi = \frac{1}{2}$ so that the lower component has zero electric charge. The upper component has one unit of positive charge, and henceforth we will write Eq. (16.2) as

$$\phi \equiv \begin{pmatrix} \phi_+ \\ \phi_0 \end{pmatrix} : (2, \frac{1}{2}), \qquad (16.19)$$

where the subscripts on the components denote their electric charges.

In passing, we should note that Eq. (16.18) looks very much like the Gell-Mann–Nishijima relation of Eq. (10.4, *p 255*). The only difference is in a factor of $\frac{1}{2}$ accompanying the hypercharge in Eq. (10.4, *p 255*). But this is inconsequential, because we could have easily defined hypercharge as $(B+S)/2$ instead of what we did in Eq. (10.9, *p 256*). Alternatively, we could have used the multiplicative arbitrariness of U(1) couplings to redefine the Y appearing in Eq. (16.18) so that it could look exactly like Eq. (10.4, *p 255*). The point is that these two equations look the same, and it is this similarity which prompted the names "weak isospin" and "weak hypercharge" for the $SU(2)_L$ part and the $U(1)_Y$ part of the gauge group.

While the formal similarity is important, it is also important to realize that the physical contents of Eqs. (10.4) and (16.18) are very different, because the properties mentioned in the two equations are different. Usual isospin symmetry (which might be called *strong isospin* in the light of the new avatar that has now appeared on the scene) applies on hadrons only, whereas leptons also transform under the weak isospin, as we will see in §16.4. The weak isospin distinguishes different chiralities of fermions, which the strong isospin does not. In other words, the formal similarity between the two formulas exists, but that's where the similarity ends.

☐ **Exercise 16.1** From Eq. (16.14), argue that the mass of the Z boson is given by

$$M_Z = \frac{1}{2}(g^2 + g'^2)^{1/2} v. \qquad (16.20)$$

[**Note :** *Remember that the Z boson is a real field, as opposed to the W boson which is complex. Therefore the mass term for the Z boson should have an extra factor of $\frac{1}{2}$.*]

☐ **Exercise 16.2** ♪ Use Eq. (16.18) to answer the following questions.

a) Which one, among ν_L and e_L, should be the upper (i.e., $T_3 = +\frac{1}{2}$) component of the $SU(2)_L$ doublet?

b) What should be the weak hypercharge of the left-chiral lepton doublet?

c) What should be the weak hypercharge of e_R?

16.3.2   Couplings of photon

Now that we have fixed the value of Y_ϕ, we know that the combination of neutral gauge bosons that acquires a mass is given by

$$Z_\mu \propto -gW_\mu^3 + g'B_\mu \, . \tag{16.21}$$

Normalizing properly, we can write

$$Z_\mu = \cos\theta_W \, W_\mu^3 - \sin\theta_W \, B_\mu \, , \tag{16.22}$$

where θ_W, called *Weinberg angle*, is defined by the relation

$$\tan\theta_W = \frac{g'}{g} \, . \tag{16.23}$$

Photon will then be given by the orthogonal combination of W_μ^3 and B_μ, i.e.,

$$A_\mu = \sin\theta_W \, W_\mu^3 + \cos\theta_W \, B_\mu \, . \tag{16.24}$$

Let us now check the couplings of the photon. Consider, for example, a fermion field ψ which has a $U(1)_Y$ quantum number Y and whose T_3 eigenvalue is T_3. (We are using the same notation for the operator and the eigenvalue, and assuming that it will be understood from the context which one is used in any particular formula.) Its coupling with the neutral gauge bosons will then be given by

$$\mathscr{L}_{\text{neutral}} = \overline{\psi}\gamma^\mu \left(-gT_3 W_\mu^3 - g'Y B_\mu \right)\psi \, . \tag{16.25}$$

Inverting Eqs. (16.22) and (16.24), we can write

$$\mathscr{L}_{\text{neutral}} = -\overline{\psi}\gamma^\mu \left(gT_3 \sin\theta_W + g'Y \cos\theta_W \right) A_\mu\psi + Z_\mu\text{-coupling.} \tag{16.26}$$

From Eq. (16.23), we see that $g\sin\theta_W = g'\cos\theta_W$. Hence, combining the two terms in the photon coupling, we obtain

$$\begin{aligned}
\mathscr{L}_{\text{em}-\text{int}} &= -g\sin\theta_W \, \overline{\psi}\gamma^\mu \left(T_3 + Y \right) A_\mu\psi \\
&= -g\sin\theta_W \, Q\overline{\psi}\gamma^\mu A_\mu\psi \, ,
\end{aligned} \tag{16.27}$$

using Eq. (16.18) in the last step. This is exactly the QED interaction provided we identify the gauge coupling constant of QED by the relation

$$e = g\sin\theta_W \, . \tag{16.28}$$

☐ **Exercise 16.3** Take a scalar particle and write its interaction with the neutral gauge bosons of $SU(2)_L \times U(1)_Y$. Make the identification of Eq. (16.28) and show that the resulting interactions with the photon are exactly the same as those in scalar QED.

□ **Exercise 16.4** Show that

$$M_W = M_Z \cos \theta_W . \tag{16.29}$$

□ **Exercise 16.5** Use Eqs. (16.16) and (16.28) to express the vacuum expectation value v in terms of M_W, e and $\sin \theta_W$. Using $\sin^2 \theta_W = 0.23$, show that $v = 246\,\text{GeV}$.

□ **Exercise 16.6** Suppose the symmetry is broken by a number of multiplets obtaining non-zero vacuum expectation values. The I-th of these multiplets transforms like a $2T_I + 1$ dimensional representation of the SU(2)$_L$ part of the gauge group, and has weak hypercharge equal to Y_I. If the neutral component obtains a VEV v_I, show that the mass relation between the gauge bosons is given by

$$\frac{M_W^2}{M_Z^2 \cos^2 \theta_W} = \frac{\sum_I \left(T_I(T_I + 1) - Y_I^2 \right) v_I^2}{\sum_I 2 Y_I^2 v_I^2} . \tag{16.30}$$

16.3.3   Gauge fixing

In §15.4, we have discussed the question of gauge fixing in great detail. We can therefore cut down on the explanation and write the gauge-fixing terms that we are going to use:

$$\mathscr{L}_{\text{gf}} = -\frac{1}{\xi_W} \left| \partial_\mu W_+^\mu + i \xi_W M_W w_+ \right|^2$$
$$- \frac{1}{2\xi_Z} (\partial_\mu Z^\mu + \xi_Z M_Z z)^2 - \frac{1}{2\xi} (\partial_\mu A^\mu)^2 . \tag{16.31}$$

Here, ξ_W, ξ_Z and ξ_A are gauge parameters. In principle they can be all different, and they define the most general R_ξ gauge for the SU(2)$_L$ × U(1)$_Y$ gauge theory. The symbols w^+ and z stand for the scalar fields that are eaten up by the W^+ and the Z boson in the process of symmetry breaking. A little reflection shows that w^+ must be ϕ^+, since there is no other positively charged scalar field in the model. And z must be the imaginary part of the uncharged component of ϕ since we have taken the vacuum expectation value to be real. In other words, the quantum fields after symmetry breaking are being represented in the following notation:

$$\phi = \begin{pmatrix} w_+ \\ \frac{1}{\sqrt{2}} (v + H + iz) \end{pmatrix} . \tag{16.32}$$

It should be noted that the gauge-fixing term for the photon field is the same as that introduced in Eq. (4.106, p 82). The photon propagator, therefore, should have the form given in Eq. (4.148, p 92). The propagators for the W and the Z bosons can easily be seen to have the form given in Eq. (15.60, p 456), and the unphysical Higgs bosons should have propagators as given in Eq. (15.62, p 456). For the sake of convenience, we summarize all these propagators in Fig. 16.1.

$W^\pm$	**Feynman rule:** $$\frac{i}{k^2 - M_W^2}\left(-g_{\mu\nu} + \frac{k_\mu k_\nu}{M_W^2}\right) - \frac{k_\mu k_\nu}{M_W^2}\left(\frac{i}{k^2 - \xi_W M_W^2}\right)$$
Z	**Feynman rule:** $$\frac{i}{k^2 - M_Z^2}\left(-g_{\mu\nu} + \frac{k_\mu k_\nu}{M_Z^2}\right) - \frac{k_\mu k_\nu}{M_Z^2}\left(\frac{i}{k^2 - \xi_Z M_Z^2}\right)$$
A	**Feynman rule:** $$\frac{i}{k^2}\left(-g_{\mu\nu} + (1 - \xi_A)\frac{k_\mu k_\nu}{k^2}\right)$$
$w^\pm$	**Feynman rule:** $$\frac{i}{k^2 - \xi_W M_W^2}$$
z	**Feynman rule:** $$\frac{i}{k^2 - \xi_Z M_Z^2}$$

Figure 16.1: Propagators of gauge bosons and unphysical Higgs bosons in R_ξ gauge. The momentum has been taken to be k for each line.

In practice, one scarcely uses the freedom of choosing different values for the gauge parameters ξ_W, ξ_Z and ξ_A. A convenient gauge is obtained by taking all of them to be equal to 1, which is called the *'t Hooft–Feynman gauge* for the present model.

□ **Exercise 16.7** *In the terms of the Lagrangian shown in Eq. (16.3), put ϕ as given in Eq. (16.32). Identify the terms which contain only a vector field and the derivative of a scalar field. Show that these terms pair with some of the terms from the gauge-fixing Lagrangian of Eq. (16.31) to produce total derivative terms which are irrelevant.*

16.4   Gauge interaction of fermions

We have already discussed the couplings of fermions with the photon. Let us now look at the couplings of fermions with the other three gauge bosons, i.e. of the $W^\pm$ and Z.

First of all, let us write the representations of the leptonic fields under the gauge group. Their SU(2) properties have already been discussed, and their U(1) quantum number can be fixed by Eq. (16.18). This gives the following representations:

$$\Psi_{\rm L} \equiv \begin{pmatrix} \nu_{\rm L} \\ e_{\rm L} \end{pmatrix} : \left(2, -\frac{1}{2}\right),$$
$$e_{\rm R} : (1, -1). \tag{16.33}$$

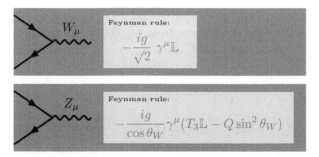

Figure 16.2: Feynman rules for fermion interactions with W and Z bosons. In the upper diagram, the incoming particle can be either the electron or the neutrino, and the outgoing particle would be the other one of the two. The Feynman rule is the same. The notation for the coupling with the Z boson has been explained in connection with Eq. (16.39).

Thus, the gauge covariant derivatives would act on these multiplets as follows:

$$D_\mu \Psi_{\mathrm L} = \left(\partial_\mu + ig\frac{\tau^a}{2}W_\mu^a + ig'(-\frac{1}{2})B_\mu\right)\Psi_{\mathrm L}\,,$$
$$D_\mu e_{\mathrm R} = \left(\partial_\mu + ig'(-1)B_\mu\right)e_{\mathrm R}\,. \tag{16.34}$$

For each fermion multiplet Ψ, right-chiral or left-chiral, the gauge covariant kinetic energy term is of the form $\overline{\Psi}i\gamma^\mu D_\mu \Psi$. The ordinary derivative term present in D_μ includes the kinetic term for the field. The rest are interactions with the gauge fields. Let us write these interaction terms for the leptonic fields.

$$\mathscr{L}_{\mathrm{int}} = -\frac{1}{2}\left(\overline{\nu}_{\mathrm L}\quad \overline{e}_{\mathrm L}\right)\gamma^\mu \begin{pmatrix} gW_\mu^3 - g'B_\mu & g(W_\mu^1 - iW_\mu^2) \\ g(W_\mu^1 + iW_\mu^2) & -gW_\mu^3 - g'B_\mu \end{pmatrix}\begin{pmatrix} \nu_{\mathrm L} \\ e_{\mathrm L} \end{pmatrix}$$
$$+\overline{e}_{\mathrm R}g'\gamma^\mu B_\mu e_{\mathrm R}\,. \tag{16.35}$$

This expression contains the interaction of the fermions with the photon, which we have already discussed. In addition, it contains the interaction of fermions with the $W^\pm$ and the Z bosons. Let us first write the interactions with the charged gauge bosons, the $W^\pm$. These terms are called the *charged-current interaction* terms. Using Eq. (16.15), these terms can be written as

$$\mathscr{L}_{\mathrm{cc}} = -\frac{g}{\sqrt{2}}\left(\overline{\nu}_{\mathrm L}\gamma^\mu W_\mu^+ e_{\mathrm L} + \overline{e}_{\mathrm L}\gamma^\mu W_\mu^- \nu_{\mathrm L}\right). \tag{16.36}$$

Note that the charged currents are purely left-chiral. This is because the right-chiral lepton field is an SU(2) singlet. The charged gauge bosons are purely SU(2) gauge bosons, and do not interact with anything that is an SU(2) singlet.

Let us now turn to the *neutral current interactions*, i.e., interactions of the Z boson. We can extract them from Eq. (16.35) by writing W_μ^3 and B_μ in terms of the mass eigenstates Z_μ and A_μ by using Eqs. (16.22) and (16.24). This gives

$$\mathscr{L}_{\rm nc} = \sqrt{g^2 + g'^2}\Big(-\frac{1}{2}\overline{\nu}_{\rm L}\gamma^\mu\nu_{\rm L}$$
$$+(\frac{1}{2} - \sin^2\theta_W)\overline{e}_{\rm L}\gamma^\mu e_{\rm L} - \sin^2\theta_W\overline{e}_{\rm R}\gamma^\mu e_{\rm R}\Big)Z_\mu. \quad (16.37)$$

Noting that $\sqrt{g^2 + g'^2} = g/\cos\theta_W$ and using identities involving the projection operators $\mathbb{L}$ and $\mathbb{R}$, we can rewrite this part of the Lagrangian as

$$\mathscr{L}_{\rm nc} = -\frac{g}{\cos\theta_W}\Big(\frac{1}{2}\overline{\nu}\gamma^\mu\mathbb{L}\nu - \overline{e}\gamma^\mu(\frac{1}{2}\mathbb{L} - \sin^2\theta_W)e\Big)Z_\mu. \quad (16.38)$$

Feynman rules for both charged and neutral current vertices of fermions have been shown in Fig. 16.2.

☐ **Exercise 16.8** Consider a fermion field ψ which has a charge Q. Its right-chiral projection is a singlet under SU(2) and the left-chiral projection has an eigenvalue T_3 under the neutral SU(2) gauge boson.

  a)   What is the weak hypercharge of $\psi_{\rm R}$?

  b)   Show that the interaction of ψ with the Z boson is given by

$$\mathscr{L}_{\rm nc} = -\frac{g}{\cos\theta_W}\overline{\psi}\gamma^\mu\Big(T_3\mathbb{L} - Q\sin^2\theta_W\Big)\psi Z_\mu. \quad (16.39)$$

☐ **Exercise 16.9** Show that the gauge interactions of the standard model, both charged and neutral currents,

  a)   Violate parity invariance

  b)   Violate charge conjugation invariance

  c)   Conserve CP

We can now discuss how to include other generations of leptons into the model. In fact, it is straightforward. All generations behave exactly the same way under the gauge group. We can write, in a notation generalized from that of Eq. (16.33), the leptonic multiplets as:

$$\Psi_{\ell{\rm L}} \equiv \begin{pmatrix} \nu_{\ell{\rm L}} \\ \ell_{\rm L} \end{pmatrix} : (2, -\frac{1}{2}),$$
$$\ell_{\rm R} : (1, -1), \quad (16.40)$$

where ℓ stands for either the electron or the muon or the tau. Obviously, interactions of fermions in other generations are also expressed by Eqs. (16.36) and (16.38) with trivial changes in notation of the fermion fields.

16.5 Yukawa sector

The model, so far, has gauge bosons, leptons and scalars. We have discussed the interactions of the gauge bosons with both leptons and scalars. Now we note that there are also gauge invariant interactions involving leptons and scalars only. Such interactions are generically called *Yukawa interactions* or *Yukawa couplings* because they were inspired by Yukawa's theory of strong interactions, described in §8.8.

Let us write down the scalar-lepton interactions in the present model. They are given by the following terms:

$$\mathscr{L}_Y = -\sum_\ell \left(h_\ell \overline{\Psi}_{\ell\mathrm{L}} \phi \ell_\mathrm{R} + h_\ell^* \overline{\ell}_\mathrm{R} \phi^\dagger \Psi_{\ell\mathrm{L}} \right) . \tag{16.41}$$

The Lagrangian can contain such terms because they are gauge invariant. To see this, consider the first term on the right hand side. Since $\Psi_{\ell\mathrm{L}}$ and ϕ are both doublets of SU(2), there is a combination of the two that is an SU(2) singlet, which is the combination written here. The field ℓ_R is an SU(2) singlet anyway, so overall the term written is an SU(2) singlet. As far as the U(1) part of the gauge group is concerned, we see that $\Psi_{\ell\mathrm{L}}$ has $Y = -\frac{1}{2}$, so $\overline{\Psi}_{\ell\mathrm{L}}$ has $Y = +\frac{1}{2}$. Adding the weak hypercharge of ϕ and ℓ_R as given in Eqs. (16.19) and (16.40), we see that the total weak hypercharge of the combination is zero, which means that the combination of operators present in the term does not violate weak hypercharge. Hence the interaction is invariant under the gauge group SU(2)$_\mathrm{L}$ × U(1)$_\mathrm{Y}$.

□ **Exercise 16.10** *In Eq. (16.41), there are two terms on the right hand side. Show that one is the hermitian conjugate of the other.*

The coupling constant h that appears in Eq. (16.41) can be taken to be real without loss of generality. What we mean is that whatever the phase of the constant h, it can always be absorbed by a redefinition of the multiplet ϕ or the field ℓ_R or the multiplet $\Psi_{\ell\mathrm{L}}$. Henceforth we will take h to be real.

On the face of it, the terms in Eq. (16.41) are cubic interaction terms. However, in the broken symmetry state, the multiplet ϕ can be written in terms of the quantum fields through the expression given in Eq. (16.32). Once we put this expression in, we find that there are also terms which are quadratic in fields and proportional to v:

$$\mathscr{L}_Y = -\frac{v}{\sqrt{2}} \sum_\ell h_\ell \left(\overline{\ell}_\mathrm{L} \ell_\mathrm{R} + \overline{\ell}_\mathrm{R} \ell_\mathrm{L} \right) + \cdots , \tag{16.42}$$

where the dots indicate all other terms. As shown in Eq. (14.24, p417), these are the mass terms for the charged leptons. The masses are given by

$$m_\ell = \frac{h_\ell v}{\sqrt{2}} . \tag{16.43}$$

This is a very interesting aspect of the model. The left-chiral part of the electron field is part of an SU(2) doublet and the right-chiral is an SU(2) singlet.

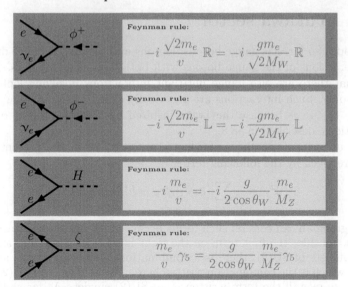

Figure 16.3: Feynman rules for fermion interactions with the unphysical and physical Higgs bosons.

Together, they cannot form a singlet, and so the gauge invariant Lagrangian cannot contain a mass term for the electron. The mass is generated through symmetry breaking.

However, even with symmetry breaking, there is no mass term for the neutrino, because there is no right-chiral component of the neutrino. Thus, in the standard model, the neutrino is massless. Experiments indicate tiny masses of neutrinos. In Ch. 22 we will discuss how to modify the standard model to take neutrino masses into account.

Of course, Eq. (16.41) also contains interaction terms. These can be written easily by using the matrix representations of the lepton doublet Ψ_L and the scalar doublet ϕ. Using the representation of ϕ given in Eq. (16.32), we obtain the following terms for the electron field:

$$\mathscr{L}_{\text{int}} = -\frac{\sqrt{2}m_e}{v}\left(\overline{\nu}_L e_R w_+ + \overline{e}_R \nu_L w_-\right) - \frac{m_e}{v}\left(\overline{e}eH + i\overline{e}\gamma_5 ez\right). \quad (16.44)$$

There are similar terms for interactions involving other charged leptons. Note that in writing this expression, we have eliminated the coupling constant h_e by using Eq. (16.43). We have also used identities involving the chirality projection operators, which were introduced earlier in §14.2.

We see that the physical Higgs boson H has intrinsic scalar couplings with fermions, and the coupling is proportional to the fermion mass. The unphysical Higgs boson couplings can be written in a slightly different manner

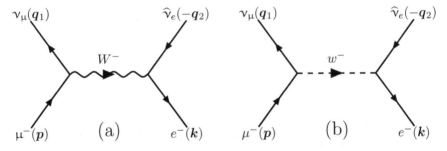

Figure 16.4: Tree-level diagrams for muon decay.

using the expressions for the gauge boson masses:

$$\mathscr{L}_{w,z} = -\frac{gm_e}{\sqrt{2}M_W}\left(\overline{\nu}_{\mathrm{L}}e_{\mathrm{R}}w_+ + \overline{e}_{\mathrm{R}}\nu_{\mathrm{L}}w_-\right) - \frac{gm_e}{2M_Z\cos\theta_W}i\overline{e}\gamma_5 ez. \quad (16.45)$$

It can be shown that, irrespective of the nature of the scalar multiplet that is responsible for symmetry breaking, the unphysical Higgs boson couplings with fermions must be of this form. This statement will be explained in Ch. 17 in a somewhat more general setting. The Feynman rules of all coupling discussed in this section have been summarized in Fig. 16.3.

One might wonder why we did not include cross terms like $\overline{\Psi}_{\ell \mathrm{L}}\phi\ell'_{\mathrm{R}}$, where $\ell' \neq \ell$, in the Yukawa Lagrangian of Eq. (16.41). Of course such terms would have been gauge invariant just as the $\ell' = \ell$ terms, because all generations transform the same way under the gauge group. Our excuse is that there is really no need to write the cross terms. The different left-chiral multiplets all transform the same way under the gauge group. Therefore, any linear combination of them would also transform the same way. The same can be said about the right-chiral fields. We can always make suitable combinations such that all cross couplings vanish. That is what we have done in writing Eq. (16.41). In other words, we have not compromised generality in writing Eq. (16.41). We have merely used the freedom available to write it in a convenient basis.

16.6 Connection with Fermi theory

Fermi interactions, as described in detail in Ch. 14, involve four fermionic field operators. In §14.9, we showed how such interactions can arise in the low energy limit of a theory with vector boson interchange between fermions. Here we discuss such low energy limits of various processes involving leptons that occur in the standard model.

16.6.1 Charged-current induced processes

We first discuss muon decay. The muon and the muon-neutrino will have the same kind of interactions that the electron and the electron-neutrinohas. Muon decay can occur at the tree level through the diagrams of Fig. 16.4.

The tau decay occurs through similar diagrams, and need not be discussed separately.

Let us write the amplitude of the W-mediated diagram in the 't Hooft–Feynman gauge. We maintain the notation for different 4-momenta that was used in §14.6. Using the couplings and the propagator deduced earlier, we obtain

$$i\mathcal{M} = \left(-\frac{ig}{\sqrt{2}}\right)^2 \left[\overline{u}_{q_1}\gamma^\lambda \mathbb{L}u_p\right] \frac{-ig_{\lambda\rho}}{(p-q_1)^2 - M_W^2} \left[\overline{u}_k \gamma^\rho \mathbb{L}v_{q_2}\right]. \quad (16.46)$$

We have put the momenta as the subscript of the spinors. The notations for momenta introduced in Fig. 16.4 tell us which mass should be involved in each spinor.

Note that the term M_W^2 in the propagator is accompanied by $(p-q_1)^2$, which can be evaluated in the muon rest-frame, taking the neutrino to be massless:

$$(p-q_1)^2 = m_\mu^2 - 2m_\mu E_{\nu_\mu}. \quad (16.47)$$

Clearly, the value of this expression is smaller than m_μ^2, and therefore negligibly small compared to M_W^2. Ignoring the momentum-dependent term, we can write

$$\mathcal{M} = -\frac{g^2}{2}\left[\overline{u}_{q_1}\gamma^\lambda \mathbb{L}u_p\right]\frac{1}{M_W^2}\left[\overline{u}_k \gamma_\lambda \mathbb{L}v_{q_2}\right]. \quad (16.48)$$

Compare this expression with Eq. (14.101, *p 431*). Remembering that $\mathbb{L} = \frac{1}{2}(1-\gamma_5)$, we see that in the limit of low energy, the standard model produces Fermi interactions, with the Fermi constant identified by

$$\frac{G_F}{\sqrt{2}} = \frac{g^2}{8M_W^2}. \quad (16.49)$$

The resulting differential energy spectrum and total decay rate have already been discussed in §14.6, and there is no need to repeat them here.

The exercise is not quite complete yet, because we have not shown that this result is gauge invariant. This is easy to see. In a general gauge, we will obtain two types of extra contributions over what has been presented already. First of all, there are extra terms in the gauge boson propagator. All these terms contain the factor $q_\lambda q_\rho$, where q is the 4-momentum of the virtual W line. Thus, the contribution coming from these terms will contain the factors

$$\left[\overline{u}_{q_1}\gamma^\lambda \mathbb{L}u_p\right] q_\lambda q_\rho \left[\overline{u}_k \gamma^\rho \mathbb{L}v_{q_2}\right]. \quad (16.50)$$

Since $q = k + q_2$, we find

$$q_\rho \left[\overline{u}_k \gamma^\rho \mathbb{L}v_{q_2}\right] = \left[\overline{u}_k (\not{k} + \not{q}_2)\mathbb{L}v_{q_2}\right]. \quad (16.51)$$

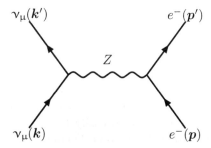

Figure 16.5: Tree level diagram for elastic scattering of muon-neutrinos off electrons. Unlike the case of muon decay, there is no additional diagram mediated by neutral scalars, because the neutrinos do not couple to neutral scalars.

We can now use the Dirac equation for the spinors to show that this expression contains a factor of the electron masses. Since the electron mass is negligible compared to the muon mass, these contributions have to be neglected.

The second type of extra contribution to the amplitude comes from the diagram of Fig. 16.4b, where a charged scalar boson is mediated. But couplings of fermions with charged scalars are proportional to fermion masses, as shown in Fig. 16.3. Therefore, this contribution is also negligible for the same reason.

☐ **Exercise 16.11** *Suppose we do not neglect the electron mass. The gauge-dependent terms must still cancel in the amplitude. Show that this is indeed true for the diagrams of Fig. 16.4.*

☐ **Exercise 16.12** *Give an example of a scattering process that occurs through charged current only.*

16.6.2   Neutral-current induced processes

Note that only the charged current contributes to the amplitude of muon decay. There cannot be any neutral current contribution, because for there to be one, the muon and the electron will have to be involved in the same vertex, which is impossible, since they belong to different multiplets of the gauge group. Gauge interactions can change only one member of a multiplet to another of the same multiplet, as emphasized earlier.

For the same reason, there cannot be any charged current that connects the electron and the muon-neutrino. The partner of the electron in the SU(2) doublet is, by definition, the electron-neutrino or ν_e. The muon-neutrino or the ν_μ is the partner of the muon in the same way. If only charged currents existed, there could not have been any interaction between the electron and the muon-neutrino at the tree level.

With the presence of neutral currents, the situation is different. In Fig. 16.5, we have shown how the elastic scattering process,

$$\nu_\mu(k) + e^-(p) \to \nu_\mu(k') + e^-(p'), \tag{16.52}$$

can be mediated by the Z boson. In this case, there is no diagram mediated by the neutral scalars, because neutrinos do not couple to them. The amplitude of this process can be written down easily by using the couplings of the Z boson derived earlier. Using the 't Hooft–Feynman gauge, we obtain

$$i\mathcal{M} = \left(-\frac{ig}{\cos\theta_W}\right)^2 \left[\overline{u}_{p'}\gamma^\lambda\left(-\frac{1}{2}\mathbb{L} + \sin^2\theta_W\right)u_p\right]$$
$$\times \frac{-ig_{\lambda\rho}}{(p-q_1)^2 - M_Z^2}\left[\overline{u}_{k'}\gamma^\rho\frac{1}{2}\mathbb{L}u_k\right]. \tag{16.53}$$

At low energies, when all momenta can be neglected in comparison with M_Z, this expression reduces to

$$\mathcal{M} = -\left(\frac{g^2}{2M_Z^2\cos^2\theta_W}\right)\left[\overline{u}_{p'}\gamma^\lambda\left(-\frac{1}{2}\mathbb{L} + \sin^2\theta_W\right)u_p\right]$$
$$\left[\overline{u}_{k'}\gamma_\lambda\mathbb{L}u_k\right]. \tag{16.54}$$

Using the mass relation between the W and the Z bosons, given in Eq. (16.29), we can write this expression in the form

$$\mathcal{M} = -\frac{4G_F}{\sqrt{2}}\left[\overline{u}_{p'}\gamma^\lambda\left(-\frac{1}{2}\mathbb{L} + \sin^2\theta_W\right)u_p\right]\left[\overline{u}_{k'}\gamma_\lambda\mathbb{L}u_k\right], \tag{16.55}$$

where we have introduced the Fermi constant through Eq. (16.49). This is exactly the form taken in Eq. (14.79, *p 426*), where we used two parameters for the electron bilinear part. Those parameters can now be identified as

$$c_V = -\frac{1}{2} + 2\sin^2\theta_W, \qquad c_A = -\frac{1}{2}. \tag{16.56}$$

The consequences of this amplitude have already been described in §14.4.

16.6.3 Processes induced by both types of currents

There can also be processes where both neutral and charged currents contribute. As an example, consider the elastic scattering of electron-neutrinos with electrons. The diagrams have been shown in Fig. 16.6. As we see, diagram (a) is the neutral current contribution, whereas diagram (b) is the charged current contribution.

Let us write the amplitudes of the two diagrams in the limit where the gauge boson masses are much larger compared to the 4-momenta of all other particles. Employing the 't Hooft–Feynman gauge, we find that the neutral

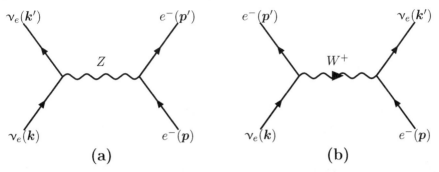

Figure 16.6: Gauge-boson mediated tree-level diagrams for elastic scattering of electron-neutrinos off electrons.

current contribution to the amplitude, $\mathscr{M}_a$, is equal to the expression given in Eq. (16.54). And the charged current contribution is given by

$$\mathscr{M}_b = \frac{g^2}{2M_W^2}\left[\overline{u}_{k'}\gamma^\lambda \mathrm{L} u_p\right]\left[\overline{u}_{p'}\gamma_\lambda \mathrm{L} u_k\right]. \tag{16.57}$$

Note that the expression looks very similar to that in Eq. (16.48). There are, of course, some notational differences. But more importantly, there is a difference in the overall sign of the expression. This is because, as far as the external lines are concerned, the two diagrams of Fig. 16.6 differ by an exchange of a pair of lines.

The total amplitude is the sum of $\mathscr{M}_a$ and $\mathscr{M}_b$. Before adding them up, we should notice that the orderings of the spinors are not the same in Eqs. (16.54) and (16.57), and they can be made the same if we perform a Fierz transformation on any one of them. In Eq. (14.75, *p425*), we noticed that the $V - A$ combinations are invariant under Fierz transformations. Since the charged current amplitude has a $V - A$ form, it is more convenient to apply Fierz transformation on it. It has to be remembered that in §14.3 we dealt with field operators, for which an extra minus sign appears while performing the Fierz transformation on it. For spinors, whose components are ordinary numbers, this sign does not appear, so the result of Fierz transformation will be

$$\left[\overline{u}_1\gamma^\mu \mathrm{L} u_2\right]\left[\overline{u}_3\gamma_\mu \mathrm{L} u_4\right] = -\left[\overline{u}_1\gamma^\mu \mathrm{L} u_4\right]\left[\overline{u}_3\gamma_\mu \mathrm{L} u_2\right], \tag{16.58}$$

where $u_1, \cdots, u_4$ are arbitrary spinors. Applying this, we can write

$$\mathscr{M}_b = -\frac{g^2}{2M_W^2}\left[\overline{u}_{p'}\gamma^\lambda \mathrm{L} u_p\right]\left[\overline{u}_{k'}\gamma_\lambda \mathrm{L} u_k\right]. \tag{16.59}$$

Adding this with the neutral current contribution, we obtain

$$\mathscr{M} = -\frac{4G_F}{\sqrt{2}}\left[\overline{u}_{p'}\gamma^\lambda\left(\frac{1}{2}\mathrm{L} + \sin^2\theta_W\right)u_p\right]\left[\overline{u}_{k'}\gamma_\lambda \mathrm{L} u_k\right]. \tag{16.60}$$

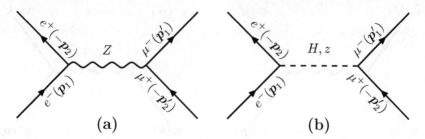

Figure 16.7: Diagram for muon–antimuon production mediated by the Z boson and the neutral scalars.

Again, this is of the form taken in Eq. (14.79, *p 426*) whose consequences have already been discussed, but with the identifications

$$c_V = \frac{1}{2} + 2\sin^2\theta_W\,, \qquad c_A = \frac{1}{2}\,. \tag{16.61}$$

16.7   Forward-backward asymmetry

In §16.6, we considered scattering and decay processes which are purely weak interaction processes. Now we discuss an interesting class of phenomena for which the dominant contribution is electromagnetic, but where weak interaction effects induce qualitative changes in the final effect.

The process that we consider is muon pair production. The electromagnetic contribution to the rate of this process was discussed in §5.4.2. The angular distribution of the scattering, given in Eq. (5.94, *p 133*), has the interesting feature of being forward-backward symmetric, as was commented during the discussion.

In the electroweak theory, there are other tree-level diagrams which contribute to this process: they have been shown in Fig. 16.7. We will assume that the energies involved are much higher than the muon mass, so that the masses of all external particles can be neglected. Since the couplings with the scalar fields are proportional to fermion masses, we can disregard the diagrams of Fig. 16.7b altogether. The amplitude is then the sum of two contributions. One of them is the photon contribution that was written in Eq. (5.87, *p 132*), and is quoted here for the sake of convenience:

$$\mathscr{M}_\gamma = \frac{e^2}{s}[\bar{v}_{\boldsymbol{p}_2}\gamma^\mu u_{\boldsymbol{p}_1}][\bar{u}_{\boldsymbol{p}_1'}\gamma_\mu v_{\boldsymbol{p}_2'}]\,. \tag{16.62}$$

The other is the contribution from the Z-mediated diagram. Using Eq. (16.28), let us write the Z boson interaction with charged leptons ℓ in the form

$$\mathscr{L}_{\text{int}} = -\frac{e}{\sin 2\theta_W}\bar{\ell}\gamma^\mu(c_V + c_A\gamma_5)\ell\,, \tag{16.63}$$

with c_V and c_A given by Eq. (16.56). Then the amplitude for the Z boson mediated diagram is given by

$$\mathscr{M}_Z = \frac{e^2 f_Z}{\mathsf{s}} [\bar{v}_{\boldsymbol{p}_2} \gamma^\mu (c_V + c_A \gamma_5) u_{\boldsymbol{p}_1}][\bar{v}_{\boldsymbol{p}_1'} \gamma^\mu (c_V + c_A \gamma_5) u_{\boldsymbol{p}_2'}], \quad (16.64)$$

where

$$f_Z = \frac{\mathsf{s}}{(\mathsf{s} - M_Z^2) \sin^2 2\theta_W}. \quad (16.65)$$

The expression for the cross-section will contain the quantity

$$\overline{|\mathscr{M}|^2} = \frac{1}{4} \sum_{\text{spins}} \left| \mathscr{M}_\gamma + \mathscr{M}_Z \right|^2 = \overline{|\mathscr{M}_\gamma|^2} + \overline{|\mathscr{M}_Z|^2} + \overline{\mathscr{M}_\gamma^* \mathscr{M}_Z} + \text{c.c.} \quad (16.66)$$

The calculation for determining $\overline{|\mathscr{M}_\gamma|^2}$ was shown in some detail in §5.4.2. Since here we are neglecting the mass of the muon, Eq. (5.93, *p 133*) reduces to

$$\overline{|\mathscr{M}_\gamma|^2} = e^4 (1 + \cos^2 \theta), \quad (16.67)$$

where θ is the scattering angle. Going through similar-looking steps, we obtain

$$\overline{|\mathscr{M}_Z|^2} = e^4 f_Z^2 \left[(c_V^2 + c_A^2)^2 (1 + \cos^2 \theta) + 8 c_V^2 c_A^2 \cos \theta \right]. \quad (16.68)$$

The cross terms can also be evaluated easily, and we obtain

$$\overline{\mathscr{M}_\gamma^* \mathscr{M}_Z} + \text{c.c.} = 2e^4 f_Z \left[c_V^2 (1 + \cos^2 \theta) + 2 c_A^2 \cos \theta \right]. \quad (16.69)$$

Adding all contributions, we obtain

$$\overline{|\mathscr{M}|^2} = e^4 \left[(1 + a_1)(1 + \cos^2 \theta) + a_2 \cos \theta \right], \quad (16.70)$$

where

$$\begin{aligned} a_1 &= f_Z^2 (c_V^2 + c_A^2)^2 + 2 f_Z c_V^2, \\ a_2 &= 8 f_Z^2 c_V^2 c_A^2 + 4 f_Z c_A^2. \end{aligned} \quad (16.71)$$

Note that the a_1 term has the same angular distribution as the original QED contribution. The a_2 term, on the other hand, generates a forward-backward asymmetry. Formally, the amount of forward-backward asymmetry can be defined as the difference between the scattering in the forward hemisphere ($0 < \theta < \pi/2$) and in the backward hemisphere ($\pi/2 < \theta < \pi$), normalized by the sum of the two quantities.

$$\mathcal{A} = \frac{\int_0^{\pi/2} d\theta \sin \theta \frac{d\sigma}{d\Omega} - \int_{\pi/2}^{\pi} d\theta \sin \theta \frac{d\sigma}{d\Omega}}{\int_0^{\pi} d\theta \sin \theta \frac{d\sigma}{d\Omega}}. \quad (16.72)$$

It is straightforward to check that with the angular dependence given by Eq. (16.70), one obtains

$$\mathcal{A} = \frac{3}{8}\frac{a_2}{1+a_1}\,. \tag{16.73}$$

The forward-backward asymmetry, therefore, is induced by a non-zero value of a_2. Looking at the expression for a_2, we find that it vanishes when $c_A = 0$. Hence, it is the axial vector coupling which is responsible for forward-backward asymmetry.

We might wonder whether there is a symmetry of the amplitude that ensures forward-backward asymmetry. In Ex. 6.7 *(p 158)*, we pointed out that the simultaneous presence of polar and axial vector currents violates parity and charge conjugation symmetries. But forward-backward asymmetry cannot be caused by a violation of either of these symmetries, because the asymmetric term is proportional to $\cos\theta$, i.e., $\boldsymbol{p}_1\cdot\boldsymbol{p}'_1$, which is invariant under both of these symmetries. Looking back at the electromagnetic amplitude of the process given in Eq. (16.62), we find that it has polar vector bilinears of the electron spinors and the muon spinors. Now suppose we consider a transformation

$$u_{(e)} \to v_{(e)}\,, \tag{16.74}$$

where the subscripted letter denotes that the transformation is applied only to the electron spinors and not to the muon spinors. This can be seen as a charge conjugation on the electron field only, leaving all other fields unchanged. Under this transformation, the electron spinor bilinear changes as follows:

$$\overline{v}_{\boldsymbol{p}_2}\gamma^\mu u_{\boldsymbol{p}_1} \to \overline{u}_{\boldsymbol{p}_2}\gamma^\mu v_{\boldsymbol{p}_1}\,. \tag{16.75}$$

But

$$\overline{u}_{\boldsymbol{p}_2}\gamma^\mu v_{\boldsymbol{p}_1} = v_{\boldsymbol{p}_2}^\top\,\mathbb{C}^{-1}\gamma^\mu\gamma_0\,\mathbb{C}u_{\boldsymbol{p}_1}^* \tag{16.76}$$

because of the conjugation relations between the u- and the v-spinors, as given in Eq. (F.80, *p 746*). Since the object is a number, we can take the transpose of the right hand side and obtain

$$\overline{u}_{\boldsymbol{p}_2}\gamma^\mu v_{\boldsymbol{p}_1} = u_{\boldsymbol{p}_1}^\dagger\,\mathbb{C}\gamma_0^\top\,(\gamma^\mu)^\top\mathbb{C}^{-1}v_{\boldsymbol{p}_2}\,, \tag{16.77}$$

using the antisymmetry of the matrix $\mathbb{C}$. Using the defining property of $\mathbb{C}$, Eq. (F.17, *p 738*), we can now write

$$\overline{u}_{\boldsymbol{p}_2}\gamma^\mu v_{\boldsymbol{p}_1} = -\,\overline{u}_{\boldsymbol{p}_1}\gamma^\mu v_{\boldsymbol{p}_2}\,. \tag{16.78}$$

Since the Mandelstam variable $\mathfrak{s}$ is symmetric under the interchange of p_1 and p_2, we find that, under the transformation of Eq. (16.74), the electromagnetic amplitude changes by an overall minus sign, which has no effect in the cross-section.

Let us now see what is the effect of this statement on forward-backward asymmetry. Eq. (16.74) basically implies interchanging the electron with the positron. Under this transformation, p_1 will change to p_2, i.e., to $-p_1$ in the CM frame, whereas p_1' and p_2' will remain unaffected. The quantity $p_1 \cdot p_1'$ will then change sign, which means that $\cos\theta$ will change sign. Thus, if Eq. (16.74) is a symmetry of the cross-section formula, there cannot be any linear term in $\cos\theta$ in the angular distribution. This guarantees forward-backward symmetry.

The symmetry is lost if there is an axial vector contribution to the amplitude, because, under the transformation of Eq. (16.74), the axial vector current transforms as

$$\overline{u}_{p_2}\gamma^\mu\gamma_5 v_{p_1} = +\overline{u}_{p_1}\gamma^\mu\gamma_5 v_{p_2}\,, \qquad (16.79)$$

which can be easily checked. Thus, the amplitude does not change by an overall sign in presence of both polar and axial vector currents, and forward-backward asymmetry is generated.

☐ **Exercise 16.13** Follow steps similar to those used for obtaining Eq. (16.78) to prove Eq. (16.79).

Chapter 17

Electroweak interaction of hadrons

In Ch. 16, we have introduced the standard electroweak theory, discussing only processes involving leptons. In this chapter, we want to extend the discussion to processes involving hadrons as well. Since hadrons are made of quarks, we start with the status of quarks in the standard electroweak model.

17.1 Quarks in standard model

Like leptons, quarks also come in three generations. As mentioned earlier in the book, there are three quarks with charge $+\frac{2}{3}$, which are represented by the letters u, c and t. And then there are the three quarks d, s and b, each of which carries a charge $-\frac{1}{3}$.

Like leptons, we will take the left-chiral quarks in doublets, and the right-chiral quarks in singlets of the $SU(2)_L$ part of the standard model gauge group. The weak hypercharges of these multiplets can be determined from Eq. (16.18, $p\,465$). These representations of the quarks under the gauge group are summarized in this equation:

$$
\begin{pmatrix} u_L \\ d'_L \end{pmatrix}, \begin{pmatrix} c_L \\ s'_L \end{pmatrix}, \begin{pmatrix} t_L \\ b'_L \end{pmatrix} \quad : \quad (2, \tfrac{1}{6}),
$$

$$
u_R, \qquad c_R, \qquad t_R \quad : \quad (1, \tfrac{2}{3}), \tag{17.1}
$$

$$
d'_R, \qquad s'_R, \qquad b'_R \quad : \quad (1, -\tfrac{1}{3}).
$$

Note that we have added primes while writing the down-type quarks, i.e., quarks with charges $-\frac{1}{3}$, because we do not want to commit at this stage who the partners are in the doublets. As we will see, this is an important issue.

There is nothing new about the gauge bosons or the scalar fields: this part of the model has already been described in Ch. 16. The mechanism of symmetry breaking, the masses of gauge bosons, are all still given by the expressions in that chapter. The main difference occurs in the Yukawa sector, i.e., of the interaction of quarks with the scalar multiplet. Using the gauge properties of the quark fields, we find that the following interactions are allowed:

$$\mathscr{L}_Y = -\sum_{A,B} \left(h^{(d)}_{AB} \overline{q}_{A\mathrm{L}} \phi d'_{B\mathrm{R}} + h^{(u)}_{AB} \overline{q}_{A\mathrm{L}} \widetilde{\phi} u_{B\mathrm{R}} \right) + \text{h.c.} \tag{17.2}$$

We owe the reader a lot of explanation regarding the notation used in this equation. The indices A, B are generation indices which run from 1 to 3. The quantities $h^{(d)}_{AB}$ and $h^{(u)}_{AB}$, for each value of A and B, represent a coupling constant. The notation q_{L} represents a left-chiral quark doublet. Along with the generational index, it can stand for one of the three doublets shown in Eq. (17.1). For example, $q_{1\mathrm{L}}$ will be the first doublet shown there, and so on. Similarly, $u_{B\mathrm{R}}$ for $B = 1$ is the first right-chiral up-type quark, i.e., quark of charge $+\frac{2}{3}$. By the same token, $d'_{2\mathrm{R}}$ means s'_{R}, whereas $d'_{3\mathrm{R}}$ means b'_{R}. And finally, we need to explain what is ϕ. As mentioned in §8.7.2, for any doublet ψ of an SU(2), the object $\varepsilon \psi^*$ also transforms as a doublet, where ε is the completely antisymmetric 2×2 matrix. Thus, from the scalar doublet ϕ given in Eq. (16.19, $p\,465$), we can define the object

$$\widetilde{\phi} \equiv \begin{pmatrix} 0 & 1 \\ -1 & 0 \end{pmatrix} \begin{pmatrix} \phi_+ \\ \phi_0 \end{pmatrix}^* = \begin{pmatrix} \phi_0^* \\ -\phi_- \end{pmatrix}, \tag{17.3}$$

which will also be a doublet under the SU(2) part of the gauge group. Because it involves a conjugation, its properties under the U(1) part of the gauge group will be opposite to that of ϕ, i.e., the representation of $\widetilde{\phi}$ will be given by

$$\widetilde{\phi} \equiv \begin{pmatrix} \phi_0^* \\ -\phi_- \end{pmatrix} : (2, -\tfrac{1}{2}). \tag{17.4}$$

It can now easily be checked that the interactions given in Eq. (17.2) are gauge invariant.

As we have seen for the case of leptons, spontaneous symmetry breaking produces mass terms for fermions. This is also true for quarks. Using the vacuum expectation value of the component ϕ_0, we find the following quadratic terms in the quark fields that arise out of the gauge invariant Yukawa interactions:

$$\mathscr{L}_{\text{mass}} = -\frac{v}{\sqrt{2}} \sum_{A,B} \left(h^{(d)}_{AB} \overline{d}'_{A\mathrm{L}} d'_{B\mathrm{R}} + h^{(u)}_{AB} \overline{u}_{A\mathrm{L}} u_{B\mathrm{R}} \right) + \text{h.c.} \,. \tag{17.5}$$

In the generation indices, $h^{(u)}$ and $h^{(d)}$ are matrices. We can choose a basis in which one of these matrices, say $h^{(u)}$, is diagonal. By adjusting the phases of the fields $u_{A\mathrm{R}}$, we can also make sure that the diagonal entries of this matrix are real and positive. (This choice does not affect any other part of the theory since the right-chiral quarks are stand-alone singlet fields, which means that their definitions are independent of the definitions of anything else.) Once this is done, we obtain the mass terms for the three up-type quarks.

But there is no guarantee that $h^{(d)}$ will be diagonal in this basis. In fact, there is no reason for it to be so. Therefore, for the down-type quarks, the mass terms that we obtain are of the form

$$-\sum_{A,B} \overline{d}'_{A\mathrm{L}} M_{AB} d'_{B\mathrm{R}} + \text{h.c.},\tag{17.6}$$

where

$$M_{AB} = h^{(d)}_{AB} v/\sqrt{2}.\tag{17.7}$$

In the generation space, this is a matrix. It can be called the *mass matrix* for down-type quarks in this basis. Because this matrix is not diagonal, there are cross terms of the form $\overline{d}'_{\mathrm{L}} s'_{\mathrm{R}}$ and so on, which should not be there if we are dealing with fields of physical particles.

And this is precisely the point of putting the primes on the down-type fields in Eq. (17.1): the fields d', s' and b' do not correspond to physical particles. To identify the physical fields, we have to diagonalize the matrix M_{AB}.

Usually, we use similarity transformations involving unitary matrices in order to diagonalize Hamiltonians in quantum mechanics. This is a special case, which works for the so-called normal matrices, i.e., matrices which commute with their hermitian conjugates: $[M, M^\dagger] = 0$. This class includes hermitian matrices, unitary matrices, and many others.

In the case at hand, however, there is no guarantee that the matrix is normal, so we need to find a method of diagonalization that works in more general cases. And to this end, we use the following theorem:

Theorem 1 *For any matrix A, we can find two unitary matrices U_L and U_R such that $U_L^\dagger A U_R$ is diagonal, with real non-negative entries along the diagonal.*

The proof is simple. For any matrix A, the matrix $A^\dagger A$ must be hermitian. Therefore, there exists a unitary matrix U such that

$$U^\dagger A^\dagger A U = D^2,\tag{17.8}$$

where D^2 is diagonal. The diagonal elements are given by

$$\left(D^2\right)_{aa} = \sum_b \left|\left(AU\right)_{ba}\right|^2,\tag{17.9}$$

and are therefore real and non-negative. Let us now define a matrix D, any element of which is the positive square root of the corresponding element of D^2, and define the matrix

$$H \equiv U D U^\dagger.\tag{17.10}$$

Clearly, H is hermitian, and

$$H^2 = U D^2 U^\dagger = A^\dagger A,\tag{17.11}$$

using Eq. (17.8) in the last step. From this relation, it is easy to see that AH^{-1} is a unitary matrix, which we will call U'. Then $A = U'H = U'UDU^\dagger$. Renaming $U'U = U_L$ and $U = U_R$, we can rewrite this relation as

$$U_L^\dagger A U_R = D,\tag{17.12}$$

which proves the theorem. In the derivation, we have assumed that the matrix H has an inverse. If not, i.e., if H is singular, the result is still true although the proof is a little more complicated.

The diagonalization involves two different unitary matrices at the two ends, and is therefore called a bi-unitary transformation. Physically, it means changing the basis for left-chiral and right-chiral fields by different amounts, so that the matrix that they sandwich is diagonal. More explicitly, identifying M to be the matrix A in Eq. (17.12), we can write

$$M = U_L D U_R^\dagger. \tag{17.13}$$

The mass terms for the down-type quarks, given in Eq. (17.6), can then be rewritten as

$$-\sum_{A,B} \overline{d}'_{AL} \left(U_L D U_R^\dagger \right)_{AB} d'_{BR} + \text{h.c.}, = -\sum_{A} \overline{d}_{AL} D_{AA} d_{AR} + \text{h.c.}, \tag{17.14}$$

where we have defined a new set of fields, dubbed d_A, by the relations

$$d_{AL} = \left(U_L^\dagger \right)_{AB} d'_{BL}, \qquad d_{AR} = \left(U_R^\dagger \right)_{AB} d'_{BR}. \tag{17.15}$$

These unprimed fields would then be the mass eigenstates, because their mass terms do not mix. This also means that the partners of u, c and t quarks in the $SU(2)_L$ doublets shown in Eq. (17.1) are not mass eigenstates, which is why we started by marking them with primes.

It should be obvious that there is nothing sacred about the up-type quarks that we do not have to use primed fields for them. We could have started with linear combinations of the doublets shown in Eq. (17.1) such that the down-type quarks appearing the doublets were the mass eigenstates, whereas the up-type quarks were superpositions of mass eigenstates. It is only a matter of convention or a choice of basis in the generation space. We could have also started with a most general basis where neither up-type nor down-type quarks in a doublet would have been a mass eigenstate. Then we would have needed to diagonalize both up-type and down-type mass matrices by bi-unitary transformations. We would have gained nothing by doing this: we would only have to wrestle more with the notations.

17.2 Gauge interaction of quarks

Let us now discuss the interaction of quarks with the electroweak gauge bosons. These come from the terms containing the action of the gauge co-variant derivative on the fermion fields. For leptons, the covariant derivatives were given in Eq. (16.34, *p 469*). For quarks, we can write a similar set of equations, substituting the appropriate values of the weak hypercharges:

$$D_\mu q_{AL} = \left(\partial_\mu + ig \frac{\tau^a}{2} W_\mu^a + ig'(-\frac{1}{6}) B_\mu \right) q_{AL},$$

$$D_\mu u_{AR} = \left(\partial_\mu + ig'(+\frac{2}{3}) B_\mu \right) u_{AR},$$

$$D_\mu d'_{AR} = \left(\partial_\mu + ig'(-\frac{1}{3}) B_\mu \right) d'_{AR}. \tag{17.16}$$

Note that the same formulas apply for fields in all generations.

Let us first look at the interactions of the charged gauge bosons, which come from the gauge covariant derivative of q_{L} only. These terms will be given by

$$\mathscr{L}_{\mathrm{cc}} = -\frac{g}{\sqrt{2}} \sum_A \left(\overline{u}_{A\mathrm{L}} \gamma^\mu W_\mu^+ d'_{A\mathrm{L}} + \overline{d}'_{A\mathrm{L}} \gamma^\mu W_\mu^- u_{A\mathrm{L}} \right). \qquad (17.17)$$

As mentioned earlier, the partners of the u_A quarks in the doublets are not eigenstates of mass. It would be more useful to write these interactions in terms of fields of physical particles. This can be easily done by using Eq. (17.15), and the result is

$$\mathscr{L}_{\mathrm{cc}} = -\frac{g}{\sqrt{2}} \sum_{A,B} \left(\overline{u}_{A\mathrm{L}} \gamma^\mu W_\mu^+ V_{AB} d_{B\mathrm{L}} + \overline{d}_{B\mathrm{L}} V_{AB}^* \gamma^\mu W_\mu^- u_{A\mathrm{L}} \right), \qquad (17.18)$$

where $V = U_L$. This shows that the charged current couples any up-type quark to a down-type quark of any generation. This phenomenon is known as *quark mixing*, and was first discussed by Cabibbo. Later, Kobayashi and Maskawa noticed that this matrix can be responsible for CP violation, a subject that will be discussed in Ch. 21. Using the initials of these three scientists, the quark mixing matrix is often called the *CKM matrix*.

The phrase *quark mixing* implies that the charged current gauge interactions mix between quarks of different generations. As a result, there cannot be any generational quantum number for quarks which is conserved. Quantum numbers like strangeness, which are associated with a single flavor of quark, are also violated because of the charged current interactions.

Let us look at the neutral current interactions now. It is easy to see that the matrices U_L and U_R will not appear in the final expressions in this sector. The reason is easy to understand. Consider the gauge interactions of the field $d'_{A\mathrm{R}}$. These are,

$$\frac{1}{3} g' \sum_A \overline{d}'_{A\mathrm{R}} \gamma^\mu B_\mu d'_{A\mathrm{R}}. \qquad (17.19)$$

But this can also be written as

$$\frac{1}{3} g' \sum_A \overline{d}_{A\mathrm{R}} \gamma^\mu B_\mu d_{A\mathrm{R}}, \qquad (17.20)$$

because the extra factors of the matrix U_R that appear through the relation of Eq. (17.15) disappear due to unitarity of the matrix U_R. The same thing happens while we deal with the gauge interaction terms of the left-chiral fields. So finally, the interactions of the photon appear in the standard form, with the appropriate charges of the quarks, and the interactions of the Z boson also have the form given in Fig. 16.2 *(p 469)*. Neutral current interactions are therefore flavor diagonal, i.e., do not change quark flavor. This is called the absence of *flavor-changing neutral current*, or *FCNC* for short. Of course, the statement about the absence of FCNC pertains only to the tree level. FCNC can arise from loop interactions, as we will see later in §17.9.5.

17.3 CKM matrix and its parametrization

It is important to reassess the results that we have obtained. In the leptonic sector, we found that the gauge interactions contain only two parameters, the gauge coupling constants g and g'. In the quark sector, this is no more the case. The gauge interactions involve, apart from the gauge coupling constants, the elements of the unitary *CKM matrix* V that appears from the diagonalization of the quark masses. The elements of this matrix can be parametrized as follows:

$$V = \begin{pmatrix} c_1 & -s_1 c_3 & -s_1 s_3 \\ s_1 c_2 & c_1 c_2 c_3 - s_2 s_3 e^{i\delta} & c_1 c_2 s_3 + s_2 c_3 e^{i\delta} \\ s_1 s_2 & c_1 s_2 c_3 + c_2 s_3 e^{i\delta} & c_1 s_2 s_3 - c_2 c_3 e^{i\delta} \end{pmatrix}. \tag{17.21}$$

where $c_a \equiv \cos\theta_a$ and $s_a \equiv \sin\theta_a$. It should be realized that the parametrization is not unique: other useful ones will be discussed in Ch. 21. The angle θ_1 is called the *Cabibbo angle*, because Cabibbo introduced it in 1963 to describe mixing between two generations of quarks. The other two angles and the phase δ were necessary because of Kobayashi and Maskawa's conjecture about the existence of a third generation of quarks.

A most general unitary 3×3 matrix can have more parameters according to our counting in §3.4.1, but those other parameters can all be absorbed in the definitions of the quark fields, an issue that will be taken up in detail in Ch. 21. The interesting thing is that, even after utilizing all phase redefinitions, one phase cannot be removed. This can be responsible for CP violation, as was noticed by Kobayashi and Maskawa.

The phase need not be restricted to the lower-right 2×2 block of the CKM matrix, as it does in the original parametrization of Kobayashi and Maskawa presented in Eq. (17.21). By redefining the phases of the quark fields, we can move the phase around within the matrix. Here is another parametrization which has proved to be quite useful:

$$V = \begin{pmatrix} c_{12}c_{13} & s_{12}c_{13} & s_{13}e^{-i\delta_0} \\ -s_{12}c_{23} - c_{12}s_{23}s_{13}e^{i\delta_0} & c_{12}c_{23} - s_{12}s_{23}s_{13}e^{i\delta_0} & s_{23}c_{13} \\ s_{12}s_{23} - c_{12}c_{23}s_{13}e^{i\delta_0} & -c_{12}s_{23} - s_{12}c_{23}s_{13}e^{i\delta_0} & c_{23}c_{13} \end{pmatrix}. \tag{17.22}$$

Many other parametrizations are possible.

The elements of the CKM matrix modify the strength of charged current interactions, as seen from Eq. (17.18). Studying various charged current processes, it is therefore possible to estimate the magnitudes of the elements. We summarize such estimates here:

$$|V| = \begin{pmatrix} 0.97418 \pm 0.00027 & 0.2255 \pm 0.0019 & (3.93 \pm 0.36) \times 10^{-3} \\ 0.230 \pm 0.011 & 1.04 \pm 0.06 & (41.2 \pm 1.1) \times 10^{-3} \\ (8.1 \pm 0.6) \times 10^{-3} & (38.7 \pm 2.3) \times 10^{-3} & > 0.74 \end{pmatrix}. \tag{17.23}$$

Note that the matrix is almost the unit matrix, in the sense that the diagonal elements are almost equal to unity and the off-diagonal elements are all small. Moreover, the off-diagonal elements involving the first two generations are much larger than all other off-diagonal elements. More explicitly, in the parametrization of Eq. (17.22), we can write $s_{13} \ll s_{23} \ll s_{12} \ll 1$. Wolfenstein proposed to make this hierarchy explicit in the notation by introducing a parameter $\lambda = s_{12}$ to denote the smallness of all elements of the mixing matrix. Then s_{23}, which should be another order of smallness down from s_{12}, can be written as $A\lambda^2$. The element V_{ub} can be called $A\lambda^3(\rho - i\eta)$: cubic in λ because s_{13} is smaller than s_{23}, and the factor $\rho - i\eta$ to ensure that this element is complex. The rest of the elements can now be found from unitarity of the matrix. Of course this parametrization is approximate, so we can retain terms only up to $\mathcal{O}\left(\lambda^2\right)$, except where the leading term of any element is $\mathcal{O}\left(\lambda^3\right)$. The form of the matrix is

$$
V = \begin{pmatrix} 1 - \frac{1}{2}\lambda^2 & \lambda & A\lambda^3(\rho - i\eta) \\ -\lambda & 1 - \frac{1}{2}\lambda^2 & A\lambda^2 \\ A\lambda^3(1 - \rho - i\eta) & -A\lambda^2 & 1 \end{pmatrix}. \tag{17.24}
$$

Note that we have four parameters in this parametrization, as is there in any other one such as the matrices in Eq. (17.21) and Eq. (17.22). The advantage of this phenomenological parametrization is that all CP-violating effects, characterized by the parameter η without which the matrix would have been real, arise from the 13 and the 31 elements of the matrix. This is a big contrast with the parametrization of Eq. (17.22) from which the Wolfenstein parametrization has been derived. As we see, five of the nine elements of the matrix of Eq. (17.22) contain the CP-violating phase δ_0. However, note for example the 22 element. Here, the phase factor $e^{i\delta_0}$ is multiplied by $s_{12}s_{23}s_{13}$. This product of the three sines is of order λ^6, and hence this term is too small to be of any phenomenological importance. Similar arguments apply to 21, 23 and 32 elements, leaving only two complex elements.

□ **Exercise 17.1** *Verify that the phase δ_0 does not contribute to the 21, 23 and 32 elements of the CKM matrix at $\mathcal{O}\left(\lambda^3\right)$.*

There was no analog of the CKM matrix in the leptonic sector because the standard model did not have any right-handed neutrino field. Therefore, there was only one kind of mass terms, viz., those of the charged leptons, and we could work in a basis of generations in which these terms were diagonal. For quarks, this freedom was lost because there is no fundamental reason to guarantee that we can take a basis in which both up-type and down-type quark masses are diagonal.

Let us now look at the flip side of the issue. We had to introduce two matrices in Eq. (17.15), U_L and U_R, in the process of applying the bi-unitary transformation to diagonalize quark matrices. Of these, U_L appeared in the charged current. But U_R seems to have vanished! It has not occurred anywhere in the currents. The reason is that we started with an unnecessarily complicated notation. The $d'_\mathbb{R}$, $s'_\mathbb{R}$ and $b'_\mathbb{R}$, mentioned in Eq. (17.1), all transform the same way under the gauge group, so any linear combination of them

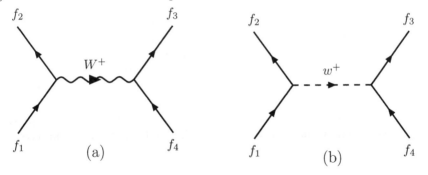

Figure 17.1: Tree-level diagrams for the process of Eq. (17.25).

would also transform the same way. Thus, we could have started with the fields d_R, s_R and b_R in that equation, the fields that would finally appear in the diagonal mass terms. That way, we never would have encountered the matrix U_R. This says that the matrix U_R is unphysical: it should not appear in any physical formula. The matrix U_L, or V, on the other hand, cannot be dispensed with even while writing the representations of the quark fields. If we choose to write the doublets using d_L, s_L and b_L, the upper components of the doublets would be superpositions of physical up-type quark fields, and this superposition would again involve the same matrix V.

Be it as it may, there are only a few parameters in the quark interactions, and all properties of hadrons should be expressible in terms of these few parameters in principle. In fact, in this chapter, we will further reduce the number of parameters by taking $\delta = 0$ and postponing the discussion of all CP-violating effects until Ch. 21. Even then, the main difficulty lies in evaluating the hadronic matrix elements. One has to make approximations and depend on symmetries only. The procedures introduce form factors in the calculation. We will see examples later in the chapter.

17.4 Yukawa interaction of quarks

As mentioned in Ch. 16, Yukawa interactions mean the interaction of fermions with spinless particles. There is really only one physical scalar in the standard model: all other three degrees of freedom present in the complex doublet are eaten up by the gauge bosons. Nevertheless, one often performs calculations in the gauges where the physical spectrum is not apparent, and in such gauges one has to deal with the unphysical modes as well. The propagators of the unphysical bosons have been given in Fig. 16.1 *(p 468)*. Here, we discuss the coupling of the unphysical bosons, as well as the physical Higgs boson, with the quarks.

We start with the unphysical Higgs bosons. Their propagators are determined by gauge invariance, as shown in §16.3. Hence, it might be suspected that their couplings should also be somehow governed by gauge invariance. And indeed, they are. To see this in some generality, consider two chiral fermions f_1 and f_2 which belong to the same multiplet of some gauge group, so that there is a coupling with f_1 coming in and $f_2 W$ going out of a vertex, W being a gauge boson. Suppose the Feynman rule for this coupling is ia^μ_{12}, which, for any value of the Lorentz index μ, is a matrix. The same applies to the two fermions f_3 and f_4, for which the Feynman rule for the coupling is ia^μ_{34}. Then there would be a W boson mediated diagram for the process

$$f_1 + f_4 \to f_2 + f_3 \,, \tag{17.25}$$

which has been shown in Fig. 17.1a. The amplitude of this diagram can be written as

$$i\mathcal{M}_a = \left[\bar{u}_2 ia^\mu_{12} u_1\right] iD^{(W)}_{\mu\nu}(k) \left[\bar{u}_3 ia^\nu_{34} u_4\right], \tag{17.26}$$

where $D^{(W)}_{\mu\nu}$ denotes the propagator for the gauge boson. In this formula, u_1 signifies the positive energy spinor with 4-momentum p^μ_1 of the particle f_1 whose mass is m_1, and so on for u_2, u_3 and u_4. The momentum of the gauge boson is obviously given by

$$k = p_1 - p_2 = p_3 - p_4 \tag{17.27}$$

through energy-momentum conservation. Now, suppose we use the general R_ξ-gauge propagator for the W boson, as given in Fig. 16.1 *(p 468)*. In this propagator, there are terms with the denominator $k^2 - M^2_W$, which have no dependence on the gauge parameter ξ. We don't worry about them. But then there is one part of the propagator that has a denominator $k^2 - \xi M^2_W$. Let us write the part of $\mathcal{M}_a$ that contains this part of the propagator:

$$\mathcal{M}^{(\xi)}_a = \left[\bar{u}_2 a^\mu_{12} u_1\right] \frac{k_\mu k_\nu}{M^2_W} \frac{1}{k^2 - \xi M^2_W} \left[\bar{u}_3 a^\nu_{34} u_4\right]. \tag{17.28}$$

This is, of course, a gauge-dependent contribution, and must vanish in the final amplitude. The cancellation must occur against the contribution of Fig. 17.1b, mediated by the unphysical charged Higgs boson w^+. Using the propagator of w^+ given in Fig. 16.1 *(p 468)*, this latter contribution can be written as

$$\mathcal{M}_b = -\left[\bar{u}_2 b_{12} u_1\right] \frac{1}{k^2 - \xi M^2_W} \left[\bar{u}_3 b_{34} u_4\right], \tag{17.29}$$

where b_{12} and b_{34} denotes the coupling of the w with the spinors. Looking at these formulas, we see that the cancellation is achieved if, between the spinors,

$$b_{12} = \frac{k_\mu}{M_W} a^\mu_{12}, \tag{17.30}$$

and a similar formula involving the other pair of fermions.

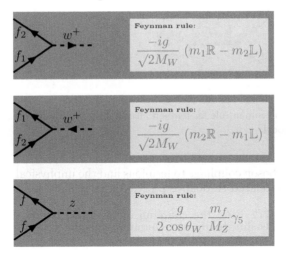

Figure 17.2: Feynman rules for fermion interactions with the unphysical Higgs bosons. As explained in the text, f_1 and f_2 are two fermion fields whose left-chiral projections form a doublet of $SU(2)_L$, and f denotes either one of them.

This is the general formula which applies for the coupling of an unphysical Higgs boson in any spontaneously broken gauge theory. For the $SU(2)_L \times U(1)_Y$ theory, $a^\mu_{12} = -(g/\sqrt{2})\gamma^\mu L$. Note that Eq. (17.27) implies

$$\left[\bar{u}_2\gamma^\mu L u_1\right]k_\mu = \bar{u}_2(\slashed{p}_1 - \slashed{p}_2)L u_1 = \bar{u}_2(R\slashed{p}_1 - \slashed{p}_2 L)u_1$$
$$= \bar{u}_2(m_1 R - m_2 L)u_1\,, \tag{17.31}$$

by the use of the Dirac equation for the spinors at the last step. The resulting Feynman rule for the $f_1 \to f_2 w^+$ vertex is presented in Fig. 17.2. Couplings of the unphysical neutral Higgs boson can also be determined through a similar argument. They also appear in Fig. 17.2.

A few comments. First, the couplings are really determined up to an overall sign. Had we defined b_{12} to be the negative of what we have shown in Eq. (17.30), the ξ-dependence in $\mathcal{M}_a + \mathcal{M}_b$ would have still canceled. This ambiguity is true, and is irrelevant. The field that is called w^+ could also be called $-w^+$, thus reversing the signs of all its couplings.

Secondly, we see that the rules given for leptons in Fig. 16.3 *(p 472)* are special cases of these general rules, as can be easily seen by putting $m_1 = 0$ and $m_2 = m_e$. And thirdly, it should be acknowledged that we have not invoked any details of the scalar sector in deriving these couplings. We have not assumed that there is one doublet of scalars in our model. More than one doublet may develop VEV and induce spontaneous symmetry breaking. There may be other kinds of multiplets responsible for driving the symmetry breaking. Whatever be the mechanism of symmetry breaking, we have shown

Figure 17.3: Feynman rule for fermion interactions with the physical Higgs boson of the standard model.

that the gauge boson couplings to fermions and the unphysical Higgs couplings to fermions are related through Eq. (17.30).

☐ **Exercise 17.2** Find the coupling of fermions to the unphysical neutral Higgs boson, taking the coupling with the Z boson from Fig. 16.2 (p 469).

☐ **Exercise 17.3** In deriving the Feynman rules for the unphysical Higgs bosons, we assumed that the coupling with the gauge boson W is purely left-chiral. This is not a necessity in a general gauge theory. If there are fermions whose right-chiral fields also transform non-trivially under the $SU(2)_L$ part of the gauge group, the gauge coupling can have the more general form $i\gamma^{\mu}(a\mathbb{L} + b\mathbb{R})$. Find the corresponding coupling for the unphysical charged Higgs boson.

The coupling of the physical Higgs boson H, on the other hand, is not necessarily determined by the gauge coupling and the fermion masses. In a model with a more elaborate choice of scalar multiplets, there can be other parameters that would enter this coupling. What we have shown in Fig. 17.3 is the coupling in the standard model with one Higgs doublet which we have been considering since Ch. 16.

Our results show that all Yukawa couplings in the standard model involve a ratio of a fermion mass and a gauge boson mass. This means that while computing physical amplitudes, the contributions of diagrams involving Yukawa couplings can be neglected in comparison with those involving gauge couplings, so far as we are dealing with fermions which are much lighter compared to the gauge bosons. In fact, this condition is satisfied for all known fundamental quarks and leptons except the top quark. So, in the processes mentioned in the rest of the chapter and in forthcoming chapters as well, we will often not even mention the Higgs boson mediated diagrams.

17.5   Leptonic decays of mesons

The main problem with hadronic processes, as described above, is the evaluation of the hadronic matrix elements. The difficulty increases with the number of hadrons in a process. So we first discuss the simplest possible processes containing only one hadron, the remaining particles being leptons.

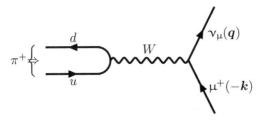

Figure 17.4: Quark-level Feynman diagram for charged pion decay. The fermion lines on the left side of the diagram correspond to one incoming u quark line and an incoming d antiquark, which together represent an incoming π^+.

17.5.1 Decay of charged pions

As discussed in §10.9, the charged pion is the lightest hadron that decays through weak interactions. The dominant decay mode of π^+ is given by

$$\pi^+(p) \to \mu^+(k)\nu_\mu(q)\,, \tag{17.32}$$

where we have put a notation for the 4-momenta of all particles that will be used in the calculations below.

Clearly, at the quark level, the process of Eq. (17.32) means

$$u\widehat{d} \to \mu^+\nu_\mu\,. \tag{17.33}$$

The tree-level diagram for the process is shown in Fig. 17.4. If we disregard the momentum dependence of the W propagator, the effective four-fermion interaction can be written as

$$\mathscr{L}_{\text{eff}} = \frac{G_F}{\sqrt{2}} V_{ud}^* \left[\overline{d}\gamma^\lambda(1-\gamma_5)u\right]\left[\overline{\nu}_\mu\gamma_\lambda(1-\gamma_5)\mu\right]. \tag{17.34}$$

Note that the CKM matrix element V_{ud} occurs in this effective Lagrangian, because the quark current between the d and the u quark is multiplied by that element in the interaction Lagrangian, as shown in Eq. (17.18).

We now have to take the matrix element of this effective Lagrangian between the initial and the final states of the process given in Eq. (17.32). For the leptonic part, this is trivial. Let us denote the leptonic current appearing in the second square bracket of Eq. (17.34) by $j_\lambda(x)$, where the co-ordinate dependence comes from the field operators. This dependence can be factored out in the form of an exponential, as shown in §5.7.1. This exponential does not enter the Feynman amplitude, but rather contributes to an energy-momentum conserving delta function. Once this factor is taken out, we are left with the matrix element of $j_\lambda(0)$, and for this, we can just use the u-spinors and v-spinors, as prescribed in Table 4.1 *(p 91)*. This gives

$$\left\langle \mu^+(k)\nu_\mu(q) \left| j_\lambda(0) \right| 0 \right\rangle = \overline{u}_{(\nu_\mu)}(q)\gamma_\lambda(1-\gamma_5)v_{(\mu)}(k)\,. \tag{17.35}$$

Note that in writing this matrix element, we have put the initial state to be the vacuum state. This is because the initial state contains no leptons, and is therefore the same as the vacuum state as far as the leptons are concerned.

Let us now look at the quark current that appears in a square bracket in Eq. (17.34), which we will call $J^\lambda(x)$. The co-ordinate dependence gives another exponential factor, as before, and we are faced with the task of finding the matrix element of $J^\lambda(0)$ between an initial pion state and a final vacuum state. The vacuum state appears in this discussion for the same reason it appeared in the context of leptons: as far as hadrons are concerned, the final state is no different from the vacuum state.

The problem is in finding the matrix element of this current, because the initial state does not contain free quarks: it contains a meson. The matrix element can be calculated only if we could solve the field theory with interacting quarks exactly, which would tell us how the quarks bind within the pion. But we don't know how to do that. So the best thing that we can do is to parametrize the matrix element.

We note that the current transforms like a vector under Lorentz transformations. The hadronic matrix element should then also transform as a vector, because the pion is a scalar state (i.e., the pion field transforms like a scalar), and so is the vacuum. Moreover, the matrix element should depend on the parameters of the states, and the only 4-vector relevant to the state is p. Thus, we can write

$$\langle 0 | J^\lambda(0) | \pi^+(p) \rangle = -\sqrt{2} i f_\pi p^\lambda \,, \tag{17.36}$$

where f_π is a constant. It is called the *pion decay constant.*

□ **Exercise 17.4** What is the mass dimension of f_π? [**Hint :** *Recall the dimension of one-particle states from §4.8.*]

Combining Eqs. (17.35) and (17.36), we can now write the amplitude for the pion decay process:

$$\mathcal{M} = -G_F V_{ud}^* f_\pi p^\lambda \, \overline{u}_{(\nu_\mu)}(q) \gamma_\lambda (1 - \gamma_5) v_{(\mu)}(k) \,. \tag{17.37}$$

Using the momentum conservation equation $p = q + k$ and the Dirac equations for the spinors, we can write it as

$$\mathcal{M} = -G_F V_{ud}^* f_\pi \, \overline{u}_{(\nu_\mu)}(q) \Big(\slashed{q} + \slashed{k} \Big)(1 - \gamma_5) v_{(\mu)}(k)$$
$$= G_F V_{ud}^* f_\pi m_\mu \, \overline{u}_{(\nu_\mu)}(q)(1 + \gamma_5) v_{(\mu)}(k) \,. \tag{17.38}$$

This gives

$$\overline{|\mathcal{M}|^2} = 8 G_F^2 |V_{ud}|^2 f_\pi^2 m_\mu^2 \, q \cdot k \,. \tag{17.39}$$

From kinematics, $q \cdot k = \frac{1}{2}(m_\pi^2 - m_\mu^2)$. So, putting the expression for $\overline{|\mathcal{M}|^2}$ into the formula for the two-body decay rate of Eq. (4.168, *p 97*), we obtain

$$\Gamma = \frac{1}{4\pi} G_F^2 |V_{ud}|^2 f_\pi^2 m_\mu^2 m_\pi \left(1 - \frac{m_\mu^2}{m_\pi^2} \right)^2 \,. \tag{17.40}$$

From the measurement of charged pion lifetime, one can find the value of the pion decay constant f_π.

□ **Exercise 17.5** *The measured lifetime for charged pion is 2.60×10^{-8} s. Using the value of the Fermi constant and the masses of the pion and the muon and assuming $V_{ud} \approx 1$, show that*

$$f_\pi = 93 \,\text{MeV} . \tag{17.41}$$

This is a cautionary note to point out that there are different conventions of defining f_π which differ by a factor of $\sqrt{2}$. Some people do not put the factor of $\sqrt{2}$ on the right hand side of Eq. (17.36), so for them f_π turns out to be about 130 MeV. The factor of $\sqrt{2}$ appears arbitrary and whimsical in the definition of the pion decay constant. Later in Eq. (18.38, $p\,536$), we will write this definition in an alternative form which will make the factor of $\sqrt{2}$ look more natural.

The factor of i that appears in Eq. (17.36) is just a convention, adopted so that the Feynman amplitude if Eq. (17.38) does not have any explicit factor of i.

The phenomenological part of the analysis is done, but we want to spend some time on some issues in the derivation. First, look at the definition of f_π in Eq. (17.36). The current J^λ has a polar vector part and an axial vector part, as seen in Eq. (17.34). Now, the pion is a pseudoscalar, i.e., its intrinsic parity is -1, whereas the vacuum state must have intrinsic parity equal to $+1$. The polar vector current behaves like any polar vector, so its spatial part has a negative intrinsic parity. The matrix element of this part between the pion and the vacuum should have positive intrinsic parity. But the expression on the right hand side involves p^λ, whose spatial part has negative intrinsic parity. This argument shows that the polar vector current does not contribute to the matrix element at all. It is only the axial vector current which does, and the matrix element is given in Eq. (17.36).

To get to the second interesting point, let us mention that the decay

$$\pi^+(p) \to e^+(k) \nu_e(q) \tag{17.42}$$

is also kinematically allowed. In fact, since the electron is much lighter than the muon, the available phase space to this channel is much larger than that to the channel $\mu^+ \nu_\mu$. The analysis for the process of Eq. (17.42) is exactly similar, with electron mass occurring in place of muon mass in the final formula. Thus we obtain

$$\frac{\Gamma(\pi^+ \to e^+ \nu_e)}{\Gamma(\pi^+ \to \mu^+ \nu_\mu)} = \frac{m_e^2}{m_\mu^2} \left(\frac{m_\pi^2 - m_e^2}{m_\pi^2 - m_\mu^2} \right)^2 , \tag{17.43}$$

which is about 1.3×10^{-4}. Why such a small branching ratio for a channel which is kinematically favored? The reason is the occurrence of the factor proportional to the charged lepton mass in the amplitude, which produces the suppression factor m_e^2/m_μ^2 in the branching ratio. The factor can be understood by helicity arguments discussed in §14.2.1. We recall that the helicity of both the final state fermions should be the same. Since the neutrino helicity is always -1, the helicity of the charged antilepton, μ^+ or e^+, should also be the same. However, the gauge current has only left-chiral fields. It can create a left-chiral neutrino if it annihilates a left-chiral charged lepton. Turning things around, we can say that if it creates a left-chiral neutrino, it can also create a right-chiral antilepton with it. We thus reach an apparent impasse: helicity of the created antilepton has to be negative, but its chirality has to be right, i.e., positive. In the massless limit, helicity coincides with chirality, so this is not possible. It becomes possible if the charged lepton is massive, through the mismatch of helicity and chirality. This mismatch is bigger if the mass is bigger, so the amplitude is proportional to the charged lepton mass.

□ **Exercise 17.6** Use the helicity argument to conclude that the rate for the decay $\pi^0 \to \nu\widehat{\nu}$ should vanish if the neutrinos are massless. Verify this by writing the matrix element and evaluating it in the manner shown for charged pion decay.

□ **Exercise 17.7** The neutral pion decays almost always into two photons, as mentioned in §10.9. It also has some rare decay modes, shown below with the branching ratio of each mode:

$$\mathcal{B}(\pi^0 \to e^+ e^-) = 6.2 \times 10^{-8},$$
$$\mathcal{B}(\pi^0 \to e^+ e^- \gamma) = 1.2 \times 10^{-2},$$
$$\mathcal{B}(\pi^0 \to e^+ e^- e^+ e^-) = 3.1 \times 10^{-5}. \tag{17.44}$$

All these decays are purely electromagnetic. Explain qualitatively their relative magnitudes.

17.5.2   Decay of charged kaons

Charged kaons, just like charged pions, can decay leptonically through the channels

$$K^+ \to \mu^+ \nu_\mu,$$
$$K^+ \to e^+ \nu_e. \tag{17.45}$$

Take the first process, for instance. It is governed by the quark-level process

$$u\widehat{s} \to \mu^+ \nu_\mu. \tag{17.46}$$

The analysis is more or less the same as that for charged pion decay, except that the CKM matrix element that appears in the amplitude is V_{us}, and that the analog of Eq. (17.36) is now

$$\langle 0 | \bar{s} \gamma^\lambda (1 - \gamma_5) u | K^+(p) \rangle = \sqrt{2} i f_K p^\lambda, \tag{17.47}$$

where f_K, like f_π, is a phenomenological constant which is predictably called the *kaon decay constant*. From the observed decay rate of charged kaons into this channel, one can find the value of this constant:

$$f_K = 110 \, \text{MeV}. \tag{17.48}$$

☐ **Exercise 17.8** Check the value of f_K quoted above, given that $|V_{us}| = 0.225$. The lifetime of the K^+, the branching ratio to the said channel, the masses of various particles, and the Fermi constant can all be found in various appendices.

☐ **Exercise 17.9** Find the numerical value for $\Gamma(K^+ \to e^+\nu_e)/\Gamma(K^+ \to \mu^+\nu_\mu)$.

But we must remember a very important difference between the leptonic K decays and leptonic π decays. For kaons, there are also purely hadronic decay modes, which are not possible for the pion because of its low mass. For example, K^+ can decay to $\pi^+\pi^0$, which has a substantial branching ratio.

17.5.3 Related processes

We can, of course, discuss the decays of many other mesons in the same way. The arguments are similar, so we need not get into details. Rather, here we discuss another class of decay processes in which only one meson is involved.

We are talking about the decay of the tau. The purely leptonic decay modes of the tau were discussed in §14.6. We found that these decay modes comprise about 35% of the total decay width of the tau. The rest involve hadrons in the final state. The tau lepton is heavier than the pion, the kaon and many other mesons, so these can appear in the final state of tau decay. In particular, there are modes which involve only one meson in the final state, e.g., $\tau \to \pi^-\nu_\tau$ or $\tau \to K^-\nu_\tau$. The quark-level processes contributing to such decays can be obtained by invoking crossing symmetry on the processes shown in Eqs. (17.33) and (17.46), with the muon replaced by tau and the muon-neutrino replaced by the tau-neutrino. The calculation of the matrix element is therefore similar to what we have done for charged pion and kaon decay. The expression for $\overline{|\mathcal{M}|^2}$ would now have an extra factor of $\frac{1}{2}$ because of averaging over initial spin, and also the quantity $q \cdot k$ appearing in Eq. (17.39) would be equal to $\frac{1}{2}(m_\tau^2 - m_\pi^2)$ from kinematics, so that we would obtain

$$\Gamma(\tau^- \to \pi^-\nu_\tau) = \frac{G_F^2}{8\pi}|V_{ud}|^2 f_\pi^2 m_\tau^3 \left(1 - \frac{m_\pi^2}{m_\tau^2}\right)^2. \qquad (17.49)$$

The calculation for the decay mode to $K^-\nu_\tau$ should be similar, with f_π and m_π replaced by f_K and m_K respectively. This means that

$$\frac{\Gamma(\tau^- \to K^-\nu_\tau)}{\Gamma(\tau^- \to \pi^-\nu_\tau)} = \left|\frac{V_{us}}{V_{ud}}\right|^2 \frac{f_K^2}{f_\pi^2} \left(\frac{m_\tau^2 - m_K^2}{m_\tau^2 - m_\pi^2}\right)^2. \qquad (17.50)$$

Thus, the decay mode into kaons is weaker because V_{us} is small. Experimentally, the ratio of the two branching ratios is about 0.065.

17.6   Spin and parity of hadronic currents

It might be worthwhile to have some general discussion of hadronic currents like those which appeared in the discussion of decays of pions and kaons. We restrict our discussion to the gauge currents which contain polar and axial vector bilinears of fermion fields. We will denote the two types of currents generically by $\mathcal{V}_\mu$ and $\mathcal{A}_\mu$, respectively.

Under parity, a vector current transforms as

$$\mathcal{V}_0 \to \mathcal{V}_0, \qquad \boldsymbol{\mathcal{V}} \to -\boldsymbol{\mathcal{V}}, \tag{17.51}$$

as has been shown in Eq. (6.26, *p 156*). So far as spatial rotations are concerned, $\mathcal{V}_0$ behaves like a scalar, whereas $\boldsymbol{\mathcal{V}}$ behaves like a vector. Hence, we can summarize the spin and parity properties of a polar vector current by writing

$$J^P(\mathcal{V}_0) = 0^+, \qquad J^P(\boldsymbol{\mathcal{V}}) = 1^-. \tag{17.52}$$

For axial vector currents, the rotation properties are the same but parity properties are opposite, i.e.,

$$J^P(\mathcal{A}_0) = 0^-, \qquad J^P(\boldsymbol{\mathcal{A}}) = 1^+. \tag{17.53}$$

One can use these properties to find the general form for the matrix elements of quark currents between different kinds of states. For example, consider matrix elements of the type $\langle 0|\mathcal{V}_\mu|M\rangle$, where M is any meson in the 0^- octet. The time component of this matrix element should have $J^P = 0^-$, since the vacuum is a 0^+ state. On the other hand, the J^P of the spatial components of the matrix element should transform like a combination of 0^+ (for the vacuum), 1^- (for the current) and 0^- (for the meson), i.e., should have $J^P = 1^+$. So the matrix element should be some quantity whose time and space components transform like 0^- and 1^+ respectively. The matrix element can depend only on the meson 4-momentum p_μ. One cannot construct any quantity that depends only on one 4-vector and transforms like $J^P = (0^-, 1^+)$. Hence this matrix element must be zero.

For the axial vector current, the same arguments show that the time and space components of the matrix element should have $J^P = (0^+, 1^-)$. Indeed, the components of p_μ itself have the same properties. Therefore the matrix element can be proportional to p_μ, and that is exactly the form in which we had written down the matrix element in Eq. (17.36), for example. Analysis of parity properties of the currents provided us with the extra information that in leptonic decays of 0^- mesons, only the axial vector current contributes: the polar vector current does not. In summary, Eq. (17.36) and its analogs for kaon etc can equivalently be written as

$$\langle 0|\mathcal{V}^\lambda(0)|M(p)\rangle = 0,$$
$$\langle 0|\mathcal{A}^\lambda(0)|M(p)\rangle = \sqrt{2}if_M p^\lambda, \tag{17.54}$$

where f_M is the decay constant of the meson M.

If we want to consider decay of one meson into another meson plus leptons, the hadronic matrix elements will be of the form $\langle M_b(k)|\mathcal{J}_\mu|M_a(p)\rangle$, where we take both M_a and M_b to be 0^- mesons, and $\mathcal{J}_\mu$ stands for either $\mathcal{V}_\mu$ or $\mathcal{A}_\mu$. In this case, the time and space components of the matrix element of the polar vector current should transform like $(0^+, 1^-)$, which is the way 4-momentum vectors transform. So we can write

$$\langle M_b(k) |\mathcal{V}_\mu(0)| M_a(p)\rangle = f_1 p_\mu + f_2 k_\mu \,, \tag{17.55}$$

where f_1 and f_2 are form factors. On the other hand,

$$\langle M_b(k) |\mathcal{A}_\mu(0)| M_a(p)\rangle = 0 \,, \tag{17.56}$$

which can be proved with very little effort. Matrix elements involving three 0^- mesons can also be parametrized through similar arguments.

There is an important difference between f_π that appears in Eq. (17.54) and the objects f_1 and f_2 that appear in Eq. (17.55). From the Lorentz transformation property of both sides of these equations, we deduce that all of them should be Lorentz scalars. For the pion decay, the only scalar that can be constructed from the parameters of the problem is p^2, which is equal to m_π^2, i.e., is a constant. On the other hand, in the matrix element involving two mesons, we can define three Lorentz invariants, viz., p^2, k^2 and $k \cdot p$. The first two are constants, related to the masses of the two mesons. The third one, which can be traded with $q^2 = (p - k)^2$, is a dynamical variable. Therefore, f_1 and f_2 that appear in Eq. (17.55) are two form factors which depend on the momentum transfer between the mesons, i.e., on q^2.

□ **Exercise 17.10** If M_a, M_b and M_c are all 0^- mesons, show that

$$\langle M_b(k)M_c(q) |\mathcal{A}_\mu| M_a(p)\rangle = f_a p_\mu + f_b k_\mu + f_c q_\mu \,,$$
$$\langle M_b(k)M_c(q) |\mathcal{V}_\mu| M_a(p)\rangle = f \varepsilon_{\mu\alpha\beta\gamma} p^\alpha k^\beta q^\gamma \,, \tag{17.57}$$

represent the most general parametrization consistent with angular momentum and parity symmetries.

We now show that there is an interesting property of the charged current involving the u and the d quarks which also helps constrain some matrix elements. This involves G-parity, that was introduced in §8.5. Note that the results of a half-rotation around the I_2 axis on the up and the down quarks are the following:

$$u \to d \,, \qquad d \to -u \,. \tag{17.58}$$

Charge conjugation changes each quark field to its conjugate field, with the prescription given in Eqs. (6.48) and (6.49). Thus, the effect of G-parity transformation on these quark fields is given by

$$Gu(x)G^{-1} = \gamma_0 \mathbb{C} d^*(x) \,, \qquad Gd(x)G^{-1} = -\gamma_0 \mathbb{C} u^*(x) \,. \tag{17.59}$$

It can then be easily shown that

$$G\left(\overline{u}Fd\right)G^{-1} = -\overline{u}\mathbb{C}F^{\top}\mathbb{C}^{-1}d\,, \tag{17.60}$$

where F is any numerical 4×4 matrix, e.g., a Dirac matrix. For any of the sixteen basic 4×4 matrices that we introduced in Eq. (4.93, $p\,79$), the effect of G-parity is to change it to $-F_C$, where F_C was tabulated in Eq. (6.68, $p\,163$). From the table, it is seen that the polar vector current is unchanged by G-parity, whereas the axial vector current changes by a sign.

This helps explain why some processes are unobserved, or have very small rates. For example, let us ask whether the decay

$$\tau \to \eta\pi^-\nu_\tau \tag{17.61}$$

is possible. The hadronic part of the matrix element involves two 0^- meson states. A little variation of the argument leading to Eq. (17.56) would show us that the axial current cannot contribute to this matrix element. The matrix element of the polar vector current also vanishes in the case in question, which can be best seen from a consideration of G-parity. The η particle has positive G-parity, the pions have negative G-parity, so together they have negative G-parity. The polar vector current, which is positive under G-parity, therefore cannot create $\eta\pi$ from the hadronic vacuum. Gauge currents therefore cannot induce this decay. If such a decay is observed, it would imply the existence of some *second-class current*, involving other type of fermion field bilinears which have different G-parity properties from the vector currents. Any bilinear that does not contain γ_5 and has the same G-parity property as the polar vector current is called a *first-class current*. The same name applies to currents which contain γ_5 and is negative under G-parity, just as the axial vector current is.

□ **Exercise 17.11** *Find the matrix for* $\exp(i\pi I_2)$ *in the* 2×2 *representation. Apply it on the isospin doublet* $\binom{u}{d}$ *and verify Eq.* (17.58).

□ **Exercise 17.12** *Prove Eq.* (17.60).

When we introduced G-parity in Ch. 8, we mentioned that it is a symmetry of the strong interaction only, and is not obeyed by weak, or even electromagnetic, interactions. One might wonder why then we invoke it in the discussion of weak decays? The reason is that only the polar vector current between the up and the down quarks can be responsible for the said process, and this part of the interaction, taken in isolation, has a well-defined G-parity, as shown in Eq. (17.60). It is true that by 'weak interactions', we do not mean just this one term but many others which do not have well-defined G-parity properties. However, to the leading order in perturbation theory, they are irrelevant for the process in question.

We now discuss matrix elements involving vector mesons such as the ρ-mesons. The simplest matrix elements would be of the form $\langle 0 | J^\mu | M^* \rangle$, where M^* denotes a generic vector meson and J^μ can be either the polar or the axial vector current. We can write

$$\langle 0 | J^\mu(0) | M^*(p, \epsilon_r) \rangle = \langle 0 | J^\mu(0)a_r^\dagger(\boldsymbol{p}) | 0 \rangle\,, \tag{17.62}$$

where ϵ_r denotes the polarization vector of the meson, and $a_r^\dagger$ is the creation operator for this polarization. The creation operator appears in the Fourier expansion of the vector meson field $\Phi^\mu(x)$:

$$\Phi^\mu(x) = \sum_r \int \frac{d^3k}{\sqrt{(2\pi)^3 2E_k}} \left(a_r^\dagger(k)\epsilon_r^{\mu*}(k)e^{ik\cdot x} + \cdots \right), \qquad (17.63)$$

where the other term contains the annihilation operator and is not relevant for the present discussion. From this, it is straightforward to see that

$$a_r^\dagger(p) = \sqrt{\frac{2E_p}{(2\pi)^3}}\; \epsilon_r^\mu(p) \int d^3x\; e^{-ip\cdot x}\Phi_\mu(x). \qquad (17.64)$$

If we put this into Eq. (17.62), we encounter an expression involving the vacuum expectation value of $\int d^3x\; e^{-ip\cdot x}\, \mathcal{J}^\mu(0)\Phi^\nu(x)$. The most general possible form for this vacuum expectation value can be written as

$$\int d^3x\; e^{-ip\cdot x}\, \langle 0\,|\mathcal{J}^\mu(0)\Phi^\nu(x)|\,0\rangle = Fg^{\mu\nu} + F'p^\mu p^\nu, \qquad (17.65)$$

where F and F' are Lorentz invariants.

Consider now any particular combination of the Lorentz indices μ and ν on both sides of Eq. (17.65). The vector meson is odd under parity. According to the definition of parity of vector fields introduced in Eq. (6.14, *p 153*), this statement means that if ν happens to be a spatial index, the intrinsic parity of the field is negative. Suppose μ is the time index. Then the right hand side of Eq. (17.65) reduces to $F'E_p p^i$, which is negative under parity. Therefore, on the left hand side we should have a current such that $\mathcal{J}^0$ is even under parity. The axial vector current does not have this property, and so

$$\langle 0\,|\mathcal{A}^\mu|\,M^*\rangle = 0. \qquad (17.66)$$

The entire matrix element comes from the polar vector current. Putting the parametrization of Eq. (17.65) back into Eq. (17.62) and using the fact that $\epsilon^\mu p_\mu = 0$, we obtain

$$\langle 0\,|\mathcal{V}^\mu(0)|\,M^*(p,\epsilon_r)\rangle = \sqrt{2}f_{M^*}\epsilon_r^\mu. \qquad (17.67)$$

The quantity f_{M^*} can be called the decay constant of the vector meson. Note that it has the dimension of squared mass. We have kept a factor of $\sqrt{2}$ to mimic the corresponding definition for the pseudoscalar mesons.

☐ **Exercise 17.13** Consider the leptonic decay of a charged vector meson of mass M. Use the form of the hadronic matrix element given in Eqs. (17.66) and (17.67) to show that the decay rate is given by

$$\Gamma = \frac{1}{2\pi}\, G_F^2 f^2 M, \qquad (17.68)$$

neglecting the masses of the charged lepton and the neutrino. [**Hint :** Use Eq. (4.41, *p 69*) for the polarization sum.]

Finally, we discuss the general form of matrix elements of currents within two baryon states. We only discuss spin-$\frac{1}{2}$ baryons, all of which have the same intrinsic parity, viz., $+1$. When the current is a polar vector current, the matrix element should transform like a polar vector. In other words, with two baryons b and b', the matrix element should have the form

$$\langle b'(p') | \mathcal{V}^\mu | b(p) \rangle = \overline{u}'(p') \Gamma^\mu u(p) \tag{17.69}$$

where u and u' denote the spinors of the baryon fields, and Γ^μ is something carrying a Lorentz index such that the combination on the right-hand side is a polar vector. There can at most be four different independent 4-vectors, but for expressions in the form that appears here, Gordon identity provides one relation between them. Therefore, only three independent combinations are possible, and one takes them in the following form:

$$\langle b'(p') | \mathcal{V}^\mu | b(p) \rangle = \overline{u}'(p') \Big[f_1 \gamma^\mu + i f_2 \sigma^{\mu\nu} q_\nu + f_3 q^\mu \Big] u(p), \tag{17.70}$$

where $q = p - p'$, and f_1, f_2 and f_3 are functions of q^2. Similarly, with the axial vector current, the most general form of the matrix element is

$$\langle b'(p') | \mathcal{A}^\mu | b(p) \rangle = \overline{u}'(p') \Big[\widetilde{f}_1 \gamma^\mu + i \widetilde{f}_2 \sigma^{\mu\nu} q_\nu + \widetilde{f}_3 q^\mu \Big] \gamma_5 u(p), \tag{17.71}$$

where $\widetilde{f}_1$, $\widetilde{f}_2$ and $\widetilde{f}_3$ are another set of form factors which are functions of q^2.

17.7 Selection rules for charged currents

Some quick rules of thumb can be derived by considering the nature of hadronic currents in the standard model. Suppose we are considering some strangeness-changing process. It can change only through charged current weak interaction, i.e., by W exchange, since strong and electromagnetic interactions, and even neutral current weak interactions, conserve strangeness. Strangeness is a property that is carried only by the s quark and the corresponding antiquark, and by convention the values of strangeness assigned to them are -1 and $+1$ respectively.

The charged current can change the s quark to any up-type quark, none of which carries any strangeness. So, in this way, strangeness can go up from -1 to zero, i.e., we have a change of strangeness $\Delta S = 1$. But, at the same time, the electric charge in the quarks has gone up from $-\frac{1}{3}$ to $+\frac{2}{3}$, i.e., by one unit. Of course electric charge is conserved: this increase means that a W^- has carried off the rest. But the W^- is not a hadron. If we consider only the charge in hadrons and call it Q_H, we can write

$$\Delta S = \Delta Q_H. \tag{17.72}$$

Of course, this rule is not absolute. If the W^-, produced as a virtual particle, produces quarks in the final state, ΔQ_H in the entire process must vanish, and therefore the selection rule of Eq. (17.72) will not hold. The selection rule will work if the W^- produces leptons in the final state. For leptonic and semi-leptonic decays of mesons, this rule will hold as long as the process uses only charged current interactions. Examples of this rule can be seen in the leptonic decays of kaons and semi-leptonic decays of the τ lepton given earlier, and in the semi-leptonic decays of kaons given later in Eq. (17.74).

Similar rules also hold for hadrons containing heavier quarks. For the charm quark, the same argument will tell us that

$$\Delta C = \Delta Q_H, \tag{17.73}$$

where C, or charm, is defined to be $+1$ for the c-quark, -1 for the corresponding antiquark, and zero for everything else. Assignment of the bottom quantum number is similar to that for the strange quark, i.e., -1 for the b-quark, and again a similar rule holds.

17.8   Semileptonic decays of mesons

We can now go one step further and consider processes involving two mesons. These would include semileptonic decay modes of the kaon and other heavier mesons. The K^+, for example, has the following semileptonic decay modes:

	mode	branching ratio
K^+	$\rightarrow$ $\pi^0 e^+ \nu_e$	4.98%
K^+	$\rightarrow$ $\pi^0 \mu^+ \nu_\mu$	3.32%

$$\tag{17.74}$$

The first one is usually called K^+_{e3} decay and the second one $K^+_{\mu 3}$ decay. These processes are driven by the quark level process

$$\hat{s} \rightarrow \hat{u} \ell^+ \nu_\ell, \tag{17.75}$$

where ℓ stands for either the muon or the electron. The leptonic part of the matrix element has the same form as that used for purely leptonic final state processes discussed in §17.5. But the hadronic part of the matrix element is $\langle \pi^0 | \bar{s} \gamma^\mu \mathbb{L} u | K^+ \rangle$. As shown in Eqs. (17.55) and (17.56), this kind of matrix element can be parametrized with two form factors, e.g.:

$$\langle \pi^0(p) | \bar{s} \gamma^\mu \mathbb{L} u | K^+(k) \rangle = f_+ (k+p)^\mu + f_- (k-p)^\mu. \tag{17.76}$$

Denoting $q = k - p$, one can argue that the two form factors are functions of the dynamical variable q^2 only. The argument is identical to that given in §5.7.1 for electromagnetic form factors, and need not be repeated here.

Many other processes can be analyzed through the same machinery. These include the pion beta decay ($\pi^+ \rightarrow \pi^0 e^+ \nu_e$), as well as decay modes of the tau lepton containing two mesons in the final state, e.g., $\tau \rightarrow \pi^0 \pi^- \nu_\tau$, $\tau \rightarrow \pi^0 K^- \nu_\tau$, $\tau \rightarrow K^0 K^- \nu_\tau$. Of course, the form factors will be different for different meson states.

☐ **Exercise 17.14** What is the mass dimension of the form factors appearing in Eq. (17.76)?

☐ **Exercise 17.15** Find the absolute square of the matrix element for $K_{\ell 3}^+$ decay and verify that the mass suppression that occurs for purely leptonic decay modes does not occur in this case.

17.9 Neutral kaons

For many reasons, historical and coincidental, neutral kaons occupy a very prominent position in the field of particle physics. In this section, we will give some introduction to the physics issues connected with neutral kaons.

17.9.1 Eigenstates in the neutral kaon sector

When neutral kaons are produced in strong interactions, they are produced either as K^0 or as $\widehat{K}^0$. As discussed in Ch. 10, the quark contents of these two states are $d\widehat{s}$ and $\widehat{d}s$ respectively. Thus, K^0 carries $+1$ unit of strangeness, whereas $\widehat{K}^0$ carries -1. These are strangeness eigenstates. Since strangeness is conserved in strong interactions, any particle created from a state of definite strangeness must be an eigenstate of strangeness.

Strangeness is not violated by electromagnetic interactions as well. In fact, even neutral current weak interactions are flavor conserving, as we have noted earlier. However, charged current weak interactions contain terms where the strange quark couples through the W boson to the up-type quarks, none of which carry any strangeness. These terms definitely violate strangeness. Thus, it is expected that when the effects of these interactions are included, the eigenstates of the Hamiltonian will not be the states $|K^0\rangle$ and $|\widehat{K}^0\rangle$, but some linear combinations of them. This is a phenomenon called K^0-$\widehat{K}^0$ mixing.

To discuss the phenomenon in a quantitative manner, let us consider a state

$$|\psi(t)\rangle = a(t)\,|K^0\rangle + b(t)\,\big|\widehat{K}^0\big\rangle \,, \tag{17.77}$$

and represent it through a column vector by writing

$$|\psi(t)\rangle = \begin{pmatrix} a(t) \\ b(t) \end{pmatrix} . \tag{17.78}$$

The time evolution of $|\psi(t)\rangle$ is governed by the Schrödinger equation,

$$i\frac{d}{dt}\,|\psi(t)\rangle = \mathbb{H}\,|\psi(t)\rangle \,, \tag{17.79}$$

where $\mathbb{H}$ is some effective Hamiltonian appropriate to this two-state system whose basis states have been taken as $|K^0\rangle$ and $|\widehat{K}^0\rangle$. Remember that, according to Eq. (4.136, *p 87*), these states are not normalized to unity. We can define normalized states

$$|1\rangle = \frac{|K^0\rangle}{\sqrt{2m_K \mathcal{V}}} , \qquad |2\rangle = \frac{|\widehat{K}^0\rangle}{\sqrt{2m_K \mathcal{V}}} , \qquad (17.80)$$

so that the elements of the effective Hamiltonian can be written as

$$\mathbb{H}_{ij} = \langle i | \mathbb{H} | j \rangle . \qquad (17.81)$$

We might be tempted to think that the Hamiltonian is hermitian, and therefore its diagonal elements are real and the off-diagonal ones are complex conjugates of each other. Indeed, in introductory texts on quantum mechanics, we always discuss Hamiltonians which are hermitian. With a hermitian H, the time evolution operator $\exp(-iHt)$ is unitary, which ensures that the probability of obtaining the particle in the entire space does not change with time. This means that the procedure is applicable for stable particles. If, on the other hand, a particle is not stable but rather decays into something else, the probability of finding it throughout the space should decrease with time as $|\psi|^2 \propto \exp(-\Gamma t)$, where Γ is the decay rate. This means that the Hamiltonian should contain a non-hermitian part. This part can be taken to be skew-hermitian without any loss of generality, since any matrix can be written as a sum of a hermitian matrix and a skew-hermitian one. For the 2×2 case at hand, we can therefore write

$$\mathbb{H} = \mathbb{M} - \frac{i}{2}\Gamma , \qquad (17.82)$$

where $\mathbb{M}$ and Γ are both 2×2 hermitian matrices.

□ **Exercise 17.16** *Using Eq. (17.79), show that*

$$\frac{d}{dt} \langle \psi(t) | \psi(t) \rangle = - \langle \psi(t) | \Gamma | \psi(t) \rangle , \qquad (17.83)$$

which tells us that the part Γ is responsible for the decay of the kaons, which depletes the probability of observing either of the two neutral kaon states.

It should be understood that this is no deviation from the basic tenets of quantum mechanics. The point is that if one starts with a state, say, $|K^0\rangle$, the time evolution of this state will not only contain the states $|K^0\rangle$ and $|\widehat{K}^0\rangle$, but also the decay products of the neutral kaons. Had we considered basis states to span these decay products as well and written the Hamiltonian in that space, the Hamiltonian would have come out to be hermitian. Here we are specializing into more restrictive questions where no attention is paid to the decay products: only K^0 and $\widehat{K}^0$ are measured. This is the reason we got a non-hermitian Hamiltonian, which should be treated as an effective Hamiltonian of the neutral kaon system.

The effective Hamiltonian $\mathbb{H}$ of the neutral kaon sector need not be hermitian, but we should find out what constraints does CPT invariance put on its elements, since we assume throughout that CPT is a good symmetry. For this, first we see what are the elements of $\mathbb{M}$ and Γ. If weak interactions did not exist, K^0 and $\widehat{K}^0$ would have been eigenstates of the Hamiltonian. They would have been stable, implying $\Gamma = 0$, and their masses would have been equal by CPT invariance. Let us call this common mass m_0. In the presence of the weak interaction Hamiltonian H_w, we can write, up to terms in the second order in perturbation theory, the expressions

$$\mathbb{M}_{ij} = m_0 \delta_{ij} + \langle i | H_w | j \rangle + \sum_n \mathcal{P} \frac{\langle i | H_w | n \rangle \langle n | H_w | j \rangle}{E_n - m_0} ,$$

$$\Gamma_{ij} = 2\pi \sum_n \delta(E_n - m_0) \langle i | H_w | n \rangle \langle n | H_w | j \rangle , \qquad (17.84)$$

where $\mathcal{P}$ denotes the *principal value*.

Let us now see how CPT transformation affects these elements. Suppose we choose the phases such that

$$\Theta\left|K^0\right\rangle = \left|\widehat{K}^0\right\rangle, \qquad \Theta\left|\widehat{K}^0\right\rangle = \left|K^0\right\rangle, \qquad (17.85)$$

where $\Theta \equiv \mathcal{CPT}$. Using Eq. (7.45, *p 196*) which shows how matrix elements transform under CPT, we can write, e.g.,

$$\left\langle \widehat{K}^0 \left| H_w \right| \widehat{K}^0 \right\rangle = \left\langle K^0 \left| H_w^\dagger \right| K^0 \right\rangle. \qquad (17.86)$$

But of course H_w is hermitian, so we can omit the dagger sign on it. This equation shows that, so long as only the first order perturbation term is concerned, we obtain $\mathbb{M}_{11} = \mathbb{M}_{22}$.

We now try the same equation on the off-diagonal elements. Here, we obtain

$$\left\langle \widehat{K}^0 \left| H_w \right| K^0 \right\rangle = \left\langle \Theta K^0 \left| H_w \right| \Theta \widehat{K}^0 \right\rangle = \left\langle K^0 \left| H_w \right| \widehat{K}^0 \right\rangle^*. \qquad (17.87)$$

This is an identity because of the hermiticity of H_w. So we get no restriction on the off-diagonal elements at first order in perturbation. Continuing the exercise to second order, the same result repeats, i.e., we obtain

$$\mathbb{H}_{11} = \mathbb{H}_{22} \qquad (17.88)$$

whereas there is no restriction on $\mathbb{H}_{12}$ and $\mathbb{H}_{21}$. This means that we can parametrize the matrix $\mathbb{H}$ as

$$\mathbb{H} = a\mathbb{1} + b\begin{pmatrix} 0 & p^2 \\ q^2 & 0 \end{pmatrix}. \qquad (17.89)$$

The parameters p and q can be taken dimensionless, with

$$|p|^2 + |q|^2 = 1, \qquad (17.90)$$

and b can be taken to be real without any loss of generality. The other parameters, a, p and q are complex. It is easy to see that the eigenvalues of this Hamiltonian are

$$a \pm bpq, \qquad (17.91)$$

and the corresponding eigenvectors

$$\begin{pmatrix} p \\ \pm q \end{pmatrix}, \quad \text{i.e.,} \quad p\left|K^0\right\rangle \pm q\left|\widehat{K}^0\right\rangle, \qquad (17.92)$$

which have the same normalization as the states $\left|K^0\right\rangle$ and $\left|\widehat{K}^0\right\rangle$. The eigenvalues are in general complex. Their real parts give the energy, and imaginary parts give the decay rates.

□ **Exercise 17.17** Verify Eq. (17.88) on the second order terms in perturbation theory.

So far, we have not assumed anything except CPT symmetry. From now on, let us ignore CP violating effects. If CP is conserved, the states $\left|K^0\right\rangle$ and $\left|\widehat{K}^0\right\rangle$ can be defined as CP conjugates of each other, i.e.,

$$\mathscr{C}\mathscr{P}\left|K^0\right\rangle = \left|\widehat{K}^0\right\rangle, \qquad \mathscr{C}\mathscr{P}\left|\widehat{K}^0\right\rangle = \left|K^0\right\rangle. \tag{17.93}$$

The Hamiltonian is CP invariant, i.e., it commutes with the operator CP: $(\mathscr{C}\mathscr{P})^{-1}\mathbb{H}(\mathscr{C}\mathscr{P}) = \mathbb{H}$. Therefore,

$$\mathbb{H}_{12} \equiv \left\langle K^0 \left|\mathbb{H}\right| \widehat{K}^0\right\rangle = \left\langle K^0 \left|(\mathscr{C}\mathscr{P})^{-1}\mathbb{H}(\mathscr{C}\mathscr{P})\right| \widehat{K}^0\right\rangle. \tag{17.94}$$

But Eq. (17.93), along with the unitarity of the operator CP, implies that

$$\left\langle K^0 \left|(\mathscr{C}\mathscr{P})^{-1}\mathbb{H}(\mathscr{C}\mathscr{P})\right| \widehat{K}^0\right\rangle = \left\langle \widehat{K}^0 \left|\mathbb{H}\right| K^0\right\rangle = \mathbb{H}_{21}. \tag{17.95}$$

We have thus proved that

$$\mathbb{H}_{12} = \mathbb{H}_{21} \tag{17.96}$$

provided CP is conserved. For the parametrization presented in Eq. (17.89), it means that $p = q$. So the eigenstates of the Hamiltonian are

$$K^0_{(\pm)} \equiv \frac{1}{\sqrt{2}}\left(\left|K^0\right\rangle \pm \left|\widehat{K}^0\right\rangle\right). \tag{17.97}$$

Note that a can still be complex, giving complex eigenvalues in Eq. (17.91), signifying that the two eigenstates can decay even if there is no CP violation, something that will be discussed in §17.9.2.

We warn the reader a little bit about the notation. We have used $K^\pm$ to denote kaons carrying a positive or a negative unit of electric charge. Here in Eq. (17.97), the plus or minus sign in the subscript does not say anything about the electric charges of the particles. The electric charge is zero: remember that we are talking of neutral kaons. The subscripted signs in the states defined in Eq. (17.97) represent the CP eigenvalue of the state, which can be verified easily by using Eq. (17.93). The eigenstates of the Hamiltonian are CP eigenstates in this case: something that we should have expected since the operator CP commutes with the Hamiltonian.

Even with the assumption of CP invariance, the two eigenvalues are not degenerate, as is shown by the expression for the eigenvalues in Eq. (17.91). Thus, there will be two different eigenstates in the neutral kaon sector, with a mass difference between them. The amount of this mass difference will be estimated in §17.9.4.

□ **Exercise 17.18** ♪ Eq. (17.93) assumes a certain phase relation between the states $\left|K^0\right\rangle$ and $\left|\widehat{K}^0\right\rangle$. More generally, both states can have some extra phase associated with them, so that one obtains a relation of the form

$$\mathscr{CP} \left| K^0 \right\rangle = e^{i\zeta} \left| \widehat{K}^0 \right\rangle . \tag{17.98}$$

Show that this implies

$$\mathscr{CP} \left| \widehat{K}^0 \right\rangle = e^{-i\zeta} \left| K^0 \right\rangle . \tag{17.99}$$

As for the conditions of CP invariance, show that Eq. (17.96) changes, in this convention, to

$$\mathbb{H}_{12} = e^{2i\zeta} \mathbb{H}_{21} . \tag{17.100}$$

17.9.2   Decays of neutral kaons into pions

The two eigenstates have very different properties regarding decays. To see this, consider the possibility of a neutral kaon decaying into two pions. Because of electric charge conservation, the two pions in the final state can be either $\pi^+ \pi^-$ or $\pi^0 \pi^0$. For the ensuing argument, it does not matter which one of these combinations we are talking about. The two pions must be produced in a $L = 0$ state, because the decaying kaon as well as the final pions are all spinless particles. Thus, the orbital parity of the final state is $+1$. The intrinsic parity of two pions is also $+1$, making the final state a parity eigenstate with eigenvalue $+1$. The initial kaon, however, has a negative intrinsic parity. Therefore, parity must be violated in the process.

This is no big deal, because weak processes violate parity maximally. But let us now look at the charge conjugation property of the final state. Whether the final state is $\pi^+ \pi^-$ or $\pi^0 \pi^0$, it contains a spinless boson and its antiparticle. As stated in Ex. 6.20 *(p 173)*, this will have to be an eigenstate of charge conjugation, with eigenvalue $(-1)^L$ in general, L being the orbital angular momentum. In the present case, $L = 0$, so the final state has an eigenvalue $+1$ of charge conjugation.

Consider the combined symmetry CP now. The final state of two pions, by being an eigenstate of parity as well as charge conjugation, must be an eigenstate of CP as well. And the eigenvalue will be the product of its parity eigenvalue and charge conjugation eigenvalue, i.e., $+1$. If CP is conserved, it can occur only in the decay of a CP eigenstate with the same eigenvalue. Thus, if CP is conserved, only $K^0_{(+)}$ can decay into two pions; $K^0_{(-)}$ cannot.

Of course, both $K^0_{(+)}$ and $K^0_{(-)}$ can decay into three pions. For $K^0_{(-)}$, this will be the dominant decay mode. But notice that the mass of the neutral kaon is 497 MeV whereas the mass of pions is around 140 MeV. Thus, with three pions in the final state, there is very little phase space for the decay products. As a result, the decay will be much slower compared to the decay into two pions. So $K^0_{(-)}$ will be long-lived, whereas $K^0_{(+)}$ will be much shorter-lived.

In reality, CP is violated, so the eigenstates will be superpositions of $K^0_{(\pm)}$. CP violation will be discussed in Ch. 21. We only want to mention here that the effects of CP violations are much smaller than the usual weak interaction effects, so the eigenstates will not change drastically. We will still have one combination that will be predominantly $K^0_{(-)}$, and it will be much longer-lived than the other. This combination will be called K^0_L, and the other eigenstate K^0_S (here, 'L' is for 'long', and 'S' for 'short'). Experimental values of their lifetimes show the huge difference that the two-pion decay mode makes for one of them:

$$\Gamma^{-1}(K^0_S) = 0.89 \times 10^{-10}\,\text{s}\,,$$
$$\Gamma^{-1}(K^0_L) = 5.11 \times 10^{-8}\,\text{s}\,. \tag{17.101}$$

17.9.3   Kaon oscillations

Because of the small difference in mass and large difference in the lifetimes of the two neutral kaon eigenstates, very interesting phenomena can be observed with kaons. Consider a neutral kaon produced in strong interactions. This would be an eigenstate of strangeness since strong interactions conserve strangeness. Let us assume that we have produced a beam of K^0 particles.

The beam now moves, and we see its time evolution. The best way to describe time evolution is to write K^0 in terms of the eigenstates, which can be done by inverting Eq. (17.92). If we ignore CP violation, this relation will be

$$\left|K^0\right\rangle = \frac{1}{\sqrt{2}}\left(\left|K^0_L\right\rangle + \left|K^0_S\right\rangle\right). \tag{17.102}$$

After some time t, this beam would evolve to

$$\left|K^0(t)\right\rangle = \frac{1}{\sqrt{2}}\left(e^{-im_L t - \gamma_L t}\left|K^0_L\right\rangle + e^{-im_S t - \gamma_S t}\left|K^0_S\right\rangle\right), \tag{17.103}$$

where m_L and m_S are the masses of the two eigenstates, and γ_L, γ_S are their decay rates. We have assumed that the particles in the beam are non-relativistic, so that the contribution of the kinetic energy can be neglected.

We notice that the state $\left|K^0(t)\right\rangle$ is not the same as the original state. The relative proportion of K^0_L and K^0_S varies with time. Thus, at any given time after the production, the beam will contain a mixture of K^0 and $\widehat{K}^0$.

It is easy to detect the presence of $\widehat{K}^0$'s in the beam. If a target made of ordinary matter is placed in the path of the beam, the particles in the beam would interact with the particles in the target. Protons in the target can produce Λ particles through the reaction

$$p + \widehat{K}^0 \rightarrow \Lambda + \pi\,, \tag{17.104}$$

which can be mediated by strong interactions. On the other hand, the K^0 cannot produce Λ or any other strange baryon from the proton, because it has the wrong value of strangeness. Thus, only the elastic scattering channel

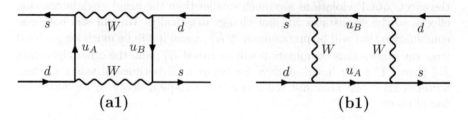

Figure 17.5: K^0-$\widehat{K}^0$ transition diagrams arising at the one-loop level in the standard model. The fermion lines to the left of each diagram represent an incoming d-quark and an incoming $\widehat{s}$, i.e., an incoming K^0. The fermion lines to the right represent an outgoing $\widehat{K}^0$.

is open for the K^0, and the cross-section is much smaller than that for the $\widehat{K}^0$. From the cross-section of scattering, we can then infer the amount of $\widehat{K}^0$ present in the beam at any time. It is seen that it oscillates with time, because of the mass difference between K_L^0 and K_S^0. The overall probability of detecting any kind of neutral kaon also goes down with time because the kaons decay.

☐ **Exercise 17.19** Neglect CP violation, so that the eigenstates are given by the combinations given in Eq. (17.97). Show that, from a beam that was originally purely K^0, the probability of finding a $\widehat{K}^0$ at a time t is given by

$$\mathrm{Prob}(\widehat{K}^0, K^0(t)) = \frac{1}{4}\left[e^{-\gamma_L t} + e^{-\gamma_S t}\right.$$
$$\left. -2e^{-\frac{1}{2}(\gamma_L+\gamma_S)t}\cos(\Delta m_K\, t)\right], \quad (17.105)$$

where Δm_K is the difference in mass of the two eigenstates.

17.9.4 K_L^0-K_S^0 mass difference

The mass eigenvalues for the neutral kaons have been given in Eq. (17.91). The formula shows that the mass difference between K_L^0 and K_S^0 arises from the off-diagonal terms in the Hamiltonian of Eq. (17.89), which was written with K^0 and $\widehat{K}^0$ as the basis states.

Weak interactions violate strangeness. So, weak interactions can induce some effective interactions that can result in transitions between K^0 and $\widehat{K}^0$. Such transition diagrams have been shown in Fig. 17.5, which are called *box diagrams* for obvious reasons. The fermion lines at the left sides of these diagrams represent an incoming d quark and an outgoing s quark. The latter can also be interpreted as an incoming antiquark $\widehat{s}$, so that the two lines at the left side stand for an incoming K^0. Similarly, the two lines at the right sides of the diagrams represent an outgoing $\widehat{K}^0$. The amplitude of such diagrams would then provide the off-diagonal elements of the Hamiltonian of Eq. (17.89). Our task here would be to estimate these amplitudes.

Let us take the diagram of Fig. 17.5a1 first. It gives rise to an effective $[\bar{s}d][\bar{s}d]$ interaction. The effective Lagrangian for this interaction is

$$i\mathscr{L}_{\text{eff}}^{(a1)} = \left(-\frac{ig}{\sqrt{2}}\right)^4 \sum_{A,B} \int \frac{d^4l}{(2\pi)^4}\left[\bar{s}V_{A2}^*\gamma^\mu \mathbb{L}\frac{i(\slashed{l}+m_{u_A})}{l^2-m_{u_A}^2}V_{A1}\gamma^\nu \mathbb{L}d\right]$$
$$\times iD_{\mu\rho}^{(W)}(l)iD_{\nu\lambda}^{(W)}(l)\left[\bar{s}V_{B2}^*\gamma^\lambda \mathbb{L}\frac{i(\slashed{l}+m_{u_B})}{l^2-m_{u_B}^2}V_{B1}\gamma^\rho \mathbb{L}d\right]. \quad (17.106)$$

Let us explain what we have written. We are looking for effective operators involving quark fields which can have non-zero matrix elements between an incoming K^0 and an outgoing $\widehat{K}^0$ state, or vice versa. We start with the working assumption that we will be able to obtain simple operators which contain no derivatives. This is the reason why we took all external quark lines at zero momentum. Once this is assumed, all internal lines carry the same loop momentum, l, which has been integrated over. The CKM matrix elements come through the coupling to the W boson. In denoting the elements of the CKM matrix, we have used the subscript 2 for the s quark and 1 for the d quark.

We now notice that the quark mass terms in the numerator of the propagators appearing in Eq. (17.106) do not contribute to the effective Lagrangian because they contain a string like $\mathbb{L}\gamma^\alpha \mathbb{L}$, which is zero. Using the 't Hooft–Feynman gauge for the W boson propagators, we then obtain

$$i\mathscr{L}_{\text{eff}}^{(a1)} = \frac{g^4}{4}\left[\bar{s}\gamma^\mu\gamma^\alpha\gamma^\nu \mathbb{L}d\right]\left[\bar{s}\gamma_\nu\gamma_\beta\gamma_\mu \mathbb{L}d\right]\sum_{A,B}V_{A2}^*V_{A1}V_{B2}^*V_{B1}$$
$$\times \int \frac{d^4l}{(2\pi)^4}\frac{l_\alpha l^\beta}{(l^2-M_W^2)^2(l^2-m_{u_A}^2)(l^2-m_{u_B}^2)}. \quad (17.107)$$

Further simplification can be obtained by transforming the integral to an invariant integral times δ_α^β via Eq. (G.10, *p 758*). In that case, all Lorentz indices are contracted between the Dirac matrices appearing in the two fermion field bilinears. Using the formula for the string of three Dirac matrices, Eq. (F.50, *p 742*), one finds

$$\left[\bar{s}\gamma^\mu\gamma^\alpha\gamma^\nu \mathbb{L}d\right]\left[\bar{s}\gamma_\nu\gamma_\alpha\gamma_\mu \mathbb{L}d\right] = 4\left[\bar{s}\gamma^\lambda \mathbb{L}d\right]\left[\bar{s}\gamma_\lambda \mathbb{L}d\right]. \quad (17.108)$$

Putting this back, we obtain

$$i\mathscr{L}_{\text{eff}}^{(a1)} = \frac{g^4}{4}\left[\bar{s}\gamma^\lambda \mathbb{L}d\right]\left[\bar{s}\gamma_\lambda \mathbb{L}d\right]\sum_{A,B}V_{A2}^*V_{A1}V_{B2}^*V_{B1}$$
$$\times \int \frac{d^4l}{(2\pi)^4}\frac{l^2}{(l^2-M_W^2)^2(l^2-m_{u_A}^2)(l^2-m_{u_B}^2)}. \quad (17.109)$$

Notice one thing now. Suppose, for the argument's sake, that all up-type quark masses are zero. In that case, the integral is independent of the quark masses, and can be taken outside the summation over the indices A, B. But in this case, the sum over A, B vanishes because of the unitarity of the CKM matrix V. In fact, the vanishing occurs even if we put all masses on one of the internal quark lines to be zero, without paying attention to the other line. This means that in the integrand we can make the replacement

$$\frac{1}{l^2 - m_{u_A}^2} \longrightarrow \frac{1}{l^2 - m_{u_A}^2} - \frac{1}{l^2}, \tag{17.110}$$

since the extra term will not contribute. The same argument applies to the other line. Hence we can write

$$
\begin{aligned}
i\mathscr{L}_{\text{eff}}^{(a1)} &= \frac{g^4}{4}\left[\bar{s}\gamma^\lambda \mathbb{L}d\right]\left[\bar{s}\gamma_\lambda \mathbb{L}d\right]\sum_{A,B} V_{A2}^* V_{A1} V_{B2}^* V_{B1} \\
&\quad \times \int \frac{d^4 l}{(2\pi)^4} \frac{m_{u_A}^2 m_{u_B}^2}{l^2(l^2 - M_W^2)^2(l^2 - m_{u_A}^2)(l^2 - m_{u_B}^2)} \\
&= \frac{g^4}{4M_W^2}\left[\bar{s}\gamma^\lambda \mathbb{L}d\right]\left[\bar{s}\gamma_\lambda \mathbb{L}d\right]\sum_{A,B} V_{A2}^* V_{A1} V_{B2}^* V_{B1} r_{u_A} r_{u_B} \\
&\quad \times \int \frac{d^4 l'}{(2\pi)^4} \frac{1}{l'^2(l'^2 - 1)^2(l'^2 - r_{u_A})(l'^2 - r_{u_B})}, \tag{17.111}
\end{aligned}
$$

where in the last step we have used a new integration variable defined by $l'^\mu = l^\mu / M_W$, and introduced the notation

$$r \equiv m^2 / M_W^2 \tag{17.112}$$

for any particle with mass m.

The implication of this argument is this. If all $r_{u_A} \ll 1$, i.e., all up-type quark masses are small compared to the only other mass present in the integral, M_W, then the integral should be suppressed by these mass ratios. This happens, as we have shown, because of the fact that the r-independent terms in the integral cancel out. This is called *GIM* cancellation, after the names of Glashow, Iliopoulos and Maiani, who first noticed that such cancellation takes place provided the up-type quarks and the down-type quarks come in pairs, and the mixing matrix is unitary. We will talk more about it in §17.9.5, in a different context.

There is a related issue. In writing the expression of Eq. (17.109), we have used the 't Hooft–Feynman gauge. In order that our calculation is consistent and gauge invariant, we should also take into account diagrams where one or both of the W lines of Fig. 17.5 are replaced by the unphysical charged Higgs. These extra diagrams corresponding to Fig. 17.5a1 have been shown in Fig. 17.6. If the masses of the quarks circulating in the loops were very small and could be neglected altogether, the diagrams containing the unphysical

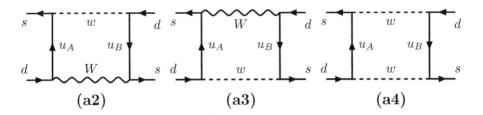

Figure 17.6: K^0-$\widehat{K}^0$ transition diagrams containing the unphysical charged Higgs boson. These diagrams correspond to Fig. 17.5a1. There is a similar set of diagrams which corresponds to Fig. 17.5b1.

Higgs boson would not have contributed, because the couplings would have vanished in the limit of vanishing quark masses. But we know that the W mediated diagrams vanish in the same limit because of GIM cancellation, and the surviving terms of the gauge boson mediated diagrams contain factors of r anyway, and can therefore be naively of the same order as the unphysical Higgs mediated diagrams.

Let us see the general nature of these extra diagrams. To be specific, we take the diagram of Fig. 17.6a2, where the upper W line of Fig. 17.5 *(p 510)* is replaced by the unphysical Higgs boson w. Couplings of quarks with the w were given in Fig. 17.2 *(p 491)*. In the present case, since we have taken the external 4-momenta to be zero, we should take the down-type quarks masses to be zero as well for the sake of consistency. The effective Lagrangian arising out of this diagram can then be written as

$$
i\mathscr{L}_{\text{eff}}^{(a2)} = \left(\frac{ig}{\sqrt{2}}\right)^4 \sum_{A,B} \int \frac{d^4 l}{(2\pi)^4} \left[\bar{s}V_{A2}^* \frac{m_{u_A}}{M_W} \mathbb{R} \frac{i(\slashed{l} + m_{u_A})}{l^2 - m_{u_A}^2} V_{A1}\gamma^\nu \mathbb{L}d\right]
$$
$$
\times i\Delta^{(w)}(l)iD_{\nu\lambda}^{(W)}(l)\left[\bar{s}V_{B2}^*\gamma^\lambda \mathbb{L}\frac{i(\slashed{l} + m_{u_B})}{l^2 - m_{u_B}^2}V_{B1}\frac{m_{u_B}}{M_W}\mathbb{L}d\right]
$$
$$
= \frac{g^4}{4M_W^2}\left[\bar{s}\gamma^\nu \mathbb{L}d\right]\left[\bar{s}\gamma^\lambda \mathbb{L}d\right]\sum_{A,B} V_{A2}^* V_{A1} V_{B2}^* V_{B1}r_{u_A}r_{u_B}
$$
$$
\times \int \frac{d^4 l'}{(2\pi)^4} \frac{1}{(l'^2-1)^2(l'^2 - r_{u_A})(l'^2 - r_{u_B})}. \tag{17.113}
$$

Note that the same combination of mixing matrix elements has appeared in this expression, and also that the field bilinears have the same form as that obtained from the W mediated diagram. The same comments apply to other diagrams mediated by the unphysical Higgs, and also to the diagram of Fig. 17.5b *(p 510)* plus all its variants involving unphysical Higgs bosons.

We can omit the details of all these other diagrams and summarize the results in the following way. As described in Appendix G, the integrals can be transformed by performing Wick rotation of the integration variable, and

subsequently performing the angular integrations through Eq. (G.26, p 761).
This gives

$$\int \frac{d^4 l'}{(2\pi)^4} \, F(l'^2) = \frac{i}{16\pi^2} \int_0^\infty dy \; yF(-y) \,, \qquad (17.114)$$

where F is any function, and y is the square of Euclidean version of the loop
momentum variable. Thus, the final result can be written as

$$\mathscr{L}_{\text{eff}} = \frac{g^4}{64\pi^2 M_W^2} \left[\bar{s}\gamma^\lambda \mathbb{L} d\right] \left[\bar{s}\gamma_\lambda \mathbb{L} d\right] \sum_{A,B} \Lambda_A \Lambda_B r_{u_A} r_{u_B} f(r_{u_A}, r_{u_B}), \qquad (17.115)$$

where

$$\Lambda_A = V_{A2}^* V_{A1} \,, \qquad (17.116)$$

and $f(r_{u_A}, r_{u_B})$ summarizes the result of the integrations over the magnitude
of the loop momentum from all diagrams.

It does not make much sense to go through the tedious calculation of all
the diagrams here. We present the result of the final integration by giving the
form of the function f:

$$f(x,y) = \frac{g(x) - g(y)}{x - y} - \frac{3}{4} \frac{1}{(1-x)(1-y)} \,,$$

$$g(x) = \left(\frac{1}{4} + \frac{3}{2}\frac{1}{1-x} - \frac{3}{4}\frac{1}{(1-x)^2}\right) \ln x \,. \qquad (17.117)$$

For $x = y$, we can take the limit $y \to x$ in $f(x,y)$ and obtain

$$f(x,x) = \frac{1}{x}\left[\frac{1}{4} + \frac{9}{4}\frac{1}{1-x} - \frac{3}{2}\frac{1}{(1-x)^2}\right] - \frac{3}{2}\frac{x}{(1-x)^3} \ln x \,. \qquad (17.118)$$

Because of the factors of r that appear in Eq. (17.115), contributions from
the u quark internal lines will be very small. We can neglect them safely.
Despite the large top quark mass, $m_t \approx 175\,\text{GeV}$, contributions involving the
top quark are also quite small because the relevant CKM elements are very
small. The dominant contribution comes from charm quark internal lines.
Since $r_c \approx 2.5 \times 10^{-4}$, we can take only the leading term in $f(x,x)$, which is
$1/x$. Thus

$$\mathscr{L}_{\text{eff}} = \frac{G_F^2 M_W^2}{2\pi^2} \, \Lambda_c^2 r_c \left[\bar{s}\gamma^\lambda \mathbb{L} d\right] \left[\bar{s}\gamma_\lambda \mathbb{L} d\right]. \qquad (17.119)$$

To evaluate the mass difference Δm_K of the two eigenstates, we first use
Eq. (17.91) to write

$$\Delta m_K = 2\,\text{Re}\,\mathbb{H}_{12} \qquad (17.120)$$

in the CP conserving case, where the off-diagonal elements are equal. The
elements of the CKM matrix, and hence the combinations Λ defined in Eq.

(17.116), can be taken to be real. The effective Hamiltonian is related to the effective Lagrangian by simply a change in sign.

Recall that $\mathscr{H}$ denotes the Hamiltonian density really, whereas $\mathbb{H}$ has the dimension of the total Hamiltonian. We have considered the effective transition element at zero momentum, so that in the co-ordinate space it is a constant everywhere. Thus, to obtain the relevant element of the total Hamiltonian, we can simply multiply the expression of $\mathscr{H}$ appearing in Eq. (17.79) by $\mathscr{V}$, the volume of the system in which we are performing all our calculations. Putting this into Eq. (17.81) and using the normalization condition of Eq. (17.80), we obtain

$$\Delta m_K = \frac{G_F^2 M_W^2}{2\pi^2} \, \Lambda_c^2 r_c \, \frac{1}{m_K} \left\langle \widehat{K}^0 \left| \left[\overline{s}\gamma^\lambda \mathbb{L}d\right]\left[\overline{s}\gamma_\lambda \mathbb{L}d\right] \right| K^0 \right\rangle . \quad (17.121)$$

Note that the arbitrary volume $\mathscr{V}$ has dropped out.

We are now left with the task of evaluating the matrix element of the quark field combinations within the states K^0 and $\widehat{K}^0$. Clearly, this cannot be done analytically, because we do not know the exact details of strong interactions: how exactly a quark and an antiquark are combined into a meson. So we need to take recourse to some approximations or wild guesses.

One thing we can do, without making any compromise, is to insert a complete set of states in between, and sum over them. But even the relevant matrix elements cannot be determined, so we make a drastic assumption: the contribution of the vacuum state dominates the sum. This is the basis of what is called the *vacuum saturation method*.

There are two ways that we can obtain a non-zero matrix element involving the vacuum. Note that we have not thought about quark color so far in this context, because we did not have to. Only weak gauge bosons and Higgs bosons are being exchanged in the diagrams, none of which can change the color of a quark. This means that the same color flows along any fermion line in the diagrams. Thus, if we care to put the color indices on the quarks, we should write the field bilinear combinations in the effective operator as $\left[\overline{s}_\alpha \gamma^\lambda \mathbb{L}d_\alpha\right]\left[\overline{s}_\beta \gamma_\lambda \mathbb{L}d_\beta\right]$, where α, β are color indices. But Fierz transformation tells us that the expression can equally well be written as $\left[\overline{s}_\alpha \gamma^\lambda \mathbb{L}d_\beta\right]\left[\overline{s}_\beta \gamma_\lambda \mathbb{L}d_\alpha\right]$. When we make the vacuum insertion, we need to consider both these possibilities. Thus we will write

$$\left\langle \widehat{K}^0 \left| \left[\overline{s}\gamma^\lambda \mathbb{L}d\right]\left[\overline{s}\gamma_\lambda \mathbb{L}d\right] \right| K^0 \right\rangle = \left\langle \widehat{K}^0 \left| \overline{s}_\alpha \gamma^\lambda \mathbb{L}d_\alpha \right| 0 \right\rangle \left\langle 0 \left| \overline{s}_\beta \gamma_\lambda \mathbb{L}d_\beta \right| K^0 \right\rangle$$
$$+ \left\langle \widehat{K}^0 \left| \overline{s}_\alpha \gamma^\lambda \mathbb{L}d_\beta \right| 0 \right\rangle \left\langle 0 \left| \overline{s}_\beta \gamma_\lambda \mathbb{L}d_\alpha \right| K^0 \right\rangle .$$
$$(17.122)$$

How do we find the matrix elements that appear in this equation? Recall the definition of the charged kaon decay constant from Eq. (17.47). Since the isospin symmetry is very well respected, this definition also implies the relation

$$\langle 0 |\bar{s}\gamma^\lambda(1-\gamma_5)d| K^0(p)\rangle = \sqrt{2}if_K p^\lambda \tag{17.123}$$

to a very good approximation. If we want to add the color indices of the quarks, this should be written as

$$\langle 0 |\bar{s}_\alpha\gamma^\lambda(1-\gamma_5)d_\beta| K^0(p)\rangle = \frac{\sqrt{2}i}{3}\delta_{\alpha\beta}f_K p^\lambda , \tag{17.124}$$

which would give Eq. (17.123) on summing over the three colors, and a vanishing matrix element when the colors do not match. The matrix elements involving the $\widehat{K}^0$ will be obtained by applying CPT transformation on the result above. Putting now this definition into Eq. (17.122) and remembering that the definition of $\mathbb{L}$ has a factor of $\frac{1}{2}$, we obtain

$$\left\langle \widehat{K}^0 \left| \left[\bar{s}\gamma^\lambda \mathbb{L}d\right]\left[\bar{s}\gamma_\lambda \mathbb{L}d\right] \right| K^0 \right\rangle = \frac{1}{2}f_K^2 m_K^2 \left(1 + \frac{1}{9}\delta_{\alpha\beta}\delta_{\alpha\beta}\right)$$
$$= \frac{2}{3}f_K^2 m_K^2 , \tag{17.125}$$

where the factor of m_K^2 comes from the square of the 4-momentum. Thus, the matrix element of the effective transition Lagrangian of Eq. (17.115) between the K^0 and the $\widehat{K}^0$ states gives

$$\Delta m_K = \frac{G_F^2 M_W^2}{3\pi^2} f_K^2 m_K \Lambda_c^2 r_c . \tag{17.126}$$

Putting in the charm quark mass 1.25 GeV, we obtain $r_c = 2.5 \times 10^{-4}$. Using the value of Cabibbo angle from Eq. (17.23), of f_K from Eq. (17.48), and of the Fermi constant and M_W, we obtain

$$\frac{\Delta m_K}{m_K} = 4.2 \times 10^{-15} . \tag{17.127}$$

If we take the experimental values, with

$$m_K \equiv \left[m(K_L^0) \text{ or } m(K_S^0)\right] = (497.6 \pm 0.024)\,\text{MeV} ,$$
$$\Delta m_K \equiv m(K_L^0) - m(K_S^0) = (3.483 \pm 0.006) \times 10^{-12}\,\text{MeV} , \tag{17.128}$$

the ratio comes out to be

$$\left.\frac{\Delta m_K}{m_K}\right|_{\text{exp}} = 7.2 \times 10^{-15} . \tag{17.129}$$

The good news is that the theoretically calculated value is of the same order of magnitude as the experimentally measured one. The bad news is that they do not agree. There is of course an error bar for the experimental number, but it cannot possibly be near a value that would make Eqs. (17.127) and (17.128) consistent with each other.

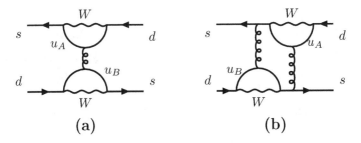

Figure 17.7: More diagrams for K^0-$\widehat{K}^0$ transition.

Is this a disaster, that the theoretical estimate falls short of the experimental value by a considerable amount? Not really. There are many things that have not been taken into account in the calculation that we have described above. Consider the first stage of the procedure, viz., that of determining the quark-level transition operator. We can include extra diagrams by adding gluon lines connecting different quark lines in the box diagrams of Fig. 17.5 *(p 510)* and Fig. 17.6 *(p 513)*. In addition, there can be extra diagrams involving gluon exchanges, such as those shown in Fig. 17.7. The first of these diagrams is called the double-*penguin diagram*. These are of course higher order in perturbation theory if we count the loops. For example, the penguin diagram has two loops whereas the box diagrams have only one. Thus, it might naively seem that these extra diagrams provide only minuscule corrections to the box diagrams. But this need not be true because the extra features present in these diagrams are gluons. At low energies, their interaction is not in the perturbative domain. The important suppression factors are the SU(2) gauge coupling constant and inverse powers of the W mass, and it is easily seen that in this respect, the diagrams of Fig. 17.7 are no different from the diagrams of Fig. 17.5 *(p 510)*: any diagram of each of these two sets contains four weak vertices and two W propagators. So the results of the two sets should be comparable. For the charm quark intermediate diagrams, it has been estimated that these strong corrections can increase the quark-level operator by a factor of about 2.

Let us now consider the second stage of the calculation, i.e., finding the matrix element of the quark-level operator between the relevant meson states. The vacuum saturation method is a poor consolation which has no reason to be very accurate. Clearly, there are other intermediate states that will contribute to the matrix element, e.g., the $\pi\pi$ states. Since K_S can decay to two pions, and K_S can be written as a superposition of K^0 and $\widehat{K}^0$, it follows that both K^0 and $\widehat{K}^0$ have non-zero matrix elements with the two-pion states. There is no reason to assume that these contributions are negligible. One can bypass the introduction of intermediate states by devising more sophisticated techniques for evaluating the matrix element, e.g., the *bag model*, which involves some educated guess about the way the quarks bind in a hadron, but these are guesses in any case, and are not supported by a solution of the QCD interactions.

Considering all these factors, we should feel happy that the simple calculation that we have outlined gives an estimate of Δm_K that has the right order of magnitude.

17.9.5 Leptonic decays of neutral kaons

Like the charged kaons, the neutral kaons can also decay leptonically or semi-leptonically. Let us first discuss semi-leptonic decays involving pions in the final state. Both K_S^0 and K_L^0 can decay to $\pi^+ e^- \widehat{\nu}_e$ and $\pi^- e^+ \nu_e$. These are called K_{e3}^0 decays, analogous to the K_{e3}^+ decay for the charged kaons. Such decay modes arise at the tree level and their analysis is done in the manner shown in §17.5.2. The rates are comparable to those of the charged kaons.

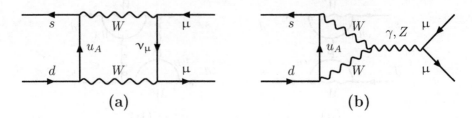

Figure 17.8: Diagrams for $K_L^0 \to \mu^+\mu^-$ arising at the one-loop level in the standard model.

However, the branching ratio of the K_{e3}^0 decay is very high, a little more than 40%. In contrast, the K_{e3}^+ decay has a branching ratio of 5.08%, and $K_{\mu 3}^+$ has 3.35%. The reason lies in the fact that the purely leptonic decay modes of the neutral kaons are extremely suppressed, whereas for the K^+, the dominant decay is into $\mu^+\nu_\mu$. Analogous decays for the neutral kaons would have to be either into $\nu\widehat{\nu}$ channels, or into $\ell^+\ell^-$ channels where ℓ is the electron or the muon. The amplitude for the $\nu\widehat{\nu}$ channels should vanish in the standard model for reasons explained in connection with the decay $\pi^+ \to e^+\nu_e$ in §17.5. For the same reason, the rate for a neutral kaon decaying to e^+e^- should be more suppressed compared to the $\mu^+\mu^-$ mode. All of these expectations are consistent with experimental results.

What was a riddle, in the early days of quark models when only three quarks were assumed to exist, was the minuscule rate for $K_L^0 \to \mu^+\mu^-$. At the quark level, this involves a process $d\widehat{s} \to \mu^+\mu^-$ or $s\widehat{d} \to \mu^+\mu^-$, i.e., there is flavor change in the quark sector without any change of electric charge. This is flavor-changing neutral current in the quark sector. As discussed earlier in §17.2, such currents are absent at the tree level. So the process has to go through loops, like those shown in Fig. 17.8. If one assumes only three flavors of quarks as Cabibbo did when he first postulated quark mixing, only the up quark can appear as the intermediate quark line. From the vertices, the factors $\cos\theta_C \sin\theta_C$ would appear in the amplitude, where θ_C is the Cabibbo angle. Loop integrations give factors of the order of $1/(16\pi^2)$. There does not seem to be any other big suppression factor compared to tree-level diagrams. And just these factors cannot take us anywhere near the experimentally observed branching ratio,

$$\mathscr{B}(K_L^0 \to \mu^+\mu^-) = (6.84 \pm 0.11) \times 10^{-9}. \qquad (17.130)$$

The solution to this riddle lies in the existence of other quarks, and in the GIM cancellation mechanism discussed earlier. In fact, this is the riddle which was addressed by Glashow, Iliopoulos and Maiani, and they showed that the riddle disappears if one assumes the existence of a fourth flavor of quark. This was in 1970, a few years before the charm quark was actually discovered experimentally. Subsequently, Gaillard and Lee showed that the same mechanism can explain the suppression of the K_L-K_S mass difference, something we have already discussed in §17.9.4.

The argument need not be given in detail for the present case, since we have already encountered it in the context of the K_L-K_S mass difference. The point is that the mixing matrix is unitary, so the contributions from different internal lines cancel in the limit of vanishing quark masses. The dominant contribution should therefore have a factor of m_c^2/M_W^2 in the amplitude, and that accounts for the suppression of the rate.

17.10 Processes involving baryons

Baryon-baryon scattering is dominated by strong interactions. Scattering of baryons against charged leptons, discussed in Ch. 13, is dominated by electromagnetic interactions. None of these processes falls under the subject matter of this chapter, which is weak interactions. Scattering of neutrinos off baryons must happen through weak interactions, because neutrinos have no other interaction. So, here we take up the process of deep inelastic scattering of neutrinos from protons. In particular, the process that we have in mind is

$$\nu_\mu + p \rightarrow \mu^- + X\,. \tag{17.131}$$

As in the notation used in §13.2 where deep inelastic scattering with electrons was discussed, X means anything that is possible in the final state. In this sense, this is also an example of inclusive process.

In Ch. 13, we discussed the content of the proton. At the quark level, the process of Eq. (17.131) can be induced by the process

$$\nu_\mu + d \rightarrow \mu^- + u\,. \tag{17.132}$$

However, there are also antiquarks in the proton, and so the process

$$\nu_\mu + \widehat{u} \rightarrow \mu^- + \widehat{d} \tag{17.133}$$

can also take place. In fact, the proton also contains strange quarks and antiquarks, so there can also be quark-level processes involving them. However, the amplitude of such processes would involve off-diagonal elements of the CKM matrix, and therefore would be small. We will neglect them in what follows. In order to maintain consistency, we will take the diagonal elements to be equal to 1.

Whatever the underlying interaction might be, the final state parton from this interaction would undergo hadronization along with the other partons which were present in the original proton, resulting in hadrons in the final hadronic state that we had called X. In order to analyze the process of Eq. (17.131), let us first analyze the fundamental processes of Eqs. (17.132) and (17.133).

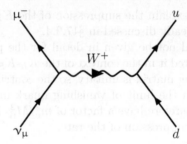

Figure 17.9: Tree-level diagrams for quark-level processes that contribute to neutrino–proton scattering.

The Feynman diagram for both these processes is given by Fig. 17.9, where the line marked d with an incoming arrow can be either interpreted as a d quark coming in or a d-type antiquark going out. The first alternative, with similar interpretation for the line marked u, gives the reaction of Eq. (17.132), whereas the second alternative gives the reaction of Eq. (17.133). We will neglect the masses of the quarks and also of the muon. For this reason, the diagram mediated by the unphysical charged Higgs boson w^+ does not contribute.

We assume that we are interested in finding the scattering cross-section for a momentum transfer in the range $m_p^2 \ll |q^2| \ll M_W^2$ so that we can neglect the proton mass and also use the 4-fermion approximation. It is not necessary to get into a detailed derivation of the Feynman amplitude for the reaction in Eq. (17.132). With minor changes in notation, the derivation is exactly similar to what we had done in §14.5. Denoting the lepton momenta in the initial and final states by k and k', and the quark momenta by $\breve{p}$ and $\breve{p}'$, we obtain

$$\overline{|\mathscr{M}_1|^2} = 64G_F^2 k \cdot \breve{p} \, k' \cdot \breve{p}' = 16G_F^2 \breve{s}^2 \,, \tag{17.134}$$

where $\mathscr{M}_1$ denotes the Feynman amplitude of the process in Eq. (17.132). For the other process, we will denote the Feynman amplitude by $\mathscr{M}_2$. The only difference in its evaluation is that v-spinors would appear in place of u-spinors. Because of this, the term involving the Levi-Civita tensor will change its sign corresponding to whatever is written in Eq. (14.95, *p 430*). When the contraction is done, one obtains

$$\overline{|\mathscr{M}_2|^2} = 64G_F^2 k \cdot \breve{p}' \, \breve{p} \cdot k' \,. \tag{17.135}$$

In §13.4, we had introduced two Lorentz invariant kinematical variables called x and y. Let us recall them now. We note that

$$y = \frac{p \cdot q}{p \cdot k} = 1 - \frac{p \cdot k'}{p \cdot k} \,, \tag{17.136}$$

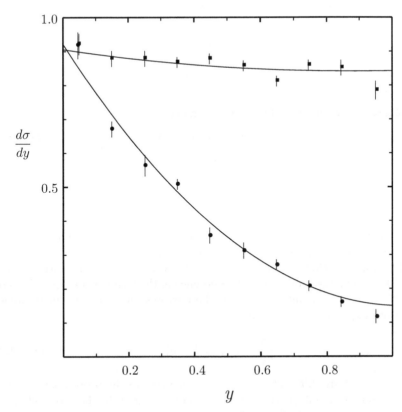

Figure 17.10: Differential cross-section for neutrino (box marks) and antineutrino (bullet marks) scattering off protons. (Redrawn from the article by F. Eisele, *Rep. Prog. Phys.* **49** (1986) 233. © IOP Publishing. With kind permission of IOP Publishing and the author.)

and that $p \cdot k = \frac{1}{2}\check{s}$ in the limit of negligible masses. Thus, $p \cdot k' = \frac{1}{2}(1-y)\check{s}$, and in the massless limit, $k \cdot p'$ is the same. Thus,

$$\overline{|\mathcal{M}_2|^2} = 16 G_F^2 \check{s}^2 (1-y)^2 . \tag{17.137}$$

Using Eq. (4.215, *p 108*), we can then write

$$\frac{d\sigma}{dt}(\nu_\mu d \rightarrow \mu^- u) = \frac{G_F^2}{\pi} ,$$

$$\frac{d\sigma}{dt}(\nu_\mu \widehat{u} \rightarrow \mu^- \widehat{d}) = \frac{G_F^2}{\pi}(1-y)^2 . \tag{17.138}$$

If we now want to write the Feynman amplitude squared for the neutrino–proton scattering, we should add contributions of Eqs. (17.134) and (17.137) with suitable weights which are the parton distribution functions. The method of combining these results has been discussed in §13.7. The result is

$$\frac{d\sigma}{dx\,dy} = \frac{G_F^2\,\mathfrak{s}}{\pi}\left[xf_d(x) + xf_{\hat{u}}(x)(1-y)^2\right],\qquad(17.139)$$

where in this formula, $\mathfrak{s}$ stands for the Mandelstam variable of the electron–proton scattering, different from $\check{\mathfrak{s}}$ that pertains to the quark-level process.

□ **Exercise 17.20** Show that, for the reaction

$$\widehat{\nu}_\mu + p \rightarrow \mu^+ + X,\qquad(17.140)$$

the formula corresponding to Eq. (17.139) is

$$\frac{d\sigma}{dx\,dy} = \frac{G_F^2\,\mathfrak{s}}{\pi}\left[xf_u(x)(1-y)^2 + xf_{\hat{d}}(x)\right].\qquad(17.141)$$

We therefore see how the knowledge of the parton distribution functions can help in calculating the cross-section of neutrino or antineutrino scattering against protons. However, even without knowing the parton distributions explicitly, we can see some dramatic difference in the formulas in Eqs. (17.139) and (17.141). Imagine integrating out these expressions over the entire range of x. Both expressions will be of the form

$$\frac{d\sigma}{dy} = A + B(1-y)^2,\qquad(17.142)$$

where A and B are different for the two processes. In the proton, the probability of finding an antiquark should be small compared to the probability of finding a u or a d quark. Thus, if we suppose that, to a first approximation, the antiquarks are not at all there in a proton, we find that $d\sigma/dy$ is constant for neutrino–proton interaction, whereas it varies as $(1-y)^2$ for antineutrino–proton interaction. There is a slight deviation from such behavior, as seen in Fig. 17.10, which confirms the presence of antiquarks by small amounts.

Chapter 18

Global symmetries of standard model

The standard model is based on a gauge symmetry, which is what we have discussed in Ch. 12, Ch. 16 and Ch. 17. This symmetry governs the dynamics of the model. However, there are many other symmetries, or near-symmetries, in particle interactions, some of which we have discussed in earlier chapters in the book. Such symmetries are often very helpful in understanding various properties of hadrons and leptons. In this chapter, we discuss how some of these symmetries can be understood from the standard model Lagrangian.

18.1 Accidental symmetries

We have discussed the basics of the standard model in two chapters. In Ch. 16, we discussed the electroweak interaction of leptons, and in Ch. 17, that of quarks. We can wonder how it was possible to discuss the two sectors independent of each other. The answer is simple. In the standard model Lagrangian, there is no interaction term which contains a quark field *and* a lepton field. Indeed, the two sectors are really separated.

For one thing, this means that we cannot change a quark to a lepton, or vice versa, through the interactions of the standard model. Let us look at the Yukawa interactions of the standard model, for example. In Eq. (17.2, *p483*), we see that a barred quark field and an unbarred quark field appear. So, whenever a quark is annihilated by the unbarred operator, the barred counterpart creates a quark or annihilates an antiquark. With either possibility, the net quark number, defined as the number of all quarks minus the number of all antiquarks, does not change. This number is usually divided by 3 and called *baryon number*. Thus, a proton has baryon number equal to one, whereas antiproton has the negative of that. Our statement above implies that baryon number is conserved, provided all particles other than quarks and antiquarks are assumed to carry zero baryon numbers.

The same thing can be said about *lepton number*, which is defined to be the number of electrons, muons, taus and all three kinds of neutrinos, minus the number of their antiparticles. But in the leptonic sector, there are more symmetries. If we look at all the interaction terms that appear in Ch. 16, we will see that there isn't any term that connects two leptonic fields of different generations. Thus, we can even define *generational lepton numbers*, e.g.,

$$L_e \equiv (\text{number of electrons and } \nu_e\text{'s})$$
$$-(\text{number of their antiparticles}), \qquad (18.1)$$

and the number will be conserved by all interactions of the standard model. The same way one can define L_μ and L_τ, using particles of the second and third generations respectively.

Note that none of these symmetries was necessary in constructing the standard model Lagrangian. In other words, we did not put these symmetries as constraints on the Lagrangian. We wanted our Lagrangian to be Poincaré invariant, of course. We also put in a gauge symmetry, which is an internal symmetry, and demanded that our Lagrangian should be invariant under that symmetry. Then how did we get a Lagrangian that conserves baryon number or lepton number?

Let us consider a simple example with two numerical variables x and y. Suppose we want to write a quadratic expression involving these two variables which will be invariant under the interchange $x \leftrightarrow y$, and also under the change of sign of one of them, $x \to -x$. Both are Z_2 symmetries, so the overall symmetry is $Z_2 \times Z_2$. It is easy to see that the most general quadratic expression consistent with this symmetry is an arbitrary multiple of $x^2 + y^2$. But this is invariant under arbitrary rotations in the x-y plane. Such transformations, as we have seen earlier, form an O(2) group. This group contains the $Z_2 \times Z_2$ symmetry that we started with, but is much bigger than that: it is a continuous symmetry. We did not impose it: it appeared accidentally in our expression. Such symmetries are called accidental symmetries.

Note that we have, in a sense, planned the accident. Had we not imposed the condition that the expression will have to be quadratic in the two variables, the O(2) symmetry would not have appeared. Alternatively, suppose we wanted to impose the $Z_2 \times Z_2$ symmetry on quadratic terms involving variables x, y and two more variables, w and z. The behavior of all these variables under each Z_2 is summarized below:

Variable	Result of application of	
	First Z_2	Second Z_2
x	y	$-x$
y	x	y
z	w	$-z$
w	z	w

$$(18.2)$$

In this case, the most general expression consistent with the rules is $a(x^2 + y^2) + b(w^2 + z^2) + c(xz + yw)$, which does not have the O(2) symmetry in the variables x and y.

It is the same thing with symmetries like baryon number and lepton number in the standard model. While writing the Lagrangian of the standard model, we put two kinds of restrictions on ourselves. First, we decide on a certain repertoire of fields that should enter the Lagrangian. The collection includes fields of different particles which have been experimentally observed (e.g., the electron), as well as fields that may be necessary for fulfilling some internal consistency of the theory. Second, we want the Lagrangian to be renormalizable, so we allow only field combinations which have mass dimension less than or equal to four. If any one of these extra conditions is sacrificed, an accidental symmetry may be lost. Let us use some examples to illustrate this comment.

The standard model contains left-handed quark and lepton doublets which transform as follows under the gauge group $SU(3)_c \times SU(2)_L \times U(1)_Y$:

$$q_L : (3, 2, \frac{1}{6}), \qquad \ell_L : (1, 2, -\frac{1}{2}). \tag{18.3}$$

Each generation has such multiplets: we have suppressed the generation index. Consider now the combination of multiplets

$$q_L q_L q_L \ell_L. \tag{18.4}$$

Never mind how the different components of the multiplets couple to one another. These will be determined by numerical factors which are not necessary for our argument. These factors will dictate that the three quark fields would make a color singlet, and that the four $SU(2)_L$ doublets would combine to form an $SU(2)_L$ singlet. The weak hypercharge quantum number of the combination is easily seen to be zero, so that the combination does not change the weak hypercharge of any state that it operates on. In short, we can make an invariant of the standard model gauge group from the four multiplets given above.

Such a combination would not be renormalizable because there are four fermion fields involved, which have a combined mass dimension of 6. But if we did not have the requirement of renormalizability, such a term would have been allowed in the Lagrangian. And it is obvious that this term would have violated both baryon number and lepton number symmetries. In fact, it would have violated both these numbers by one unit.

There is in fact a very easy argument to prove that baryon number violating dimension-6 operators would violate lepton number as well. Only quarks carry baryon numbers in the standard model, and each quark field violates baryon number by $\frac{1}{3}$. Since all baryons carry integral baryon numbers, we need three quark fields to violate baryon number. With three quark fields only, angular momentum is not conserved, so we need another fermion field. All other fermion fields available in the standard model carry lepton numbers. Hence the violation of lepton number.

☐ **Exercise 18.1** With the standard model multiplets, there are six dimension-6 B-violating operators in total, not counting the differences between different generations of fermions. Eq. (18.4) represents two of these six, because both the first two and the last two fields can be joined into a singlet or a triplet of the $SU(2)_L$ group.

Identify the remaining four and show that in each case, lepton number is violated by the same amount.

Consider next what would have happened if we suddenly discovered a multiplet of scalar bosons which transformed like $(3, 1, \frac{1}{3})$ of the standard model gauge group. If we really did, we should have included this multiplet in writing the Lagrangian of the standard model. In other words, the Lagrangian would have contained terms involving this multiplet, Δ (no connection with the Δ-baryons). In particular, there would have been terms like the following:

$$\mathscr{L}_\Delta = \left(h_1 q_{\mathrm{L}} q_{\mathrm{L}} \Delta + h_2 q_{\mathrm{L}} \ell_{\mathrm{L}} \Delta^\dagger \right) + \text{h.c.}, \tag{18.5}$$

where h_1 and h_2 are coupling constants. These two terms, taken together, would violate baryon number as well as lepton number symmetry.

It should be realized that any of these two terms, in absence of the other, cannot violate baryon number. Take, e.g., $h_2 = 0$ in Eq. (18.5). The remaining term contains two quark fields, which have a combined baryon number of $\frac{2}{3}$. Therefore, if we assign the Δ boson a baryon number equal to $-\frac{2}{3}$, the interaction term does not violate baryon number. Similarly, if we consider $h_1 = 0$, we can assign a baryon number and a lepton number to Δ that would ensure conservation of baryon and lepton numbers. But when both are nonzero, we cannot have a consistent assignment of baryon and lepton numbers for the field Δ, and it is then that both these numbers are violated.

One might wonder why we are entitled to assign baryon number and lepton number arbitrarily to field multiplets in order to save these symmetries. Shouldn't baryons and leptons be fermions? The answer is positive so far as we have only the fields of the standard model. If we have to introduce more fields in the Lagrangian for whatever reason, we cannot be sure about the answer from the outset. Leaving aside the question of whether only fermions should be included in the club of baryons, let us try to see what are the symmetries of a given Lagrangian. For example, if $h_1 = 0$ in Eq. (18.5), we find that the interaction term is invariant under two independent global U(1) symmetries. Under one of them, q_{L} and Δ carry equal charges whereas ℓ_{L} is neutral. Under the other, ℓ_{L} and Δ carry equal charges whereas q_{L} is neutral. Unquestionably, these symmetries are present in the Lagrangian. If someone has reservations against a boson being called a baryon or a lepton, we can give some other names to these symmetries. But there is no point really in proliferating names. The first U(1) symmetry is the same as the baryon number symmetry so far as all standard model fields are concerned, so it is much easier to continue with the name 'baryon number' for this symmetry. For the same reason, the second symmetry can be called 'lepton number'.

Of course, none of the possibilities discussed here is part of the standard model. The standard model does not have non-renormalizable terms in the Lagrangian, neither does it have extra scalar fields which are color triplets. So baryon number and lepton number are indeed accidental symmetries of the standard model. This statement should be taken with a grain of salt, and will be reanalyzed in §18.4.

18.2 Approximate symmetries

In addition to the absolutely conserved quantum numbers that appear accidentally in the standard model, there are other symmetries which are only approximate. We have discussed many of these in earlier chapters of this book.

Consider, for example, parity or space inversion. It is a symmetry so far as strong and electromagnetic interactions are concerned, because the theories for these interactions are vector-like. For QCD, this means that left- and right-chiral fields transform the same way under its gauge group. QED has the same property: left- and right-chiral components of the same fermion have the same electric charge. Consequently, gauge interactions contain only polar vector bilinears of fermions. Through our experience with photon interactions, we know that with only polar vector currents, parity is conserved. However, under the electroweak gauge group, left- and right-chiral fermion fields transform differently, and therefore fermion currents obtained after adding the left-chiral and the right-chiral currents are not purely polar vector type: there are axial vector currents as well. This leads to parity violation, which is a property of weak interactions only. If we do not make such subdivisions in the interactions and consider all standard model interactions at the same time, all we can say is that parity is an approximate symmetry. What it means is that a parity conserving amplitude would be larger than a parity violating amplitude in a general sense. The same thing can be said about charge conjugation invariance.

The case of time reversal invariance is somewhat different. Polar and axial vector currents behave the same way under time reversal, as seen from Eq. (7.32, p 193). Thus, if gauge currents constituted the only interactions, time reversal symmetry would have been exact. However, other interactions in the standard model violate time reversal invariance by a small amount. Because of CPT invariance, it also implies a minute violation of CP invariance. This will be discussed in detail in Ch. 21.

There are more symmetries which are respected by both strong and electromagnetic interactions, but violated by weak interactions. An example is the property called strangeness, which came up in the discussion of Ch. 10 many times. A little thought shows that the same statement can be made about any flavor of particles. For example, consider an electron. If we define a quantum number that is +1 for the electron, −1 for the positron and zero for everything else in the world, this number is not violated at the electromagnetic interaction vertex involving a photon. There is no question of it being violated through strong interactions, because strong interactions are completely oblivious of the existence of electrons. The quantum number so defined can change only in weak interactions, e.g., in beta decay. For quarks, the argument is the same so far as electromagnetic interactions are concerned. For strong interactions, we only need notice that gluons, by construction of the theory, can change only quark color: they do not do anything to the flavor.

We have also discussed symmetries like isospin, which are respected only by strong interactions. The flavor SU(3), discussed in Ch. 10, also falls in this category. Isospin would have been an exact symmetry of the Lagrangian if the up quark and the down quark were degenerate. Similarly, if the u, d and the s quarks were all degenerate, flavor SU(3) symmetry would have been exact. But the quark masses are unequal, so these symmetries are only approximate ones. Some of the flavor SU(3) predictions have quite large discrepancies with experimental results, because the mass of the strange quark is much larger than those of the up and the down quarks.

And that brings in the question: why is isospin symmetry so good? The quark masses, as mentioned in Eq. (13.98, $p\,409$), show that the down quark is probably about twice as heavy as the up quark, and can even be quite a bit heavier than that. And yet, the predictions of isospin symmetry are respected to within a percent or better. For example, the neutron-proton mass difference is about a thousandth of their average mass, as given in Eq. (8.1, $p\,202$). Why such fantastic agreement?

The answer is that the masses of the up and the down quarks, whatever may be the ratio between the two of them, are both very small compared to the mass of all hadrons. Thus, to a first approximation, we can take both masses to be zero, and in this limit isospin is conserved. We can even take both to have the same mass, and isospin is still conserved. A dimensionless measure of isospin breaking would then comprise the difference $|m_d - m_u|$ divided by a typical energy or mass scale for hadrons. As we just said, all such scales are much heavier than $|m_d - m_u|$, which explains why isospin violations are so small.

The strange quark mass is, however, much larger, as seen in Eq. (13.99, $p\,409$). Differences such as $|m_s - m_d|$ are therefore not that small compared to hadron masses, resulting in considerable discrepancies between the predictions of flavor SU(3) and experimental results.

18.3   Chiral symmetries

18.3.1   Symmetries of massless Dirac Lagrangian

Let us now consider a different kind of flavor symmetry of the Lagrangian, which is realized approximately as well. To motivate this, consider the Lagrangian of a free Dirac particle. It has the symmetry

$$\psi(x) \rightarrow \psi'(x) = e^{i\theta}\psi(x)\,. \qquad (18.6)$$

The conserved Noether charge for this symmetry is just the number of ψ particles. Each term in the free Dirac Lagrangian annihilates a particle but creates one at the same time, so that the total number does not change.

In §14.2, we discussed how any Lagrangian involving fermion fields can be written using chiral projections only. The free Dirac Lagrangian can be written in the form

$$\mathscr{L} = \overline{\psi}_L i\gamma^\mu \partial_\mu \psi_L + \overline{\psi}_R i\gamma^\mu \partial_\mu \psi_R - m(\overline{\psi}_L \psi_R + \overline{\psi}_R \psi_L). \tag{18.7}$$

The mass term involves conversion from one chirality to another. If the mass happens to vanish, the kinetic terms will conserve the numbers of ψ_L and ψ_R separately. The Lagrangian in the massless limit therefore has more symmetries than what appears in Eq. (18.6). Obviously, the symmetry can be written as

$$\psi_L(x) \to \psi_L'(x) = e^{i\theta_1} \psi_L(x), \qquad \psi_R(x) \to \psi_R'(x) = e^{i\theta_2} \psi_R(x). \tag{18.8}$$

Such transformations which act differently on different chiralities of fermion fields are called *chiral transformations*, and symmetries corresponding to these transformations are called *chiral symmetries*.

□ **Exercise 18.2** *Show that the transformation of Eq. (18.8) can also be written as*

$$\psi'(x) = e^{i(\alpha + \beta\gamma_5)} \psi(x), \tag{18.9}$$

where α and β are related to θ_1 and θ_2. Find these relations.

Whether we write the symmetry in terms of the chiral fields as in Eq. (18.8) or in terms of the total field as in Eq. (18.9), it is clear that the symmetry operation now involves two parameters which can be chosen independently, so that there are two U(1) symmetries. In other words, the massless Dirac Lagrangian has the symmetry U(1) × U(1). One of these symmetries, connected with the parameter α in Eq. (18.9), is a vector-like symmetry, in the sense that it does not distinguish between left and right chiralities. The other symmetry, corresponding to the parameter β, is chiral.

Let us now consider the case where there are N massless Dirac fields. Quite obviously, the Lagrangian of these fields will have a flavor symmetry U(N) × U(N). Since the group U(N) is identical to the group SU(N) × U(1), as shown in Eq. (3.9, *p 39*), we can also say that the flavor symmetry for N massless Dirac fields is

$$SU(N) \times SU(N) \times U(1) \times U(1). \tag{18.10}$$

One of the U(1) factors corresponds to a vectorial symmetry under which all fields transform the same way. For N flavors of quarks, this symmetry is the same as the baryon number symmetry. The other U(1) symmetry will be discussed in §18.4. The rest of the symmetry group is SU(N) × SU(N), which is what we are going to discuss for the moment.

This symmetry can at best be an approximate one. For example, since the masses of the up and the down quarks are very small, we can think of the corresponding SU(2) × SU(2) symmetry being an approximate symmetry of the standard model Lagrangian to a very good extent.

There is a big advantage of considering this approximate symmetry. Suppose this symmetry is spontaneously broken by some means to the diagonal vectorial SU(2) symmetry, which is the isospin symmetry. $SU(2) \times SU(2)$ symmetry has six generators, whereas the diagonal SU(2) symmetry has three. The spontaneous symmetry breaking, in this case, breaks three of the six generators, and we should expect three Goldstone bosons as a result.

As discussed in §15.2, a Goldstone boson should be massless. So, does that mean that we should expect three massless bosons? Not really, because the symmetry $SU(2) \times SU(2)$ was not exact to begin with. It was an approximate symmetry, realized in the limit $m_u = m_d = 0$. This means that, if m_u and m_d were really zero, we would have obtained three massless scalars from the spontaneous breaking of chiral symmetry,

$$SU(2) \times SU(2) \to SU(2). \tag{18.11}$$

But since m_u and m_d are not really zero, we should not expect to see some massless scalars. Rather, we should find three particles whose masses are small, since m_u and m_d are small.

There are, indeed, three such particles that we find in the hadronic spectrum whose masses are very small. These are the three pions. They are much lighter compared to any other hadron. We can find a rationale of the smallness of their masses through chiral symmetry: they are the Goldstone bosons of the symmetry breaking shown in Eq. (18.11). They have some mass because the symmetry that is being broken here was not an exact one to begin with: the mass terms of the quarks do not respect this symmetry. The statement will be quantified in §18.3.6.

18.3.2 Sigma model

To understand the nature of this symmetry breaking, let us forget about quarks and gauge bosons for a while, and discuss a model with nucleons and pions. In the Yukawa picture, the pions mediated a force between nucleons, and the nature of this interaction was discussed in §8.8. In order to make our point regarding chiral symmetries, we need to introduce another scalar field. This field is usually denoted by $\sigma(x)$ and therefore the model is called the sigma model.

The Lagrangian of this model is given by

$$\mathscr{L} = \overline{N} i \gamma^\mu \partial_\mu N + \frac{1}{2}(\partial^\mu \sigma)(\partial_\mu \sigma) + \frac{1}{2}(\partial^\mu \pi_a)(\partial_\mu \pi_a)$$
$$+ g\overline{N}\Big(\sigma + i\tau_a \gamma_5 \pi_a\Big) N - V(\sigma, \pi), \tag{18.12}$$

where the τ_a's are the Pauli matrices, and

$$V(\sigma, \pi) = \frac{1}{2}\mu^2(\sigma^2 + \pi_a \pi_a) + \frac{1}{4}\lambda(\sigma^2 + \pi_a \pi_a)^2 \tag{18.13}$$

represents the potential involving the scalar fields. If the field σ is considered an isosinglet, then this Lagrangian definitely has the isospin symmetry, under which the nucleons transform as a doublet and the pions as a triplet. More explicitly, under this symmetry, the infinitesimal transformations are

$$\delta N = -i\frac{\tau_a}{2}\alpha_a N \,,$$
$$\delta \pi_a = \varepsilon_{abc}\alpha_b \pi_c \,,$$
$$\delta \sigma = 0 \,, \tag{18.14}$$

where α_a are infinitesimal parameters, all independent of x.

This is an $SU(2)$ symmetry no doubt, but there is some additional symmetry in this Lagrangian. This can be made explicit by using three infinitesimal parameters β_a:

$$\delta \sigma = -\beta_a \pi_a \,,$$
$$\delta \pi_a = \beta_a \sigma \,,$$
$$\delta N = -i\frac{\tau_a}{2}\beta_a \gamma_5 N \,. \tag{18.15}$$

Note that because of the presence of γ_5 in the transformation rule for nucleons, this is a chiral symmetry. It inflicts different transformations on the left-chiral and right-chiral components of the nucleons.

The vectorial and the chiral symmetries do not commute. However, we can rewrite the transformations using the two chiralities of nucleons, left and right. Each chirality of nucleons will also transform like $SU(2)$ doublets. What is more, the two types of transformations commute, so that the total symmetry is $SU(2) \times SU(2)$: one factor corresponding to the left-chiral and the other corresponding to the right-chiral part of the nucleon. These comments will be explained in more detail in §18.3.3.

☐ **Exercise 18.3** Show that the transformations given in Eqs. (18.14) and (18.15), for infinitesimal α_a's and β_a's, are indeed symmetries of the Lagrangian.

☐ **Exercise 18.4** Show that the Noether currents corresponding to the vectorial $SU(2)$ transformations are

$$J_a^\mu = \overline{N}\frac{\tau_a}{2}\gamma^\mu N + \varepsilon_{abc}\pi_b \partial_\mu \pi_c \,, \tag{18.16}$$

and those corresponding to the chiral $SU(2)$ transformations are

$$\widetilde{J}_a^\mu = \overline{N}\frac{\tau_a}{2}\gamma^\mu \gamma_5 N + (\partial^\mu \sigma)\pi_a - (\partial^\mu \pi_a)\sigma \,. \tag{18.17}$$

☐ **Exercise 18.5** The Lagrangian of Eq. (18.12) is also invariant under parity. Show that the intrinsic parity of the σ-field is positive, whereas that of the pion fields is negative.

Now suppose that the parameter μ^2 appearing in the potential is in fact negative. Then the minimum of the potential will not occur for vanishing values of the four scalar fields. Rather, it will happen at the values satisfying

$$\sigma^2 + \pi_a \pi_a = \frac{-\mu^2}{\lambda} \equiv v^2 \,. \tag{18.18}$$

Suppose we find the system in the vacuum where $\langle\sigma\rangle = v$, and the expectation value of the π fields are zero. Since the field σ was invariant under the isospin transformations, the isospin symmetry does not know about this field, and is therefore intact. However, the chiral SU(2) symmetry must be broken, since σ transforms non-trivially under this symmetry. So we see a concrete example where symmetry breaking of the type mentioned in Eq. (18.11) can be realized.

We now expect three Goldstone bosons. To identify them, we note that σ cannot be regarded as a quantum field now. We can rewrite the Lagrangian in terms of the quantum fields $\tilde{\sigma} \equiv \sigma - v$ and the π_a's. Once we do this, it is easily seen that the π_a's are the Goldstone bosons.

□ **Exercise 18.6** *Find the mass of the σ-particle after symmetry breaking.*

□ **Exercise 18.7** *Is the parity symmetry broken spontaneously by the vacuum expectation value of the σ-field?*

18.3.3 Currents and charges

Let us now come back to quarks. The free Lagrangian of the quarks can be written as

$$\mathscr{L}_{\text{free}} = \sum_q \overline{q}_\alpha i\gamma^\mu \partial_\mu q_\alpha - (\text{mass terms}). \tag{18.19}$$

The sum over q runs over all quark flavors, and three colors of each flavor. The index α on the quarks is a color index, which is assumed to be summed over. The color symmetry is SU(3), which is gauged by the introduction of the gluon fields, resulting in the theory of quantum chromodynamics or QCD. Here, we are interested in the flavor symmetries. If we consider the limit in which the up and the down quarks are massless, there is an SU(2) × SU(2) flavor symmetry. Denoting the isospin doublet by

$$\Psi \equiv \begin{pmatrix} u \\ d \end{pmatrix}, \tag{18.20}$$

these flavor symmetry transformations can be written as

$$\Psi \to \Psi' = \exp\left(-i\frac{\tau_a}{2}\alpha_a\right)\Psi \tag{18.21}$$

and

$$\Psi \to \Psi' = \exp\left(-i\frac{\tau_a}{2}\beta_a\gamma_5\right)\Psi. \tag{18.22}$$

The first one is a vector-like symmetry, the second one is chiral. The generators of SU(2) in the doublet representation are the Pauli matrices, which have been denoted by τ_a.

We can easily find the Noether currents corresponding to these symmetries. For the vectorial SU(2) symmetry given in Eq. (18.21), the Noether currents are

$$J_a^\mu(x) = \overline{\Psi}(x)\gamma^\mu \frac{T_a}{2}\Psi(x)\,, \tag{18.23}$$

whereas for the chiral SU(2) symmetry of Eq. (18.22), the currents are

$$\widetilde{J}_a^\mu(x) = \overline{\Psi}(x)\gamma^\mu \gamma_5 \frac{T_a}{2}\Psi(x)\,. \tag{18.24}$$

The conserved charges are therefore given by

$$Q_a = \int d^3x\, J_a^0(x) = \int d^3x\, \Psi^\dagger(x)\frac{T_a}{2}\Psi(x)\,,$$
$$\widetilde{Q}_a = \int d^3x\, \widetilde{J}_a^0(x) = \int d^3x\, \Psi^\dagger(x)\gamma_5\frac{T_a}{2}\Psi(x)\,. \tag{18.25}$$

We can now try to see what are the commutation relations between the charges. First, the commutator between two vectorial charges. With explicit indices, we can write $\Psi^\dagger(x)\tau_a\Psi(x) = \Psi_A^\dagger(x)(\tau_a)_{AB}\Psi_B(x)$, where A, B take the values 1 and 2 to denote which flavor of the doublet Ψ is implied. Then,

$$[Q_a, Q_b] = \left(\frac{T_a}{2}\right)_{AB}\left(\frac{T_b}{2}\right)_{CD}\int d^3x \int d^3y \left[\Psi_A^\dagger(x)\Psi_B(x), \Psi_C^\dagger(y)\Psi_D(y)\right]. \tag{18.26}$$

Since the charges Q_a are conserved, we can take both charges at the same time, so that the right hand side of this equation involves an equal-time commutator involving field variables. The commutator can be reduced by using the identity

$$[PQ, RS] = P[Q, R]_+S - PR[Q, S]_+ + [P, R]_+SQ - R[P, S]_+Q\,, \tag{18.27}$$

which is valid for any set of objects P, Q, R and S whose multiplication is associative. This way, we obtain equal-time anticommutators. Using the anticommutation rule given in Eq. (4.128, *p 86*) which, in more explicit notation, is given by

$$\left[\psi_A(x), \psi_B^\dagger(y)\right]_+\bigg|_{x_0=y_0} = \delta_{AB}\delta^3(\boldsymbol{x}-\boldsymbol{y})\,, \tag{18.28}$$

where A, B denote the components of the spinor fields, we obtain

$$[Q_a, Q_b] = \left(\frac{T_a}{2}\right)_{AB}\left(\frac{T_b}{2}\right)_{CD}\int d^3x \left(\delta_{BC}\Psi_A^\dagger(x)\Psi_D(x) - \delta_{AD}\Psi_C^\dagger(x)\Psi_B(x)\right)$$
$$= \int d^3x\, \Psi^\dagger(x)\left[\frac{T_a}{2}, \frac{T_b}{2}\right]\Psi(x)\,. \tag{18.29}$$

Using now the commutation relation of the Pauli matrices, we find that

$$\left[Q_a, Q_b\right] = i\varepsilon_{abc}Q_c\,, \tag{18.30}$$

which is an SU(2) algebra. However, the chiral charges do not form an SU(2) algebra. A similar exercise gives

$$\left[\widetilde{Q}_a, \widetilde{Q}_b\right] = i\varepsilon_{abc}Q_c\,, \tag{18.31}$$

with Q, rather than $\widetilde{Q}$, on the right hand side. And finally, the commutation between two different kinds of charges is

$$\left[Q_a, \widetilde{Q}_b\right] = i\varepsilon_{abc}\widetilde{Q}_c\,. \tag{18.32}$$

We said that the flavor symmetry is $SU(2) \times SU(2)$. Obviously, neither the Q_a's nor the $\widetilde{Q}_a$'s are generators of either of these two SU(2) factors, because the generators of the two factors should commute. It is easily seen that if we define the combinations

$$Q_a^{(\mathrm{L})} = \frac{1}{2}\left(Q_a - \widetilde{Q}_a\right)\,, \qquad Q_a^{(\mathrm{R})} = \frac{1}{2}\left(Q_a + \widetilde{Q}_a\right)\,, \tag{18.33}$$

then the $Q_a^{(\mathrm{L})}$'s can be the generators of one of the SU(2) factors, and $Q_a^{(\mathrm{R})}$'s can be the generators of the other SU(2). Moreover, any of the $Q^{(\mathrm{L})}$ generators commutes with any of the $Q^{(\mathrm{R})}$ generators, so the two SU(2) factors are independent, and the symmetry can be called $SU(2) \times SU(2)$.

☐ **Exercise 18.8** Verify that the charges defined in Eq. (18.33) have the following commutators:

$$\left[Q_a^{(\mathrm{L})}, Q_b^{(\mathrm{L})}\right] = i\varepsilon_{abc}Q_c^{(\mathrm{L})}\,,$$

$$\left[Q_a^{(\mathrm{R})}, Q_b^{(\mathrm{R})}\right] = i\varepsilon_{abc}Q_c^{(\mathrm{R})}\,,$$

$$\left[Q_a^{(\mathrm{L})}, Q_b^{(\mathrm{R})}\right] = 0\,. \tag{18.34}$$

The relations in Eq. (18.34) are commutation relation between various charges, and can be called the *charge algebra*. In a similar manner, we can find out the commutators between a charge and the time component of a current. For example, following the steps that lead to Eq. (18.30), we can obtain the result

$$\left[Q_a, J_b^0(x)\right] = i\varepsilon_{abc}J_c^0(x)\,. \tag{18.35}$$

There are similar analogs of the other commutators, which are easy to guess. Going one step further, we can find the commutator between the time components of currents. For example, we obtain

$$\left[J_a^0(x), J_b^0(y)\right]\Big|_{x_0=y_0} = i\varepsilon_{abc}J_c^0(x)\delta^3(\boldsymbol{x} - \boldsymbol{y})\,. \tag{18.36}$$

These relations, along with similar relations involving the time component of the axial vector currents, comprise what is called the *current algebra*. One can also include similar commutators involving the spatial parts of the currents, but there is some complication in it which we need not discuss.

Why are these relations useful? Consider the Lagrangian to be divided into two parts, $\mathcal{L}_0$ and $\mathcal{L}_1$, such that the chiral symmetry is an exact symmetry of $\mathcal{L}_0$. We can find the Noether currents of the symmetry, as we have done in Eqs. (18.23) and (18.24). We can also find the commutators, as we have done. Now, what are the effects of $\mathcal{L}_1$ on all these exercises? We can still follow the prescription of Eq. (4.109, *p 82*) to define a current. If $\mathcal{L}_1$ does not contain any derivative term in the fields, we will still obtain the same expression for the currents. Of course, these currents will not be conserved since $\mathcal{L}_1$ does not have the chiral symmetry. But, any expression involving the currents will still be valid, provided it has not used the current conservation at any stage. The charge algebra and the current algebra are precisely expressions of this sort. So, the physical consequences derived from these commutators would be valid even if the chiral symmetry is not exact.

18.3.4 Chiral symmetry breaking

The $SU(2) \times SU(2)$ symmetry was established in Eq. (18.33). If this symmetry is somehow broken spontaneously down to an $SU(2)$ subgroup, we should expect three Goldstone bosons. We have already seen, in Eq. (18.30), that the Q_a's satisfy an $SU(2)$ algebra among themselves. So it is tempting to guess that the remnant $SU(2)$ symmetry will be spanned by these generators.

However, in order to break a symmetry spontaneously, one needs some vacuum expectation value that transforms non-trivially under that symmetry. In all examples of spontaneous symmetry breaking discussed so far, we have relied on the VEV of some scalar field. That cannot work here. There is no boson in the Lagrangian of Eq. (18.19). However, in §15.2, we mentioned that the strength of Goldstone's theorem lies in the fact that even in the absence of bosons in the fundamental Lagrangian, we will obtain three massless bosonic states in the spectrum of states of the theory.

We must therefore need some composite operator, composed of quark fields, to develop a VEV. The operator should transform like a scalar, because otherwise we would break the proper Lorentz symmetry. The simplest such operators are of the form $\bar{q}q$ for any quark field q. Such an operator will be able to create a quark-antiquark pair from the vacuum. Thus, the statement that such operators have a VEV is equivalent to the statement that the vacuum state contains quark-antiquark pairs.

There are quite a few reasons to believe that this statement is true. If the quarks are really massless, it does not cost energy to create quark-antiquark pairs from the vacuum, so it is natural to suppose that such pairs will remain in the ground state of the system. After all, it is known that the VEV of Cooper pairs of electrons gives rise to the phenomenon of superconductivity.

Secondly, we know from earlier chapters that

$$\overline{q}q = \overline{q}_\text{L} q_\text{R} + \overline{q}_\text{R} q_\text{L} \,. \tag{18.37}$$

Thus, if $\overline{q}q$ develops a VEV, it would imply that the vacuum can take in a right-chiral quark and give back a left-handed quark, and vice versa. In other words, the vacuum can exchange a right-chiral quark for a left-chiral one, change the chirality of a quark. This is also what happens with the mass term in the Dirac equation. Thus, the vacuum in this case would emulate a mass term for the quarks, some sort of an induced mass. This can explain why a quark in a hadron has a large constituent mass, as we saw in §13.10.

Clearly, since the vacuum can change a left-chiral quark to a right-chiral one, the numbers of quarks of left and right chiralities are not separately conserved. It means that the vacuum does not respect chiral symmetries of quarks. Thus, with this vacuum, the chiral SU(2) generators $\tilde{Q}_a$ will be broken. The vectorial generators Q_a will remain unbroken, and will constitute the vectorial SU(2) symmetry that is isospin. There will be three Goldstone bosons corresponding to the three broken generators $\tilde{Q}_a$. These should be the pions.

18.3.5   PCAC and soft pion theorem

There are many ramifications of the idea that the pions are Goldstone bosons of chiral symmetry breaking. We will discuss some of them here.

In §17.5, we introduced the matrix element of quark currents between the pion and the vacuum states. The quark current in the standard model Lagrangian has polar and axial vector parts. We remarked that only the axial part contributes to the matrix element. So the definition of the pion decay constant, given in Eq. (17.36, p 494), can also be written as

$$\left\langle 0 \left| \tilde{J}_a^\mu(x) \right| \pi_b(p) \right\rangle = i f_\pi \delta_{ab} p^\mu e^{-ip \cdot x} \,, \tag{18.38}$$

where we have also generalized the relation to include the neutral pion and the neutral current as well, and noting that isospin invariance would imply the Kronecker delta on the right hand side.

The matrix element of the divergence of the axial current can be written easily now, using the translation operator to relate $\tilde{J}^\mu(x)$ to $\tilde{J}^\mu(0)$, as shown in Eq. (5.115, p 139). This gives

$$\left\langle 0 \left| \partial_\mu \tilde{J}_a^\mu(0) \right| \pi_b(p) \right\rangle = f_\pi \delta_{ab} m_\pi^2 \,, \tag{18.39}$$

where we have used the relation $p^\mu p_\mu = m_\pi^2$. This is an interesting relation which says that if the axial current is conserved, the pion is massless.

☐ **Exercise 18.9** Take the definition of the pion decay constant from Eq. (18.38), with $\tilde{J}_a^\mu$ defined through Eq. (18.24). The charged pion fields are defined as $\left|\pi^{+}\right\rangle = \left(\left|\pi_1\right\rangle + i\left|\pi_2\right\rangle\right)/\sqrt{2}$. Use this to show that Eq. (18.38) is indeed equivalent to the definition given in Eq. (17.36, p 494).

So far, this is just the definition of f_π. We can go a bit further if we express the Kronecker delta on the right hand side of these equations in the form

$$\langle 0 \left|\phi_a(x)\right| \pi_b(p)\rangle = e^{-ip\cdot x}\delta_{ab}\,, \tag{18.40}$$

where ϕ_a represents the pion field. Eq. (18.39) now becomes

$$\left\langle 0 \left|\partial_\mu \tilde{J}_a^\mu(0)\right| \pi_b(p)\right\rangle = f_\pi m_\pi^2 \langle 0 \left|\phi_a(0)\right| \pi_b(p)\rangle\,, \tag{18.41}$$

where both sides of the equation contain matrix elements between the same set of states. There is a conjecture that the relation is valid also at the operator level, i.e., one can write operator relation

$$\partial_\mu \tilde{J}_a^\mu(x) = f_\pi m_\pi^2 \phi_a(x)\,. \tag{18.42}$$

This conjecture goes by the name *partially conserved axial current*, better known by the acronym *PCAC*.

□ **Exercise 18.10** The Lagrangian of the sigma model was given in Eqs. (18.12) and (18.13). Define a new theory by adding a term $a\sigma$ to this Lagrangian. This violates the chiral $SU(2)$ symmetry explicitly, but the vectorial $SU(2)$ is intact because the σ field is invariant under it. Use the expression of the chiral $SU(2)$ currents given in Eq. (18.17), take its divergence, and use the equations of motion to show that the divergence is indeed of the PCAC form.

With the help of PCAC, we can deduce a class of relations for matrix elements of arbitrary operators within states containing pions. To deduce such relations, let us introduce a very standard machinery of quantum field theory called the Lehmann–Symannzik–Zimmermann (or LSZ) reduction formula. The aim of this reduction is to express any matrix element in terms of the vacuum expectation value of some operator. If we have an initial state containing one pion and any number of other particles which are collectively called α, and a final state containing some particles which we collectively call β, then for an arbitrary operator $O(x)$, the reduction formula reads

$$\langle \beta \left|O(0)\right| \alpha\pi_a(p)\rangle = i \int d^4x\, e^{-ip\cdot x}(\Box + m_\pi^2) \left\langle \beta \left|\left(O(0)\phi(x)\right)_T\right| \alpha\right\rangle. \tag{18.43}$$

The matrix element on the right hand side of this equation is of a *time-ordered product*, which will be defined shortly. Without getting into the definition, it is clear that, applying similar relations on the other particles in the initial and the final states, we can ultimately obtain a vacuum expectation value, as mentioned earlier. But here we are not interested in reaching that goal. Rather, we first integrate by parts, transferring the derivative onto the exponential factor, and obtain

$$\langle \beta \,|O(0)|\, \alpha \pi_a(p)\rangle = i(-p^2 + m_\pi^2) \int d^4x \, e^{-ip \cdot x} \left\langle \beta \left| \left(O(0)\phi(x) \right)_T \right| \alpha \right\rangle . \tag{18.44}$$

Then we use the PCAC relation, Eq. (18.42), to rewrite the matrix element as

$$\langle \beta \,|O(0)|\, \alpha \pi_a(p)\rangle = \frac{i(-p^2 + m_\pi^2)}{f_\pi m_\pi^2}$$
$$\times \int d^4x \, e^{-ip \cdot x} \left\langle \beta \left| \left(O(0)\partial_\mu \tilde{J}_a^\mu(x) \right)_T \right| \alpha \right\rangle . \tag{18.45}$$

We have not had an occasion to introduce time-ordered products of operators since we did not derive the results of quantum field theory in a formal manner. For any two operators $A(x)$ and $B(y)$, the time-ordered product is defined as

$$\left(A(x)B(y) \right)_T \equiv \Theta(x_0 - y_0)A(x)B(y) + \Theta(y_0 - x_0)B(y)A(x) , \tag{18.46}$$

where Θ denotes the unit step function, first introduced in Eq. (4.158, *p 96*). In other words, in the time-ordered product of two operators, the operator at a later time always sits to the left of the other one. If we now take a spatial derivative with respect to any of the co-ordinates, the result will also be a time-ordered product. For example,

$$\frac{\partial}{\partial x^i} \left(A(x)B(y) \right)_T = \left(\frac{\partial A(x)}{\partial x^i} B(y) \right)_T . \tag{18.47}$$

But if we take a derivative with respect to time, we need to consider derivatives of the step function as well, which are delta functions. Therefore we obtain

$$\frac{\partial}{\partial x^0} \left(A(x)B(y) \right)_T = \left(\frac{\partial A(x)}{\partial x^0} B(y) \right)_T + \delta(x_0 - y_0)\left[A(x), B(y) \right] . \tag{18.48}$$

Using this result, we can write

$$\left\langle \beta \left| \left(O(0)\partial_\mu \tilde{J}_a^\mu(x) \right)_T \right| \alpha \right\rangle = \partial_\mu \left\langle \beta \left| \left(O(0)\tilde{J}_a^\mu(x) \right)_T \right| \alpha \right\rangle$$
$$- \delta(x_0)\left[\tilde{J}_a^\mu(x), O(0) \right] . \tag{18.49}$$

Putting this back into Eq. (18.45) and performing some partial integrations, we obtain

$$\langle \beta \,|O(0)|\, \alpha \pi_a(p)\rangle = \frac{i(-p^2 + m_\pi^2)}{f_\pi m_\pi^2} \int d^4x \, e^{-ip \cdot x}$$
$$\times \left\langle \beta \left| ip_\mu \left(O(0)\tilde{J}_a^\mu(x) \right)_T - \delta(x_0)\left[\tilde{J}_a^\mu(x), O(0) \right] \right| \alpha \right\rangle . \tag{18.50}$$

Suppose now we consider this result in the limit $p^\mu \to 0$. This is usually called the *soft limit*. On the right hand side, the term containing the time-ordered product now vanishes. In the other term, the integration over x can be performed to yield the axial charge. Thus we obtain

$$\lim_{p^\mu \to 0} \langle \beta \,|O(0)|\, \alpha \pi_a(p)\rangle = -\frac{i}{f_\pi} \left\langle \beta \left| \left[\tilde{Q}_a, O(0) \right] \right| \alpha \right\rangle . \tag{18.51}$$

This is called the *soft pion theorem*. On the left hand side, we have a matrix element within two states, one of which contains a pion. In the soft pion limit, this matrix element is equal to another matrix element where the pion has been removed from the external state, and the operator is a commutator involving the axial charge. There is, of course, a similar result where the removed pion is in the final state.

18.3.6   Quark masses

So far, we have considered the consequences of neglecting the masses of the u and the d quarks. We can go further and consider the s quark to be massless as well. In that case, the global flavor symmetry would be SU(3) × SU(3), apart from the two U(1) factors one of which is baryon number symmetry and the other will be discussed in §18.4. The global symmetry implies polar and axial vector currents. If the axial SU(3) currents are spontaneously broken, then there will be eight Goldstone bosons. Three of them are the three pions. The others are naturally the other members of the lightest meson octet that was discussed in Ch. 10, containing also the kaons and the eta. We will represent all these particles collectively by the uppercase letter Π in what follows, and their fields by $\Phi(x)$. Generalization of the formulas derived above for pions is then obvious.

Quark mass terms in the Lagrangian are not invariant under chiral symmetries, as has been said before. Because of these terms, the chiral symmetry is not really a symmetry of the Lagrangian, and the members of the pseudoscalar meson octet should not be massless. The mass of these mesons must then somehow depend on the quark masses. This fact entails crucial information regarding quark masses, as we will now see.

The mass terms in the Lagrangian are

$$\mathscr{L}_{\mathrm{mass}} = -\left(m_u \bar{u}u + m_d \bar{d}d + m_s \bar{s}s \right). \tag{18.52}$$

The corresponding terms in the Hamiltonian would have opposite signs, and will be denoted by $\mathscr{H}'$. We can treat these terms as perturbations and estimate meson masses. In first order in perturbation theory, we should write

$$m_a^2 = \langle \Pi_a | \mathscr{H}' | \Pi_a \rangle, \tag{18.53}$$

where the index a can take values from 1 to 8. We can use the soft pion theorem to remove the mesons from the external states. The result will be valid only in the soft limit, and would read

$$m_a^2 f_a^2 = \left\langle 0 \left| \left[\tilde{Q}_a, \left[\tilde{Q}_a, \mathscr{H}' \right] \right] \right| 0 \right\rangle, \tag{18.54}$$

where there is no summation on the repeated indices on any side. If we take non-hermitian generators, then one of the $\tilde{Q}_a$'s should be replaced by its hermitian conjugate.

Evaluating the commutators present in Eq. (18.54) is easy. For SU(3) ×
SU(3) symmetry, the axial charges are given by

$$\tilde{Q}_a = \int d^3x \ \Psi^\dagger(x) \frac{\lambda_a}{2} \gamma_5 \Psi(x) \,, \qquad (18.55)$$

where now Ψ denotes a column matrix with u, d and s quarks as entries, and
λ_a's are the SU(3) generators in the fundamental representation, presented in
Eq. (10.13, *p 259*). On the other hand, $\mathscr{H}'$ can be written as

$$\mathscr{H}' = \Psi^\dagger \gamma_0 \mathbb{M} \Psi \,, \qquad (18.56)$$

where

$$\mathbb{M} = \text{diag}(m_u, m_d, m_s) \,. \qquad (18.57)$$

Evaluation of commutators now follows more or less the same steps as those
obtained while evaluating charge commutators in §18.3.3. We can use Eq.
(18.27), just as we did there. The difference is that now there is a Dirac
matrix γ_5 in one factor and a γ_0 in the other, which anticommute. This gives
an extra minus sign between the two terms, and we end up with

$$\left[\tilde{Q}_a, \mathscr{H}'\right] = \overline{\Psi} \left[\frac{\lambda_a}{2}, \mathbb{M}\right]_+ \gamma_5 \Psi \,. \qquad (18.58)$$

The evaluation of the second commutator is similar, and it gives, finally,

$$\left[\tilde{Q}_a, \left[\tilde{Q}_a, \mathscr{H}'\right]\right] = \overline{\Psi} \left[\frac{\lambda_a}{2}, \left[\frac{\lambda_a}{2}, \mathbb{M}\right]_+\right]_+ \Psi \,. \qquad (18.59)$$

For example, if we consider $\tilde{Q}_3$, we would obtain

$$\left[\tilde{Q}_3, \left[\tilde{Q}_3, \mathscr{H}'\right]\right] = \overline{\Psi} \begin{pmatrix} m_u & 0 & 0 \\ 0 & m_d & 0 \\ 0 & 0 & 0 \end{pmatrix} \Psi = m_u \bar{u}u + m_d \bar{d}d \,. \quad (18.60)$$

When we put this into Eq. (18.54), we will obtain vacuum expectation values
of $\bar{u}u$ and $\bar{d}d$. In the first order of perturbation, we can neglect the difference
between such vacuum expectation values and write all of them as $\langle \bar{q}q \rangle$. Thus
we obtain

$$f_\pi^2 m_\pi^2 = (m_u + m_d) \langle \bar{q}q \rangle \,, \qquad (18.61)$$

since $\tilde{Q}_3$ is part of a triplet under isospin, and so is the pion. Using the same
procedure with $\tilde{Q}_4$ instead of $\tilde{Q}_3$, we would obtain a similar relation involving
the isodoublet kaons:

$$f_K^2 m_K^2 = (m_s + m_d) \langle \bar{q}q \rangle \,. \qquad (18.62)$$

And with $\widetilde{Q}_8$, we obtain the mass formula for the isosinglet in the SU(3) octet, which is the η:

$$f_\eta^2 m_\eta^2 = \frac{1}{3}(m_u + m_d + 4m_s)\langle \bar{q}q \rangle \,. \tag{18.63}$$

We have written one mass relation for each isomultiplet, tacitly taking a viewpoint that isospin symmetry is not broken. This can be a consistent viewpoint only if $m_u = m_d$. Using this, and taking the decay constants to be equal as well, we obtain Eq. (10.82, *p 280*), the Gell-Mann–Okubo mass formula for pseudoscalar mesons. Moreover, we obtain the relation

$$\frac{m_\pi^2}{2m_K^2 - m_\pi^2} = \frac{m_0}{m_s}, \tag{18.64}$$

where m_0 is the common mass of the up and the down quarks in this limit. Using the experimentally known values of the pion and kaon masses, we obtain

$$\frac{m_0}{m_s} \approx \frac{1}{26}. \tag{18.65}$$

This shows that the up and the down quarks are substantially lighter than the strange quark, implying that the isospin is a much better symmetry than the SU(3) flavor symmetry.

We can also introduce isospin breaking effects coming from electromagnetic interactions, which will depend only on the charges of the quarks and not their flavors. So the corrections for π^+ and for K^+ would be equal. We can then interpret the previously obtained equations as the ones which are valid for the neutral mesons, and write the formulas for the charged mesons as

$$\begin{aligned} f^2 m_{\pi^+}^2 &= (m_u + m_d)\langle \bar{q}q \rangle + \Delta \,, \\ f^2 m_{K^+}^2 &= (m_s + m_u)\langle \bar{q}q \rangle + \Delta \,, \end{aligned} \tag{18.66}$$

using a common value of the decay constants. This gives

$$\frac{m_{K^+}^2 - m_{\pi^+}^2 - m_{K^0}^2}{m_{K^0}^2 - m_{K^+}^2 + m_{\pi^+}^2 - 2m_{\pi^0}^2} = \frac{m_d}{m_u}. \tag{18.67}$$

Note that the electromagnetic correction drops out of the ratio. Putting in the masses of the mesons, we obtain

$$\frac{m_d}{m_u} \approx 1.8. \tag{18.68}$$

This is the result that we mentioned in §18.2: the down quark is much heavier than the up quark.

The ratios mentioned above do not determine any of the masses. However, we can take a hint from the baryon decuplet, for which any two successive isospin multiplets has a mass difference of about 150 MeV. If we assign this difference to the mass of a strange quark, Eq. (18.65) tells us that the average mass of the up and the down quarks should be about 6 MeV. Using Eq. (18.68) now, we can obtain

$$m_u \approx 4 \, \text{MeV}, \qquad m_d \approx 7.5 \, \text{MeV}. \tag{18.69}$$

Quark masses similar to these were quoted in §13.10.

Contrary to the constituent quark masses which appear in the formulas for magnetic moments etc, these are really the quark masses that appear in the Lagrangian, because these are responsible for the fact that chiral symmetry is not a symmetry of the Lagrangian. Since the quarks cannot be seen as free particles, their masses have to be estimated in such indirect manner. In addition, quark masses are parameters in the Lagrangian which, like the coupling constants, depend on the momentum scale. This increases the confusion with the exact values of the quark masses. All in all, the estimates have a lot of arbitrariness. For example, the masses quoted in Ch. 13 are somewhat smaller than the values mentioned here.

18.3.7 Chiral Lagrangians

In §15.3, we discussed that the interactions of Goldstone bosons are always derivative interactions, i.e., they vanish in the limit that the 4-momentum of the Goldstone boson goes to zero. If pions are thought of as Goldstone bosons of chiral symmetry breaking, the same should apply to them. We showed that we can write the fields in a "polar" form in which this is explicit. For a U(1) symmetry, we presented this polar form in Eq. (15.36, *p 451*). In a similar fashion, we can construct a matrix

$$U \equiv \exp\left(i\frac{\tau_a}{2}\pi_a/f_\pi\right), \tag{18.70}$$

that contains the pion fields. Since the pions are supposed to be the Goldstone bosons of the SU(2) × SU(2) flavor symmetry, the Lagrangian should contain only derivatives of U if this symmetry is exact in the Lagrangian.

To construct a Lagrangian with the derivatives, we need to know how U transforms under the SU(2) × SU(2) flavor symmetry. For this, it is easiest to introduce the nucleons in the Lagrangian. The infinitesimal transformation on the nucleon fields were given in Eqs. (18.14) and (18.15). We can use these to see that the finite transformations on the left and right chiralities of the nucleon are of the form

$$N_{\text{L}} \to N_{\text{L}}' = V_L N_{\text{L}}, \qquad N_{\text{R}} \to N_{\text{R}}' = V_R N_{\text{R}}, \tag{18.71}$$

where V_L and V_R are both 2 × 2 matrices, of the form

$$V_L = \exp\left(-i\frac{\tau_a}{2}\theta_a^{(L)}\right), \tag{18.72}$$

and likewise for V_R. Since we are talking of global transformations here, the parameters $\theta_a^{(L)}$ and $\theta_a^{(R)}$, or equivalently the matrices V_L and V_R, should be independent of the spacetime co-ordinates.

To see how these transformations affect the pion fields, let us consider the pion–nucleon interaction term of Eq. (18.12). To the lowest order in the pion field, this can be written as

$$gf_\pi \left(\overline{N}_{\mathbb{L}} U N_{\mathbb{R}} + \overline{N}_{\mathbb{R}} U^\dagger N_{\mathbb{L}} \right) . \tag{18.73}$$

It is now clear that, in order that this is invariant under SU(2) × SU(2), the transformation of U should read like

$$U \to U' = V_L U V_R^\dagger . \tag{18.74}$$

The effective Lagrangian involving the pions can now be written down as something that contains only derivatives of U, and is invariant under the transformation of Eq. (18.74). This Lagrangian has to be of the form

$$\mathscr{L}_{\text{eff}} = f_\pi^2 \, \text{Tr} \left((\partial_\mu U^\dagger)(\partial^\mu U) \right) + \cdots , \tag{18.75}$$

where the dots indicate terms with more factors of U. It is easy to see that one obtains the expected kinetic terms for the pions if one expands the exponential in U. The first non-vanishing term would be quadratic in pion fields:

$$\mathscr{L}_{\text{eff}}^{(2)} = \frac{1}{4} (\partial_\mu \pi_a)(\partial^\mu \pi_b) \, \text{Tr} \left(\tau_a \tau_b \right) . \tag{18.76}$$

According to the normalization of the SU(2) generators in the fundamental representation that was decided upon in Eq. (11.54, *p 308*), $\text{Tr} \left(\tau_a \tau_b \right) = 2\delta_{ab}$ since the generators are really $\tau_a/2$. Thus we obtain

$$\mathscr{L}_{\text{eff}}^{(2)} = \frac{1}{2} (\partial_\mu \pi_a)(\partial^\mu \pi_a) , \tag{18.77}$$

which is the kinetic term for the pion fields. The interactions come from higher order terms in the pion field.

It should be remembered that the SU(2) × SU(2) symmetry is only an approximate symmetry of the Lagrangian. Thus, the effective Lagrangian involving pions can contain also some terms which are not invariant under this symmetry. Consider, for example, the following Lagrangian:

$$\mathscr{L}_{\text{eff}} = f_\pi^2 \, \text{Tr} \left((\partial_\mu U^\dagger)(\partial^\mu U) \right) + m_\pi^2 f_\pi^2 \, \text{Tr}(U + U^\dagger) . \tag{18.78}$$

The first term is the same as that given in Eq. (18.75), and is invariant under the SU(2) × SU(2) flavor symmetry. The second term is not invariant. Expanding U in a power series of the pion field, we see that indeed the quantity m_π present in this term is the pion mass. Also note that the new term, though not invariant under SU(2) × SU(2), is invariant under the diagonal subgroup

SU(2) under which $V_L = V_R$. This is the isospin subgroup, and because it is intact, the neutral and the charged pions come out with the same mass.

The pion interaction with photons can be incorporated into this effective Lagrangian easily. The trick, as always, is to change the ordinary derivative into a gauge covariant derivative,

$$D_\mu = \partial_\mu + ieQA_\mu. \tag{18.79}$$

The value of Q should be taken as ± 1 for $\pi^\pm$, and zero for π^0. The electromagnetic interaction terms will also not be invariant under the $SU(2) \times SU(2)$ flavor symmetry. From such terms, and also higher order terms in the matrix U, low-energy properties of pions can be calculated.

18.4 Anomalies

We proposed to postpone the discussion about the U(1) factors in the flavor symmetry group. Now we take it up.

As mentioned in Eq. (18.10), there are two U(1) symmetries. The vectorial one is baryon number symmetry. The axial U(1) is the one under which all quark fields transform as

$$q(x) \rightarrow e^{i\beta\gamma_5} q(x), \tag{18.80}$$

with the same value of β. This symmetry would also be broken spontaneously by the $\langle \bar{q}q \rangle$ VEVs. If quarks were massless, one would have expected a Goldstone boson. Of course, the quark mass terms in the Lagrangian break this symmetry explicitly, so the Goldstone boson will not be massless. Naively, one should expect a particle which would be roughly as heavy as the pions if only the up and down quarks are involved. If also the strange quark is involved, the mass can be expected to be roughly the same as the mass of the kaons or the eta. And yet, there exists no meson which fits this description. The η', which is a singlet of flavor SU(3), has a mass of 958 MeV, almost twice as heavy as the η. If η' is identified as the Goldstone boson of the broken chiral U(1) symmetry, we have reasons to suspect that the symmetry is more badly broken in the Lagrangian than the chiral SU(3). This turns out to be the case indeed, for reasons that we will explain in this section.

18.4.1 Failure of Noether's theorem

Noether's theorem tells us that, corresponding to a continuous symmetry of the Lagrangian, there exists a 4-vector, called the Noether current, whose divergence is zero. The theorem was originally proved in the context of classical field theory. We now show that, when quantum corrections are included, there are cases where the theorem no longer holds, i.e., the divergence of the Noether current does not vanish.

a)   A simple example

To see this, we need not go through the elaborate framework where a number of flavors of quark fields are involved. We can just take QED, with one fermion field $\psi(x)$ interacting with the photon field, and ask ourselves whether the axial vector current,

$$\widetilde{J}^\mu = \overline{\psi}\gamma^\mu\gamma_5\psi\,, \tag{18.81}$$

is conserved in the limit that the fermion is massless. From the QED Lagrangian given in Eq. (5.7, *p 113*) and the resulting equations of motion, we expect

$$\partial_\mu \widetilde{J}^\mu = 2im\overline{\psi}\gamma_5\psi \equiv 2im\widetilde{J}\,, \tag{18.82}$$

where $\widetilde{J}$ stands for the pseudoscalar bilinear of the fermion field. This would imply that the axial vector current has vanishing divergence in the massless limit, as is expected from Noether's theorem and invariance of the Lagrangian under the axial symmetry

$$\psi(x) \rightarrow e^{i\beta\gamma_5}\,\psi(x)\,. \tag{18.83}$$

□ **Exercise 18.11** Verify Eq. (18.82).

Let us now define the following matrix element of creating two photons from the vacuum:

$$\int d^4x\, e^{-iq\cdot x} \left\langle \gamma(k_1)\gamma(k_2)\left|\widetilde{J}^\lambda(x)\right|0\right\rangle = E_{\mu\nu}\mathscr{T}^{\mu\nu\lambda}(k_1,k_2)\,, \tag{18.84}$$

where $E_{\mu\nu}$ is just a shorthand for the expression

$$E_{\mu\nu} = (2\pi)^4\delta^4(k_1 + k_2 - q)\epsilon_\mu^*(\boldsymbol{k}_1)\epsilon_\nu^*(\boldsymbol{k}_2)\,, \tag{18.85}$$

and $\mathscr{T}^{\mu\nu\lambda}(k_1,k_2)$ is the quantity whose properties will be the focus of our discussion. If we contract both sides of Eq. (18.84) by q_λ, we obtain

$$\begin{aligned}
E_{\mu\nu}\,q_\lambda\mathscr{T}^{\mu\nu\lambda} &= \int d^4x \left(i\partial_\lambda e^{-iq\cdot x}\right)\left\langle \gamma(k_1)\gamma(k_2)\left|\widetilde{J}^\lambda(x)\right|0\right\rangle \\
&= -i\int d^4x\, e^{-iq\cdot x}\left\langle \gamma(k_1)\gamma(k_2)\left|\partial_\lambda\widetilde{J}^\lambda(x)\right|0\right\rangle\,, \tag{18.86}
\end{aligned}$$

by performing integration by parts. A vanishing divergence of the axial vector current would then imply the relation

$$q^\lambda\mathscr{T}_{\mu\nu\lambda}(k_1,k_2) = 0\,. \tag{18.87}$$

□ **Exercise 18.12** *If the mass of the fermion is not neglected, show that the corresponding equation is*

$$q^\lambda\mathscr{T}_{\mu\nu\lambda}(k_1,k_2) = 2m\mathscr{T}_{\mu\nu}(k_1,k_2)\,, \tag{18.88}$$

where $\mathscr{T}_{\mu\nu}$ is defined through a relation that is very similar to Eq. (18.84), except that the pseudoscalar current is involved instead of the axial vector current.

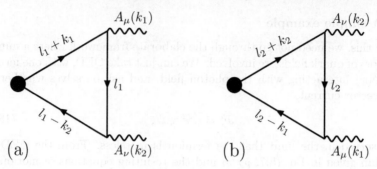

Figure 18.1: Triangle diagrams for the 3-point function involving two photons and an axial vector current. The blob on the left vertex in each diagram represents an axial vector coupling.

We can now try to see whether Eq. (18.87) is respected once we take loop corrections into account. In particular, we can evaluate the left hand side in the massless limit and check whether it really vanishes. At the one-loop level, diagrams for the quantity $\mathcal{T}_{\mu\nu\lambda}$ are presented in Fig. 18.1, where the blob represents the axial coupling of the fermion that results from $\widetilde{J}^\mu$. The contributions from the two diagrams can be written down, using the usual Feynman rules. With our notation for the momenta for different legs in the loop given in the figure, we obtain

$$i\mathcal{T}_{\mu\nu\lambda} = -(-ieQ)^2 \int \frac{d^4 l}{(2\pi)^4}\ \mathrm{Tr}\left(\frac{i}{\slashed{l}_1}\gamma_\mu \frac{i}{\slashed{l}_1 + \slashed{k}_1} i\gamma_\lambda \gamma_5 \frac{i}{\slashed{l}_1 - \slashed{k}_2}\gamma_\nu \right.$$
$$\left. + \frac{i}{\slashed{l}_2}\gamma_\nu \frac{i}{\slashed{l}_2 + \slashed{k}_2} i\gamma_\lambda \gamma_5 \frac{i}{\slashed{l}_2 - \slashed{k}_1}\gamma_\mu \right), \qquad (18.89)$$

where the electric charge of the fermion is eQ, and the overall minus sign is for a closed fermion loop. Note that, in writing Eq. (18.89), l_1 and l_2 have not been identified with the loop integration variable l, for reasons that will be clear soon. In general, l can be the momentum of any of the internal lines, or even some combination involving the external momenta.

If we now contract the expression by q^λ and use relations like

$$\slashed{q}\gamma_5 = (\slashed{l}_1 + \slashed{k}_1 - \slashed{l}_1 + \slashed{k}_2)\gamma_5 = (\slashed{l}_1 + \slashed{k}_1)\gamma_5 + \gamma_5(\slashed{l}_1 - \slashed{k}_2), \qquad (18.90)$$

we arrive at the one-loop relation

$$q^\lambda \mathcal{T}_{\mu\nu\lambda} = -ie^2 Q^2 \int \frac{d^4 l}{(2\pi)^4}\ \mathrm{Tr}\left(\frac{1}{\slashed{l}_1}\gamma_\mu \frac{1}{\slashed{l}_1 - \slashed{k}_2}\gamma_\nu \gamma_5 - \frac{1}{\slashed{l}_1}\gamma_\mu \frac{1}{\slashed{l}_1 + \slashed{k}_1}\gamma_\nu \gamma_5 \right.$$
$$\left. + \frac{1}{\slashed{l}_2}\gamma_\nu \frac{1}{\slashed{l}_2 - \slashed{k}_1}\gamma_\mu \gamma_5 - \frac{1}{\slashed{l}_2}\gamma_\nu \frac{1}{\slashed{l}_2 + \slashed{k}_2}\gamma_\mu \gamma_5 \right). \qquad (18.91)$$

Performing the traces now and reorganizing the terms, we can write

$$q^\lambda \mathcal{T}_{\mu\nu\lambda} = 4e^2 Q^2 \varepsilon_{\mu\nu\alpha\beta} \int \frac{d^4l}{(2\pi)^4} \left(\frac{l_2^\alpha k_2^\beta}{l_2^2(l_2+k_2)^2} - \frac{l_1^\alpha k_2^\beta}{l_1^2(l_1-k_2)^2} \right.$$
$$\left. - \frac{k_1^\alpha l_2^\beta}{l_2^2(l_2-k_1)^2} + \frac{k_1^\alpha l_1^\beta}{l_1^2(l_1+k_1)^2} \right). \tag{18.92}$$

Naively, it seems that this expression vanishes, as one should expect from Noether's theorem. If we identify both l_1 and l_2 with the loop integral variable l, and replace l by $l+k_2$ in the integrand of the first term, the first two terms of the last equation cancel each other. Likewise, the cancellation of the other two terms can be seen by suitable shift of the integral variable. There is a catch, however: all four terms in the integral are linearly divergent, and it is dangerous to shift the integration variable arbitrarily in such integrals. In other words, one cannot afford to be naive in this situation.

To understand the problem with linearly divergent integrals, let us see an example of an integration over just one variable, of the form

$$I(a) = \int_{-\infty}^{+\infty} dx \left[f(x+a) - f(x) \right], \tag{18.93}$$

where a is a constant, and $f(x)$ is a function with the property

$$\lim_{x\to\pm\infty} f(x) = \text{constant}, \tag{18.94}$$

i.e., at positive or negative infinity, the function does not vanish but all of its derivatives do. We can expand $f(x+a)$ in a Taylor series, which gives

$$I(a) = \int_{-\infty}^{+\infty} dx \left[af'(x) + \frac{1}{2}a^2 f''(x) + \cdots \right]. \tag{18.95}$$

This will give

$$I(a) = a\left[f(+\infty) - f(-\infty) \right], \tag{18.96}$$

and there is no reason that this must be zero.

□ **Exercise 18.13** To see a concrete example, consider $f(x) = \tanh x$. The indefinite integral of "$\tanh(x+a) - \tanh x$" can be performed exactly. Put the limits and show that the result agrees with the expression obtained in Eq. (18.96).

It is now easy to derive a similar result for integrals on more than one variable. Consider integrals of the form

$$I_\alpha(a) = \int d^D x \left[f_\alpha(x+a) - f_\alpha(x) \right]. \tag{18.97}$$

Here, x and a are vectors in a D-dimensional Euclidean space, and we take a function which itself carries a vector index, because this is the kind of function that we will encounter. Taylor expansion now gives

$$I(a) = \int d^D x \left[a^\mu \partial_\mu f_\alpha(x) + \cdots \right]. \tag{18.98}$$

Only the first derivative term will not automatically vanish for a linearly divergent integral, because using Gauss theorem we can write it as

$$I_\alpha(a) = a^\mu \int dS_D \, n_\mu f_\alpha(x), \tag{18.99}$$

where dS_D denotes an element of the $(D-1)$-dimensional surface of the entire space, and n^μ is the unit normal on it. Obviously, $dS_D = R^{D-1} d\Omega_D$ for a surface of radius R, and $n_\mu = R_\mu/R$. The integration over the angular variables implied in Ω_D has been shown in Eq. (G.26, p 761). Using the result, we obtain

$$I_\alpha(a) = a^\mu \frac{2\pi^{D/2}}{\Gamma(D/2)} \lim_{R\to\infty} R_\mu R^{D-2} f_\alpha(R).$$
(18.100)

We will encounter integrands that have the form

$$f_\alpha(R) = R_\alpha/R^D$$
(18.101)

for large R. For such functions, we convert $R_\mu R_\alpha$ to $\frac{1}{D} R^2 g_{\mu\alpha}$ in the integrand, as indicated in Eq. (G.10, p 758). This gives

$$I_\alpha(a) = a_\alpha \frac{2\pi^{D/2}}{D\Gamma(D/2)}.$$
(18.102)

For integrals over Minkowski spaces, there will be an additional factor of i because of Wick rotation, described in Appendix G. Specializing for $D = 4$, we can write

$$I_\alpha(a) = i\frac{\pi^2}{2} a_\alpha.$$
(18.103)

We, therefore, need to be careful about the linear divergence while evaluating the expression of Eq. (18.92). Before actually evaluating the integral, let us see what would happen if we contract the expression of Eq. (18.89) by one of the photon momenta, e.g., k_1. The result of the contraction for this case would be

$$k_1^\mu \mathcal{T}_{\mu\nu\lambda} = -ie^2 Q^2 \int \frac{d^4l}{(2\pi)^4} \text{Tr}\left(\frac{1}{\slashed{l}_1} \slashed{k}_1 \frac{1}{\slashed{l}_1 + \slashed{k}_1} \gamma_\lambda \gamma_5 \frac{1}{\slashed{l}_1 - \slashed{k}_2} \gamma_\nu \right.$$
$$\left. + \frac{1}{\slashed{l}_2} \gamma_\nu \frac{1}{\slashed{l}_2 + \slashed{k}_2} \gamma_\lambda \gamma_5 \frac{1}{\slashed{l}_2 - \slashed{k}_1} \slashed{k}_1\right).$$
(18.104)

Again, using relations like

$$\frac{1}{\slashed{l}_1} \slashed{k}_1 \frac{1}{\slashed{l}_1 + \slashed{k}_1} = \frac{1}{\slashed{l}_1}\left(\slashed{l}_1 + \slashed{k}_1 - \slashed{l}_1\right)\frac{1}{\slashed{l}_1 + \slashed{k}_1} = \frac{1}{\slashed{l}_1} - \frac{1}{\slashed{l}_1 + \slashed{k}_1},$$
(18.105)

we obtain

$$k_1^\mu \mathcal{T}_{\mu\nu\lambda} = -ie^2 Q^2 \int \frac{d^4l}{(2\pi)^4} \text{Tr}\left[\left(\frac{1}{\slashed{l}_1 - \slashed{k}_2} \gamma_\nu \frac{1}{\slashed{l}_1} - \frac{1}{\slashed{l}_2} \gamma_\nu \frac{1}{\slashed{l}_2 + \slashed{k}_2}\right.\right.$$
$$\left.\left. + \frac{1}{\slashed{l}_2 - \slashed{k}_1} \gamma_\nu \frac{1}{\slashed{l}_2 + \slashed{k}_2} - \frac{1}{\slashed{l}_1 - \slashed{k}_2} \gamma_\nu \frac{1}{\slashed{l}_1 + \slashed{k}_1}\right)\gamma_\lambda\gamma_5\right],$$
(18.106)

where we have used the cyclic property of traces to keep $\gamma_\lambda\gamma_5$ at the extreme right for each term. After the traces are evaluated, this becomes

$$k_1^\mu \mathcal{T}_{\mu\nu\lambda} \propto \epsilon_{\lambda\nu\alpha\beta} \int d^4l \left(\frac{l_1^\alpha k_2^\beta}{l_1^2(l_1 - k_2)^2} - \frac{l_2^\alpha k_2^\beta}{l_2^2(l_2 + k_2)^2}\right.$$
$$\left. + \frac{l_2^\alpha(k_1 + k_2)^\beta}{(l_2 - k_1)^2(l_2 + k_2)^2} - \frac{l_1^\alpha(k_1 + k_2)^\beta}{(l_1 - k_2)^2(l_1 + k_1)^2}\right),$$
(18.107)

keeping only the linearly divergent terms in the integral, because other terms
do not contribute.

In order to proceed, we now need to be more specific about l_1 and l_2. Let
us take

$$l_1 = l + b_1 k_1 + b_2 k_2\,, \tag{18.108}$$

keeping the numerical constants b_1 and b_2 unspecified at this stage. This way,
we remain as general as we can. For example, if $b_1 = b_2 = 0$, the loop integral
variable is the momentum of the rightmost side of the triangle in Fig. 18.1
*(p 546)*a. If $b_1 = 1$ and $b_2 = 0$, the loop integral variable is the momentum of
the upper left side, and if $b_1 = 0$ and $b_2 = -1$, it is the momentum of the
lower left side. For any other combination of values of b_1 and b_2, the loop
momentum cannot be identified with the momentum on any of the legs. In
any case, Eq. (18.108) tells us that for the diagram in Fig. 18.1 *(p 546)*b, we
must take

$$l_2 = l + b_1 k_2 + b_2 k_1\,, \tag{18.109}$$

because the expression in Eq. (18.89) must obey Bose symmetry, i.e., should
be unchanged if we interchange the two identical photons in the outer lines,
i.e., make the changes

$$k_1 \leftrightarrow k_2\,, \qquad \mu \leftrightarrow \nu\,. \tag{18.110}$$

Thus,

$$l_1 - l_2 = (b_1 - b_2)(k_1 - k_2)\,. \tag{18.111}$$

We now note that the first two terms of Eq. (18.107) can be written in the
form

$$\varepsilon_{\lambda\nu\alpha\beta}\, k_2^\beta \int d^4l \left[f_\alpha(l_1) - f_\alpha(l_2 + k_2) \right] \tag{18.112}$$

where

$$f_\alpha(l) = \frac{l_\alpha}{l^2(l - k_2)^2}\,. \tag{18.113}$$

Using Eq. (18.103), the result of the integral of these two terms can therefore
be written as

$$i\frac{\pi^2}{2}\varepsilon_{\lambda\nu\alpha\beta}\, k_2^\beta (l_1 - l_2 - k_2)^\alpha = i\frac{\pi^2}{2}(b_1 - b_2)\varepsilon_{\lambda\nu\alpha\beta}\, k_1^\alpha k_2^\beta\,. \tag{18.114}$$

Similarly, the last two terms of Eq. (18.107) can be written in the form

$$\varepsilon_{\lambda\nu\alpha\beta}\, (k_1 + k_2)^\beta \int d^4l \left[f_\alpha(l_2 - k_1) - f_\alpha(l_1 - k_2) \right] \tag{18.115}$$

for a suitably defined function f_α. The value of this integral will be

$$i\frac{\pi^2}{2}2(b_2 - b_1 - 1)\varepsilon_{\lambda\nu\alpha\beta}k_1^\alpha k_2^\beta. \tag{18.116}$$

The total contribution can be obtained by adding these two contributions. But the photon couples to the gauge current, which must be conserved. This implies that the tensor $\mathscr{T}_{\mu\nu\lambda}$ should satisfy the relations

$$k_1^\mu T_{\mu\nu\lambda} = 0, \qquad k_2^\nu T_{\mu\nu\lambda} = 0. \tag{18.117}$$

This shows that b_1 and b_2 cannot be completely arbitrary: they must obey the relation

$$b_2 - b_1 = 2. \tag{18.118}$$

With this knowledge, let us now look at Eq. (18.92). The algebraic steps are very similar to those encountered already, and we do not give the details. Using Eq. (18.118), we obtain

$$q^\lambda \mathscr{T}_{\mu\nu\lambda} = -\frac{ie^2Q^2}{2\pi^2}\varepsilon_{\mu\nu\alpha\beta}k_1^\alpha k_2^\beta \tag{18.119}$$

when the fermion inside the loop is massless. If we perform the same calculation with a mass m of the fermion in the loop, we would obtain this extra term in addition to the mass-dependent contribution shown in Eq. (18.88), i.e.,

$$q^\lambda \mathscr{T}_{\mu\nu\lambda}(k_1, k_2) = 2m\mathscr{T}_{\mu\nu}(k_1, k_2) - \frac{ie^2Q^2}{2\pi^2}\varepsilon_{\mu\nu\alpha\beta}k_1^\alpha k_2^\beta. \tag{18.120}$$

The important part is the new addition on the right hand side compared to Eq. (18.88), i.e., the loop contribution shown in Eq. (18.119), which implies that the axial current is not conserved even in the massless limit, and there is a failure of Noether's theorem. Such cases, where a symmetry of the classical Lagrangian is broken by the quantum corrections, is called *anomaly*. Classical symmetries which have anomaly are called *anomalous symmetries*.

One can also express the result in co-ordinate space. For this, one needs to multiply both sides of Eq. (18.120) by the polarization vectors of the two photons and take the inverse Fourier transform. This would yield

$$\partial_\mu \tilde{J}^\mu = 2im\tilde{J} + \frac{e^2Q^2}{8\pi^2}F^{\alpha\beta}\tilde{F}_{\alpha\beta}, \tag{18.121}$$

where

$$\tilde{F}_{\alpha\beta} = \frac{1}{2}\varepsilon_{\alpha\beta\mu\nu}F^{\mu\nu}, \tag{18.122}$$

is the dual field-strength tensor, introduced in Eq. (5.143, *p 144*).

   □ **Exercise 18.14** Go back to the diagrams of Fig. 18.1 (p 546) and write the amplitudes without neglecting the fermion mass. Go through the procedures mentioned in the text to show how the mass dependent term of Eq. (18.120) appears through direct evaluation of the diagrams.

b) Variations

In our discussion so far, the anomalous term comes from the electromagnetic field-strength tensor. If the fermions coupling to the axial current couple to gluons as well, there will be a similar term involving the field-strength tensor of QCD. In general, for any non-abelian symmetry, the Feynman rules for the vertices would involve not only just the coupling constant but also a matrix that is a representation of a generator of the group. If the lines connected to the triangle diagrams correspond to the currents $\widetilde{J}_\lambda^A$, J_μ^b and J_ν^c where b, c are gauge indices, the formula corresponding to Eq. (18.121) would be:

$$\partial_\mu \widetilde{J}_A^\mu = \frac{g_b g_c}{16\pi^2} \operatorname{Tr}\left(\mathsf{T}_A \left[T_b, T_c\right]_+\right) F_b^{\alpha\beta} \widetilde{F}_{\alpha\beta}^c + (\text{mass terms}), \quad (18.123)$$

where g_b and g_c are the gauge coupling constants that accompany the generators T_b and T_c, and F is the field-strength tensor for this gauge symmetry.

It should be noted that we have used a different kind of index for the axial current, and a different-looking letter for the accompanying generator. The reason is that this current can carry some index other than the gauge index. In questions of interest, the currents in fact do. For example, the flavor currents that we are considering carry a flavor index, whereas they are neutral under color. The trace appearing in the anomaly expression should be taken over all indices, gauge and any others.

It is easy to see why the anticommutator of generators occurs in Eq. (18.123). Imagine the diagrams of Fig. 18.1 *(p 546)* with these non-abelian currents at the vertices. If we pick up the generators $\mathsf{T}_A T_b T_c$ by following the opposite direction of the arrows on the fermion line in the first diagram, we will pick up $\mathsf{T}_A T_c T_b$ from the second diagram.

We can also contemplate chiral theories in which the left-chiral and right-chiral projections of fermions transform like different representations of the group. The two chiral projectors differ by a sign of the γ_5 term, so the contribution of the two chiralities to the triangle diagrams would come with opposite signs. Therefore, in such theories, the factor involving the generators in the anomalous term should be written as

$$A_{Abc} \equiv \operatorname{Tr}\left(\mathsf{T}_A \left[T_b, T_c\right]_+\right)_R - \operatorname{Tr}\left(\mathsf{T}_A \left[T_b, T_c\right]_+\right)_L, \quad (18.124)$$

where the two traces are over the representations of right and left chiral projections of fermions, as indicated by the subscripts. A chiral current $\widetilde{J}_A^\mu$ is non-anomalous if the *anomaly co-efficient* A_{Abc} is zero for any choice of b and c. If the chiral current in question is a gauge current, all anomaly co-efficients associated with it must vanish in order that the theory is renormalizable.

18.4.2 QCD anomaly of $U(1)_A$

Let us now consider which of the flavor currents can have a QCD anomaly. This means that we consider Eq. (18.123) when both b and c are color indices.

The flavor group commutes with the color group. In other words, the color of a quark does not depend on its flavor in any way. Thus, in this case, the trace decomposes into a flavor trace and a color trace. The flavor trace contains just the trace of T_A. If we consider a flavor current belonging to the SU(N) × SU(N) part of the flavor group, this trace is zero, as shown in §10.3. Thus, all currents corresponding to this part of the flavor group are free from QCD anomaly.

In more pedestrian terms, we can look at what happens to a flavor current of the form $\overline{u}\gamma_\mu\gamma_5 u - \overline{d}\gamma_\mu\gamma_5 d$. The color interactions are flavor-blind. So the mass-independent part of the contribution coming from u quarks in the loop would be the same as those coming from d quarks in the loop, except that the flavor current couples with opposite signs to the two flavors. Hence, the contributions from up quarks and down quarks cancel.

Such cancellation does not take place only for the axial U(1) current. All quarks contribute with the same magnitude and same sign to this current, implying that the generator is the diagonal unit matrix. Thus, the trace over the flavor space produces a factor N_F, the number of quark flavors. In the color sector, we can use the cyclicity of the trace to write the relevant trace in the form $2\,\mathrm{Tr}(T_b T_c)$. According to the normalization convention chosen for the generators in §11.2, this trace should be equal to $\delta_{b\overline{c}}$ since each flavor of quark is in the fundamental representation of the color group. So we obtain,

$$\partial_\mu \widetilde{J}^\mu = \frac{g_s^2 N_F}{16\pi^2} G^{\mu\nu}_{\overline{a}} \widetilde{G}^a_{\mu\nu} + (\cdots)\,, \tag{18.125}$$

where G denotes the QCD field-strength tensor. The terms omitted on the right hand side include fermion masses, and also possibly an anomaly term involving the QED field-strength tensor.

In the preamble for §18.4, we mentioned the problem with the mass of the η' meson. We now see why this ceases to be mysterious once anomalies are known. The point is that, even in the absence of quark masses, the chiral U(1) symmetry is broken by anomalies. In other words, even if we neglect quark masses, the η' cannot be treated like a Goldstone boson. No wonder then that the mass of the η' is much larger than the masses of the mesons in the octet.

18.4.3   Decay of neutral pions

The SU(N) flavor currents do not have any QCD anomaly, as we have argued. However, they can have QED anomalies. The basic reason is that the electromagnetic U(1) group does not commute with the flavor groups. In other words, different flavors of fermions carry different electric charges. This is crucial in the explanation of several physical phenomena, most striking of which is the decay of the neutral pion.

The neutral pion decays into two photons. The relevant amplitude should be represented by $\langle \gamma(k_1)\gamma(k_2) | \pi^0(p) \rangle$. Using the LSZ reduction formula, we can write

$$\langle \gamma(k_1)\gamma(k_2) | \pi^0(p) \rangle = i \int d^4x \, e^{-ip\cdot x} (\Box + m_\pi^2) \, \langle \gamma(k_1)\gamma(k_2) | \phi_0(x) | 0 \rangle \,,$$

$$(18.126)$$

where $\phi_0(x)$ is the field operator for the neutral pion. Performing integration by parts, this can be written as

$$\langle \gamma(k_1)\gamma(k_2) | \pi^0(p) \rangle = i(-p^2 + m_\pi^2) \int d^4x \, e^{-ip\cdot x} \, \langle \gamma(k_1)\gamma(k_2) | \phi_0(x) | 0 \rangle \,.$$

$$(18.127)$$

Sutherland and Veltman then used the PCAC relation, Eq. (18.42), to claim that the amplitude is proportional to

$$\int d^4x \, e^{-iq\cdot x} \left\langle \gamma(k_1)\gamma(k_2) \left| \partial_\mu \tilde{J}_3^\mu(x) \right| 0 \right\rangle \,, \qquad (18.128)$$

where the subscript '3' on the current signifies that it transforms like the third, or the neutral generator of the flavor SU(2). The matrix element present in this equation is exactly of the form that appears on the right hand side of Eq. (18.86), and should therefore have the form given in that equation, which involves the tensor $\mathcal{T}_{\mu\nu\lambda}(k_1,k_2)$. Sutherland and Veltman tried to write the most general expression for this tensor subject to the condition that it should obey Bose symmetry, and should be transverse to the momentum of any of the photons, in the sense given in Eq. (18.117), when the photons are on-shell, i.e.,

$$k_1^2 = k_2^2 = 0 \,. \qquad (18.129)$$

This expression is of the form

$$\mathcal{T}_{\mu\nu\lambda}(k_1,k_2) = k_1^\alpha k_2^\beta \left[\varepsilon_{\mu\nu\alpha\beta} q_\lambda T_1(p^2) + \left(\varepsilon_{\mu\lambda\alpha\beta} k_{2\nu} - \varepsilon_{\nu\lambda\alpha\beta} k_{1\mu} \right) T_2(p^2) \right.$$
$$\left. + \left((\delta_\mu^\rho k_{1\nu} - \delta_\nu^\rho k_{2\mu}) \varepsilon_{\rho\lambda\alpha\beta} - g_{\alpha\beta} \varepsilon_{\mu\nu\lambda\rho} (k_1^\rho - k_2^\rho) \right) T_3(p^2) \right] \,,$$

$$(18.130)$$

where T_1, T_2 and T_3 are three Lorentz invariant form factors. This gives

$$q^\lambda \mathcal{T}_{\mu\nu\lambda}(k_1,k_2) = \varepsilon_{\mu\nu\alpha\beta} k_1^\alpha k_2^\beta p^2 \left(T_1(p^2) + T_3(p^2) \right) \,. \qquad (18.131)$$

In the soft pion limit, $p^2 = 0$, this expression can be non-zero only if T_1 or T_3 has a term that goes like $1/p^2$ for small p^2. Such terms can arise only from propagators of massless particles. There is no such known massless particle. Hence, the amplitude must vanish. This is the important and shocking result by Sutherland and Veltman, that the neutral pion decay amplitude vanishes in the soft pion limit.

But the neutral pion decays nevertheless! And we now understand what goes wrong in this argument. The PCAC relation, Eq. (18.42), implies that the axial current is conserved in the limit when the pion mass is zero. We realize that this must be modified. The effect of anomaly must be there. The correct relation should therefore be

$$\partial_\mu \tilde{J}_A^\mu(x) = f_\pi m_\pi^2 \phi_A(x) + \frac{e^2}{8\pi^2} \operatorname{Tr}\left(\mathsf{T}_A Q^2\right) F^{\alpha\beta}(x)\tilde{F}_{\alpha\beta}(x), \quad (18.132)$$

where F is the electromagnetic field-strength tensor. The extra term added here is exactly what we wrote in a more general notation in Eq. (18.123). For electromagnetic interactions, the gauge coupling constant is e, and the charge Q plays the role of the generator.

The $\pi^0 \to 2\gamma$ amplitude now contains not only the contribution shown in Eq. (18.128), but also another contribution from the anomaly term, which is

$$\frac{e^2 \operatorname{Tr}\left(\mathsf{T}_3 Q^2\right)}{8\pi^2} \frac{-p^2 + m_\pi^2}{f_\pi m_\pi^2} \int d^4x\, e^{-ip\cdot x} \left\langle \gamma(k_1)\gamma(k_2) \left| F^{\alpha\beta}(x)\tilde{F}_{\alpha\beta}(x) \right| 0 \right\rangle.$$

$$(18.133)$$

In the soft limit, this amplitude is the same as that coming from an effective interaction term

$$\mathscr{L}_{\text{eff}} = K\phi_0(x)F^{\alpha\beta}(x)\tilde{F}_{\alpha\beta}(x), \quad (18.134)$$

where

$$K = \frac{e^2 \operatorname{Tr}\left(\mathsf{T}_3 Q^2\right)}{8\pi^2 f_\pi} = \frac{\alpha \operatorname{Tr}\left(\mathsf{T}_3 Q^2\right)}{2\pi f_\pi}. \quad (18.135)$$

This gives a decay rate

$$\Gamma = \frac{K^2 m_\pi^3}{4\pi}. \quad (18.136)$$

In order to obtain a numerical value, we need to evaluate $\operatorname{Tr}\left(\mathsf{T}_3 Q^2\right)$. Only the u and the d quarks have non-zero values of T_3, and the values are $+\frac{1}{2}$ and $-\frac{1}{2}$ respectively. Using the values of charges of these quarks, we obtain

$$\operatorname{Tr}\left(\mathsf{T}_3 Q^2\right) = N_c\left(\frac{1}{2}\times\frac{4}{9} - \frac{1}{2}\times\frac{1}{9}\right) = \frac{1}{6}N_c, \quad (18.137)$$

where N_c is the number of quark colors. Putting this into the expression for the decay rate written down above, we obtain

$$\Gamma = (N_c/3)^2 \times 1.11 \times 10^{16}\,\text{s}^{-1}. \quad (18.138)$$

The measured value of this lifetime,

$$\tau(\pi^0) = 8.52 \times 10^{-17}\,\mathrm{s}\,, \tag{18.139}$$

corresponds to $N_c = 3$, an experimental confirmation of the fact that quarks come in three colors.

□ **Exercise 18.15** *Starting from the effective interaction given in Eq. (18.134), deduce Eq. (18.136) by using the following steps.*

    a) *Show that the Feynman amplitude is given by*

$$\mathcal{M} = 4iK\varepsilon_{\mu\nu\lambda\rho}k_1^\mu k_2^\lambda \epsilon_1^\nu \epsilon_2^\rho\,, \tag{18.140}$$

       where ϵ_1 and ϵ_2 are the polarization vectors for the photon with momenta k_1 and k_2 respectively.

    b) *Take the absolute square of the Feynman amplitude and sum over all photon polarizations. Use Eq. (4.33, p 68) for evaluating the polarization sums.*

    c) *Now integrate over the phase space factors, remembering that only half of the total solid angle is available for a single photon because the two photons are identical.*

There is another interesting aspect of this result. Suppose, instead of working with quarks, we treat the nucleons as fundamental particles that couple to the pions through Yukawa interactions. We can perform the same calculation. All arguments leading to Eq. (18.136) would still be valid, with K still given by Eq. (18.135). The only difference would be that the triangular loops will now contain the nucleons. However, note that the neutron has $Q = 0$, so it does not contribute to K. The proton has $T_3 = \frac{1}{2}$, so that we obtain $\mathrm{Tr}\left(T_3Q^2\right) = \frac{1}{2}$, which is the same as that obtained with quarks for $N_c = 3$. This means that we would obtain the same value for the neutral pion decay rate if we ignore the quark degrees of freedom and work with nucleons. This is ensured by the fact that

$$\mathrm{Tr}\left(T_3Q^2\right)_{\mathrm{nucleons}} = \mathrm{Tr}\left(T_3Q^2\right)_{\mathrm{quarks}}\,, \tag{18.141}$$

provided there are three colors of quarks. This is called the *anomaly matching* condition, and has to be satisfied by any substructure of hadrons in order that one obtains the correct value of the neutral pion decay rate, which is driven by anomaly.

18.4.4 Anomaly cancellation for gauge currents

We have been discussing anomalies for chiral currents. The standard electroweak model has chiral currents. So, are they also *not conserved* in the unbroken phase of the model? If the answer to this question turns out to be 'yes', that would jeopardize the renormalizability of the standard model.

Fortunately, the answer is 'no'. This answer derives not from any fundamental structure of the standard model gauge group, but from the representations of the fermions under the gauge group. Let us try to understand the answer in a somewhat oblique way. We will take the representation of the quark and lepton fields under the $SU(3)_c$ and $SU(2)_L$ groups, and examine to what extent can the vanishing of triangle anomalies involving various combinations of gauge currents determine the weak hypercharges of these multiplets. Thus, we start with undefined weak hypercharges Y_l and Y_q for the lepton and the quark doublets, and Y_e, Y_u and Y_d for the right-chiral $SU(2)$ singlets of the particles denoted by the subscripts. Fermions in the second and third generations are not considered, because they will produce identical contributions. First, from the $\boxed{331}$ triangles, i.e., triangles where two of the vertices couple to the $SU(3)_c$ gauge bosons and the other to the $U(1)_Y$ gauge boson of the standard model, we obtain the condition

$$2Y_q = Y_u + Y_d\,, \tag{18.142}$$

which needs to be satisfied in order that the relevant anomaly vanishes. From the $\boxed{221}$ triangles, we obtain

$$3Y_q + Y_l = 0\,, \tag{18.143}$$

remembering that quarks can come in three colors. Triangles like $\boxed{311}$ or $\boxed{211}$ will vanish identically because of the tracelessness of the non-abelian generators. Thus the only other non-trivial condition comes from the $\boxed{111}$ triangles, which give

$$2Y_l^3 - Y_e^3 + 6Y_q^3 - 3Y_u^3 - 3Y_d^3 = 0\,. \tag{18.144}$$

We can obtain further constraints on the weak hypercharges by demanding that the left- and right-chiral components of a fermion must have the same electric charge. Since the electric charge is a linear combination of the two neutral generators of the electroweak group, we can write a relation of the form

$$aQ = T_3 + bY\,, \tag{18.145}$$

where a and b are constants. As emphasized after Eq. (16.18, $p\,465$), we can take $a = b = 1$ by using the multiplicative arbitrariness in defining $U(1)$ quantum numbers. Using this, we obtain the relations

$$Y_q = Y_u - \frac{1}{2} = Y_d + \frac{1}{2}\,, \qquad Y_l = Y_e + \frac{1}{2}\,. \tag{18.146}$$

Eq. (18.142) becomes redundant in the light of these equations. The five relations in Eqs. (18.143), (18.144) and (18.146) have the solutions for the weak hypercharges that exactly coincide with the values given for the standard model in Eqs. (16.40) and (17.1). We have thus shown that all gauge currents in the standard model are anomaly free, and that the charges of the particles are unique once the choice $a = b = 1$ is made in Eq. (18.145).

18.4.5   Non-anomalous global symmetry

In §18.1, we mentioned that baryon number is an accidental global symmetry of the standard model. So are the generational lepton numbers. Now that we have learned about anomalies, let us check whether these statements are true.

Let us take the case of baryon number first. We can consider triangle diagrams where one vertex couples like the baryon number. If the other two vertices are QCD vertices, the corresponding anomaly co-efficient must vanish, since there will be no chiral coupling in such diagrams. Let look at the case where the other two vertices couple to SU(2) currents, e.g., with T_3. The anomaly co-efficient for a single generation of fermions will be

$$\mathrm{Tr}\left(BT_3^2\right) = 3 \times 2 \times \frac{1}{3} \times \frac{1}{4} = \frac{1}{2}. \tag{18.147}$$

The factor 3 comes from the number of colors, the factor 2 comes from two different flavors, and then $\frac{1}{3}$ is the value of B for any of the quarks, and $\frac{1}{4}$ is the value of T_3^2 for a quark of any color or flavor. The bottom line is that baryon number has a non-vanishing anomaly co-efficient with the gauge currents, and its current therefore has a non-vanishing divergence. The same is true for the lepton number current.

□ **Exercise 18.16** *Show that, for fermions from a single generation,*

$$\mathrm{Tr}\left(LT_3^2\right) = \frac{1}{2}, \tag{18.148}$$

where L stands for lepton number.

□ **Exercise 18.17** *Calculate the anomaly co-efficients for baryon number and lepton number with two U(1)$_Y$ currents.*

Strictly speaking, then, baryon number and lepton number are not global symmetries of the standard model. They are violated by the anomaly. However, this is the only source of violation of baryon number and lepton number. The Lagrangian, being a local object, i.e., being a function of fields and their derivatives at a given spacetime point, does not have the information of this violation. Hence the Feynman rules derived from the Lagrangian will not have this information either, so perturbative calculations will never reveal any effect of this violation. Only non-perturbative effects such as instantons can violate baryon number and lepton number. But Eqs. (18.147) and (18.148) also show that the anomaly co-efficients vanish for the combination $B - L$. This is therefore a non-anomalous symmetry.

Chapter 19

Bosons of standard model

The electroweak sector of the standard model has four gauge bosons: W^+, W^-, Z and the photon. In addition, the symmetry breaking mechanism leaves a neutral scalar as a physical state, which is called the *Higgs boson*. In earlier chapters, we have discussed only processes where these bosons appear as intermediate virtual particles. But they can also occur in the initial or final state of a physical process. In this chapter, we discuss some processes of this kind.

19.1 Interactions among bosons

The standard model Lagrangian has many terms which signify interactions involving the gauge bosons and the spinless bosons. Here, "spinless bosons" mean the physical Higgs boson, as well as the unphysical degrees of freedom that vanish in the unitary gauge but can be present as internal lines in general gauges. In this section, we examine these interactions.

19.1.1 Self-interaction among gauge bosons

One of the most distinguishing features of non-abelian gauge theories is the self-coupling of gauge bosons. In Ch. 11, we derived the general rules for cubic and quartic couplings of gauge bosons in a general gauge theory. Such couplings exist in the standard electroweak model because of the SU(2) factor in its gauge group. Instead of taking the real fields W_μ^1, W_μ^2 and W_μ^3, we take the SU(2) gauge bosons in the form of $W_\mu^\pm$ which were defined in Eq. (16.15, p 464), and denote W_μ^3 by W_μ^0. The generators associated with $W_\mu^\pm$ and W_μ^0 are $T_\pm$ and T_0, where, in terms of the hermitian generators $T_{1,2,3}$,

$$T_\pm = \frac{1}{\sqrt{2}}(T_1 \pm iT_2), \qquad T_0 \equiv T_3. \qquad (19.1)$$

The commutation relations between these generators are:

$$[T_+, T_-] = T_0, \qquad [T_0, T_\pm] = \pm T_\pm. \qquad (19.2)$$

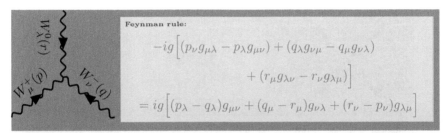

Figure 19.1: The only basic cubic interaction of gauge bosons in the electroweak sector of the standard model. The momentum associated with each leg is given in the parentheses.

This means that the non-zero structure constants of the SU(2) group, with this choice of generators, are

$$f_{+-0} = -i\,, \quad f_{0++} = -i\,, \quad f_{0--} = i\,, \tag{19.3}$$

and the relations obtained by using the antisymmetry of the first two indices.

☐ **Exercise 19.1** *Verify that the structure constants of the SU(2) group given in Eq. (19.3) satisfy the symmetry relations given in Eq. (11.53, p 308).*

Armed with the structure constants, we now find the gauge boson couplings. The cubic interaction among gauge bosons in a general non-abelian gauge theory was given in Eq. (11.74, *p 312*). Letting the gauge indices take the 'values' $+$, $-$ and 0, and putting in the structure constants from Eq. (19.3), we obtain the cubic gauge interaction terms for the SU(2) group as follows:

$$\mathcal{L}_{\text{cubic}} = -ig\left[W_+^\alpha W_-^\beta \partial_{[\alpha} W_{\beta]}^0 + W_0^\alpha W_+^\beta \partial_{[\alpha} W_{\beta]}^- - W_0^\alpha W_-^\beta \partial_{[\alpha} W_{\beta]}^+\right], \tag{19.4}$$

where we have used a shorthand notation

$$\partial_{[\alpha} V_{\beta]} \equiv \partial_\alpha V_\beta - \partial_\beta V_\alpha\,. \tag{19.5}$$

The Feynman rule for the cubic vertex can be read from Eq. (19.4), or directly from Fig. 11.2 *(p 313)*. This is shown in Fig. 19.1.

To interpret the Feynman rule in terms of physical bosons, we need to recall that W_μ^0 is not an eigenstate of the Hamiltonian. The eigenstates are the photon and the Z_μ, and the relation is

$$W_\mu^0 = \cos\theta_W\, Z_\mu + \sin\theta_W\, A_\mu\,. \tag{19.6}$$

Thus, the Feynman rule for the WWZ vertex will have an extra factor of $\cos\theta_W$ in it, whereas the WW-photon vertex will have an extra factor of

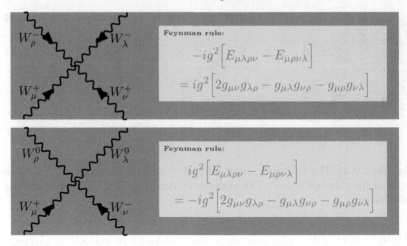

Figure 19.2: Feynman rules for quartic interactions of gauge bosons in the standard electroweak model. The tensor E with four indices has been defined in Eq. (11.71, *p 311*).

$\sin \theta_W$. Recalling the relation $g \sin \theta_W = e$ which was shown in Eq. (16.28, *p 466*), we can also say that for the vertex involving the photon, the factor g appearing in Fig. 19.1 will have to be replaced by e.

We now talk about quartic couplings. The SU(2) gauge group can have two kinds of quartic couplings: a coupling with $W^+W^+W^-W^-$ at a vertex, and another type with $W^+W^-W^0W^0$. There can be only one more combination of SU(2) gauge bosons that would conserve electric charge at the vertex, viz., $W^0W^0W^0W^0$. But a look at the general vertex given in Fig. 11.1 (*p 312*) quickly confirms that a quartic vertex cannot be built from four copies of the same gauge boson: the associated structure constants of the group vanish.

The Feynman rule for the coupling of four charged gauge bosons can be easily read from the general form of the quartic coupling given in Fig. 11.1 (*p 312*) and putting in the proper structure constants. Of course it has to be remembered, as for the case of cubic couplings, that W^0 is not a physical boson. The $W^+W^-W^0W^0$ coupling shown in Fig. 19.2 therefore entails three kinds of quartic vertices involving neutral gauge bosons: W^+W^-ZZ, W^+W^-ZA and W^+W^-AA. Along with the usual factor of g^2 that appears in the Feynman rule of quartic couplings, there will be an extra factor of $\cos \theta_W$ for each Z boson in the vertex and a factor of $\sin \theta_W$ for each photon in the vertex. The Feynman rules have been given with these explicit factors in Appendix H, and are not repeated here.

There is an interesting point about the WW-photon coupling that is worth mentioning here. In Ch. 5, we derived photon couplings with any other field by the method of minimal substitution, i.e., by taking the free Lagrangian of this other field and replacing the ordinary derivatives in that Lagrangian by the gauge covariant derivative

$$D_\mu = \partial_\mu + ieQA_\mu \,, \tag{19.7}$$

where eQ is the charge of the particle involved. For fermion fields, this method gave us the trilinear coupling of QED, shown in Eq. (5.17, *p 116*). For charged scalar fields, we got both trilinear and quadrilinear couplings, as shown in Fig. 5.7 *(p 135)*.

Let us see what we would have obtained if we had followed the same procedure for the W bosons. The free Lagrangian for the W bosons is given by

$$\begin{aligned}
\mathscr{L}_{\text{free}} &= -\frac{1}{2}(\partial_\mu W_\nu^+ - \partial_\nu W_\mu^+)(\partial^\mu W_-^\nu - \partial^\nu W_-^\mu) \\
&= -(g^{\alpha\mu}g^{\beta\nu} - g^{\alpha\nu}g^{\beta\mu})(\partial_\mu W_\nu^+)(\partial_\alpha W_\beta^-) \,.
\end{aligned} \tag{19.8}$$

The procedure of minimal substitution would yield

$$\mathscr{L}_{\text{ms}} = -(g^{\alpha\mu}g^{\beta\nu} - g^{\alpha\nu}g^{\beta\mu})\Big((\partial_\mu + ieA_\mu)W_\nu^+\Big)\Big((\partial_\alpha - ieA_\alpha)W_\beta^-\Big). \tag{19.9}$$

The quartic interaction term present in Eq. (19.9) is

$$\begin{aligned}
\mathscr{L}_{\text{ms4}} &= -e^2(g^{\mu\nu}g^{\lambda\rho} - g^{\nu\lambda}g^{\mu\rho})A_\mu A_\nu W_\lambda^+ W_\rho^- \\
&= -\frac{1}{2}e^2(2g^{\mu\nu}g^{\lambda\rho} - g^{\nu\lambda}g^{\mu\rho} - g^{\mu\lambda}g^{\nu\rho})A_\mu A_\nu W_\lambda^+ W_\rho^- \,, \tag{19.10}
\end{aligned}$$

where in the last step we have taken the factors of the metric tensors to be symmetric in the exchange of the indices μ and ν since the expression multiplies $A_\mu A_\nu$. This interaction is the same as that obtained in the pure gauge Lagrangian of the Yang-Mills theory.

This is not the case for the cubic couplings involving the photon. Eq. (19.9) contains the following cubic terms:

$$\mathscr{L}_{\text{ms3}} = -ie(g^{\alpha\mu}g^{\beta\nu} - g^{\alpha\nu}g^{\beta\mu})\Big[A_\mu W_\nu^+(\partial_\alpha W_\beta^-) - (\partial_\mu W_\nu^+)A_\alpha W_\beta^-\Big]. \tag{19.11}$$

Certainly these terms are contained in Eq. (19.4), but that equation contains the extra cubic terms

$$-ieW_+^\alpha W_-^\beta F_{\alpha\beta} \,, \tag{19.12}$$

where $F_{\alpha\beta}$ is the field-strength tensor for electromagnetism. It therefore has to be remembered that the electromagnetic interactions of the W bosons cannot be obtained from the prescription of minimal substitution. The terms obtained from minimal substitution have to be augmented by the term shown in Eq. (19.12) in order to obtain the photon interactions that come out of the pure gauge Lagrangian of the standard model.

In passing, we want to note that a coupling of the form shown in Eq. (19.12) gives a magnetic moment of the charged vector boson. This can be inferred from the analogy with the case of fermions, where we saw that an effective interaction of the form $\overline{\psi}\sigma_{\mu\nu}\psi F^{\mu\nu}$ indicates a coupling of the spin to the magnetic field. If we put in the matrix indices to make the pattern more explicit, we would see that this effective interaction is of the form

$$(\overline{\psi})_a (\sigma_{\mu\nu})_{ab} (\psi)_b F^{\mu\nu}, \tag{19.13}$$

where a, b are matrix indices. If we mimic the same thing for vector bosons, we should substitute $\sigma_{\mu\nu}$'s by the generators of the Lorentz group in the vector representation, i.e., should write

$$W_\lambda^+ (S_{\mu\nu})^{\lambda\rho} W_\rho^- F^{\mu\nu}, \tag{19.14}$$

because this would reduce to the interaction $\boldsymbol{S}\cdot\boldsymbol{B}$ in the non-relativistic limit. Using the form of these matrices given in Eq. (3.57, p 54), one easily obtains the form given in Eq. (19.12).

19.1.2   Self-interaction among scalars

The scalar potential of the standard model was given in Eq. (16.5, p 463). In the Lagrangian, the potential comes with a negative sign. Apart from a constant that is not important, this part of the Lagrangian can be written as

$$\mathscr{L}_{\text{sc}} = -\lambda(\phi^\dagger\phi - v^2/2)^2. \tag{19.15}$$

The quantum fields in the multiplet can be represented in the form

$$\phi = \begin{pmatrix} w_+ \\ \frac{1}{\sqrt{2}}(v + H + iz) \end{pmatrix}, \tag{19.16}$$

as shown in Eq. (16.32, p 467). Putting this in, we obtain

$$\mathscr{L}_{\text{sc}} = -\lambda \left[w^+ w^- + \frac{1}{2}(2vH + H^2 + z^2) \right]^2. \tag{19.17}$$

The only mass term is $-\lambda v^2 H^2$, which means

$$M_H^2 = 2\lambda v^2. \tag{19.18}$$

We can, therefore, use M_H^2 instead of λ as the independent parameter in the scalar potential. The VEV of the Higgs multiplet, v, can also be exchanged in favor of the W boson mass, which is

$$M_W = \frac{1}{2}gv \tag{19.19}$$

as shown in Eq. (16.16, p 464). Thus, the scalar potential terms in the Lagrangian can be written as

$$\mathscr{L}_{\text{sc}} = -\frac{g^2 M_H^2}{8 M_W^2} \left[w^+ w^- + \frac{2M_W}{g} H + \frac{1}{2}H^2 + \frac{1}{2}z^2 \right]^2. \tag{19.20}$$

All cubic couplings present in this expression have at least one leg of the physical Higgs boson H. There is no coupling such as the zzz or the $w^+ w^- z$.

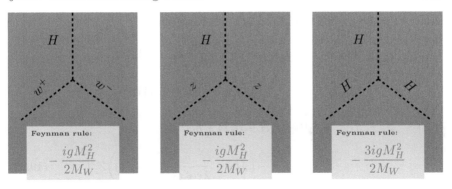

Figure 19.3: Feynman rules for cubic interactions of fundamental scalars in the standard model.

Three kinds of cubic vertices are present in the expression: w^+w^-H, zzH and HHH. The corresponding Feynman rules are shown in Fig. 19.3 and repeated in Appendix H. Note that the Feynman rule for the zzH vertex contains not only the numerical co-efficient of zzH in Eq. (19.20) but also an extra factor of $2! = 2$ because of the two identical field operators z. Similarly, there is an extra factor of $3! = 6$ in the Feynman rule for the HHH vertex.

The quartic couplings involve the squares of the three quadratic terms in Eq. (19.20), and their products of one with another. The Feynman rules for these vertices are given in Appendix H, and are not repeated here.

19.1.3 Gauge-scalar interactions

All interactions involving gauge fields and scalar fields come from the covariant derivative term of the scalar doublet. Using the representation matrices of the doublet representation of SU(2) as well as the value of the weak isospin, the covariant derivative acting on the multiplet ϕ, given in Eq. (16.4, *p 463*), can be written as

$$-iD_\mu\phi = -i\partial_\mu\phi + \frac{1}{2}\begin{pmatrix} gW^0_\mu + g'B_\mu & \sqrt{2}gW^+_\mu \\ \sqrt{2}gW^-_\mu & -gW^0_\mu + g'B_\mu \end{pmatrix}\phi, \quad (19.21)$$

where W^0_μ is the neutral gauge boson of SU(2), written as W^3_μ in Ch. 16. We can replace W^0_μ and B_μ by the eigenstates Z_μ and A_μ through the defining relations in Eqs. (16.22) and (16.24). Then we obtain

$$
\begin{aligned}
\left|D_\mu \phi\right|^2 = & \left| -i\partial_\mu w^+ + \left(\frac{g \cos 2\theta_W}{2\cos\theta_W} Z_\mu + eA_\mu\right)w^+ \right. \\
& \left. + \left(M_W + \frac{g}{2}(H+iz)\right)W_\mu^+ \right|^2 \\
& + \frac{1}{2}\left| (-i\partial_\mu H + \partial_\mu z) + gw^+ W_\mu^- \right. \\
& \left. - \left(M_Z + \frac{g}{2\cos\theta_W}(H+iz)\right)Z_\mu \right|^2 .
\end{aligned}
\tag{19.22}
$$

There are many kinds of terms in this expression. First, we see that there are the mass terms of the W and the Z bosons, which we identified earlier in §16.3. Then there are terms which contain one gauge boson and the derivative of one scalar field. These are the pathological terms which we first encountered in §15.4: they clutter the interpretation of particle states and are removed by gauge fixing. The rest of the terms are interactions. There are cubic and quartic vertices.

Let us take a look first at the terms involving the photon field. There are quartic terms involving the w^+, i.e., $w^+ w^- A_\mu A^\mu$, and the co-efficient of this term is exactly what one would expect from scalar electrodynamics involving the field w^+. There are also cubic vertices involving $w^+ w^- A$, and those couplings are also exactly what one would expect from scalar QED. But then we notice that there are also interaction terms like

$$
eM_W A^\mu w^+ W_\mu^- + \text{h.c.}
\tag{19.23}
$$

The Feynman rule of the resulting vertex is shown in Fig. 19.4.

Such terms, involving two gauge bosons and one scalar field, were not discussed in Ch. 11. The reason is that they do not exist unless the theory is spontaneously broken, as the presence of the factor M_W indicates in no uncertain terms. There are such terms involving other pairs of gauge bosons, as we will see shortly.

The term in Eq. (19.23) is perplexing from another point of view. In earlier chapters, on many occasions we have made the comment that the photon cannot change one particle to another. And now, here we see that a W boson can become a w line through the emission or absorption of a photon. The two statements are reconciled when we recall that the fields $w^\pm$ are not physical, and that they can be removed from the Lagrangian by a suitable choice of the gauge, as described in §15.4.4.

The cubic couplings involving the W and the Z bosons can also be read from Eq. (19.22), and they look very similar to the photon coupling, except for an overall multiplicative factor. There are also trilinear couplings involving the physical Higgs boson H, all of which are shown in Fig. 19.4. Note the absentee list: there is no coupling of this kind involving the z, the unphysical neutral Higgs boson.

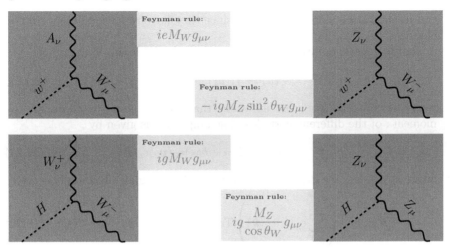

Figure 19.4: Feynman rules for vertices involving two gauge bosons and one scalar boson.

In the category of quartic couplings involving two gauge bosons, there is no absentee. The combinations w^+w^-, zz and HH appear in all possible combinations that conserve electric charge. The Feynman rules for these vertices are easily readable from Eq. (19.22) and are given in Appendix H. We don't repeat them here.

□ **Exercise 19.2** *Starting from the Lagrangian terms given in Eq. (19.22), derive the Feynman rules shown in Fig. 19.4.*

19.2 Decay of gauge bosons

The W and the Z bosons, being massive, can decay into lighter particles. In particular, their coupling to fermions can induce decays into two fermions. For example, we can have

$$W^+ \to e^+ + \nu_e \tag{19.24}$$

or

$$Z \to e^+ + e^-. \tag{19.25}$$

To obtain the decay rates of such processes, we notice that the fermion interactions with gauge bosons are generically of the form

$$\mathscr{L}_{\text{int}} = -\overline{f}_1 \gamma^\mu (a - b\gamma_5) f_2 V_\mu, \tag{19.26}$$

where V represents either the W or the Z, and the fermions f_1 and f_2 might be the same field or two different fields. In the case of W interactions, the two fermion fields must be different since they have to have different electric charges.

In the generic notation of Eq. (19.26), the gauge boson V decays as

$$V(k) \to f_1(p_1) + \widehat{f}_2(p_2), \qquad (19.27)$$

where we have denoted the notations that we are going to adopt for the momenta of the different particles. The amplitude is given by

$$\mathcal{M} = -\overline{u}_1(\boldsymbol{p}_1)\gamma^\mu (a - b\gamma_5)v_2(\boldsymbol{p}_2)\epsilon_\mu(\boldsymbol{k}), \qquad (19.28)$$

where $\epsilon_\mu(\boldsymbol{k})$ denotes the polarization vector for the gauge boson. This gives

$$\overline{|\mathcal{M}|^2} = \frac{1}{3}\left(\sum_{\text{pol}} \epsilon_\mu(\boldsymbol{k})\epsilon_\nu^*(\boldsymbol{k})\right) \text{Tr}\left[\not{p}_1 \gamma^\mu (a - b\gamma_5)\not{p}_2 \gamma^\nu (a - b\gamma_5)\right]. \quad (19.29)$$

Note that we are neglecting fermion masses. Also note that the averaging over initial state polarizations imply division by 3, since the massive vector boson has three independent polarization states.

We now use the polarization sum from Eq. (4.41, p 69) and evaluate the traces of the Dirac matrices. Note that the terms proportional to ab do not contribute to this result because they are antisymmetric in the Lorentz indices, whereas the polarization sum is symmetric. So we obtain

$$\begin{aligned}
\overline{|\mathcal{M}|^2} &= \frac{4}{3}(a^2 + b^2)\left(-g_{\mu\nu} + \frac{k_\mu k_\nu}{M^2}\right)\left[p_1^\mu p_2^\nu + p_1^\nu p_2^\mu - g^{\mu\nu} p_1 \cdot p_2\right] \\
&= \frac{4}{3}(a^2 + b^2)\left[p_1 \cdot p_2 + \frac{2k \cdot p_1\, k \cdot p_2}{M^2}\right]. \qquad (19.30)
\end{aligned}$$

Kinematical equations imply $k \cdot p_1 = k \cdot p_2 = p_1 \cdot p_2 = \frac{1}{2}M^2$, so that

$$\overline{|\mathcal{M}|^2} = \frac{4}{3}(a^2 + b^2)M^2. \qquad (19.31)$$

The decay rate of the vector boson can now be written by using Eq. (4.168, p 97):

$$\Gamma = \frac{M}{12\pi}(a^2 + b^2). \qquad (19.32)$$

To get a feel for the numerical values, consider the decay of the W^+ boson into a charged antilepton, ℓ^+, and the associated neutrino, ν_ℓ. In this case

$$a = b = \frac{g}{2\sqrt{2}}. \qquad (19.33)$$

Using Eq. (16.28, p 466), we can write

$$\Gamma(W^+ \to \ell^+ \nu_\ell) = \frac{g^2 M_W}{48\pi} = \frac{\alpha M_W}{12 \sin^2 \theta_W}. \qquad (19.34)$$

Experimentally, the total decay width of the W boson is obtained to be

$$\Gamma_{\text{tot}}(W) = (2.141 \pm 0.041)\,\text{GeV}, \qquad (19.35)$$

and the branching ratio to the $e^+\nu_e$ mode, for example, is

$$\mathscr{B}(W^+ \to e^+\nu_e) = 10.75\%. \qquad (19.36)$$

Combining the two pieces of data, we obtain

$$\Gamma(W^+ \to e^+\nu_e) = 230\,\text{MeV}. \qquad (19.37)$$

If we use the experimentally measured mass of the W boson, $M_W = 80.4\,\text{GeV}$, and the value $\sin^2\theta_W = 0.23$ obtained from neutrino scattering experiments, Eq. (19.34) gives a decay rate of 211 MeV if we use the low energy value $\alpha = 1/137$. However, the important point to notice here is that the low energy value should not be used. As explained in §12.2, the couplings depend on the momentum scale of the interaction. In this case, we should use the value of α at a scale equal to the W mass. As mentioned in Ex. 12.2 *(p 332)*, this value is close to about $1/128$. If we use this value, we obtain the decay rate to be about 228 MeV, in excellent agreement with experimental results.

For the Z boson decay, Eq. (19.32) is still applicable, only the values of a and b are different. Looking back at Eq. (16.38, *p 470*), we find

$$a = \frac{g}{2\cos\theta_W}\left(T_{3L} - 2Q\sin^2\theta_W\right), \qquad b = \frac{g}{2\cos\theta_W}T_{3L}, \qquad (19.38)$$

where T_{3L} is the value of the diagonal generator of weak isospin for the left-chiral projection of the particle. The right-chiral fields, as we have seen, are all singlets under weak isospin in the standard model.

It is to be noted that the decay rate is very sensitive to the properties of the final particle-antiparticle pair. For example, for the charged leptons, $T_{3L} = -\frac{1}{2}$ and $Q = -1$, so a turns out to be very small. The branching ratio of the Z to $\ell^+\ell^-$ pairs comes out to be very low, about 3.3% for each charged lepton. For the quarks, the branching ratios are much larger because a and b are both appreciable. For example, the branching ratios to $b\widehat{b}$ and $c\widehat{c}$ pairs are about 15.6% and 12.0% respectively.

- □ **Exercise 19.3** *Show that the relative branching ratios of Z boson decay to $\ell^+\ell^-$, $c\widehat{c}$ and $b\widehat{b}$ channels are consistent with the formulas given in Eqs. (19.32) and (19.38).*

- □ **Exercise 19.4** *Calculate the decay rate for the process $Z \to \nu\widehat{\nu}$. Take $\sin^2\theta_W = 0.23$ and neglect neutrino mass. Put in $M_Z = 91.2\,\text{GeV}$ and verify that the rate comes out to be 165 MeV.*

- □ **Exercise 19.5** ♩ *Why can't a Z boson decay into two photons?*

The Z boson can be produced as a resonance in e^+e^- collisions, and the width of the resonance has been found to be

$$\Gamma_{\text{tot}}(Z) = (2.4952 \pm 0.0023)\,\text{GeV}. \qquad (19.39)$$

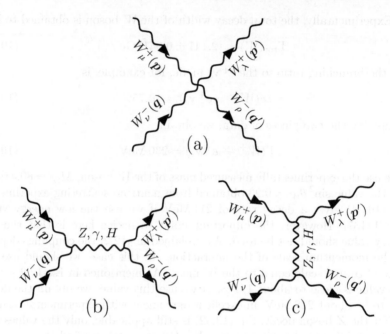

Figure 19.5: Tree-level diagrams for W^+W^- elastic scattering in the standard model.

The decay rates to the charged leptons and quarks account only for about 80% of this width. The remaining 20% of the width must come from Z boson decays to $\nu\bar{\nu}$ pairs. Indeed, with three neutrinos each contributing to 165 MeV of width, one can account for the remaining decay width with very good accuracy.

19.3   Scattering of gauge bosons

In a U(1) gauge theory such as QED, the gauge boson does not have any self-interaction. Thus, it can interact only through loop effects, and such interactions are therefore very much suppressed. A photon-photon scattering can take place through a 1-loop diagram involving four vertices, and the amplitude is therefore of order e^4, or the cross-section of order α^4. For non-abelian gauge bosons, this is not the case, Even at tree level they can interact, yielding an amplitude that is $\mathcal{O}\left(g^2\right)$ where g is the gauge coupling constant. For example, tree-level diagrams of W^+W^- elastic scattering have been shown in Fig. 19.5.

The Feynman amplitude of the scattering process can be written as

$$\mathcal{M} = \epsilon_\mu(p)\epsilon_\nu(q)\epsilon^*_\lambda(p')\epsilon^*_\rho(q')\mathcal{M}^{\mu\nu\lambda\rho}, \tag{19.40}$$

where $\epsilon^\mu(p)$, for example, is the polarization vector of the incoming W^+ boson with momentum p. The contribution to $\mathscr{M}^{\mu\nu\lambda\rho}$ is different from different diagrams. The diagram containing the quartic coupling gives

$$\mathscr{M}_{(a)}^{\mu\nu\lambda\rho} = g^2 (2g^{\mu\rho}g^{\lambda\nu} - g^{\mu\lambda}g^{\nu\rho} - g^{\mu\nu}g^{\rho\lambda}). \tag{19.41}$$

The diagrams with intermediate photon line give the following contributions to $\mathscr{M}_{\mu\nu\lambda\rho}$:

$$\mathscr{M}_{(b\gamma)}^{\mu\nu\lambda\rho} = -\frac{e^2 g_{\alpha\beta}}{\mathfrak{s}} \left[(p-q)^\alpha g^{\mu\nu} - (q+2p)^\nu g^{\alpha\mu} + (p+2q)^\mu g^{\nu\alpha} \right]$$
$$\times \left[(-p' + q')^\beta g^{\lambda\rho} + (q' + 2p')^\rho g^{\beta\lambda} + (-p' - 2q')^\lambda g^{\rho\beta} \right]$$

$$\mathscr{M}_{(c\gamma)}^{\mu\nu\lambda\rho} = -\frac{e^2 g_{\alpha\beta}}{\mathfrak{t}} \left[(p+p')^\alpha g^{\mu\lambda} + (p' - 2p)^\lambda g^{\alpha\mu} + (p - 2p')^\mu g^{\lambda\alpha} \right]$$
$$\times \left[(q+q')^\beta g^{\nu\rho} + (q' - 2q)^\rho g^{\beta\nu} + (q - 2q')^\nu g^{\rho\beta} \right], \tag{19.42}$$

where $\mathfrak{s}$ is one of the Mandelstam variables. Contributions from the Z-mediated diagrams are similar: only that e has to be replaced by $g\cos\theta_W$, and the propagator denominator will contain the Z boson mass, i.e., $\mathfrak{s}$ and $\mathfrak{t}$ in the denominators have to be replaced by $(\mathfrak{s} - M_Z^2)$ and $(\mathfrak{t} - M_Z^2)$ respectively in the 't Hooft–Feynman gauge. The diagrams mediated by the Higgs boson H will be discussed shortly.

□ **Exercise 19.6** We have used the Z propagators in the 't Hooft–Feynman gauge. In a general R_ξ gauge, there will be more terms in the propagator. Check that these terms do not contribute to the amplitude.

In order to obtain the cross-section for completely unpolarized W bosons, one has to average over initial polarization states and sum over final polarization states, as we have done many times for fermion spins. There is however one aspect of this scattering that has to be specially mentioned. Since this aspect pertains to the longitudinal polarization states, we might as well consider the case where all the W bosons in initial as well as in final states are longitudinally polarized.

The longitudinal polarization vectors have been mentioned in Eq. (4.40, p 69). If we are considering scattering at energies much higher compared to the mass of a massive vector boson, the energy can be taken to be equal to the magnitude of 3-momentum and we can write

$$\epsilon_l^\mu(k) \approx \frac{k^\mu}{M}. \tag{19.43}$$

If we put this form into Eq. (19.40), the resulting expression looks problematic at the first sight. For example, consider the contribution of the quartic vertex to the Feynman amplitude. It is

$$\mathscr{M}_{(a)} = \frac{g^2}{4M_W^4} \left(2\mathfrak{u}^2 - \mathfrak{s}^2 - \mathfrak{t}^2 \right), \tag{19.44}$$

where we have used the Mandelstam variables, with

$$\mathfrak{s} = (p+q)^2, \qquad \mathfrak{t} = (p-p')^2, \qquad \mathfrak{u} = (p-q')^2. \tag{19.45}$$

If we calculate the contribution of this term alone to the cross-section, we will find a contribution that goes like $g^4 \mathfrak{s}^3/M_W^8$. In §14.8, we discussed that such behavior causes problem with partial wave unitarity, since it is clear that only very few partial waves contribute to this scattering.

Fortunately, there is nothing to worry about. Although individual terms show this troublesome behavior, the potentially offending terms all cancel when we sum them up. In fact, in this approximation where the gauge boson masses are completely neglected everywhere except in the denominator of the polarization vector, we find

$$\mathcal{M}_{(b\gamma+bZ)} = \frac{g^2}{4M_W^4}(\mathfrak{t}^2 - \mathfrak{u}^2),$$

$$\mathcal{M}_{(c\gamma+cZ)} = \frac{g^2}{4M_W^4}(\mathfrak{s}^2 - \mathfrak{u}^2). \tag{19.46}$$

The sum of these two contributions cancels $\mathcal{M}_{(a)}$. Thus we have shown that there is no term in the amplitude which goes like $1/M_W^4$ and violates partial wave unitarity constraint.

The Higgs boson-mediated diagrams have been kept out of this argument because the WWH coupling is proportional to M_W, as can be seen from Fig. 19.4 (p 565). With two such couplings and the factors of $1/M_W$ from each polarization vector, the Higgs-mediated diagrams have at best an overall factor of $1/M_W^2$. Of course, if the total amplitude contains a term with M_W^2 in the denominator, the cross-section will contain a contribution that goes like $g^4\mathfrak{s}/M_W^4$. This will also violate the partial wave unitarity constraint. Therefore, if the theory has to make sense, these terms need to cancel as well. Checking this cancellation is however a much more complex task. There are many terms. The form of the longitudinal polarization vector given in Eq. (19.43) should now be modified to include corrections; the Mandelstam variables contain W mass terms; and even in propagators we should write

$$\frac{1}{\mathfrak{s}-M_Z^2} \approx \frac{1}{\mathfrak{s}} + \frac{M_Z^2}{\mathfrak{s}^2} + \cdots \tag{19.47}$$

and use the relation $M_W = M_Z\cos\theta_W$. Finally, after all these operations, it is found that the $1/M_W^2$ term does not cancel in the amplitude, but its co-efficient does not grow with energy either. In fact, the co-efficient is proportional to M_H^2. We will discuss these terms in §19.4.

19.4   Equivalence theorem

It is to be noted that in §19.3, we have not really calculated the cross-section for the cross-section of longitudinally polarized W^+-W^- scattering. We have only indicated some features of the result at high energies. The calculation itself is complicated, as the expressions for different contributions to the Feynman amplitude suggest.

Fortunately, there is an easy way out. At energies much higher than the masses, any amplitude involving longitudinally polarized gauge bosons can be obtained by substituting the external gauge boson lines by the corresponding unphysical Higgs boson and using the couplings of these would-be Goldstone modes to calculate the diagrams. This statement is called the *equivalence theorem*.

Let us use this theorem to calculate the cross-section of W^+-W^- scattering with longitudinally polarized W bosons. The diagrams are obtained by replacing the external $W^\pm$ lines with $w^\pm$ lines, and are not shown here. Using the Feynman rules given earlier in this chapter and in Appendix H, we obtain

$$i\mathcal{M}'_{(a)} = -\frac{ig^2 M_H^2}{2M_W^2},$$

$$i\mathcal{M}'_{(bH)} = \left(\frac{-igM_H^2}{2M_W}\right)^2 \frac{i}{\mathsf{s} - M_H^2}$$

$$i\mathcal{M}'_{(cH)} = \left(\frac{-igM_H^2}{2M_W}\right)^2 \frac{i}{\mathsf{t} - M_H^2}. \qquad (19.48)$$

Addition of these contributions gives

$$\mathcal{M}'_{(a+bH+cH)} = -\frac{g^2 M_H^2}{4M_W^2}\left[\frac{\mathsf{s}}{\mathsf{s} - M_H^2} + \frac{\mathsf{t}}{\mathsf{t} - M_H^2}\right]. \qquad (19.49)$$

These are the only terms with M_W in the denominator. Thus, we already see the simplification in relation to working with the gauge bosons themselves: the $1/M_W^4$ terms, which canceled between different diagrams of Fig. 19.5 *(p 568)*, do not appear in this procedure at all.

We are still left with the photon and the Z boson-mediated diagrams. These contributions are easily calculated, yielding

$$\mathcal{M}'_{(bZ+cZ)} = -\left(\frac{g\cos 2\theta_W}{2\cos\theta_W}\right)^2\left[\frac{(p-k)\cdot(p'-k')}{\mathsf{s} - M_Z^2} + \frac{(p+p')\cdot(k+k')}{\mathsf{t} - M_Z^2}\right]$$

$$= -\left(\frac{g\cos 2\theta_W}{2\cos\theta_W}\right)^2\left[\frac{\mathsf{u}-\mathsf{t}}{\mathsf{s} - M_Z^2} + \frac{\mathsf{s}-\mathsf{u}}{\mathsf{t} - M_Z^2}\right],$$

$$\mathcal{M}'_{(b\gamma+c\gamma)} = -e^2\left[\frac{\mathsf{u}-\mathsf{t}}{\mathsf{s}} + \frac{\mathsf{s}-\mathsf{u}}{\mathsf{t}}\right]. \qquad (19.50)$$

All these terms are obtained even if we use the W bosons in the outer legs. In addition, some other terms also appear, but they all vanish in the limit $M_W^2/\mathsf{s} \to 0$. This is a demonstration of the equivalence theorem.

Right when we introduced the idea of the Higgs mechanism in Ch. 15, we said that in a spontaneously broken gauge theory, the unphysical Higgs modes are 'unphysical' as the name suggests, and they cannot appear as external lines of any diagram. One might wonder whether the algorithm of the equivalence theorem is insane because it clearly advises us to use the unphysical Higgs modes as external legs. However, there is really no problem. The point is that the equivalence is obtained only in the limit $M_W \to 0$. Since $M_W = \frac{1}{2}gv$, one can think of approaching this limit in two different ways: as $g \to 0$ or as $v \to 0$. If we consider the $g \to 0$ limit, the theory is not a gauge theory at all. It only has global symmetries. Symmetry breaking in this case would produce real Goldstone bosons, as discussed in §15.1. Therefore, the procedure makes sense. On the other hand, if we consider the limit $v \to 0$, there is a gauge symmetry but it remains unbroken. So, even this limit makes sense because the gauge bosons do not eat up any scalar degrees of freedom, and all scalar modes remain physical.

□ **Exercise 19.7** For the process $WZ \to WZ$ involving longitudinally polarized W and Z bosons, use the equivalence theorem to find the terms in the amplitude that contain the factor M_H^2/M_W^2.

Why does the equivalence theorem work? Notice that the theorem prescribes replacement of longitudinally polarized gauge boson by the corresponding unphysical Higgs *only* on the external lines. The propagators and other factors coming from internal lines would be the same in both ways of evaluating the amplitude. If we use the gauge bosons on the external lines, there would be a factor of ϵ_l^μ in the amplitude, because of the Feynman rule for external vector boson lines. In Eq. (19.43), we showed that this factor is roughly equal to k^μ/M, where k^μ is the 4-momentum and M is the mass of the gauge boson, provided the energy is much larger than M. Obviously, this factor would be absent if we work with the alternative formulation where the would-be Goldstone mode is considered to be the external line. Thus, the amplitude calculated in two ways can be equal only if the same factors appear in the other formalism for some other reason.

The only other difference between the two diagrams would be in the vertex where this external line meets other parts of the Feynman diagram. If we use gauge boson in the external line, we will have to use the gauge couplings. With an unphysical Higgs boson on the external line, we need to use its couplings. However, recall from the discussion of §17.4, and in particular from Eq. (17.30, *p490*), that the latter coupling is precisely k^μ/M times the gauge couplings. Thus, in the regime where the longitudinal polarization vector can be written as k^μ/M, the gauge boson diagrams would give the same results as the diagrams with the unphysical Higgs bosons in the external lines.

□ **Exercise 19.8** Earlier in this chapter, we have derived the Feynman rule for the WWZ coupling. Use it, and the diagram of Fig. 19.5 (*p568*) with Z and z intermediate lines. Writing the propagators in the R_ξ gauge, show that the WWz coupling vanishes from the requirement that the amplitude is independent of ξ.

□ **Exercise 19.9** Consider the process $We \to A\nu_e$ where A is a photon. Draw the tree-level diagrams of this process mediated by the W boson and the unphysical Higgs boson w. Take the WW-photon coupling from the discussion of §19.1. Work in the R_ξ-gauge and find the Ww-photon coupling from the fact that the amplitude should be independent of ξ. Check that your final result agrees with the expression given in Fig. 19.4 (*p565*).

19.5 Custodial symmetry

In describing the phenomenon of spontaneous symmetry breaking in the standard electroweak theory, we said that the gauge symmetry $SU(2)_L \times U(1)_Y$ breaks down to $U(1)$, the latter being the symmetry of QED. This statement does not seem to be an accurate description of the symmetry breaking phenomenon if we look only at the scalar potential which is responsible for spontaneous symmetry breaking. The reason is that this part of the Lagrangian possesses symmetry that is larger than the gauge symmetry. In other words, the scalar potential has some accidental symmetry.

To appreciate the point, we recall that the potential, as shown in Eq. (16.5, *p 463*), is a function of $\phi^\dagger \phi$. Here ϕ is a doublet of the $SU(2)$ part of the gauge group, i.e., it has two components. We can write the doublet in the form

$$\phi \equiv \begin{pmatrix} a_1 + ib_1 \\ a_2 + ib_2 \end{pmatrix}, \tag{19.51}$$

where a_1, a_2, b_1, b_2 are real fields. Then,

$$\phi^\dagger \phi = a_1^2 + a_2^2 + b_1^2 + b_2^2. \tag{19.52}$$

Since the potential is a function of this expression, the potential remains unaffected by any orthogonal transformation involving the variables a_1, a_2, b_1, b_2. In other words, the potential is invariant under rotations in this four-parameter space, and therefore has an $O(4)$ symmetry. Using very similar steps that we had used in §3.6 to show that the proper Lorentz group, $SO(3,1)$, is equivalent to $SU(2) \times SU(2)$, we can show that $SO(4)$ symmetry is equivalent to an $SU(2) \times SU(2)$ symmetry. Obviously, this is bigger than the $SU(2) \times U(1)$ gauge symmetry that we wanted to implement, implying that we have some accidental symmetry in the potential.

So far, we have been talking about the symmetry in the unbroken phase, i.e., when the doublet ϕ has no VEV. Let us now shift our attention to the situation where ϕ has a VEV. As discussed earlier, the VEV can be taken in any one direction in this four-parameter space. If we take $\langle a_2 \rangle \neq 0$ as we did in Ch. 16, the $O(4)$ symmetry is broken. The direction of a_2 has been singled out. The remaining three directions however still remain equivalent, which means that there is an unbroken $O(3)$ symmetry. Again, $SO(3)$ is the same as $SU(2)$ so far as the group algebra is concerned, so we can describe the symmetry breaking process as

$$SU(2) \times SU(2) \to SU(2). \tag{19.53}$$

This left-over symmetry is called *custodial symmetry*.

What would be the physical implication of this custodial symmetry? It would mean that the quanta of the fields a_1, b_1 and b_2 would be degenerate. If we have only the scalar fields in the theory, this would be a trivial statement because the symmetry would have been global, and the aforementioned particles would have been the Goldstone bosons of the symmetry breaking process. Obviously, they would be degenerate, because they are all massless.

When we introduce the gauge bosons and talk about a gauge theory, there are non-trivial ramifications of the custodial symmetry. These ramifications are clear and dramatic if we consider a doublet of scalar fields, and put it in a gauge theory with a gauge symmetry SU(2). The scalar potential would have the same form even in this case, and all arguments given earlier in this section would hold. The three scalar fields a_1, b_1 and b_2 would transform like a triplet of the custodial SU(2) symmetry in the scalar potential. Because it is unbroken in the scalar potential, the mass terms for these three fields would be the same for any choice of the gauge-fixing term. Accordingly, the gauge bosons that eat them up would also have equal masses after the symmetry breaking, owing to the custodial SU(2) symmetry.

> ☐ **Exercise 19.10** *Take an* SU(2) *gauge theory, along with a doublet of scalar field. If the doublet develops a VEV, show that all three gauge bosons acquire equal masses.*

Now consider what would happen if the gauge symmetry were SU(2) × U(1). First consider the limit in which g', the coupling constant of the U(1) part of the gauge group, is vanishingly small. In this case, there is really no U(1) gauge symmetry, and the custodial symmetry would guarantee that the three gauge bosons of the SU(2) part of the gauge group are degenerate after the doublet of scalars develop a VEV. In particular, this will be the mass of the charged gauge bosons, so we call this contribution to the mass-squared values by M_W^2. In the limit $g' \to 0$, this will also be the mass of the neutral gauge boson belonging to the SU(2) part of the gauge group.

We now want to consider the effects of the fact that g' is non-zero in the real world. In the W_μ^0-B_μ basis, the mass matrix of the neutral gauge bosons must be of the form

$$\begin{pmatrix} M_W^2 & M'^2 \\ M'^2 & M_B^2 \end{pmatrix} \tag{19.54}$$

for some M' and M_B. The upper left element, the direct mass contribution for the W^0 boson, would be equal to the mass of the charged W bosons because of the custodial symmetry. The two off-diagonal elements would be equal because the mass matrix should be hermitian. Diagonalization of this matrix would give one massless gauge boson, viz., the photon, which implies that the determinant of the matrix is zero, i.e., $M_B^2 = M'^4/M_W^2$. Further, if we define the photon through Eq. (16.24, *p466*), i.e., define the Weinberg angle by saying that the photon is a superposition of W^0 and B with the relative weights $\sin\theta_W$ and $\cos\theta_W$, then that implies that, in Eq. (19.54), one should have $M_W^2/M'^2 = -\cot\theta_W$. Utilizing these relations, we can rewrite the matrix of Eq. (16.24, *p466*) in the form

$$\begin{pmatrix} M_W^2 & -M_W^2 \tan\theta_W \\ -M_W^2 \tan\theta_W & M_W^2 \tan^2\theta_W \end{pmatrix}. \tag{19.55}$$

The square of the Z boson mass would be the only non-zero eigenvalue of this matrix, i.e., $M_Z^2 = M_W^2(1 + \tan^2\theta_W)$, or

$$M_W^2 = M_Z^2 \cos^2\theta_W . \tag{19.56}$$

This relation between the masses of the W boson and the Z boson was first encountered in Eq. (16.29, *p467*), where it was derived as a tree-level relation. We now see that the relation is a consequence of the custodial SU(2) symmetry, which means that it will be obeyed to all orders in perturbation theory provided one does not include the effects of breaking of this custodial SU(2). As we said, the accidental larger symmetry SU(2) × SU(2) exists only in the Higgs potential. Thus, inclusion of effects of the gauge interactions or fermion mass or Yukawa coupling would produce effects that violate the custodial SU(2), and thereby violate the mass relation of Eq. (19.56). Of course, these effects can only come through loop diagrams, and will therefore be small.

19.6 Loop corrections

Loop corrections to weak interaction processes are important for several reasons. First, precision tests have been performed near the Z mass scale at the LEP and in later machines. In order to test the experimentally obtained results against their theoretical predictions, one needs accuracy of order of one part in a thousand or better. At this level, certainly one-loop corrections are relevant. Second, in view of the astounding success of the standard model, it is clear that if there is any physics beyond the standard model, its effects would be small. In order to discover any such effect, one must make sure that the standard model effects have properly been taken into account. Third, some approximate symmetries are violated at the loop level, and it is important to study the effects of their violation. One such symmetry is the custodial SU(2) symmetry, which we have discussed in §19.5.

Before embarking on a discussion on the loop corrections, it is important to note that at the tree level, there are equivalent definitions of some physical parameters which are violated by loop corrections. Take, for example, the definition of the Weinberg angle θ_W. It appears in the relation between different gauge coupling constants, as in Eq. (16.23, *p466*) or in Eq. (16.28, *p466*). It also appears in the mass relation between gauge bosons, Eq. (16.29, *p467*). These relations are equivalent at the tree level, but need not be so once loop corrections are included. We should be careful about which definition should be carried over to the discussion of higher order effects. The answer to this question is guided by experimental data. In the gauge sector of the standard model, three parameters are measured with very high accuracy. These are the fine-structure constant α, the Fermi constant G_F and the Z mass, M_Z. Interestingly, there is a tree-level relation that connects these three parameters

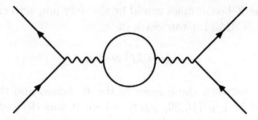

Figure 19.6: A typical diagram at the one-loop level that contributes to fermion–fermion scattering.

to the Weinberg angle:

$$\sin^2 2\theta_W = \frac{4\pi\alpha}{\sqrt{2G_F M_Z^2}}. \tag{19.57}$$

It is therefore appropriate to take this equation to be the definition of the Weinberg angle to all orders. This means that the mass relation such as Eq. (16.29, p 467) might get modified at the loop level. Indeed, since the mass relation follows from the custodial symmetry, a symmetry that is violated by various terms of the Lagrangian, it is expected to break down at the loop level. One therefore defines the ratio

$$\rho \equiv \frac{M_W^2}{M_Z^2 \cos^2 \theta_W}, \tag{19.58}$$

which is called the ρ parameter. At the tree level in the standard model, one should obtain $\rho = 1$. This is a result of the symmetry being broken by the doublet. For more general schemes of symmetry breaking, the value of ρ is obtained through Eq. (16.30, p 467).

19.6.1   Oblique parameters

Let us consider loop corrections to the scattering of two light fermions. The scattering will be mediated by gauge bosons. On the intermediate line, there can be self-energy loops for the gauge bosons, as shown in Fig. 19.6. In the corresponding tree-level diagram without the loop in the middle, the internal line would contribute to the Feynman amplitude only through its propagator. In the one-loop diagram, we will have two propagators, sandwiched by the self-energy of the gauge boson. Any self-energy function, with incoming and outgoing momentum equal to q, will contain two terms in general: one proportional to the metric tensor and another to $q_\mu q_\nu$. Propagators of gauge bosons have also the same general Lorentz structure. Thus, these terms can be written symbolically in the form

$$\underbrace{\left(ag_{\mu\alpha} + bq_{\mu}q_{\alpha}\right)}_{\text{propagator}} \underbrace{\left(\Pi(q^2)g^{\alpha\beta} + \Delta(q^2)q^{\alpha}q^{\beta}\right)}_{\text{self-energy}} \underbrace{\left(ag_{\nu\beta} + bq_{\nu}q_{\beta}\right)}_{\text{propagator}} . \quad (19.59)$$

The momentum terms in the propagators will contract with the Dirac matrices coming from the fermion vertices, and will be proportional to the masses of the external fermions when the Dirac equation for the spinors is used. For high energies and light external fermions, we can neglect these terms. We then see that the term in self-energy containing $q^{\alpha}q^{\beta}$ can also be neglected for the same reason. We therefore find that the only corrections will come through the invariant function $\Pi(q^2)$ that appears in the self-energy of gauge bosons. These corrections are called *oblique corrections*, which means that they are not corrections to the fermion vertex directly. Rather, they appear through the self-energy function of gauge bosons somewhat indirectly or in an oblique manner.

In fact, there is not a single function $\Pi(q^2)$. There will be different functions, depending on what the gauge boson lines are at the two ends. There are four gauge bosons in the standard model. Since the W bosons are charged, there can be four possible self-energy functions, WW, ZZ, $Z\gamma$ and $\gamma\gamma$. The Π-functions for these different self-energies can be written as

$$\Pi_{WW}(q^2) = \overline{\Pi}_{WW} + q^2 \overline{\Pi}'_{WW} + \cdots , \quad (19.60a)$$

$$\Pi_{ZZ}(q^2) = \overline{\Pi}_{ZZ} + q^2 \overline{\Pi}'_{ZZ} + \cdots , \quad (19.60b)$$

$$\Pi_{Z\gamma}(q^2) = q^2 \overline{\Pi}'_{Z\gamma} + \cdots , \quad (19.60c)$$

$$\Pi_{\gamma\gamma}(q^2) = q^2 \overline{\Pi}'_{\gamma\gamma} + \cdots , \quad (19.60d)$$

where Π' denotes the derivative of Π with respect to q^2, and the bars on Π and Π' imply that the said quantity has to be evaluated at $q^2 = 0$. The dots denote higher order terms in q^2 which do not concern us. Note that $\Pi_{\gamma\gamma}(q^2)$ as well as $\Pi_{Z\gamma}(q^2)$ must vanish at $q^2 = 0$ because otherwise the photon would obtain a mass, something that is not allowable by the electromagnetic U(1) symmetry which is not broken.

There are thus six parameters in Eq. (19.60) which need to be considered, the values of the two $\overline{\Pi}$'s and four $\overline{\Pi}'$'s. There are three parameters which are very well-measured, viz., the fine-structure constant α, the Fermi constant G_F and the Z mass, M_Z. Three of the six parameters obtained in Eq. (19.60) can be absorbed in the definition of these three measured parameters. The other three can then be used to test the standard model, e.g., to check whether there exist unknown particles which, even if not seen directly, will nevertheless affect the values of the self-energy functions through their contributions in loops.

To define the combination of the parameters that will be convenient for this purpose, it is better to carry out the discussion in terms of the currents of the form $J_a^{\mu} = \overline{\psi}\gamma^{\mu}T_a\psi$ for the hermitian generators T_a of the SU(2), and

the electromagnetic current J_Q^μ. The currents that couple to W and Z can be
written as

$$J_W^\mu = \frac{g}{\sqrt{2}}(J_1^\mu \pm i J_2^\mu)\,,$$

$$J_Z^\mu = \frac{g}{c_W}(J_3^\mu - s_W^2 J_Q^\mu)\,, \tag{19.61}$$

using Eqs. (16.36) and (16.38) and writing $c_W = \cos\theta_W$ and $s_W = \sin\theta_W$ for
the sake of brevity. The notation J_Q stands for the electromagnetic current.
If we denote the current-current correlation functions as Π_{11} etc, then

$$\Pi_{WW} = \frac{1}{2}g^2\Big(\Pi_{11} + \Pi_{22}\Big) = g^2\Pi_{11}\,, \tag{19.62a}$$

$$\Pi_{ZZ} = \frac{g^2}{c_W^2}\Big(\Pi_{33} - 2s_W^2\Pi_{3Q} + s_W^4\Pi_{QQ}\Big)\,, \tag{19.62b}$$

$$\Pi_{Z\gamma} = \frac{ge}{c_W}\Big(\Pi_{3Q} - s_W^2\Pi_{QQ}\Big)\,, \tag{19.62c}$$

and

$$\Pi_{\gamma\gamma} = e^2\Pi_{QQ} = g^2 s_W^2 \Pi_{QQ}\,. \tag{19.62d}$$

Of course the relations in Eq. (19.62) are valid definitions for all q^2. They can
also be written in terms of the values of these functions at $q^2 = 0$, and the
derivatives at $q^2 = 0$, and so on.

The oblique corrections are parametrized by the following quantities:

$$\alpha S = 4e^2(\overline{\Pi}'_{33} - \overline{\Pi}'_{QQ})\,, \tag{19.63a}$$

$$\alpha T = \frac{g^2}{M_W^2}(\overline{\Pi}_{11} - \overline{\Pi}_{33})\,, \tag{19.63b}$$

$$\alpha U = 4e^2(\overline{\Pi}'_{11} - \overline{\Pi}'_{33})\,, \tag{19.63c}$$

which are called the *oblique parameters* or alternatively *Peskin-Takeuchi parameters* after the people who initiated this kind of analysis. Note that the
parameters S, T and U are all dimensionless.

☐ **Exercise 19.11** *Show that, in terms of the self-energies in the basis
of physical gauge bosons, the oblique parameters are defined as*

$$\alpha S = 4s_W^2 c_W^2\left[\overline{\Pi}'_{ZZ} - \frac{c_W^2 - s_W^2}{s_W c_W}\overline{\Pi}'_{Z\gamma} - \overline{\Pi}'_{\gamma\gamma}\right]\,,$$

$$\alpha T = \frac{\overline{\Pi}_{WW}}{M_W^2} - \frac{\overline{\Pi}_{ZZ}}{M_Z^2}\,,$$

$$\alpha U = 4s_W^2\left[\overline{\Pi}'_{WW} - c_W^2\overline{\Pi}_{ZZ} - 2s_W c_W\overline{\Pi}'_{Z\gamma} - s_W^2\overline{\Pi}'_{\gamma\gamma}\right]\,. \tag{19.64}$$

The custodial symmetry would imply $\Pi_{aa} = \Pi_{bb}$ (no summation implied
on the indices) where the a and b stand for SU(2) indices. Thus, custodial

symmetry would imply that $T = 0$ and $U = 0$, although S can be non-zero. On the other hand, if the weak isospin symmetry, i.e., the SU(2) part of the gauge group, remains intact, that would also imply $T = U = 0$.

To demonstrate how these parameters express loop corrections to tree-level relations, let us consider the ρ parameter. The tree-level contributions to the W and Z boson masses were given in Eqs. (16.16) and (16.29). After adding the self-energy terms, the 2-point function for the W boson will be

$$(k^2 - \frac{1}{4}g^2v^2)g_{\mu\nu} - \Pi_{WW}g_{\mu\nu} + (q_\mu q_\nu \text{ terms}). \qquad (19.65)$$

The $q_\mu q_\nu$ terms include gauge-fixing terms, which are not important. The inverse of the 2-point function is the propagator, whose pole gives the mass of the concerned particle. Taking the inverse of the expression given Eq. (19.65), we find that the mass of the W boson is given by

$$M_W^2 = \frac{1}{4}g^2v^2 + \overline{\Pi}_{WW}. \qquad (19.66)$$

Similarly, the mass of the Z boson will be given by

$$M_Z^2 = \frac{1}{4}(g^2 + g'^2)v^2 + \overline{\Pi}_{ZZ}. \qquad (19.67)$$

Thus,

$$\begin{aligned}
\rho &= \left(1 + \frac{4\overline{\Pi}_{WW}}{g^2v^2}\right) \Big/ \left(1 + \frac{4\overline{\Pi}_{ZZ}}{(g^2 + g'^2)v^2}\right) \\
&\approx 1 + \frac{4\overline{\Pi}_{WW}}{g^2v^2} - \frac{4\overline{\Pi}_{ZZ}}{(g^2 + g'^2)v^2}.
\end{aligned} \qquad (19.68)$$

Neglecting higher order corrections, this expression can be written as

$$\rho = 1 + \frac{\overline{\Pi}_{WW}}{M_W^2} - \frac{\overline{\Pi}_{ZZ}}{M_Z^2} = 1 + \alpha T. \qquad (19.69)$$

Experimental measurements of the ρ parameter therefore places bounds on the oblique parameter T.

19.6.2 Evaluation of T

As an example of loop corrections affecting tree-level results, we present here in some detail the calculations of one-loop corrections to the ρ parameter. As seen from Eq. (19.69), it means calculation of the oblique parameter T. As commented earlier, T can be non-zero by effects which break the custodial symmetry. Such effects are present in the fermion sector. So we calculate the contribution to the oblique parameter T that comes from fermion loops in the self-energy diagram of gauge bosons.

The couplings to the W bosons are purely left-chiral. The Z boson couples to both left and right chiralities of fermions. The calculation is facilitated by first evaluating self-energy functions with purely chiral couplings of fermions to gauge bosons, and with arbitrary masses m_1 and m_2 in the two fermion lines. For example, if both vertices have left-chiral couplings, the corresponding self-energy tensor is denoted by $\Pi^{\mu\nu}_{(LL)}(m_1, m_2, q)$. Explicitly, using the momentum notation defined in Fig. 19.7, we can write

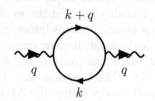

Figure 19.7: Self energy diagram for gauge bosons, with fermions in the loop.

$$i\Pi^{\mu\nu}_{(LL)}(m_1, m_2, q) = -\int \frac{d^4k}{(2\pi)^4} \, \text{Tr} \left[i\gamma^\mu \text{L} \frac{i(\slashed{k} + m_2)}{k^2 - m_2^2} \right.$$
$$\left. \times i\gamma^\nu \text{L} \frac{i(\slashed{k} + \slashed{q} + m_1)}{(k+q)^2 - m_1^2} \right], \quad (19.70)$$

where the minus sign outside the integral comes from the closed fermion loop, as mentioned in §4.10.3. Combining the denominators by the trick shown in Eq. (G.2, *p 756*), we obtain

$$i\Pi^{\mu\nu}_{(LL)} = -\int_0^1 dx \int \frac{d^4k}{(2\pi)^4} \frac{\text{Tr}\left(\gamma^\mu \slashed{k} \gamma^\nu (\slashed{k} + \slashed{q}) \text{R}\right)}{\left(k^2 + 2xk \cdot q + xq^2 - X_{12}\right)^2}, \quad (19.71)$$

where, for the sake of brevity, we define the shorthand

$$X_{12} = xm_1^2 + (1-x)m_2^2. \quad (19.72)$$

We now rewrite the expression in terms of a shifted momentum $l = k + xq$. There will be three kinds of objects carrying the Lorentz indices: $l^\mu l^\nu$, $g^{\mu\nu}$ and $q^\mu q^\nu$. The last type of terms are irrelevant for our purpose, as explained earlier. We are only interested in the co-efficient of $g^{\mu\nu}$. The $l^\mu l^\nu$ term also contributes to this co-efficient, because within the integrand, one can replace $l^\mu l^\nu$ by $\frac{1}{4}l^2 g^{\mu\nu}$, a result that has been demonstrated in Appendix G, as Eq. (G.10, *p 758*). After evaluating the trace, we therefore have to merely identify the co-efficient of $g^{\mu\nu}$, which is

$$i\Pi_{(LL)} = \int_0^1 dx \int \frac{d^4l}{(2\pi)^4} \frac{l^2 - 2x(1-x)q^2}{\left(l^2 + x(1-x)q^2 - X_{12}\right)^2}. \quad (19.73)$$

This integral is divergent in 4-dimensional spacetime, so we need to perform the integral by going over to a D-dimensional spacetime. The result of such integration has been given in Eq. (G.34, *p 762*). In Eq. (19.63b), we need only the values of self-energies at $q^2 = 0$, which comes out to be

$$\overline{\Pi}_{(LL)}(m_1, m_2) = \frac{2\Gamma(2 - {}^D\!/_2)}{(4\pi)^{D/_2}} \int_0^1 dx \left(X_{12}\right)^{-1+D/_2}. \tag{19.74}$$

Note that this quantity depends only on the two masses in the loop, as we have explicitly indicated on the left hand side of this equation.

In §G.6 of Appendix G, we have discussed how the divergent part of such integrals can be separated from the finite part. Following the prescription presented in Eq. (G.49, *p 764*), we obtain

$$\overline{\Pi}_{(LL)}(m_1, m_2) = \frac{1}{8\pi^2} \int_0^1 dx \, X_{12} \left(\frac{1}{\varepsilon'} - \ln X_{12}\right), \tag{19.75}$$

where

$$\varepsilon' = \frac{1}{2 - {}^D\!/_2} - \text{(some constants)}, \tag{19.76}$$

as presented in Eq. (G.46, *p 764*) or Eq. (G.48, *p 764*). The remaining integration on the Feynman parameter x is trivial, and it yields the result

$$\overline{\Pi}_{(LL)}(m_1, m_2) = \frac{1}{16\pi^2} \left[(m_1^2 + m_2^2)\frac{1}{\varepsilon'} + \frac{1}{2}(m_1^2 + m_2^2) \right.$$
$$\left. - \frac{m_1^4 \ln m_1^2 - m_2^4 \ln m_2^2}{m_1^2 - m_2^2} \right]. \tag{19.77}$$

It might be unpleasant seeing logarithms of dimensionful quantities like m_1^2 or m_2^2. In §G.6 of Appendix G, we explain why such monstrosities appear, and how things can look sane by introducing an arbitrary mass scale in the calculations. However, we assure that the final result will be devoid of such monstrosities so we might be better off by just ignoring the awkward looks of the intermediate steps.

Since the fermion to W boson coupling has a factor of $g/\sqrt{2}$, it is quite easy to see that

$$\overline{\Pi}_{WW}(m_1, m_2) = \frac{1}{2}g^2\overline{\Pi}_{(LL)}(m_1, m_2), \tag{19.78}$$

where m_1 and m_2 denote masses of the partners in a fermion doublet. On the other hand, the Feynman rule for the coupling of a fermion with the Z boson, given in Fig. 16.2 *(p 469)*, can be written in the form

$$-\frac{ig}{\cos\theta_W}\gamma^\mu\left((T_3 - Q\sin^2\theta_W)\mathbb{L} - Q\sin^2\theta_W\mathbb{R}\right), \tag{19.79}$$

so that we can write

$$\Pi_{ZZ} = \left(\frac{g}{\cos\theta_W}\right)^2 \Big[(T_3 - Q\sin^2\theta_W)^2 \Pi_{(LL)} $$
$$ - (T_3 - Q\sin^2\theta_W)Q\sin^2\theta_W \left(\Pi_{(LR)} + \Pi_{(RL)}\right) $$
$$ + (Q\sin^2\theta_W)^2 \Pi_{(RR)} \Big]. \tag{19.80}$$

In the evaluation of the traces of Dirac matrices during the calculation of $\Pi_{(LL)}$, the terms with a single factor of γ_5 do not contribute at all. Hence, if we change the sign of these terms, the result will not be affected. This means that

$$\Pi_{(LL)} = \Pi_{(RR)}. \tag{19.81a}$$

For the same reason, we have

$$\Pi_{(LR)} = \Pi_{(RL)}. \tag{19.81b}$$

For the self-energy of the Z boson, we will always have the same fermion on both internal lines. Thus, we need the value of $\overline{\Pi}_{(LL)}$ after putting $m_1 = m_2 = m$ in Eq. (19.77). This can be done by taking the limit $m_2 \to m$ in Eq. (19.77), or, more easily by putting the two masses to be equal in Eq. (19.75). The result is

$$\overline{\Pi}_{(LL)}(m,m) = \frac{m^2}{8\pi^2}\left(\frac{1}{\varepsilon'} - \ln m^2\right). \tag{19.82}$$

Evaluation of the LR loops can also be performed. For equal masses in the loop, the result is

$$\overline{\Pi}_{(LR)}(m,m) = -\frac{m^2}{8\pi^2}\left(\frac{1}{\varepsilon'} - \ln m^2\right) = -\overline{\Pi}_{(LL)}(m,m). \tag{19.83}$$

Because of Eqs. (19.81) and (19.83), the expression in Eq. (19.80) reduces to the form

$$\overline{\Pi}_{ZZ} = \left(\frac{g}{\cos\theta_W}\right)^2 T_3^2 \overline{\Pi}_{(LL)}(m,m) \tag{19.84}$$

for a fermion of mass m in the loop. We need to add the contributions from two fermions, of masses m_1 and m_2, which are partners in a doublet. This gives

$$\overline{\Pi}_{ZZ} = \frac{g^2}{32\pi^2\cos^2\theta_W}\left[m_1^2\left(\frac{1}{\varepsilon'} - \ln m_1^2\right) + m_2^2\left(\frac{1}{\varepsilon'} - \ln m_2^2\right)\right]. \tag{19.85}$$

We now take $\overline{\Pi}_{WW}$ from Eqs. (19.78) and (19.77) and $\overline{\Pi}_{ZZ}$ from Eq. (19.85), and put them into Eq. (19.63b). Since the result contains an explicit factor of g^2, we can put $M_W = M_Z \cos\theta_W$ in the rest of the factors, thereby obtaining

$$\alpha T = \frac{g^2}{32\pi^2 M_W^2}\left[\frac{1}{2}(m_1^2 + m_2^2) - \frac{m_1^2 m_2^2}{m_1^2 - m_2^2}\ln\frac{m_1^2}{m_2^2}\right]$$
$$= \frac{G_F}{4\sqrt{2}\pi^2}\left[\frac{1}{2}(m_1^2 + m_2^2) - \frac{m_1^2 m_2^2}{m_1^2 - m_2^2}\ln\frac{m_1^2}{m_2^2}\right]. \tag{19.86}$$

Notice that, as promised, the awkward logarithms have disappeared from this expression, as they should because it is a measurable quantity. Also notice that the divergent terms have all disappeared from this expression.

The expression in Eq. (19.86) is valid for one doublet of fermions. Contributions like this one occur for each doublet. In particular, for quark doublets, there can be three different colors of quarks circulating in the loop, so there will be a color factor of 3.

The implication of this result is profound. It shows that the two members of a fermion doublet cannot be arbitrarily separated in terms of their mass eigenvalues. Because, if they do, they would contribute too much to the ρ parameter. This argument was in fact very useful for obtaining an upper bound of the top quark mass even before the top quark was discovered. We will outline the argument in §20.4.

□ **Exercise 19.12** *Suppose we have a doublet of fermions whose partners have equal mass. Weak isospin is not violated by these mass terms, and therefore these fermions should not contribute to the oblique parameter T. Verify that indeed the contribution of Eq. (19.86) vanishes if $m_1 = m_2$.*

19.7 Higgs boson

The Higgs boson is the only spinless elementary particle, according to the standard model. It was also the last particle in the standard model for which experimental evidence was obtained. In fact, it created a record of a sort. The standard model was proposed in 1967, and the discovery of the Higgs boson was announced in 2012, i.e., a long 45 years later. It remained elusive for almost half a century.

19.7.1 Elusiveness

Why is the Higgs boson so elusive? One part of the answer is that it is heavy. However, that cannot be the whole story. The t quark, in fact, is much heavier than the Higgs boson. And yet, the t quark was discovered in the mid-1990s whereas the Higgs boson defied detection for almost two more decades.

The other big reason for the elusiveness of the Higgs boson is its coupling to fermions. In §16.5, we saw that the coupling of the Higgs boson to any fermion is proportional to the mass of the fermion, and is in fact given by

$$h_f = \frac{\sqrt{2}m_f}{v} = \frac{gm_f}{\sqrt{2}M_W}. \tag{19.87}$$

Our experimental detectors are made out particles in the first generation of fermions because they are constituents of stable material. More often than not, the experimental probes are also made out of the same. The masses of the first generation fermions are much smaller compared to the masses of fermions in the other generations. Accordingly, couplings of these fermions with the Higgs boson are also small. For the electron, Eq. (19.87) tells us that the coupling is of order 10^{-7}. With the up and the down quarks, the couplings are a little bigger, maybe by an order of magnitude. But even then, these are minuscule couplings. This is the basic reason why the Higgs boson is hard to produce and to detect.

There is a theoretical part of the elusiveness. Eq. (19.18) gives the value of the Higgs boson mass in terms of the parameters in the standard model Lagrangian. Using Eq. (19.19), we can write this mass formula in the form

$$M_H^2 = \frac{8\lambda M_W^2}{g^2} = \frac{\sqrt{2}\lambda}{G_F} . \tag{19.88}$$

We know the value of the Fermi constant G_F quite well, but there is also the factor λ sitting in the mass formula. This quantity, λ, does not appear in any interaction involving gauge bosons and/or fermions. This is a totally independent coupling constant that occurs in the theory: we have no idea about its magnitude from all experimental facts that we know about fermions and gauge bosons.

The experimentalists therefore had to start searching the Higgs boson from very small masses. With machines of increasing energy, they could gradually rule out higher and higher masses of the Higgs boson until they found some indication of the particle.

19.7.2   Theoretical bounds on mass

Since the Higgs boson mass is not predicted by the standard model, as explained in connection with Eq. (19.88), we can start by finding out whether theoretical arguments can at least put some bounds on the Higgs boson mass. There are a few such arguments.

a)   Unitarity bound

In §14.8, we discussed how partial wave unitarity constraints must be obeyed by scattering cross-sections. The argument can be applied to WW scattering. At energies much higher than the W boson mass, the dominant part of the amplitude comes from the longitudinally polarized W bosons. The Feynman amplitude for this scattering was calculated in §19.4, and it has terms that grow with the Higgs boson mass. Arguments using partial wave unitarity will say that this growth cannot be unlimited, thereby putting an upper bound on the Higgs boson mass.

Since the bound is going to be relevant for large values of M_H, we look at the terms in the Feynman amplitude that grow most quickly with M_H. These are the terms given in Eq. (19.49). At energies much higher than the Higgs boson mass as well, we find that

$$|\mathcal{M}| \approx \frac{g^2 M_H^2}{2 M_W^2} . \qquad (19.89)$$

This part of the amplitude is angle-independent, i.e., has only the zeroth partial wave. Therefore, from Eq. (14.147, *p 439*), we find that

$$\frac{1}{64\pi^2 \mathsf{s}} \left(\frac{g^2 M_H^2}{2 M_W^2} \right)^2 < \frac{4}{\mathsf{s}} , \qquad (19.90)$$

which is equivalent to

$$M_H < \frac{4\sqrt{2\pi} M_W}{g} = 2v \sqrt{2\pi} . \qquad (19.91)$$

Putting in the known value of v, we obtain

$$M_H < 1.2 \, \text{TeV} . \qquad (19.92)$$

Other processes yield similar upper bounds on M_H.

☐ **Exercise 19.13** Find the unitarity bound for M_H using the $WZ \to WZ$ scattering amplitude studied in Ex. 19.7 (p 572).

☐ **Exercise 19.14** Do the same thing for $ZZ \to ZZ$ elastic scattering. [**Note :** *Only the Higgs boson can mediate this process at the tree-level.*]

b)   Triviality bound

In §12.2, we discussed that the gauge coupling constants run, i.e., their values depend on the momentum scale at which they are measured. The same is true for other coupling constants in the theory. In particular, for the self-coupling λ of the Higgs multiplet, the evolution equation is given by

$$\frac{d\lambda}{d\ln(\mu^2)} = \frac{3\lambda^2}{4\pi^2} \qquad (19.93)$$

to the one-loop approximation. Integrating this equation, one obtains

$$\lambda(\mu) = \frac{\lambda(\mu_0)}{1 - \frac{3\lambda(\mu_0)}{4\pi^2} \ln \frac{\mu^2}{\mu_0^2}} , \qquad (19.94)$$

where μ_0 is some fixed scale where the value of λ is $\lambda(\mu_0)$. This equation, taken literally, implies that the value of λ becomes infinite at a scale

$$\mu_*^2 = \mu_0^2 \exp \left(\frac{4\pi^2}{3\lambda(\mu_0)} \right) . \qquad (19.95)$$

This scale is called the *Landau pole*. The only way to avoid this catastrophe at any scale is to have $\lambda = 0$, i.e., a trivial theory.

We cannot have $\lambda = 0$, because then the Higgs potential cannot have a non-zero minimum. Thus, we need λ at least at the weak scale, characterized by the VEV of the Higgs multiplet v. We can then ask, what should be the value of $\lambda(v)$ at the weak scale so that, within a reasonable domain of validity of our theory, the value of λ remains finite everywhere?

If the upper boundary of this "reasonable" domain is taken to be Λ, then we want $\lambda(\Lambda)$ to remain finite everywhere up to Λ. This happens if the denominator of Eq. (19.94) does not become zero even at Λ, i.e., if

$$\lambda(v) < \frac{4\pi^2}{3\ln(\Lambda^2/v^2)}\,. \tag{19.96}$$

The mass of the Higgs boson is therefore bounded by

$$M_H^2 = 2\lambda(v)v^2 < \frac{8\pi^2 v^2}{3\ln(\Lambda^2/v^2)}\,. \tag{19.97}$$

This produces an upper limit for the Higgs boson mass, and the value depends on the value of Λ that we consider reasonable. Some examples follow:

$$\Lambda = 10^{16}\,\text{GeV} \Rightarrow M_H < 160\,\text{GeV}\,,$$
$$\Lambda = 10^{19}\,\text{GeV} \Rightarrow M_H < 144\,\text{GeV}\,. \tag{19.98}$$

Why do we consider these values of Λ to be reasonable? As we said, the coupling constants run. If we consider the running of the gauge coupling constants of the electroweak SU(2) and U(1), and also of the color SU(3), we find that they cross one another at about an energy scale of 10^{16} GeV. At higher energies, presumably a new state of affairs will prevail, as discussed briefly in Ch. 23. This is the reason to believe that we should not extrapolate any consideration with the standard model to energy scales larger than about 10^{16} GeV.

In discussing particle interactions, we have always neglected the effects of gravitation. Gravitational interactions have an inherent energy scale given by Newton's constant. This energy scale is characterized by Planck mass, which was mentioned in Ex. 1.7 (p 13). Its value comes out to be of order 10^{19} GeV. Thus, above the Planck mass, gravitational effects are bound to become very important and all our arguments should be reconsidered.

Remember that in deducing the bound on M_H, we have assumed that the evolution of the coupling constant λ depends only on the value of λ itself. This is not true. The evolution of λ should depend on the gauge coupling constants as well, and also on the Yukawa couplings. Although most of the Yukawa couplings are minuscule, as discussed in connection with Eq. (19.87), the top quark mass is larger than the W boson mass by more than a factor of 2 and therefore its Yukawa coupling cannot be neglected. Including its effect as well as the effects of gauge couplings, the evolution equation for λ at the one-loop level should read

$$\frac{d\lambda}{d\ln(\mu^2)} = \frac{1}{16\pi^2}\Big[12\lambda^2 + 12\lambda h_t^2 - 12h_t^4$$
$$-\frac{3}{2}\lambda(3g_2^2 + g_1^2) + \frac{3}{16}(2g_2^4 + (g_2^2 + g_1^2)^2)\Big]\,, \tag{19.99}$$

where we have used the notations g_2 and g_1, rather than g and g' used in Ch. 16, for the $SU(2)_L$ and $U(1)_Y$ parts of the gauge groups in order to make the notations more obvious. Considerations of these other terms change the numerical values given in Eq. (19.98), but not by huge amounts.

□ **Exercise 19.15** For each term on the right hand side of Eq. (19.99), identify at least a single one-loop diagram that can be responsible for it.

c) Stability bound

There is also an argument to assert that the Higgs mass cannot be very small. Notice that the expression for the Higgs boson mass contains the VEV v and the scalar self-coupling λ. The value of the VEV is known to be 246 GeV from the known values of the W boson mass and gauge coupling constant. So, the Higgs boson mass can be small if λ is small. Now, the expression for the VEV, as given in Eq. (16.7, *p 463*), shows that the parameter μ^2 in the Lagrangian must have a small magnitude. If the magnitude is too small, its sign may be overturned by quantum corrections. If that happens, there is no symmetry breaking, and no standard electroweak theory. Hence a lower bound on Higgs boson mass.

In order to set a numerical value for the bound, we need to calculate quantum corrections to the Higgs potential. Rather than attempting it, we provide here a quick and dirty way of having a taste of the result. For small λ, the dominant terms of Eq. (19.99) can be written as

$$\frac{d\lambda}{d\ln(\mu^2)} = \frac{1}{16\pi^2}\left[-12h_t^4 + \frac{3}{16}(2g_2^4 + (g_2^2 + g_1^2)^2)\right]. \tag{19.100}$$

Note that the contribution coming from the top quark Yukawa coupling is negative. It is now known that h_t is almost equal to unity, so it can be the most dominant term and drive λ down. For some mass scale, λ will become negative. That would be a disaster because the scalar potential will then not have any minimum. This problem can be avoided only if λ is not very small so that the terms shown in Eq. (19.100) can never dictate the dominant running for λ. The lower bound on λ, thus obtained, implies a lower bound on the Higgs boson mass:

$$M_H > 134\,\text{GeV}, \tag{19.101}$$

if we do not want λ to be negative anywhere below 10^{16} GeV.

d) Bound from oblique parameters

In §19.6.2, we showed how the mass difference between partners of an $SU(2)_L$ doublet contributes to the oblique parameter T. The calculation we performed there pertained to a fermion doublet. However, similar calculation can be performed for a scalar multiplet as well. The contribution to T comes out

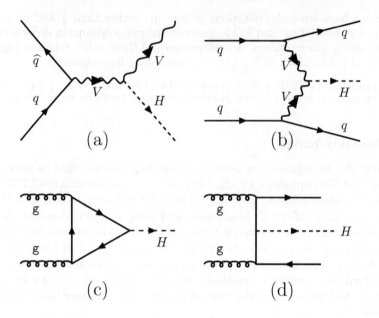

Figure 19.8: Feynman diagrams for main processes that can be responsible for Higgs boson production at hadron colliders. Each vector boson line marked V can represent either the W or the Z boson. Remember that, according to the arrow convention that we have employed, the outgoing arrow on the line marked $\hat{q}$ in diagram (a) represents an incoming antiquark.

to be proportional to $\ln(M_H/M_Z)$. In principle, one can put a bound on the Higgs boson mass from the experimental limits on the ρ parameter. But, since the dependence on the mass is logarithmic, this bound is not very useful.

19.7.3   Production

In order to detect the Higgs boson experimentally, one has to produce the particle through some interaction. As we emphasized earlier, the couplings of the Higgs boson to all first generation fermions are very small. So, direct production through the coupling to the electron or to light quarks would be a hopeless proposal.

Instead, one should explore the possibilities of producing the Higgs boson through its gauge coupling, or coupling to the top quark. Depending on the initial particles, different production mechanisms might dominate. For example, if one tries to produce the Higgs boson in hadronic colliders such as the LHC, the initial state will contain quarks, antiquarks and gluons. From our discussion in §13.9, recall that at very high energies such as what LHC possesses, there is a substantial component of ocean quarks and gluons in the proton, so that even a *pp* collider like the LHC will have lots of antiquarks and gluons in the colliding beams, apart from the valence quarks that are present in the proton. Thus, the Higgs boson might be created from quark and antiquarks or from gluons. In Fig. 19.8, we show some processes that are expected to dominate.

If, instead, we are thinking of the production of the Higgs boson in a e^+e^- collider machine, all processes involving quarks and gluons in the initial state are irrelevant for obvious reasons. Instead, the processes that can dominate the production have Feynman diagrams like those in Fig. 19.8a and Fig. 19.8b, but with all quarks replaced by electrons and all antiquarks by positrons. The gauge bosons V shown in these diagrams will have to be Z bosons. The suggested modifications of Fig. 19.8a would give rise to a process

$$e^+e^- \to ZH \,, \tag{19.102}$$

whereas that of Fig. 19.8b would imply a process

$$e^+e^- \to e^+e^-H \,. \tag{19.103}$$

In the LEP machine described in Ch. 9, at a CM energy of 205 GeV, signature of the process in Eq. (19.102) was searched for, and nothing was found. Since the couplings are known, the null result can only mean that the Higgs boson mass is higher than the available energy that is the CM energy minus the Z boson mass, or 91 GeV. This imposes an experimental lower bound

$$M_H > 114 \, \text{GeV} \,. \tag{19.104}$$

19.7.4 Decay

The Higgs boson will have a small lifetime, so, once it is created, one cannot possibly detect it through the direct observation of its track. Rather, one has to look for decay channels of the Higgs bosons in order to look for resonances in these channels. Therefore we need to know what could be the dominant decay channels for the Higgs bosons. Not surprisingly, the answer to this question depends on the mass of the Higgs boson. Let us identify several regions of Higgs mass and discuss what might be the dominant decays in those regions.

The coupling of the Higgs boson to any fermion is proportional to the mass of that fermion. Therefore, heavier fermions have larger couplings to the Higgs boson and therefore the matrix element of Higgs boson decay would be larger for heavier fermions. For example, the Higgs boson would prefer to decay to a $b\widehat{b}$ pair rather than to a $c\widehat{c}$ pair. The decays to an even lighter quark-antiquark pair, i.e., to $u\widehat{u}$, $d\widehat{d}$ and $s\widehat{s}$, are so small that we will not talk about them at all. For the same reason, we will not talk about decays into leptons, with the only exception of decays into the $\tau^+\tau^-$.

If the Higgs boson mass M_H were smaller than M_Z, then, as said above, decay to $b\widehat{b}$ pair and subsequently into bottom-carrying hadrons would have

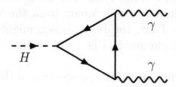

Figure 19.9: Feynman diagram for the process $H \to \gamma\gamma$. The loop can contain any charged fermion.

been the most dominant. Subdominant processes would include $c\bar{c}$ pair and $\tau^+\tau^-$ pair. Decays to all other fermions would be negligible. However, the Higgs boson could decay into a photon pair, $\gamma\gamma$. The Higgs boson is electrically uncharged, so it does not couple to the photon directly. The simplest way that it can decay into two photons is through the one-loop graph shown in Fig. 19.9. There are contributions from all charged fermions in the loop because they all have couplings with the Higgs boson. However, the largest contribution comes from the top quark in the loop, because its coupling with the Higgs boson is large. The couplings with the photon are gauge couplings, so we know how large they are. Thus the only suppression in the diagram is coming from typical numerical factors that appear in loop integration. That is why the branching ratio of the process can be quite large for small values of M_H, as seen in Fig. 19.10 *(p 591)*.

If the Higgs boson is heavier, new channels, involving heavier particles, gradually gain importance. The qualitative features regarding the importance of different channels should be understandable by looking at Fig. 19.10. For example, the $t\hat{t}$ channel has vanishing contribution below $2m_t$. For $M_H > 2m_t$, the top contribution rises very fast and overpowers all other fermionic decay modes because the top coupling to the Higgs boson is much larger than the coupling of any other fermion to the Higgs boson.

The case of the decay to two weak gauge bosons is also qualitatively understandable from the figure. For $M_H > 2M_W$, decay is possible only to virtual WW or ZZ pairs, and the rate is low. For $M_H = 2M_W$, there is a resonance, which is seen as a hump in the curve for the WW final state. Remember that the graph uses logarithmic scales, so that the hump is less pronounced than it would be on a linear scale, but is still discernible. At this mass, the branching ratios to all other channels dip sharply, as is seen in the figure. Again, it will have to be remembered that a dip looks much more dramatic on a logarithmic scale than it does on the linear scale.

☐ **Exercise 19.16** *The coupling of the Higgs boson to WW as well as ZZ pairs is given in Fig. 19.4 (p 565). Use them to find the ratio $\Gamma(H \to ZZ)/\Gamma(H \to WW)$ in the regime of high Higgs mass where the*

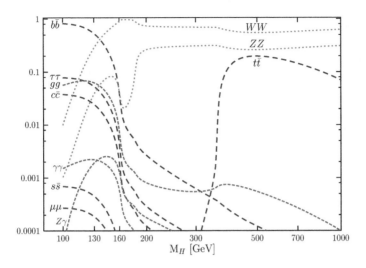

Figure 19.10: Branching ratio for the Higgs decay into various channels for a range of values of the Higgs boson mass M_H. (Adapted from A. Djouadi, hep-ph/0503172, with permission from the author.)

W and the Z masses can be neglected. See whether your result agrees with that inferred from Fig. 19.10.

It is not just the branching ratios that are important. One needs to consider, along with them, how good is the detection for each channel. For Higgs masses somewhat larger than 100 GeV, the two-photon mode may be the best. Although its branching ratio is substantially smaller than some other modes, the final state signal is clear. Consider, on the contrary the $b\widehat{b}$ or $\tau^+\tau^-$ modes, which have much larger branching ratios. However, the b or the τ is not directly detected: detection is only possible through identifying their decay products. This makes the analysis less reliable than that of a two-photon state.

If the Higgs mass is substantially higher, WW or ZZ channels should be better for the detection of the particle. Although the WW branching ratio is higher, this channel is less convenient for detection than the ZZ channel. The reason is that if either of the W's decays leptonically, neutrinos will be produced, which will not be detected, and therefore one will not be able to reconstruct the energy and momentum. Alternatively, the W's can decay hadronically, but such decays are always difficult to analyze because the effects of hadronization cannot be quantified. On the other hand, if the Higgs decay produces a ZZ pair, there is some probability that both Z's will decay into charged leptons only, producing two charged leptons and corresponding antileptons for the detector. This would constitute the best possibility for

Chapter 19. Bosons of standard model

detection of a heavy Higgs boson, and the final mode just described is often called the *gold-plated mode*.

The first announcement of observation of a Higgs boson by two different experiments at the LHC was based on these two decay modes described above. The mass of the Higgs boson was found to be about 125 GeV.

Chapter 20

Hadrons involving heavy quark flavors

We have so far discussed hadrons that involve the up, down and the strange quark flavors. In a sense all these quarks can be called light: they are all lighter than the nucleon, or even lighter than the QCD scale parameter Λ_{QCD}. There are, however, quarks which are heavier than Λ_{QCD}, and even heavier than the nucleon. Three such flavors of quark are known: charm (c), bottom (b) and top (t). In this chapter, we discuss hadrons involving them.

There is a reason for deferring the discussion of these quarks for so long. In earlier chapters, we have taken some symmetry or some kind of interaction as the central theme and brought in different particles to provide examples on the theme. Here, we will take a different approach. Now that we have introduced all basic interactions and the major symmetries, we can discuss both strong and weak interactions of fermions with heavy flavors. There is only one kind of phenomenon that we will leave out here: CP violation in heavy quark systems will be treated along with CP violation in light quark systems in Ch. 21.

20.1 Charm quark and charmed hadrons

20.1.1 Prediction of charm quark

The charm quark was discovered in 1974. But it was conjectured on theoretical grounds a few years before that.

When Cabibbo put forward the idea of quark mixing, only three types of quarks were known: the three light quarks u, d and s. Cabibbo, therefore, conjectured that the weak charged current connects the up quark to a superposition of the down and the strange quarks. In modern language, we can say that

$$\begin{pmatrix} u \\ d\cos\theta_c + s\sin\theta_c \end{pmatrix} \tag{20.1}$$

593

constitutes a doublet of the $SU(2)_L$ part of the electroweak gauge group. If this were the case, only the up quark could have been in the internal quark line for the process $K_L \to \mu^+\mu^-$ shown in Fig. 17.8 *(p 518)*. As explained in §17.9.5, the extremely small branching ratio of this process could not have been explained this way. Glashow, Iliopoulos and Maiani realized that if there is a fourth quark that also couples to both d and s quarks through charged currents, the leading term coming from this quark in the loop cancels the leading contribution coming from up quarks in the loop. This is the GIM cancellation mechanism discussed earlier. And this extra quark, the fourth flavor, was named the charm quark or the c quark.

The same argument can be given from many other processes. For example, consider the calculation of the K_L-K_s mass difference, presented in §17.9.4. Suppose there is no third generation, and there is no charm quark either. The small factor $r_c \equiv m_c^2/M_W^2$ that appears in Eq. (17.126, *p 516*) would not have been present in that case. Instead, there would be a factor of order 1, coming from the integration of the loop diagrams with only the u quarks in it. The calculated value of Δm_K would have been much larger than the experimental value in that case. It is the GIM cancellation that ensures that the leading terms, of order 1, cancel among themselves, so that the surviving terms contain a suppression factor and produces a theoretical result that is comparable with the experimental values. Because of such rare processes, the need for the charm quark was felt as soon as the GIM paper came out.

20.1.2   Discovery of charmonium

Then in 1974, there were two experiments which saw a *charmonium* state. The generic name indicates a bound state of the c quark and its antiquark, just like positronium is the name of a bound state of positron and its antiparticle, the electron. One of these experiments was done near the eastern coast of the USA at the Brookhaven National Laboratory under the leadership of Samuel Ting. The other was performed at Stanford, near the opposite coast, under the leadership of Burton Richter. They were different experiments. We will start by describing the method used by Richter's group.

Consider what happens to the inelastic cross-section of e^+e^- collision as we increase the CM energy. For CM energies less than twice the muon mass, the inelastic cross-section is zero: the collision is totally elastic because no inelastic channel is available at such energies. Once the CM energy goes above twice the muon mass, the process $e^+e^- \to \mu^+\mu^-$ becomes possible. For energies just a little bigger than $2m_\mu$, there are some threshold effects involving the muon mass, as shown in Eq. (5.95, *p 133*). As the energy is increased further, these contributions drop out and we obtain

$$\sigma(e^+e^- \to \mu^+\mu^-) = \frac{4\pi\alpha^2}{3s}. \tag{20.2}$$

If the CM energy is high enough, hadrons can also be produced. Let us define the ratio

$$R \equiv \frac{\sigma(e^+e^- \to \text{hardons})}{\sigma(e^+e^- \to \mu^+\mu^-)}. \tag{20.3}$$

The ratio has to be taken for the same value of the CM energy of the initial e^+e^- pair, and will depend on this energy. Equivalently, we can say that the ratio depends on the Mandelstam variable $\mathfrak{s}$.

At the basic level, hadron production should be governed by quark-antiquark production. The cross-section for the production of a free quark-antiquark pair of a certain flavor can be calculated in exactly the same manner as $\mu^+\mu^-$ production. The only difference is that the quark charge Q_q is not the same as the muon charge, and there are three different colors of quarks corresponding to a certain flavor. Taking these into account, we can write

$$\sigma(e^+e^- \to q\widehat{q}) = 3Q_q^2\,\sigma(e^+e^- \to \mu^+\mu^-), \tag{20.4}$$

neglecting the masses of the final-state fermions. Thus, at any given value of $\mathfrak{s}$, the ratio R should be given by

$$R = 3\sum_q Q_q^2, \tag{20.5}$$

where the sum is over all flavors of quarks which can be created along with its antiquark at the given value of $\mathfrak{s}$, i.e., whose mass is below the CM energy of the incoming electron.

Of course the quark-level process may be more complicated. For example, it is possible that hadrons are formed from the products of the basic reaction $e^+e^- \to q\widehat{q}g$. The cross-section for this process will have an extra factor of α_3. Thus, a term proportional to α_3 will add to the naïve contribution of Eq. (20.5). However, at high energy the strong coupling constant is small, as discussed in Ch. 12, so these effects would be small and R will be given by the simple expression of Eq. (20.5).

If only hadrons involving the u, d and s quarks are produced in the final state, the ratio R will have the value $\frac{4}{3}+\frac{1}{3}+\frac{1}{3} = 2$. Fig. 20.1 shows that a little above 1 GeV, this is more or less the value of R obtained from experiments. But then, above 3 GeV, the ratio increases. The increase can be explained if there is a fourth quark of charge $Q = \frac{2}{3}$, which can contribute an extra $\frac{4}{3}$ to R. This rise in R clearly indicates that, around $\sqrt{\mathfrak{s}} = 3$ GeV, one crosses a threshold for producing a new quark.

But then what happens exactly at the threshold? Shouldn't we see a hump, as explained in §9.6.1? The answer is of course 'yes', but in this case the hump was more like a sharp spike, i.e., very narrow, so much so that it was missed at first. The first such spike is encountered at a mass of 3097 MeV, and its width is only about 93 keV. This is the lightest bound state of the charm quark and its antiquark, and is called the J/ψ particle. The width is

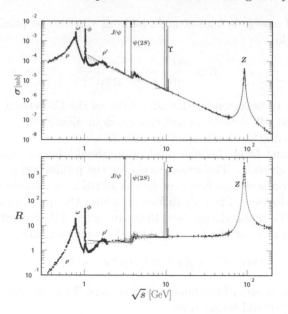

Figure 20.1: The ratio R as a function of the CM energy $\sqrt{s}$. [Reprinted with permission from: Particle Data Group, Phys. Rev. **D 86** (2012) 010001; © (2012) by the American Physical Society.]

so small that the rise and the fall of the ratio R at the resonance cannot be shown in the scale of the plot in Fig. 20.1. Hence a vertical line is drawn at the relevant energy to indicate the position of the spike.

The name of the particle is somewhat strange. It almost looks like the experts could not agree whether to call it J or call it ψ, and hence left both names with a slash mark in between. In fact, that is exactly what happened historically. As mentioned earlier, two groups found the particle roughly at the same time. One group wanted to call it J, the other preferred ψ. In the ensuing discussion, we will use the symbol Ψ, showing both letters in a superposed configuration.

At this point, let us comment on the method that Ting's group had used to obtain the same particle. They used a fixed-target experiment, bombarding a beryllium target by a proton beam with momentum 28.5 GeV. Of course the target contained protons, and there were reactions of the form

$$p + p \rightarrow e^+ + e^- + X \,, \tag{20.6}$$

where X means 'anything'. They looked at the CM energies of the e^+e^- pair to find out whether they are products of the decay of an intermediate particle formed in the reaction. In other words, they tried to see whether the process shown in Eq. (20.6) can occur through the production of a short-lived particle

$$p + p \rightarrow \Psi + X \,, \tag{20.7}$$

followed by the decay of this particle:

$$\psi \rightarrow e^+ + e^- \,. \tag{20.8}$$

In §9.6.2, we described how the presence of such intermediate short-lived particles can be ascertained by reconstructing events. That is exactly what they did, and found a peak for the e^+e^- invariant energy at $3.1\,\text{GeV}$.

There are of course other bounds states of $c\hat{c}$. Fig. 20.1 shows the spike corresponding one such state that is marked $\psi(2S)$. It has a mass of $3686\,\text{MeV}$ and total width of $317\,\text{keV}$, and was discovered within about two weeks from the discovery of the state mentioned earlier. As its name indicates, the c quark and its antiquark in this meson have a spatial wavefunction corresponding to a 2S state. The earlier state mentioned, the lightest $c\hat{c}$ state, has a 1S configuration. Other states have also been discovered, with different spatial and spin configurations of the c quark and its antiquark. All $c\hat{c}$ states are collectively called charmonium states.

20.1.3   Flavor SU(4)

With three flavors of quarks, we discussed the flavor SU(3) symmetry in Ch. 10. With the addition of the charm quark, we can think of extending the group to SU(4), for which the u, d, s and c quarks form the fundamental representation. We will denote this group by writing $SU(4)_F$, where the subscript 'F' stands for flavor. By the same token, the SU(3) symmetry discussed in Ch. 10 can be denoted by $SU(3)_F$.

The advantage of considering the flavor group is that the possible meson and baryon combinations can be identified easily from the representations of the group. Some of the simplest irreducible representations of SU(4) are as follows:

$$
\begin{array}{llcccc}
\text{Dimension} & : & 4 & 6 & 10 & 15 \\
\text{Young tableaux} & : & \square & \begin{array}{c}\square\\\square\end{array} & \square\square & \begin{array}{c}\square\square\\\square\end{array}
\end{array} \tag{20.9}
$$

The 15-dimensional representation is the adjoint representation of SU(4).

□ **Exercise 20.1** Using the method of finding the complex conjugate representation of any Young tableaux described in Fig. 10.4 (p 267), determine which of the representations given in Eq. (20.9) are real.

To understand the $SU(4)_F$ multiplets in terms of the multiplets of $SU(3)_F$, we have to consider the decomposition of $SU(4)_F$ representations under the subgroup $SU(3)_F \times U(1)_X$, where the generator of X commutes with all generators of $SU(3)_F$. The normalization of the U(1) quantum number X is arbitrary, and we can set it in a way such that the fundamental representation of $SU(4)_F$ decomposes as

$$4 \longmapsto \left(3, -\frac{1}{3}\right) + (1, 1)\,. \tag{20.10}$$

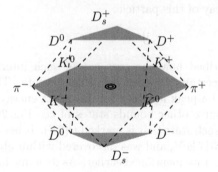

Figure 20.2: Mesons belonging to the $15+1$ representations of $SU(4)_F$. The middle level contains the $SU(3)_F$ octet. The states at the center are not labeled. They have been discussed in the text.

The $SU(3)_F$ triplet consists of the u, d and s flavors of quark, as discussed in Ch. 10. The singlet of $SU(3)_F$ is the charm quark. So we see that the value of X is $+1$ for the charm quark, and $-\frac{1}{3}$ for all lighter quarks. If we introduce the *charm quantum number* C which is $+1$ for the c quark, -1 for its antiquark and zero for everything else, we can write

$$X = -B + \frac{4}{3}C, \qquad (20.11)$$

where B is baryon number, which is equal to $\frac{1}{3}$ for any quark.

a) Mesons

If we combine a quark and an antiquark into a meson, the combination will transform as

$$4 \times 4^* = 1 + 15. \qquad (20.12)$$

The decomposition of the adjoint representation can be calculated from Eqs. (20.12) and (20.10). The result is

$$15 \longrightarrow (8,0) + \left(3, -\frac{4}{3}\right) + \left(3^*, \frac{4}{3}\right) + (1,0). \qquad (20.13)$$

Since $B = 0$ for mesons, we see from Eq. (20.11) that the members of the octet of $SU(3)_F$ have $C = 0$, i.e., they do not contain any charm quark. The 3^* representation of $SU(3)_F$ has $C = 1$, so its members must contain the charm quark, plus the antiquark of either u, or d, or s. The 3 representation of $SU(3)_F$ has $C = -1$, so it will have $\hat{c}$ along with one of the lighter quarks u, or d, or s.

The spin and parity of the lowest mass mesons are given by $J^P = 0^-$, as discussed in Ch. 10. Let us first discuss such mesons. The $SU(3)_F$-octet of mesons contained in the **15** of SU(4)$_F$ comprises the pions, the kaons and the eta. All of these mesons are made up of the u, d, s quarks and their antiquarks, and were presented in Fig. 10.1 *(p 256)*. The $C = +1$ mesons are called D mesons:

$$|D^+\rangle = c\widehat{d}, \qquad |D^0\rangle = c\widehat{u}, \qquad |D_s^+\rangle = c\widehat{s}. \tag{20.14}$$

These three mesons transform like a **3** representation of SU(3)$_F$. The conjugate mesons, which contain the $\widehat{c}$ and form a $\overline{\mathbf{3}}$ representation of SU(3)$_F$, are called D^-, $\widehat{D}^0$ and D_s^- respectively. All these mesons have been observed, and their masses are around 1900 MeV. The D^+ and the D^0 form a doublet of isospin, and their masses are very close, around 1865 MeV. The mass of D_s^+ is larger by about 100 MeV, a difference that we encountered between different isomultiplets within an SU(3)$_F$ multiplet while studying the SU(3)$_F$ multiplets in Ch. 10. More accurate values of the masses have been given in Table B.4 *(p 721)*.

Finally, we discuss the mesons that occupy the center of the figure shown in Fig. 20.2. There can be four states at the center, corresponding to the combinations $u\widehat{u}$, $d\widehat{d}$, $s\widehat{s}$ and $c\widehat{c}$. The actual mesons are linear superpositions of these combinations. Three such combinations, the π^0, the η and the η', have been discussed in Ch. 10. The fourth state is a $c\widehat{c}$ state, and is called the η_c. Its mass is 2980 MeV.

The $J^P = 1^-$ mesons have the same representation under the flavor group, and therefore Fig. 20.2 applies to the vector mesons as well, with only changes in the names of the particles. The SU(3)$_F$ octet plus the SU(3)$_F$ singlet, in this case, are considered together as a nonet of particles, as mentioned in §10.7. This nonet includes the isotriplet ρ mesons, the K^* mesons, plus the ω and the ϕ. The mesons that take the place of the D mesons are called the D^* mesons. The only other state is a $c\widehat{c}$ state, and this is the charmonium state Ψ.

b) Baryons

Let us now discuss baryons. Since they have three quarks, their transformation under SU(4)$_F$ would be like $4 \times 4 \times 4$. Moreover, we argued in §10.11 that the quark wavefunctions are antisymmetric in color. Thus, in the lowest lying states for which the spatial wavefunction is symmetric, the spin-flavor part of the wavefunction must also be symmetric. In analogy with the spin-flavor group SU(6) that we considered in §10.11.1, here we need to consider the spin-flavor SU(8) whose fundamental representation would consist of four different flavors of quarks with spins up or down. The completely symmetric part of the product of three fundamentals is

$$(\mathbf{8} \times \mathbf{8} \times \mathbf{8})_{\mathrm{symm}} = \mathbf{120}. \tag{20.15}$$

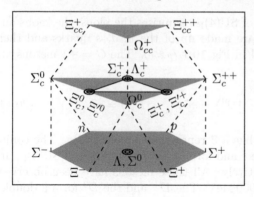

Name	Quark content	Mass in MeV
Λ_c^+	$(ud - du)c$	2286
Σ_c	qqc	2454
Ω_c^0	ssc	2697
Ξ_c	qsc	2469
Ξ_c'	qsc	2577
Ξ_{cc}	qcc	
Ω_{cc}^+	scc	

Figure 20.3: Spin-$\frac{1}{2}$ baryons belonging to the $\mathbf{20}_m$ representation of $SU(4)_F$. The gray polygon at the lowest and the highest levels denote the octet and the antitriplet of $SU(3)_F$ that appear in Eq. (20.17). The middle level contains the sextet (in gray) and the triplet (in white). Concentric elliptical mark at any place indicates that there are two states at that place. In the table, q means either u or d quark.

If we decompose the $\mathbf{120}$ in terms of the subgroup $SU(4)_F \otimes SU(2)$ where $SU(2)$ is the rotation group, we obtain

$$\mathbf{120} \implies (\mathbf{20}_s, \mathbf{4}) + (\mathbf{20}_m, \mathbf{2}). \tag{20.16}$$

The first thing on the right hand side represents spin-$\frac{3}{2}$ baryons, the second thing spin-$\frac{1}{2}$ baryons. Notice that, as far as the $SU(4)_F$ group is concerned, both types of baryons transform like a 20-dimensional representation. However, the two are not the same representation. The representation denoted by $\mathbf{20}_s$ is completely symmetric in the $SU(4)_F$ group indices, i.e., it has a Young tableaux ⊞⊞⊞. The other one is the representation ⊟⊟, which means that it has mixed symmetry.

It might seem peculiar that two different representations of the same group have the same dimension. If it does, then that's because the most familiar non-abelian group, SU(2), does not have this property. But there is nothing strange about it. For abelian groups, this is the rule rather than exception since all irreducible representations are 1-dimensional. For discrete non-abelian groups also such representations are quite common. For SU(2), however, all representations of the same dimension are equivalent. There is no reason to expect the same thing for SU(4) or any other group.

□ **Exercise 20.2** In fact, apart from the two different irreducible 20-dimensional representations mentioned already, $SU(4)$ has yet another of the same dimension, given by the Young tableaux ⊟⊟. Verify this statement.

The two different 20-dimensional representations decompose differently under the $SU(3)_F \times U(1)_X$ subgroup.

$$\mathbf{20}_s \implies (\mathbf{10}, -1) + (\mathbf{6}, \frac{1}{3}) + (\mathbf{3}, \frac{5}{3}) + (\mathbf{1}, 3),$$

$$20_m \longrightarrow (8,-1) + (6,\frac{1}{3}) + (3,\frac{1}{3}) + (3^*,\frac{5}{3}) . \qquad (20.17)$$

Let us first look at the 20_m, which are spin-$\frac{1}{2}$ baryons according to Eq. (20.16). The X quantum number of the octet of $SU(3)_F$ implies that it has $C = 0$, since $B = 1$ for these three-quark states. Indeed, this is the charmless octet that contains the nucleon, and that was discussed in Ch. 10. Calculation of the charm quantum number of the other $SU(3)_F$ multiplets shows that the baryons transforming as the 6 and 3 of $SU(3)_F$ contain one c quark, whereas those transforming as the 3^* of $SU(3)_F$ contain two c quarks. These states have been shown in Fig. 20.3, where the number of c quarks is denoted by a subscript.

The lettered names of the baryons appear in the figures. The naming scheme can be explained as follows. The two lightest quarks are the u and the d. If both of them occur in a baryon along with some third flavor of quark, the combination can be a isospin singlet or a triplet. The isospin singlet is called Λ and the triplet is collectively called Σ. Of course, when we say a quark 'occurs' in a baryon, we mean that it is one of the valence quarks. The ocean quarks do not matter in this discussion. When only one out of the three valence quarks is either u or d, the baryon must be a isospin doublet, and is denoted by the letter Ξ. If a baryon does not contain any u or d quark, it is represented by the letter Ω. Such baryons are clearly isosinglets.

Of course just this much is not enough, it will lead to the same name for many different baryons. To remove such degeneracies, two things are done. A superscript is added denoting the electric charge of the baryon. which helps distinguish between different members of the same isomultiplet. And a subscript is added, which tells us the flavors of the quarks which are neither u nor d. For example, if a Σ particle has a c quark in addition to the u and d quarks, it is called Σ_c. If the additional quark happens to be a strange quark, the subscript is omitted. Thus, in Ch. 10, the baryons that we had encountered in Fig. 10.6 *(p 274)* and Fig. 10.7 *(p 275)*, there is no subscript on any of the baryons. But in Fig. 20.3 and Fig. 20.4, subscripts appear with charmed baryons.

Note that there are some points on the figure where two baryons appear. We encountered this phenomenon while discussing $SU(3)_F$ as well. Take, for example, the udc baryons. There is one such baryon where the u and the d quarks form an isospin singlet. This is called the Λ_c^+, according to the naming scheme explained above. There is another which would be part of an isotriplet, along with uuc and ddc baryons, which is the Σ_c^+. Similarly, there is a dsc baryon that is a U-spin singlet, and one that is part of a U-spin triplet. In the first one, the d and the s quarks are in an antisymmetric combination, whereas in the second one they are in a symmetric combination. An usc baryon can similarly be a V-spin singlet or triplet. Since neither U-spin nor V-spin is a very good symmetry, the real particles are superpositions of these two states which are appropriate for members of isospin doublets. With these two uncharged and two singly charged states, there are two isodoublets: one

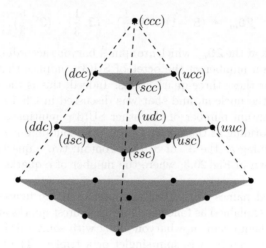

Figure 20.4: Spin-$\frac{3}{2}$ baryons in SU(4). The lowest level contains the decuplet of baryons of $SU(3)_F$, which have not been labeled here.

whose members are denoted by Ξ_c, and the other whose members are called Ξ'_c. The masses of these isodoublets are presented in Fig. 20.3.

The $\mathbf{20}_s$ of $SU(4)_F$ contains baryons with spin-$\frac{3}{2}$. The $\mathbf{10}$-plet of $SU(3)_F$ has $C = 0$ through Eq. (20.11). Obviously, this is the $SU(3)_F$ decuplet of baryons that was discussed in Ch. 10. Then there is a sextet of $SU(3)_F$ with $C = 1$, i.e., these baryons have one c quark and two other quarks which are u, d or s. There is also a triplet of $SU(3)_F$ with $C = 2$, baryons which contain two c quarks. And finally, the $SU(3)_F$ singlet with $C = 3$ is a baryon composed of three c quarks. All these members of the $\mathbf{20}_s$ of $SU(4)_F$ are shown in Fig. 20.4.

Having discussed the advantage of considering the $SU(4)_F$ group, it should be added that this symmetry is very badly broken, which is why the states in the same multiplet have a wide range of masses. But this is only to be expected! The isospin symmetry is respected very well, at the per cent level. The $SU(3)_F$ symmetry is much worse, as we have discussed in Ch. 10, because the strange quark mass is much larger than that of the u and the d quarks. And the $SU(4)_F$ must be even worse because the charm quark is much heavier than the s quark, as is obvious from the masses of the charmed hadrons.

20.1.4   Discovery of other charm-containing hadrons

The charmonium state ψ has a mass of about 3100 MeV. It is a $J^P = 1^-$ state, as we have mentioned. The charmonium state with $J^P = 0^-$ is called η_c, and is lighter than the ψ by more than 100 MeV. Despite being heavier, the ψ state was discovered before the η_c. One might wonder, why wasn't the η_c observed first, at a lower energy?

The answer to this question lies in the spin of the state. The experiments were done with colliding e^+-e^- beams. In such experiments, the initial particles would have to go through a virtual photon, which in turn would produce the $c\hat{c}$ state. Thus, so far as the hadronic states are concerned, it is a transition from the vacuum state to the meson state through the virtual photon, which has spin 1. Since the vacuum does not carry any angular momentum, this is not possible unless the meson also has spin 1. Thus the ψ can be produced, but not the η_c. One might notice in this connection that Fig. 20.1 *(p 596)* has peaks for the ω and ϕ mesons, which are members of the 1^- nonet, but not for the π^0 or the η or the η' which are spinless. The reason is the same. Of course the argument applies only for the case of a single photon exchange. If more photons are exchanged, the process is possible, but then also the process is higher order in e and therefore much weaker. Or alternatively, the η_c can be produced along with some other particle so that the above argument does not apply. Historically, the η_c was not discovered through its direct production. Rather, it was found from the decay of the ψ:

$$\psi \to \eta_c + \gamma. \tag{20.18}$$

Charmonium states like ψ and η_c do not carry any charm quantum number. The quantum number cancels between the c quark and its antiquark. The D mesons, on the other hand, have non-zero charm quantum number. As discussed before, the masses of the D mesons are a little below 2 GeV. Hence they are much lighter than the ψ, or even than the η_c. And yet, they were discovered later because from e^+e^- collisions, one cannot produce a single D meson which carries charm: one must have to produce also a $\hat{D}$ meson along with it. Hence the total CM energy required in the e^+e^- beam would be twice that of the mass of a D meson, i.e., close to 4 GeV. In 1976, when the energy could be increased to cross this mark, it was observed that in the products, $K^-\pi^+$ CM energies go through a hump at about 1.9 GeV. According to the method described in §9.6.2, this indicates the presence of a particle that decays into $K^-\pi^+$. This is the D^0. Other D mesons were discovered in similar ways.

☐ **Exercise 20.3** *From the spin and parity of the states, show that the ψ is a $^3S_1$ state of the charm quark and its antiquark, whereas the η_c is a $^1S_0$ state.*

20.1.5 Decay of mesons with charm quark

We mentioned earlier that the charmonium states are very narrow. Through the time-energy uncertainty relation, it implies that their lifetimes are quite large. The charmonium state ψ at 3097 MeV has a width of about 93 keV, as already mentioned. To understand how small it is, compare it with the fact that the ω meson, with mass 782 MeV, has a width of 8.5 MeV. The charmonium ψ, despite being about four times heavier, has a width which is about 90 times smaller.

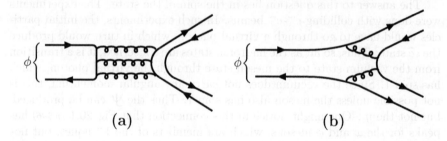

Figure 20.5: The diagram in the left shows how the $s\hat{s}$ meson can decay to two mesons, none of which carries any strangeness. The diagram on the right shows decay where the decay particles carry strangeness. The diagram to the left is OZI suppressed with respect to the diagram to the right. The text explains the reason.

If we consider the decay of ψ to e^+e^- or any other final state containing leptons, the discrepancy is easily explained. Leptonic final states must have to come through electromagnetic interactions, for which the rate must be smaller compared to typical strong interaction rates. But the ψ decays mostly hadronically. In fact, branching ratio of hadronic final states is 87.7%. There are multi-pion final states, and also final states involving the rho meson, kaons, etc. Why are these decays much slower than typical strong interaction decays?

The answer lies in the OZI rule that we had mentioned in Ch. 10, but did not really explain. Let us now try to understand the reason for the OZI suppression, using the ϕ meson decay as a paradigm. As we said in Ch. 10, the ϕ meson is almost purely an $s\hat{s}$ state. If it has to decay to non-strange particles, the diagram has to look like Fig. 20.5a. It is important to note that there have to be at least three gluons exchanged. A single gluon exchange is impossible since the quark-antiquark pair in the initial state are in a color singlet combination, and therefore cannot couple to a gluon. Two gluons are also not possible, because the ϕ meson is odd under charge conjugation, whereas a state with two gluons is even. Thus, at least three gluons are necessary, as shown in Fig. 20.5a, and the amplitude will therefore contain a factor g_3^6, i.e., the rate will contain α_3^6. The diagram in Fig. 20.5b, on the other hand, has a rate proportional to α_3^4. Moreover, remember from Ch. 12 that the value of α_3 depends quite strongly on the energy scale. In Fig. 20.5, the virtual gluons should have 4-momenta of the order of the ϕ meson mass. On the contrary, if we consider a decay of the form Fig. 20.5b where the final state contains the s and the $\hat{s}$ in two different particles, much smaller momentum transfer would be involved. At low momentum transfers, the QCD coupling constant is large, and it falls sharply as the momentum transfer increases. Hence, each factor of α_3 appearing in Fig. 20.5a is much smaller than any α_3 in Fig. 20.5b. This is the reason why $\phi \to \rho\pi$ decay is suppressed with respect to $\phi \to KK$ decay.

Coming back to the question of ψ decay, we see that the situation here is worse compared to the ϕ decay. Here, there is no possible decay mode that is not OZI suppressed. In other words, the ψ cannot decay into a particle containing charm quantum number $C = +1$ and another one containing $C = -1$. There simply isn't any charmed meson with mass less than half the mass of the ψ, or even the $\psi(3686)$. Thus, the c and the $\hat{c}$ both have to be absent in the final states, and the final state must be OZI suppressed. This is the reason the decays are slow, and the resonance is narrow.

□ **Exercise 20.4** Using G-parity arguments, show that the ψ cannot decay to two pions through strong interactions.

Let us now discuss the decay of the D mesons. These carry charm quantum numbers. The decay products cannot, because there are no charmed particles lighter than the D mesons. So the decays must be weak decays. One consequence of this fact is that the lifetimes are quite long, of the order of 10^{-15} s. It also has important implication toward the decay products. The weak neutral current cannot change flavor, as discussed in Ch. 17. So the charged current, which contains the CKM matrix elements, must be responsible for the decay. The largest CKM matrix element involving charm quark is V_{cs} (in fact, its magnitude is close to unity), so the c quark will decay more readily to the s quark than to the d quark. Therefore, the dominant decay channels of the D mesons contain strangeness. For example, 54.7% of D^0 decays contains K^- in the final state.

There is a related phenomenon, viz., that the branching ratio of D^0 decays containing a K^+ is only about 3.4%. The reason for this discrepancy between K^+ and K^- decay modes can easily be understood by recalling that K^+ contains $u\hat{s}$ whereas K^- contains $s\hat{u}$. A D meson contains the c quark which can easily be converted to the s quark, as explained above. An $\hat{u}$ will then appear as a member of a quark-antiquark pair that will be produced, as seen on the right end of Fig. 20.5b. That way, a K^- is obtained in the final state. In order to obtain a K^+, the original c quark must go to a d quark which is highly suppressed because the relevant CKM element is small, and then somehow lots of quark-antiquark pairs have to be produced, among which there will be a u and an $\hat{s}$.

□ **Exercise 20.5** Draw a quark-level diagram to show how a K^+ might appear in the final state of a D^0 decay.

□ **Exercise 20.6** We said that the lifetimes of the D mesons are of the order of 10^{-15} s. What are the widths in energy units?

Decays of D^* mesons are very different. They can decay strongly to a D meson, accompanied by a pion or a photon. If the accompanying particle is a pion, the decay can be mediated by strong interactions, and therefore the width is large, of the order of a few MeV.

20.1.6   Decay of baryons with charm quark

Let us start with the $J^P = \frac{1}{2}^+$ baryons. The Σ_c baryons have masses around 2455 MeV whereas the Λ_c^+ mass is 2286 MeV. There is thus enough mass difference for the decay

$$\Sigma_c \rightarrow \Lambda_c^+ + \pi \qquad\qquad\qquad (20.19)$$

to take place. As it happens, this is the only mode for which a strong decay is allowed. Therefore, the Σ_c baryons decay almost fully through this channel.

> □ **Exercise 20.7** *If the isospin symmetry were exact, show that one should expect* $\Gamma(\Sigma_c^{++} \rightarrow \Lambda_c^+ + \pi^+) = \Gamma(\Sigma_c^+ \rightarrow \Lambda_c^+ + \pi^0)$.

The Λ_c^+ is the lightest charmed baryon, and therefore cannot decay strongly. Thus it has a much larger lifetime, roughly 2×10^{-13} s. The weak decays preferentially have final states containing strange particles, as explained earlier in connection with the decay of the charmed mesons. Thus, the dominant decay modes are of two types. One type consists of proton and strange mesons in the decay products, like $p\widehat{K}^0$ or $pK^-\pi^+$. In the other type, the Λ baryon carries the strangeness, and some non-strange mesons are produced along with it.

The Ξ_c and Ξ_c' baryons carry both charm and strangeness, and are the lightest baryons which have this property. So they have to decay via weak interactions. As explained in the context of mesons, the charm quark likes to decay to a strange quark. Add to it the strange quark that was already there in the initial state, and one gets two units of strangeness in the final state. Of course, other values of strangeness are possible, but such final states are CKM suppressed.

Among the charmed baryons with $J^P = \frac{3}{2}^+$, all but one should be able to decay strongly. The exception is Ω_{ccc}^{++}, which has to decay via weak interactions.

20.2   Bottom quark

Was there any prediction of the bottom quark? The answer is both 'yes' and 'no'. Once the charm quark was discovered, the existence of two generations of quarks was confirmed. The first generation comprised the u and the d quarks, the second comprised the c and the s quark. The GIM cancellation between these two generations explained the smallness of flavor-changing decays or the K_L-K_S mass difference. So, in a sense, there was no compelling reason to look beyond the two generations of quarks.

Except one, perhaps. Kobayashi and Maskawa discovered that gauge interaction of fermions cannot explain CP violation unless the number of fermion generations is at least three. In Ch. 21, we will explain this statement in more detail. Since CP violation was discovered in 1964, this could have been taken as a serious hint for third generation quarks. But historically, despite this hint, the idea of third generation quarks was not taken very seriously, and many people thought that CP violation might be explained by some other means like an extension of the scalar sector of the standard model.

20.2.1   Discovery of bottom-containing hadrons

And so, it was a mild surprise when, in 1977, a team led by Lederman discovered a new resonance in Fermilab while studying the energy of $\mu^+\mu^-$ pairs in a fixed target experiment of proton beam hitting a beryllium target. A year later, the same resonance was produced in e^+e^- collision in DESY, and it was proved that the resonance is a meson. Such a heavy meson could not be explained with the four quarks that were known at that time, so the hypothesis of the existence of a new quark was necessary. The rise in the R value above this resonance showed that the electric charge of this quark has the same magnitude as that of the d and s quarks. In analogy with the name 'down', this quark came to be known as the 'bottom' quark, or the b quark. Some people prefer to think that the name is not 'bottom' but 'beauty', in the same line as 'strangeness' or 'charm'. Either way, it is the b quark.

The resonance seen was a $b\hat{b}$ meson. It has mass of 9.460 GeV. Like the ψ, it is a state with spin 1. It came to be called the upsilon (Υ), or more precisely $\Upsilon(1S)$. It is one member of the family of particles which are tagged by the generic name *bottomonium*, bound states of $b\hat{b}$. As for the case of charmonium, other bottomonium states have also been observed, like the $\Upsilon(2S)$ at 10.023 GeV and the $\Upsilon(3S)$ at 10.355 GeV.

There have been searches for hadrons containing the bottom quantum number, and they have been very successful. The B^+ meson, with mass 5.279 GeV and quark content $u\hat{b}$, as well as its antiparticle B^+, has been widely studied. Neutral B mesons like B^0 ($d\hat{b}$) and B_s^0 ($s\hat{b}$), as well as their antiparticles $\hat{B}^0$ ($b\hat{d}$) and $\hat{B}_s^0$ ($b\hat{s}$), have also been discovered and studied extensively. Charged particles carrying both bottom and charm quantum numbers have also been discovered. They are called B_c^+ ($c\hat{b}$) and B_c^- ($b\hat{c}$), and have a mass of 6.277 GeV. Some bottom-containing baryons have also been discovered, like the Λ_b^0 whose valence quarks are udb, and the Ξ_b^0 and Ξ_b^- whose valence quarks are usb and dsb respectively.

20.2.2   Decay of bottom-containing hadrons

The bottomonium state $\Upsilon(1S)$ is a very narrow resonance, the width being only about 54 keV. For the charmonium, the reason for the narrowness is OZI suppression. For the bottomonium, the reason is the same, with the additional comment that the Υ is much heavier than ψ, and so the QCD coupling constant is even smaller at this scale and the decay rate is smaller. The width of the $\Upsilon(2S)$ and $\Upsilon(3S)$ states are also comparable to that of $\Upsilon(1S)$. However, $\Upsilon(4S)$ has a mass of 10.58 GeV, and is heavy enough to decay into $B\hat{B}$ mesons, i.e., one meson containing a b quark and another

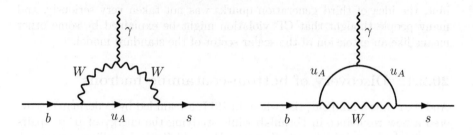

Figure 20.6: One-loop diagrams for the quark-level process $b \to s + \gamma$. In any renormalizable gauge, there are extra diagrams in which one or both of the internal W bosons are replaced by the unphysical charged Higgs boson.

containing a $\widehat{b}$. Such decays are not OZI suppressed, and so the $\Upsilon(4S)$ has a large width, 20.5 MeV.

The mesons carrying bottom quantum number turn out to be much longer lived than the mesons carrying charm quantum number. The reason is that the charm quark decays to a strange quark, and the corresponding CKM matrix element is almost equal to unity. On the other hand, the bottom quark decays to a charm quark, and the relevant CKM matrix element, V_{cb}, is quite small, of order λ^2 in the Wolfenstein parametrization of the CKM matrix. Because of this, the amplitudes are suppressed, and the lifetimes of bottomed mesons are typically of order 10^{-13} s.

An important process that deserves some attention is the quark-level transition $b \to s\gamma$. At the hadronic level, one might think that the simplest process to show the effect of this transition would be $B^+ \to K^+\gamma$ or $B^0 \to K^0\gamma$. But that is not correct. These processes are in fact absolutely forbidden since they would involve a $J = 0$ to $J = 0$ radiative transition. However, the final state can contain an extra pion. Or else, the final state might contain K^*, members of the $J^P = 1^-$ nonet of mesons, instead of the K mesons. The resulting final state is a two-body state, and hence can be very accurately studied. The quark-level process can go through one-loop diagrams shown in Fig. 20.6. Effectively, the transition constitutes a flavor-changing neutral current, and the GIM cancellation operates here. The intermediate t quark lines contribute appreciably because the top mass is large.

20.3 Neutral meson–antimeson systems

We have discussed the physics of neutral kaons in Ch. 17. With the discovery of the third generation of quarks, there arose the possibility of studying other similar systems consisting of two mesons which are antiparticles of each other. We will denote the meson states with well-defined quark flavors by M^0 and $\widehat{M}^0$. For example, M^0 might be B^0, in which case $\widehat{M}^0$ would be $\widehat{B}^0$, and we will be studying the physics of the B^0-$\widehat{B}^0$ system. We can also consider the system composed of B_s^0 and $\widehat{B}_s^0$, or even of the charmed hadrons D^0 and $\widehat{D}^0$. In essence, all such systems are similar to the K^0-$\widehat{K}^0$ system. Since we have already described the neutral kaon system in great detail, we will not get into similar details of these mesons. We will only point out the similarities and differences with the neutral kaon sector qualitatively.

First, the similarities. The quark structure is similar, as we have already commented. In the approximation that CP violation is neglected, one combination of M^0 and $\widehat{M}^0$ is a CP eigenstate with eigenvalue $+1$ and another with eigenvalue -1. The two eigenstates are not exactly degenerate: there is a tiny mass difference between them, and also a difference in their decay rates, both induced by weak interaction effects. When CP violation is included, one can obtain CP-violating phenomena involving such mesons, as will be discussed in Ch. 21.

Let us now look at the differences with the neutral kaon sector. The biggest difference lies in the fact that the neutral kaons are not very much heavier than the combined mass of three pions, so that the three-pion decay mode is quite suppressed: the decay rate of K_L^0 is about two orders of magnitude smaller than that of K_S^0. The B mesons, on the other hand, are much heavier, so the decay rates into the two-pion and three-pion channels are almost equal. Because of this difference, it makes little sense to call the two eigenstates by the names 'short-lived' and 'long-lived'. Instead, one usually denotes the heavier eigenstate by M_H^0 and the lighter one by M_L^0. In the calculation of the K_L^0-K_S^0 mass difference in the standard model, we found that almost the entire contribution comes from c quark intermediate lines in the box diagrams. The t quark contributions were negligible, despite the large t quark mass, because the mixing angles were very small. If we use Wolfenstein parametrization of the CKM matrix, we find that $V_{ts}V_{td} \sim \lambda^5$, whereas $V_{cs}V_{cd} \sim \lambda$. For the B mesons, the relative magnitudes of the mixing matrix elements will not be similar. Looking at Eq. (17.24, *p 488*), we find that for B_s mesons, the relevant combinations of mixing matrix elements are $V_{ts}V_{tb} \sim \lambda^2$, and $V_{cs}V_{cb}$ also goes like the same power of λ. For B_d mesons, the relevant magnitudes are $V_{td}V_{tb} \sim \lambda^3$ which is smaller than the corresponding quantity for the B_s mesons, but then $V_{cd}V_{cb}$ is also of the same order. The conclusion is that both c quark and t quark intermediate lines have to be considered in the calculation. Since the t quark is more than 100 times heavier than the c quark, the contributions to the B^0-$\widehat{B}^0$ or to the B_s^0-$\widehat{B}_s^0$ are large:

$$\Delta m_{B^0} = (3.337 \pm 0.033) \times 10^{-10} \,\mathrm{MeV}$$
$$= (0.507 \pm 0.004) \times 10^{12} \,\mathrm{s}^{-1}, \tag{20.20a}$$
$$\Delta m_{B_s^0} = (116.4 \pm 0.5) \times 10^{-10} \,\mathrm{MeV}$$
$$= (17.69 \pm 0.08) \times 10^{12} \,\mathrm{s}^{-1}. \tag{20.20b}$$

Note that the mass difference is larger for the B_s^0-$\widehat{B}_s^0$ system, a fact that can be understood from the differences in the power of the Wolfenstein parameter

λ appearing in the two expressions, explained earlier in the paragraph. For the D^0-$\widehat{D}^0$ system, the mass difference is much smaller:

$$\Delta m_{D^0} = (1.44^{+0.48}_{-0.50}) \times 10^{10}\,\text{s}^{-1}. \tag{20.21}$$

The main reason for the smaller value is that the predominant contribution comes from the s quark mass, which is much smaller than the t quark mass.

20.4   Top quark

When the b quark was discovered, it was tempting to assume that there was also another quark, which has the same electric charge as u and c quarks. People who thought b as bottom called this new quark the 'top' quark. People who thought b as beauty wanted to call it by the name 'truth'. So the name t quark was universally accepted.

However, mere analogy with the other two generations cannot be a proof of existence of the t quark. Direct detection had to wait till 1994 because the top quark turned out to be much heavier than the bottom quark. However, even before the direct detection, it was clearly known that the bottom quark cannot be a lone quark in the third generation of fermions. The demonstration came from Z decay. The point was that the branching fraction of Z decaying to hadrons containing b quarks was measured to be 15.12%. The branching ratio of $Z \to e^+e^-$ is 3.363%, which means that

$$\frac{\Gamma(Z \to b\widehat{b})}{\Gamma(Z \to e^+e^-)} \approx 4.5. \tag{20.22}$$

The decay rate of the Z boson to fermion-antifermion pair was calculated in §19.2. Using the results given in Eqs. (19.32) and (19.38), we obtain

$$\frac{\Gamma(Z \to b\widehat{b})}{\Gamma(Z \to e^+e^-)} = 3 \times \frac{\left(T_{3L}^{(b)} + \frac{2}{3}\sin^2\theta_W\right)^2 + \left(T_{3L}^{(b)}\right)^2}{\left(-\frac{1}{2} + 2\sin^2\theta_W\right)^2 + \frac{1}{4}}, \tag{20.23}$$

where the factor 3 in the numerator is the color factor, which takes into account the fact that there are three different colors of b that the decay of Z can produce.

For a rough estimate, we can put $\sin^2\theta_W = \frac{1}{4}$. If the b quark were a singlet of the SU(2) part of the gauge group, the ratio of the decay rates would have been equal to $\frac{1}{3}$, nowhere close to the experimental result. If, on the other hand, the b quark is assumed to have $T_{3L} = -\frac{1}{2}$, the ratio comes out to be $4\frac{1}{3}$, which seems the right choice. With a more accurate value of $\sin^2\theta_W$, the analytical result agrees perfectly with the experimental one.

> □ **Exercise 20.8** In §16.7, we calculated the forward-backward asymmetry in the reaction e^+e^- going to a fermion-antifermion pair. Show that this asymmetry would vanish if the b quark happened to be a singlet of SU(2)$_L$.

Thus, even before the discovery of the t quark, it was known that the b quark has $T_{3L} = -\frac{1}{2}$, and therefore it must have a partner in the third generation of quarks. The actual discovery, however, had to wait quite a while. The reason is that the ratios of masses of the two quarks in each of the first two generations lie within a factor of 10 or so. Since the bottom quark mass is about 4.5 GeV, naively it was expected that the top quark could not be much heavier than 45 GeV or so. This naive expectation was jolted when the Z boson was not found to decay into $t\widehat{t}$, signifying that the top quark has to be heavier than $\frac{1}{2}M_Z$, leaving hardly any room for its mass to be within an order of magnitude of the bottom mass. The lower limit was steadily increased as more and more data was accumulated.

There were upper limits as well. These limits came from the effect of virtual top quarks in various processes. We have commented that the contribution coming from top quark internal lines is negligible in the calculation of the K_L^0-K_S^0 mass difference. However, for the B^0-$\widehat{B}^0$ system, the contribution of top quark internal lines is dominant, and the measured value of the mass difference of the neutral B mesons provides an upper limit of the top quark mass. The ratio of the masses of the W and the Z boson also receives one-loop corrections from top quark loops, and here also the measured value of the ρ parameter can be used to estimate an upper bound of the top mass. These estimates indicated an upper limit of roughly 200 GeV.

There was also the limit that came from the measurement of the W to Z boson mass ratio, or, more precisely, of the ρ parameter defined in Eq. (19.58, p 576). In Eq. (19.86, p 583), we showed the contribution to this parameter that comes from a doublet of fermions. If $m_1 \gg m_2$, the dominant term in this contribution is

$$\frac{3G_F m_1^2}{8\sqrt{2}\pi^2} = 3.1 \times 10^{-3} \times \left(\frac{m_1}{100\,\text{GeV}}\right)^2, \qquad (20.24)$$

including an explicit color factor. The experimental bounds on the ρ parameter indicated that the top mass has to be less than 200 GeV or so.

As the experimental lower limit on top mass was pushed beyond the Z mass, it was clear that such a heavy top would predominantly decay to a real W-boson and a b quark:

$$t \to W^+ + b. \qquad (20.25)$$

This channel was searched for in the Tevatron which started operating in 1987. As mentioned in Ch. 9, it is a proton-antiproton collider. A top quark is produced in such colliders through strong interactions, and is therefore produced in the pair $t\widehat{t}$. The $\widehat{t}$ decays into $W^- + \widehat{b}$. Thus, the production of a $t\widehat{t}$ pair can be identified from the decay mode containing $W^+W^-b\widehat{b}$. Searching from channels containing at least two leptons which arise from the decay of the W's or the b's, the top quark was finally discovered in 1994.

□ **Exercise 20.9** Ignoring the mass of the b quark, show that the rate of the top decay process shown in Eq. (20.25) is given by

$$\Gamma = \frac{\alpha m_t}{16 \sin^2 \theta_W} \frac{m_t^2}{M_W^2} \left(1 - \frac{M_W^2}{m_t^2}\right)^2 \left(1 + \frac{2M_W^2}{m_t^2}\right). \tag{20.26}$$

There is one aspect in which the top quark is very different from any other quark. The top quark mass is about 173 GeV. Putting it into Eq. (20.26) along with the values of M_W, α and $\sin^2 \theta_W$, one obtains that the lifetime is roughly $(1.5\,\text{GeV})^{-1}$. This is significantly smaller than $\Lambda_{\text{QCD}}^{-1}$, since $\Lambda_{\text{QCD}} \approx 200\,\text{MeV}$. The reason why this comparison is important is the following. Once a quark is produced in a high energy reaction, it takes some time to hadronize. The hadronization time is of the order of $\Lambda_{\text{QCD}}^{-1}$. Thus, we find that a top quark produced in a high energy collision process decays before it can hadronize.

20.5   Quark masses

We have discussed the masses of the light quarks — u, d and s — in §18.3.6. Here, we discuss the masses of the heavy quarks c, b and t. We call these three quarks "heavy" not only because they are heavier than the other three, but because their masses are much bigger than the QCD scale Λ_{QCD}. This means that for discussing processes involving these quarks, the coupling constant can be taken to be in the perturbative region.

A very rough estimate of the mass of a particular quark can be obtained by determining the mass of a meson that contains the quark and its antiquark, and then dividing by two. But this cannot be a good algorithm. For example, take the charm quark. The $c\bar{c}$ bound state η_c has a mass of 2980 MeV, whereas the ψ mass is about 3100 MeV. The excess amount of mass for the ψ can be thought of as the energy of an excited state of its constituents. Even if we go by the lightest meson, we need to remember that a considerable amount of the mass of a hadron comes from the ocean quarks and gluons. To account for these things, and also to fit the effects of these masses when these quarks appear as virtual lines, one usually takes

$$m_c = 1.27\,\text{GeV}. \tag{20.27}$$

For the case of the bottom quark, the exercise is simpler because its mass is much higher:

$$m_b = 4.18\,\text{GeV}. \tag{20.28}$$

And for the top quark, the mass is so large that one easily concludes

$$m_t = 173.5\,\text{GeV} \tag{20.29}$$

from the minimum energy required for observing top events.

There is one important point to be mentioned in connection with quark masses. Earlier in Ch. 12, we argued that the physical coupling constant is scale-dependent. The mass of any particle, being also a parameter in the Lagrangian, is also scale-dependent for the same reason. We have never mentioned this point so far, because there is a very natural definition of mass: it is the parameter that appears in the kinematical or dynamical equations for the relevant particle. For example, if we want to mention the mass of the electron, we can just see how much it bends in a given magnetic field, and deduce the mass from that. It is true that the mass parameter that appears in the Lagrangian will be scale-dependent, but there is an on-shell value of the mass, which is what we accept as the physical mass.

For quarks, we cannot use this on-shell definition because there are no free quarks. Thus, the values quoted for the quark masses pertain to some particular energy scale. We have not mentioned the scales because the variation with scale is not of much importance in our discussions. However, the variations are not insignificant.

20.6 Heavy quark effective theory

In Ch. 18, we saw that certain properties of hadrons are easily understandable by considering the massless limits of the light quarks. Similarly, there are properties of hadrons containing one or more heavy quarks which can be understood by considering the limit that the heavy quark masses are infinite. This is the topic that we discuss in this section.

20.6.1 Symmetries

We will denote heavy quarks by $\mathscr{Q}$, the calligraphic capital Q. We consider hadrons with only one heavy quark, the remaining quarks or antiquarks being light. In such systems, even if momentum is exchanged between the heavy quark and other quarks, the change of the velocity 4-vector will be given by $\delta u^\mu = \delta p^\mu / m_\mathscr{Q}$, which vanishes in the limit $m_\mathscr{Q} \to \infty$. Thus, the velocity 4-vector of the heavy quark can be taken to be a conserved quantity, characteristic of the system. This is reminiscent of the standard treatment of the hydrogen atom, where we can consider the proton to be at rest without any loss of consistency, because the proton is much heavier than the electron.

It is therefore convenient to deal with quark fields characterized by a velocity 4-vector u^μ. To this end, we rewrite the heavy quark field $\mathscr{Q}(x)$ in the form

$$\mathscr{Q}(x) = \exp\left(-im_\mathscr{Q} u \cdot x\right)\left(\mathscr{Q}_u(x) + \mathfrak{Q}_u(x)\right), \tag{20.30}$$

where

$$\mathscr{Q}_u(x) = \exp\left(im_\mathscr{Q} u \cdot x\right) \frac{1 + \slashed{u}}{2}\, \mathscr{Q}(x),$$

$$\mathcal{Q}_{\alpha}(x) = \exp\left(im_{\mathcal{Q}}v \cdot x\right) \frac{1 - \slashed{v}}{2} \, \mathcal{Q}(x)\,. \tag{20.31}$$

☐ **Exercise 20.10** Show that $P_{\pm} \equiv \frac{1}{2}(1 \pm \slashed{v})$ are orthogonal projection matrices, i.e.,

$$P_{+}^{2} = P_{+}\,, \qquad P_{-}^{2} = P_{-}\,, \qquad P_{+}P_{-} = P_{-}P_{+} = 0\,. \tag{20.32}$$

☐ **Exercise 20.11** Verify that

$$\slashed{v}\mathcal{Q}_{\alpha}(x) = \mathcal{Q}_{\alpha}(x)\,, \qquad \slashed{v}\mathfrak{Q}_{\alpha}(x) = -\mathfrak{Q}_{\alpha}(x)\,. \tag{20.33}$$

☐ **Exercise 20.12** Show that

$$\frac{1 + \slashed{v}}{2}\,\gamma^{\mu}\,\frac{1 + \slashed{v}}{2} = \frac{1 + \slashed{v}}{2}\,v^{\mu}\,\frac{1 + \slashed{v}}{2}\,. \tag{20.34}$$

Find the similar equation with a Dirac matrix flanked by $\frac{1}{2}(1 - \slashed{v})$.
[**Note :** *In fact, it is not necessary to put down two factors of the projection matrix on the right hand side of Eq. (20.34).*]

The significance of the projection matrices appearing in Eq. (20.31) is straightforward. Note that

$$\frac{1 \pm \slashed{v}}{2} = \frac{m \pm \slashed{p}}{2m} \tag{20.35}$$

for any particle with mass m. In the appendix, in §F.2.7, we argue that these matrices project out the positive energy and negative energy spinors, i.e., spinors corresponding to particles and antiparticles. Therefore, the field $\mathcal{Q}_{\alpha}(x)$ corresponds to quarks and $\mathfrak{Q}_{\alpha}(x)$ to antiquarks. It can then be easily seen that the free Lagrangian of the heavy quark can be rewritten in the form

$$\mathscr{L}_{0} = \overline{\mathcal{Q}}\, i\gamma^{\mu}\partial_{\mu}\mathcal{Q} - m_{\mathcal{Q}}\overline{\mathcal{Q}}\mathcal{Q} = \overline{\mathcal{Q}}_{v}\, iv^{\mu}\partial_{\mu}\mathcal{Q}_{v} + \cdots\,, \tag{20.36}$$

using the identity of Eq. (20.34) and omitting the antiquark terms, which will be suppressed in the heavy mass limit. This shows that the Lagrangian is independent of the quark mass. Therefore, the physical consequences of the theory should be independent of the heavy quark mass as well. In other words, if we consider two different systems, one containing a heavy quark $\mathcal{Q}_{1}$ and the other a different heavy quark $\mathcal{Q}_{2}$, plus other lighter quarks which are the same in both systems, properties of both systems should be identical to the extent that corrections involving the inverse of the heavy quark masses can be ignored. This is called the *heavy quark flavor symmetry*.

Once again, we can go back to the analogy of the hydrogen atom. An ordinary hydrogen atom has one proton as its nucleus. Suppose we compare it with the deuterium and the tritium, whose nuclei contain one and two neutrons respectively. The energy levels of the electron in these atoms will be the same as those of the ordinary hydrogen atom, because we consider the nuclei to be infinitely heavy in all three cases. If we consider the three nuclei as different 'flavors', this example constitutes a perfect analog of the heavy quark flavor symmetry for atomic systems.

There are more symmetries in the heavy quark limit. To see them, let us try to find the propagator of a heavy quark. Starting from the free term seen in Eq. (20.36), if we follow the method of finding the propagator that was detailed in Ch. 4, we would obtain the propagator of the $\mathscr{Q}_v$ field as $i/(v \cdot k)$. Usually, is presented in the form

$$\frac{1}{2}(1 + \slashed{v}) \frac{i}{v \cdot k} . \tag{20.37}$$

The projection matrix, $\frac{1}{2}(1 + \slashed{v})$, appears here only to remind us that the associated field contains a projection matrix. It would not matter at all in the evaluation of any Feynman amplitude involving these fields, because the vertices involving these fields will contain the projection matrix as well, and powers of a projection matrix is the matrix itself.

The important point of Eq. (20.37) is that, apart from the projection matrix that is superfluous, the propagator is just a multiple of the unit matrix. Similarly, the interaction of the quarks with some gauge bosons V_μ^a can also be written as

$$g\overline{\mathscr{Q}}\gamma^\mu T_a \mathscr{Q} V_\mu^a = g\overline{\mathscr{Q}} v^\mu T_a \mathscr{Q} V_\mu^a , \tag{20.38}$$

by using Eq. (20.34). Here also the Dirac matrix reduces to a unit matrix. Thus, there is nothing in the vertices or in the propagators of heavy quark fields that can change a particular component of the Dirac field to a different one. This means that the physical consequences of this theory is independent of the spin of the heavy quarks. This is called the *heavy quark spin symmetry*.

20.6.2 Hadron fields

Heavy quark spin symmetry would connect different particles which appear in different representations of the Lorentz group, e.g., have different spins. For example, consider a pseudoscalar meson consisting of a heavy quark $\mathscr{Q}$ and a light antiquark $\widehat{q}$. The spin of $\widehat{q}$ should be opposite to the spin of $\mathscr{Q}$ since the total spin of the meson is zero. Now suppose we apply heavy quark spin rotation. In course of the rotation, the spin of $\mathscr{Q}$ can be in the same direction as that of $\widehat{q}$. In this situation, the total spin of the system cannot possibly be zero. In fact, it will be 1, so that we will encounter a vector meson. Thus the pseudoscalar and vector mesons are related through heavy quark spin symmetry. In general, the spin of a hadron, $\boldsymbol{J}$, can be written in the form

$$\boldsymbol{J} = \boldsymbol{S}_h + \boldsymbol{S}_l , \tag{20.39}$$

where $\boldsymbol{S}_h$ is the contribution of the heavy quarks and $\boldsymbol{S}_l$ of the light quarks in the hadron, and the sum has to be performed according to the standard methods of addition of angular momenta. The point is that unless s_l, the quantum number corresponding to the eigenvalue for $\boldsymbol{S}_l^2$, is zero, different values of j are obtained for the same $\boldsymbol{S}_h$, and hadrons with these different values of j therefore belong to the same irreducible multiplet under the heavy quark spin symmetry. For example, as described above, the $J^P = 0^-$ and $J^P = 1^-$ mesons will belong to the same multiplet.

It is therefore convenient to express the $J^P = 0^-$ and $J^P = 1^-$ mesons through the same field. Suppose there is a pseudoscalar meson M containing a particular heavy quark and some light antiquark. With the same combination of the heavy quark and the light antiquark, there is a vector meson M^*, where the asterisk stands for an excited state of the same constituents. We will denote the corresponding fields by $\phi(x)$ and $\Phi^\mu(x)$. The field combining these two, which will be used in the context of heavy quark spin symmetry, is written in the form

$$\mathfrak{M}_\alpha(x) = \frac{1 + \not{v}}{2}\left[\gamma_\mu \Phi_\alpha^\mu(x) + i\phi_\alpha(x)\gamma_5\right]. \tag{20.40}$$

The fields $\phi_\alpha(x)$ etc are defined in the same manner as $\mathscr{Q}_\alpha$ was defined in Eq. (20.31).

Let us explain different aspects of the definition that occurs in Eq. (20.40). First, notice that we have put in the projection matrix, $(1+\not{v})/2$, to ensure that only the meson involving a heavy quark is relevant here, and not its antiparticle with a heavy antiquark. Second, we have used the factor of i with the pseudoscalar, so that both terms in the square bracket have the same kind of hermiticity property. In particular, if any of these two terms is denoted by A, we have $A^\dagger = \gamma_0 A \gamma_0$. Third, note that the field $\mathfrak{M}_\alpha(x)$ is in the form of a 4×4 matrix. This has been done so that the rows of this matrix can transform as doublets of $\boldsymbol{S}_l$ whereas the columns as doublets of $\boldsymbol{S}_h$. In other words,

$$\left[\boldsymbol{S}_h, \mathfrak{M}_\alpha\right] = \frac{1}{2}\boldsymbol{\Sigma}\mathfrak{M}_\alpha,$$

$$\left[\boldsymbol{S}_l, \mathfrak{M}_\alpha\right] = -\frac{1}{2}\mathfrak{M}_\alpha\boldsymbol{\Sigma}, \tag{20.41}$$

where $\boldsymbol{\Sigma}$ represents the 4×4 generators of rotation which were discussed in some detail in §14.2.1. Therefore, Eq. (20.39) implies that under a spatial rotation by an amount θ, the change of the field $\mathfrak{M}_\alpha(x)$ is given by

$$\delta\mathfrak{M}_\alpha = i\left[\boldsymbol{\theta} \cdot \boldsymbol{J}, \mathfrak{M}_\alpha\right] = \frac{i}{2}\left[\boldsymbol{\theta} \cdot \boldsymbol{\Sigma}, \mathfrak{M}_\alpha\right]. \tag{20.42}$$

If we now consider the meson in the rest frame so that $\not{v} = \gamma_0$, it is easy to see that $\delta M(x) = 0$, since both γ_0 and γ_5 commute with all components of $\boldsymbol{\Sigma}$. This shows that $\phi(x)$ indeed is a spinless field.

   □ **Exercise 20.13** From Eq. (20.42), show that, in the rest frame where $\not{v} = \gamma_0$, the change of the field $\boldsymbol{\Phi}(x)$ under rotation is given by

$$\delta\boldsymbol{\Phi}(x) = \boldsymbol{\theta} \times \boldsymbol{\Phi}(x), \tag{20.43}$$

which is the expected transformation property of a vector field.

   □ **Exercise 20.14** Show that the field $\mathfrak{M}_\alpha(x)$ satisfies the identities

$$\not{v}\mathfrak{M}_\alpha(x) = \mathfrak{M}_\alpha(x), \qquad \mathfrak{M}_\alpha(x)\not{v} = -\mathfrak{M}_\alpha(x). \tag{20.44}$$

[**Hint :** *For proving the second identity, one needs to use the fact that the spin-1 particle M_*^μ should have a polarization vector orthogonal to its momentum, as mentioned for example in Eq. (4.28, p 67). In the present case, we can write the condition as $v_\mu M_*^\mu = 0$, which implies $\not{M}_* \not{v} = -\not{v}\not{M}_*$.*]

For spin-$\frac{1}{2}$ baryons containing a single heavy quark, the formalism is much less complicated. The reason is that now the light degrees of freedom have total spin zero. So a baryon alone constitutes an irreducible representation of the heavy quark spin symmetry, and is described by a spinor field $\mathcal{B}_v(x)$ that satisfies the constraint

$$\not{v}\,\mathcal{B}_v(x) = \mathcal{B}_v(x)\,. \tag{20.45}$$

This equation is equivalent to the equation of motion for the spinors, as in Eq. (4.52, p 71). The spinor solutions of this equation, characterized by the velocity vector v^μ, can be normalized by the relation

$$\bar{u}(v,s)\gamma^\mu u(v,s) = 2v^\mu\,, \tag{20.46}$$

which is a special case of the Gordon identity, Eq. (F.123, p 752).

20.6.3   Consequences

There are many consequences of these heavy quark symmetries. We give a few examples here to convey an idea of the operations involved and of the results obtained.

a)   Decay of pseudoscalar and vector mesons

In §20.6.2, we mentioned that a pseudoscalar and a vector meson consisting of the same heavy quark and the same light antiquark belong to the same irreducible representation under the heavy quark spin symmetry. For example, both D^+ meson and D_*^+ meson have the quark content $c\hat{d}$. Decay rates of two such mesons should therefore be related in the heavy quark limit.

   In §17.6, we discussed how the decay constants for pseudoscalar as well as vector mesons can be defined. We will now show that if we take a pseudoscalar and a vector meson, both containing the same heavy quark Q and the same light antiquark $\hat{q}$, their decay constants are related by heavy quark spin symmetry. The hadronic part of the matrix element is of the form $\langle 0|J^\mu|\text{meson}\rangle$, where the meson is either the pseudoscalar M or the vector M^*. And the current is of the form $\bar{q}\Gamma^\lambda Q$, where Γ^λ can be either γ^λ or $\gamma^\lambda\gamma_5$. Note that we can write

$$\bar{q}\Gamma^\lambda Q = \text{tr}\left(\Gamma^\lambda Q\bar{q}\right)\,. \tag{20.47}$$

   Replacing the field operator of Q by the field Q_v in the heavy quark limit, we can contemplate how the current will look if we try to express it using the irreducible meson field introduced in Eq. (20.40). Whatever is the most general expression possible, it should contain only one factor of the field $\mathfrak{M}_\alpha$, and it should transform the same way under heavy quark spin symmetry. Of course, $\mathfrak{M}_\alpha$ itself transforms like $\mathfrak{Q}\bar{q}$ under this symmetry, so the most general possible form for the current would be

$$\bar{q}\Gamma^\lambda \mathfrak{Q}_\alpha = \text{tr}\left(X\Gamma^\lambda \mathfrak{M}_\alpha\right), \tag{20.48}$$

where X is invariant under the heavy quark spin symmetry. Thus, X can only contain α, and the most general form for it would be

$$X = a_0 \mathbb{1} + a_1 \not{\alpha}. \tag{20.49}$$

Recalling the form for the field $\mathfrak{M}_\alpha$ from Eq. (20.40), we find

$$\text{tr}\left(X\Gamma^\lambda \mathfrak{M}_\alpha\right) = (a_0 - a_1)\,\text{tr}\left(\Gamma^\lambda \mathfrak{M}_\alpha\right), \tag{20.50}$$

using cyclic property of trace and Eq. (20.44). Now note that

$$\text{tr}\left(\gamma^\lambda \mathfrak{M}_\alpha\right) = 2\Phi_\alpha^\lambda,$$
$$\text{tr}\left(\gamma^\lambda \gamma_5 \mathfrak{M}_\alpha\right) = -2i\alpha^\lambda \phi_\alpha. \tag{20.51}$$

Therefore, under heavy quark symmetry, we find that

$$\langle 0 |\bar{q}\gamma^\lambda \mathfrak{Q}| M^*(p,\epsilon)\rangle = 2\langle 0 |\Phi_\alpha^\lambda| M^*(p,\epsilon)\rangle = 2(a_0 - a_1)N\epsilon^\lambda,$$
$$\langle 0 |\bar{q}\gamma^\lambda \gamma_5 \mathfrak{Q}| M(p)\rangle = -2i\alpha^\lambda \langle 0 |\phi_\alpha| M(p)\rangle = -2i\alpha^\lambda (a_0 - a_1)N, \tag{20.52}$$

where N depends on the normalization on the states. Comparing these equations with the equations that define the decay constants for the mesons, and using $\alpha^\lambda = p^\lambda/m_M$ in the heavy quark limit, we find the relation

$$f_{M^*} = m_M f_M. \tag{20.53}$$

b)   The decay $\Lambda_c \to \Lambda e^+ \nu_e$

As mentioned earlier in this chapter, the Λ_c baryon has the quark content cud. The Λ, on the other hand, has an s quark in place of the c. Thus, at the quark level, the decay $\Lambda_c \to \Lambda e^+ \nu_e$ implies a process

$$c \to s e^+ \nu_e. \tag{20.54}$$

Only weak interactions can inflict such a process. In writing the Feynman amplitude, there is no trouble with the leptonic part. In fact, exactly a similar piece appeared when we wrote the amplitude for the decay rate of the charged pion. The problem lies with the hadronic part of the Feynman amplitude, where we encounter a matrix element of the form

$$\langle \Lambda(p') | \bar{s}\gamma^\mu (1 - \gamma_5) c | \Lambda_c(p) \rangle \,, \tag{20.55}$$

that comes from charged currents of weak interactions. As we saw in §17.6, the most general parametrization of the matrix elements involves six form factors.

Since the charm quark is a heavy quark, let us check whether the heavy quark symmetries can bring about any simplification. We replace the field of the charm quark by $c_{\mathscr{u}}$, as described earlier in this section. The matrix elements should now be of the form

$$\langle \Lambda(p') | \bar{s}\Gamma^\mu c | \Lambda_c(p) \rangle = \bar{u}' X \Gamma^\mu u \,, \tag{20.56}$$

where Γ^μ can be either γ^μ or $\gamma^\mu \gamma_5$, and X depends on the velocity 4-vector $\mathscr{u}^\mu$. The most general matrix that can be built up from the vector $\mathscr{u}^\mu$ is of the form

$$X = A + B\mathscr{\not{u}} \,. \tag{20.57}$$

It is thus easily seen that both f_1 and $\tilde{f}_1$ contain a contribution that is equal to A. As regards the B term, let us note that since the charm quark is very heavy compared to the other quarks in Λ_c, its momentum dominates the momentum of the entire hadron. In other words, $p^\mu \approx m_{\Lambda_c} \mathscr{u}^\mu$. Then, for example,

$$\bar{u}' \mathscr{\not{u}} \gamma^\mu u = \frac{1}{m_{\Lambda_c}} \bar{u}' \left(\not{p}' + \not{q} \right) \gamma^\mu u = \frac{1}{m_{\Lambda_c}} \bar{u}' \left(m_\Lambda \gamma^\mu + \not{q}\gamma^\mu \right) u \,, \tag{20.58}$$

using the Dirac equation for the Λ baryon. But

$$\not{q}\gamma^\mu = q_\alpha \gamma^\alpha \gamma^\mu = q_\alpha \left(g^{\alpha\mu} - i\sigma^{\alpha\mu} \right) \,. \tag{20.59}$$

There are similar equations involving an extra factor of γ_5. Putting these back into Eq. (20.56), we find that the six form factors introduced in Eqs. (17.70) and (17.71) are all given in terms of A and B as follows:

$$f_1 = \tilde{f}_1 = A + \frac{m_\Lambda}{m_{\Lambda_c}} B \,;$$
$$f_2 = f_3 = \tilde{f}_2 = \tilde{f}_3 = \frac{m_\Lambda}{m_{\Lambda_c}} B \,. \tag{20.60}$$

Needless to say, this is a huge simplification of the hadronic matrix element involved in the decay.

Chapter 21

CP violation

21.1   CP violation and complex parameters

We have discussed CP symmetry in §6.11. We found that if the only interactions in a model are the gauge interactions of fermions, CP cannot be
violated. This puts CP in a special position so far as the standard model is
concerned. Parity and charge conjugation are both broken by gauge interactions, as we have commented in Ch. 16, in particular through Ex. 16.9 *(p 470)*.
The combined symmetry CP cannot be broken in this manner.

In the standard model, there are other interactions, of course. There is a
scalar multiplet which has gauge interactions as well as self interactions. In
addition, there are Yukawa interactions between fermions and scalars.

Let us start with the Yukawa interactions and check if they can be CP-
violating. We start with a general form of such interactions, given by

$$\mathscr{L}_{\text{int}} = \overline{\psi}_1(a + b\gamma_5)\psi_2\phi + \overline{\psi}_2(a^* - b^*\gamma_5)\psi_1\phi^\dagger \,, \tag{21.1}$$

where the second term is the hermitian conjugate of the first. We now take
the CP conjugate of the first term. Using Eq. (6.165, *p 186*), we obtain

$$(\mathscr{C}\mathscr{P})\overline{\psi}_1(a + b\gamma_5)\psi_2\phi(\mathscr{C}\mathscr{P})^{-1} = \eta_{CP}^{(1)*}\eta_{CP}^{(2)}\eta_{CP}^{(\phi)}\overline{\psi}_2(a - b\gamma_5)\psi_1\phi^\dagger \,. \tag{21.2}$$

The Lagrangian of Eq. (21.1) will be invariant under CP if this CP conjugate
of the first term is equal to the second term appearing in that equation, i.e.,
if

$$\frac{a^*}{a} = \frac{b^*}{b} = \eta_{CP}^{(1)*}\eta_{CP}^{(2)}\eta_{CP}^{(\phi)} \,. \tag{21.3}$$

This shows that if a and b are real, we can choose the intrinsic CP phases of
all fermion and scalar fields to be 1 and then CP will be conserved. There
must be some complex parameter in the Yukawa couplings in order that CP
can be violated.

However, it must also be noted that presence of complex parameters, while
a necessary condition for CP violation, is not sufficient. For example, suppose

we have just two fermion fields in the model and their Yukawa interaction as given in Eq. (21.1). If a and b are complex but have the same phases, i.e., if $a^*/a = b^*/b$, we can take the intrinsic phase of the scalar field to be equal to this value and the intrinsic CP phases of the fermion fields to be 1, and satisfy Eq. (21.3) this way. One must therefore have complex coupling parameters with different phases in order that the Lagrangian is CP-violating.

We should warn the reader that the presence of complex parameters with different phases is a necessary condition for CP violation, but not sufficient. Some phases might be absorbed in the definitions of various fields. An example of this kind of redefinition will be discussed in §21.2.

We can now understand, in a different way, why gauge interactions do not violate CP. From the definition of the gauge coupling through the covariant derivative, it follows that the gauge coupling must be real. So, even gauge interactions involving scalar particles cannot violate CP.

Finally, there are self-interactions of the scalar multiplet in the standard model. Since there is only one scalar multiplet, it can be easily seen that the quartic coupling term $\lambda(\phi^\dagger\phi)^2$ implies that the coupling constant λ should be real. So, to summarize, we found that the Yukawa interactions must, somehow or other, be responsible for CP violation in the standard model.

21.2 Kobayashi–Maskawa theory of CP violation

For fermions in the first and the second generations, the Yukawa couplings are very small. For example, let us look at Eq. (16.43, p471). Since the vacuum expectation value of the Higgs field is 246 GeV and the electron mass is 0.51 MeV, the Yukawa coupling of the electron is about 2×10^{-6}. For quarks in the first generation, the number would be roughly ten times larger. With such small couplings, any effect of CP violation would be virtually impossible to observe.

Kobayashi and Maskawa realized that there is a way that the complex parameters of the Yukawa sector sneak into the gauge sector. The link between the two sectors is provided by the fact that the Yukawa couplings are responsible for the fermion masses in the standard model. In general, the mass terms connect any left-chiral quark field to any right-chiral one, so that mass matrices are not diagonal. The mass matrices can be diagonalized through unitary transformations. If now one writes the gauge interactions in terms of the eigenstates, we find that the charged current gauge interactions contains a unitary matrix, called the CKM matrix, which was introduced in §17.1.

In §6.11, we showed that CP is conserved if the coupling constants in the gauge interactions are all real. Now, we find that the gauge couplings involve the CKM matrix, which is a unitary matrix and can therefore contain complex elements. Therefore, the gauge currents can be CP-violating. However, we should also bear in mind the warning issued in §21.1, viz., that complex parameters are not necessarily CP-violating. Let us check whether the mixing matrix indeed contains any complex parameter that can violate CP.

In order to obtain maximum possible insight from this analysis, we do not fix the number of fermion generations at first. Let the number of generations be N. The CKM matrix is therefore an $N \times N$ unitary matrix. In §3.4.1, it was shown that such matrices contain, in general, N^2 real parameters. If only real values were allowed for the elements of the mixing matrix, the matrix would describe a rotation in an N-dimensional space, which requires $\frac{1}{2}N(N-1)$ parameters. The remaining parameters in the unitary matrix are therefore phases, and there are $\frac{1}{2}N(N+1)$ of them.

But not all of these phases have observable effects. The reason is that the fields involved in the interaction Lagrangian can be defined up to some arbitrary phase. Thus, by adjusting the phases of the u_A fields, we can absorb one phase from each row of the mixing matrix. Similarly, we can absorb one phase from each column by adjusting the phases of the down-type quark fields. It therefore seems that we can absorb $2N$ phases from the mixing matrix. But this is not really correct, because if the phases of all up-type and all down-type quarks are changed and changed by the same amount, there will be no effect on the mixing matrix. Counting this one exception out, we can now write the number of physically observable phases in the CKM matrix to be

$$\frac{1}{2}N(N+1) - (2N-1) = \frac{1}{2}(N-1)(N-2). \qquad (21.4)$$

This is an interesting result. It shows that if there were one or even two generations of fermions, gauge currents could not have violated CP. CP violation through gauge currents is possible only if the number of fermion generations is three or more. This was the crucial observation made by Kobayashi and Maskawa, at a time when there was no experimental signature for the third generation. Experimental proof of the existence of CP violation was obtained in 1964. In a sense, Kobayashi and Maskawa indicated that these observations can be explained if there is a third generation of quarks. With three generations, Eq. (21.4) tells us that there is only one CP-violating parameter in the mixing matrix. In §17.3, we discussed different ways of representing this one CP-violating phase, as well as the three mixing angles. The Kobayashi–Maskawa theory of CP violation assumes that all CP-violating effect observed in nature owe their origin to this one parameter in the mixing matrix.

☐ **Exercise 21.1** *There is only one CP-violating combination of the parameters in the Lagrangian of Eq. (21.1). Identify it.*

☐ **Exercise 21.2** *Redo Ex. 6.25 (p 184), this time allowing for the possibility that a and b might be complex. Verify that the tree-level result does not show any effect of CP violation, i.e., the CP-violating parameter identified in Ex. 21.1 does not appear in the result.*

21.3 Rephasing invariant formulation

While counting CP-violating parameters in §21.2, we mentioned that some of the phases occurring in the mixing matrix can be absorbed in the redefinition of the fields in the charged current interactions. Since these phases are unphysical, it would be elegant and helpful if we can build up the theory of CP violation explicitly in a form that does not depend on these phases.

The charged current interaction between quark fields was given in Eq. (17.18, *p 486*). Physical implications of this Lagrangian should not change if we inflict the following redefinition of the quark fields:

$$u_{A\mathbb{L}} \to e^{i\alpha_A} u_{A\mathbb{L}}, \qquad d_{A\mathbb{L}} \to e^{i\beta_A} d_{A\mathbb{L}}. \tag{21.5}$$

Clearly, under these redefinitions, the elements of the mixing matrix change by the rule

$$V_{AB} \to e^{i(\alpha_A - \beta_B)} V_{AB}. \tag{21.6}$$

The matrix V contains phases, which undergo changes because of this redefinition. Note that, according to our summation convention announced in §2.1, there is no implied summation if indices like A or B are repeated.

In principle one can think of a much more elaborate redefinition scheme, in which the different quark fields with the same electric charge are also mixed, i.e., transformations of the form

$$u_{A\mathbb{L}} \to \sum_B U_{AB} u_{B\mathbb{L}}, \tag{21.7}$$

and a similar one for the down-type quarks, where U is a unitary matrix. Such transformations should also not affect any physical consequence of the theory. However, general transformations like this are awkward because then one has to work with fields which are not eigenstates of the Hamiltonian, and also in the end one does not find anything more than what we are going to describe.

We should now look for combinations of the matrix elements of V which are invariant under phase transformations of the type shown in Eq. (21.6). The obvious ones are the absolute squares of the elements, $|V_{AB}|^2$. But these are all real, and therefore are not responsible for CP violation. Complex numbers can be obtained in quartic combinations of the form

$$T_{ABA'B'} \equiv V_{AB} V_{A'B'} V_{AB'}^* V_{A'B}^* \qquad \text{(no sum)}, \tag{21.8}$$

which are invariant under rephasing. Note that, although there are N^4 such combinations for N generations of quarks, not all of them are complex. If $A = A'$ or $B = B'$, the corresponding T is real. There are also properties like

$$T_{ABA'B'} = T_{AB'A'B}^* = T_{A'BAB'}^* = T_{A'B'AB}, \tag{21.9}$$

which cuts down on the number of independent complex numbers. Finally, there are also the transitive relations

$$T_{ABA'B'} T_{A'BA''B'} = \left| V_{A'B} V_{A'B'} \right|^2 T_{ABA''B'}. \tag{21.10}$$

Taking all these into account, it can be argued that all independent complex combinations can be taken as

$$T_{AB1N}, \qquad \text{with } A \le B, \ A \neq 1, \ B \neq N. \qquad (21.11)$$

It can be easily checked that the number of such combination is exactly equal to the number of CP-violating phases counted in Eq. (21.4). The imaginary parts of these combinations are rephasing invariant CP-violating parameters.

For three generations of quarks, as is known experimentally, there is only one such rephasing invariant. It is called the *Jarlskog invariant* and denoted by J. According to Eq. (21.11), we can write $J = \text{Im}(T_{2213})$. In the convention of Kobayashi and Maskawa, shown in Eq. (17.21, *p487*), we obtain

$$J = \text{Im}(T_{2213}) = \text{Im}\left(V_{22}V_{13}V_{23}^*V_{12}^*\right) = c_1 c_2 c_3 s_1^2 s_2 s_3 \sin\delta. \qquad (21.12)$$

As we said earlier, the mixing angles are all small, so that their cosines are close to 1. Thus, the CP-violating parameter turns out to be

$$J \approx s_1^2 s_2 s_3 \sin\delta. \qquad (21.13)$$

It is a small parameter because it contains four powers of sine functions of the different mixing angles. Putting in the experimentally known values of the parameters in the CKM matrix, we get

$$J \approx 3 \times 10^{-5}. \qquad (21.14)$$

This parameter will appear in the calculation of all CP-violating observables. This does not necessarily mean that all CP-violating phenomena are suppressed to the level of one part in 10^5. In order to obtain a suppression factor for CP violation, a CP-violating phenomenon has to be compared with a CP-conserving one. The CP-conserving observable might also be suppressed by some powers of the different sine functions appearing in the CKM matrix. The ratio of the two will vary from process to process. For example, we will see later in this chapter that in the definition of the neutral kaon eigenstates, there is a CP-violating parameter that is of order 10^{-3}, being a ratio of a CP-violating matrix element that is of order J and a CP-conserving matrix element that contains s_1^2. There can be even larger values of CP-violating parameters if the corresponding CP-conserving parameter is more suppressed.

☐ **Exercise 21.3** Verify that, if the mixing angles are defined through Eq. (17.22, *p487*), the Jarlskog invariant turns out to be $c_{12}s_{12}c_{23}s_{23}c_{13}^2s_{13} \sin\delta$.

☐ **Exercise 21.4** Verify that in the Wolfenstein parametrization of the CKM matrix, the Jarlskog invariant is given by $A^2\lambda^6\eta$.

The expression in Eq. (21.12) shows another interesting feature of CP violation. If any of the mixing angles vanishes, CP violation disappears. This is expected since with one vanishing angle, there is effectively mixing between the two remaining generations, and Eq. (21.4) shows that there cannot be any CP violation in this case.

It should be noted that if quarks from two different generations were degenerate for some reason, the mixing angle between those two generations would become meaningless. The same argument then applies, and implies that CP violation should vanish in this case. Therefore, some authors prefer to define the CP-violating parameter in the following way:

$$J' = F_u F_d J \,, \tag{21.15}$$

where J is the Jarlskog invariant given above, and

$$F_u = \frac{(m_c^2 - m_u^2)(m_t^2 - m_u^2)(m_t^2 - m_c^2)}{M_W^6} \,, \tag{21.16}$$

with a similar definition of F_d in terms of the down-type quark masses. The conceptual point can be appreciated, but this parameter will not be very useful in our subsequent analysis.

21.4 CP-violating decays of kaons

CP violation was first observed in neutral kaon decays. In this section, we discuss in some detail the phenomenology associated with such decays.

21.4.1 Observables and experimental results

In §17.9.2, we argued that if CP is conserved, the CP-odd combination of neutral kaon cannot decay into two pions. The CP-even combination can, and therefore it becomes short-lived. Clearly then, if one observes that the long-lived eigenstate experiences the decay mode $K_L^0 \to 2\pi$, that will signal CP violation. Indeed, this was how CP violation was first discovered by Fitch, Cronin, Christenson and Turley in 1964. Dimensionless measurables for CP-violating 2π decay modes can be taken to be the amplitude ratios

$$\eta_{+-} \equiv \frac{\langle \pi^+ \pi^- | \mathbb{T} | K_L^0 \rangle}{\langle \pi^+ \pi^- | \mathbb{T} | K_S^0 \rangle}, \qquad \eta_{00} \equiv \frac{\langle \pi^0 \pi^0 | \mathbb{T} | K_L^0 \rangle}{\langle \pi^0 \pi^0 | \mathbb{T} | K_S^0 \rangle}. \tag{21.17}$$

The symbol $\mathbb{T}$ appearing in the middle of each matrix element signifies the *T-matrix*, whose matrix element between two states give the Feynman amplitude of passing from one state to the other. Experimentally, one obtains the results

$$|\eta_{+-}| = (2.232 \pm 0.011) \times 10^{-3} \,,$$
$$|\eta_{00}| = (2.220 \pm 0.011) \times 10^{-3} \,,$$
$$\arg(\eta_{+-}) = (43.51 \pm 0.05)^\circ \,,$$
$$\arg(\eta_{00}) = (43.52 \pm 0.05)^\circ \,. \tag{21.18}$$

There can be other decays that will signal CP violation. For example, the K_L^0 decays to the channels $\pi^+\ell^-\widehat{\nu_\ell}$ and $\pi^-\ell^+\nu_\ell$ where ℓ is either the muon or the electron. These two final states are CP conjugates of each other. If CP is conserved, K_L^0 would be a CP eigenstate, and therefore its decay rate into these two semileptonic modes would be equal. Thus the ratios

$$\delta_{(\ell)} \equiv \frac{\Gamma(K_L^0 \to \pi^-\ell^+\nu_\ell) - \Gamma(K_L^0 \to \pi^+\ell^-\widehat{\nu_\ell})}{\Gamma(K_L^0 \to \pi^-\ell^+\nu_\ell) + \Gamma(K_L^0 \to \pi^+\ell^-\widehat{\nu_\ell})} \qquad (21.19)$$

would constitute measures of CP violation. Experiments show that

$$\delta_{(e)} \approx \delta_{(\mu)} = (3.32 \pm 0.06) \times 10^{-3} \,. \qquad (21.20)$$

21.4.2 Identifying sources of CP violation

In §17.9, we showed that, with the assumption of CPT invariance, the eigenstates of the Hamiltonian of the K^0-$\widehat{K}^0$ system are the combinations

$$\left|K_S^0\right\rangle = p\left|K^0\right\rangle + q\left|\widehat{K}^0\right\rangle \,, \qquad \left|K_L^0\right\rangle = p\left|K^0\right\rangle - q\left|\widehat{K}^0\right\rangle \,, \qquad (21.21)$$

normalized with the relation $|p|^2 + |q|^2 = 1$. The parameters p and q were defined in Eq. (17.89, *p 506*) through the effective Hamiltonian of the neutral kaon system.

We took the phase convention of the states $\left|K^0\right\rangle$ and $\left|\widehat{K}^0\right\rangle$ such that

$$\mathscr{CP}\left|K^0\right\rangle = \left|\widehat{K}^0\right\rangle \,. \qquad (21.22)$$

In this convention, we showed that CP conservation implies $p = q$. The relative phase between p and q can be changed by redefining the states $\left|K^0\right\rangle$ and $\left|\widehat{K}^0\right\rangle$, as discussed in Ex. 21.7 below. The absolute values cannot be changed this way, so that if we define the quantity

$$\delta \equiv |p|^2 - |q|^2 \,, \qquad (21.23)$$

any non-zero value of δ will be a signature of CP violation. This is called *CP violation from mixing* by some people, and *indirect CP violation* by some others.

It should be noticed that Eq. (21.21) implies $\left\langle K_S^0 \middle| K_L^0 \right\rangle = \delta$, i.e., the two eigenstates are not orthogonal if $\delta \neq 0$. They are not expected to be so, because the effective Hamiltonian of Eq. (17.89, *p 506*) does not commute with its hermitian conjugate unless $|p| = |q|$. In other words, for $|p| \neq |q|$, the effective Hamiltonian is not a normal matrix.

In general, for a matrix M one can define the right and left eigenstates by the equations

$$M\left|a\right\rangle = \alpha\left|a\right\rangle \,, \qquad \left\langle b\right| M = \beta\left\langle b\right| \,, \qquad (21.24)$$

where α and β are numbers. If a matrix is not normal, its left eigenstates are not the same as its right eigenstates. What is guaranteed, however, is that, the right and the left eigenstates can be made to satisfy orthogonality conditions of the form

$$\left\langle b_i \middle| a_j \right\rangle = \delta_{ij} \qquad (21.25)$$

by choosing proper normalization of the states.

□ **Exercise 21.5** Show that the Hamiltonian matrix given in Eq. (17.89, p 506) is a normal matrix only if $\delta = 0$, where δ is defined in Eq. (21.23).

□ **Exercise 21.6** Find the left eigenstates of the matrix $\mathbb{H}$ given in Eq. (17.89, p 506) and show that the orthogonality conditions of Eq. (21.25) are indeed satisfied.

□ **Exercise 21.7** Suppose, instead of taking the phase convention of Eq. (21.22), we choose the phases of the states such that

$$\mathscr{CP}\left|K^0\right\rangle = e^{i\varsigma}\left|\widehat{K}^0\right\rangle . \tag{21.26}$$

Show that this implies

$$\mathscr{CP}\left|\widehat{K}^0\right\rangle = e^{-i\varsigma}\left|K^0\right\rangle . \tag{21.27}$$

In this convention, show that CP invariance implies the relation

$$q = p\, e^{i\varsigma} . \tag{21.28}$$

There may be other sources of CP violation. To identify them, let us consider the decay amplitudes of the neutral kaon states to a certain final state f and to its CP conjugate state $\widehat{f}$. The amplitudes will in general be different depending on whether the initial state is $\left|K^0\right\rangle$ or $\left|\widehat{K}^0\right\rangle$. Let us denote these amplitudes by

$$A_f \equiv \left\langle f\left|\mathbb{T}\right|K^0\right\rangle , \qquad \widehat{A}_f = \left\langle f\left|\mathbb{T}\right|\widehat{K}^0\right\rangle ,$$
$$A_{\widehat{f}} \equiv \left\langle \widehat{f}\left|\mathbb{T}\right|K^0\right\rangle , \qquad \widehat{A}_{\widehat{f}} = \left\langle \widehat{f}\left|\mathbb{T}\right|\widehat{K}^0\right\rangle . \tag{21.29}$$

If the Lagrangian is CP-conserving, the T-matrix will also be CP-conserving. From this, it is easy to deduce that if CP is conserved, then, in our phase convention implied in Eq. (21.22),

$$A_f = \widehat{A}_{\widehat{f}}, \qquad \widehat{A}_f = A_{\widehat{f}}. \tag{21.30}$$

More generally, without committing ourselves to any particular phase convention, we can write the consequences of CP conservation as

$$\left|A_f\right| = \left|\widehat{A}_{\widehat{f}}\right|, \qquad \left|\widehat{A}_f\right| = \left|A_{\widehat{f}}\right|. \tag{21.31}$$

If these relations are not satisfied by the amplitudes, then also CP will be violated. Some people call it *direct CP violation*, and some call it simply *CP violation from amplitudes*.

□ **Exercise 21.8** If we use the definition of the phases of the states as in Eq. (21.26), show that CP invariance implies the relations

$$A_f = e^{i\varsigma}\widehat{A}_{\widehat{f}}, \qquad A_{\widehat{f}} = e^{i\varsigma}\widehat{A}_f , \tag{21.32}$$

from which Eq. (21.31) follows. [**Hint :** *Follow the derivation of Eq. (17.96, p 507).*]

There is a very important difference between between direct and indirect CP violation. The indirect violation comes from the complexity of the off-diagonal element of the effective Hamiltonian matrix in the K^0-$\widehat{K}^0$ basis. Since K^0 has strangeness $S = +1$ and $\widehat{K}^0$ has the opposite, this kind of CP violation requires a $|\Delta S| = 2$ effective operator. On the other hand, since kaons are the lightest hadrons carrying strangeness, their decay products should all be non-strange, and therefore the transition matrix elements should induce $|\Delta S| = 1$. We will see later how such operators arise in the standard model.

In the direct sector, if we want to define convenient dimensionless parameters which signal CP violation, we should know what kind of final state we are talking about. We will have two kinds of final states in our mind: the 2π decays and the semileptonic $K_{\ell 3}$ decays, and we will take them up shortly.

Finally, we want to note that there is another possible way that CP can be violated in decays of mesons. Consider, for example, the ratio

$$\lambda_f \equiv \frac{q}{p} \frac{\widehat{A}_f}{A_f}, \tag{21.33}$$

where the final state f is a CP eigenstate. It is to be noted that such ratios have a very interesting property. Suppose we change the phase of the state $|\widehat{K}^0\rangle$ by an arbitrary amount α, without making any change in the phase of the state $|K^0\rangle$. Off-diagonal elements of the effective Hamiltonian matrix will now change. The elements would now be given by $p'^2 = e^{i\alpha} p^2$ and $q'^2 = e^{-i\alpha} q^2$. Thus, $q'/p' = e^{-i\alpha} q/p$. But the relative phases of A_f and $\widehat{A}_f$ would also change, and in fact would change exactly in a way that the effect is canceled in the ratio λ_f. Thus, λ_f is independent of the relative phase factor in the definition of the states $|K^0\rangle$ and $|\widehat{K}^0\rangle$, and there lies its importance.

The value of λ_f can therefore be obtained in any phase convention between $|K^0\rangle$ and $|\widehat{K}^0\rangle$. In the phase convention that we have been using all along, CP conservation implies Eqs. (17.96) and (21.30), i.e., $\lambda_f = 1$. It must be true then that $\lambda_f \neq 1$ is a signal of CP violation no matter what phase convention we use. Note that, for an arbitrary phase convention, absence of direct and indirect CP violation merely ensures that $|\lambda_f| = 1$. If $\lambda_f \neq 1$ although $|\lambda_f| = 1$, we say that it is a phenomenon of *interference CP violation*.

21.4.3   CP violation in semileptonic decays

It is easy to relate the leptonic CP asymmetry, defined in Eq. (21.19), to the parameters in the effective Hamiltonian of the neutral kaon sector. Notice that the $\Delta S = \Delta Q_H$ rule, proved in §17.7, tells us that

$$\mathfrak{a}(K^0 \to \pi^+ \ell^- \widehat{\nu}_\ell) = 0 , \qquad \mathfrak{a}(\widehat{K}^0 \to \pi^- \ell^+ \nu_\ell) = 0 . \tag{21.34}$$

Thus, the amplitudes for the decays involved can be written as

$$\mathfrak{a}(K_L^0 \to \pi^- \ell^+ \nu_\ell) = p\, \mathfrak{a}(K^0 \to \pi^- \ell^+ \nu_\ell)\,,$$
$$\mathfrak{a}(K_L^0 \to \pi^+ \ell^- \widehat{\nu}_\ell) = -q\, \mathfrak{a}(\widehat{K}^0 \to \pi^+ \ell^- \widehat{\nu}_\ell)\,, \tag{21.35}$$

because of the way K_L^0 is defined, Eq. (21.21). Further, note that CPT invariance would imply

$$\mathfrak{a}(K^0 \to \pi^- \ell^+ \nu_\ell) = \mathfrak{a}(\widehat{K}^0 \to \pi^+ \ell^- \widehat{\nu}_\ell)\,. \tag{21.36}$$

Thus we obtain, from the definitions in Eqs. (21.19) and (21.23), the result

$$\delta_{(\ell)} = \frac{|p|^2 - |q|^2}{|p|^2 + |q|^2} = \delta\,. \tag{21.37}$$

Measurement of this asymmetry, Eq. (21.20), tells us right away that the indirect CP violation is small compared to usual weak interaction strengths which guides each of the decays involved in defining the asymmetry parameter $\delta_{(\ell)}$.

21.4.4 CP violation in decays into two pions

a) The CP-violating parameters ϵ and ϵ'

Kaons and pions have no spin. So the two pions in the final state must be in a state of zero orbital angular momentum. This state is symmetric in an exchange of the two pions. Because of the generalized Pauli principle, the two pions should also be in a symmetric state with respect to isospin. Since pions have $I = 1$, the combination of two pions can have $I = 0, 1, 2$. Of these, the states $I = 1$ are antisymmetric, and hence are not allowed in this case because of Bose symmetry. There are, therefore, basically four transition matrix elements, with either K_L^0 or K_S^0 in the initial state, and the two different isospin states of two pions in the final state. We will denote the final states by $(\pi\pi)_I$, with $I = 0$ or 2. Of course, by $|(\pi\pi)_2\rangle$, we mean the state with $I = 2$ and $I_3 = 0$, because only this state will be electrically neutral and can be the final state in the decay of a neutral kaon.

Dimensionless quantities can be defined by taking ratios of matrix elements. There will be three independent ratios, and we take them as follows:

$$\omega \equiv \frac{\langle(\pi\pi)_2|\mathbf{T}|K_S^0\rangle}{\langle(\pi\pi)_0|\mathbf{T}|K_S^0\rangle}\,, \qquad \epsilon_I \equiv \frac{\langle(\pi\pi)_I|\mathbf{T}|K_L^0\rangle}{\langle(\pi\pi)_0|\mathbf{T}|K_S^0\rangle}\,, \tag{21.38}$$

with $I = 0$ or 2. The parameter ω quantifies the relative strength of transition to the $I = 2$ state with respect to the $I = 0$ state, and need not signal CP violation. The other two are both indicators of CP violation. The quantity ϵ_0 is often shortened to just ϵ, without any subscript.

Let us now see how these ratios can be expressed in terms of the quantities defined in Eq. (21.29). As argued earlier, the two-pion state is a CP eigenstate, so f and $\widehat{f}$ that appears in the general notation of Eq. (21.29) are identical in this case. However, because the two-pion state can be in two different isospin combinations, it is useful to define the matrix elements specific to isospin eigenstates, viz.,

$$A'_I = \left\langle (\pi\pi)_I \left| \mathbb{T} \right| K^0 \right\rangle , \qquad \widehat{A}'_I = \left\langle (\pi\pi)_I \left| \mathbb{T} \right| \widehat{K}^0 \right\rangle . \qquad (21.39)$$

CPT invariance would imply some relation between these amplitudes. To find it, we assume that the CPT transformation properties of K^0 and $\widehat{K}^0$ are given by Eq. (17.85, p 506). Then, using Eq. (7.45, p 196) which shows how matrix elements transform under CPT, we can write,

$$\left\langle (\pi\pi)_I \left| \mathbb{T} \right| K^0 \right\rangle = \left\langle \widehat{K}^0 \left| \mathbb{T} \right| \Theta(\pi\pi)_I \right\rangle , \qquad (21.40)$$

since the $\mathbb{T}$-matrix is hermitian, a property that follows from the hermiticity of the Lagrangian. Obviously we have used the notation

$$|\Theta(\pi\pi)_I\rangle \equiv \Theta \, |(\pi\pi)_I\rangle \qquad (21.41)$$

in writing the relation above. The question now is, what is this quantity?

We have already argued that a neutral two-pion state is an eigenstate of CP with eigenvalue $+1$. Therefore, the action of Θ, or CPT, on the state is the same as the action of the time-reversal operator. The time-reversal operator turns an outgoing state into an incoming state and vice versa. There is strong interaction between two pions, so that even with just two pions coming in, the state going out finally into infinity is not exactly the same state, but rather related by a phase factor:

$$\left| (\pi\pi)_I^{\text{out}} \right\rangle = e^{2i\delta_I} \left| (\pi\pi)_I^{\text{in}} \right\rangle . \qquad (21.42)$$

In Eq. (21.40), the two-pion state on the right hand side of the equation appears as an in-state whereas that on the left hand side in an out-state. More explicitly, we can write

$$\begin{aligned}
\left\langle (\pi\pi)_I^{\text{out}} \left| \mathbb{T} \right| K^0 \right\rangle &= \left\langle \widehat{K}^0 \left| \mathbb{T} \right| (\pi\pi)_I^{\text{in}} \right\rangle \\
&= e^{2i\delta_I} \left\langle \widehat{K}^0 \left| \mathbb{T} \right| (\pi\pi)_I^{\text{out}} \right\rangle \\
&= e^{2i\delta_I} \left\langle (\pi\pi)_I^{\text{out}} \left| \mathbb{T} \right| \widehat{K}^0 \right\rangle^* .
\end{aligned} \qquad (21.43)$$

It is the out-state of the two pions that is relevant in the definition of Eq. (21.39). So Eq. (21.43) tells us that the consequence of CPT invariance is given by

$$A'_I = e^{2i\delta_I} \widehat{A}'^*_I . \qquad (21.44)$$

There is a notationally neater way of saying the same thing. We define

$$A_I \equiv e^{-i\delta_I} A'_I , \qquad \widehat{A}_I \equiv e^{-i\delta_I} \widehat{A}'_I , \qquad (21.45)$$

so that the condition for CPT invariance reads

$$A_I^* = \widehat{A}_I \, . \tag{21.46}$$

In a sense, this notation has also some conceptual advantage because the matrix elements given in Eq. (21.39) are now denoted by

$$\langle (\pi\pi)_I \, | \mathbf{T} | \, K^0 \rangle = A_I e^{i\delta_I} \, , \qquad \langle (\pi\pi)_I \, | \mathbf{T} | \, \widehat{K}^0 \rangle = \widehat{A}_I e^{i\delta_I} \, . \tag{21.47}$$

This, by the way, is not a way to express a complex number, the amplitude, by its magnitude and phase. In fact there is nothing so far in our discussion that tells us that even these newly defined objects, A_I and $\widehat{A}_I$, are not complex. They are, in general. Eq. (21.45) merely separates out a part of the phase of the amplitude. For two-pion elastic scattering in which we have two pions both in the initial and in the final states, the phase shift is $2\delta_I$, as defined in Eq. (21.42). Here we are discussing a process in which the two pions appear only in the final state. The phase shift due to the final-state interactions between these pions is half the phase shift that appears in Eq. (21.42), which is what has been separated out in the definitions of the amplitudes in Eq. (21.47), which allows us to deal with the rest of the amplitude much more easily.

☐ **Exercise 21.9** With the definitions of the CPT and CP transformation properties given in Eqs. (17.85) and (17.93), show that both A_I and $\widehat{A}_I$ should be real if CP and CPT are both conserved.

☐ **Exercise 21.10** Eq. (21.46) expresses the constraint of CPT invariance in the phase convention chosen in Eq. (17.85, p 506). More generally, one can take

$$\Theta \, | K^0 \rangle = e^{i\nu} \, | \widehat{K}^0 \rangle \, , \tag{21.48}$$

where $\Theta \equiv \mathscr{CPT}$. Show that this implies

$$\Theta \, | \widehat{K}^0 \rangle = e^{i\nu} \, | K^0 \rangle \, . \tag{21.49}$$

In this convention, find the equivalent of Eq. (21.46).

☐ **Exercise 21.11** Show that, irrespective of the phase factor defined in Eq. (21.48), the following relations hold if CPT invariance holds:

$$A_0^* A_2 = \widehat{A}_0 \widehat{A}_2^* \, , \qquad A_0 \widehat{A}_2^* = \widehat{A}_0^* A_2 \, , \tag{21.50}$$

and, of course, $|A_I| = |\widehat{A}_I|$.

Using Eq. (21.21), we can then immediately write

$$\langle (\pi\pi)_I \, | \mathbf{T} | \, K_L^0 \rangle = (p A_I - q \widehat{A}_I) e^{i\delta_I} \, ,$$
$$\langle (\pi\pi)_I \, | \mathbf{T} | \, K_S^0 \rangle = (p A_I + q \widehat{A}_I) e^{i\delta_I} \, . \tag{21.51}$$

Hence, using the shorthand

$$\delta_S \equiv \delta_2 - \delta_0, \tag{21.52}$$

we obtain from the definitions given in Eq. (21.38) the following relations:

$$\epsilon \equiv \epsilon_0 = \frac{1 - \lambda_0}{1 + \lambda_0}, \qquad \epsilon_2 = \frac{1 - \lambda_2}{1 + \lambda_0}\frac{A_2}{A_0}e^{i\delta_S}, \tag{21.53}$$

and

$$\omega = \frac{1 + \lambda_2}{1 + \lambda_0}\frac{A_2}{A_0}e^{i\delta_S} \tag{21.54}$$

where the λ_I's are the quantities λ_f, defined in Eq. (21.33), for the two-pion final states, i.e.,

$$\lambda_I \equiv \frac{q}{p}\frac{\widehat{A}_I}{A_I}. \tag{21.55}$$

Arguments given earlier confirm that no matter how we define the phase of the state $\left|K^0\right\rangle$, we obtain the same value for the λ_I's as well as the ratio A_2/A_0. The difference of the phase shifts, denoted by δ_S, is a measurable parameter, free from any arbitrariness. Consequently, the definitions of ϵ_I and ω are also free from any phase convention.

In the next step, we note that the isospin states $\left|(\pi\pi)_0\right\rangle$ and $\left|(\pi\pi)_2\right\rangle$ are superpositions of the states $\left|\pi^+\pi^-\right\rangle$ and $\left|\pi^0\pi^0\right\rangle$, connected by the Clebsch–Gordan co-efficients. Inverting these equations, we can obtain equations of the type

$$\left|\pi^+\pi^-\right\rangle = a\left|(\pi\pi)_0\right\rangle + b\left|(\pi\pi)_2\right\rangle,$$
$$\left|\pi^0\pi^0\right\rangle = c\left|(\pi\pi)_0\right\rangle + d\left|(\pi\pi)_2\right\rangle. \tag{21.56}$$

for known co-efficients a, b, c, d. Then we can write

$$\left\langle\pi^+\pi^-\left|\mathbf{T}\right|K_L^0\right\rangle = a\left\langle(\pi\pi)_0\left|\mathbf{T}\right|K_L^0\right\rangle + b\left\langle(\pi\pi)_2\left|\mathbf{T}\right|K_L^0\right\rangle$$
$$= a(pA_0 - q\widehat{A}_0)e^{i\delta_0} + b(pA_2 - q\widehat{A}_2)e^{i\delta_2}, \tag{21.57}$$

and similarly other matrix elements that appear in the definitions of the quantities η_{+-} and η_{00}, defined in Eq. (21.17). With their help, we find

$$\eta_{+-} = \frac{a(pA_0 - q\widehat{A}_0) + b(pA_2 - q\widehat{A}_2)e^{i\delta_S}}{a(pA_0 + q\widehat{A}_0) + b(pA_2 + q\widehat{A}_2)e^{i\delta_S}}$$
$$= \frac{a(pA_0 - q\widehat{A}_0) + b(pA_2 - q\widehat{A}_2)e^{i\delta_S}}{(a + b\omega)(pA_0 + q\widehat{A}_0)}$$
$$= \frac{a}{a + b\omega}\epsilon + \frac{b}{a + b\omega}\epsilon_2. \tag{21.58}$$

Now comes another important point. Kaons have $I = \frac{1}{2}$. Thus, going to the $I = 0$ two-pion state involves $|\Delta I| = \frac{1}{2}$, whereas going to $I = 2$ state involves $|\Delta I| = \frac{3}{2}$. Standard model interactions can easily inflict the former kind of transitions because an s quark, for example, can change to a u quark through charged current interaction. But $|\Delta I| = \frac{3}{2}$ has to be more complicated, and therefore suppressed. So both ϵ_2 and ω should be much smaller than unity. Thus we can write

$$\frac{1}{a+b\omega} = \frac{1}{a}\left(1+\frac{b}{a}\omega\right)^{-1} \approx \frac{1}{a}\left(1-\frac{b}{a}\omega\right) \tag{21.59}$$

and neglect terms which involve $\epsilon_2\omega$. This allows us to write

$$\eta_{+-} = \epsilon + \frac{b}{a}(\epsilon_2 - \epsilon\omega). \tag{21.60}$$

We can write it as

$$\eta_{+-} = \epsilon + \epsilon' \tag{21.61}$$

by defining a new parameter ϵ'. The ratio η_{00} would also involve the combination $(\epsilon_2 - \epsilon\omega)$, and will therefore have a term proportional to ϵ'. Calculation of η_{00}, keeping the correct Clebsch–Gordan co-efficients, yields

$$\eta_{00} = \epsilon - 2\epsilon'. \tag{21.62}$$

□ **Exercise 21.12** *Clebsch–Gordan co-efficients tables give*

$$|(\pi\pi)_2\rangle = \frac{1}{\sqrt{6}}\left(|(\pi^+\pi^-)\rangle - 2|\pi^0\pi^0\rangle + |(\pi^-\pi^+)\rangle\right),$$

$$|(\pi\pi)_0\rangle = \frac{1}{\sqrt{3}}\left(|(\pi^+\pi^-)\rangle + |\pi^0\pi^0\rangle + |(\pi^-\pi^+)\rangle\right). \tag{21.63}$$

Note that here distinction has been made between $\left|(\pi^+\pi^-)\right\rangle$ and $\left|(\pi^-\pi^+)\right\rangle$, where the first one means that some arbitrarily assigned particle 1 is π^+ and the other one is π^-, where in the ket $\left|(\pi^-\pi^+)\right\rangle$ the tags are just the opposite. Experimentally, there is no difference between the two states: you get one π^+ and one π^-. So it is best to use

$$|\pi^+\pi^-\rangle \equiv \frac{1}{\sqrt{2}}\left(|(\pi^+\pi^-)\rangle + |(\pi^-\pi^+)\rangle\right), \tag{21.64}$$

which does not refer to these arbitrary tags. Use this to find the values of the co-efficients a, b, c, d defined in Eq. (21.56) and hence derive Eq. (21.62) from a definition of ϵ' that fixes Eq. (21.61).

b)   Experimental determination of parameters

We find that the CP-violating observables can be expressed in terms of ϵ and ϵ'. We note from Eqs. (21.61) and (21.62) that the quantities η_{+-} and η_{00} would be equal if $\epsilon' = 0$. Indeed, Eq. (21.18) tells us that these two quantities differ by only a very small amount, indicating that $|\epsilon'| \ll |\epsilon|$. Thus, taking only first order effects in the ratio ϵ'/ϵ, we can write

$$|\eta_{+-}| = |\epsilon|\left(1 + \mathrm{Re}(\epsilon'/\epsilon)\right),$$
$$|\eta_{00}| = |\epsilon|\left(1 - 2\,\mathrm{Re}(\epsilon'/\epsilon)\right). \tag{21.65}$$

Comparing these expressions with the experimental values given in Eq. (21.18), one obtains

$$|\epsilon| = (2.228 \pm 0.011) \times 10^{-3},$$
$$\mathrm{Re}(\epsilon'/\epsilon) = (1.66 \pm 0.23) \times 10^{-3}. \tag{21.66}$$

In the original experiment that discovered CP violation, only ϵ could be measured. In other words, the results were consistent with $\epsilon' = 0$ within the error bars. The value of ϵ'/ϵ has been measured much later, in the 1990's, with the increase of precision.

The phase of ϵ can be derived from experimentally measured rates by neglecting ϵ' once again. We will see presently that this implies neglecting direct CP violation, and in this case the decay asymmetry defined in Eq. (21.23) is given by

$$\delta_{(\ell)} = \frac{2\,\mathrm{Re}\,\epsilon}{1 + |\epsilon|^2}. \tag{21.67}$$

From the measured value of $\delta_{(\ell)}$ given in Eq. (21.20), we therefore obtain $\mathrm{Re}\,\epsilon \approx 1.66 \times 10^{-3}$. Comparing this result with the value of $|\epsilon|$, one can obtain the value of $\arg(\epsilon)$. The value quoted in the literature, taking error bars into account, is

$$\arg(\epsilon) = (43.52 \pm 0.05)^\circ. \tag{21.68}$$

☐ **Exercise 21.13** *In absence of direct CP violation when Eq. (21.30) holds, show that*

$$\frac{q}{p} = \frac{1 - \epsilon}{1 + \epsilon}, \tag{21.69}$$

and use it to derive Eq. (21.67), using the smallness of ϵ.

It is also important to discuss the values of the CP conserving parameters which are relevant for two-pion decays of neutral kaons. If CP were conserved, the amplitudes A_I would have been real, as noted in Ex. 21.9 *(p 631)*. Their values could then be estimated from the CP-conserving decay $K_S \to \pi\pi$. The general formula for two-body decays, given in Eq. (4.168, p 97), gives

$$\Gamma(K_S^0 \to \pi\pi) = \frac{1}{16\pi m_K}\sqrt{1 - \frac{4m_\pi^2}{m_K^2}}\left|\langle \pi\pi \left| \mathbb{T} \right| K_S^0 \rangle\right|^2, \qquad (21.70)$$

where the final state, denoted by $\pi\pi$, can contain either charged pions or neutral pions. For the present purpose, we can ignore CP-violating effects and take K_S to be the CP-even state defined in Eq. (17.97, p 507). Then,

$$\langle \pi\pi \left| \mathbb{T} \right| K_S^0 \rangle = \frac{1}{\sqrt{2}}\left(\langle \pi\pi \left| \mathbb{T} \right| K^0 \rangle + \langle \pi\pi \left| \mathbb{T} \right| \widehat{K}^0 \rangle\right). \qquad (21.71)$$

Using the Clebsch–Gordan co-efficients a, b, c, d defined in Eq. (21.56), we obtain

$$\langle \pi^+\pi^- \left| \mathbb{T} \right| K_S^0 \rangle = \frac{1}{\sqrt{2}}\left(a(A_0 + \widehat{A}_0)e^{i\delta_0} + b(A_2 + \widehat{A}_2)e^{i\delta_2}\right),$$

$$\langle \pi^0\pi^0 \left| \mathbb{T} \right| K_S^0 \rangle = \frac{1}{\sqrt{2}}\left(c(A_0 + \widehat{A}_0)e^{i\delta_0} + d(A_2 + \widehat{A}_2)e^{i\delta_2}\right). \qquad (21.72)$$

The quantities A_I and $\widehat{A}_I$ are related through Eq. (21.46) because of CPT invariance. Therefore, remembering the small mass difference between charged and neutral pions, we can write

$$\frac{\Gamma(K_S^0 \to \pi^+\pi^-)}{\Gamma(K_S^0 \to \pi^0\pi^0)} = \sqrt{\frac{m_K^2 - 4m_{\pi^+}^2}{m_K^2 - 4m_{\pi^0}^2}}\left|\frac{a\,\mathrm{Re}\,A_0 + be^{i\delta_S}\,\mathrm{Re}\,A_2}{c\,\mathrm{Re}\,A_0 + de^{i\delta_S}\,\mathrm{Re}\,A_2}\right|^2. \quad (21.73)$$

Recall that the parameter ω is not CP-violating. Thus, in its definition of Eq. (21.38), if we put the CP-conserving values $p = q = 1/\sqrt{2}$, we get

$$\omega = \frac{A_2 + \widehat{A}_2}{A_0 + \widehat{A}_0}e^{i\delta_S} = \frac{\mathrm{Re}\,A_2}{\mathrm{Re}\,A_0}e^{i\delta_S}. \qquad (21.74)$$

Therefore we can write

$$\frac{\Gamma(K_S^0 \to \pi^+\pi^-)}{\Gamma(K_S^0 \to \pi^0\pi^0)} = \sqrt{\frac{m_K^2 - 4m_{\pi^+}^2}{m_K^2 - 4m_{\pi^0}^2}}\left|\frac{a + b\omega}{c + d\omega}\right|^2. \qquad (21.75)$$

Eq. (21.74) tells us that, apart from small corrections that might come from CP violation, the phase of ω is equal to δ_S. This phase is measured in pion-pion elastic scattering, and its value is obtained to be

$$\delta_S = (-41.4 \pm 8.1)°. \qquad (21.76)$$

Now, using the experimentally measured branching ratios,

$$\mathscr{B}(K_S^0 \to \pi^+\pi^-) = (69.20 \pm 0.05)\%,$$
$$\mathscr{B}(K_S^0 \to \pi^0\pi^0) = (30.69 \pm 0.05)\%, \qquad (21.77)$$

we can find $|\omega|$. After that, we can use this value and the measured decay rate of K_S^0 decays, to obtain $|\operatorname{Re} A_0|$. Since CP violation is small, we can also say that $|A_0| \approx |\operatorname{Re} A_0|$.

□ **Exercise 21.14** Do the arithmetic. Show that

$$|\omega| = 0.045 \,, \tag{21.78}$$
$$|A_0| = 4.71 \times 10^{-4} \,\text{MeV} \,. \tag{21.79}$$

Use Table B.4 (p 721) for the masses of the kaons and pions.

c)　Identifying direct and indirect CP violation

Let us try to see whether it is direct or it is indirect CP violation that gives rise to ϵ and ϵ'. First, check ϵ', which is equal to $\epsilon_2 - \epsilon\omega$ apart from an overall factor. Using the definitions of these quantities and amplitudes of the form given in Eq. (21.51), we find

$$\epsilon_2 - \epsilon\omega = \frac{2pq(\widehat{A}_0 A_2 - A_0 \widehat{A}_2)e^{i\delta_S}}{(pA_0 + q\widehat{A}_0)^2} \,. \tag{21.80}$$

From Eq. (21.30) (or more generally from Eq. (21.32)), we see that this expression vanishes if there is no direct CP violation, irrespective of whether there is any other kind of CP violation. Therefore, ϵ' is a measure of direct CP violation.

Let us now look at ϵ. Even if direct CP violation vanishes, this quantity can still be non-zero provided there is indirect CP violation, i.e., $p \neq q$. It is therefore a measure of indirect CP violation. Of course, it can also contain effects of direct CP violation, but from the smallness of ϵ'/ϵ, it is clear that such effects must be small. The dominant contribution to ϵ comes from indirect CP violation.

d)　Indirect CP violation and the effective Hamiltonian

Indirect CP violation must be expressible in terms of the elements of the effective Hamiltonian matrix. For this, we note that the definition of p and q in Eq. (17.89, p 506) implies

$$\frac{q}{p} = \sqrt{\frac{\mathbb{H}_{21}}{\mathbb{H}_{12}}} = \sqrt{\frac{\mathbb{M}_{12}^* - \frac{i}{2}\Gamma_{12}^*}{\mathbb{M}_{12} - \frac{i}{2}\Gamma_{12}}} \,, \tag{21.81}$$

where in writing the last step, we have used the definition in Eq. (17.82, p 505), and the fact that both $\mathbb{M}$ and Γ are hermitian matrices.

Let us now write

$$\mathbb{M}_{12} - \frac{i}{2}\Gamma_{12} \equiv \mu_1 + i\mu_2 \,. \tag{21.82}$$

The break-up on the right hand side is not into real and imaginary parts.
Rather, we take

$$\mu_1 = \operatorname{Re} \mathbb{M}_{12} - \frac{i}{2} \operatorname{Re} \Gamma_{12}, \qquad \mu_2 = \operatorname{Im} \mathbb{M}_{12} - \frac{i}{2} \operatorname{Im} \Gamma_{12}, \qquad (21.83)$$

so that both μ_1 and μ_2 are in general complex. Clearly,

$$\mathbb{M}_{12}^* - \frac{i}{2} \Gamma_{12}^* = \mu_1 - i\mu_2. \qquad (21.84)$$

Note that the condition of CP invariance was the equality of the off-diagonal
elements of the effective Hamiltonian, Eq. (17.96, *p 507*). Since the matrices
$\mathbb{M}$ and Γ are hermitian by definition, this condition is obtained if both $\mathbb{M}_{12}$
and Γ_{12} are real, i.e., if μ_2 vanishes. Any non-zero value of μ_2 signals CP
violation. Since CP violation is small, we can write

$$\frac{q}{p} = \sqrt{\frac{1 - i\mu_2/\mu_1}{1 + i\mu_2/\mu_1}} \approx 1 - i\frac{\mu_2}{\mu_1}, \qquad (21.85)$$

taking the square root that produces the value unity in the CP-conserving
limit. Also, let us write

$$A_0 = |A_0| e^{i\xi_0}, \qquad (21.86)$$

so that, using the consequence of CPT conservation from Eq. (21.46), we can
write

$$\frac{\widehat{A}_0}{A_0} = e^{-2i\xi_0}. \qquad (21.87)$$

We now go back to Eq. (21.53). Since $\lambda_0 = 1$ if CP is conserved, and since
CP violation is small, we can write

$$\epsilon = \frac{1}{2}(1 - \lambda_0). \qquad (21.88)$$

Using the definition of λ_0 from Eq. (21.55), we now obtain

$$\epsilon = \frac{1}{2}\left(1 - (1 - i\frac{\mu_2}{\mu_1})(1 - 2i\xi_0)\right) \approx \frac{i}{2}\left(\frac{\mu_2}{\mu_1} + 2\xi_0\right). \qquad (21.89)$$

Using the definitions of μ_1 and μ_2, we can now write

$$\epsilon = \frac{i}{2}\left(\frac{\operatorname{Im} \mathbb{M}_{12} - \frac{i}{2} \operatorname{Im} \Gamma_{12}}{\operatorname{Re} \mathbb{M}_{12} - \frac{i}{2} \operatorname{Re} \Gamma_{12}} + 2\xi_0\right). \qquad (21.90)$$

Obviously, the ξ_0 term represents the contribution of direct CP violation to
ϵ. The rest is the contribution from indirect CP violation.

Putting in more phenomenological knowledge, we can simplify Eq. (21.90)
further. First, we should write the denominator in terms of observable quan-
tities. Ignoring CP violation, $\mathbb{M}_{12}$ and Γ_{12} are real, and the eigenstates of the
Hamiltonian are the following.

1. The CP-even state, $K_{(+)}^0 \approx K_S^0$, with eigenvalue $\mathbb{H}_{11} + \mathbb{H}_{12}$.

2. The CP-odd state, $K_{(-)}^0 \approx K_L^0$, with eigenvalue $\mathbb{H}_{11} - \mathbb{H}_{12}$.

The experimental numbers given in Eq. (17.101, p 509) and Eq. (17.128, p 516) show that the K_L^0 is the heavier state and also it has a smaller decay rate. Thus, if we define the absolute values of the mass difference and decay rate difference of the two eigenvalues by Δm_K and $\Delta \gamma_K$, we get

$$\mathrm{Re}\, \mathbb{M}_{12} = -\frac{1}{2}\Delta m_K\,, \qquad \mathrm{Re}\, \Gamma_{12} = \frac{1}{2}\Delta \Gamma_K\,. \tag{21.91}$$

Next we note that the experimental values quoted also show that

$$\Delta \Gamma_K \approx 2\Delta m_K\,. \tag{21.92}$$

Since the mass difference as well as the dominant decay modes come from CP-invariant physics, we can write

$$\mathrm{Re}\, \mathbb{M}_{12} - \frac{i}{2}\,\mathrm{Re}\, \Gamma_{12} = \frac{1}{2}\left(-\Delta m_K - \frac{i}{2}\Delta \Gamma_K\right) \approx -\frac{e^{i\pi/4}}{\sqrt{2}}\Delta m_K\,. \tag{21.93}$$

We now turn our attention to $\mathrm{Im}\, \Gamma_{12}$. For this, we look back at Eq. (17.84, p 505). For kaons, the contribution of the $I = 0$ two-pion intermediate state is so overwhelming compared to that of any other state that, as a first approximation, one can ignore all other intermediate states. With this assumption, one obtains

$$\frac{\Gamma_{12}}{\Gamma_{11}} = \frac{\langle K^0|H_w|(\pi\pi)_0\rangle\langle(\pi\pi)_0|H_w|\widehat{K}^0\rangle}{\langle K^0|H_w|(\pi\pi)_0\rangle\langle(\pi\pi)_0|H_w|K^0\rangle} = \frac{\widehat{A}_0}{A_0} = e^{-2i\xi_0}\,, \tag{21.94}$$

using Eq. (21.87) in the last step. To a first approximation, we can use the CP-positive combination of the neutral kaon fields, shown in Eq. (17.97, p 507), to evaluate Γ_{11}. Then we can use

$$\left|\langle(\pi\pi)_0|H_w|K^0\rangle\right|^2 \approx \frac{1}{2}\left|\langle(\pi\pi)_0|H_w|K_S^0\rangle\right|^2\,, \tag{21.95}$$

giving $\Gamma_{11} \approx \frac{1}{2}\Gamma_S \approx \frac{1}{2}\Delta\Gamma_K$, since $\Gamma_L \ll \Gamma_S$. So we obtain

$$\Gamma_{12} \approx \frac{1}{2}e^{-2i\xi_0}\Delta\Gamma_K\,. \tag{21.96}$$

Noting that ξ_0 must be small since CP violation is small, we can keep only up to the first order term in it and write

$$\mathrm{Im}\, \Gamma_{12} \approx -\xi_0\Delta\Gamma_K \approx -2\xi_0\Delta m_K\,. \tag{21.97}$$

Putting these things back into Eq. (21.90), we obtain

$$\epsilon = \frac{e^{i\pi/4}}{\sqrt{2}}\left(-\frac{\mathrm{Im}\, \mathbb{M}_{12}}{\Delta m_K} + \xi_0\right)\,. \tag{21.98}$$

From this, it seems that $\arg(\epsilon) = \pi/4$ assuming the combination in parentheses to be positive. This is a result of the relation between Δm_K and $\Delta\Gamma_K$ given in Eq. (21.92). Since that relation is only approximately obeyed by experimental data, the phase of ϵ is a little away from $\pi/4$, about 43.5°.

□ **Exercise 21.15** *Show that, if one does not assume Eq. (21.92), one obtains*

$$\epsilon = \frac{\exp\left(i\tan^{-1}r\right)}{\sqrt{1+r^2}}\left(-\frac{\mathrm{Im}\,M_{12}}{\Delta m_K} + \xi_0\right), \tag{21.99}$$

where

$$r = \frac{2\Delta m_K}{\Delta\Gamma_K}. \tag{21.100}$$

e) Direct CP-violating parameter

We can start from the expression of Eq. (21.80), and multiply it by b/a, where a and b were defined in Eq. (21.56), to obtain ϵ'. This gives

$$\epsilon' = \frac{\sqrt{2}pq(\widehat{A}_0 A_2 - A_0 \widehat{A}_2)e^{i\delta_S}}{(pA_0 + q\widehat{A}_0)^2} = \frac{\sqrt{2}e^{i\delta_S}}{(1+\lambda_0)^2}\frac{A_2}{A_0}(\lambda_0 - \lambda_2), \tag{21.101}$$

according to the definition of λ_I in Eq. (21.55). Note that the same definitions imply $A_2\lambda_2/A_0 = \widehat{A}_2\lambda_0/\widehat{A}_0$. We can use this relation to eliminate λ_2 and write

$$\epsilon' = \frac{\sqrt{2}\lambda_0 e^{i\delta_S}}{(1+\lambda_0)^2}\left(\frac{A_2}{A_0} - \frac{\widehat{A}_2}{\widehat{A}_0}\right) = \frac{\sqrt{2}\lambda_0 e^{i\delta_S}}{(1+\lambda_0)^2}\left(\frac{A_2}{A_0} - \frac{A_2^*}{A_0^*}\right), \tag{21.102}$$

using Eq. (21.50) in the last step. CP-violating effects are small, so we can take them only to first order. In other words, we substitute the CP conserving values for every factor that is non-zero even if CP is conserved. This gives

$$\epsilon' = \frac{ie^{i\delta_S}}{\sqrt{2}}\,\mathrm{Im}(A_2/A_0). \tag{21.103}$$

□ **Exercise 21.16** *Show that ω can be given through an expression quite similar to that in Eq. (21.103),*

$$\omega = e^{i\delta_S}\,\mathrm{Re}(A_2/A_0), \tag{21.104}$$

neglecting CP-violating effects.

We thus see that the phase of ϵ' is related to δ_S. As for the magnitude, we can write

$$\left|\frac{\epsilon'}{\epsilon}\right| = \frac{\mathrm{Im}(A_2 A_0^*)}{\sqrt{2}\,|\epsilon A_0^2|} = \frac{\mathrm{Im}\,A_2\,\mathrm{Re}\,A_0 - \mathrm{Re}\,A_2\,\mathrm{Im}\,A_0}{\sqrt{2}\,|\epsilon A_0^2|}. \tag{21.105}$$

Since CP violation is small, we should expect $|\mathrm{Im}\,A_I| \ll |\mathrm{Re}\,A_I|$, which implies that $\mathrm{Re}\,A_I \approx |A_I|$, so that we obtain

$$\left|\frac{\epsilon'}{\epsilon}\right| = \frac{\mathrm{Im}\,A_2 - |\omega|\,\mathrm{Im}\,A_0}{\sqrt{2}\,|\epsilon A_0|}. \tag{21.106}$$

This shows that the imaginary parts of the amplitudes are responsible for direct CP violation, in accordance with the result indicated in Ex. 21.9 *(p 631)*. Putting in the experimentally determined values of $|\omega|$, $|\epsilon|$ and $|A_0|$, we obtain

$$\left|\frac{\epsilon'}{\epsilon}\right| = \frac{\text{Im}\, A_2 - 0.045\,\text{Im}\, A_0}{1.48 \times 10^{-6}\,\text{MeV}}. \tag{21.107}$$

21.4.5 Standard model estimates

a) Estimating ϵ

From Eq. (21.98), we understand that the task of estimating ϵ reduces to the task of estimating $\text{Im}\, M_{12}$, since ξ_0 pertains to direct CP violation, which is much smaller than indirect CP violation. This result can be easily read off the calculations that we had performed in §17.9.4. In particular, let us look at Eq. (17.115, *p 514*). For the purpose of estimating the mass difference, we assumed the CKM matrix elements to be real. All we have to do now is to relax this assumption and take the imaginary part of the result. The terms involving m_u can still be neglected safely because of the extreme smallness of m_u. For the other terms, we use the fact that all mixing angles are small, so we put their cosines equal to unity. Then, in terms of the parametrization of the CKM matrix given in Eq. (17.21, *p 487*), we obtain

$$\text{Im}(V_{cs}^* V_{cd})^2 \approx 2s_1^2 s_2 s_3 s_\delta, \tag{21.108a}$$

$$\text{Im}(V_{ts}^* V_{td})^2 \approx 2s_1^2 s_2^2 s_3 s_\delta (s_2 + s_3 s_\delta), \tag{21.108b}$$

$$\text{Im}(V_{cs}^* V_{cd} V_{ts}^* V_{td}) \approx s_1^2 s_2 s_3 s_\delta, \tag{21.108c}$$

where $s_\delta \equiv \sin\delta$. The expression on the left hand side of Eq. (21.108c) is J, the Jarlskog parameter. Factoring it out from other terms as well, we can write

$$\text{Im} \sum_{A,B} \Lambda_A \Lambda_B r_{u_A} r_{u_B} f(r_{u_A}, r_{u_B}) = 2J\Bigg[r_c^2 f(r_c, r_c) - r_c r_t f(r_c, r_t)$$

$$- s_2(s_2 + s_3 c_\delta) r_t^2 f(r_t, r_t) \Bigg]. \tag{21.109}$$

As mentioned in §17.9.4, each term in the square bracket should in fact be multiplied with a strong interaction correction factor.

The important thing to note is that the expression in Eq. (21.109) has an overall factor of $s_1^2 s_2 s_3 s_\delta$. The real part of the same sum was discussed in §17.9.4, and was given to a very good approximation by

$$\text{Re} \sum_{A,B} \Lambda_A \Lambda_B r_{u_A} r_{u_B} f(r_{u_A}, r_{u_B}) \approx \Lambda_c^2 r_c^2 f(r_c, r_c)$$

$$\approx s_1^2 r_c^2 f(r_c, r_c). \tag{21.110}$$

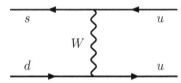
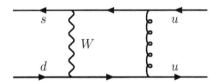

Figure 21.1: Tree diagram for $du \to su$ which contributes $K^0 \to 2\pi$ transition.

Figure 21.2: Example of a diagram that adds to the contribution of Fig. 21.1.

Therefore, from Eq. (21.98), we obtain

$$\epsilon \approx \frac{e^{i\pi/4}}{\sqrt{2}} s_2 s_3 s_\delta \left[1 - r_t f(r_c, r_t) - s_2(s_2 + s_3 c_\delta) \frac{r_t^2}{r_c} f(r_t, r_t) \right], \quad (21.111)$$

using $f(r_c, r_c) \approx 1/r_c$. This shows why CP violation is small. The CP-violating phase δ need not be small. The point is that the CP-violating parameter ϵ contains the product of two small angles. Because of this, the magnitude of ϵ is of order 10^{-3}, as shown in Eq. (21.66).

□ **Exercise 21.17** Using the Wolfenstein parametrization, show that the expression on the left hand side of Eq. (21.109) contains an overall factor $A^2 \lambda^6 \eta$.

b) Estimating ϵ'/ϵ

From Eq. (21.103), we see that the contributions to ϵ' come from the decay amplitudes, which have $|\Delta S| = 1$. At the quark level, this involves a $d + X \to s + X$ transition, where X stands from other quarks and antiquarks which are constituents of the final state pions.

The simplest diagram for a $dX \to sX$ type of transition has been shown in Fig. 21.1. This is, in fact, a $du \to su$ transition. If we neglect the momentum dependence in the W boson propagator, the amplitude of this diagram is given by

$$i\mathcal{M}_1 = \left(\frac{-ig}{\sqrt{2}} \right)^2 \frac{i}{M_W^2} V_{us}^* V_{ud} \left[\overline{s} \gamma^\mu \mathsf{L} u \right] \left[\overline{u} \gamma_\mu \mathsf{L} d \right], \quad (21.112)$$

or

$$\mathcal{M}_1 = -\frac{G_F}{\sqrt{2}} \Lambda_u \mathcal{O}_1, \quad (21.113)$$

where $\Lambda_u = V_{us}^* V_{ud}$, and $\mathcal{O}_1$ is the quark-level operator

$$\mathcal{O}_1 = \left[\overline{s} \gamma^\mu (1 - \gamma_5) u \right] \left[\overline{u} \gamma_\mu (1 - \gamma_5) d \right]. \quad (21.114)$$

However, the contribution of this simplest diagram to A_0 and A_2 cannot induce CP violation because the CKM matrix elements appearing in this amplitude are real, as seen from the parametrization of the CKM matrix given in Eq. (17.21, *p 487*) or Eq. (17.22, *p 487*). More generally, one can say that the co-efficient involves effects of only two generations, and that is not enough to inflict CP violation. So we need to look for other diagrams.

There are of course QCD corrections, through gluon exchanges, to the simple diagram. An example has been shown in Fig. 21.2. There are similar diagrams involving gluon exchange between other fermion lines. It should be noted that there is a qualitative difference between this diagram and the tree-level diagram given earlier. The point is that, in Fig. 21.1, the quark color cannot change on any of the two fermion lines. So, in Eq. (21.114), the color indices are the same for the quark fields within the same bilinear. In Fig. 21.2, the gluon can induce color change, so that the quark-level operator obtained from this diagram would contain the bilinears

$$\left[\bar{s}_\alpha \gamma^\mu (1 - \gamma_5)\lambda^a_{\alpha\beta} u_\beta\right]\left[\bar{u}_{\alpha'} \gamma_\mu (1 - \gamma_5)\lambda^a_{\alpha'\beta'} d_{\beta'}\right], \qquad (21.115)$$

where the primed and unprimed α, β are color indices, and the index a runs over the eight gluons. The λ^a's appear from the gluon vertices. They satisfy the completeness relation given in Eq. (10.18, *p 259*), which can be used to express the operator of Eq. (21.115) as $-\frac{2}{3}\mathcal{O}_1 + 2\mathcal{O}_2$, where

$$\mathcal{O}_2 = \left[\bar{s}_\alpha \gamma^\mu (1 - \gamma_5)u_\beta\right]\left[\bar{u}_\beta \gamma_\mu (1 - \gamma_5)d_\alpha\right]. \qquad (21.116)$$

However, this diagram and similar ones with gluon exchanges between other pairs of quark lines do not contribute to ϵ' either, for the same reason as that mentioned in context of Fig. 21.1.

There are other diagrams, of course. One such diagram is shown in Fig. 21.3a. These are called *penguin diagrams* because, with some small reorientations, the diagram can be made to look like a cartoon of a bird. [**Hint :** *Rotate the diagram clockwise by 90°, think of the d and s lines as the wings, the gluon line as the torso, and use some imagination.*] This also justifies the name of double-penguin diagrams shown in Fig. 17.7 *(p 517)*.

Note that the external fermions line attached to the gluon in Fig. 21.3 have been non-committally marked q, because it can be any flavor of quark. Secondly, there are self-energy diagrams shown in Fig. 21.3b which must be added with the penguin diagram in order to obtain finite results. The amplitude of these diagrams involves the following operators:

$$\mathcal{O}_3 = \left[\bar{s}\gamma^\mu (1 - \gamma_5)d\right]\sum_q \left[\bar{q}\gamma_\mu (1 - \gamma_5)q\right],$$

$$\mathcal{O}_4 = \left[\bar{s}_\alpha \gamma^\mu (1 - \gamma_5)d_\beta\right]\sum_q \left[\bar{q}_\beta \gamma_\mu (1 - \gamma_5)q_\alpha\right],$$

$$\mathcal{O}_5 = \left[\bar{s}\gamma^\mu (1 - \gamma_5)d\right]\sum_q \left[\bar{q}\gamma_\mu (1 + \gamma_5)q\right],$$

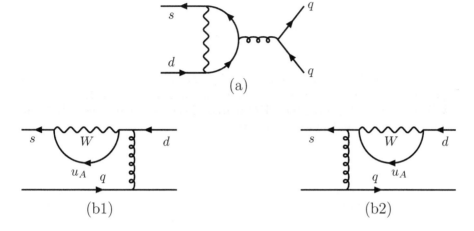

(a)

(b1)   (b2)

Figure 21.3: (a) Gluonic penguin diagram; (b) Self-energy diagrams with gluon exchange which should be added to the gluonic penguin diagram in order to obtain a gauge invariant result.

$$\mathcal{O}_6 = \left[\bar{s}_\alpha \gamma^\mu (1 - \gamma_5) d_\beta\right] \sum_q \left[\bar{q}_\beta \gamma_\mu (1 + \gamma_5) q_\alpha\right] . \qquad (21.117)$$

Note that now there are quark-level bilinears with $V + A$ structure as well. The reason is easily understood if one notes that the gluon couples to quarks vectorially, and a vector bilinear can be written as a sum of a $V - A$ and a $V + A$ bilinear. The penguin diagrams can contribute to CP-violating amplitudes, because the loop can involve up-type quarks from all three generations.

There are other diagrams where the gluon is replaced by a photon or a Z boson. These are called electroweak penguin diagrams. Since the electromagnetic coupling is proportional to the quark charge, and the Z boson coupling also has a term proportional to the quark charge, these diagrams introduce amplitudes with the following new operators:

$$\mathcal{O}_7 = \left[\bar{s}\gamma^\mu (1 - \gamma_5) d\right] \sum_q Q_q \left[\bar{q}\gamma_\mu (1 - \gamma_5) q\right] ,$$

$$\mathcal{O}_8 = \left[\bar{s}_\alpha \gamma^\mu (1 - \gamma_5) d_\beta\right] \sum_q Q_q \left[\bar{q}_\beta \gamma_\mu (1 - \gamma_5) q_\alpha\right] ,$$

$$\mathcal{O}_9 = \left[\bar{s}\gamma^\mu (1 - \gamma_5) d\right] \sum_q Q_q \left[\bar{q}\gamma_\mu (1 + \gamma_5) q\right] ,$$

$$\mathcal{O}_{10} = \left[\bar{s}_\alpha \gamma^\mu (1 - \gamma_5) d_\beta\right] \sum_q Q_q \left[\bar{q}_\beta \gamma_\mu (1 + \gamma_5) q_\alpha\right] , \qquad (21.118)$$

where the electric charge of the quark q is eQ_q.

Combining all contributions, we can write the matrix elements in the form

$$A_I e^{i\delta_I} = -\frac{G_F}{\sqrt{2}} \sum_{n=1}^{10} \left[\Lambda_u z_n^{(u)} - \Lambda_t z_n^{(t)} \right] \langle (\pi\pi)_I | \mathscr{O}_n | K^0 \rangle . \quad (21.119)$$

Note that the factor Λ_u appeared in Eq. (21.113) from the vertices. The quantity Λ_t occurs the same way. From the diagrams in Fig. 21.1 *(p 641)* as well as in Fig. 21.2 *(p 641)*, the CKM matrix elements would come only in the combination Λ_u, and so we can say that

$$z_1^{(t)} = z_2^{(t)} = 0 \qquad\qquad\qquad (21.120)$$

in Eq. (21.119). Contributions from all other diagrams would involve charm and top quarks in the loops, and would therefore contain terms proportional to Λ_c and Λ_t. However, $\Lambda_u + \Lambda_c + \Lambda_t = 0$ because of unitarity, which can be used to eliminate any one of the three. In writing Eq. (21.119), we have chosen to eliminate Λ_c.

The co-efficients z_n appearing in Eq. (21.119) contain, first of all, factors that come from the evaluation of the diagrams shown. Moreover, they also contain results from QCD corrections to those diagrams, obtained by adding extra gluon lines, which should be comparable in magnitude in the low energy limit where the QCD coupling constant is not small. Results of these corrections are incorporated in multiplicative constants which are called *Wilson co-efficients*. They are estimated through lattice calculations. We will not discuss the details of the procedure in this book.

By now it should be clear why the estimation of ϵ' is so much more difficult than that of ϵ. There are ten different quark level operators, and we will have to take their matrix elements between an initial state K^0 and a final state of two pions. The final state pions can be in two different states of total isospin, and the matrix elements would be different for the two states. There is some consolation in noting that

$$\langle (\pi\pi)_2 | \mathscr{O}_n | K^0 \rangle = 0 \qquad \text{for } n = 3, 4, 5, 6 , \qquad (21.121)$$

since the $q\widehat{q}$ pair produced in the QCD penguins is produced from a vertex with gluons, which can produce them only in an $I = 0$ state. Still, there is a huge number of hadronic matrix elements to be computed, compared to only one for the case of ϵ. And the difference is not just in the volume of work to be performed. As mentioned during the estimation of ϵ, the matrix elements of quark-level operators cannot be determined very reliably. So, if each matrix element is uncertain by some amount, the overall result must be very uncertain.

Despite these uncertainties, we should at least try to make some estimate and try to ascertain that $|\epsilon'|$ comes out to be much smaller than $|\epsilon|$, consistent with experimental results presented in Eq. (21.66). For this, it is convenient to rewrite Eq. (21.105) in the form

$$\left|\frac{\epsilon'}{\epsilon}\right| = \frac{\text{Im}\left(A_2\Lambda_u^*(A_0\Lambda_u^*)^*\right)}{\sqrt{2}\,|\epsilon A_0^2\Lambda_u^2|} \approx \frac{\text{Im}(A_2\Lambda_u^*) - |\omega|\,\text{Im}(A_0\Lambda_u^*)}{\sqrt{2}\,|\epsilon A_0\Lambda_u|}. \quad (21.122)$$

Then we multiply Eq. (21.119) by Λ_u^* and take the imaginary parts of both sides. On the right hand side, there will be some terms which will contain $\text{Im}\langle(\pi\pi)_I|\mathscr{O}_n|K^0\rangle$. These are the parts which give rise to the phase shift, and should be equated with the term involving $\sin\delta_I$ on the left hand side. The other terms give the CP-violating parts of the amplitude, and are as follows:

$$\text{Im}(A_I\Lambda_u^*)\cos\delta_I = \frac{G_F}{2|\epsilon A_0\Lambda_u|}\,\text{Im}(\Lambda_t\Lambda_u^*)\,\text{Re}\,\langle(\pi\pi)_I\,|\mathscr{O}_n|\,K^0\rangle. \quad (21.123)$$

Note that the Jarlskog parameter is given by

$$J = \text{Im}(\Lambda_t\Lambda_u^*) \quad (21.124)$$

according to the definition given in §21.3. So we obtain

$$\left|\frac{\epsilon'}{\epsilon}\right| = \frac{G_F J}{|\epsilon A_0\Lambda_u|}\sum_n z_n^{(t)}\left[\frac{\text{Re}\,\langle(\pi\pi)_2|\mathscr{O}_n|K^0\rangle}{\cos\delta_2} - \frac{|\omega|\,\text{Re}\,\langle(\pi\pi)_0|\mathscr{O}_n|K^0\rangle}{\cos\delta_0}\right]. \quad (21.125)$$

To obtain a rough estimate for the matrix elements, we use a factorization ansatz:

$$\langle\pi(k_1)\pi(k_2)\,|\mathscr{J}_1\mathscr{J}_2|\,K^0(p)\rangle = \langle\pi(k_1)\,|\mathscr{J}_1|\,0\rangle\,\langle\pi(k_2)\,|\mathscr{J}_2|\,K^0(p)\rangle, \quad (21.126)$$

where $\mathscr{J}_1$ and $\mathscr{J}_2$ are two bilinears of quark field operators, and the division of the operators $\mathscr{O}_n$ into $\mathscr{J}_1$ and $\mathscr{J}_2$ has to be done in all possible ways. For example, if we consider $\mathscr{O}_1$, we can write

$$\langle\pi(k_1)\pi(k_2)\,|\mathscr{O}_1|\,K^0(p)\rangle = \langle\pi(k_2)\,|\bar{s}\gamma^\mu(1-\gamma_5)u|\,K^0(p)\rangle$$
$$\times\,\langle\pi(k_1)\,|\bar{u}\gamma_\mu(1-\gamma_5)d|\,0\rangle. \quad (21.127)$$

These matrix elements can be written down in the general forms given in §17.6. The contributing parts of the matrix elements are

$$\langle\pi(k_1)\,|\bar{u}\gamma^\mu\gamma_5 d|\,0\rangle = \sqrt{2}if_\pi k_1^\mu,$$
$$\langle\pi(k_2)\,|\bar{s}\gamma_\mu u|\,K^0(p)\rangle = f_1 p_\mu + f_2 k_{2\mu}. \quad (21.128)$$

Using $p\cdot k_1 = \tfrac{1}{2}m_K^2$ and $k_1\cdot k_2 = \tfrac{1}{2}m_K^2 - m_\pi^2$, we can then write

$$\langle\pi(k_1)\pi(k_2)\,|\mathscr{O}_1|\,K^0(p)\rangle \sim f_\pi m_K^2. \quad (21.129)$$

Matrix elements of other operators are expected to be of the same order of magnitude. Folding in all factors together, we obtain

$$\left|\frac{\epsilon'}{\epsilon}\right| \lesssim \frac{G_F J}{|\epsilon A_0 \Lambda_u|} f_\pi m_K^2, \qquad (21.130)$$

ignoring the factors $z_n^{(t)}$ in this expression. There are estimates for the $z_n^{(t)}$'s, and they are all small. Hence the less than sign in the above expression.

Putting in the known values of the quantities that appear in this estimate, we obtain the value of $|\epsilon'/\epsilon|$ to be about 0.035. This already shows that the value should be small. As said earlier, this is as much as we can do without having a reliable way of estimating the matrix elements.

21.5 Other signals of CP violation

21.5.1 Decays of mesons involving heavier quarks

We have discussed in detail signatures of CP violation in neutral kaon decays. CP violation has been observed in decays of other mesons as well, and we discuss the main ideas here.

a) Similarities and dissimilarities with kaon sector

Neutral kaons are either $d\widehat{s}$ or $\widehat{d}s$. Similar mesons involving the bottom quark are called $B_q^0 \equiv \widehat{b}q$ or $\widehat{B}_q^0 \equiv b\widehat{q}$, where q means either the d quark or the s quark. Because of quark flavor violating amplitudes such as those which arise from the box diagrams analogous to the ones shown in Fig. 17.5 *(p 510)*, the physical eigenstates will be linear combinations of B_q^0 and $\widehat{B}_q^0$. If we neglect the effects of CP violation, one of the eigenstates will be CP-even, and the other CP-odd. A signature of CP violation would consist of two-pion decays of the eigenstate that is predominantly CP-odd.

The analysis is very similar to that for neutral kaon decays, and need not be repeated. We only point out the differences between the kaon case and the neutral B meson case. First, the eigenstates in the neutral kaon sector were earmarked by their decay rates, one of them being much longer lived than the other. It should be realized that such huge differences in decay rates occur in the neutral kaon sector because, barring the small effect of CP violation, the CP-odd eigenstate can decay only into three pions, and the masses of the pions and the kaons are such that the phase space is very small for this decay. This is a peculiarity of the kaons that is not expected to be present for more massive mesons. For B mesons which are much heavier, there are many channels available for decay, so the lifetimes for both eigenstates are comparable. For this reason, the eigenstates of the neutral B mesons are tagged not by their lifetimes, but by their masses. Thus, for the non-strange mesons, the eigenstates are called B_H^0 and B_L^0, where the subscript 'L' now stands for 'light' and not 'long-lived' as in the case for kaons, whereas the subscript 'H' signifies 'heavy'. Experimentally, one obtains

$$m(B_H^0) - m(B_L^0) = (3.337 \pm 0.033) \times 10^{-10}\,\text{MeV}\,. \qquad (21.131)$$

For the neutral B mesons carrying strangeness, the corresponding value is

$$m(B_{sH}^0) - m(B_{sL}^0) = (117.0 \pm 0.8) \times 10^{-10}\,\text{MeV}\,. \qquad (21.132)$$

☐ **Exercise 21.18** *From the experimental values given in Eqs. (17.128), (21.131) and (21.132), note that*

$$m(B_{sH}^0) - m(B_{sL}^0) \gg m(B_H^0) - m(B_L^0) \gg m(K_L^0) - m(K_S^0)\,.(21.133)$$

Explain this hierarchy qualitatively by looking at Eq. (17.126, p 516) and making educated guesses for similar equations for the other mesons.

The second characteristic of the kaon sector that need not be identical to other neutral mesons is the fact that the predominantly CP-odd eigenstate is heavier in mass. As shown in §17.9.4, the mass difference between two neutral eigenstates depends on a combination of elements of the CKM matrix. The sign of the mass difference will depend on the signs of these CKM elements.

Another characteristic of the kaon sector, already mentioned in connection with the first one, is that neutral kaons can decay hadronically only into pions. In case of heavier mesons, many other decay modes are possible, some of which, like the two-pion mode, consist of a CP eigenstate in the final state. Therefore, CP violation can also be studied by looking at the decay of the two neutral meson eigenstates into such channels. Examples of such channels will be given shortly.

The analysis of CP violation for different final states will have some obvious differences. For the case of kaon decay into two pions, we have considered tree diagrams as well as penguin diagrams while estimating the direct CP violation. Both type of diagrams need not exist for an arbitrary final state. In some cases, only one or the other type might exist. In some others, neither might exist, and CP violation would come only from the box diagrams. In Ex. 21.19, we give some examples of different kinds of processes.

☐ **Exercise 21.19** *For neutral kaons decaying into two pions, there are tree diagrams and penguin diagrams. For each of the following decays of the neutral meson B_d^0 (i.e., the bound state $\widehat{b}d$), identify the quark-level process and argue that only the mentioned combinations of tree and qcd-penguin diagrams are possible.*

    a) $B_d^0 \to K^+\pi^-$: Both tree and qcd-penguin diagrams possible.

    b) $B_d^0 \to \widehat{D}^0\pi^0$: No penguin diagram, only tree diagrams.

    c) $B_d^0 \to K^0\widehat{K}^0$: No tree diagram, only qcd-penguin diagram.

    d) $B_d^0 \to K^-\pi^+$: Neither penguin nor tree diagram.

There can be similar studies involving the D mesons. The neutral D mesons contain $c\hat{u}$ and $\hat{c}u$. However, the effects are expected to be much smaller than for kaon or B meson systems. Let us try to understand the reason by considering the box diagrams. We recall that the expression of Eq. (21.109) for the kaon system has a sum over the two intermediate quark lines. The amplitude of the diagram, apart from overall constant factors and the function f that comes from loop integration, contains factors of the form $F_q^{(K)} \equiv V_{qd}V_{qs}^* m_q^2$ apart from where q is an up-type quark. In fact, there are two such factors, coming from the two internal quark lines. For the B meson system, similarly, there are two factors of the form $F_q^{(B)} \equiv V_{qd}V_{qb}^* m_q^2$. On the other hand, for the neutral D meson system, this factors are $F_q^{(D)} \equiv V_{cq}V_{uq}^* m_q^2$, where now q is a down-type quark, because these are the quarks now appearing in the loop. Let us now compare these quantities, using the Wolfenstein parametrization of the CKM matrix that was presented in Eq. (17.24, $p\,488$):

q belongs to	Approximate value of		
	$F_q^{(K)}$	$F_q^{(B)}$	$F_q^{(D)}$
1st generation	λm_d^2	$\lambda^3 m_u^2$	λm_d^2
2nd generation	λm_s^2	$\lambda^3 m_c^2$	λm_s^2
3rd generation	$\lambda^5 m_t^2$	$\lambda^3 m_t^2$	$\lambda^5 m_b^2$

$$(21.134)$$

Recall that λ is the sine of the Cabibbo angle, and its value is about 0.22. The masses of the various quarks have been given in §18.3 and §20.5. It is clearly seen that the values of F_q are much smaller for the D meson system compared to the K or B meson systems. Thus, according to known physics, the effects in the D meson system must be very small.

b)   Time-dependent CP violation effects

However, in B meson decays, CP violation need not be searched for in channels with only pions in the decay products. There can be other strategies for seeing CP-violating signals. One such strategy involves study of the decay of a neutral meson and its antiparticle to the same final state f which is an eigenstate of CP. In order to keep the notation as general as possible, we will denote the meson by M^0 and its antiparticle by $\widehat{M^0}$. Thus, M^0 can be either B^0 (i.e., B_d^0) or B_s^0. In a collision process, an M^0-$\widehat{M^0}$ pair is produced together. For example, an B^0-$\widehat{B}^0$ pair can be produced in the decay of an $\Upsilon(4S)$ meson. If one of the members of the pair is identified by observing its decay into a semi-leptonic mode, we know for sure what the other member is. For example, suppose we find that one member decays into a charged antilepton, ℓ^+, and something else, which we call X. The B^0 meson contains the antiquark $\hat{b}$, which can go to $\hat{c} + W^+$, and the W^+ can decay to ℓ^+ and a neutrino. If the meson contain the quark b, such decays would have produced a charged lepton, ℓ^-. Thus, if we obtain ℓ^+ in the decay mode of one member of the pair B^0-$\widehat{B}^0$, we can be sure that it is the B^0, and therefore the other member must be $\widehat{B}^0$. The evolution of this other member in time can then be studied. This procedure of identifying a particle, through the decay of another particle in a well-defined pair, is called *tagging*.

In order to develop a formalism and identify CP-violating parameters, we start from the definitions of the eigenstates:

$$\left|M_H^0\right\rangle = p\left|M^0\right\rangle + q\left|\widehat{M}^0\right\rangle ,$$
$$\left|M_L^0\right\rangle = p\left|M^0\right\rangle - q\left|\widehat{M}^0\right\rangle . \tag{21.135}$$

It resembles Eq. (21.21), though of course the parameters p and q will be different for different systems. We do not put explicit indications of the system in the notations for the parameters p and q.

Inverting these equations, we obtain

$$\left|M^0\right\rangle = \frac{1}{2p}\left(\left|M_H^0\right\rangle + \left|M_L^0\right\rangle\right) ,$$
$$\left|\widehat{M}^0\right\rangle = \frac{1}{2q}\left(\left|M_H^0\right\rangle - \left|M_L^0\right\rangle\right) . \tag{21.136}$$

Thus, if a particle is produced as M^0 at $t = 0$, after a time t its state will become

$$\left|M^0(t)\right\rangle = \frac{1}{2p}\left(e^{-im_H t - \gamma_H t}\left|M_H^0\right\rangle + e^{-im_L t - \gamma_L t}\left|M_L^0\right\rangle\right), \quad (21.137)$$

where m_H, m_L are the masses and γ_H, γ_L the lifetimes of the eigenstates. Using Eq. (21.135), this state can be rewritten in the form

$$\left|M^0(t)\right\rangle = \mathscr{R}_+(t)\left|M^0\right\rangle - \frac{q}{p}\mathscr{R}_-(t)\left|\widehat{M}^0\right\rangle , \tag{21.138}$$

where

$$\mathscr{R}_\pm = \frac{1}{2}\left(e^{-im_H t - \gamma_H t} \pm e^{-im_L t - \gamma_L t}\right) . \tag{21.139}$$

Therefore, for a certain final state f, the matrix element for transition from the state $M^0(t)$ would be given by

$$\left\langle f\left|\mathsf{T}\right|M^0(t)\right\rangle = \mathscr{R}_+(t)A_f - \frac{q}{p}\mathscr{R}_-(t)\widehat{A}_f , \tag{21.140}$$

with the obvious definitions for A_f and $\widehat{A}_f$. The similar equation for a state that originated as $\widehat{M}^0$ is given by

$$\left\langle f\left|\mathsf{T}\right|\widehat{M}^0(t)\right\rangle = \mathscr{R}_+(t)\widehat{A}_f - \frac{p}{q}\mathscr{R}_-(t)A_f . \tag{21.141}$$

The absolute square of these amplitudes contain the quantities

$$\left|\mathscr{R}_\pm\right|^2 = \frac{1}{2}e^{-\gamma t}\left[\cosh\frac{1}{2}\Delta\gamma\, t \pm \cos\Delta m t\right] ,$$
$$\mathscr{R}_\pm^*\mathscr{R}_\mp = -\frac{1}{2}e^{-\gamma t}\left[\sinh\frac{1}{2}\Delta\gamma\, t \pm i\sin\Delta m t\right] , \tag{21.142}$$

where

$$\gamma = \frac{1}{2}(\gamma_H + \gamma_L), \qquad \Delta\gamma = \gamma_H - \gamma_L. \qquad (21.143)$$

□ **Exercise 21.20** Verify Eq. (21.141).

For the rest of the discussion, we will assume that $\Delta\gamma = 0$, in view of the fact that the lifetimes are nearly equal, as commented earlier. Then,

$$\left| \langle f | \mathbb{T} | M^0(t) \rangle \right|^2 \propto |A_f|^2 \left[1 + C_f \cos\Delta mt - S_f \sin\Delta mt \right],$$

$$\left| \langle f | \mathbb{T} | \widehat{M}^0(t) \rangle \right|^2 \propto \frac{|\widehat{A}_f|^2}{|\lambda_f|^2} \left[1 - C_f \cos\Delta mt + S_f \sin\Delta mt \right], \qquad (21.144)$$

with the same constant of proportionality, and with

$$C_f = \frac{1 - |\lambda_f|^2}{1 + |\lambda_f|^2}, \qquad S_f = \frac{2\,\mathrm{Im}\,\lambda_f}{1 + |\lambda_f|^2}. \qquad (21.145)$$

Clearly, any non-zero value of either C_f or S_f signals CP violation, since CP conservation requires $\lambda_f = 1$, as argued in §21.4.2. The important point is that even if there is no indirect CP violation so that $|\widehat{A}_f| = |\lambda_f A_f|$, we can still obtain CP-violating effects. This is an example of interference CP violation. The magnitude of CP asymmetry in the decays will be given by

$$\mathscr{A}_{\mathrm{CP}} = \frac{\left| \langle f | \mathbb{T} | M^0(t) \rangle \right|^2 - \left| \langle f | \mathbb{T} | \widehat{M}^0(t) \rangle \right|^2}{\left| \langle f | \mathbb{T} | M^0(t) \rangle \right|^2 + \left| \langle f | \mathbb{T} | \widehat{M}^0(t) \rangle \right|^2}$$

$$= C_f \cos\Delta mt - S_f \sin\Delta mt. \qquad (21.146)$$

□ **Exercise 21.21** Evaluate the left hand sides of Eq. (21.144) and find the constant of proportionality that has been omitted in writing both expressions.

Evidences for such CP violation have been found for a large number of final states. An example of such final state is $\psi + K_S^0$. In absence of CP violation, K_S^0 is a CP-even eigenstate. The charmonium state ψ, as mentioned in Ex. 20.3 *(p 603)*, is a $^3S_1$ state of the charm quark and its antiquark, and is therefore CP-odd, according to the parity and charge conjugation properties of fermion-antifermion bound states expounded in §6.7. So the overall state is an eigenstate of CP, with eigenvalue -1. Measurements of CP-violating S or C parameters, for B^0 or $\widehat{B}^0$ decaying to this and other CP eigenstates, are summarized in Table 21.1. It is to be noted that, unlike the parameters δ, ϵ and ϵ' encountered in the kaon sector, these measures of CP violation are not very small.

Table 21.1: Time-dependent CP-violating effects in B^0 and $\widehat{B}^0$ decays. Note that the list is not exhaustive: only some sample channels are shown.

Final state	Parameter	Experimental value
ψK_S^0	S	0.679 ± 0.020
$\psi \pi^0$	S	-0.93 ± 0.15
$\pi^+ \pi^-$	C	-0.36 ± 0.06
$\eta' K_S^0$	S	0.59 ± 0.07
$D^+ D^-$	S	-0.98 ± 0.17

21.5.2   CP-violating correlations

Certain correlations between momentum and/or spin in various processes can signal CP violation. Obviously, one needs to consider correlations in two different processes involving the CP conjugates of each other. We have already discussed such an example in Ex. 7.6 *(p 198)* concerning the decay of μ^- and μ^+. The neutrinos and antineutrinos emitted in the decays cannot be detected, so we can only measure the spin of the decaying $\mu^\pm$ in their rest frame, and the spin and the momentum of the $e^\pm$ produced in the decay. With these measurable quantities, the most general parametrization of the differential decay rates of μ^- and μ^+ was written in Eq. (7.56, *p 198*). CP conservation implies some relations between the parameters. Since momentum changes sign under parity and angular momentum doesn't, CP transformation would convert an electron's 3-momentum to a positron's 3-momentum in the opposite direction. With such arguments, we can easily show that CP conservation would imply the relations

$$R_0^+ = R_0^-, \quad A^- = -A^+, \quad B^- = -B^+, \quad C^- = C^+, \quad D^- = -D^+,$$
$$(21.147)$$

for the parameters that appear in Eq. (7.56, *p 198*).

Of course this example involves leptons, and cannot occur at any appreciable level unless there is some source of CP violation in the leptonic sector, a topic that will be discussed in Ch. 22. Similar correlations can in principle be devised for processes involving hadrons. However, carrying out such an experiment to see any signature of CP violation is not easy since the correlations would involve polarized beams, and making such beams is not easy since the lifetimes of baryons are very short.

21.5.3   Electric dipole moment

In §6.9.3 and §7.7, we argued that the presence of non-zero electric dipole moment (EDM) of a fermion signals violation of parity and time-reversal

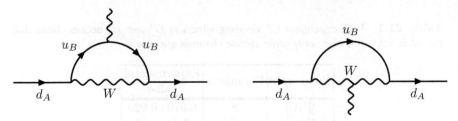

Figure 21.4: 1-loop diagram that generates an effective coupling of fermions with the photon, involving weak vertices in the loop.

symmetries. If CPT is assumed to be conserved, violation of time-reversal symmetry would be equivalent to CP violation. Therefore, a non-zero EDM of a particle can be taken as a signature of CP violation, provided of course that CPT is conserved.

Pure electromagnetic interactions do not violate CP. So, the electric dipole moment must come from diagrams that has some weak vertices as well. The simplest coupling with the photon that involves an internal weak vertex is the one-loop diagram shown in Fig. 21.4.

However, it is easy to see that this one-loop diagram cannot produce any CP-violating effect. The reason is simple. Imaginary part in the amplitude can come only from the CKM matrix elements. However, with the fermions shown in the figure, one vertex will contain a factor of V_{BA} whereas the other will have V_{BA}^*. Thus the amplitude will contain $|V_{BA}|^2$ and will be real. It has been argued that even two-loop diagrams cannot produce a CP-violating for factor assuming that the CKM matrix is the only source for CP violation. Thus, contributions to electric dipole moments can come from three-loop diagrams. This means that such contributions must be very much suppressed, not only because CP violation is small, but also by loop factors and the coupling constants that appear in these high-loop diagrams.

Of course we do not observe free quarks. But the point is that if the u or the d quark has an electric dipole moment, it can contribute to the electric dipole moment of the neutron. Very sensitive experiments have been performed to find the electric dipole moment of the neutron. They are all null experiments, and produce only an upper bound:

$$d_E^{(\text{neutron})} \leq 2.9 \times 10^{-26} \, e \, \text{cm} \,. \qquad (21.148)$$

However, because only high-order loops might be responsible for a non-zero value, the theoretical estimate is smaller than this upper bound by many orders of magnitude. So, unless there is some other source of CP violation, it does not look probable that the electric dipole moment of the neutron will be found experimentally in the near future.

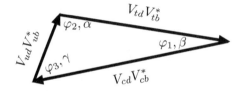

Figure 21.5: A schematic representation of the unitarity triangle. There are two kinds of notations used for the angles of the triangle. Both have been shown in the figure.

21.6   Unitarity triangle

There is another way of representing CP violation, which works well because there are three generations of quarks. Unitarity of the CKM matrix elements implies that different rows (or columns) of the matrix must be orthogonal. For example, we can write

$$V_{ud}V_{us}^* + V_{cd}V_{cs}^* + V_{td}V_{ts}^* = 0. \tag{21.149}$$

On the complex plane, if we draw arrows corresponding to the complex numbers $V_{ud}V_{us}^*$ etc, and arrange the tip of one arrow to coincide with the base of another, the three arrows will form a triangle. Such triangles are called *unitarity triangle*s.

   If there is no CP violation, all elements of the CKM matrix can be taken to be real. In this case, each of the three terms on the left hand side of Eq. (21.149) would be real, and the only way they can add up to zero is that one of them has the opposite sign compared to the other two, and has the magnitude equal to the sum of the other two. In other words, the name unitarity triangle would not have been appropriate in this case, because it would have been a collapsed triangle with vanishing area. Because of CP violation, this does not happen: the three sides of the triangle are really different, and the triangle encompasses a finite area. Thus, the area of the unitarity triangles is a measure of CP violation.

   Eq. (21.149), as it stands, is the statement of orthogonality of the first two rows of the CKM matrix. One can consider other pairs of rows, or even of columns, to obtain other relations of the same type. These will give different unitarity triangles. However, the area of all such triangles are equal. In fact, it can be easily shown that the area is half of the Jarlskog invariant introduced in Eq. (21.12).

   The shape of the unitarity triangle, as dictated by present data, has been shown in Fig. 21.5, using the orthogonality of the first and the third columns of the CKM matrix. There are two different notations for the angles that exist in the literature, and we present both, along with their expressions in terms of the CKM matrix elements in the Wolfenstein parametrization.

$$\beta \equiv \varphi_1 = \arg\left(-\frac{V_{cd}V_{cb}^*}{V_{td}V_{tb}^*}\right) \approx \arg\left(\frac{1}{1-\rho-i\eta}\right), \qquad (21.150a)$$

$$\alpha \equiv \varphi_2 = \arg\left(-\frac{V_{td}V_{tb}^*}{V_{ud}V_{ub}^*}\right) \approx \arg\left(-\frac{1-\rho-i\eta}{\rho+i\eta}\right). \qquad (21.150b)$$

$$\gamma \equiv \varphi_3 = \arg\left(-\frac{V_{ud}V_{ub}^*}{V_{cd}V_{cb}^*}\right) \approx \arg\left(\rho+i\eta\right). \qquad (21.150c)$$

Note that the angles do not depend on the parameters λ and A that appear in the Wolfenstein parametrization.

□ **Exercise 21.22** Show that, if we rescale and reorient the axes in the complex plane such that the vertices containing the angles φ_3 and φ_1 have co-ordinates $(0,0)$ and $(1,0)$ respectively, the co-ordinates of the vertex containing the angle φ_2 should be (ρ, η).

□ **Exercise 21.23** We have made a statement in the text that the area of any unitarity triangle is given by $\frac{1}{2}J$, where J is the Jarlskog invariant. Prove the statement.

The angles of the unitarity triangle can be estimated in a variety of ways. Take, for example, the angle φ_2. It can be estimated from the decay of B^0 (i.e., B_d^0) and $\widehat{B}^0$ to $\pi^+\pi^-$. By definition, the parameter λ for this decay will be given by

$$\lambda_{(B^0 \to \pi^+\pi^-)} = \frac{q}{p}\frac{\langle \pi^+\pi^-|\mathbb{T}|\widehat{B}^0\rangle}{\langle \pi^+\pi^-|\mathbb{T}|B^0\rangle}, \qquad (21.151)$$

where q and p are the parameters in the effective Hamiltonian matrix for the B^0-$\widehat{B}^0$ system. The value of this ratio can be read from Eq. (21.81). Recalling that $|\Gamma_{12}| \ll |\mathbb{M}_{12}|$ for the B system, we can write

$$\frac{q}{p} = \sqrt{\frac{\mathbb{M}_{12}^*}{\mathbb{M}_{12}}}. \qquad (21.152)$$

Contributions to $\mathbb{M}_{12}$ come, as for the kaon system, through box diagrams. For the B system, the box diagrams with internal t quark lines should dominate because of the huge mass of the top quark, and so the contribution should be proportional to $(V_{td}^*V_{tb})^2$. Hence $q/p \sim (V_{td}V_{tb}^*)/(V_{td}^*V_{tb})$. On the other hand, the decay amplitudes are dominated by tree diagrams. For example, the decay $\widehat{B}^0 \to \pi^+\pi^-$ involves $b \to d u\widehat{u}$ at the quark level, which occurs through W boson exchange with a factor $V_{ub}V_{ud}^*$. Similarly, the $B^0 \to \pi^+\pi^-$ comes with a factor $V_{ub}^*V_{ud}$. Combining all factors, we obtain

$$\lambda_{(B^0 \to \pi^+\pi^-)} \sim \frac{V_{td}V_{tb}^*}{V_{td}^*V_{tb}}\frac{V_{ub}V_{ud}^*}{V_{ub}^*V_{ud}} \approx \frac{V_{td}}{V_{td}^*}\frac{V_{ub}}{V_{td}^*}, \qquad (21.153)$$

using the fact that all diagonal elements of the CKM matrix are very close to unity. Thus,

$$\arg \lambda_{(B^0 \to \pi^+ \pi^-)} = \arg\left(\frac{V_{td}}{V_{ub}^*}\right) + \arg\left(\frac{V_{ub}}{V_{td}^*}\right) = 2\arg\left(\frac{V_{td}}{V_{ub}^*}\right), (21.154)$$

which is related to φ_2 since, if we put the diagonal CKM elements to be equal to unity, Eq. (21.150b) says that $\varphi_2 = \arg(-V_{td}/V_{ub}^*)$. Similarly, φ_1 can be estimated from the decays $B^0 \to \psi + K_S$ and $\widehat{B}^0 \to \psi + K_S$.

There is one reason why the unitarity triangles provide a very convenient way of detecting and measuring CP violation. The absolute values of the elements of the CKM matrix are obtainable from the rates of CP conserving processes. If we collect data of the absolute values of the elements that appear, e.g., in Eq. (21.149) and find the absolute values of the three terms in that equation, we obtain the lengths of the three sides of the triangle. From elementary geometry, we know that the sides of a triangle determine the triangle uniquely, apart from its orientation in a plane. In particular, the area of a triangle can be found from the length of its sides. Thus, information about CP violation can be obtained by measuring CP conserving processes only.

21.7 CP violation and T violation

We have always assumed CPT invariance in our analysis. With this assumption, CP violation is equivalent to time-reversal (or T) violation. Experimentally, however, one cannot take any principle to be sacrosanct. We should therefore be critical about whether results described in this chapter are measurements of CP violation really, or of T violation.

As already emphasized, presence of a non-zero electric dipole moment of any fermion is really a signature for T violation. On the other hand, results obtained from various decay processes are definitely signatures of CP violation: they do not say anything directly about T violation since it is not possible to perform an experiment in the opposite direction, where two or more particles come together and fuse into one particle.

Other tests of T violation have been suggested in analogy with the CP violation tests described in §21.5.1(b) involving time-dependent effects in B meson decays. In that section, we looked at the time evolution of the flavor states M^0 and $\widehat{M}^0$, which have well-defined bottom quantum number, i.e., contains either a bottom quark or its antiquark. The bottomness was determined by looking at the decay of its partner which was produced in the same reaction, and which must have the opposite value of bottomness.

While discussing CP violation in §21.5.1(b), we used tagging the mesons by their leptonic decay modes. Exactly in a similar way, we can tag by other properties as well. This realization opens up the possibility of a lot of other tests for checking different discrete symmetries. The idea is explained in Fig. 21.6 for T violation. Consider the left panel. At $t = t_0$, a B^0-$\widehat{B}^0$ pair

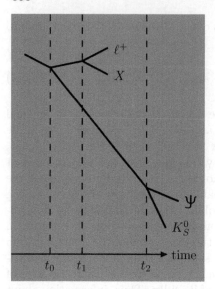 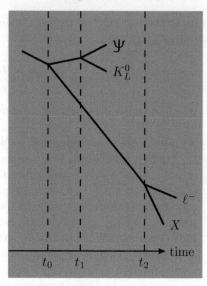

Figure 21.6: Schematic diagram for test of T violation. Time flows to the right of each diagram, as indicated by the arrow. The dashed lines correspond to some fixed values of time. Similar strategy can be employed for testing CP or CPT symmetries, as described in the text.

is produced, let us say from $\Upsilon(4S)$ decay. At time $t = t_1$, we see a decay to $\ell^+ X$, i.e., a charged antilepton and anything else. As explained in §21.5.1(b), this means that the decay occurs to a B^0. The other particle that survives is therefore a $\widehat{B}^0$. Let us suppose that we detect it finally at time $t = t_2$ through the decay channel ψK_S^0. As discussed earlier, apart from small CP-violating effects in the kaon sector, this decay product is an eigenstate of CP with eigenvalue -1. For this reason, we can call the B state that decays into it as $B_{(-)}$. This chain of events has been indicated by the entry of the first column of the first line in Table 21.2 *(p 657)*.

   If we are interested in CP violation, we would compare this process with the process where the decay at $t = t_1$ produced $\ell^- X$, the rest being equal. This was the discussion of §21.5.1(b). Now, if we are interested in T violation, we would look for the time-reversed process of a $B_{(-)}$ transforming into a $\widehat{B}^0$. In this case, we want to look at a process in which the decay at $t = t_1$ occurred for the state orthogonal to $B_{(-)}$, which is a CP-even eigenstate, barring small CP-violating effects in the kaon sector. Thus, it would decay into ψK_L^0. The surviving $B_{(-)}$, after traveling for a time $t_2 - t_1$, would have a $\widehat{B}^0$ component in it, which will be identified by the decay product $\ell^- X$. In a similar way, we can also test CPT symmetry by choosing suitable final states, as indicated in Table 21.2.

Table 21.2: Examples of different kinds of taggings required for tests of different discrete symmetries. In each case, we show in square brackets the decay products by which the first particle is tagged, and the decay product that identifies the final particle.

Original process	Conjugated process under		
	T	CP	CPT
$\widehat{B}^0 \;\; [\ell^+ X]$ $\downarrow$ $B_{(-)} \;\; [\psi K_S]$	$B_{(-)} \;\; [\psi K_L]$ $\downarrow$ $\widehat{B}^0 \;\; [\ell^- X]$	$B^0 \;\; [\ell^- X]$ $\downarrow$ $B_{(-)} \;\; [\psi K_S]$	$B_{(-)} \;\; [\psi K_S]$ $\downarrow$ $B^0 \;\; [\ell^- X]$
$B_{(+)} \;\; [\psi K_S]$ $\downarrow$ $B^0 \;\; [\ell^+ X]$	$B^0 \;\; [\ell^- X]$ $\downarrow$ $B_{(+)} \;\; [\psi K_L]$	$B_{(+)} \;\; [\psi K_S]$ $\downarrow$ $\widehat{B}^0 \;\; [\ell^- X]$	$\widehat{B}^0 \;\; [\ell^+ X]$ $\downarrow$ $B_{(+)} \;\; [\psi K_L]$

Following the analysis presented in §21.5.1(b), we can easily deduce the expression for matrix elements of the form $\langle B_{(\beta)}|\mathsf{T}|B_{(\alpha)}\rangle$, where $B_{(\alpha)}$ and $B_{(\beta)}$ are states like B^0, $\widehat{B}^0$, $B_{(+)}$, $B_{(-)}$, which appear in the argument given above. If we assume $\Delta\gamma = 0$, i.e., $\gamma_H = \gamma_L \equiv \gamma$, the general form of such matrix elements is given by

$$\langle B_{(\beta)}|\mathsf{T}|B_{(\alpha)}\rangle \propto e^{-\gamma t}\left[1 + C_{\alpha,\beta}\cos\Delta mt + S_{\alpha,\beta}\sin\Delta mt\right]. \quad (21.155)$$

One can find out the co-efficients $C_{\alpha,\beta}$ and $S_{\alpha,\beta}$ for different initial and final states by looking at the decay products. If the results for these co-efficients are different in a process and its T-conjugated process, that would indicate T violation. For example, if

$$C_{\widehat{B}^0,B_{(-)}} \neq C_{B_{(-)},\widehat{B}^0}, \quad (21.156)$$

it would signal T violation. Similar inequality with the S co-efficients would also signal the same thing. The *BaBar* collaboration has reported evidence of T violation through such data in 2012.

21.8   Strong CP problem

21.8.1   Effective θ-parameter

We have focused on weak interactions while discussing CP violation. The reason is simple: CP-violating effects are very small. In fact, they are weaker than usual weak interactions, as the smallness of CP-violating parameters like ϵ indicates. Naively, one should expect such effects to come out of weak interactions.

However, in §12.8, we pointed out that the QCD Lagrangian admits of a term that violates parity as well as time reversal. This term, shown in Eq. (12.204, *p 376*), can be written as

$$\mathscr{L}_{\text{new}} = \frac{\theta_{\text{QCD}} g_3^2}{32\pi^2} \mathsf{G}_{\mu\nu}^{a\dagger} \widetilde{\mathsf{G}}_a^{\mu\nu} , \tag{21.157}$$

where θ_{QCD} is a parameter, g_3 is the coupling constant of QCD and G denotes the QCD field-strength tensor. The factor $g_3^2/32\pi^2$ has been put in in view of Eq. (12.230, *p 381*): if the gauge fields have a fixed winding number, the integral of this new term would be θ_{QCD} times the winding number.

It is important to realize that one cannot get rid of this term by simply postulating that the QCD Lagrangian conserves parity or time reversal. The reason for this can be understood from our discussion of chiral symmetries in Ch. 18. In particular, in Eq. (18.125, *p 552*), we showed that, even if all quarks are massless, the axial U(1) current is not conserved. It has an anomaly given by

$$\partial_\mu \widetilde{J}^\mu = \frac{g_3^2 N_F}{16\pi^2} \mathsf{G}_{\mu\nu}^{a\dagger} \widetilde{\mathsf{G}}_a^{\mu\nu} , \tag{21.158}$$

where N_F is the number of quark flavors. Thus, if we perform an axial rotation by an amount β on each quark field, as shown in Eq. (18.80, *p 544*), the Lagrangian changes by an amount $(\beta/N_F)\partial_\mu \widetilde{J}^\mu$ according to Eq. (4.108, *p 82*), which means that a term like that in Eq. (21.157) will be induced by the axial transformation.

This is not merely an esoteric point. In §17.1, we described how, starting from doublets of the SU(2) part of the electroweak gauge group, a bi-unitary transformation is needed to obtain the physical quark fields. As explained there, a bi-unitary transformation means different transformations on the left-chiral and right-chiral fields, which must involve an axial transformation. In order to identify the U(1) part of the axial transformation involved in this process, let us go back to Eq. (17.15, *p 485*). As described in §3.3, the diagonalizing matrices U_L and U_R can be written in the form

$$U_C = U_C' X_C , \tag{21.159}$$

where $C = L, R$, with U_C' being a matrix of unit determinant and X_C a multiple of the unit matrix. The former matrix inflicts transformations on the flavor SU(N) group for N generations of fermions, whereas the latter one inflicts U(1) transformations. Clearly,

$$X_C = e^{i\Delta_C/N} \, \mathbb{1} , \tag{21.160}$$

where $e^{i\Delta_C}$ is the determinant of U_C. This means that the quantity β mentioned after Eq. (21.158) is $\Delta_R - \Delta_L$.

To express this amount using the mass matrix, we can use Eq. (21.159) to rewrite Eq. (17.13, *p 485*) in the form

$$M = U_L' X_L D X_R^\dagger U_R'^\dagger . \tag{21.161}$$

Since U'_L and U'_R have unit determinants, this implies

$$\det M = \det D \cdot \det X_L / \det X_R = \det D \cdot e^{i(\Delta_L - \Delta_R)}. \quad (21.162)$$

Recall that the matrix D is a diagonal matrix whose elements are the physical masses of the down-type quarks, so its determinant is real. Thus we obtain

$$\arg \det M = \Delta_L - \Delta_R, \quad (21.163)$$

which is the amount of chiral rotation on the quarks.

In the discussion so far, we have used formulas from §17.1 where we had assumed that we have taken a basis of the doublets such that the mass matrix of the up-type quarks is diagonal. More generally, we can start from an arbitrary basis where both up-type and down-type quarks have non-diagonal Yukawa couplings, then there will be an extra contribution to the chiral rotation coming from the up-type quark matrices. Denoting the up-type and down-type quark matrices by $M^{(u)}$ and $M^{(d)}$ in an arbitrary basis, we can write the chiral U(1) rotation amount as $\arg \det M^{(u)} + \arg \det M^{(d)}$, or, more succinctly, as $\arg \det(M^{(u)} M^{(d)})$. Adding this to the QCD contribution, we find that the effective value of the θ parameter is given by

$$\theta_{\text{eff}} = \theta_{\text{QCD}} + \arg \det(M^{(u)} M^{(d)}). \quad (21.164)$$

The mass matrices of the quarks have to be complex in order to accommodate weak CP violation, so the second term on the right hand side cannot be zero. Clearly, the effective θ cannot be zero in that case, because two effects coming from two very different sectors of the theory cannot possibly cancel each other. The effective θ parameter will give rise to CP-violating effects, which we discuss next.

21.8.2   Physical effects of θ

Let us consider the CP-violating effect which we have discussed in most detail in this chapter, viz., CP violation in the neutral kaons. Let us consider, e.g., the parameter ϵ corresponding to indirect CP violation, and ask ourselves the question: can it have a contribution from the θ parameter?

The answer is 'no', and the reason is the following. The parameter ϵ appears in the superposition of K^0 and $\widehat{K}^0$ in the mass eigenstates. Both K^0 and $\widehat{K}^0$ have the same parity properties, hence parity properties do not change because of the superposition. Thus, the superposition is CP-violating because it is C-violating. In short, the parameter ϵ violates C and T, but not P. The θ parameter, on the other hand, is the co-efficient of an operator that violates P and T, but not C.

Physical effects of θ must therefore be sought for in measurables which violate P and T, but not C. Such a parameter is the electric dipole moment (EDM) of a particle. The parameter θ can contribute to the EDM of strongly interacting particles.

Let us try to estimate the contribution to the EDM of a nucleon. To the lowest order, the contribution must be linear in θ. The contribution must also have one power of the QED coupling constant e because of the coupling to the photon. If chiral symmetry were exact in the Lagrangian, a chiral rotation would have been physically inconsequential, and so θ would have been an irrelevant parameter. However, if chiral symmetry were exact in the Lagrangian and spontaneously broken, the pion mass would have been zero. So, the explicit breakdown of chiral symmetry is connected to the value of m_π^2, and the contribution should involve this as a factor. Combining everything, we can write

$$d_{(N)} \sim \frac{e\theta_{\text{eff}} m_\pi^2}{m_N^3}, \qquad (21.165)$$

where m_N is the nucleon mass, which is the only factor that can make the dimensions right. The experimentally known upper bound on the EDM of neutron, given in Eq. (21.148), then implies that the effective θ parameter is constrained by

$$|\theta_{\text{eff}}| \lesssim 10^{-9}. \qquad (21.166)$$

This is the best empirical bound on the parameter.

21.8.3   Can θ be irrelevant?

Thus, if θ exists, its value will have to be very tiny. Very tiny parameters are always uncomfortable from a theoretical point of view, because it is hard to expect that two contributions from very different sectors of the theory could cancel to such a good accuracy. Besides, even if one insists on such a cancellation at the lowest level, it is not clear that higher order contributions would not destabilize it. Such problems are called *fine tuning* problems in a theory.

It would be helpful therefore if there is some symmetry that would guarantee the cancellation and will imply $\theta_{\text{eff}} = 0$ exactly. As mentioned in §21.8.2, chiral symmetry can do the job. However, the standard model does not have chiral symmetry: the Yukawa couplings do not obey this symmetry. However, instead of the standard model which has only one doublet of scalar fields, let us consider a model with two doublets called ϕ_u and ϕ_d, and Yukawa couplings given by

$$\mathscr{L}_Y = -\sum_{A,B} \left(h_{AB}^{(d)} \bar{q}_{A\mathbb{L}} \phi_d d_{B\mathbb{R}} + h_{AB}^{(u)} \bar{q}_{A\mathbb{L}} \tilde{\phi}_u u_{B\mathbb{R}} \right) + \text{h.c.} \qquad (21.167)$$

Notice that in this case, the mass of the up-type quarks would come from the VEV of ϕ_u whereas the mass of the down-type quarks would come from the VEV of ϕ_d. The notable thing is that these Yukawa couplings are invariant under the following transformations:

$$q_{\mathrm{L}} \to q_{\mathrm{L}}, \qquad d_{\mathrm{R}} \to e^{i\beta} d_{\mathrm{R}}, \qquad u_{\mathrm{R}} \to e^{i\beta} u_{\mathrm{R}},$$
$$\phi_d \to e^{-i\beta}\phi_d, \qquad \phi_u \to e^{i\beta}\phi_u. \tag{21.168}$$

We have omitted the generation indices, implying that the same transformations should be applied to quarks belonging to all generations. Clearly, the left-chiral and the right-chiral quarks transform differently, so this involves an axial symmetry; and the symmetry is U(1), because it is flavor-blind. This symmetry is called the *Peccei–Quinn symmetry* after the people who first talked about it.

Obviously, the problem with the strong CP violation disappears with a Lagrangian that possesses Peccei–Quinn symmetry, because chiral transformations render the θ-term irrelevant. However, a new problem arises from the fact that when ϕ_u and ϕ_d develop VEVs to break the gauge symmetry and to give masses to fermions, the Peccei–Quinn symmetry breaks spontaneously as well, and that would imply that there should be a Goldstone boson in the particle spectrum. Because of its link with the axial symmetry, this Goldstone boson is called the *axion*.

To be precise, the axion would not be exactly massless. The reason is that the axial U(1) is anomalous, as we have discussed in §18.4.2. The same reason that makes η' heavier than the mesons in the flavor octet would give the axion some small mass. An estimate of the mass is not important for our discussion.

In §15.3, we showed that the interactions of any Goldstone boson are suppressed by inverse powers of the VEV responsible for the symmetry breaking that gives rise to the Goldstone boson. The VEVs of the multiplets ϕ_u and ϕ_d will both contribute to the W boson mass, and instead of the expression of Eq. (16.16, *p464*), we will obtain

$$M_W^2 = \frac{1}{4}g^2(v_u^2 + v_d^2) \tag{21.169}$$

for the model with two doublets. This will imply that $\sqrt{v_u^2 + v_d^2} = 246\,\mathrm{GeV}$, so that the VEVs are in the range of a couple of hundred GeVs at the most. From this, one can estimate their couplings, and hence the rate for decay of a hadron, or even of a nuclear excited state, by axion emission. Experiments were carried out to look for such processes, and they were not found.

In fact, the kind of processes alluded to can take place in the core of a star as well, and the resulting axion would then escape from the star, causing energy loss of the star. From known bounds on stellar energy loss and lifetimes, one can put an upper bound on the couplings of such light scalars. The result is that, if the Peccei–Quinn symmetry is broken by a VEV v_{PQ}, the couplings can be small enough if

$$v_{\mathrm{PQ}} \gtrsim 10^9\,\mathrm{GeV}. \tag{21.170}$$

Therefore, if the Peccei–Quinn idea has to be implemented to get rid of the strong CP problem, one needs to break the Peccei–Quinn symmetry at a very high scale, consistent with Eq. (21.170). This can be done by VEVs of extra scalars which carry axial U(1) charges, and which are standard model singlets so that the VEVs do not break the electroweak gauge group at the high scale. There are various models with this general idea. We do not give the details since so far there has not been any experimental indication favoring any of these ideas.

Chapter 22

Neutrino mass and lepton mixing

For a long time it was believed that the neutrinos are massless fermions. Experiments were consistent with zero mass of neutrinos, although each experiment had its error bar. Around 1970 or so, astrophysical and cosmological motivations for neutrino mass were felt. Around the turn of the millennium, the existence of neutrino masses was firmly established, although the mass values are not quite known yet. In this chapter, we will discuss theoretical and experimental ramifications of neutrino mass.

While discussing the standard electroweak theory in Ch. 16, we assumed the neutrinos to be massless. One can ask whether the conclusions reached in that chapter are reliable since the assumption has been proved wrong experimentally. The answer is in the positive: although neutrinos have mass, the masses are so small that in most situations they can be neglected. They become important only because there are some phenomena which would be impossible without non-zero neutrino masses. Some such phenomena will be discussed in this chapter. And since the neutrinos are massless in the standard model, we need to go beyond the standard model in these discussions.

22.1   Simple extension of standard model

In §17.1, we discussed how quarks obtain masses in the standard model. Why can't the neutrinos obtain masses in the same way?

The answer is simple. It takes a left-chiral and a right-chiral field to construct a mass term. For each flavor of quark, we have the left-chiral component in an $SU(2)_L$ doublet and the right-chiral component in a singlet. The same is true for charged leptons, as seen from Eq. (16.40, p 470), and indeed the charged leptons also obtain masses the same way the quarks do, i.e., through the Yukawa coupling after symmetry breaking. But the same equation shows us that we did not put in a right-chiral neutrino. So, no wonder that the neutrinos remain massless in the standard model.

This was not an oversight on the part of the people who proposed the standard model. Neutrino helicity was measured in different experiments, as indicated in §14.2, and the result was found to be consistent with -1. In other words, there was no experimental indication that a neutrino can be right-handed. In gauge theories, one has to specify the gauge transformation properties of right-chiral and left-chiral fermion fields. If one admits only left-chiral neutrinos and not right-chiral ones, that would imply that the neutrinos are massless, and then helicity and chirality would mean the same thing, as explained in §14.2. Thus, absence of right-chiral neutrinos would explain why neutrinos cannot be found with right-handed helicity in any experiment. For this reason, right-chiral neutrinos were not considered part of the physical reality and not introduced in the standard model in its original version.

There were, however, some indirect indications of neutrino mass from astrophysics and cosmology. In addition, there was the æsthetic point that the fermion content of the standard model looks lopsided without a right-chiral neutrino. We can rectify the situation by introducing a right-chiral neutrino field for each generation of fermions. Like the right-handed quarks and charged leptons, they can be assumed to be $SU(2)_L$ singlets. The electric charge formula given in Eq. (16.18, *p 465*) then tells us that its weak hypercharge should be zero, i.e., it should be invariant under the $U(1)_Y$ part of the gauge group as well. There is thus no extra gauge interaction because of the introduction of these fields.

But there will be new Yukawa interactions. Within a single generation, in addition to the Yukawa terms shown in Eq. (16.41, *p 471*), the following extra terms are allowed by gauge invariance:

$$\mathscr{L}'_Y = -h'\overline{\psi}_L\widetilde{\phi}\nu_R - h'^*\overline{\nu}_R\widetilde{\phi}^\dagger\psi_L\,, \tag{22.1}$$

where $\widetilde{\phi}$ was defined in Eq. (17.4, *p 483*). When the scalar doublet obtains a VEV, the neutrino obtains a mass in a way that is exactly similar to the way that quarks and charged leptons obtain masses.

When we consider multiple generations of fermions, we obtain one extra feature. In the case of quarks, we found that the up-type and the down-type quark mass matrices are not diagonalized by the same transformation, which led to the phenomenon of quark mixing. Similarly, there is no reason why the neutrino mass matrix and the charge lepton mass matrix should be diagonalized by the same transformations. This implies *lepton mixing*.

The field that shares space in a doublet with a charged lepton ℓ can still be called ν_ℓ, i.e., the partner of e_L in the $SU(2)_L$ can still be called ν_e. The point is that these fields will not be eigenstates of the Hamiltonian. Let us denote the eigenstates by ν_A for $A = 1, 2, 3$. These two sets of states should be related through a unitary matrix:

$$\nu_{\ell L} = \sum_A U_{\ell A}\nu_{AL}\,. \tag{22.2}$$

Since $\nu_{\ell L}$ and ℓ_L form a doublet, the interaction of the W bosons with the leptons can be written as

$$\mathscr{L}_{\rm cc} = -\frac{g}{\sqrt{2}} \sum_\ell \bar{\ell}_{\rm L} \gamma^\mu W_\mu^- \nu_{\ell \rm L} + \text{h.c.} \tag{22.3}$$

Using Eq. (22.2), we can rewrite this equation as

$$\mathscr{L}_{\rm cc} = -\frac{g}{\sqrt{2}} \sum_{\ell,A} \left(\bar{\ell}_{\rm L} \gamma^\mu W_\mu^- U_{\ell A} \nu_{A\rm L} + \bar{\nu}_{A\rm L} \gamma^\mu W_\mu^+ U_{\ell A}^* \ell_{\rm L} \right). \tag{22.4}$$

The matrix U, similar to the CKM matrix for quarks, is called the *PMNS matrix* in the context of leptons. The acronym is derived from the names of the scientists Pontecorvo, Maki, Nakagawa and Sakata, who were the early proponents of the possibility of lepton mixing.

Neutrino mass and lepton mixing therefore go hand in hand. We are not saying that the Yukawa couplings described in Eq. (22.1) constitute the full story of neutrino mass. In fact, in §22.6, we will argue that it is more likely that neutrinos obtain masses from some other mechanism. But, no matter how the masses are obtained, the source is different from the source of masses of charged leptons, and therefore there is no reason that the mass matrices of charged leptons and of neutrinos should be diagonalized by the same unitary matrix. This means that there would be lepton mixing.

22.2 Neutrino oscillation

The only observed consequence of lepton mixing is flavor oscillation of neutrinos, often abbreviated to *neutrino oscillation*. In this section, we discuss the theoretical framework for understanding the phenomenon and the experimental findings.

22.2.1 Theoretical analysis

Suppose we have produced a neutrino in an experiment through a charged current interaction. We know that a particular flavor of charged lepton ℓ was annihilated in the process or its antiparticle has been produced. Through Eq. (22.3), it means that the state $\nu_{\ell L}$ has been produced. If there is lepton mixing, this is not an eigenstate of the Hamiltonian. Rather, this is a superposition of different eigenstates, as shown in Eq. (22.2). We can call each such superposition a *flavor state*, meaning that each one couples to a particular flavor of charged lepton through the charged current. If this state now evolves in time, at time t the state will be

$$|\nu_\ell(t)\rangle = \sum_A U_{\ell A} e^{-iE_A t} |\nu_A\rangle , \tag{22.5}$$

where E_A is the energy of the eigenstate ν_A. If the neutrinos are emitted with a momentum of magnitude p,

$$E_A = \sqrt{p^2 + m_A^2}, \qquad (22.6)$$

where m_A is the mass of ν_A. Since the energies of different eigenstates are not equal, the state at time t is not, in general, the same state as that given in Eq. (22.2). So, at time t, the state will be a different superposition of the eigenstates, which will contain other flavor states as well. The situation is similar to kaon oscillations described in §17.9.3: strangeness eigenstates K^0 and $\widehat{K}^0$ are produced in strong interaction, but neither of them is an eigenstate of the Hamiltonian.

To be more quantitative, let us make a few simplifying assumptions. First, we suppose that there are only two generations of fermions, and denote the neutrino flavors by ν_e and ν_μ. In this case, the PMNS matrix is of the form

$$U = \begin{pmatrix} \cos\theta & -\sin\theta \\ \sin\theta & \cos\theta \end{pmatrix}. \qquad (22.7)$$

Second, we assume that the neutrino masses are very small compared to the momenta. If the magnitude of neutrino momentum is p, we can write

$$E_A = p + \frac{m_A^2}{2p}, \qquad (22.8)$$

ignoring higher order corrections in mass. If the neutrino is produced in the state $|\nu_e\rangle$, then, after a time t, the state will evolve to

$$|\nu_e(t)\rangle = e^{-ipt}\left[e^{-im_1^2 t/2p}\cos\theta\,|\nu_1\rangle - e^{-im_2^2 t/2p}\sin\theta\,|\nu_2\rangle\right]. \qquad (22.9)$$

Once again, we emphasize that this is not the state $|\nu_e\rangle$. This is the state in which an initial $|\nu_e\rangle$ state has evolved after time t. The probability of finding the state $|\nu_e\rangle$ in this beam is

$$P_{\nu_e\nu_e}(t) \equiv \left|\langle\nu_e|\nu_e(t)\rangle\right|^2 = 1 - \sin^2 2\theta \sin^2\left(\frac{\Delta m^2}{4p}t\right), \qquad (22.10)$$

where

$$\Delta m^2 \equiv m_2^2 - m_1^2. \qquad (22.11)$$

This can be called the *survival probability*. The opposite would be the conversion probability, i.e., the probability of finding a ν_μ from this beam. Obviously, this would be

$$P_{\nu_e\nu_\mu}(t) = 1 - P_{\nu_e\nu_e}(t). \qquad (22.12)$$

Because each of the probabilities shown in Eqs. (22.10) and (22.12) has an oscillating factor in t, the phenomenon is called *neutrino oscillation*.

□ **Exercise 22.1** Consider more than two generations of neutrinos and allow for phases in the mixing matrix. Show that the probability of finding a neutrino flavor $\nu_{\ell'}$ at a time t after producing a neutrino flavor ν_ℓ is given by

$$P_{\nu_\ell \nu'_\ell}(t) = \sum_{A,B} \left| C_{\ell\ell' AB} \right| \cos\left(\frac{(m_A^2 - m_B^2)t}{2p} - \arg C_{\ell\ell' AB} \right), \quad (22.13)$$

where

$$C_{\ell\ell' AB} = U_{\ell A} U_{\ell' B} U^*_{\ell' A} U^*_{\ell B}. \quad (22.14)$$

In a real experiment, neutrinos cannot be monochromatic. There should be some distribution of the momentum values. Let us say that the fraction of neutrinos having momentum values between p and $p + dp$ is $\Phi(p)dp$, so that

$$\int dp\, \Phi(p) = 1. \quad (22.15)$$

Then the survival probability of a neutrino beam of a pure flavor would be

$$P_{\text{surv}}(t) = 1 - \sin^2 2\theta \int dp\, \Phi(p) \sin^2\left(\frac{\Delta m^2}{4p} t \right). \quad (22.16)$$

Using the dimensionless variable

$$r \equiv \frac{\Delta m^2}{4\langle p \rangle} t \quad (22.17)$$

where $\langle p \rangle$ is the mean value of p, Eq. (22.16) can be rewritten as

$$P_{\text{surv}}(t) = 1 - \frac{1}{2} \sin^2 2\theta \int dp\, \Phi(p)\left[1 - \cos(2r\langle p \rangle /p) \right]. \quad (22.18)$$

The corresponding formula for the conversion probability should be obvious. If $r \gg 1$, the cosine term will fluctuate wildly with p, so that its integral will vanish and we will obtain

$$P_{\text{surv}}(t) = 1 - \frac{1}{2} \sin^2 2\theta \quad (22.19)$$

irrespective of the detailed nature of $\Phi(p)$. In the other extreme, if $r \ll 1$, we can keep only the first non-trivial dependence in r in the cosine term, which would give

$$P_{\text{surv}}(t) = 1 - Kr^2 \sin^2 2\theta, \quad (22.20)$$

where K is independent of t.

Experiments are of two basic kinds. In one kind, one tries to measure the survival probability, i.e., produces a beam of a certain flavor and tries to find, after some time, whether the probability of finding that flavor is still

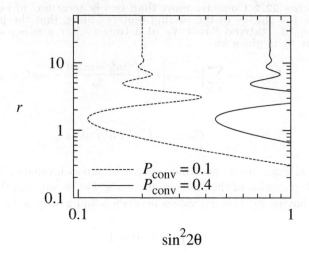

Figure 22.1: Contours of equal probability for two different values of the conversion probability. A Gaussian energy spectrum has been assumed, with a standard deviation of $0.2 \langle p \rangle$.

unity. Such experiments are called *disappearance experiments*. The other experiments where one looks for a flavor of neutrino that was not there in the original beam, are called *appearance experiments*. In either kind, the data are analyzed by using the momentum distribution function $\Phi(p)$ from the design of the experiment, the probability is measured, and it is found out which set of values of Δm^2 and θ is consistent with the data.

As an illustrating example, let us consider a Gaussian distribution of momenta in an incoming beam of neutrinos. The nature of the equi-probability contours for this case have been shown in Fig. 22.1. Thus, for example, if a disappearance experiment finds that the disappearance probability at a certain distance lies between the two values marked in the figure, the region between the two lines is allowed. Using the values of t and $\langle p \rangle$, this allowed region can be mapped into a region in the parameters Δm^2 and θ.

□ **Exercise 22.2** Show that $K = \langle p^2 \rangle \langle 1/p^2 \rangle$ in Eq. (22.20).

□ **Exercise 22.3** In Fig. 22.1, check that for large r, the lines of equal probability are consistent with Eq. (22.19), and that the slopes of the lines for $r \ll 1$ are consistent with Eq. (22.20).

Inspired by the theory of neutrino oscillation, some authors raised the question of whether there can be a similar oscillation phenomenon with charged leptons. This does not make much sense. We talk about flavor oscillation of neutrinos in situations where we know which charged lepton (or antilepton) has been produced in a reaction, but do not know which neutrino (or antineutrino) eigenstate has been produced. Without a knowledge of the neutrino eigenstate, all we can say is that the produced state is a superposition of eigenstates which matches with its charged counterpart. By the same token, in order to talk about a charged lepton oscillation experiment, we need to imagine a situation where we know that a particular eigenstate of neutrino has been produced but do not know which charged lepton accompanies it. If the neutrinos and the charged leptons had comparable masses, such a situation would not have been inconceivable. However, the neutrino masses are much smaller than the masses of the charged leptons. With the amount of precision that would be necessary to identify a particular eigenstate of neutrino, it is virtually impossible not to know which charged lepton accompanies it. So, the idea of charged lepton oscillation is an unreal idea.

22.2.2 Experimental data

We classify the experimental data depending on the source of neutrinos. Thus, when we talk of *solar* neutrino data, it means the measurements were done on neutrinos that were produced in the sun and detected on the earth. For *terrestrial* searches, the source and the detector are both earth bound. Although terrestrial experiments can be the most controlled ones, a lot about neutrino oscillation was learned from extra-terrestrial sources of neutrinos. In fact, the experiments with extra-terrestrial sources gave data on neutrino oscillations much before any terrestrial experiment did.

a) Solar neutrinos

The sun produces electron-neutrinosin nuclear reactions. In the first stage of nuclear reactions taking place in the sun or any comparable star, two protons fuse to form a deuteron:

$$p + p \rightarrow d + e^+ + \nu_e \,. \tag{22.21}$$

The resulting neutrino, being only weakly interacting, does not experience rescattering within the sun and comes out. There are also neutrinos produced from reactions involving nuclei with higher mass numbers. The nuclear physics of these neutrino production mechanisms are well-studied in terrestrial experiments, and the cross-sections of such reactions are known to a good accuracy. Based on such data, one can calculate numerically the number of neutrinos produced from the sun and their energy distribution. Since the 1960s, experiments were set up to detect these *solar neutrinos*. The earliest experiment was led by Davis at the Homestake mines in South Dakota, USA. They detected neutrinos through the inverse beta decay reaction

$$\nu_e + {}^{37}\text{Cl} \rightarrow e^- + {}^{37}\text{Ar} \,, \tag{22.22}$$

and found that they were getting about one-third of the number of neutrinos that was predicted from the solar models. After that, the Kamiokande group in Japan tried detection of solar neutrinos through elastic neutrino–electron scattering. They also did not find as many neutrinos as expected. Interestingly, their result was about half the expected flux, different from the depletion observed in the Homestake experiment. Later, various other experiments were set up, and the experimental results always fell short of the solar model expectation. The ratio of observed flux and expected flux was different in different experiments. The solar model was suspected for a while, but as data from different experiments accumulated, it became clear that changes in the solar model cannot possibly explain the results of all experiments that detected solar neutrinos.

The other possibility was that the neutrinos have some property beyond the standard model. In particular, if neutrinos are massive and they mix, there can be flavor oscillations. The sun can produce only ν_e's: the reactions going on in the solar interior are fusion reactions, like that shown in Eq. (22.21), which are nuclear processes involving binding energies of the order of a few MeV, and these energies are not high enough to produce a ν_μ or a ν_τ with the associated charged lepton. With neutrino mixing, the ν_e would oscillate on its way from the sun to the earth, and the beam would not contain as many ν_e's as are produced in the sun. This can explain why the experiments detected fewer neutrinos than expected from a theory in which neutrino mass and mixing had not been taken into account. Further, since the oscillation probability depends on the neutrino energy, such a scenario would also explain why different experiments, which were set up to detect neutrinos in different energy ranges, showed different depletion of ν_e flux.

The question of depletion arose because the detection processes employed by the various experiments were biased toward electron-neutrinos.In the case of the Homestake experiments, only ν_e's could be detected, because it was not possible to produce a muon or a tau through an inverse beta decay reaction with the energies carried by the neutrinos. Such was the case of many other detectors as well, which used *radiochemical methods of detection*, i.e., inverse beta decay processes involving some nucleus or other. For the electron-scattering detectors like the Kamiokande, muon-neutrinos could also be detected, but the scattering cross-section of ν_μ's with the electron is much smaller than the cross-section of ν_e-e scattering. Hence there was a bias here as well.

The matter was settled by the SNO (Sudbury Neutrino Observatory) group in Canada, whose detector contained heavy water. They detected neutrinos through a number of reactions, including some of the reactions used by earlier groups. However, they had an extra channel of detection, through the decomposition of deuteron:

$$\nu + d \rightarrow \nu + n + p \,. \tag{22.23}$$

This reaction takes place through neutral current weak interaction, and therefore has the same cross-section for all neutrino flavors. Thus, even if the ν_e oscillates to some other flavor of neutrino like the ν_μ or the ν_τ, this reaction should be able to detect it. Indeed, they found depletion in electron-scattering and other channels where the ν_e's are preferentially detected, but the flux agreed with the solar model calculations when the neutral current detection channel was used.

b) Atmospheric neutrinos

Atmospheric neutrinos also presented some anomaly. By *atmospheric neutrinos*, we mean neutrinos that are produced by cosmic rays hitting the atmosphere. Cosmic ray particles are predominantly protons. When they hit nuclei in the atmosphere, pions are produced. Charged pions decay, almost entirely into muons and associated neutrinos. Subsequently, the muons decay. The chain of these reactions can be depicted as follows:

$$p + X \rightarrow \pi^{\pm} + Y$$
$$\hookrightarrow \mu^{\pm} + \text{``}\nu_{\mu}\text{''} \tag{22.24}$$
$$\hookrightarrow e^{\pm} + \text{``}\nu_e\text{''} + \text{``}\nu_{\mu}\text{''}$$

where X and Y denote nuclei. We have put quotation marks around the neutrinos to indicate that the particle might be either a neutrino or an antineutrino: the distinction is not made in the detection process. We see that twice as many muon-type neutrinos will be produced compared to the number of electron-type neutrinos. This ratio can be modified a bit because of other secondary interactions, and can also depend on the energy of the neutrinos. Such effects can be calculated to a fairly good degree of accuracy. Experiments were set up, first by the Kamiokande group, to detect these neutrinos. It was observed that the flux of muon-type neutrinos was less than what was expected. Again, this can be explained if the muon-type neutrinos oscillate into some other neutrino flavor.

c) Terrestrial searches

Not surprisingly, all early terrestrial experiments gave null results. Of course, within the error bars, neutrino oscillation was allowed, but all data was also consistent with no oscillation. First terrestrial evidence of neutrino oscillation came from the KamLAND experiment in Japan. This is a disappearance experiment, in which researchers used electron-type antineutrino ($\widehat{\nu}_e$) beams from several reactors at distances between 150 and 200 km from the detector. The energies of these antineutrinos are in the range of a few MeVs, peaking at about 3.5 MeV. In the detector, they looked for the survival probability of the $\widehat{\nu}_e$'s and found a depletion from the original number. Their results have been shown in Fig. 22.2, where L is the distance between the production point and the detection point of the neutrinos.

It should be noted that the results show data points for values of L/E spanning almost a factor of 5. This is achieved because the experiment could measure energies of the antineutrinos. Depletion rates have been calculated for antineutrinos with different energies, giving a wide range of values of L/E. The oscillatory behavior of the survival probability as a function of L/E is obvious from the figure. Fig. 22.2 shows the best fit values for Δm^2 and θ corresponding to the KamLAND data only. We don't present these values here.

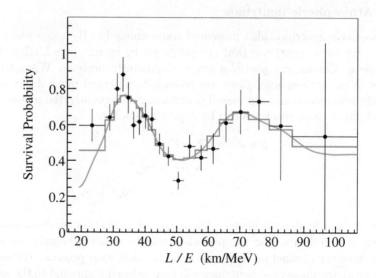

Figure 22.2: KamLAND result. The blobs are the data points, after making subtractions for the background and from geo-neutrinos. The fit, with the best values suggested by KamLAND, has been shown with a histogram and with a continuous curve. [Reprinted with permission from: The KamLAND collaboration, Phys. Rev. Lett. **100** (2008) 221803; © (2008) by the American Physical Society.]

In §22.2.4, we will present a summary of results from all different experiments taken together.

The KamLAND experiment, along with other experiments on solar and atmospheric neutrinos, helped determine two of the three mixing angles of the PMNS matrix, and it turned out that both these angles are quite large. The remaining angle θ_{13} that governs the 13 element of the PMNS matrix is small. Until very recently, all experimental data was consistent with the 13 element being zero. The situation changed when three different experiments published their results almost at the same time in 2012: the Daya Bay experiment based in China, the RENO experiment in Korea, and the double Chooz experiment in France. These are all disappearance experiments which used $\widehat{\nu}_e$'s from reactors, and a combination of near and far detectors to infer a depletion of the flux of the antineutrinos. The value of θ_{13} was found to be approximately $9°$. We will mention the result more accurately in §22.2.4 when we summarize all results regarding neutrino mass and mixing.

d)   Order-of-magnitude analysis

Eq. (22.18) gives the expression for the survival probability in a neutrino oscillation experiment. In a given experiment, the time t can be replaced, in the natural units, by the source-to-detector distance x which is known. Also known is the energy distribution of the neutrinos used. Clearly then, a given experiment is most efficient in detecting neutrino oscillations if the value of Δm^2 is such that $r \sim 1$, i.e., if $\Delta m^2 \sim \overline{m}^2$, where

$$\overline{m}^2 = \frac{4 \langle \mathbf{p} \rangle}{x}. \tag{22.25}$$

This quantity is usually called the *figure of merit* of a neutrino oscillation experiment. Putting in relevant conversion factors, we can write

$$\frac{\overline{m}^2}{1\,\mathrm{eV}^2} = 7.92 \times 10^{-4} \left(\frac{\langle \mathbf{p} \rangle}{1\,\mathrm{MeV}} \right) \times \left(\frac{x}{1\,\mathrm{km}} \right)^{-1}. \tag{22.26}$$

For the KamLAND experiment, using the average energy and path length as mentioned, we find that $\overline{m}^2 \sim 10^{-5}\,\mathrm{eV}^2$, and indeed, the experiment found some allowed region in the Δm^2 vs θ parameter space around that value of Δm^2. For atmospheric neutrinos, energies are of order GeV or so, whereas the path length is a few thousand kilometers for neutrinos coming through the earth after being produced in the atmosphere near the diametrically opposite point. Thus, evidence of atmospheric neutrino oscillations would provide an allowed region around $\Delta m^2 \sim 10^{-3}\,\mathrm{eV}^2$. For solar neutrinos, however, the same analysis gives a figure of merit of order $10^{-10}\,\mathrm{eV}^2$. However, there are other allowed regions, because our analysis so far has neglected one aspect of neutrino oscillations that becomes very important for solar neutrinos. We discuss it next.

22.2.3 Matter effects

Neutrinos are produced mainly in the core region of the sun, where the temperature is high. On the way out of the sun, they must pass through regions of high density of matter. Wolfenstein pointed out that, when neutrinos travel through a medium, coherent forward scattering with the particles forming the medium can change the effective neutrino Hamiltonian. To quantify this contribution, let us consider a normal medium consisting of electrons, protons and neutrons. Consider first the interactions with the electrons in the medium. The effective Lagrangian for neutrino–electron interactions was given in Eq. (14.78, *p 426*). When the electrons belong to the medium, we need to average over them in order to obtain the contribution that is quadratic in the neutrino fields. This term is

$$\mathscr{L}_{\mathrm{mat}} = -\sqrt{2}G_F \left\langle \overline{\psi}_{(e)} \gamma^\mu (c_V - c_A \gamma_5) \psi_{(e)} \right\rangle \left[\overline{\psi}_{(\nu)} \gamma_\mu \mathbb{L} \psi_{(\nu)} \right], \tag{22.27}$$

where the angular brackets indicate the averaging. Let us assume the electrons in the medium are non-relativistic, an excellent approximation for the sun whose core temperature is a few keVs. In this limit, the spatial parts of the axial vector currents average to spin. If we assume that the medium does not have any overall spin, this part is zero. The temporal part of the axial vector current is negligible. The spatial part of the polar vector currents is proportional to the average velocity, which should also be negligible. So we are left with only the temporal part of the polar vector current, which gives the number density. Thus,

$$\mathscr{L}_{\text{mat}} = -\sqrt{2}G_F n_e c_V \left[\overline{\psi}_{(\nu)} \gamma_0 \mathbb{L}\psi_{(\nu)}\right]. \tag{22.28}$$

If we augment the free Lagrangian of neutrinos by this quadratic term and write the equation of motion of the neutrino field, the effective energy of the neutrino comes out to be

$$E = \sqrt{\mathbf{p}^2 + m^2} + \sqrt{2}G_F n_e c_V. \tag{22.29}$$

□ **Exercise 22.4** Add the term shown in Eq. (22.28) to the free Dirac Lagrangian and show that the plane wave solutions of the resulting equation obey the dispersion relation of Eq. (22.29).

This is for one flavor of neutrino. Suppose now we consider the time evolution of a state involving two flavors of neutrinos, e.g., the ν_e and the ν_μ. The effective 2×2 Hamiltonian of the system should have a density-dependent term for each flavor, with the appropriate value of c_V. Reading the values of c_V from Eqs. (16.56) and (16.61), we find that the effective Hamiltonian has a contribution proportional to n_e which is given by

$$\sqrt{2}G_F n_e \begin{pmatrix} \frac{1}{2} + 2\sin^2\theta_W & 0 \\ 0 & -\frac{1}{2} + 2\sin^2\theta_W \end{pmatrix} \tag{22.30}$$

in the ν_e-ν_μ basis.

There are also terms that depend on the number densities of protons and neutrons in the medium. These arise from neutral current interactions, and are therefore equal for ν_e and ν_μ. In the 2×2 basis used above, these contributions will produce a term proportional to the unit matrix. Adding these, and also writing some part of the n_e term as a multiple of the unit matrix, we can write the effective Hamiltonian as

$$\widetilde{H} = H + a\mathbb{1} + \begin{pmatrix} \sqrt{2}G_F n_e & 0 \\ 0 & 0 \end{pmatrix}, \tag{22.31}$$

where H is the Hamiltonian in absence of all matter effects, i.e., the Hamiltonian for neutrinos traveling through the vacuum, the case that was discussed in §22.2.1. All contributions to a, the co-efficient of the unit matrix $\mathbb{1}$, have not been calculated really. The reason is that they are irrelevant for our purpose, as we will see shortly.

What is H? The eigenvalues were given in Eq. (22.8), in the approximation that we have employed. If we take the eigenstates as basis states, the Hamiltonian matrix would have been diagonal, with the eigenvalues as diagonal entries. But Eq. (22.31) has been written by using ν_e and ν_μ as basis states. Using Eq. (22.2), we find that in this basis, the Hamiltonian matrix should be given by

$$H = U \begin{pmatrix} E_1 & 0 \\ 0 & E_2 \end{pmatrix} U^\dagger. \tag{22.32}$$

Using the form of U given earlier, we obtain

$$\widetilde{H} = a' \mathbb{1} + \frac{1}{4\mathrm{p}} \begin{pmatrix} -\Delta m^2 \cos 2\theta + 2A & \Delta m^2 \sin 2\theta \\ \Delta m^2 \sin 2\theta & \Delta m^2 \cos 2\theta \end{pmatrix}, \tag{22.33}$$

where

$$A = 2\sqrt{2} G_F n_e \mathrm{p}, \tag{22.34}$$

and a' contains not only a but also some other terms, like the term p and terms proportional to the sum of the mass squares, which are also of the form of a multiple of the unit matrix. This matrix can be diagonalized by similarity transformation involving a matrix $\widetilde{U}$ which looks like the matrix U of Eq. (22.7), with the angle θ replaced by $\widetilde{\theta}$, given by

$$\tan 2\widetilde{\theta} = \frac{2\widetilde{H}_{12}}{\widetilde{H}_{22} - \widetilde{H}_{11}} = \frac{\Delta m^2 \sin 2\theta}{\Delta m^2 \cos 2\theta - A}. \tag{22.35}$$

This is part of the reason why the exact expression of a' is not necessary: the effective mixing angle is independent of this contribution. The other part of the reason has to do with the fact that the eigenvalues of the matrix $\widetilde{H}$ are given by

$$a' + \frac{1}{4\mathrm{p}} \left[A \pm \sqrt{\left(\Delta m^2 \cos 2\theta - A\right)^2 + \left(\Delta m^2 \sin 2\theta\right)^2} \right], \tag{22.36}$$

so that the difference of the eigenvalues, which is responsible for oscillation phenomena, is also independent of a'.

The important point is that the eigenvalues as well as the mixing angle depend on the density of ambient matter through A. In the sun, the neutrinos are produced near the core, where the mixing angle can be much different from the mixing angle outside the sun because of these matter effects. This can affect the survival and conversion probabilities significantly. To get a quantitative feeling for the effect, suppose that the mixing angle is $\widetilde{\theta}_0$ at the point where the neutrinos are created. At this point, the eigenstates are, say, $\widetilde{\nu}_1$ and $\widetilde{\nu}_2$, with

$$\nu_e = \widetilde{\nu}_1 \cos \widetilde{\theta}_0 + \widetilde{\nu}_2 \sin \widetilde{\theta}_0. \tag{22.37}$$

Thus, there is a probability of $\cos^2 \widetilde{\theta}_0$ that the produced ν_e is in eigenstate 1, and a probability of $\sin^2 \widetilde{\theta}_0$ that it is in eigenstate 2. If adiabatic conditions prevail, the eigenstate 1 (corresponding to the eigenvalue with a negative sign before the square root in Eq. (22.36)) remains eigenstate 1, although with the change in ambient density it becomes a different superposition of ν_e and ν_μ because the mixing angle changes. Finally, the neutrinos are detected on the surface of the earth, where matter density is negligible, the mixing angle is equal to its vacuum value θ. This means that the ν_e is the combination

$$\nu_e = \nu_1 \cos\theta + \nu_2 \sin\theta, \tag{22.38}$$

in terms of the vacuum eigenstates ν_1 and ν_2. Now, if we have a detection system that is sensitive only to ν_e, we will have a probability of $\cos^2\theta$ of detection if the neutrino is in eigenstate 1, and $\sin^2\theta$ if the neutrino is in eigenstate 2. Combining the probabilities at production and detection points, we obtain that the survival probability of neutrinos is given by

$$P_{\text{surv}}^{(\text{ad})} = \cos^2\widetilde{\theta}_0 \cos^2\theta + \sin^2\widetilde{\theta}_0 \sin^2\theta = \frac{1}{2}\left[1 + \cos 2\theta_0 \cos 2\theta\right]. \tag{22.39}$$

There are two assumptions that go into this formula. First, there is a wide spectrum of neutrino energies in the beam, so that the oscillating terms average out when we integrate over the energies. This is the reason we added probabilities instead of adding amplitudes and then squaring the sum. The second one is that adiabatic conditions prevail, which is indicated by the declaration 'ad' in the formula.

If the second assumption does not hold, i.e., if there is a probability X of jumping from one eigenstate to another, the survival probability would be given by

$$P_{\text{surv}} = (1-X)P_{\text{surv}}^{(\text{ad})} + X P_{\text{conv}}^{(\text{ad})} = \frac{1}{2}\left[1 + (1-2X)\cos 2\theta_0 \cos 2\theta\right]. \tag{22.40}$$

The jumping probability X can be obtained once the nature of variation of density is known, but the expression is not necessary for the ensuing discussion.

The important point is that the conversion from one neutrino flavor to another can be much more efficient in the background of a changing density. To see an example, consider that the neutrino was produced in a region of infinite density, so that $\widetilde{\theta}_0 = \pi/2$. Eq. (22.39) then gives a survival probability of $\sin^2\theta$. In the vacuum background, Eq. (22.19) tells us that the energy-averaged survival probability is $1 - \frac{1}{2}\sin^2 2\theta$. The difference is dramatic if θ is small.

The reason for this big difference is that the neutrinos on their way out of the sun cross a region where the electron number density satisfies the equation $\Delta m^2 \cos 2\theta = A$, so that the effective mixing angle becomes $\pi/4$. The effective mass-squared difference also attains a minimum at this point. This is a resonance. The phenomenon can be called resonant oscillation At and near this point, oscillations are very efficient and therefore the original ν_e beam converts largely into another flavor.

The resonant oscillation phenomenon will occur as long as the denominator of the right side of Eq. (22.35) goes through the value zero. On the earth, where the neutrino is detected, we have $A \approx 0$ and therefore the denominator is positive. Thus, in order that the denominator passes through zero, matter density at the point of production of neutrinos should be high enough so that $A_0 > \Delta m^2 \cos 2\theta$, A_0 being the value of A at the point of production. Using the fact that the matter density near the solar core is about $150\,\mathrm{g/cm}^3$ and that typical energies of neutrinos coming out of the sun are of the order of a few MeV, this means that

$$\Delta m^2 \cos 2\theta \lesssim 10^{-4}\,\mathrm{eV}^2\,. \tag{22.41}$$

Indeed, solutions consistent with all solar neutrino experiments are obtained for values of Δm^2 close to this upper bound, as we will summarize presently.

22.2.4 Summary

Result of analysis of data from terrestrial, atmospheric and solar neutrino experiments are shown in Fig. 22.3. Here, we summarize the allowed values of different mass and mixing parameters. First, from the solar neutrino data, one obtains

$$\Delta m_\odot^2 = (7.50 \pm 0.20) \times 10^{-5}\,\mathrm{eV}^2\,, \tag{22.42a}$$
$$\sin^2 2\theta_\odot = 0.857 \pm 0.024\,. \tag{22.42b}$$

The atmospheric neutrino data suggest the following values:

$$\Delta m_{\mathrm{atm}}^2 = (2.32^{+0.12}_{-0.08}) \times 10^{-3}\,\mathrm{eV}^2\,, \tag{22.42c}$$
$$\sin^2 2\theta_{\mathrm{atm}} > 0.95\,. \tag{22.42d}$$

Finally, the terrestrial $\widehat{\nu}_e$-disappearance experiments described in §22.2.2(c) imply

$$\sin^2 2\theta_{13} = 0.098 \pm 0.013\,. \tag{22.42e}$$

Obviously, the Δm^2 values dictated by atmospheric neutrinos and solar neutrinos cannot pertain to the same pair of neutrino eigenstates. This means that the neutrino oscillations responsible for these two phenomena do not occur between the same two pairs of neutrino flavors. For the case of solar neutrino, we know that the sun produces ν_e only, so this must be one of the flavors involved in the oscillation phenomenon. On the other hand, for atmospheric neutrinos, the two flavors involved must be ν_μ and ν_τ, because at the relevant value of Δm^2, there is no appreciable oscillation of ν_e's, as has been confirmed by terrestrial experiments.

The neutrino eigenstates responsible for the depletion of solar ν_e's can be called ν_1 and ν_2, with the convention that $m_2 > m_1$, so that we can identify the parameters of the solar neutrino oscillation as $\Delta m_\odot^2 = \Delta m_{21}^2 \equiv m_2^2 - m_1^2$ and $\theta_\odot = \theta_{12}$. The atmospheric neutrino data then provides information about the absolute value of the mass-squared difference Δm_{13}^2 or Δm_{23}^2 and

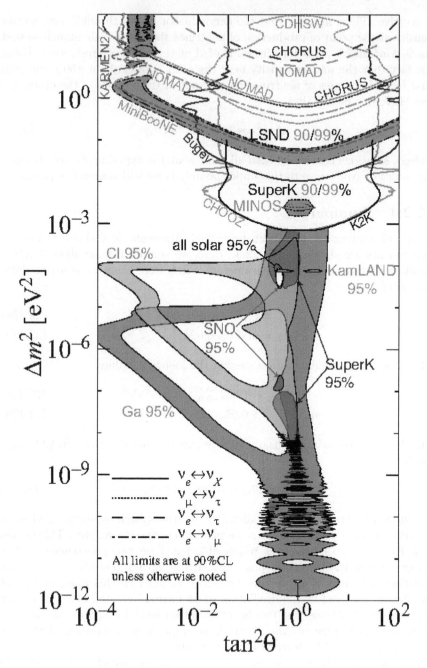

Figure 22.3: Summary of solutions of data coming from terrestrial, atmospheric and solar neutrino experiments. For each experiment, the data has been analyzed with the assumption that oscillation occurs only between two neutrino flavors. [Reprinted with permission from: Particle Data Group, Phys. Rev. **D 86** (2012) 010001; © (2012) by the American Physical Society.]

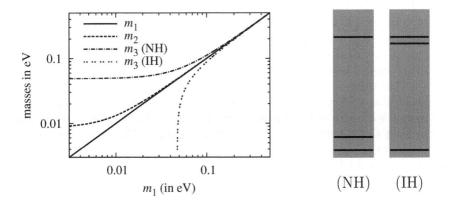

Figure 22.4: Two possible hierarchies for neutrino masses. In the plot on the left side, we show the masses as functions of m_1, where there are two possible solutions for m_3, one for normal hierarchy (NH), the other for inverted hierarchy (IH). On the right side, we show schematically the arrangement of eigenvalues in the two hierarchies.

mixing angle θ_{23}. The allowed values of these parameters are given in Eq. (22.42).

With the three mixing angles determined, we can therefore write the PMNS matrix if CP violating phases can be ignored. Using the central values of the results shown in Eq. (22.42) for θ_{12} and θ_{13}, and using $\theta_{23} = \pi/4$, this matrix is given by

$$U = \begin{pmatrix} 0.820 & 0.551 & 0.159 \\ -0.487 & 0.524 & 0.698 \\ 0.301 & -0.649 & 0.698 \end{pmatrix}. \tag{22.43}$$

Let us now turn to the information on mass eigenvalues. Because of the smallness of the allowed value of Δm_{21}^2, the masses m_2 and m_1 must be very close together. The other mass, m_3, must be somewhat apart from the first two. Since the atmospheric neutrino experiments give only the absolute value of Δm_{13}^2 or Δm_{23}^2, it is not clear whether this third eigenstate is heavier or lighter than the former two. Thus, there can be two possible patterns of neutrino mass eigenvalues. The first one, with $m_3 > m_2, m_1$, is called *normal hierarchy* because naively one expects that smaller masses should involve smaller mass-squared differences. The alternative pattern, with a small m_3, is called *inverted hierarchy*. Both possibilities have been shown in Fig. 22.4.

Naively, if we assume that each mass squared difference is dominated by only one eigenvalue, we see that one mass eigenvalue should be of order 10^{-1} eV, another around $10^{-2.5}$ eV, and the third one much smaller than that. It is, however, not impossible that in either hierarchy pattern, the mass values are in fact much larger than these benchmark values, and the values of Δm^2 turn out to be small because the neutrinos are *nearly degenerate*, as depicted by the right end of the graph in Fig. 22.4. However, even in this case, there are cosmological and astrophysical bounds which imply that none of the known neutrinos can be heavier than about $1\,\mathrm{eV}$.

This raises a question: why are neutrinos so light? Consider the first generation of fermions. The electron, the up and the down quarks all have masses within an order of magnitude of $1\,\mathrm{MeV}$. The electron neutrino mass is at least six or seven orders of magnitude smaller. The same story continues in other generations. Of course one can add right-chiral neutrino fields in the standard model and produce neutrino masses through the VEV of the Higgs field in much the same way as one produces masses from charged leptons and quarks. In this case, the smallness of neutrino masses can be attributed to the smallness of the relevant Yukawa couplings. But it is difficult to understand why these couplings have to have such minuscule values compared to all other couplings. This is one of the puzzles of neutrino mass.

The second puzzle is neutrino mixing. As seen from oscillation experiments, two out of three mixing angles are very large. In the quark sector, the mixing angles are all very small. The phenomenon of lepton mixing must then have some characteristics which are very different from those of quark mixing.

Such oddities are not totally unexpected given the fact that the neutrinos differ from all other elementary fermions in one important aspect. They do not have any electric charge. Also, they do not carry color. Thus, they do not transform under the unbroken gauge group of fundamental interactions, $\mathrm{SU}(3)_\mathrm{c} \times \mathrm{U}(1)_\mathrm{em}$. Interactions of this unbroken gauge group therefore cannot distinguish a neutrino from its antiparticle. This opens up a new possibility that the neutrinos might be their own antiparticles.

22.3   Majorana fermions

If a fermion is its own antiparticle, it is called a *Majorana fermion*. In this section, we discuss the formalism associated with Majorana fermions.

22.3.1   Definition

In §4.4, we tried to construct quadratic invariants of a fermion field of the form $\psi^\dagger M \psi$, where M is some fixed matrix. We found that the choice $M = \gamma_0$ gives us an invariant, and we used this invariant in the Dirac Lagrangian. Since this term is quadratic in fields and does not contain any derivative, it was identified as the mass term for the fermion field ψ.

We disregarded the fact that there can be other kinds of quadratic invariants involving fermion fields. For scalar fields, both ϕ^2 and $\phi^\dagger \phi$ are invariant quadratics. For fermion fields, we seemed to have tried the second kind, but not the first one. Of course, a fermion field carries a spinor index, which must be contracted between the two fields in order that we obtain a Lorentz invariant. Thus we can ask whether we can make any invariant of the form $\psi_a A_{ab} \psi_b$, which can be written as $\psi^\top A \psi$ in matrix notation, where A is a constant matrix.

Using the Lorentz transformation property of a fermion field from Eq. (4.73, *p 76*), we find

$$\psi'^\top(x') A \psi'(x') = \psi^\top(x) \exp\left(-\frac{i}{4} \omega^{\mu\nu} \sigma_{\mu\nu}^\top\right) A \exp\left(-\frac{i}{4} \omega^{\mu\nu} \sigma_{\mu\nu}\right) \psi(x).$$

$$(22.44)$$

Examining the first order terms in the transformation parameters $\omega^{\mu\nu}$, we see that invariance will be achieved if the matrix A satisfies the condition

$$\sigma_{\mu\nu}^\top A + A \sigma_{\mu\nu} = 0,$$

$$(22.45)$$

i.e., if

$$\sigma_{\mu\nu}^\top = -A \sigma_{\mu\nu} A^{-1}.$$

$$(22.46)$$

Comparing this with Eq. (6.56, *p 161*), we find that we can choose

$$A = \mathbb{C}^{-1},$$

$$(22.47)$$

i.e., $\psi^\top \mathbb{C}^{-1} \psi$ is a Lorentz invariant.

We might ask ourselves why we did not try to incorporate this kind of invariant earlier. The answer does not come from Lorentz invariance. Hermiticity also cannot be an impediment: although the term by itself is not hermitian, we are free to add its hermitian conjugate in the Lagrangian. The problem is that, if the field ψ is charged, the invariant $\psi^\top \mathbb{C}^{-1} \psi$ can annihilate two units of this charge, and would therefore defy any charge conservation. If we deal with electrons, e.g., we cannot put in this kind of invariant just because of electric charge conservation. For quarks, the same argument holds.

But how about neutrinos? They do not have electric charge. So, can we consider this new kind of invariant for them? If lepton number is conserved, we cannot, because that is one charge that the neutrinos carry. But we discussed in Ch. 18 that there is nothing sacred about lepton number: it is just an accidental symmetry of the standard model, and can easily cease to be so if there are extra fields in the model. So, for neutrinos, it is worthwhile to consider the possibility of this kind of term.

We mentioned that for scalars there can be two kinds of non-derivative quadratic terms, ϕ^2 (and its hermitian conjugate $\phi^\dagger \phi^\dagger$) and $\phi^\dagger \phi$. The two types of quadratic terms are equivalent if $\phi = \phi^\dagger$, i.e, if the field $\phi(x)$ is a real scalar field. We can ask what is the corresponding condition for fermion fields. In all earlier chapters, we have dealt with quadratic terms of the form $\overline{\psi}\psi$. Now we encounter $\psi^\top \mathbb{C}^{-1}\psi$. Both are Lorentz invariant. However, the field operator $\psi^\top \mathbb{C}^{-1} \psi$ is not hermitian, so it might be accompanied with a phase factor. So the equality of two kinds of mass terms requires

$$\overline{\psi} = e^{i\alpha}\, \psi^\top \mathbb{C}^{-1}\,. \qquad (22.48)$$

Taking the transpose of both sides and using the antisymmetry of the matrix $\mathbb{C}$ which was first seen in Eq. (6.60, *p 162*) and proved in Appendix F, we can write it in the form

$$\widehat{\psi} = e^{i\alpha}\psi\,, \qquad (22.49)$$

where $\widehat{\psi} = \gamma_0 \mathbb{C}\psi^*$, as defined in Eq. (6.49, *p 160*). Fermion fields satisfying this condition are called *Majorana fermion* fields.

☐ **Exercise 22.5** Deduce Eq. (22.49) from Eq. (22.48).

☐ **Exercise 22.6** Argue that the choice of A given in Eq. (22.47), although made by comparing only the first order terms in $\omega^{\mu\nu}$, ensures the invariance of $\psi^\top(x)A\psi(x)$ to all orders in $\omega^{\mu\nu}$.

A real scalar field can likewise be defined by the relation

$$\phi^\dagger = \phi\,. \qquad (22.50)$$

Using this condition on the plane wave expansion of a general real field given in Eq. (4.15, *p 65*), we can easily find that $a(\boldsymbol{p}) = \widehat{a}(\boldsymbol{p})$, which reduces the plane wave expansion to that of a real scalar field given in Eq. (4.12, *p 64*). It means that the particle and the antiparticle are annihilated by the same operator. Similarly, they are created by the same operator. In short, it means that the particle is the same as its antiparticle.

☐ **Exercise 22.7** A more general definition of a real scalar field can be taken as

$$\phi^\dagger = e^{i\alpha}\phi\,. \qquad (22.51)$$

With this definition, verify that the annihilation operators for "particles" and "antiparticles" differ by a phase, which means that a so-called "antiparticle" state with a certain momentum is nothing but the particle state with the same momentum, multiplied by an extra constant phase factor.

Eq. (22.49) implies that a Majorana fermion is its own antiparticle. This means that such a fermion cannot carry any conserved charge, and that is why bilinears of the form $\psi^\top \mathbb{C}^{-1}\psi$ can be allowed in the Lagrangian, as discussed earlier.

22.3.2 Feynman rules

Looking back at the plane wave expansion of a Dirac field given in Eq. (4.65, *p 72*) and imposing the Majorana condition Eq. (22.49) on it, we find that the plane wave expansion of a Majorana field is of the form

$$\psi(x) = \sum_s \int D^3p \left(d_s(\boldsymbol{p})u_s(\boldsymbol{p})e^{-ip\cdot x} + \lambda^* d_s^\dagger(\boldsymbol{p})v_s(\boldsymbol{p})e^{+ip\cdot x} \right), \qquad (22.52)$$

Table 22.1: Feynman rules for external Majorana fermion lines.

Operator responsible	Feynman rule for incoming	outgoing
ψ	$u_{\boldsymbol{p},s}$	$\lambda^* v_{\boldsymbol{p},s}$
$\overline{\psi}$	$\lambda \overline{v}_{\boldsymbol{p},s}$	$\overline{u}_{\boldsymbol{p},s}$

noting that the u-spinors and the v-spinors satisfy the conjugation relations

$$v_s = \gamma_0 \mathbb{C} u_s^* , \qquad u_s = \gamma_0 \mathbb{C} v_s^* , \tag{22.53}$$

which have been proved in Appendix F. Comparison with Eq. (22.49) shows that

$$\lambda = e^{i\alpha} . \tag{22.54}$$

The quantity λ^* is called the *creation phase factor* by some authors because it appears along with the creation operator. But this name makes no sense, since the phase could just as well be put with the other term. It is better to call α the *intrinsic Majorana phase* in view of its appearance in the definition of Eq. (22.49), or $e^{i\alpha}$ the *intrinsic Majorana phase factor*.

It should be noted that the intrinsic Majorana phase is not a physical quantity. It can always be taken to be zero by transferring the phase into the field. To be precise, instead of taking $\psi(x)$ that appears in Eq. (22.52) as the field, if we take $e^{i\alpha/2}\psi(x)$ as the field, the Majorana phase would be zero, i.e., the phase factor would be unity. However, the freedom with the phase is sometimes useful, so we keep it in our discussion, at least to show that it does not contribute to physical amplitudes.

Looking at Eq. (22.52), we find that the field operator $\psi(x)$ can annihilate as well as create a Majorana particle. So can $\overline{\psi}(x)$. The Feynman rules for Majorana particles are therefore a bit more elaborate than those for Dirac fields. When we first wrote the Feynman rules in Table 4.1 *(p 91)*, we assumed the fermions to be Dirac particles, so that there was one set of rules for fermions and one set for antifermions. Since a Majorana fermion is the same as its antiparticle, both sets of rules would apply for a Majorana fermion. In other words, an incoming fermion line can contribute either $u_s(\boldsymbol{p})$ or $\overline{v}_s(\boldsymbol{p})$ to the Feynman amplitude. An outgoing Majorana fermion line, similarly, can contribute either $\overline{u}_s(\boldsymbol{p})$ or $v_s(\boldsymbol{p})$. The contribution depends on which field operator is annihilating or creating the fermion: ψ or $\overline{\psi}$, as elaborated in Table 22.1.

To see a concrete example, consider the decay of a boson to two neutral fermions, with 3-momenta $\boldsymbol{p}_1$ and $\boldsymbol{p}_2$. Suppose the interaction Lagrangian contains the bilinear $\overline{\psi}F\psi$, where F is some numerical matrix. It is possible that the field operator $\overline{\psi}$ has created the particle with momentum $\boldsymbol{p}_1$ and

ψ has created the other particle with momentum $\boldsymbol{p}_2$; but it can also be the other way around. Reading the Feynman rules for these two possibilities from Table 22.1, we conclude that the Feynman amplitude for the process will be of the form

$$\mathcal{M} = \lambda^* \Big[\overline{u}(\boldsymbol{p}_1) F v(\boldsymbol{p}_2) - \overline{v}(\boldsymbol{p}_1) F u(\boldsymbol{p}_2) \Big] \mathcal{M}_0 \,, \tag{22.55}$$

where $\mathcal{M}_0$ is a factor that comes from the field operator of the initial state particle. The minus sign between the two terms in Eq. (22.55) appears for the usual reason: taking two fermions in opposite order. If the final state were the particle-antiparticle pair of a Dirac fermion, the amplitude would have contained only one of these two terms, depending on which one is particle and which one is antiparticle.

Eq. (22.55) can be written in a more useful form by using Eq. (22.53) to deduce

$$\overline{v}(\boldsymbol{p}_1) F u(\boldsymbol{p}_2) = u^{\top}(\boldsymbol{p}_1) \mathbb{C}^{-1} F \gamma_0 \mathbb{C} v^*(\boldsymbol{p}_2) \,. \tag{22.56}$$

Further, since the expression is ultimately a number, we can also write it as the transpose of the matrix expression on the right hand side. Using the properties of the matrix $\mathbb{C}$ to be found in Appendix F, we obtain

$$\mathcal{M} = \lambda^* \, \overline{u}(\boldsymbol{p}_1) \Big[F + \mathbb{C} F^{\top} \mathbb{C}^{-1} \Big] v(\boldsymbol{p}_2) \mathcal{M}_0 \,. \tag{22.57}$$

□ **Exercise 22.8** Show that Eq. (22.57) can be deduced from Eq. (22.55).

We can easily check explicitly how this form of the Feynman amplitude affects the decay width of the Z boson. Purely from kinematical considerations, we found that the decay rate should be proportional to $(a^2 + b^2)$ if the final particle mass is neglected, where a and b are the strengths of the polar and axial vector couplings in the interactions, as in Eq. (19.26, p 565). The associated algebra was performed in §19.2 for a Dirac particle-antiparticle pair in the final state. If, instead, the final state particles are Majorana neutrinos, the Feynman amplitude derived from the same interaction Lagrangian would be different because of the extra term in Eq. (22.57). When we add this extra term, we find that the polar vector term does not contribute to the Feynman amplitude at all because of the definition of the matrix $\mathbb{C}$ appearing in Eq. (6.57, p 161). Due to the same definition, the axial vector term gets a contribution two times bigger in the Feynman amplitude. In other words, instead of Eq. (19.28, p 566), the amplitude turns out to be

$$\mathcal{M} = 2b \, \overline{u}_1(\boldsymbol{p}_1) \gamma^\mu \gamma_5 v_2(\boldsymbol{p}_2) \epsilon_\mu(\boldsymbol{k}) \,, \tag{22.58}$$

where $\epsilon_\mu(\boldsymbol{k})$ denotes the polarization vector of the Z boson. In the square of the amplitude, instead of the factor $(a^2 + b^2)$, we should obtain a factor $4b^2$. There is also an extra factor of $1/2!$ since we now have two identical particles

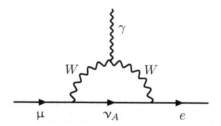

Figure 22.5: One-loop diagram for the process $\mu \to e + \gamma$. In any renormalizable gauge, there are extra diagrams in which one or both of the internal W bosons are replaced by the unphysical charged Higgs boson.

in the final state, as explained right after Eq. (5.45, *p 123*) in the context of electron–electron scattering. Thus, the decay rate into two Majorana neutrinos can be obtained from Eq. (19.32, *p 566*) by replacing the factor $(a^2 + b^2)$ by $2b^2$.

Now, here comes the punch line. Neutrino coupling to the Z boson is purely left-chiral, i.e., $a = b$. Therefore, $(a^2 + b^2)$ is equal to $2b^2$. The intrinsic Majorana phase does not appear in the expression for the rate, which confirms our earlier comment that this phase is unphysical. So, after this entire exercise, we see that the Dirac or Majorana nature of the neutrino does not make any difference in the Z boson decay rate. This is a statement that is correct only if the neutrino mass is zero. For non-zero masses, there will be corrections in the expression for the decay rate, and the rate will depend on the Dirac or Majorana nature of the neutrino. However, for practical purposes this difference is useless since the corrections would contain the factor m_ν/M_Z, which is tiny. We therefore have to look for other kind of signatures in order to distinguish between the Dirac or Majorana nature of neutrinos, a topic that will be discussed in §22.5.

22.4 Consequences of lepton mixing

We have already discussed neutrino oscillations, which are consequences of neutrino mixing. Here, we discuss more processes which might occur because of lepton mixing. None of these processes has been observed yet.

22.4.1 Lepton flavor violation

Lepton mixing implies that the generational lepton numbers are not conserved. One can think of many such processes. The most easily observable would be processes in which the initial and final states do not contain any neutrino. No such process has been observed so far. We discuss some processes which are expected, and discuss what sort of rates to expect for them.

a)   Radiative decays

For example, the muon might decay into the electron with the emission of a photon:

$$\mu \to e + \gamma. \tag{22.59}$$

Of course it cannot happen at the tree level. Fig. 22.5 shows how the process might occur at the one-loop level. There can be similar decays of the τ lepton, where the final state can contain either the muon or the electron.

Experimentally, only an upper bound is known for the branching ratio of the process in Eq. (22.59):

$$\mathscr{B}(\mu \to e + \gamma) < 2.4 \times 10^{-12}. \tag{22.60}$$

Let us see how much we can guess about the amplitude and rate for the process without actually evaluating the diagrams. In §5.7.1, we found the most general form for the electromagnetic vertex involving the same fermion in the initial and final states. The result was shown in Eq. (5.121, *p 140*). The same analysis can be carried out for the case when the initial and the final fermions are different. The outcome of the analysis is that the vertex function will be of the form

$$\Gamma_\lambda = \Big(F + F_5 \gamma_5\Big)\sigma_{\lambda\rho} q^\rho, \tag{22.61}$$

where q is the photon momentum, and F and F_5 are form factors, both of which can be functions of q^2 if we consider the general case where the photon need not be on-shell.

- ☐ **Exercise 22.9** *Prove that the most general form of the electromagnetic vertex function is given by Eq. (22.61) when two different fermions are involved at the vertex.* [**Hint** : *Start from the most general form allowed by Lorentz invariance, and use the gauge invariance condition of Eq. (5.120, p 140).*]

- ☐ **Exercise 22.10** *From the expression of the vertex function, show that the decay rate of the process $f \to f' + \gamma$ is given by*

$$\Gamma = \frac{(m^2 - m'^2)^3}{8\pi m^3}\left(|F|^2 + |F_5|^2\right). \tag{22.62}$$

For the physical process of radiative decay as shown in Eq. (22.59), we have $q^2 = 0$ of course. So we need the values of $F(0)$ and $F_5(0)$. Since these are constants, it is easy to take inverse Fourier transform and go back to the co-ordinate space representation, which immediately tells us that both $F(0)$ and $F_5(0)$ have mass dimension -1.

With this in mind, let us see what are the factors that will definitely occur in the amplitude. Since internal W lines are involved, their propagators and couplings should provide factors of g^2/M_W^2, i.e., a factor of order G_F. The charged current Lagrangian of Eq. (22.4) shows that there should also be the factors $U_{A\mu}U_{Ae}^*$ from the two weak vertices. The photon vertex gives a factor of e.

Moreover, Eq. (22.61) shows that the effective operator contains the bilinears $\overline{\psi}_{(e)}\sigma_{\lambda\rho}\psi_{(\mu)}$ and $\overline{\psi}_{(e)}\sigma_{\lambda\rho}\gamma_5\psi_{(\mu)}$, which connect a left-chiral field to a right-chiral field, as seen in Eq. (14.26, *p 417*). Thus, either the initial muon has to be right-chiral and the final electron left-chiral, or just the other way around. However, only left-chiral fields participate in charged current gauge interactions. The only term in the Lagrangian that can change the chirality of a fermion is the mass term. The implication is that the amplitude must contain a factor of the mass of the external fermions. If we neglect the electron mass compared to the muon mass, only the muon mass can appear. Thus, if $m_e = 0$, the initial state contains a right-chiral muon, which turns left-chiral just because chirality is not conserved for massive particles. This left-chiral muon then participates in charged current weak interactions. In the effective operator, we must then have $F(0) = F_5(0)$ so that we can start with a right-chiral muon. Combining all these arguments, we see that the amplitude must be of the form

$$F = F_5 = eG_F m_\mu \sum_A U_{\mu A} U_{eA}^* \, f(M_W, m_{\nu_A}) \,, \qquad (22.63)$$

where f denotes some function which depends on the masses of the internal particles, and whose form can be obtained only by evaluating the diagrams carefully.

From the dimensional argument given earlier, it follows that the function f has to be dimensionless. Therefore, instead of two separate parameters M_W and m_{ν_A}, it can depend only on their ratio. Since each end of the neutrino line comes with the left-chiral projector $\mathbb{L}$ from the vertex, the mass term in the numerator of the neutrino propagator cannot contribute. The only dependence on the neutrino mass comes from the denominator, where the mass appears as squared. In summary, the function f can depend only on the ratio $(m_{\nu_A}/M_W)^2$. The ratio is bound to be small, so we can think of expanding the function f as a power series in it. The constant term will not contribute at all in the amplitude: because of the unitarity of the mixing matrix, the sum over A will vanish for the constant term. This is the phenomenon of GIM cancellation that we have encountered in case of quarks earlier. It occurs for the same reason here: although the matrix is not the quark mixing matrix, it is a unitary matrix.

The linear order terms in the power series expansion do not cancel, so we obtain

$$F = F_5 \sim eG_F m_\mu \sum_A U_{\mu A} U_{eA}^* \left(\frac{m_{\nu_A}}{M_W}\right)^2 \,. \qquad (22.64)$$

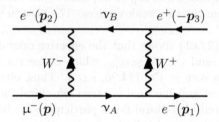

Figure 22.6: One-loop box diagram for the process $\mu \to 3e$. Recall our convention of arrows described in §5.2, by which the inward going arrow on the positron line with momentum $-p_3$ implies an outgoing positron with momentum p_3. In any renormalizable gauge, there are extra diagrams in which one or both of the internal W bosons are replaced by the unphysical charged Higgs boson. In addition, there is another set of diagrams in which the two outgoing electron momenta are interchanged.

Using Eq. (22.62) and comparing with the usual muon decay mode given in Eq. (14.117, *p 434*), we obtain

$$\mathscr{B}(\mu \to e + \gamma) \sim 192\pi^3 \alpha \left| \sum_A U_{\mu A} U_{eA}^* \left(\frac{m_{\nu_A}}{M_W} \right)^2 \right|^2. \qquad (22.65)$$

For any acceptable value of neutrino masses, this branching ratio is well below 10^{-20}, and therefore is far from the presently known limit shown in Eq. (22.60). This explains why the process has not been seen so far. In fact, if it is observed in the near future, it will necessitate some new and unusual physics beyond just neutrino mass and mixing.

Processes like $\tau \to \mu + \gamma$ and $\tau \to e + \gamma$ are also expected to occur because of neutrino mixing. They are also expected to be very suppressed for exactly the same reason.

b) Purely leptonic decays

We can also contemplate decay processes involving charged leptons and antileptons only, e.g.,

$$\mu^- \to e^- e^- e^+ \qquad (22.66)$$

and its charge conjugate, as well as similar decays of the τ lepton where the final state can contain both muons and electrons. The process written in Eq. (22.66) is often dubbed $\mu \to 3e$, knowing fully well that whoever reads or hears about it would know that one of the final state particles must be oppositely charged from the other two because of electric charge conservation. This process can occur at the one-loop level. One possible diagram is obtained by attaching an outgoing $e^+ e^-$ pair to the photon line of Fig. 22.5 *(p 685)*, whereby the photon line itself becomes an internal virtual line. There are other possibilities shown in Fig. 22.6.

c) Processes involving neutrinos

There can be flavor-changing processes involving neutrinos as well. For example, all but the lightest neutrino should decay to a lighter neutrino and a photon:

$$\nu_A \to \nu_B + \gamma. \tag{22.67}$$

The diagrams will be similar to that in Fig. 22.5 *(p 685)*, except that the charged lepton lines and the neutrino lines are interchanged. Because the charged lepton appears as an internal line, the photon can also come out of that line instead of the W boson line. The analysis is very similar to that of the process $\mu \to e + \gamma$. The rates are very small, not only because of GIM cancellation, but also because the neutrino masses are very small.

d) Processes involving hadrons

Hadronic decay products can also reveal lepton flavor violation. No such process has been observed. We give here some examples of such decay channels, along with the experimental upper bound for the branching ratio of each channel. The list is, of course, not exhaustive.

$$\mathcal{B}(\pi^0 \to \mu^+ e^-) < 3.8 \times 10^{-10}; \tag{22.68a}$$

$$\mathcal{B}(\eta \to \mu^+ e^-) < 6 \times 10^{-6}; \tag{22.68b}$$

$$\mathcal{B}(K^+ \to \pi^+ \mu^+ e^-) < 1.3 \times 10^{-11}; \tag{22.68c}$$

$$\mathcal{B}(K^0_L \to \mu^\pm e^\mp) < 4.7 \times 10^{-12}. \tag{22.68d}$$

22.4.2 CP violation in leptonic sector

In §21.2, we found that the mixing matrix in the quark sector can be responsible for CP violation. In the same way, the leptonic mixing matrix can contain observable phases which can induce CP violation.

The number of such phases would be exactly the same as that in the quark sector if the neutrinos are Dirac fermions. If the neutrinos are Majorana particles, the story would be different. In §21.2, we mentioned that, for N generations of fermions, we can remove N phases from the mixing matrix by redefining the phases of the up-type quark fields, and N more by redefining the phases of the down-type quark fields, with one constraint. In the leptonic sector, if the neutrinos are Majorana particles, we cannot redefine their phases. Thus, only N phases can be removed, and we are left with

$$\frac{1}{2}N(N+1) - N = \frac{1}{2}N(N-1) \tag{22.69}$$

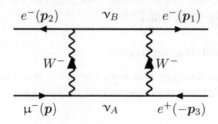

Figure 22.7: Extra box diagrams for the process $\mu \to 3e$ if the neutrinos are Majorana particles. There are several accompanying diagrams, as described for Fig. 22.6.

phases which can inflict CP violation. In other words, compared to the Dirac case, there are $N - 1$ extra phases. These extra phases are often called Majorana phases for CP violation. The presence of such phases would mean, among other things, that even with two generations of fermions one can observe CP-violating effects. For three generations of fermions, we should expect three CP-violating phases of which two would be Majorana phases, provided the neutrinos are Majorana fermions.

For a while, it was believed that the Majorana phases can only be observable in lepton number-violating processes, since these appear only if the neutrinos are Majorana particles, and the existence of Majorana particles requires lepton number violation. It was later realized that it need not be so. For example, consider $\mu \to 3e$, which can proceed through the diagram of Fig. 22.6. However, with Majorana neutrinos, diagrams like the one in Fig. 22.7 are also possible. Lepton number is broken on each of the solid lines in this diagram, so the CP-violating Majorana phases will enter the amplitude. But lepton number is violated by equal and opposite amounts on the two solid lines, so the overall diagram conserves lepton number.

As for the case of quarks, CP violation in the leptonic sector can also be described in terms of rephasing invariant parameters. If neutrinos are Dirac particles, the number and structure of these rephasing invariants are exactly the same as those in the quark sector, except that it is the lepton mixing matrix which should be used to evaluate their magnitude. But if the neutrinos are Majorana fermions, there will be more CP-violating phases, and consequently more rephasing invariants.

This can be seen most easily by taking $\alpha = 0$ in the definition of Majorana fermions given in Eq. (22.49). In this case, the phase of the Majorana field cannot be arbitrarily changed, and we cannot use the analogs of phase redefinitions shown in Eq. (21.5, *p 623*) for the neutrinos. The charged lepton fields can be changed by giving them phase transformations

$$\ell_{\mathrm{L}} \to e^{i\theta_\ell}\ell_{\mathrm{L}}\,. \tag{22.70}$$

This inflicts the transformation

$$U_{\ell A} \to e^{-i\theta_\ell} U_{\ell A} \,, \tag{22.71}$$

and we need to find which combinations of the mixing matrix elements are invariant under these transformations. Clearly, the lowest order combinations that satisfy this criterion are of the form

$$S_{AB\ell} \equiv U_{\ell A} U_{\ell B}^* \,, \qquad \text{(no sum)} \,. \tag{22.72}$$

For $A = B$, these are just the absolute squares of mixing matrix elements, which do not contain information about CP violation, as argued in the context of quarks in §21.3. The CP-violating independent parameters can be taken to be

$$\mathrm{Im}\left(S_{AB\ell}\right) \qquad \text{with } A \leq B \,. \tag{22.73}$$

Clearly, the number of such parameters agrees with the number of CP-violating phases determined in Eq. (22.69).

☐ **Exercise 22.11** *One can also form the quartic invariants, of the type shown in Eq. (21.8, p 623), for the leptonic sector. Show that they can be expressed in terms of the quadratic invariants given in Eq. (22.72) and hence can be disregarded in the list of independent CP-violating parameters.*

22.5   Lepton number violation

The processes mentioned in §22.4.1 are expected to take place at some level or other, irrespective whether neutrinos are Dirac or Majorana particles. If neutrinos happen to be Majorana particles, the Lagrangian should contain lepton number-violating terms, and therefore one should also expect processes which violate lepton number. In this section, we discuss some such possibilities.

22.5.1   Neutrinoless double beta decay

It is known that there are some nuclei that cannot undergo beta decay because it is energetically forbidden. For example, consider the selenium isotope $^{82}_{34}\mathrm{Se}$. Normal beta decay would have produced $^{82}_{35}\mathrm{Br}$ from it, but it cannot happen because the ground state of the bromine isotope is higher than that of the selenium isotope. However, the element with the next higher atomic number, krypton, has a ground state that is much lower than the selenium ground state, so the decay

$$^{82}_{34}\mathrm{Se} \to {}^{82}_{36}\mathrm{Kr} + 2e^- + 2\widehat{\nu}_e \tag{22.74}$$

is possible. Such processes have been observed and are called *double beta decay*.

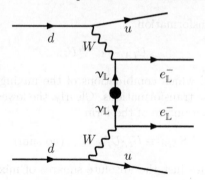

Figure 22.8: Quark-level diagram of $\beta\beta_{0\nu}$ process induced by Majorana neutrino mass.

If lepton number is violated, there is also the possibility of observing decays where there is no neutrino in the final state:

$$X(A, Z) \rightarrow Y(A, Z + 2) + 2e^- \tag{22.75}$$

where X and Y are two nuclei whose mass numbers and atomic numbers are shown in parentheses. At the nucleon level, this will mean transitions of the type

$$n + n \rightarrow p + p + e + e, \tag{22.76}$$

and at the quark level,

$$d + d \rightarrow u + u + e + e. \tag{22.77}$$

Fig. 22.8 shows how this process can occur at the tree level provided the neutrino is a Majorana particle. Such processes are given the name *neutrinoless double beta decay*, and sometimes abbreviated to $\beta\beta_{0\nu}$.

We now argue that the amplitude of the process must contain a factor of the neutrino mass. For this, it is enough to consider just the vertices involving the neutrino and the neutrino propagator, and neglect neutrino mixing for the time being. This factor is then

$$\mathcal{M}_{\lambda\rho} = \Big[\bar{u}_e(p_1)\gamma_\lambda \mathsf{L}\Big]_a \Big[\bar{u}_e(p_2)\gamma_\rho \mathsf{L}\Big]_b \Big[S_{(\nu\nu)}(q)\Big]_{ab} - (p_1 \leftrightarrow p_2). \tag{22.78}$$

Here, a, b are Dirac indices, and $S_{(\nu\nu)}(q)$ denotes the neutrino propagator with momentum q. In a matrix notation, we can write this expression as

$$\mathcal{M}_{\lambda\rho} = \Big[\bar{u}_e(p_1)\gamma_\lambda \mathsf{L}\Big] \Big[S_{(\nu\nu)}(q)\Big] \Big[\bar{u}_e(p_2)\gamma_\rho \mathsf{L}\Big]^\top - (p_1 \leftrightarrow p_2). \tag{22.79}$$

Note that, using Eq. (6.57, *p 161*), we can write

$$\Big[\bar{u}_e(p_2)\gamma_\rho \mathsf{L}\Big]^\top = \mathsf{L}^\top \gamma_\rho^\top \gamma_0^\top u_e^*(p_2) = \mathbb{C}^{-1}\mathsf{L}\gamma_\rho\gamma_0 \mathbb{C} u_e^*(p_2)$$

$$= \mathbb{C}^{-1}\mathsf{L}\gamma_\rho v_e(p_2) = \mathbb{C}^{-1}\gamma_\rho \mathbb{R} v_e(p_2), \tag{22.80}$$

using the relationship between the u-spinors and the v-spinors, Eq. (22.53), in the last step.

Let us now look at the propagator. We have denoted it by $S_{(\nu\nu)}$, and let us begin by explaining the parenthetical subscript. At the two vertices, interaction terms of the type $\bar{e}\gamma^\mu \nu W_\mu$ produce the two electrons, and two neutrino field operators are left over. None of them is $\bar{\nu}$. Rather, both are just the field ν, which is why we have put the subscripts on the propagator. To obtain the propagator, we should express the free Lagrangian of the Majorana neutrino field in terms of the ν fields at both ends, i.e., should not use $\bar{\nu}$. This can be easily done by using Eq. (22.48), and we can write the free Dirac Lagrangian for a Majorana field as

$$\mathscr{L} = \frac{1}{2}\nu^\top \mathbb{C}^{-1}\left(i\gamma^\mu \partial_\mu - m\right)\nu, \tag{22.81}$$

where the overall factor of $\frac{1}{2}$ is used for self-conjugate fields so that one can combine them to make complex fields, as has been shown for scalar fields in Eq. (4.101, *p81*). We can now follow the procedure of §4.10.2 to find the propagator. Without the factor of $\mathbb{C}^{-1}$, we would have obtained $S_F(q)$, as in Eq. (4.145, *p92*). Because of the $\mathbb{C}^{-1}$, we get an extra factor of $\mathbb{C}$ after $S_F(q)$ while taking the inverse. Thus, finally, we can write

$$\mathscr{M}_{\lambda\rho} = \bar{u}_e(p_1)\gamma_\lambda \mathbb{L}\,\frac{\slashed{q}+m_\nu}{q^2-m_\nu^2}\,\gamma_\rho \mathbb{R} v_e(p_2) - (p_1 \leftrightarrow p_2). \tag{22.82}$$

It is now clear that the $\slashed{q}$-term in the numerator of the propagator does not contribute to the amplitude due to the chiral projection operators. Neglecting the neutrino mass term in the denominator, we can write

$$\mathscr{M}_{\lambda\rho} = \frac{m_\nu}{q^2}\,\bar{u}_e(p_1)\gamma_\lambda\gamma_\rho \mathbb{R} v_e(p_2) - (p_1 \leftrightarrow p_2), \tag{22.83}$$

which shows the proportionality with neutrino mass.

It is easy to include the effect of neutrino mixing. Let us consider that the internal neutrino line is the eigenstate ν_A. There will be two factors of mixing matrix elements. The factor of m_ν in Eq. (22.83) will thus be replaced by

$$\langle m_\nu \rangle \equiv \sum_A m_{\nu_A}\left(U_{eA}\right)^2. \tag{22.84}$$

Thus, m_ν replaced by this combination, $\langle m_\mu \rangle$, will give the leptonic part of the Feynman amplitude. The other part involves the matrix element involving the initial and the final nuclei. There is a considerable amount of uncertainty in estimating this part of the amplitude due to nuclear effects. However, a non-zero measurement would definitely tell us that there is lepton number violation, and will determine the combination of neutrino masses and mixings, shown in Eq. (22.84), within some limits of error. A null experiment will put an upper bound on this combination. There has been one claim of a non-null signal, but the consensus is that till now, there is only an upper bound,

$$\left| \langle m_\nu \rangle \right| \lesssim 0.5 \, \text{eV} . \tag{22.85}$$

If we combine this result with the results of neutrino oscillation, we can obtain much more information about neutrino masses. Because the different mass-squared values are bounded by solar and atmospheric data, we can correlate the smallest neutrino mass eigenvalue and the value of $\langle m_\nu \rangle$. The correlated values depend on the type of mass hierarchy that exists among the neutrinos — normal or inverted — and might be useful in determining which kind of hierarchy exists in nature.

22.5.2   Lepton to antilepton conversion

One can also think of the inverse process of the neutrinoless double beta decay, viz.,

$$e^+ + X(A, Z) \rightarrow e^- + Y(A, Z + 2) \tag{22.86}$$

or its reverse. Or one can also think of such processes involving charged leptons of different generations, e.g.,

$$\mu^+ + X(A, Z) \rightarrow e^- + Y(A, Z + 2) . \tag{22.87}$$

Such processes have not been observed, but surely, their observation would indicate lepton number violation.

☐ **Exercise 22.12** *Draw a Feynman diagram for the process*

$$\mu^+ + W^- \rightarrow e^- + W^+ \tag{22.88}$$

in a model with neutrino mixing. Show that the matrix element squared can be expressed in terms of the rephasing invariants $S_{AB\ell}$ introduced in Eq. (22.72).

22.6   Models of neutrino mass

In the standard electroweak model proposed by Weinberg and Salam, neutrinos were considered massless, consistent with the experimental knowledge that was available at that time. Now we know that neutrinos have mass. So we need to modify the standard model to accommodate this fact. In this section, we discuss various ways of doing that, as well as some general strategies for achieving the goal.

22.6.1   Adding singlet neutrinos

The most obvious way of obtaining massive neutrinos is to imitate the method by which all other fermions get their masses. The idea has already been described in §22.1 in some detail. We introduce right-chiral chargeless fermion fields, $\nu_{\ell R}$, one for each generation. There can be extra Yukawa interaction terms of the form

$$\mathscr{L}_Y = -\sum_{\ell,\ell'} \tilde{h}_{\ell\ell'} \overline{\psi}_{\ell L} \tilde{\phi} \nu_{\ell' R} + \text{h.c.} \tag{22.89}$$

Once the scalar doublet obtains VEV, a mass matrix will be generated for the neutrinos:

$$m^{(D)}_{\ell\ell'} = \tilde{h}_{\ell\ell'} v / \sqrt{2}. \tag{22.90}$$

Diagonalization of this matrix will give the mass eigenvalues of neutrinos and mixing angles in the leptonic sector. The parenthetical superscript 'D' indicates that the neutrinos will be Dirac particles in this case, like all other fermions.

This explanation of neutrino masses might be straightforward, but it is unsatisfactory for many reasons. Firstly, it gives absolutely no hint to the two puzzles of neutrino mass mentioned at the end of §22.2: why the masses are so small and mixings (at least some of them) so large. Sure, the elements of the the mass matrix might happen to be such that both these features are achieved, but that's hardly an explanation!

The second reason for the dissatisfaction is that the model is incomplete. Since the right-chiral neutrinos are $SU(2)_L$ singlets, the electric charge formula given in Eq. (16.18, *p465*) implies that its weak hypercharge should be zero. Thus the electroweak gauge representation of the right-chiral neutrinos is given by:

$$\nu_{\ell R} : (1,0), \tag{22.91}$$

in the notation that was used in Eqs. (16.19) and (16.40). In other words, these fields are invariant under the entire gauge group of the standard model. Therefore, any combination made from these fields alone would be gauge invariant. In particular, we can put in the Lorentz invariant combinations

$$-\frac{1}{2} \sum_{\ell,\ell'} M_{\ell\ell'} \nu_{\ell R}^{\top} \mathbb{C}^{-1} \nu_{\ell' R} + \text{h.c.} \tag{22.92}$$

If there were no other mass terms for the neutrinos, these would have given Majorana masses to the singlet neutrinos.

This second problem mentioned above is very easy to rectify. We can just decide to consider these extra mass terms to be added to the ones shown in Eq. (22.89). Then all mass terms for neutral fermions can be summarized in an expression of the form

$$\mathscr{L}_{\text{mass}} = -\frac{1}{2} \left(\overline{\nu}_L \quad \overline{\widehat{\nu}}_L \right) \begin{pmatrix} 0 & m^{(D)} \\ (m^{(D)})^{\top} & M \end{pmatrix} \begin{pmatrix} \widehat{\nu}_R \\ \nu_R \end{pmatrix} + \text{h.c.} \tag{22.93}$$

We need to explain the notation that has been used in writing this expression. Objects like ν_L that appear in Eq. (22.93) are, for N generations of fermions, collections of N fields of a certain chirality. M and $m^{(D)}$ are both $N \times N$ matrices whose elements have been given in Eqs. (22.90) and (22.92). In order to write the Dirac mass terms, we have used

$$\overline{\nu}_L \nu_R + \overline{\nu}_R \nu_L = \frac{1}{2}\left(\overline{\nu}_L \nu_R + \overline{\nu}_R \nu_L\right) + \text{h.c.}\,, \qquad (22.94)$$

and then used the identity

$$\overline{\nu}_{\ell R}\nu_{\ell' L} = \overline{\widehat{\nu}}_{\ell' L}\widehat{\nu}_{\ell R} \qquad (22.95)$$

to rewrite the hermitian conjugate parts.

□ **Exercise 22.13** *Prove Eq. (22.95).*

□ **Exercise 22.14** *Use Eq. (22.95) and similar equations to show that the mass matrix, written in the form of Eq. (22.93), must be a symmetric matrix.*

In order to obtain the mass eigenvalues of neutral leptons, we need to diagonalize the $2N \times 2N$ mass matrix that appears in Eq. (22.93). To get a feel for what kind of things may be expected, let us consider the case of $N = 1$, for which the matrix is 2×2, and M as well as $m^{(D)}$ are numbers. Let us also assume that $M \gg m^{(D)}$. In this case, diagonalization of the mass matrix will yield two eigenvalues,

$$m_1 \approx \frac{(m^{(D)})^2}{M}\,, \qquad m_2 \approx M\,. \qquad (22.96)$$

The eigenstate with the smaller eigenvalue will be chiefly the state which is $SU(2)_L$ doublet, whereas the eigenstate with mass M will be chiefly the singlet.

There is a tricky point here that is worth noting. For the single generation case, the eigenvalues of the matrix that appears in Eq. (22.93) are

$$\mu_1 \approx -\frac{(m^{(D)})^2}{M}\,, \qquad \mu_2 \approx M\,. \qquad (22.97)$$

Note the minus sign in the first eigenvalue. Both these eigenvalues cannot be positive for real entries in the mass matrix, and therefore cannot be the physical masses of the two eigenstates. In writing Eq. (22.96), we assumed that M is positive, so it qualifies as the mass of one eigenstate. The other eigenvalue is then negative. However, for fermion fields, the sign of the mass term is just a matter of convention. Suppose we have some fermion field ψ for which the free Lagrangian is

$$\mathcal{L} = \overline{\psi}i\gamma^\mu \partial_\mu \psi + \mu\overline{\psi}\psi\,, \qquad (22.98)$$

with $\mu > 0$, i.e., same as the free Dirac Lagrangian except for the sign of the non-derivative term. We can now define

$$\widetilde{\psi} = \gamma_5 \psi\,, \qquad (22.99)$$

then it is easily seen that, in terms of this new field, the Lagrangian reads

$$\mathscr{L} = \overline{\widetilde{\psi}} i \gamma^\mu \partial_\mu \widetilde{\psi} - \mu \overline{\widetilde{\psi}} \widetilde{\psi} \,, \tag{22.100}$$

which has the right sign for the mass term. Since Eq. (22.99) implies $\widetilde{\psi}_{\text{L}} = -\psi_{\text{L}}$ and $\widetilde{\psi}_{\text{R}} = +\psi_{\text{R}}$, the multiplication by γ_5 implies that, in the interpretation of the mass term, we need to put in an extra minus sign in the left-chiral components of the fields compared to the right-chiral ones. So, to summarize, the masses of the physical particles are the absolute values of the eigenvalues, which is what we had written in Eq. (22.96).

□ **Exercise 22.15** Verify Eq. (22.100).

In a sense, the eigenvalue structure shown in Eq. (22.96) solves the other problem of neutrino mass for us. After all, the neutrinos whose masses we measure are those which take part in gauge interactions, i.e., are non-singlets. Eq. (22.96) shows that the masses of these neutrinos are much smaller than $m^{(D)}$, the mass that is obtained from Yukawa couplings with doublet Higgs. Since all charged fermions get their masses from their couplings with doublet Higgs bosons, this provides a rationale of the unusual lightness of neutrinos. The generic idea that the active neutrinos are light because some other neutral fermions are heavy goes by the name of *seesaw mechanism*.

It should be noted that we have obtained two neutrino eigenstates per generation, as is indicated by Eq. (22.96). With the same number of degrees of freedom, we got one charged lepton in a generation. This means that each of the neutrino eigenstates has half the number of degrees of freedom that a Dirac fermion possesses. This can happen if the neutrino eigenstates are Majorana particles. The seesaw mechanism of explaining the lightness of neutrino masses is therefore intimately connected to the Majorana nature of the neutrinos.

Nevertheless, the model is still not satisfactory. The whole point of gauge symmetries is to restrict and relate interactions between different particles. If we introduce gauge singlets, they do not have any gauge interaction and seem like outsiders in the gauge theory. A gauge singlet is acceptable only if it belongs to a non-trivial representation of a bigger gauge group that is valid at some higher energy. We will say more about it in Ch. 23.

22.6.2 Effective Lagrangian for neutrino mass

If introducing singlet neutrinos is not a good way of explaining neutrino masses, we will have to make do with the doublet neutrinos only. In this case, the neutrinos will have to be Majorana particles. Since Majorana masses violate lepton number, we can ask how one can violate lepton number with just the standard model multiplets.

Of course, no renormalizable operator will violate lepton number, something we have discussed in §18.1. If we look for dimension-5 operators, we find one combination that violates lepton number:

$$\mathscr{L}_5 = \frac{f}{M} \, \Psi_{\text{L}} \Psi_{\text{L}} \phi \phi \,, \tag{22.101}$$

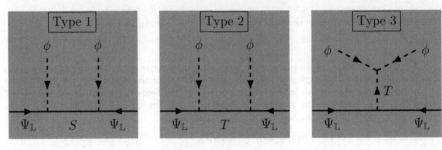

Figure 22.9: Different versions of the seesaw mechanisms for generation of neutrino mass. The labels S and T are indicative of whether the relevant particle transforms like a singlet or a triplet of the $SU(2)_L$ part of the standard electroweak gauge group.

where f is dimensionless and M has dimensions of mass. We have merely written the multiplets and not indicated how they combine to make the gauge invariant operator. But this is enough to ensure that the combination is invariant under the $U(1)_Y$ part of the gauge group. It is also trivially invariant under the color group $SU(3)_c$, since each multiplet involved in the combination is a color singlet. $SU(2)_L$ invariance can also be achieved, in fact in a number of ways, which we will discuss presently.

Clearly, from considerations of the $U(1)_Y$ part of the gauge group alone, we see that one component of the interaction of Eq. (22.101) is of the form $f\nu_L\nu_L\phi_0\phi_0/M$. When the doublet Higgs multiplet obtains a VEV $v/\sqrt{2}$ and we rewrite this term in terms of the quantum fields, one of the terms will be $(fv^2/2M)\nu_L\nu_L$. This is a Majorana mass term for the field ν_L, with mass eigenvalue fv^2/M.

We have been discussing the matter as if there is just one lepton doublet Ψ_L. In reality, there are more, one for each generation. The lepton doublets therefore carry a generation index. Accordingly, there are generation indices on f. This means that, after the Higgs doublet ϕ obtains a VEV, the mass matrix generated for the neutrinos would have the elements

$$f_{\ell\ell'}\, v^2/M. \tag{22.102}$$

If $M \gg v$, this gives small masses for neutrinos. This is also the seesaw mechanism in action, because smallness of neutrino mass owes its origin to the largeness of M.

We are of course not pleading for introducing non-renormalizable terms like the one in Eq. (22.101). What we are hinting at is that there may be heavy particles in the theory which take part in the mass-generating process as internal lines, and their propagators supply the power of inverse mass that occurs in the expression. This is exactly the way that the non-renormalizable Fermi coupling arises in the low-energy limit of the standard model, as shown in Eq. (16.49, *p474*).

Let us now enumerate different ways of generating the dimension-5 operator at the tree level from renormalizable interactions. Note that $\Psi_{\rm L}$ and ϕ are both doublets of SU(2)$_{\rm L}$. Thus, if we take the combination $\Psi_{\rm L}\phi$, it can transform either as a singlet or as a triplet of SU(2)$_{\rm L}$, with weak hypercharge equal to zero. As far as Lorentz transformations are concerned, the combination transforms like a fermion. Thus, we can make a renormalizable interaction involving $\Psi_{\rm L}\phi$ and another fermion field. This extra field will have $Y = 0$, and will transform like a singlet of SU(2)$_{\rm L}$ if $\Psi_{\rm L}\phi$ is a singlet, and triplet if $\Psi_{\rm L}\phi$ is triplet, so that the overall combination is a gauge singlet. This extra field can act as an intermediate line and produce the dimension-5 operator at low energies, where M will be the mass of this extra field. The two possibilities have been shown in Fig. 22.9 as type-1 and type-2 seesaw mechanisms.

There is a third way that seesaw mechanism might work at the tree level, that is also shown in Fig. 22.9. In this case, we take two $\Psi_{\rm L}$ operators and combine them into a triplet. The two ϕ operators can also be combined into a triplet. The message between these two pairs can be mediated by a scalar boson triplet. There is no fourth alternative in which both $\Psi_{\rm L}\Psi_{\rm L}$ and $\phi\phi$ combine into SU(2)$_{\rm L}$ singlet combinations and the scalar triplet T is replaced by a scalar singlet. Such effective operators might be generated, but they will not contribute to any neutrino mass term.

□ **Exercise 22.16** What is the weak hypercharge of the scalar triplet T that appears in Fig. 22.9?

□ **Exercise 22.17** Why can't neutrino mass be generated from operators in which the combination $\Psi_{\rm L}\Psi_{\rm L}$ appears in an SU(2)$_{\rm L}$ singlet?

A realization of the type-1 seesaw mechanism is the model described in §22.6.1. The fermion field S that appears in Fig. 22.9 represents the singlet neutrinos introduced in that model. The two couplings are precisely of the type shown in Eq. (22.89). For other two types of seesaw mechanism, one can also construct similar renormalizable models. And of course, there can be many more realizations of the dimension-5 effective operator if one considers loop diagrams for generating neutrino masses. We do not discuss any of these models here.

Chapter 23

Beyond the standard model

No matter how successful the standard model may be in describing properties of elementary particles, it is quite obvious that it cannot be the final theory of everything. For one thing, the standard model does not explain or describe gravitation: it ignores gravity for reasons explained in Ch. 1. Moreover, we know that the neutrinos have mass, a feature that was absent in the standard model. So, there is more than one reason to look beyond the standard model. There are various theoretical ideas on this general theme. In this chapter, we will try to give an outline to some such ideas. But first, we will discuss some other shortcomings of the standard model.

23.1 Shortcomings of standard model

Even if we set aside the question of neutrino mass or gravitational interactions, there are theoretical problems within the standard model. For example, let us count the number of parameters that are present in the standard model whose values are not predicted by the model: they have to be determined by experiments. We have enumerated them in Table 23.1, and see that the answer is 19. That is a huge number for anything that wants to be the final

Description	Comment	Number
Gauge coupling constants		3
Parameters in scalar potential	[See Eq. (16.5, *p 463*)]	2
Charged lepton masses		3
Quark masses		6
CKM matrix angles		3
CP-violating phase		1
Strong CP parameter	[See §21.8]	1
Total		19

Table 23.1: Counting parameters of the standard model.

or ultimate theory of interactions. If neutrino masses have to be incorporated in the model, the number increases further.

It should be noticed that we have counted physical parameters only. The CKM matrix, e.g., can have more phases in it, but we have shown in §21.2 that they will be unphysical. Only one phase of the CKM matrix is physical and is responsible for CP violation, which is what we have counted in Table 23.1.

Electric charges of all elementary particles are in simple integral ratios. In fact, if we take the charge of the down quark as the unit, all charges can be written as integral multiples of it. This is a remarkable and surprising fact. When we write the gauge transformation of a bigger group like SU(2), it comes with the generators of the group, and the eigenvalues of these generators are indeed quantized because of the non-trivial algebra of the group. But for U(1), when we specify the gauge transformation as in Eq. (5.2, *p 112*), the value of Q can be anything; there is no restriction at all. And yet, we find that in the real world, the charges are quantized. The standard model does not provide any explanation of this phenomenon of *charge quantization*.

CP violation is observed in weak interactions but not in strong interactions. And yet, if we write the most general renormalizable gauge invariant Lagrangian of the QCD part of the standard model, we find that there is one term that violates CP. This was discussed in §21.8. Of course, the coupling constant multiplying this CP-violating operator is a free parameter, as indicated in Table 23.1, and its value can be tiny. The standard model does not provide any rationale for the tiny value of this parameter.

There is also the question of the Higgs boson mass. Self-energy diagrams for scalar fields are quadratically divergent. Therefore, loop corrections to the Higgs boson mass are of order Λ^2 if we use a cut-off Λ while performing integrations over loop momenta. There does not seem to be any reason why this correction should not be very large. Since we have neglected gravitational interactions all throughout, we can say that the theory is not expected to be valid near or above the Planck mass, beyond which gravitational interactions cannot be neglected. So, if we use the Planck mass as a benchmark cut-off Λ, loop corrections to the Higgs mass squared comes out to be more than 30 orders of magnitude larger than the physical mass of the boson. We can of course say that the bare mass is also of the same order, and the two cancel to give a physical mass of the order of the weak scale. But that would require a huge cancellation. Such cancellations seem unnatural, and are referred to as *fine tuning*. This is certainly not a desirable feature of any model.

To understand why we pick only the Higgs boson mass to make the point, let us contrast it with the question of, say, the electron mass. The electron mass also obtains loop corrections, but these corrections are all proportional to the electron mass itself. In other words, if the electron mass were zero to begin with, it would have remained zero after all loop corrections. There would be a chiral symmetry, much like those described for quarks in Ch. 18, which would ensure that the mass remains zero. Moreover, the electron self-energy diagrams are only logarithmically divergent, meaning that the Λ dependence in the loop corrections would depend only on $\log \Lambda$. Even if Λ is large by many orders of magnitude compared to the electron mass, the correction term, going like $\alpha m_e \log(\Lambda/m_e)$ cannot be very large. For the Higgs boson, there is no symmetry to protect the mass, and this is what causes the fine tuning problem.

These and other problems will be on focus when we discuss ideas beyond the standard model in the rest of this chapter.

23.2   Left–right symmetric model

In the standard model, left-chiral and right-chiral fermion fields are treated very differently, which is why parity violation is inherent in the structure of the model itself. It would be more satisfying if the two kinds of chiral projections can be treated on a more similar footing, breaking parity from the dynamics of the model rather than by the assignments. In order to achieve this goal, we can consider the gauge group to include a factor $SU(2)_R$ in addition to the usual $SU(2)_L$. Just as the left-chiral fermions are doublets under the $SU(2)_L$ that is a part of the standard model gauge group, we will take the right-chiral fermions to be doublets under the $SU(2)_R$ part of an extended gauge group. Thus, instead of the transformation properties listed in Eq. (17.1, *p 482*), we now consider that the quark fields transform as follows:

$$\begin{pmatrix} u_L \\ d'_L \end{pmatrix}, \begin{pmatrix} c_L \\ s'_L \end{pmatrix}, \begin{pmatrix} t_L \\ b'_L \end{pmatrix} \quad : \quad (2,1,X_q),$$

$$\begin{pmatrix} u_R \\ d'_R \end{pmatrix}, \begin{pmatrix} c_R \\ s'_R \end{pmatrix}, \begin{pmatrix} t_R \\ b'_R \end{pmatrix} \quad : \quad (1,2,X_q). \tag{23.1}$$

In the notation for transformation properties, the first number denotes dimensionality of the representation under the $SU(2)_L$ factor of the gauge group. The second number is the dimensionality under the $SU(2)_R$ factor. Finally, the number X_q is the quantum number under a $U(1)$ factor of the gauge group. We will soon see that this number is something quite familiar to us.

If we try to make similar assignment for leptons, the first thing that strikes us is the fact that we cannot do it with only the fields present in the standard model. We must add right-chiral neutrinos to go with the right-chiral components of the charged leptons so that they can form doublets of $SU(2)_R$. In other words, we should take

$$\begin{pmatrix} \nu_{eL} \\ e_L \end{pmatrix}, \begin{pmatrix} \nu_{\mu L} \\ \mu_L \end{pmatrix}, \begin{pmatrix} \nu_{\tau L} \\ \tau_L \end{pmatrix} \quad : \quad (2,1,X_\ell),$$

$$\begin{pmatrix} \nu_{eR} \\ e_R \end{pmatrix}, \begin{pmatrix} \nu_{\mu R} \\ \mu_R \end{pmatrix}, \begin{pmatrix} \nu_{\tau R} \\ \tau_R \end{pmatrix} \quad : \quad (1,2,X_\ell). \tag{23.2}$$

Electric charge of a particle should have a formula similar to Eq. (16.18, *p 465*). Note that the expression for Q must now contain $T_{3L} + T_{3R}$ so that the left and right-chiral components of a fermion acquire the same electric charge. Also, the U(1) quantum numbers should be equal for the left and right-chiral components of the same fermion, as we have already imposed in Eqs. (23.1) and (23.2). We have earlier said many times that any U(1) quantum number has a multiplicative arbitrariness. We can fix the normalization of X in such a way that it appears without any extra numerical factor in the expression for the electric charge Q, i.e., we obtain a formula $Q = T_{3L} + T_{3R} + X$. This then immediately shows that $X_q = \frac{1}{6}$, $X_\ell = -\frac{1}{2}$, implying that for all known fermions, we can express the quantum number X as a simple combination of the baryon number B and lepton number L:

$$X = \frac{B - L}{2} \tag{23.3}$$

and the electric charge formula as

$$Q = T_{3L} + T_{3R} + \frac{B - L}{2}. \tag{23.4}$$

The extended electroweak symmetry can therefore be called $SU(2)_L \times SU(2)_R \times U(1)_{B-L}$. Interestingly, $B - L$, which was a non-anomalous global symmetry of the standard model, has been taken as part of the gauge symmetry of this model. In addition, there is an $SU(2)_R$ symmetry which means that even right-chiral fermions have charged current gauge interactions mediated by the charged gauge bosons of $SU(2)_R$, which we can call $W_R^\pm$. We will presently see why this does not disturb any of the successes of the standard model.

If there is a discrete Z_2 symmetry that interchanges the two SU(2) parts of the interactions, it immediately follows that the interactions of left-chiral and right-chiral fermions are identical and therefore parity is conserved. For this reason, this model is called the left–right symmetric model.

Of course the gauge symmetry has to be broken spontaneously. The chain of symmetry breaking should be as follows:

$$SU(2)_L \times SU(2)_R \times U(1)_{B-L} \longrightarrow SU(2)_L \times U(1)_Y \longrightarrow U(1)_{em}. \tag{23.5}$$

If the first stage of this symmetry breaking occurs at a scale much higher than the electroweak scale, three gauge bosons would obtain masses at the scale. These will include the charged gauge bosons of the $SU(2)_R$ and a combination of the neutral gauge bosons of $SU(2)_R$ and $U(1)_{B-L}$. Charged current gauge interactions involving right-chiral fermions will have an effective Fermi constant that will be given by a formula like that in Eq. (16.49, *p 474*), except that the heavy masses would come in the place of the usual W boson mass. Such interactions will therefore be suppressed, and the non-observation of right-chiral currents can be translated to a lower limit on the W_R masses.

□ **Exercise 23.1** Which representation of scalar fields is necessary to give masses to quarks?

Taking strong interactions into account as well, the gauge group of this model can be written as $SU(3)_c \times SU(2)_L \times SU(2)_R \times U(1)_{B-L}$. Under the $SU(2)$ factors of this group, we know that the transformation of fermion fields depends on chirality. Let us now summarize the transformation properties under the remaining part, i.e., $SU(3)_c \times U(1)_{B-L}$. Quarks of either chirality transform like $(3, \frac{1}{3})$. Leptons transform like $(1, -1)$. In Eq. (10.47, *p 269*), we showed how the fundamental representation of $SU(3)$ decomposes under $SU(2) \times U(1)$. Comparing with that formula, it is easy to convince oneself that the quark and lepton representations are $SU(3) \times U(1)$ decompositions of a fundamental representation of $SU(4)$. It is therefore tempting to conjecture that the gauge group of particle interactions is $SU(4) \times SU(2)_L \times SU(2)_R$, and we see only standard model gauge interactions because symmetry is broken at some high scale whose virtual effects are beyond our experimental accuracies.

This $SU(4)$ part of the gauge group, proposed by Pati and Salam, has the interesting property that fermions belonging to a fundamental representation of this group contain three quarks of three different colors, plus a lepton field. Thus this model has the distinguishing feature that there are gauge interactions involving a quark and a lepton field. Such interactions were absent in the standard model, and also in its left–right symmetric electroweak extension discussed above.

23.3   Grand unified theories

Georgi and Glashow, in 1974, suggested something that goes even further. They tried to unify all three fundamental interactions except gravitation into one simple gauge group.

Before we proceed, we should explain what a *simple group* is. For this, we first need to discuss the notion of an *invariant subgroup*. For any group $\mathcal{G}$, a subset of elements $\mathcal{H}$ form an invariant subgroup if for any $H \in \mathcal{H}$ and any $G \in \mathcal{G}$, the element GHG^{-1} is an element of $\mathcal{H}$. Obviously, for any group $\mathcal{G}$, the trivial group consisting of only the identity element is an invariant subgroup, and the entire group $\mathcal{G}$ itself is also an invariant subgroup. A simple group is a group that has no other invariant subgroup.

Thus, a group of the form $\mathcal{G}_1 \times \mathcal{G}_2$, i.e., where all elements of $\mathcal{G}_1$ commute with all elements of $\mathcal{G}_2$, can never be a simple group since both $\mathcal{G}_1$ and $\mathcal{G}_2$ are invariant subgroups. For continuous groups which are of interest to us in the context of gauge theories, a simple group can be identified as one which cannot be factorized into mutually commuting subgroups, and any of whose generators can be written as the commutator of two generators. A $U(1)$ group therefore is not simple, because it has only one generator that commutes with itself, and consequently the generator itself cannot be written as a commutator. $SU(n)$ groups for $n > 1$ are simple groups. So are $SO(n)$ groups for $n > 1$ with the exception of $SO(4)$, and some others which need not be discussed for our purpose.

□ **Exercise 23.2** Show that the $SO(4)$ algebra is equivalent to $SU(2) \times SU(2)$ algebra. [**Hint :** *See the algebra of the Lorentz group discussed in Ch. 3.*]

If a gauge group has two or more commuting factors, there can be one gauge coupling constant for each of these factors. The standard model gauge group has three gauge coupling constants corresponding to the $SU(3)_c$, $SU(2)_L$ and $U(1)_Y$ factors. If one chooses a simple group as the gauge group, there can be only one gauge coupling constant. In this sense, all gauge interactions can be unified in this one constant.

Of course, the gauge group must have $SU(3)_c \times SU(2)_L \times U(1)_Y$ as a subgroup, so that in the process of symmetry breaking we can obtain the standard model gauge group to be valid at some energy scale. This means that the group must have at least a rank of 4, since, as mentioned in §10.3, the ranks of $SU(3)$ and $SU(2)$ are 2 and 1 respectively, and the rank of $U(1)$ is obviously 1. Georgi and Glashow analyzed all rank-4 simple Lie groups and found that only $SU(5)$ fits the bill. To explain what we mean by 'fitting the bill', let us consider the decompositions of some of the lowest dimensional representations of $SU(5)$ under the standard model gauge group. The fundamental representation decomposes as

$$5 \longrightarrow (3, 1)_{-\frac{1}{3}} + (1, 2)_{\frac{1}{2}} \,. \tag{23.6}$$

We have written the $SU(3) \times SU(2)$ representations in parentheses and $U(1)$ quantum number as a subscript. The normalization of the $U(1)$ quantum number is of course arbitrary. The rank-2 antisymmetric tensor representation of $SU(5)$ is 10-dimensional, and it decomposes as

$$10 \longrightarrow (3, 2)_{\frac{1}{6}} + (3^*, 1)_{-\frac{2}{3}} + (1, 1)_1 \,. \tag{23.7}$$

Comparing with Eq. (17.1, *p 482*), it is seen that the color-triplet weak-doublet component has exactly the quantum numbers for the quark doublet. The color antitriplet appearing in this decomposition is the complex conjugate of u_R, which can be called $\widehat{u}_L$. The color singlet transforms like $\widehat{e}_L$, whose representation should be complex conjugate to that of e_R. Remember that for $SU(2)$, the complex conjugate of any representation is equivalent to the original representation, and any $U(1)$ quantum number reverses sign under complex conjugation.

Now look at the decomposition of 5 again. It has a weak doublet which is color singlet, but the $U(1)$ quantum number is the opposite of that of the lepton doublet given in Eq. (16.33, *p 468*). But then, if we consider 5^*, that will contain a weak doublet with the correct weak hypercharge. The 5^* will also contain a color antitriplet, whose representation would match exactly with that of $\widehat{d}_L$.

So let us summarize what we have obtained. We can assign the following fields to the 5^* representation of $SU(5)$:

$$\begin{pmatrix} \widehat{d}_1 \\ \widehat{d}_2 \\ \widehat{d}_3 \\ \nu_e \\ e \end{pmatrix}_L \quad : \quad 5^* \,, \tag{23.8}$$

where the subscripts 1,2,3 indicate the three colors. The overall subscript $\mathbb{L}$ indicates that each element of the column is a left-chiral field. The $\mathbf{10}$ will be an antisymmetric rank-2 tensor representation, and can be written in the form of a matrix:

$$\begin{pmatrix} 0 & \widehat{u}_3 & -\widehat{u}_2 & u_1 & d_1 \\ & 0 & \widehat{u}_1 & u_2 & d_2 \\ & & 0 & u_3 & d_3 \\ & & & 0 & \widehat{e} \\ & & & & 0 \end{pmatrix}_{\mathbb{L}} \quad : \quad \mathbf{10}. \tag{23.9}$$

We have not written the fields below the diagonal, which are obtainable by antisymmetry of this matrix. This pattern can be repeated for all fermion generations, thus obtaining the transformation property of all fermions under the gauge group.

While writing the behaviors of different fields under the standard model gauge group, we used the left-chiral and right-chiral projections of the so-called "particles". The representations of the "antiparticles" were the complex conjugates which were implied, and not written down explicitly. We can take an alternative strategy and assign only left-chiral particles and antiparticles to the representations of the gauge group. The right-chiral ones will be implied to be included in the complex conjugate representations. More explicitly, we had, for example, given the representation of u_R under the electroweak gauge group in Eq. (17.1, $p\,482$). Now we will have $\widehat{u}_L$, and the representation of u_R will be the complex conjugate one.

□ **Exercise 23.3** Use Young tableaux to show that, in SU(5),

$$\mathbf{5} \times \mathbf{5} = \mathbf{10} + \mathbf{15}. \tag{23.10}$$

Take the decomposition of $\mathbf{5}$ under the standard model gauge group from Eq. (23.6), perform Kronecker product on it by itself, extract the antisymmetric part and show that Eq. (23.7) is obtained.

Let us now look at the gauge bosons of SU(5). They constitute the adjoint representation of the group, which is 24-dimensional. Since

$$\mathbf{5} \times \mathbf{5}^* = \mathbf{24} + \mathbf{1} \tag{23.11}$$

in SU(5), the decomposition of $\mathbf{24}$ under the standard model gauge group can be easily determined by using Eq. (23.6):

$$\mathbf{24} \longrightarrow (\mathbf{8}, \mathbf{1})_0 + (\mathbf{1}, \mathbf{3})_0 + (\mathbf{1}, \mathbf{1})_0 + (\mathbf{3}, \mathbf{2})_{5/6} + (\mathbf{3}, \mathbf{2})_{-5/6}. \tag{23.12}$$

The color octet represents the gluons, the gauge bosons of the $SU(3)_c$ subgroup of SU(5). The $(\mathbf{1}, \mathbf{3})_0$ and the $(\mathbf{1}, \mathbf{1})_0$ are the representations of the gauge bosons of the SU(2) and U(1) parts of the gauge group. And then there are 12 more gauge bosons in SU(5) which are not part of the standard model. These are extra gauge bosons, and we should check what kinds of interactions can they induce.

It is enough to consider only the $(\mathbf{3}, \mathbf{2})_{5/6}$ part. The other part is its complex conjugate, and will do the same things in reverse order. Using the formula

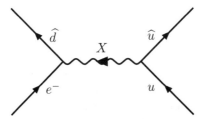

Figure 23.1: Diagram for baryon number violation induced by the exchange of virtual X bosons. Each arrow has the name of some particle written next to it, and represents either that particle going along the arrow or its antiparticle going the opposite way.

for electric charge in Eq. (16.18, *p465*), we find that this part contains three gauge bosons with $Q = \frac{4}{3}$, and three more with $Q = \frac{1}{3}$. We will refer to the first kind as X bosons and the second kind as Y bosons. In any gauge theory, there are gauge boson vertices with two members of a fermion multiplet, as shown in Fig. 11.4 *(p316)*. Thus, for example, an X boson will have vertices like $\widehat{d}\gamma^\mu e X_\mu$ (omitting the color indices) with the members of the 5*-representation. Similarly, with the members of the 10-representation, it can have interaction terms like $\bar{u}\gamma^\mu \widehat{u} X_\mu$. The interaction terms involving the Y boson can be similarly written.

Looking at the two sets of interactions of the X boson, we see that the two, taken together, must violate both baryon number and lepton number, because no consistent definition of these numbers can be given for the X bosons. Violation of baryon number and lepton number will then also be seen in processes involving ordinary fermions. For example, Fig. 23.1 shows how an effective quark-level interaction of the form

$$ud \to \widehat{u}e^+ \tag{23.13}$$

can be generated through the exchange of virtual X bosons. This can provide a decay channel for the proton:

$$p \to \pi^0 e^+. \tag{23.14}$$

The decay has not been observed yet: there is no experimental result to justify grand unified theories.

☐ **Exercise 23.4** Write the interaction terms involving the Y bosons and show that the exchange of virtual Y bosons can also give rise to proton decay channels.

Even though there is no new phenomenon to support grand unified theories, some known properties of particles, taken for granted in the standard model, find explanations in these theories. We outlined the phenomenon of charge quantization in §23.1. For a U(1) gauge group like that of QED, there is no reason why the charges would be quantized. However, now the QED gauge group is part of the grand unified gauge group, which is non-abelian. The eigenvalues of the generators of a non-abelian gauge group are quantized, the most famous example being the quantization of angular momentum components which satisfy the algebra of the rotation group. To be specific, the trace of any generator vanishes for an $SU(N)$ group. Thus, the sum of charges of the particles present in the 5^* representation shown in Eq. (23.8) should be zero. Since the color group is unbroken, the electric charges of the different colored d-quarks must be the same. If we assume that the neutrino charge is zero, it then follows that

$$Q_e - 3Q_d = 0\,,\qquad\qquad (23.15)$$

which shows why the charge of the electron has to be an exact multiple of the charge of the d quark.

How do we break the grand unified symmetry $SU(5)$ down to the manifest symmetry of particle interactions, the $SU(3)$ of color and the $U(1)$ of electromagnetism? It has to be done in two steps. In the first stage, at a high energy scale, $SU(5)$ breaks and the standard model gauge group appears:

$$SU(5) \to SU(3)_c \times SU(2)_L \times U(1)_Y\,.\qquad\qquad (23.16)$$

For this to happen, there must be a non-zero VEV of a non-trivial multiplet of $SU(5)$ along a direction that is a singlet under the standard model gauge group. Looking at Eq. (23.12), we see that the adjoint representation of $SU(5)$ has such a singlet. So, if there is a Higgs multiplet that transforms like the adjoint, it is possible that under some conditions regarding the parameters in the scalar potential, it will develop a VEV and reduce the gauge group to $SU(3)_c \times SU(2)_L \times U(1)_Y$. In the next step, the electroweak gauge group $SU(2)_L \times U(1)_Y$ has to be broken, and this can be done by the VEV of a 5-dimensional multiplet of Higgs bosons. As seen from Eq. (23.6), the decomposition of the 5-dimensional multiplet contains one part that transforms exactly like the Higgs doublet of the standard model gauge group.

The central attractive idea is that of only one gauge coupling constant. But what does that mean physically, given that the coupling constants of the $SU(3)$, the $SU(2)$ and the $U(1)$ parts of the standard model appear to be very different in measurements? The answer lies in the fact that the coupling constants depend on the energy scale, as was discussed in §12.2. From Eq. (12.46, $p\,334$), we found that the QCD coupling constant decreases with energy. From Eq. (12.39, $p\,332$), we see that the coupling constant of any $U(1)$ group should increase with energy. Since the QCD coupling constant is larger at the weak scale compared to the other ones, it is clear that at some energy, the $SU(3)_c$ and the $U(1)_Y$ coupling constant will be equal. If the $SU(2)_L$ coupling constant also meets at the same point, it would mean that all couplings will be equal at that scale, and that can be taken as the scale above which the $SU(5)$ theory would be valid with one single coupling constant.

At the time that this idea was proposed by Georgi, Quinn and Weinberg, the three gauge coupling constants of the standard model gauge group were not known to a very good accuracy. With the accuracy that was available at that time, the extrapolation of the coupling constants to high energies seemed to indicate that they would meet at around 10^{14} to 10^{15} GeV. But now, the accuracy of the data has improved a lot, and it seems that the three do not meet at the same point.

This does not necessarily mean that the idea of grand unification is wrong. It just says that if the idea has to work, the standard model gauge group cannot emerge from the grand unified group in one step, i.e., the chain of symmetry breaking indicated in Eq. (23.16) does not work in the real world. One can contemplate a bigger gauge group like SO(10). Orthogonal groups have spinor representations, the most well-known example being the spinors of the rotation group. The smallest spinor representation of SO(10) is $\mathbf{16}$-dimensional. If we consider the SU(5) subgroup of SO(10), this representation decomposes as follows:

$$\mathbf{16} \to \mathbf{10} + \mathbf{5}^* + \mathbf{1}. \tag{23.17}$$

We have already seen the decomposition of the $\mathbf{10}$ and $\mathbf{5}^*$ representations of SU(5) under the standard model gauge group and argued that they contain all left-chiral fields within one generation of fermions of the standard model. The extra singlet present in the $\mathbf{16}$ of SO(10) has to be a singlet of the standard model, and therefore a neutrino field. We mentioned in Ch. 22 that such singlet neutrinos are necessary in many models of neutrino mass.

☐ **Exercise 23.5** We said that SU(5) is a subgroup of SO(10). Show that any SU(N) group is a subgroup of SO($2N$).

We are considering the subgroup SU(5) only for the sake of convenience. We do not want to imply that the SO(10) breaks to SU(5) before further symmetry breaking takes place. In fact, that scheme would imply that the standard model interaction would first have to unify into SU(5), and then at some higher scale the SO(10) symmetry takes over — and would not work for the same reason that SU(5) does not work as a grand unified gauge group. However, SO(10) can break through many other channels to the standard model gauge group. Most interestingly, the left–right symmetric SU(4) × SU(2)$_{\text{L}}$ × SU(2)$_{\text{R}}$ containing the Pati-Salam SU(4) is a subgroup of SO(10). So, symmetry breaking can proceed through this group as an intermediate symmetry group. There are many other possible symmetry breaking chains, and other possibilities for the grand unified gauge group as well, all of which need not be discussed in the brief outline of the basic idea.

23.4   Horizontal symmetry

Grand unified models like SU(5) or SO(10) do not explain or even address the question of why fermions come in three generations. In SO(10), for example, one irreducible representation of left-chiral fields contains all fields of a single generation. The number of such multiplets, i.e., the number of generations, is taken as an empirical input. It is natural to wonder why the number of fermion generations is indeed three as we have seen. There are related questions, of course. Is it really true that there are no extra fermion generations beyond the three that we have already seen? Are there any relations between the masses of different generations of fermions, including possibly some relation with the mixing angles that appear in the CKM matrix for quarks and also the PMNS matrix for leptons? Such questions can be addressed only if we have a theory for explaining different generations of fermions.

One obvious idea to try is that of some kind of symmetry that connects different generations. For example, consider the idea that there is a SU(3) family symmetry, under which, e.g., the u quark, the c quark and the t quark form a triplet. The down-type quarks also form a triplet, and so do the leptons of the three generations. Such symmetries are often called *horizontal symmetries*. The name is a pictorial reminder to the list of quark fields given in Eq. (17.1, *p482*): transformations of this group act horizontally in this arrangement. The symmetry being SU(3), immediately we understand why the number of fermion generations is not 2 or 4: because SU(3) does not have any 2- or 4-dimensional irreducible representations. Although this argument does not explain why the number of fermion generations is not 8, at least the number of possible answers can be narrowed down considerably.

The horizontal symmetry does not have to be a continuous group. One can also contemplate discrete non-abelian groups which have 3-dimensional irreducible representations to explain the existence of three fermion generations.

The problem is that, whether continuous or discrete, an unbroken horizontal symmetry would imply that the different generations of fermions should be degenerate. This is far from what we observe in the real world. Horizontal symmetries must therefore be broken. The pattern of breaking would dictate mass relations. It is fair to say that so far it has not been possible to find any horizontal symmetry whose breaking pattern has provided any useful relation between quark masses and mixing.

There are related problems. If a horizontal symmetry is not accidental, it cannot be approximate. Then it will have to be broken spontaneously. If the symmetry happens to be global and continuous, spontaneous breaking of the symmetry would produce Goldstone bosons. The theory will therefore have an obligation to explain why these Goldstone bosons have not been observed yet despite the fact that they are massless. On the other hand, if the horizontal symmetry is a gauge symmetry, it will have to be broken at sufficiently high energy in order to explain why we have not seen the associated gauge bosons, and also flavor-changing neutral currents that such gauge bosons would mediate. In that case, it is not clear why the mass difference between, say the electron and the muon, is so much smaller than the scale of the horizontal symmetry breaking.

If there is a gauged horizontal symmetry, then we can also think of whether this symmetry is unified with the gauge symmetries of the standard model. Obviously this requires symmetries much larger than SO(10), which can accommodate fermions of a single family into its spinor representation. There are higher orthogonal groups that can be suitable for the job. For example, consider the group SO(18). Obviously it has a subgroup SO(10) × SO(8). We can take the SO(10) part as the group for single family unification, and the SO(8) part as the horizontal symmetry group. The spinor representation of the SO(18) group is 256-dimensional. Under the subgroup SO(10) × SO(8), it decomposes as follows:

$$256 \implies (16, 8_s) + (16^*, 8_c), \tag{23.18}$$

where the 8_s indicates the 8-dimensional spinor representation of SO(8), and 8_c is its complex conjugate representation. Thus, a single spinor representation of SO(18) contains all three generations of known fermions, and in fact many more. There are discussions in the literature how only three generations can have masses below the weak scale while the others will become superheavy, i.e., acquire masses at the grand unification scale. In this sense, this model explains why there are only three fermion generations. Detailed understanding of the masses of the fermions in the light generations and their mixing patterns is lacking, from SO(18) models or otherwise.

23.5   Supersymmetry

In §23.3, we talked about grand unification, which unifies the properties of particles under internal symmetries like the gauge symmetries associated with strong, weak and electromagnetic interactions. There is another kind of unification that one can contemplate. There are fermions, and there are bosons. They are distinguished by their spin angular momentum: half-integral for the former class and integral for the latter ones. If we have some kind of a symmetry operation that changes the angular momentum by a half unit, a boson would change to a fermion, and vice versa, by the action of such an operation. In that case, we could consider a fermion and a boson as different manifestations of a single kind of some superparticle. This is the basic idea of supersymmetry.

To implement this idea, one has to extend the Poincaré invariance by the addition of extra generators which behave fermionically, i.e., they have anticommutation relations among themselves. Application of such operators on a bosonic state would produce a fermionic state and vice versa. Each known particle, boson and fermion, must have a superpartner of the opposite kind in such theories. Observation of any of these superpartners would constitute a direct proof of supersymmetry.

The behavior of a particle and its superpartner must be the same under strong as well as electroweak gauge symmetry. In addition, supersymmetry also implies that the masses of any particle and its superpartner should be the same. Immediately, this means that supersymmetry cannot be an exact symmetry of nature, because if it were, then we would have known a boson with the mass and the charge of the electron, and a massless fermion as the partner of the photon, and so on. If such particles existed with such low mass, we would have discovered them long ago.

So the reason for the non-observation of the superpartners has to be the fact that they are heavy. This means that supersymmetry has to be broken. There is no consensus on how this might happen.

However, there seems to be some agreement on the amount of supersymmetry breaking. This comes from the fact that supersymmetry can explain one of the shortcomings of the standard model mentioned in §23.1, viz., the one about the Higgs boson mass. We said that the loop corrections to the Higgs boson mass are of order Λ^2 if a momentum cut-off Λ is used to evaluate the integrals. In a supersymmetric theory, however, if there is a loop with a boson circulating in it, there must also be a loop with the superpartner of the boson circulating in it. If supersymmetry is exact, the $\mathcal{O}\left(\Lambda^2\right)$ contributions from two such diagrams cancel each other exactly. In a realistic theory where supersymmetry is broken at a scale M_S, the cancellation of the two diagrams leaves a remainder of order M_S^2. Thus, as long as M_S is not very much larger than the weak scale, one does not need any fine tuning for the Higgs boson mass.

In the simplest form, supersymmetry transformation parameters are taken to be independent of the spacetime point, i.e., the transformations are considered to be global. The corresponding transformations on the ordinary spacetime are also taken to be global, and we obtain theories in the Minkowskian spacetime augmented by supersymmetric extensions of it. If instead we consider local transformations, one obtains supersymmetric extensions of theories on more general spacetime. Since Einstein's general theory of relativity is a field theory of gravity based on generalized geometry of spacetime, the supersymmetric extensions give a supersymmetric version of the theory of gravity, called *supergravity*.

23.6   Higher dimensional theories

There was a time when people believed that there were only two kinds of fundamental interactions, the gravitational and the electromagnetic. In the nineteenth century, Maxwell formulated a classical field theory for electromagnetic interactions. In the beginning of the twentieth century, Einstein did the same for gravitational interactions through his *general theory of relativity*. This latter theory was based on the independence of physical phenomena on the co-ordinate transformations in a 4-dimensional spacetime. In the 1920s, Kałuza and Klein pointed out an interesting variant of the idea of co-ordinate transformations. Suppose the spacetime were 5-dimensional. Four of these

five dimensions are the usual extended ones, but the fifth one is like a circle, i.e., the range of values that the fifth co-ordinate can take is finite, and
the same value repeats periodically if one 'travels' along the fifth dimension.
Such a dimension would then be called a *compactified dimension*, as opposed
to the other four *extended dimensions*. Since the compactified dimension is
like a circle, transformations of the fifth co-ordinate would be like U(1) electromagnetic gauge transformations. Thus, inclusion of such an extra dimension
would include the electromagnetic field equations, in addition to the gravitational field equations coming from the four extended dimensions. In this
sense, one can unify gravitational and electromagnetic interactions within a
5-dimensional geometry. This idea is called the Kałuza–Klein theory after the
name of the people who proposed it.

The obvious question that comes is this: why can't we see the circular
nature of the fifth dimension? The only reasonable answer is that the radius
of the circle is very small, i.e., one needs very high energies to probe this
dimension. If the radius of the circle is so small that the energies needed to
probe it would be larger than any energy obtained in any experiment, that
would explain why the fifth dimension has gone unnoticed in all experiments
so far.

Will there be low-energy signatures of this idea? In principle, yes. The
Klein–Gordon equation for a 5-dimensional theory would read

$$\Box\Phi - \partial_y^2\Phi + m^2\Phi = 0\,,\tag{23.19}$$

where the box stands for the 4-dimensional Klein–Gordon operator and ∂_y
denotes derivative with respect to the fifth co-ordinate, y. The field Φ should
be a function of the four usual co-ordinates, which will be denoted by x, and
of y. If the radius of the circle along the fifth dimension is R, the solutions
should have the form

$$\Phi(x,y) = \sum_n \varphi_n(x)\exp(iny/R)\tag{23.20}$$

where n is an integer, so that y and $y + 2\pi R$ can denote the same physical
point. Putting this back into Eq. (23.19), we find that $\varphi_n(x)$ satisfies the
equation

$$\Box\varphi_n + \left(m^2 + \frac{4\pi^2 n^2}{R^2}\right)\varphi_n = 0\,.\tag{23.21}$$

Thus, in four dimensions, we should see not only just one particle with mass m,
but an infinite tower of particles corresponding to all positive integer values of
n. Of course if R is small, these particles will be very heavy and therefore not
detectable directly, but effects of these particles appearing as virtual particles
in a Feynman diagram might be detectable. So far, there have been lower
bounds on the masses of these particles based on such analysis.

The 5-dimensional spacetime is good for a marriage between gravitation
and electromagnetism only. We now know that there are other interactions.

One can take the cue from the 5-dimensional example and construct theories in higher dimensions, where all but the usual four dimensions are compactified, and the symmetry groups involving co-ordinate transformations of those compact dimensions give further gauge interactions.

23.7   String theory

String theory provides a more radical approach to unification. Unlike all the previous ideas described in this chapter, this theory is not fundamentally a quantum field theory of particles. The fundamental objects considered in string theory are 1-dimensional objects or strings. Just as the Lagrangian for a point particle is the length of the worldline, and contains only one parameter, viz., the mass of the particle, the Lagrangian of a string is the area of the worldsheet swept by the string, and it also contains just one parameter, the mass per unit length of the string, also called the string tension.

Particles are the different modes of vibrations of a fundamental string. A fundamental note on a string instrument is some wave with a specific frequency and specific wavelength. In quantum theory, frequency is equivalent to energy, and wavelength (or, to be more precise, the wave vector) to 3-momentum. Thus, a normal mode corresponds to a particle with specific values of energy and momentum.

Early enthusiasm with string theory stemmed from the fact that the normal modes of a closed string theory contains a massless particle with spin 2. This can be easily identified with a *graviton*, the supposed mediator of gravitational interactions. There was a hope that string theory would therefore be able to describe all fundamental interactions including gravitation.

It was soon realized that the goal of including the other three kinds of interactions would not be straightforward or easy. The string theories have certain anomalies which depend on the number of spacetime dimensions. For ordinary strings, the spacetime dimension has to be equal to 26 in order that the theory is free from such anomalies. However, this theory is useless anyway because its vibrational modes can never contain any fermion. It also has other problems, like the presence of a tachyonic mode. Thus we need to go to what are called *superstrings*. The transition from ordinary or bosonic strings to superstrings is somewhat similar to the transition from Poincaré invariance to supersymmetry. Once this is made, one finds that a 10-dimensional spacetime is needed to lay down a consistent quantum theory of superstrings.

String theory is attractive for several reasons. First, it includes gravity, and therefore holds the promise of unifying all interactions. Second, string interactions do not give infinite results like point-particle theories: they are automatically finite at every order in perturbation theory.

There are in fact five different kinds of string theories in 10 dimensions. They differ in the possible modes of vibration of the strings. These theories have to be compactified so that finally there exist only four extended dimensions. There are many ways of compactifying any given theory, and therefore many different possibilities for the 4-dimensional theory. It is not clear whether any of them describes the real world.

Appendix A

Units and constants

Table A.1: Conversion factors in natural units ($\hbar = c = 1$). M_P is the Planck mass, defined in Eq. (1.11, *p 14*).

$$[\text{Row heading}] = [\text{Table entry}] \times [\text{Column heading}]$$
$$[\text{Column heading}]^{-1} = [\text{Table entry}] \times [\text{Row heading}]^{-1}$$

	GeV	gm	cm^{-1}	s^{-1}	M_P	erg
GeV		1.78×10^{-24}	5.06×10^{13}	1.52×10^{24}	8.18×10^{-20}	1.60×10^{-3}
gm	5.62×10^{23}		2.84×10^{37}	8.53×10^{47}	4.60×10^{4}	9.00×10^{20}
cm^{-1}	1.98×10^{-14}	3.51×10^{-38}		3.00×10^{10}	1.62×10^{-33}	3.17×10^{-17}
s^{-1}	6.58×10^{-25}	1.17×10^{-48}	3.33×10^{-11}		5.38×10^{-44}	1.05×10^{-27}
M_P	1.22×10^{19}	2.17×10^{-5}	6.17×10^{32}	1.85×10^{43}		1.95×10^{16}
erg	6.24×10^{2}	1.11×10^{-21}	3.16×10^{16}	9.49×10^{26}	5.11×10^{-17}	

Other units which do not have natural dimensions of energy or inverse energy	1 coulomb $= 1.89 \times 10^{18}$ 1 gauss $\quad = 1.96 \times 10^{-20}\,\text{GeV}^2$

Table A.2: Physical constants.

Constant	Symbol	Value
Speed of light in vacuum	c	$3.00 \times 10^{10}\,\mathrm{cm\ s^{-1}}$
Planck constant	$\hbar$	$1.05 \times 10^{-27}\,\mathrm{erg\,s}$
Fine structure constant	α	$1/137$
Fine structure constant (at scale M_Z)	$\alpha(M_Z)$	$1/128$
Fermi constant	G_F	$1.16 \times 10^{-5}\,\mathrm{GeV^{-2}}$
Strong coupling constant (at scale M_Z)	$\alpha_3(M_Z)$	0.118
Measure of weak mixing angle	$\sin^2\theta_W$	0.2316

Appendix B

Short summary of particle properties

In this appendix, we give a short summary of properties of various particles. The lists include all elementary particles of the standard model and in addition many hadrons. The following disclaimers should be kept in mind while using these tables.

- Only a few significant digits in the values of the masses, widths (or lifetimes) and branching ratios have been given.
- Errors on the values have not been shown.
- Not all decay modes have been given.
- Not all known baryons and mesons have been listed.

Table B.1: Properties of fundamental bosons.

Particle	Spin	Mass in GeV	Width in GeV	Decay mode	Decay product	$\mathscr{B}$ (%)
γ	1	0	0			
gluons	1	0	0			
W^+	1	80.39	2.09	weak	$e^+\nu$	10.75
					$\mu^+\nu$	10.57
					$\tau^+\nu$	11.25
					hadrons	67.60
Z	1	91.19	2.50	weak	e^+e^-	3.363
					$\mu^+\mu^-$	3.366
					$\tau^+\tau^-$	3.370
					hadrons	69.91
					invisible	20.00
H	0	126				

Table B.2: Properties of leptons.

Particle	Mass (in MeV)	Lifetime (second)	Decay mode	Decay product	$\mathscr{B}$ (%)
e	0.511	∞			
μ	106	2.2×10^{-6}	weak	$e\nu_\mu\widehat{\nu}_e$	100
τ	1777	2.9×10^{-13}	weak	$e\nu_\tau\widehat{\nu}_e$	17.83
				$\mu\nu_\tau\widehat{\nu}_\mu$	17.41
				$\pi^-\nu_\tau$	10.83
				$K^-\nu_\tau$	7.00

For neutrinos, the differences of mass-squared values are known:

$$\Delta m_{21}^2 = 7.50 \times 10^{-5}\,\mathrm{eV}^2$$
$$\left|\Delta m_{23}^2\right| = 2.32 \times 10^{-3}\,\mathrm{eV}^2 .$$

Values for the mixing angles have been given in Eq. (22.42, *p 677*).

Table B.3: Properties of various baryons.

[**Note :** *In the column with the heading 'Lifetime (or Width)', ordinary entries are for lifetimes, and entries given in parentheses are for widths.*]

Particle	J^P	Mass (MeV)	Lifetime (or Width)	Decay mode	Decay product	$\mathscr{B}$ (%)
p	$\frac{1}{2}^+$	938	∞			
n	$\frac{1}{2}^+$	939	$880.1\,\mathrm{s}$	weak	$pe\widehat{\nu}_e$	100
Λ	$\frac{1}{2}^+$	1115	$2.6 \times 10^{-10}\,\mathrm{s}$	weak	$p\pi^-$	63.9
					$n\pi^0$	35.8
					$n\gamma$	0.17
Σ^+	$\frac{1}{2}^+$	1189.3	$0.8 \times 10^{-10}\,\mathrm{s}$	weak	$p\pi^0$	51.57
					$n\pi^+$	48.31
					$p\gamma$	0.12
Σ^0	$\frac{1}{2}^+$	1192.6	$7.4 \times 10^{-20}\,\mathrm{s}$	em	$\Lambda\gamma$	100
Σ^-	$\frac{1}{2}^+$	1197.4	$1.48 \times 10^{-10}\,\mathrm{s}$	weak	$n\pi^-$	99.848
					$ne^-\widehat{\nu}_e$	0.102
Ξ^0	$\frac{1}{2}^+$	1314.8	$2.9 \times 10^{-10}\,\mathrm{s}$	weak	$\Lambda\pi^0$	99.522
					$\Lambda\gamma$	0.118
					$\Sigma^0\gamma$	0.333
Ξ^-	$\frac{1}{2}^+$	1321.3	$1.6 \times 10^{-10}\,\mathrm{s}$	weak	$\Lambda\pi^-$	99.887
Δ	$\frac{3}{2}^+$	1232	$(117\,\mathrm{MeV})$	strong	$N\pi$	100

(Continued on next page)

Table B.3 (Continued from previous page)

Particle	J^P	Mass (MeV)	Lifetime (or Width)	Decay mode	product	$\mathscr{B}$ (%)
Σ^{*+}	$\frac{3}{2}^+$	1382.8	(36.0 MeV)	strong		
Σ^{*0}	$\frac{3}{2}^+$	1383.7	(36 MeV)		$\Lambda\pi$	87.0
					$\Sigma\pi$	11.7
Σ^{*-}	$\frac{3}{2}^+$	1387.2	(39.4 MeV)			
Ξ^{*0}	$\frac{3}{2}^+$	1531.8	(9.1 MeV)	strong	$\Xi\pi$	100
Ξ^{*-}	$\frac{3}{2}^+$	1535.0	(9.9 MeV)	strong	$\Xi\pi$	100
Ω^-	$\frac{3}{2}^+$	1672.5	0.82×10^{-10} s	weak	ΛK^-	67.8
					$\Xi^0\pi^-$	23.6
					$\Xi^-\pi^0$	8.6
Λ_c^+	$\frac{1}{2}^+$	2286	2.0×10^{-13} s	weak	$p\widehat{K}^0$	2.3
					$pK^-\pi^+$	5.0
Σ_c^{++}	$\frac{1}{2}^+$	2454	(2.26 MeV)	strong	$\Lambda_c^+\pi^+$	$\approx$100
Σ_c^+	$\frac{1}{2}^+$	2452.9	($<$ 4.6 MeV)	strong	$\Lambda_c^+\pi^0$	$\approx$100
Σ_c^0	$\frac{1}{2}^+$	2453.7	(2.16 MeV)	strong	$\Lambda_c^+\pi^-$	$\approx$100

Table B.4: Properties of various mesons.

[**Note :** *In the column with the heading 'Lifetime (or Width)', ordinary entries are for lifetimes, and entries given in parentheses are for widths. For some heavy mesons, we do not list any decay mode because there are many of them, and none stands out.*]

Particle	J^P	Mass (MeV)	Lifetime (or Width)	Decay mode	product	$\mathscr{B}$ (%)
π^+	0^-	139.6	2.6×10^{-8} s	weak	$\mu^+\nu_\mu$	99.988
					$e^+\nu_e$	0.012
π^0	0^-	135.0	8.5×10^{-17} s	em	$\gamma\gamma$	98.823
					$e^+e^-\gamma$	1.174
K^+	0^-	493.7	1.2×10^{-8} s	weak	$\mu^+\nu_\mu$	63.54
					$\pi^0\mu^+\nu_\mu$	3.35
					$\pi^0e^+\nu_e$	5.08
					$\pi^+\pi^0$	20.68
					$\pi^+\pi^0\pi^0$	1.76
					$\pi^+\pi^+\pi^-$	5.59
K_L	0^-	497.6	5.2×10^{-8} s	weak	$\pi^0\pi^0\pi^0$	19.52
					$\pi^+\pi^-\pi^0$	12.54
					$\pi^\pm\mu^\mp\nu$	27
					$\pi^\pm e^\mp\nu$	38.8

(Continued on next page)

Table B.4 (Continued from previous page)

Particle	J^P	Mass (MeV)	Lifetime (or Width)	Decay mode	Decay product	$\mathscr{B}$ (%)
K_S	0^-	497.6	$8.9 \times 10^{-11}\,\text{s}$	weak	$\pi^+\pi^-$	68.6
					$\pi^0\pi^0$	31.4
η	0^-	547.5	$(1.30\,\text{keV})$	em	$\gamma\gamma$	39.31
					$e^+e^-\gamma$	0.69
					$\pi^0\pi^0\pi^0$	32.57
					$\pi^+\pi^-\pi^0$	22.74
					$\pi^+\pi^-\gamma$	4.60
η'	0^-	957.8	$(0.199\,\text{MeV})$	em	$\pi^+\pi^-\eta$	43.4
					$\pi^0\pi^0\eta$	21.6
					$\rho^0\gamma$	29.3
					$\omega\gamma$	2.75
					$\gamma\gamma$	2.18
$\rho^\pm, \rho^0$	1^-	775.8	$(149.1\,\text{MeV})$	strong	$\pi\pi$	≈ 100
K^{*+}	1^-	891.6	$(50.8\,\text{MeV})$	strong	$K\pi$	≈ 100
K^{*0}	1^-	896.0	$(50.3\,\text{MeV})$	strong	$K\pi$	≈ 100
					$K^0\gamma$	0.23
ω	1^-	782.7	$(8.49\,\text{MeV})$	strong	$\pi^+\pi^-\pi^0$	89.2
					$\pi^0\gamma$	8.28
					$\pi^+\pi^-$	1.53
ϕ	1^-	1019.5	$(4.26\,\text{MeV})$	strong	K^+K^-	48.9
					$K_L^0 K_S^0$	34.2
					$\rho\pi, \pi\pi\pi$	15.32
					$\eta\gamma$	1.31
D^0	0^-	1864.9	$4.1 \times 10^{-13}\,\text{s}$	weak	$K^-e^+\nu_e$	3.55
					$K^-\mu^+\nu_\mu$	3.30
					$K^-\pi^+$	3.88
D^+	0^-	1869.6	$1.04 \times 10^{-12}\,\text{s}$	weak		
D_s^+	0^-	1968.5	$5.00 \times 10^{-13}\,\text{s}$	weak		
B^+	0^-	5279.2	$1.64 \times 10^{-12}\,\text{s}$	weak		
B^0	0^-	5279.6	$1.52 \times 10^{-12}\,\text{s}$	weak		
B_s^0	0^-	5366.8	$1.50 \times 10^{-12}\,\text{s}$	weak		
B_c^+	0^-	6277	$4.53 \times 10^{-13}\,\text{s}$	weak		
$\eta_c(1S)$	0^-	2981.0	$(29.7\,\text{MeV})$		$K\widehat{K}\pi$	7.2
					$\eta\pi^+\pi^-$	4.9
ψ	1^-	3097.0	$(92.9\,\text{keV})$	em	hadrons	87.7
					e^+e^-	5.94
					$\mu^+\mu^-$	5.93
$\Upsilon(1S)$	1^-	9460.3	$(54.0\,\text{keV})$			

Table B.5: Properties of quarks. Note that quarks have not been directly observed. Their masses have been estimated from various theoretical considerations explained in different chapters.

Particle	Mass (in MeV)	I_3	Strangeness (S)	Charm quantum no. (C)	Bottom quantum no. (B)	Top quantum no. (T)
d	4.8	$-\frac{1}{2}$	0	0	0	0
u	2.3	$+\frac{1}{2}$	0	0	0	0
s	95	0	-1	0	0	0
c	1.275×10^3	0	0	$+1$	0	0
b	4.18×10^3	0	0	0	-1	0
t	173.5×10^3	0	0	0	0	$+1$

Values for the elements of the CKM matrix have been shown in Eq. (17.23, *p 487*).

Appendix C

Timeline of major advances in particle physics

1896 • Thomson discovered the electron.

1900 • Planck introduced the idea of quanta of radiation in order to explain the spectrum of blackbody radiation.

1905 • Einstein explained photoelectricity by using the idea of the photon.

1909 • Geiger and Marsden performed the alpha particle scattering experiment.

1911 • Rutherford explained the results of Geiger and Marsden by introducing the idea of the nucleus.

1912 • Hess discovered the existence of cosmic rays.

• Wilson constructed the first cloud chamber.

1928 • Dirac proposed the relativistic theory of electrons.

1930 • Pauli conjectured the existence of neutrinos.

1931 • Lawrence built the first cyclotron.

1932 • Chadwick discovered the neutron.

• Anderson discovered the positron in cosmic rays.

1933 • Fermi proposed a theory of weak interactions.

1935 • Yukawa proposed the pion.

1936 • Anderson discovered the muon while searching for the pions.

• Following an earlier cue from Heisenberg, the idea of SU(2) isospin symmetry of strong interactions was suggested by various authors including Breit, Condon, Present, Cassen, Feenberg.

1947 • The charged pions were discovered by a group involving Lattes, Occhialini, Powell.

1949 • The neutral pion was discovered.

1953 • Glaser invented the bubble chamber.

1954 • Yang and Mills proposed non-abelian gauge theories.

1955 • Segrè and Chamberlain discovered the antiproton.

1956 • Neutrinos were detected by Reines and Cowan.

• Lee and Yang suggested the possibility that weak interactions violate parity.

1957 • Parity violation in weak interactions was observed by Wu and collaborators, and also by Lederman and collaborators in two different experiments.

1961 • Gell-Mann and Ne'eman independently realized that various hadrons constitute representations of the group SU(3).

• Glashow proposed that a non-abelian gauge theory based on the group $SU(2)_L \times U(1)_Y$ might be responsible for weak interactions.

1962 • Lederman, Schwarz and Steinberger showed that the muon-neutrino is different from the electron-neutrino.

1964 • Gell-Mann and Zweig independently proposed the idea of quarks in order to explain the multitude of hadrons.

• Cabibbo introduced what is known as "Cabibbo angle" today: the fact that strangeness-changing charge weak currents are more suppressed than strangeness-conserving ones.

• CP violation was discovered by Fitch, Cronin and collaborators.

• In three different papers, Englert, Brout, Higgs, Hagen, Kibble and Guralnik proposed a mechanism whereby gauge bosons can acquire masses through spontaneous symmetry breaking.

1967 • Weinberg proposed "A model of leptons", the electroweak theory for the leptonic sector.

1968 • Salam proposed independently the model that Weinberg proposed earlier for weak interactions.

• A team led by Friedman, Kendall and Taylor performed deep inelastic scattering of electrons off protons.

1969 • Feynman, and independently Bjorken and Paschos argued that the results of deep inelastic scattering experiments imply substructure of the proton.

1970 • Glashow, Iliopoulos and Maiani showed how to extend the Weinberg–Salam model to quarks.

1971 • 't Hooft showed that spontaneously broken gauge theories are renormalizable.

1973 • Existence of weak neutral currents was discovered at CERN.

- Kobayashi and Maskawa showed that CP violation can be explained if the number of fermion generations is more than two.

1974
- The charm quark was discovered.

- A paper by Gross and Wilzcek and another one by Politzer showed that non-abelian gauge theories show the property of asymptotic freedom. QCD, a gauge theory based on SU(3), became gradually the accepted theory of strong interaction.

1975
- The tau lepton was discovered by Perl and collaborators.

1977
- The bottom quark was discovered by a group led by Lederman.

1983
- Evidence of W and Z bosons was found at the SpS collider by a team led by Rubbia.

1994
- The top quark was discovered at the Tevatron.

1999
- Direct CP violation was definitively established in the neutral kaon system.

2001
- The ν_τ was found.

2012
- The existence of a boson was announced. It seemed like the Higgs boson of the standard model.

Appendix D

Properties of spacetime

In this appendix, we summarize the basic ingredients of spacetime according to the special theory of relativity. Most (if not all) of the material presented here has already been introduced in the text. The material is summarized here for ready reference.

D.1 Metric tensor

In accordance with the special theory of relativity, we assume everywhere in this book that spacetime is Minkowskian. This means, first of all, that space and time together constitute the geometry. Points in this geometry are denoted by the co-ordinates x^μ, where $\mu = 0$ gives the time and $\mu = 1, 2, 3$ give the three directions of space. Secondly, the geometry is endowed with a metric

$$g_{\mu\nu} = \text{diag}(1, -1, -1, -1), \tag{D.1}$$

i.e., the invariant distance between two neighboring points with co-ordinates x^μ and $x^\mu + dx^\mu$ is given by

$$ds^2 = g_{\mu\nu}dx^\mu dx^\nu. \tag{D.2}$$

Since ds^2 is invariant or a scalar, and the co-ordinate differential dx^μ represents a vector, it follows that $g_{\mu\nu}$ must be a rank-2 covariant tensor. This is called the *metric tensor*.

For a co-ordinate transformation given by

$$x'^\mu = \Lambda^\mu{}_\nu \, x^\nu, \tag{D.3}$$

the invariance of ds^2 implies the relation

$$g_{\mu\nu}\Lambda^\mu{}_\alpha\Lambda^\nu{}_\beta = g_{\alpha\beta}, \tag{D.4}$$

as was shown in Eq. (2.25, p 21). The left hand side of this equation gives the expression for the components of the tensor $g_{\mu\nu}$ in the new co-ordinate system. The equation then shows that these components are the same in the old and the new co-ordinate systems, which is why the metric tensor is sometimes called an *invariant tensor*.

The inverse of the metric tensor can be defined through the relation

$$g^{\mu\alpha}g_{\alpha\nu} = \delta^\mu_\nu \,, \tag{D.5}$$

where the right hand side represents the Kronecker delta, whose value is 1 if the two indices are the same, and zero otherwise. Note that by contracting the remaining free indices, we obtain

$$g^{\mu\alpha}g_{\alpha\mu} = 4 \,. \tag{D.6}$$

The Kronecker delta, as well as the metric tensor and the inverse metric, are symmetric in their indices by definition, and we do not pay any attention to which index comes earlier and which later. In contrast, the order of the indices in, say, $\Lambda^\mu{}_\nu$ is important, since the Lorentz transformation matrices are not necessarily symmetric.

The Kronecker delta, and the inverse metric (which is also sometimes loosely called the *metric*) are also examples of invariant tensors. Because of their inter-relation, some people denote the Kronecker delta by g^μ_ν.

☐ **Exercise D.1** Show that the inverse metric and the Kronecker delta follow tensor transformation laws, and that their components are the same under co-ordinate transformations that keep ds^2 invariant.

☐ **Exercise D.2** Show that, if in one co-ordinate system we define a tensor $\delta_{\mu\nu}$ whose components are 1 if $\mu = \nu$ and zero otherwise, the components do not remain the same in another co-ordinate system. [**Note :** This is why $\delta_{\mu\nu}$, or similarly $\delta^{\mu\nu}$, cannot be defined.]

D.2 Levi-Civita symbol

It is possible to define another object whose components remain unaltered under any proper Lorentz transformation. This is called the *Levi-Civita symbol*, which has as many indices as the space on which it is defined, and is antisymmetric with respect to the interchange of any two indices. For the 4-dimensional spacetime, we can denote it by $\varepsilon_{\mu\nu\lambda\rho}$. Obviously, a component of this tensor can be non-zero only if all indices are different. We have chosen

$$\varepsilon_{0123} = +1 \,, \tag{D.7}$$

and all other non-zero components follow from it. It also follows, from the usual rules of raising and lowering indices, that

$$\varepsilon^{0123} = -1 \,. \tag{D.8}$$

We discuss various properties of this object here.

D.2.1   Contraction properties

Consider the product of two Levi-Civita symbols, i.e., an expression of the form $\varepsilon^{\mu\nu\lambda\rho}\varepsilon_{\mu'\nu'\lambda'\rho'}$. The product must be antisymmetric in the exchange of any two of the indices μ, ν, λ, ρ, and similarly for the primed indices. We know that the determinant of a matrix changes sign if any two of its rows, or any two of its columns, are interchanged. This suggests that we can put the result in the form of a matrix, each of whose rows is marked by one of the unprimed indices, and each column by a primed index. A little bit of inspection then shows that the result is necessarily of the following form:

$$\varepsilon^{\mu\nu\lambda\rho}\varepsilon_{\mu'\nu'\lambda'\rho'} = - \begin{Vmatrix} \delta^{\mu}_{\mu'} & \delta^{\mu}_{\nu'} & \delta^{\mu}_{\lambda'} & \delta^{\mu}_{\rho'} \\ \delta^{\nu}_{\mu'} & \delta^{\nu}_{\nu'} & \delta^{\nu}_{\lambda'} & \delta^{\nu}_{\rho'} \\ \delta^{\lambda}_{\mu'} & \delta^{\lambda}_{\nu'} & \delta^{\lambda}_{\lambda'} & \delta^{\lambda}_{\rho'} \\ \delta^{\rho}_{\mu'} & \delta^{\rho}_{\nu'} & \delta^{\rho}_{\lambda'} & \delta^{\rho}_{\rho'} \end{Vmatrix}, \tag{D.9}$$

where the pair of two vertical lines on two sides of the matrix indicates the determinant of the matrix.

Contracting both sides by $\delta^{\mu'}_{\mu}$, we obtain the formula

$$\varepsilon^{\mu\nu\lambda\rho}\varepsilon_{\mu\nu'\lambda'\rho'} = - \begin{Vmatrix} \delta^{\nu}_{\nu'} & \delta^{\nu}_{\lambda'} & \delta^{\nu}_{\rho'} \\ \delta^{\lambda}_{\nu'} & \delta^{\lambda}_{\lambda'} & \delta^{\lambda}_{\rho'} \\ \delta^{\rho}_{\nu'} & \delta^{\rho}_{\lambda'} & \delta^{\rho}_{\rho'} \end{Vmatrix}. \tag{D.10}$$

Another contraction yields the result

$$\varepsilon^{\mu\nu\lambda\rho}\varepsilon_{\mu\nu\lambda'\rho'} = -2! \begin{Vmatrix} \delta^{\lambda}_{\lambda'} & \delta^{\lambda}_{\rho'} \\ \delta^{\rho}_{\lambda'} & \delta^{\rho}_{\rho'} \end{Vmatrix}, \tag{D.11}$$

and yet another gives

$$\varepsilon^{\mu\nu\lambda\rho}\varepsilon_{\mu\nu\lambda\rho'} = -3!\, \delta^{\rho}_{\rho'}. \tag{D.12}$$

Finally, if all indices are contracted, we obtain

$$\varepsilon^{\mu\nu\lambda\rho}\varepsilon_{\mu\nu\lambda\rho} = -4!. \tag{D.13}$$

☐ **Exercise D.3** Verify Eqs. (D.10), (D.11), (D.12) and (D.13), starting from Eq. (D.9).

D.2.2   Transformation properties

The reader has probably noticed that despite the fact that the Levi-Civita symbol has four indices, we have never referred to it as a rank-4 tensor in this appendix. The reason should now be explained, by determining the transformation property of the Levi-Civita symbol under a transformation of co-ordinates.

Let us denote the transformation matrix by Λ, as usual, with the rows and columns marked by the indices 0, 1, 2, 3 rather than the usual 1, 2, 3, 4. The standard rule for obtaining determinant of a matrix can be summarized by writing

$$\det \Lambda = -\varepsilon^{\sigma\tau\omega\zeta}\Lambda^0{}_\sigma\Lambda^1{}_\tau\Lambda^2{}_\omega\Lambda^3{}_\zeta, \tag{D.14}$$

where the minus sign on the right side appears because of the sign convention we adopted in Eq. (D.8). Here, we take the row indices (the upper indices) in order, and the column indices in all possible orders, ensuring that the four column indices are different for each term in the determinant. We could also take the row indices in all possible orders, ensuring that each of them is different, by writing

$$\det \Lambda = -\frac{1}{4!}\varepsilon_{\mu\nu\lambda\rho}\varepsilon^{\sigma\tau\omega\zeta}\Lambda^\mu{}_\sigma\Lambda^\nu{}_\tau\Lambda^\lambda{}_\omega\Lambda^\rho{}_\zeta. \tag{D.15}$$

The factor of $1/4!$ appears in this expression to eliminate multiple occurrences of the same term. Now, multiplying both sides by $\varepsilon_{\alpha\beta\gamma\delta}$ and using Eq. (D.9), we obtain

$$\varepsilon_{\alpha\beta\gamma\delta}\det \boldsymbol{\Lambda} = -\frac{1}{4!}\varepsilon_{\mu\nu\lambda\rho}\delta^{\sigma,\tau,\omega,\zeta}_{\alpha,\beta,\gamma,\delta}\Lambda^\mu{}_\sigma\Lambda^\nu{}_\tau\Lambda^\lambda{}_\omega\Lambda^\rho{}_\zeta, \tag{D.16}$$

where $\delta^{\mu',\nu',\lambda',\rho'}_{\alpha,\beta,\gamma,\delta}$ is the type of determinant that appears on the right hand side of Eq. (D.9). This determinant will have 24 terms when expanded, and a little inspection shows that each of them will give the same result after contraction with the other indices present here. Thus we obtain

$$\varepsilon_{\alpha\beta\gamma\delta}\det \boldsymbol{\Lambda} = \varepsilon_{\mu\nu\lambda\rho}\Lambda^\mu{}_\alpha\Lambda^\nu{}_\beta\Lambda^\lambda{}_\gamma\Lambda^\rho{}_\delta. \tag{D.17}$$

Let us see what message is conveyed by this equation. If the Levi-Civita symbol were a tensor, the right hand side of this equation would have given its components in a different system of co-ordinates. The equation then shows that the components in the changed system are not necessarily equal to those in the original system, so $\varepsilon_{\mu\nu\lambda\rho}$ is not really an invariant tensor because of the presence of the determinant on the left hand side. Objects which have extra powers of the determinant of the transformation are called *tensor densities*.

If we consider proper Lorentz transformations only, then of course $\det \Lambda = 1$, and the Levi-Civita symbol is indeed a tensor. Because of this, we have sometimes loosely referred to $\varepsilon_{\mu\nu\lambda\rho}$ as a tensor in the chapters. However, it is important to remember that its transformation property is not like that of a tensor when transformations like space inversion are involved.

In fact, it holds the key to parity-violating interactions. The reader might cross-check all instances of parity violation that we have discussed in various chapters and discover that all of them involved the presence of the Levi-Civita symbol in the Lagrangian. Sometimes this presence was explicit, as in Eq. (6.148, *p 182*). More often, it came through γ_5, since this matrix is defined by using the Levi-Civita symbol, as mentioned in Eq. (F.12, *p 737*).

D.2.3 Spatial antisymmetric tensor

In many places, we need to use the spatial components of 4-vectors, and there it is useful to make a spatial reduction of the Levi-Civita symbol. This symbol should have three indices, all spatial, and we choose

$$\varepsilon_{123} = +1\,, \tag{D.18}$$

and all other components determined by the completely antisymmetric property. The corresponding object with upper indices is also defined the same way:

$$\varepsilon^{123} = +1\,. \tag{D.19}$$

Thus, the value of ε_{ijk} is equal to that of ε_{0ijk}, but the same thing cannot be said for the Levi-Civita symbols with superscripts. It should also be noticed that, because of the convention chosen, we won't be able to raise the indices of ε_{ijk} (or lower those of ε^{ijk}) by the Minkowski metric. But that would not be the end of the world since we always have to make some adjustments while dealing with vectors and tensors of the rotation group. For example, the dot product of two 4-vectors A and B is defined to be $A^\mu B_\mu$, but the dot product of two 3-vectors is not $A^i B_i$, but rather $A^i B^i$, or equivalently $A_i B_i$. The convention chosen in Eq. (D.19) allows us to write the components of the cross product of two 3-vectors in the form

$$(\boldsymbol{A} \times \boldsymbol{B})^i = \varepsilon^{ijk} A_j B_k\,. \tag{D.20}$$

Products of two such Levi-Civita symbols can be given in expressions similar to those given in §D.2.1. For example, we have

$$\varepsilon_{ijk}\varepsilon_{i'j'k'} = \left\| \begin{matrix} \delta_{ii'} & \delta_{ij'} & \delta_{ik'} \\ \delta_{ji'} & \delta_{jj'} & \delta_{jk'} \\ \delta_{ki'} & \delta_{kj'} & \delta_{kk'} \end{matrix} \right\|\,. \tag{D.21}$$

Contraction of one or more indices yields the formulas

$$\varepsilon_{ijk}\varepsilon_{ij'k'} = \delta_{jj'}\delta_{kk'} - \delta_{jk'}\delta_{kj'}\,, \tag{D.22a}$$

$$\varepsilon_{ijk}\varepsilon_{ijk'} = 2\delta_{kk'}\,, \tag{D.22b}$$

$$\varepsilon_{ijk}\varepsilon_{ijk} = 6\,. \tag{D.22c}$$

Appendix E

Clebsch–Gordan co-efficients

In §3.5.3, we mentioned that the product of the states in two irreducible representations does not constitute an irreducible representation in general. This is true for any group. Here we present the recipe for writing the irreducible representations in the product of two irreducible representations of the rotation group.

Any irreducible representation of the rotation group is characterized by a number j which is integer or half-integer. There are $2j+1$ states in the representation characterized by the number j. Each of these states is characterized by a number m whose set of admissible values starts from $-j$, increases by one unit until it reaches $+j$.

The tables given in this appendix show how to write the Kronecker products of two representations. The Kronecker product of the representations j_1 and j_2 contains all representations from $|j_1 - j_2|$ up to $j_1 + j_2$. For any j within this range, the state $|j, m\rangle$ can be written as

$$|j, m\rangle = \sum_{m_1=-j_1}^{+j_1} C(j, m|j_1, m_1; j_2, m - m_1) \, |j_1, m_1\rangle \, |j_2, m - m_1\rangle . \quad (E.1)$$

The co-efficients denoted by $C(j, m|j_1, m_1; j_2, m - m_1)$ are called the Clebsch–Gordan co-efficients. For different combinations of j_1 and j_2, we tabulate the co-efficients.

A few things need to be realized by anyone interested in using these tables. First, each co-efficient appearing in the definition of a $|j, m\rangle$ state can be multiplied by an overall phase factor. In other words, only the relative phase between two terms is physically important. The overall phase has been fixed by some arbitrary convention.

Second, the tables express properties of the algebra of SU(2). So the tables can be used in any situation where the SU(2) algebra is important, irrespective of what the SU(2) means physically. The tables apply for the rotational group for sure. They are also applicable for internal SU(2) symmetries like isospin or weak isospin. In fact, we have made use of these tables in many places in the text while talking about symmetries other than the rotation symmetry.

$j_1 = \frac{1}{2}$					
$j_2 = \frac{1}{2}$		Possible $\|j, m\rangle$ states			
m_1	m_2	$\|1,1\rangle$	$\|1,0\rangle$	$\|0,0\rangle$	$\|1,-1\rangle$
$\frac{1}{2}$	$\frac{1}{2}$	1			
$\frac{1}{2}$	$\frac{1}{2}$		$\frac{1}{2}$	$\frac{1}{2}$	
$\frac{1}{2}$	$-\frac{1}{2}$		$\frac{1}{2}$	$-\frac{1}{2}$	
$-\frac{1}{2}$	$-\frac{1}{2}$				1

Table E.1: Clebsch–Gordan co-efficients for $\frac{1}{2} \times \frac{1}{2}$. The tabulated values are the squares of the co-efficients multiplied by the sign of the co-efficient. In other words, for a co-efficient $a > 0$, we tabulate a^2. For a co-efficient $a < 0$, we tabulate $-a^2$.

$j_1 = 1$							
$j_2 = \frac{1}{2}$		Possible $\|j, m\rangle$ states					
m_1	m_2	$\|\frac{3}{2},\frac{3}{2}\rangle$	$\|\frac{3}{2},\frac{1}{2}\rangle$	$\|\frac{1}{2},\frac{1}{2}\rangle$	$\|\frac{3}{2},-\frac{1}{2}\rangle$	$\|\frac{1}{2},-\frac{1}{2}\rangle$	$\|\frac{3}{2},-\frac{3}{2}\rangle$
1	$\frac{1}{2}$	1					
1	$-\frac{1}{2}$		$\frac{1}{3}$	$\frac{2}{3}$			
0	$\frac{1}{2}$		$\frac{2}{3}$	$-\frac{1}{3}$			
0	$-\frac{1}{2}$				$\frac{2}{3}$	$\frac{1}{3}$	
-1	$\frac{1}{2}$				$\frac{1}{3}$	$-\frac{2}{3}$	
-1	$-\frac{1}{2}$						1

Table E.2: Clebsch–Gordan co-efficients for $1 \times \frac{1}{2}$. See caption of Table E.1 for conventions in tabulation.

$j_1 = 1$ $j_2 = 1$		Possible $\lvert j,m\rangle$ states								
m_1	m_2	$\lvert 2,2\rangle$	$\lvert 2,1\rangle$	$\lvert 1,1\rangle$	$\lvert 2,0\rangle$	$\lvert 1,0\rangle$	$\lvert 0,0\rangle$	$\lvert 2,-1\rangle$	$\lvert 1,-1\rangle$	$\lvert 2,-2\rangle$
1	1	1								
1	0		$\frac{1}{2}$	$\frac{1}{2}$						
0	1		$\frac{1}{2}$	$-\frac{1}{2}$						
1	-1				$\frac{1}{6}$	$\frac{1}{2}$	$\frac{1}{3}$			
0	0				$\frac{2}{3}$	0	$-\frac{1}{3}$			
-1	1				$\frac{1}{6}$	$-\frac{1}{2}$	$\frac{1}{3}$			
-1	0							$\frac{1}{2}$	$\frac{1}{2}$	
0	-1							$\frac{1}{2}$	$-\frac{1}{2}$	
-1	-1									1

Table E.3: Clebsch–Gordan co-efficients for 1×1. See caption of Table E.1 for conventions in tabulation.

Appendix F

Dirac matrices and spinors

Throughout the text, we have talked about Dirac matrices and spinors in bits and pieces, as and when the need arose. For someone who is not familiar with these objects, this might pose some problems. So, in this appendix, we give a self-contained summary of the properties of Dirac matrices and spinors.

F.1 Dirac matrices

F.1.1 Basic properties

Dirac suggested that, for spin-$\frac{1}{2}$ fermions, one should not use the Klein–Gordon equation. Instead, he introduced the Hamiltonian

$$H = \boldsymbol{\alpha} \cdot \boldsymbol{p} + \beta m \qquad \text{(F.1)}$$

in order to obtain a Schrödinger equation that is first order in both space and time derivatives. He observed that in order to reproduce the relativistic relation between energy and momentum, Eq. (4.1, p 62), the objects $\boldsymbol{\alpha}$ and β should satisfy the relations

$$\begin{aligned} \left[\alpha^i, \alpha^j\right]_+ &= 2\delta_{ij}\,, \\ \left[\alpha^i, \beta\right]_+ &= 0\,, \\ \beta^2 &= 1\,, \end{aligned} \qquad \text{(F.2)}$$

where the notation $[A, B]_+$ stands for the anticommutator of A and B, i.e., $AB + BA$. The anticommutation property cannot be satisfied if the objects $\boldsymbol{\alpha}$ and β are numbers. The relations show that we need four mutually anti-commuting matrices. We will show that these matrices must be traceless, and then they must be at least 4×4 matrices. They must be hermitian since the Hamiltonian of Eq. (F.1) has to be hermitian.

☐ **Exercise F.1** Take the square of the Hamiltonian given in Eq. (F.1), assuming that the objects $\boldsymbol{\alpha}$ and β commute with all components of momentum, but not assuming any commutation relation between themselves. Compare the resulting equation with Eq. (4.1, *p 62*) and show that the relations given in Eq. (F.2) result.

For field-theoretic purposes, it is more convenient to use the matrices

$$\gamma^i = \beta \alpha^i \qquad (F.3)$$

instead of the matrices α^i, and rename β,

$$\beta = \gamma^0, \qquad (F.4)$$

so that we can denote the set of these four matrices by a compact notation,

$$\gamma^\mu = \{\gamma^0, \gamma^i\}. \qquad (F.5)$$

We will also use the notation

$$\gamma_\mu = g_{\mu\nu}\gamma^\nu \qquad (F.6)$$

just as we do for Lorentz vectors. However, despite the notation, one should not think that the four matrices transform like vectors. They are fixed matrices. The vector index is only a reminder that bilinears of the form $\overline{\psi}\gamma^\mu\psi$ transform like vectors, as shown in §4.4.2.

It is straightforward to see that the set of relations in Eq. (F.2) can be summarized as

$$\left[\gamma_\mu, \gamma_\nu\right]_+ = 2g_{\mu\nu}\mathbb{1}. \qquad (F.7)$$

The relation is called the *Clifford algebra* of the matrices. In future, we will often omit the unit matrix that is present on the right hand side.

Obviously γ^0 is a hermitian matrix because it is equal to β. But the matrices γ^i are not hermitian. Using the hermiticity properties of α and β in conjunction with the anticommutation properties of Eq. (F.2), it is easy to see that

$$\gamma_i^\dagger = -\gamma_i. \qquad (F.8)$$

So the γ_i's are anti-hermitian whereas γ_0 is hermitian, a fact which can be written in a compact form as

$$\gamma_\mu^\dagger = \gamma_0\gamma_\mu\gamma_0. \qquad (F.9)$$

Eqs. (F.7) and (F.9) are the relations which define the Dirac matrices, alternatively called the gamma matrices. The explicit form for the matrices is not unique. In fact, it is easy to see that if one defines an alternative set of matrices by the relation

$$\widetilde{\gamma}_\mu = U\gamma_\mu U^\dagger \qquad (F.10)$$

where U is a unitary matrix, the matrices $\tilde{\gamma}_\mu$ also satisfy the relations in Eqs. (F.7) and (F.9) and can therefore serve as the Dirac matrices. Conversely, it is also true that if two sets of matrices satisfy both Eqs. (F.7) and (F.9), they are related unitarily, i.e., through a relation like that in Eq. (F.10). Because of the arbitrariness in the explicit form of the matrices, we will not introduce any explicit form of these matrices. We will just use the defining relations of Eqs. (F.7) and (F.9) to derive other properties of these matrices.

F.1.2 Associated matrices

The anticommutators of Dirac matrices are proportional to the unit matrix, as expressed in Eq. (F.7). The commutators are given the names

$$\sigma_{\mu\nu} = \frac{i}{2}\left[\gamma_\mu, \gamma_\nu\right]. \tag{F.11}$$

The importance of this definition lies in the fact that the matrices $\frac{1}{2}\sigma_{\mu\nu}$ are representations of the generators of the Lorentz group.

□ **Exercise F.2** Verify the statement, i.e., show that if in Eq. (3.56, p 54) we replace all occurrences of the Lorentz group generators $\mathscr{J}$ by the matrices $\frac{1}{2}\sigma$, it produces a valid equation.

The matrix γ_5 has been defined in Eq. (4.88, p 78). Notice that, using the anticommutation property of the Dirac matrices, this definition can be cast in the form

$$\gamma_5 = \frac{i}{4!}\varepsilon_{\mu\nu\lambda\rho}\gamma^\mu\gamma^\nu\gamma^\lambda\gamma^\rho. \tag{F.12}$$

This matrix anticommutes with all Dirac matrices:

$$\left[\gamma_\mu, \gamma_5\right]_+ = 0, \tag{F.13}$$

and therefore commutes with all sigma matrices, as expressed in Eq. (4.89, p 78). The factor i appearing in the definition ensures that

$$\left(\gamma_5\right)^2 = 1, \tag{F.14}$$

and also that

$$\gamma_5^\dagger = \gamma_5. \tag{F.15}$$

Another useful matrix is defined through Eq. (F.10). The point is that, given the anticommutation relation of Eq. (F.7), it is easy to see that the anticommutator of the matrices $-\gamma_\mu^\top$ should be the same. Thus, there must exist a unitary matrix $\mathbb{C}$,

$$\mathbb{C}^\dagger = \mathbb{C}^{-1}, \tag{F.16}$$

such that

$$\mathbb{C}^{-1}\gamma_\mu\mathbb{C} = -\gamma_\mu^\mathsf{T}\,. \tag{F.17}$$

Note that Eqs. (F.9) and (F.17) imply that

$$\gamma_\mu^* = -\mathbb{C}^{-1}\gamma_0\gamma_\mu\gamma_0\mathbb{C}\, . = -\left(\gamma_0\mathbb{C}\right)^{-1}\gamma_\mu\left(\gamma_0\mathbb{C}\right). \tag{F.18}$$

☐ **Exercise F.3** Show that

$$\mathbb{C}^{-1}\gamma_5\mathbb{C} = \gamma_5^\mathsf{T}\,. \tag{F.19}$$

F.1.3 Contraction formulas

Here we derive simplified forms for strings of Dirac matrices containing some contracted indices. Clearly,

$$\gamma^\mu\gamma_\mu = 4\,, \tag{F.20}$$

a result that follows by contracting both sides of Eq. (F.7) with $g^{\mu\nu}$. If we have another Dirac matrix between the pair that have contracted indices, we can use Eq. (F.7) to write

$$\gamma^\mu\gamma_\nu\gamma_\mu = \gamma^\mu\left(2g_{\mu\nu} - \gamma_\mu\gamma_\nu\right) = 2\gamma_\nu - \gamma^\mu\gamma_\mu\gamma_\nu\,. \tag{F.21}$$

Using Eq. (F.20) now, we finally obtain

$$\gamma^\mu\gamma_\nu\gamma_\mu = -2\gamma_\nu\,. \tag{F.22}$$

The procedure can be continued for longer strings. The results are:

$$\gamma^\alpha\gamma_\mu\gamma_\nu\gamma_\alpha = 4g_{\mu\nu}\,, \tag{F.23}$$

$$\gamma^\alpha\gamma_\mu\gamma_\nu\gamma_\lambda\gamma_\alpha = -2\gamma_\lambda\gamma_\nu\gamma_\mu\,, \tag{F.24}$$

$$\gamma^\alpha\gamma_\mu\gamma_\nu\gamma_\lambda\gamma_\rho\gamma_\alpha = 2(\gamma_\rho\gamma_\mu\gamma_\nu\gamma_\lambda + \gamma_\lambda\gamma_\nu\gamma_\mu\gamma_\rho)\,. \tag{F.25}$$

The last one can be expressed in a simpler way, mentioned in Eq. (F.51).

☐ **Exercise F.4** Derive the following contraction formulas involving the sigma matrices:

$$\sigma^{\mu\nu}\sigma_{\mu\nu} = 12\,, \tag{F.26}$$

$$\sigma^{\mu\nu}\gamma^\lambda\sigma_{\mu\nu} = 0\,, \tag{F.27}$$

$$\sigma^{\mu\nu}\sigma^{\lambda\rho}\sigma_{\mu\nu} = -4\sigma^{\lambda\rho}\,. \tag{F.28}$$

F.1.4 Trace formulas

It is easily seen that each of the γ_μ's is traceless. For this, use Eq. (F.14) to write

$$\mathrm{Tr}(\gamma_\mu) = \mathrm{Tr}(\gamma_\mu\gamma_5\gamma_5)\,. \tag{F.29}$$

This final expression can be manipulated further. First, we can use the cyclic property of traces to write it as $\text{Tr}(\gamma_5 \gamma_\mu \gamma_5)$. Then we can use the anticommutation of γ_5 and γ_μ to write it as $-\text{Tr}(\gamma_\mu \gamma_5 \gamma_5)$, which is nothing but $-\text{Tr}(\gamma_\mu)$ because of Eq. (F.14). The equality of this expression with the original one tells us that

$$\text{Tr}(\gamma_\mu) = 0 \,. \tag{F.30}$$

In a similar way, it can be proved that any string containing an odd number of Dirac matrices is also traceless, i.e.:

$$\text{Tr}(\gamma_{\mu_1} \gamma_{\mu_2} \cdots \gamma_{\mu_{2n+1}}) = 0 \,. \tag{F.31}$$

For strings with $2n$ Dirac matrices, trace is not necessarily zero. First, consider a string with $n = 0$:

$$\text{Tr}(\mathbb{1}) = 4 \,, \tag{F.32}$$

since we are dealing with 4×4 matrices. When $n = 2$, we can write

$$\text{Tr}(\gamma_\mu \gamma_\nu) = \text{Tr}(2g_{\mu\nu} \mathbb{1} - \gamma_\nu \gamma_\mu) \tag{F.33}$$

by using the basic anticommutation relation. Since $\text{Tr}(A+B) = \text{Tr}\, A + \text{Tr}\, B$, we can write it as

$$\text{Tr}(\gamma_\mu \gamma_\nu) = 2g_{\mu\nu} \, \text{Tr}(\mathbb{1}) - \text{Tr}(\gamma_\nu \gamma_\mu) \,. \tag{F.34}$$

But the last trace appearing in this equation is equal to the expression on the left hand side because of the cyclic property of traces. Using this and Eq. (F.32), we obtain

$$\text{Tr}(\gamma_\mu \gamma_\nu) = 4g_{\mu\nu} \,. \tag{F.35}$$

To obtain the trace of a string of four Dirac matrices, we first use the anticommutation relation to write

$$\begin{aligned}
\text{Tr}\left(\gamma_\mu \gamma_\nu \gamma_\lambda \gamma_\rho\right) &= 2g_{\mu\nu} \, \text{Tr}\left(\gamma_\lambda \gamma_\rho\right) - \text{Tr}\left(\gamma_\nu \gamma_\mu \gamma_\lambda \gamma_\rho\right) \\
&= 8g_{\mu\nu}g_{\lambda\rho} - \text{Tr}\left(\gamma_\mu \gamma_\lambda \gamma_\rho \gamma_\nu\right) \,,
\end{aligned} \tag{F.36}$$

using the cyclicity of traces to put the γ_ν at the end in the last term. We now use the anticommutation relation repeatedly to make the factor of γ_ν hop its way back to the position where it started from, viz., just after γ_μ. This would give

$$\text{Tr}\left(\gamma_\mu \gamma_\nu \gamma_\lambda \gamma_\rho\right) = 8(g_{\mu\nu}g_{\lambda\rho} - g_{\nu\rho}g_{\mu\lambda} + g_{\nu\lambda}g_{\mu\rho}) - \text{Tr}\left(\gamma_\mu \gamma_\nu \gamma_\lambda \gamma_\rho\right) \,. \tag{F.37}$$

Finally, transposing the remaining trace to the left hand side, we obtain the result

$$\text{Tr}\left(\gamma_\mu \gamma_\nu \gamma_\lambda \gamma_\rho\right) = 4(g_{\mu\nu}g_{\lambda\rho} - g_{\mu\lambda}g_{\nu\rho} + g_{\mu\rho}g_{\nu\lambda}) \,. \tag{F.38}$$

□ **Exercise F.5** Follow this inductive procedure to show that

$$\text{Tr}\left(\gamma_\alpha\gamma_\beta\gamma_{\mu_1}\gamma_{\mu_2}\cdots\gamma_{\mu_{2n}}\right) = g_{\alpha\beta}\,\text{Tr}\left(\gamma_{\mu_1}\gamma_{\mu_2}\cdots\gamma_{\mu_{2n}}\right)$$
$$-\sum_{r=1}^{2n}(-1)^r g_{\mu_r\beta}\,\text{Tr}\left(\gamma_\alpha\gamma_{\mu_1}\cdots\not\mu_{2n}\right)_{\not\mu_r}\quad(\text{F.39})$$

where in the last term, the crossed-out μ_r means that the corresponding index should be omitted in the trace.

Trace of six Dirac matrices will contain 15 terms, but we don't need that elaborate formula. Any such long string can be traded for a number of shorter strings by making use of Eq. (F.50).

It is also useful to find traces of strings involving γ_5. First, we should note that

$$\text{Tr}(\gamma_5) = 0\,,\qquad(\text{F.40})$$

which can be proved in much the same way that is used to prove the tracelessness of the Dirac matrices, using the fact that γ_0 anticommutes with γ_5. It is also obvious that any string containing γ_5 and an odd number of Dirac matrices is traceless, because γ_5 can be seen as a string of four Dirac matrices, making a total of an odd number of Dirac matrices in the string. Then note that

$$\text{Tr}(\gamma_\mu\gamma_\nu\gamma_5) = -\text{Tr}(\gamma_\mu\gamma_5\gamma_\nu) = -\text{Tr}(\gamma_\nu\gamma_\mu\gamma_5)\,,\qquad(\text{F.41})$$

using the cyclicity of traces in obtaining the last step. The trace must then be antisymmetric in the indices μ and ν. There is no such object that contains only two Lorentz indices and is antisymmetric in them. Thus the trace must be zero:

$$\text{Tr}(\gamma_\mu\gamma_\nu\gamma_5) = 0\,.\qquad(\text{F.42})$$

What about the trace of four Dirac matrices multiplied by γ_5? Using the anticommutation relation, we can write

$$\text{Tr}(\gamma_\mu\gamma_\nu\gamma_\lambda\gamma_\rho\gamma_5) = 2g_{\mu\nu}\,\text{Tr}(\gamma_\lambda\gamma_\rho\gamma_5) - \text{Tr}(\gamma_\nu\gamma_\mu\gamma_\lambda\gamma_\rho\gamma_5)\,.\qquad(\text{F.43})$$

The first term on the right hand side is zero because of Eq. (F.42), and so we reach the conclusion that the trace must be antisymmetric in the indices μ and ν. Through similar manipulations, it can be shown that the trace is in fact antisymmetric in the exchange of any pair of indices. Hence it must be proportional to the Levi-Civita tensor $\varepsilon_{\mu\nu\lambda\rho}$. The proportionality constant can be easily determined by taking the four different Dirac matrices and using the definition of γ_5 given in Eq. (4.88, *p 78*). This gives

$$\text{Tr}(\gamma_\mu\gamma_\nu\gamma_\lambda\gamma_\rho\gamma_5) = 4i\varepsilon_{\mu\nu\lambda\rho}\,.\qquad(\text{F.44})$$

F.1.5   Strings of Dirac matrices

In Eq. (4.93, p 79), we identified a set of 16 matrices and mentioned that they
can be used as a basis set for all 4×4 matrices. This means that any 4×4
matrix can be expressed as a linear superposition of the 16 matrices shown
in Eq. (4.93, p 79). Here we give a few examples with some strings of Dirac
matrices, with and without γ_5, expressing them as superpositions of the 16
basis matrices.

We start with a string of two Dirac matrices. Using Eq. (F.7) and Eq.
(F.11), we obtain easily the relation

$$\gamma_\mu \gamma_\nu = g_{\mu\nu} - i\sigma_{\mu\nu} \, . \tag{F.45}$$

More explicitly, the first term on the right hand side should be written as
$g_{\mu\nu} \, 1\!\!1$, indicating the unit matrix. The formula then shows that a string of
two Dirac matrices can be expressed as a superposition of the unit matrix
and the matrices $\sigma_{\mu\nu}$.

In order to get to longer strings, we first use the result

$$\varepsilon_{\alpha\beta\mu\nu} \sigma^{\mu\nu} \gamma_5 = 2i\sigma_{\alpha\beta} \, . \tag{F.46}$$

It can be easily checked by taking each possible combination of the indices
α and β and exhausting all possibilities. A more algebraic proof is outlined
in Ex. F.6. Once this is taken for granted, we can contract both sides of this
equation by $\varepsilon^{\alpha\beta\lambda\rho}$, use Eq. (D.11, p 729), and obtain

$$\sigma^{\lambda\rho} \gamma_5 = -\frac{i}{2} \varepsilon^{\lambda\rho\alpha\beta} \sigma_{\alpha\beta} \, , \tag{F.47}$$

which shows how $\sigma^{\lambda\rho}\gamma_5$ is expressed in terms of the basic 16 matrices. It is
now easy to multiply Eq. (F.45) from the right by γ_5, and use Eq. (F.47) to
obtain

$$\gamma_\mu \gamma_\nu \gamma_5 = g_{\mu\nu} \gamma_5 - \frac{1}{2} \varepsilon_{\mu\nu\alpha\beta} \sigma^{\alpha\beta} \, . \tag{F.48}$$

A special case of this equation, with μ being the time index and ν a space index, was given in Eq. (14.10, p 414) while discussing helicity. (While comparing
with Eq. (14.6, p 413), remember that $\Sigma_i = -\Sigma^i$.)

□ **Exercise F.6** *Prove Eq. (F.46) using the outline given here. First,
write γ_5 as in Eq. (F.12). The resulting expression will have a product of two Levi-Civita tensors. Use Eq. (D.9, p 729) to express this
product as strings of elements of the metric tensor. While contracting the Dirac matrices with the metric tensor, use the contraction
formulas given in §F.1.3. This will give the desired result.*

In order to deal with strings of three Dirac matrices, it is useful to start
from the expression $i\varepsilon_{\mu\nu\lambda\rho} \gamma^\rho \gamma_5$, and use the definition of γ_5 from Eq. (F.12).
Using Eq. (D.9, p 729) and various contraction formulas from §F.1.3, we obtain
the following expression:

$$i\varepsilon_{\mu\nu\lambda\rho} \gamma^\rho \gamma_5 = -\frac{1}{6}\Big(\gamma_\mu \gamma_\nu \gamma_\lambda + \gamma_\nu \gamma_\lambda \gamma_\mu + \gamma_\lambda \gamma_\mu \gamma_\nu - (\mu \leftrightarrow \nu) \Big) \, . \tag{F.49}$$

742 *Appendix F. Dirac matrices and spinors*

Using the anticommutation relations between the Dirac matrices to push γ_μ forward in every term, followed by γ_ν and γ_λ, we obtain an identity of the form

$$\gamma_\mu\gamma_\nu\gamma_\lambda = g_{\mu\nu}\gamma_\lambda - g_{\lambda\mu}\gamma_\nu + g_{\nu\lambda}\gamma_\mu - i\varepsilon_{\mu\nu\lambda\rho}\gamma^\rho\gamma_5 . \tag{F.50}$$

This is an extremely important identity because it allows one to reduce the length of any string of Dirac matrices that is more than three matrices long. For example, if one encounters a trace with six Dirac matrices, one can use this formula to reduce the result to some traces with four Dirac matrices only, with and without an extra factor of γ_5.

☐ **Exercise F.7** Show that the trace formula of Eq. (F.38) can be easily derived by using Eq. (F.50).

☐ **Exercise F.8** Use Eq. (F.50) to express the contraction formula of Eq. (F.25) in the following form involving fewer Dirac matrices on the right hand side:

$$\gamma^\alpha\gamma_\mu\gamma_\nu\gamma_\lambda\gamma_\rho\gamma_\alpha = 4(g_{\mu\nu}g_{\lambda\rho} + g_{\mu\rho}g_{\nu\lambda} - g_{\mu\lambda}g_{\nu\rho} - i\varepsilon_{\mu\nu\lambda\rho}\gamma_5) . \tag{F.51}$$

F.1.6 A property of the matrix $\mathbb{C}$

The matrix $\mathbb{C}$ was defined in Eq. (F.17). Taking the transpose of both sides of this equation, we obtain

$$\mathbb{C}^\top\gamma_\mu^\top(\mathbb{C}^{-1})^\top = -\gamma_\mu . \tag{F.52}$$

Using now Eq. (F.17) again on the left hand side of this equation, we obtain the relation

$$\left[\gamma_\mu, \mathbb{C}^\top\mathbb{C}^{-1}\right] = 0 . \tag{F.53}$$

Any 4×4 matrix can be written as a linear superposition of the 16 basis matrices given in (4.93). This way, it can be easily seen that if a matrix commutes with all four Dirac matrices γ_μ, then it must be a multiple of a unit matrix. Thus we can write

$$\mathbb{C}^\top = a\mathbb{C} , \tag{F.54}$$

where a is a scalar. Taking the transpose of this equation, we obtain $\mathbb{C} = a\mathbb{C}^\top = a^2\mathbb{C}$, so that the only possible values of a are

$$a = \pm 1 . \tag{F.55}$$

In order to find the right value of a, we rewrite Eq. (F.17) in the form

$$\gamma_\mu\mathbb{C} = -\mathbb{C}\gamma_\mu^\top = -a(\gamma_\mu\mathbb{C})^\top , \tag{F.56}$$

implying that $\gamma_\mu\mathbb{C}$ is antisymmetric if $a = +1$ and symmetric if $a = -1$. Using the definition of the σ-matrices, we can easily show that the matrices $\sigma_{\mu\nu}\mathbb{C}$

must also have the same property: antisymmetric if $a = +1$ and symmetric if $a = -1$. This shows that $a = +1$ is untenable, because that would produce ten antisymmetric matrices which are linearly independent. For 4×4, there can be at most six independent antisymmetric matrices. Thus $a = -1$, which means that

$$\mathbb{C}^\mathsf{T} = -\mathbb{C}, \qquad (\text{F.57})$$

i.e., the matrix $\mathbb{C}$ is antisymmetric.

F.2 Dirac spinors

F.2.1 Plane wave solutions of Dirac equation

The Schrödinger equation with the Dirac Hamiltonian is often called the Dirac equation:

$$i\frac{\partial \psi}{\partial t} = -i\boldsymbol{\alpha} \cdot \boldsymbol{\nabla}\psi + \beta m\psi. \qquad (\text{F.58})$$

Multiplying both sides from the left by the matrix β or γ^0, this equation can also be written in the form

$$i\gamma^\mu \partial_\mu \psi - m\psi = 0. \qquad (\text{F.59})$$

Since γ^μ's are 4×4 matrices, ψ must be a 4-component column vector. Let us therefore try solutions of the form

$$\psi(x) = u_{\boldsymbol{p}} e^{-ip\cdot x}, \qquad (\text{F.60})$$

where $u_{\boldsymbol{p}}$ is a 4-component column vector. This is called a *spinor* solution. According to the discussion after Eq. (4.7, *p 63*), the $u_{\boldsymbol{p}}$ appearing in Eq. (F.60) must be a positive energy spinor. Similarly, there will be negative energy spinors, defined by solutions of the form

$$\psi(x) = v_{\boldsymbol{p}} e^{+ip\cdot x}. \qquad (\text{F.61})$$

Obviously, there will be four linearly independent solutions altogether, which will include two of the first kind and two of the second. We will distinguish between different solutions of the same type by an extra subscript index, i.e., by writing $u_{\boldsymbol{p},s}$ and $v_{\boldsymbol{p},s}$. When we write any mathematical formula with only one subscript on the spinors, the subscript should always be understood to be the 3-momentum appearing in the plane wave solution, and the formula to be valid irrespective of which of the two different solutions we take.

The spinors satisfy the equations

$$\left(\not{p} - m\right)u_{\boldsymbol{p}} = 0, \qquad (\text{F.62a})$$

$$\left(\not{p} + m\right)v_{\boldsymbol{p}} = 0, \qquad (\text{F.62b})$$

where

$$\not{p} \equiv \gamma^\mu p_\mu. \tag{F.63}$$

These can be found by inserting Eqs. (F.60) and (F.61) into the Dirac equation, Eq. (F.59). Of course, the explicit solutions of Eq. (F.62) depend on the explicit forms of the Dirac matrices. These are representation dependent, and therefore are not of interest to us. Independent of the representation, we can make a few observations that will be helpful for future manipulations with the spinors. First, note how Eq. (F.62a) looks for $\boldsymbol{p} = 0$. In this case $p_0 = m$, so that the equation reduces to

$$(\gamma_0 - 1)u_0 = 0, \tag{F.64}$$

which means that u_0 is an eigenvector of γ_0 with eigenvalue $+1$. Similarly, Eq. (F.62b) shows that v_0 should be an eigenvector of γ_0 with eigenvalue -1.

In Eq. (F.30), we showed that the Dirac matrices are traceless. Therefore, the matrix γ_0 must have two eigenvectors corresponding to the eigenvalue $+1$, and two for the eigenvalue -1. Let us denote the normalized eigenvectors by ξ_s and χ_s, with s takes two values which label the two degenerate solutions:

$$\gamma_0\xi_s = \xi_s, \qquad \gamma_0\chi_s = -\chi_s. \tag{F.65}$$

In order to obtain a mutually orthogonal set of eigenvectors, we can take them to be simultaneous eigenstates of some other hermitian matrix that commutes with γ_0. For example, σ_{12} can fit this role, and we can take

$$\sigma_{12}\xi_s = s\xi_s, \qquad \sigma_{12}\chi_s = s\chi_s, \tag{F.66}$$

where the index s takes the values

$$s = \pm. \tag{F.67}$$

The normalization conditions on the eigenvectors can then be set as

$$\xi_s^\dagger \xi_{s'} = \delta_{ss'}, \qquad \chi_s^\dagger \chi_{s'} = \delta_{ss'}, \qquad \xi_s^\dagger \chi_{s'} = \chi_s^\dagger \xi_{s'} = 0. \tag{F.68}$$

The zero momentum solutions should be proportional to these eigenvectors. The normalization condition will be discussed shortly in §F.2.2.

The solutions for arbitrary momentum can be constructed in the following way:

$$u_{\boldsymbol{p},s} = N_p(\not{p} + m)\xi_s, \tag{F.69a}$$

$$v_{\boldsymbol{p},s} = N_p(-\not{p} + m)\chi_{-s}, \tag{F.69b}$$

where N_p is a normalizing factor. Note that these are not eigenvectors of either γ_0 or σ_{12} for any non-zero 3-momentum. In §4.4, we outlined the arguments which show that these expressions indeed satisfy Eqs. (F.62a) and (F.62b), and we do not repeat that part of the argument here.

F.2.2 Normalization of spinors

Note that Eq. (F.69) can be written in the form

$$u_{\boldsymbol{p},s} = N_p\big(E_p + m - \gamma_i p_i\big)\xi_s \,, \tag{F.70a}$$

$$v_{\boldsymbol{p},s} = N_p\big(E_p + m + \gamma_i p_i\big)\chi_{-s} \,, \tag{F.70b}$$

using Eq. (F.65). Thus, remembering that $\gamma_i^\dagger = -\gamma_i$, we can write

$$u_{\boldsymbol{p},s}^\dagger u_{\boldsymbol{p},s'} = N_p^2 \xi_s^\dagger \big(E_p + m + \gamma_i p_i\big)\big(E_p + m - \gamma_j p_j\big)\xi_{s'}$$

$$= N_p^2 \xi_s^\dagger \Big[(E_p + m)^2 - \gamma_i\gamma_j p_i p_j\Big]\xi_{s'} \,. \tag{F.71}$$

In writing the last form, we have used the relation

$$\xi_s^\dagger \gamma_i \xi_{s'} = 0 \,, \tag{F.72}$$

which can be easily shown by using Eq. (F.65) and the fact that γ_0 anticommutes with any γ_i. Next, we use the anticommutation property of the Dirac matrices to write $\gamma_i\gamma_j p_i p_j = -\boldsymbol{p}^2 = -E_p^2 + m^2$. We now put it in and use Eq. (F.68) to perform similar manipulations with the v-spinors. The results show that if we set

$$N_p = \frac{1}{\sqrt{E_p + m}} \,, \tag{F.73}$$

then our normalization condition would read

$$u_{\boldsymbol{p},s}^\dagger u_{\boldsymbol{p},s'} = 2E_{\boldsymbol{p}}\delta_{s,s'} \,, \tag{F.74a}$$

$$v_{\boldsymbol{p},s}^\dagger v_{\boldsymbol{p},s'} = 2E_{\boldsymbol{p}}\delta_{s,s'} \,. \tag{F.74b}$$

In addition, we would have

$$u_{\boldsymbol{p},s}^\dagger v_{-\boldsymbol{p},s'} = 0 \,, \tag{F.75a}$$

$$v_{\boldsymbol{p},s}^\dagger u_{-\boldsymbol{p},s'} = 0 \,. \tag{F.75b}$$

Of course the choice of N_p merely defines a convention. But we want to point out that our convention is a good one insofar as the definitions in Eq. (F.69), as well as the normalization conditions obtained in Eq. (F.74), are all valid for arbitrary masses of the fermion, including zero. An alternative way of representing the normalization conditions will be discussed in §F.3.1.

F.2.3 Conjugation properties

Now let us look at the object $\gamma_0 \mathbb{C}\xi_s^*$, where the matrix $\mathbb{C}$ has been defined in §F.1. Note that Eq. (F.17) implies that

$$\gamma_0 \mathbb{C}\xi_s^* = -\mathbb{C}\gamma_0^\top \xi_s^* = -\mathbb{C}\gamma_0^* \xi_s^* = -\mathbb{C}\xi_s^* \,, \tag{F.76}$$

using the hermiticity of γ_0, and the definition of ξ_s. On the other hand, since σ_{12} is hermitian and it commutes with γ_0, we can write

$$\sigma_{12}\gamma_0\mathbb{C}\xi_s^* = \gamma_0\sigma_{12}\mathbb{C}\xi_s^* = -\gamma_0\mathbb{C}\sigma_{12}^{\mathsf{T}}\xi_s^* = -\gamma_0\mathbb{C}\sigma_{12}^*\xi_s^* = -s\gamma_0\mathbb{C}\xi_s^*\,, \quad \text{(F.77)}$$

where we have made use of Eq. (6.56, *p 161*) on the way. These two equations show that $\gamma_0\mathbb{C}\xi_s^*$ is an eigenstate of γ_0 with eigenvalue -1, and of σ_{12} with eigenvalue $-s$. Its norm can be shown to be 1 easily, which means that $\gamma_0\mathbb{C}\xi_s^*$ is equal to χ_{-s} up to a phase. The phase factor, in fact, can be set to unity by adjusting the overall phase of the matrix $\mathbb{C}$, and we can write

$$\gamma_0\mathbb{C}\xi_s^* = \chi_{-s}\,. \quad \text{(F.78a)}$$

Using the properties of the matrix $\mathbb{C}$ deduced earlier, one can now show that this implies

$$\gamma_0\mathbb{C}\chi_s^* = \xi_{-s}\,. \quad \text{(F.78b)}$$

□ **Exercise F.9** *Prove Eq. (F.78b) from Eq. (F.78a), using properties of the matrix $\mathbb{C}$ such as those given in Eqs. (F.16) and (F.57).*

To see similar results of conjugation with spinors with arbitrary momentum, we make use of Eq. (F.18) to write

$$\gamma_0\mathbb{C}u_{\boldsymbol{p},s}^* = N_p\gamma_0\mathbb{C}(\gamma_\mu^* p^\mu + m)\xi_s^* = N_p(-\gamma_\mu p^\mu + m)\gamma_0\mathbb{C}\xi_s^*\,. \quad \text{(F.79)}$$

Now, we can use Eq. (F.78a) to discover that the right hand side is nothing but $v_{\boldsymbol{p},s}$. A similar relation is obtained by taking the complex conjugate of a v-spinor. We display both equations here:

$$\gamma_0\mathbb{C}u_{\boldsymbol{p},s}^* = v_{\boldsymbol{p},s}\,, \quad \text{(F.80a)}$$
$$\gamma_0\mathbb{C}v_{\boldsymbol{p},s}^* = u_{\boldsymbol{p},s}\,. \quad \text{(F.80b)}$$

These relations appear in the text as Eq. (6.108, *p 170*), in the context of charge conjugation properties of fermions.

F.2.4 Result of γ_0 multiplying the spinors

What happens when γ_0 multiplies the spinors? It is convenient to use the formulas in Eq. (F.70) in order to answer this question. We see that

$$\gamma_0 u_{\boldsymbol{p},s} = N_p\gamma_0\big(E_p + m - \gamma_i p_i\big)\xi_s\,. \quad \text{(F.81)}$$

Let us now try to move the factor of γ_0 to the right of the expression in the parentheses. Since γ_0 anticommutes with γ_i, we obtain

$$\gamma_0 u_{\boldsymbol{p},s} = N_p\big(E_p + m + \gamma_i p_i\big)\gamma_0\xi_s = N_p\big(E_p + m + \gamma_i p_i\big)\xi_s\,, \quad \text{(F.82)}$$

using Eq. (F.65) for the last step. This shows that on the right hand side, we have also obtained a u-spinor, but which corresponds to a 3-momentum which is reversed compared to the original one. This, and the result of the similar analysis on v-spinors can be summarized in the equations

$$\gamma_0 u_{p,s} = u_{-p,s} \, , \qquad \gamma_0 v_{p,s} = -v_{-p,s} \, . \tag{F.83}$$

These relations first appear in the text in Eq. (6.80, *p 165*).

F.2.5   Result of γ_5 multiplying the spinors

Consider $\gamma_5 \xi_s$. Since γ_5 anticommutes with γ_0, it is easy to see that $\gamma_5 \xi_s$ is an eigenstate of γ_0 with eigenvalue -1. It is also an eigenstate of σ_{12}, with eigenvalue s, since γ_5 commutes with σ_{12}. It appears to have all properties of χ_s, and it might be tempting to identify $\gamma_5 \xi_s$ with χ_s. However, that cannot be done. There can be an extra phase factor in the relation, which cannot be eliminated by choice of phases of the eigenvectors of γ_0, because we have already utilized the freedom by committing to Eq. (F.78). Thus, we should write

$$\gamma_5 \xi_s = \eta_s \chi_s \, , \tag{F.84a}$$

where η_s is a phase factor. Multiplying each side by γ_5, we obtain

$$\gamma_5 \chi_s = \eta_s^* \xi_s \, . \tag{F.84b}$$

As we said, the two phase factors η_+ and η_- cannot be arbitrarily chosen: there is a relation between them that follows from Eq. (F.78). To see this, note that Eqs. (F.78a) and (F.84b) imply the relation

$$\chi_- = \gamma_0 \mathbb{C} \xi_+^* = \gamma_0 \mathbb{C} \Big(\eta_+ \gamma_5 \chi_+ \Big)^* = \eta_+^* \gamma_0 \mathbb{C} \gamma_5^\top \chi_+^* \, , \tag{F.85}$$

using the fact that γ_5 is hermitian. At this point, we can make use of Eq. (F.19) to continue the exercise further:

$$\chi_- = \eta_+^* \gamma_0 \gamma_5 \mathbb{C} \chi_+^* = -\eta_+^* \gamma_5 \gamma_0 \mathbb{C} \chi_+^* = -\eta_+^* \gamma_5 \xi_- = -\eta_+^* \eta_- \chi_- \, , \tag{F.86}$$

using Eq. (F.84a) in the last step. This shows that we must take

$$\eta_- = -\eta_+ \, . \tag{F.87}$$

Now let us see what happens when γ_5 multiplies the spinors. Starting from the definitions of the u-spinors in Eq. (F.70a), we obtain

$$\gamma_5 u_{p,s} = N_p \gamma_5 \big(E_p + m - \gamma_i p_i \big) \xi_s = N_p \big(E_p + m + \gamma_i p_i \big) \gamma_5 \xi_s \, . \tag{F.88}$$

Using Eq. (F.84a) now and the definition of the v-spinors from Eq. (F.69b), we can write

$$\gamma_5 u_{p,s} = \eta_s v_{p,-s} \, . \tag{F.89a}$$

Similarly, one obtains

$$\gamma_5 v_{p,s} = -\eta_s^* u_{p,-s} \,. \tag{F.89b}$$

One can use these properties of the spinors to write their conjugation properties in another equivalent but useful way. Notice that

$$\begin{aligned}
\gamma_5 \mathbb{C} u_{p,s}^* &= \gamma_5 \gamma_0 v_{p,s} = -\gamma_0 \gamma_5 v_{p,s} \\
&= \eta_s^* \gamma_0 u_{p,-s} = \eta_s^* u_{-p,-s} \,,
\end{aligned} \tag{F.90}$$

using Eqs. (F.83) and (F.89) on the way. This equation, and the similar equation with v-spinors, can be written as

$$\mathbb{C}^{-1} \gamma_5 u_{p,s} = -\eta_s u_{-p,-s}^* \,, \tag{F.91a}$$

$$\mathbb{C}^{-1} \gamma_5 v_{p,s} = -\eta_s^* v_{-p,-s}^* \,. \tag{F.91b}$$

These relations were introduced as Eq. (7.59, *p 199*) in the context of time-reversal transformation, with the choice of η_s given by

$$\eta_s = s \,, \tag{F.92}$$

with s defined in Eq. (F.67), which is a particularly simple way of satisfying Eq. (F.87).

☐ **Exercise F.10** Show that

$$\mathbb{C}^{-1} u_{p,s} = v_{-p,s}^* \,, \tag{F.93a}$$

$$\mathbb{C}^{-1} v_{p,s} = -u_{-p,s}^* \,. \tag{F.93b}$$

☐ **Exercise F.11** In the Dirac-Pauli representation, one takes

$$\gamma^0 = \begin{bmatrix} 1 & 0 \\ 0 & -1 \end{bmatrix} , \qquad \gamma^i = \begin{bmatrix} 0 & \sigma^i \\ -\sigma^i & 0 \end{bmatrix} . \tag{F.94}$$

Using $\mathbb{C} = i\gamma^2\gamma^0$, show that the spinors consistent with the normalization and phase conventions given here are as follows:

$$u_+ = \begin{bmatrix} e_+ \\ \sigma \cdot n e_+ \end{bmatrix}, \qquad u_- = \begin{bmatrix} e_- \\ \sigma \cdot n e_- \end{bmatrix}, \tag{F.95a}$$

$$v_+ = -\begin{bmatrix} \sigma \cdot n e_- \\ e_- \end{bmatrix}, \qquad v_- = -\begin{bmatrix} \sigma \cdot n e_+ \\ e_+ \end{bmatrix}, \tag{F.95b}$$

where

$$e_+ = \sqrt{E_p + m} \begin{pmatrix} 1 \\ 0 \end{pmatrix} , \qquad e_- = \sqrt{E_p + m} \begin{pmatrix} 0 \\ 1 \end{pmatrix} , \tag{F.96}$$

and

$$n = \frac{p}{E_p + m} \,. \tag{F.97}$$

[**Note :** *By matrices enclosed by square brackets, we want to signify that the elements are written in a shorthand notation in which each displayed element is a block of length 2.*]

F.2.6 Spin sums

Often one needs to use expressions of the sort

$$\sum_s (u_{\boldsymbol{p},s})_a (\overline{u}_{\boldsymbol{p},s})_b \,, \tag{F.98}$$

where the subscripts indicate elements of the spinor. Obviously, the results can be written in terms of the elements of a matrix S_u, defined by

$$S_u = \sum_s u_{\boldsymbol{p},s} \overline{u}_{\boldsymbol{p},s} \,. \tag{F.99}$$

To find the matrix, we let it act on u and v spinors with the same momentum. Note that

$$S_u u_{\boldsymbol{p},s'} = \sum_s u_{\boldsymbol{p},s} \overline{u}_{\boldsymbol{p},s} u_{\boldsymbol{p},s'} \,. \tag{F.100}$$

Using Eq. (F.128a) which will be proved later, we obtain

$$S_u u_{\boldsymbol{p},s'} = 2m u_{\boldsymbol{p},s'} \,. \tag{F.101}$$

On the other hand, Eq. (F.129) shows that

$$S_u v_{\boldsymbol{p},s'} = 0 \,. \tag{F.102}$$

It should be noticed at this point that

$$(\not{p} + m) u_{\boldsymbol{p},s'} = 2m u_{\boldsymbol{p},s'} \,, \qquad (\not{p} + m) v_{\boldsymbol{p},s'} = 0 \,, \tag{F.103}$$

which follow from the defining equations of u and v spinors, Eqs. (F.62a) and (F.62b). Thus S_u and $\not{p} + m$ produce the same result on the u and v spinors. Since any 4-component column vector can be expressed as a superposition of the four u and v spinors, it means that S_u and $\not{p} + m$ produce the same result on any spinor, and therefore they must be identical. Similar arguments can be given on the sum of v spinors, and the results can be summarized as

$$\sum_s u_{\boldsymbol{p},s} \overline{u}_{\boldsymbol{p},s} = \not{p} + m \,, \tag{F.104a}$$

$$\sum_s v_{\boldsymbol{p},s} \overline{v}_{\boldsymbol{p},s} = \not{p} - m \,. \tag{F.104b}$$

F.2.7 Projection matrices on spinors

There are four spinor solutions altogether for any given momentum. Two of these are u-spinors, and two are v-spinors. The first kind satisfies Eq. (F.62a), whereas the second kind satisfies Eq. (F.62b). But how do we define the two different solutions of each kind? For many applications, it is not necessary to do so. But there are other cases when it is. In such cases, depending on the need, we have to project out one of the u-spinors or one of the v-spinors. This can be done with the help of projection matrices.

In order to obtain a projection matrix, one first needs to find a matrix Q with the property

$$Q^2 = \mathbb{1}. \tag{F.105}$$

With the help of this, we define the two matrices

$$Q_\pm \equiv \frac{1}{2}(\mathbb{1} \pm Q). \tag{F.106}$$

It is easy to see that these matrices have the following properties:

$$Q_\pm^2 = Q_\pm, \tag{F.107a}$$
$$Q_+Q_- = Q_-Q_+ = 0. \tag{F.107b}$$

Eq. (F.107a) implies that the matrices $Q_\pm$ are projection matrices, and Eq. (F.107b) says that they are mutually orthogonal.

Eq. (F.107a) implies that the eigenvalues of the projection matrices $Q_\pm$ are 0 and 1. It is easy to see that if we start with an arbitrary choice for the two u-spinors, Q_+u will be a spinor which will be an eigenstate of Q_+ with eigenvalue +1 and also of Q_- with eigenvalue zero. Similarly, Q_-u will be an eigenstate of Q_- with eigenvalue +1 and of Q_+ with eigenvalue zero. Below, we present several examples of possible choices of Q and comment on the corresponding projections on spinors.

☐ **Exercise F.12** ✫ Show that for 2×2 matrices, the most general choice for Q satisfying Eq. (F.105) is either a diagonal matrix with +1 or −1 as diagonal elements, or of the form

$$Q = \begin{pmatrix} a_1 & \dfrac{1 - a_1^2}{a_2} \\ a_2 & -a_1 \end{pmatrix}, \tag{F.108}$$

with $a_2 \neq 0$. Show that the eigenvector with eigenvalue η ($\eta = \pm 1$) is given by

$$E_\eta = \begin{pmatrix} \eta + a_1 \\ a_2 \end{pmatrix}. \tag{F.109}$$

Show also that, for an arbitrary column vector E with two elements,
$$Q_\eta E = (\text{number}) \times E_\eta. \tag{F.110}$$

a)   Chirality projection

Eq. (F.14) tells us that we have γ_5 as a choice for Q. The projection matrices that follow are usually called the chirality projection matrices, and denoted by

$$\mathbb{R} = \frac{1}{2}(1 + \gamma_5), \qquad \mathbb{L} = \frac{1}{2}(1 - \gamma_5). \tag{F.111}$$

Their properties have been discussed in detail in §14.2.

b) Helicity projection

The helicity operator for a spinor was defined in §14.2.1:

$$h \equiv \frac{\boldsymbol{\Sigma} \cdot \boldsymbol{p}}{\mathrm{p}} . \qquad (F.112)$$

It was mentioned there that $h^2 = 1$. Therefore, the helicity projection operators are

$$h_\pm = \frac{1}{2}\left(1 \pm \frac{\boldsymbol{\Sigma} \cdot \boldsymbol{p}}{\mathrm{p}}\right) . \qquad (F.113)$$

c) Spin projection

In the rest frame of a particle of mass m, the momentum 4-vector is given by $(m, \boldsymbol{0})$. In this frame, the spin of the particle would be a spatial vector. We can define a 4-vector for spin, with the prescription

$$\left. s^\mu \right|_{\text{rest frame}} = (0, \hat{\boldsymbol{s}}) , \qquad (F.114)$$

where $\hat{\boldsymbol{s}}$ is a unit spatial vector in the direction of the spin, i.e.,

$$\hat{\boldsymbol{s}} \cdot \hat{\boldsymbol{s}} = 1 . \qquad (F.115)$$

In a general frame in which the particle has energy E and 3-momentum $\boldsymbol{p}$, the spin 4-vector takes the form

$$s^\mu = \left(\frac{\boldsymbol{p} \cdot \hat{\boldsymbol{s}}}{m}, \hat{\boldsymbol{s}} + \frac{(\boldsymbol{p} \cdot \hat{\boldsymbol{s}})\boldsymbol{p}}{m(E + m)}\right) . \qquad (F.116)$$

☐ **Exercise F.13** *Derive Eq. (F.116) from Eq. (F.114) using Lorentz transformation equations for vectors.*

☐ **Exercise F.14** *Verify that*

$$s^\mu p_\mu = 0 , \qquad s^\mu s_\mu = -1 , \qquad (F.117)$$

explicitly from the form given in Eq. (F.116). [**Note :** *The results should be obvious if one goes back to the rest frame and uses Lorentz invariance.*]

We now have a choice of $\mathbb{Q}$ in the form of $\gamma_5 \not{s}$. Thus, a spin projection operator can be defined as

$$P_{\hat{\boldsymbol{s}}} = \frac{1}{2}(1 + \gamma_5 \not{s}) . \qquad (F.118)$$

d) Energy projection

Since $(\not{p})^2 = m^2$, a candidate for $\mathbb{Q}$ is $\not{p}/m$. The projection operators corresponding to this choice are given by

$$\Lambda_\pm = \frac{m \pm \not{p}}{2m} . \qquad (F.119)$$

Clearly, since the u- and v-spinors obey Eqs. (F.62a) and (F.62b),

$$\Lambda_+ u_{\boldsymbol{p}} = u_{\boldsymbol{p}}, \qquad \Lambda_+ v_{\boldsymbol{p}} = 0, \tag{F.120}$$

and so we can say that Λ_+ projects out the positive energy spinors or the u-spinors. Similarly, Λ_- projects out the v-spinors.

F.3   Bilinears

F.3.1   Gordon identity

There are inter-relations between different types of bilinears, as we show here. Consider two positive-energy spinors $u_{\boldsymbol{p}_1}$ and $u_{\boldsymbol{p}_2}$ corresponding to two fields, whose quanta have masses m_1 and m_2 respectively. And now consider the bilinear

$$\bar{u}_{\boldsymbol{p}_2}(p_1 + p_2)^\lambda (a + b\gamma_5) u_{\boldsymbol{p}_1}. \tag{F.121}$$

We can write

$$\begin{aligned}
p_1^\lambda &= (\gamma^\lambda \gamma^\rho + i\sigma^{\lambda\rho}) p_{1\rho} = \gamma^\lambda \not{p}_1 + i\sigma^{\lambda\rho} p_{1\rho}, \\
p_2^\lambda &= (\gamma^\rho \gamma^\lambda - i\sigma^{\lambda\rho}) p_{2\rho} = \not{p}_2 \gamma^\lambda - i\sigma^{\lambda\rho} p_{2\rho}.
\end{aligned} \tag{F.122}$$

Putting these identities in and using the Dirac equations for the spinors, we obtain

$$\bar{u}_{\boldsymbol{p}_2}(p_1 + p_2)^\lambda u_{\boldsymbol{p}_1} = \bar{u}_{\boldsymbol{p}_2}\Big((m_1 + m_2)\gamma^\lambda + i\sigma^{\lambda\rho} q_\rho\Big) u_{\boldsymbol{p}_1}, \tag{F.123}$$

where

$$q = p_1 - p_2. \tag{F.124}$$

This is called the *Gordon identity*.

☐ **Exercise F.15** *Prove a similar equation involving γ_5:*

$$\bar{u}_{\boldsymbol{p}_2}(p_1 + p_2)^\lambda \gamma_5 u_{\boldsymbol{p}_1} = \bar{u}_{\boldsymbol{p}_2}\Big((m_2 - m_1)\gamma^\lambda + i\sigma^{\lambda\rho} q_\rho\Big)\gamma_5 u_{\boldsymbol{p}_1}. \tag{F.125}$$

☐ **Exercise F.16** *Find the corresponding equations with v-spinors.*

As an important application of the Gordon identity, consider Eq. (F.123) for the case when both spinors contain the same mass m and the same momentum $\boldsymbol{p}$. The resulting equation is then

$$p^\lambda \bar{u}_{\boldsymbol{p}} u_{\boldsymbol{p}} = m \, \bar{u}_{\boldsymbol{p}} \gamma^\lambda u_{\boldsymbol{p}}. \tag{F.126}$$

In particular, if we consider the time component of this equation, we obtain

$$E_{\boldsymbol{p}} \bar{u}_{\boldsymbol{p}} u_{\boldsymbol{p}} = m \, \bar{u}_{\boldsymbol{p}} \gamma^0 u_{\boldsymbol{p}} = m \, u_{\boldsymbol{p}}^\dagger u_{\boldsymbol{p}}. \tag{F.127}$$

Comparing this equation and the corresponding equation with v-spinors with Eq. (F.74), we find that the normalization conditions for spinors are equivalent to the relations

$$\overline{u}_{p,s} u_{p,s'} = 2m\delta_{s,s'}, \qquad \text{(F.128a)}$$

$$\overline{v}_{p,s} v_{p,s'} = -2m\delta_{s,s'}. \qquad \text{(F.128b)}$$

Unless $m = 0$, these equations can also be used for normalization of spinors.

☐ **Exercise F.17** Show that

$$\overline{u}_{p,s} v_{p,s'} = 0, \qquad \overline{v}_{p,s} u_{p,s'} = 0. \qquad \text{(F.129)}$$

[**Hint :** *For example, one can use Eq. (F.75) along with Eq. (F.83).*]

F.3.2 Squaring amplitudes

Amplitudes for processes involving fermions contain bilinears of the form $\overline{w}_1 F w_2$, where w stands for a spinor, either u-type or v-type, and F is a 4×4 matrix. In order to calculate the rate of any process, we need to find the absolute square of the amplitude. This requires, first of all, that we find the complex conjugate of a bilinear. Since the bilinear, taken as a whole for a particular matrix F, is a number, we can as well take the hermitian conjugate: it would not make any difference. Thus,

$$\left(\overline{w}_1 F w_2\right)^\dagger = \left(w_1^\dagger \gamma_0 F w_2\right)^\dagger = w_2^\dagger F^\dagger \gamma_0 w_1 = \overline{w}_2 \gamma_0 F^\dagger \gamma_0 w_1. \quad \text{(F.130)}$$

If we introduce a symbol to write

$$\left(\overline{w}_1 F w_2\right)^\dagger = \overline{w}_2 F^\ddagger w_1, \qquad \text{(F.131)}$$

then our previous analysis shows that

$$F^\ddagger = \gamma_0 F^\dagger \gamma_0. \qquad \text{(F.132)}$$

For ready reference, let us tabulate $F^\ddagger$ for all 16 independent matrices that we chose in Eq. (4.93, *p 79*):

F	$\mathbb{1}$	γ_μ	$\sigma_{\mu\nu}$	$\gamma_\mu \gamma_5$	γ_5
$F^\ddagger$	$\mathbb{1}$	γ_μ	$\sigma_{\mu\nu}$	$\gamma_\mu \gamma_5$	$-\gamma_5$

$$\text{(F.133)}$$

In evaluating the rate of a process, we often want to sum over the spins of the fermions. Let us see then what is obtained by performing the spin sum over a bilinear and its complex conjugate.

$$\sum_{\text{spin}} \left|\overline{w}_1 F w_2\right|^2 = \sum_{\text{spin}} \left(\overline{w}_1 F w_2\right)\left(\overline{w}_2 F^\ddagger w_1\right). \qquad \text{(F.134)}$$

We write it explicitly in terms of matrix elements. Once we do this, we need not pay any attention to the ordering of different factors, and we can bring together terms which depend on the spins s_1 and s_2 which correspond to the spinors w_1 and w_2 respectively.

$$
\begin{aligned}
\sum_{\text{spin}} \left| \overline{w}_1 F w_2 \right|^2 &= (F_{ab})(F^\ddagger)_{cd} \sum_{s_1} (\overline{w}_1)_a (w_1)_d \sum_{s_2} (\overline{w}_2)_c (w_2)_b \\
&= (F_{ab})(F^\ddagger)_{cd} \sum_{s_1} (w_1 \overline{w}_1)_{da} \sum_{s_2} (w_2 \overline{w}_2)_{bc} .
\end{aligned} \tag{F.135}
$$

Spin sums such as the ones appearing in this expression have been evaluated in §F.2.6. Denoting the momenta of the spinors by p_1 and p_2 and the masses m_1 and m_2 respectively, we obtain

$$
\begin{aligned}
\sum_{\text{spin}} \left| \overline{w}_1 F w_2 \right|^2 &= (F_{ab})(F^\ddagger)_{cd} (\slashed{p}_1 + \eta_{w_1} m_1)_{da} (\slashed{p}_2 + \eta_{w_2} m_2)_{bc} \\
&= \mathrm{Tr}\left((\slashed{p}_1 + \eta_{w_1} m_1) F (\slashed{p}_2 + \eta_{w_2} m_2) F^\ddagger \right) ,
\end{aligned} \tag{F.136}
$$

where η_w is $+1$ if w happens to be a u-spinor, and -1 if it happens to be a v-spinor.

This result was used many times in the text. Moreover, it was stretched to use in places even where there is no spin sum involved. For example, in §6.9.4 or §14.7, where we worked with spin-polarized initial states, we introduced a modified F, including a spin projection operator, so that we could perform the sum over spins anyway. The projection operator in this case ensured that the contribution to the sum came from one spin polarization only.

F.3.3 Non-relativistic reduction

In many situations, the physical meaning of a fermion bilinear can be understood easily by interpreting the bilinear in the non-relativistic limit. For this purpose, we summarize here the non-relativistic limits of a number of bilinear combinations of spinors. We will perform the deductions involving u-spinors only, leaving the corresponding manipulations for the v-spinors as exercise.

In the non-relativistic limit, all components of the 3-momentum are much smaller compared to the mass of the particle. Therefore we can take p_i/m as an expansion parameter and keep only the leading order terms in it. For example, Eq. (F.70) tells us that

$$
u_{\boldsymbol{p},s} = \sqrt{2m}(1 - \gamma_i p_i / 2m)\xi_s + \mathcal{O}\left(\mathbf{p}^2\right) . \tag{F.137}
$$

In what follows, we will neglect the 3-momentum altogether, so that for any 4×4 matrix F sandwiched between two u-spinors, we write

$$
\overline{u}_{\boldsymbol{p},s} F u_{\boldsymbol{p}',s'} = u_{\boldsymbol{p},s}^\dagger \gamma_0 F u_{\boldsymbol{p}',s'} \xrightarrow{NR} 2m \, \xi_s^\dagger \gamma_0 F \xi_{s'} . \tag{F.138}
$$

Now note that the hermitian conjugate of Eq. (F.65) tells us that $\xi_s^\dagger \gamma_0 = \xi_s^\dagger$. Thus

$$\bar{u}_{\boldsymbol{p},s} F u_{\boldsymbol{p}',s'} \xrightarrow{NR} 2m\, \xi_s^\dagger F \xi_{s'}\,. \tag{F.139}$$

Let us now find out the combination $\xi_s^\dagger F \xi_{s'}$ for various choices of F.

First, the scalar bilinear matrix element. This corresponds to $F = \mathbb{1}$. So the non-relativistic limit involves

$$\xi_s^\dagger \xi_{s'} = \delta_{ss'}\,, \tag{F.140}$$

because of the way these eigenvectors were normalized.

Let us next consider the pseudoscalar matrix element, i.e., $F = \gamma_5$. Recall that $[\gamma_0, \gamma_5]_+ = 0$. Sandwiching both sides between $\xi_s^\dagger$ and $\xi_{s'}$ and using Eq. (F.65), we obtain

$$\xi_s^\dagger \gamma_5 \xi_{s'} = 0\,. \tag{F.141}$$

The same argument would also imply that

$$\xi_s^\dagger \gamma_i \xi_{s'} = 0\,. \tag{F.142}$$

This shows that the matrix element of the vector bilinear has vanishing spatial components. As for the temporal component, we can use Eq. (F.65) to write

$$\xi_s^\dagger \gamma_0 \xi_{s'} = \xi_s^\dagger \xi_{s'} = \delta_{ss'}\,. \tag{F.143}$$

If $F = \gamma_0\gamma_5$, the matrix element vanishes by the same reason that gave us Eq. (F.141). As for the spatial parts of the axial bilinear, we invoke Eq. (14.10, *p414*), or equivalently Eq. (F.48) to write

$$\xi_s^\dagger \gamma_i \gamma_5 \xi_{s'} = \xi_s^\dagger \gamma_0 \gamma_i \gamma_5 \xi_{s'} = -\frac{1}{2}\varepsilon_{ijk}\xi_s^\dagger \sigma^{jk}\xi_{s'}\,. \tag{F.144}$$

For $s = s'$, we see that the spatial parts of the axial vector bilinear are related to the spin expectation values in the non-relativistic limit.

Among the basic bilinears, we are now left with the tensor bilinears only. The bilinears involving σ_{ij}, i.e., space-space components, are the spin expectation values. The others, i.e., the ones involving σ_{0i}, vanish in this limit. This can be seen by using Eqs. (F.65) and (F.142).

Appendix G

Evaluation of loop integrals

In this appendix, we summarize various techniques which are helpful in evaluating momentum integrations that arise in the amplitudes of loop diagrams.

G.1 Introducing Feynman parameters

First, we note that

$$\int_0^1 d\zeta \, \frac{1}{\left[\zeta a_1 + (1 - \zeta)a_2\right]^2} = \frac{1}{a_1 a_2} , \qquad (G.1)$$

which can be verified very easily. This identity is often used in an integration: when the integrand contains two factors in the denominator, these two factors can be written in the form given in the left side of Eq. (G.1). The variable ζ appearing in this integration is called a *Feynman parameter*.

The formula can be generalized in a form that can be used when there are more factors in the denominator. This generalization is:

$$\Gamma(n) \left(\prod_{i=1}^n \int_0^1 d\zeta_i \right) \frac{\delta\left(1 - \sum_{i=1}^n \zeta_i\right)}{\left[\sum_{i=1}^n \zeta_i a_i\right]^n} = \prod_{i=1}^n \frac{1}{a_i} . \qquad (G.2)$$

Sometimes, even a more general formula is helpful, where each factor in the denominator appears with a power. This more general formula is this:

$$\frac{\Gamma(\sum_i \alpha_i)}{\prod_i \Gamma(\alpha_i)} \left(\prod_i \int_0^1 d\zeta_i \right) \frac{\delta\left(1 - \sum_i \zeta_i\right) \prod_i \zeta_i^{\alpha_i - 1}}{\left[\sum_i \zeta_i a_i\right]^{\sum_i \alpha_i}} = \prod_i \frac{1}{a_i^{\alpha_i}} . \qquad (G.3)$$

□ **Exercise G.1** ✿ Inductively or otherwise, prove Eqs. (G.2) and (G.3).

The formulas do not seem to be very simple and one may wonder what might one possibly gain by writing a string of factors in the denominator in terms of an integral. To appreciate the reason, let us consider an expression of the form

$$I = \int \frac{d^4k}{(2\pi)^4} \frac{f(k)}{\prod_{i=1}^{n}\left((k+p_i)^2 - A_i^2\right)}, \tag{G.4}$$

where $f(k)$ is some function of the momentum that is integrated over. Using Eq. (G.2) for the denominator, we can write

$$I = \Gamma(n)\left(\prod_{i=1}^{n}\int_0^1 d\zeta_i\right)\int \frac{d^4k}{(2\pi)^4} \frac{\delta\left(1 - \sum_{i=1}^{n}\zeta_i\right)f(k)}{\left[\sum_{i=1}^{n}\zeta_i\left((k+p_i)^2 - A_i^2\right)\right]^n}. \tag{G.5}$$

Look at the denominator now. Within the square brackets, the co-efficient of k^2 is the sum of all Feynman parameters, which is equal to unity because of the delta function appearing in the numerator. Thus we obtain

$$I = \Gamma(n)\left(\prod_{i=1}^{n}\int_0^1 d\zeta_i\right)\int \frac{d^4k}{(2\pi)^4} \frac{\delta\left(1 - \sum_{i=1}^{n}\zeta_i\right)f(k)}{\left[k^2 + \sum_{i=1}^{n}\left(2\zeta_i k \cdot p_i + \zeta_i(p_i^2 - A_i^2)\right)\right]^n}. \tag{G.6}$$

We can now shift to a new momentum integration variable defined by $k + \sum_i \zeta_i p_i$. This will eliminate the term linear in k in the denominator. Using the same letter k to write the new momentum variable, we then obtain

$$I = \Gamma(n)\left(\prod_{i=1}^{n}\int_0^1 d\zeta_i\right)\int \frac{d^4k}{(2\pi)^4} \frac{\delta\left(1 - \sum_{i=1}^{n}\zeta_i\right)f(k - \sum_i \zeta_i p_i)}{\left[k^2 - (\sum \zeta_i p_i)^2 + \zeta_i(p_i^2 - A_i^2)\right]^n}. \tag{G.7}$$

The simplification lies in this form. Notice that the denominator is now of the form of $(k^2 - B^2)^n$, where the quantity B^2, a function of the Feynman parameters, is a Lorentz invariant quantity. The momentum integration can be exactly and easily performed for an integrand of this sort, as we will show gradually in the rest of this appendix.

G.2   Reduction to invariant integrals

Let us now look at the numerator of the integral given in Eq. (G.6). We have
not specified what kind of a function $f(k)$ is. It might be a Lorentz invariant,
but it might also contain vector indices. In the latter case, the overall integral
will also bear these indices. If the indices in the integrand are carried by
any momentum other than k, these factors can simply be pulled outside the
integral. We therefore need to consider only the other case, viz., when the
Lorentz indices are carried by the integration momentum k itself. We show
that in these cases, the momentum can be easily related to a Lorentz invariant
integral.

Notice that terms with odd powers of k in the numerator vanish on integration. Thus, the simplest non-trivial example will contain two Lorentz
indices, and the integral will be of the form

$$I_{\mu\nu} = \int d^4k \; k_\mu k_\nu F(k^2) \,, \tag{G.8}$$

where $F(k^2)$ is an arbitrary function. The result must satisfy the property
$I_{\mu\nu} = I_{\nu\mu}$, and would obviously be independent of any momentum. It can
then only be proportional to the metric tensor $g_{\mu\nu}$, so that we can write

$$I_{\mu\nu} = g_{\mu\nu} I \,, \tag{G.9}$$

where I is a Lorentz invariant integral. This integral can be identified by
contracting both sides of this equation by $g^{\mu\nu}$ and using Eq. (D.6). The
result is

$$\int d^4k \; k_\mu k_\nu F(k^2) = \frac{1}{4} g_{\mu\nu} \int d^4k \; k^2 F(k^2) \,. \tag{G.10}$$

The same procedure can be applied for terms with larger number of
Lorentz indices in the integrand. For example,

$$\int d^4k \; k_\mu k_\nu k_\lambda k_\rho F(k^2) = \frac{1}{24} (g_{\mu\nu} g_{\lambda\rho} + g_{\mu\lambda} g_{\nu\rho} + g_{\mu\rho} g_{\nu\lambda})$$
$$\times \int d^4k \; (k^2)^2 F(k^2) \,. \tag{G.11}$$

□ **Exercise G.2** Prove Eq. (G.11).

At the end of the discussion of §G.1, we found that the denominator of loop
integrations can be brought to a certain form that involves only k^2. Now we
see that the whatever the numerators might be, the result of the integration
would be related to an integral of the sort

$$I_D(r, s) = \int \frac{d^D k}{(2\pi)^D} \frac{(k^2)^r}{(k^2 - B^2)^s} \,. \tag{G.12}$$

In fact, to be quite general, we have also written the integration measure as
if it applies for an integration in a D-dimensional spacetime. As we will see
in §G.6, the choice for this generalization is not merely a whim: it also helps
the evaluation of certain integrals.

G.3 Wick rotation

The form of the denominator presented in Eq. (G.12) was motivated by the expression of Eq. (G.4), where the factors in the denominator were supposed to come from various propagators that appear in an amplitude. To be very precise, the denominator of a propagator is not merely a momentum squared minus a mass squared. It contains also a vanishingly small imaginary part, something that we did not mention when we introduced the propagators in Ch. 4. For example, the Feynman rule for a internal line carrying a scalar particle of mass m should really be written as

$$i\Delta_F(p) = \frac{i}{p^2 - m^2 + i\epsilon} \tag{G.13}$$

and the limit $\epsilon \to 0$ should be taken at the end of every calculation. For most operations, this subtle point is irrelevant. This is, however, one occasion when it is not. We should remember that all these small imaginary parts from all propagators combine, so that, instead of writing the typical loop integral as in Eq. (G.12), we should write

$$I_D(r,s) = \int \frac{d^D k}{(2\pi)^D} \frac{(k^2)^r}{(k^2 - B^2 + i\epsilon)^s} . \tag{G.14}$$

This is the integral we will evaluate in the rest of this appendix.

The first step toward the momentum integration involves a change in the variable k_0. We define a new variable $\underline{k}_0$ by

$$k_0 = i\underline{k}_0 . \tag{G.15}$$

Clearly then, Eq. (G.14) can be written as

$$I_D(r,s) = i(-1)^{r+s} \int \frac{d^D \underline{k}}{(2\pi)^D} \frac{(\underline{k}^2)^r}{(\underline{k}^2 + B^2)^s} , \tag{G.16}$$

where, in this equation,

$$\underline{k}^2 = \underline{k}_0^2 + \boldsymbol{k}^2 . \tag{G.17}$$

This is how the square of a vector would have been defined if we were dealing with a Euclidean space. The transformation therefore amounts to going from the Minkowski space variables to Euclidean space variables. If we consider complex values of k_0, this can be seen as an analytic continuation in that parameter. It is a very useful tool, and has been invoked several times in the text.

There is a tricky point in going from Eq. (G.14) to Eq. (G.16) which we explain here. Naïvely, one would expect that if one makes the transformation shown in Eq. (G.15), the limits of the $\underline{k}_0$-integration would be from $-i\infty$ to $+i\infty$, since the corresponding limits are from $-\infty$ to

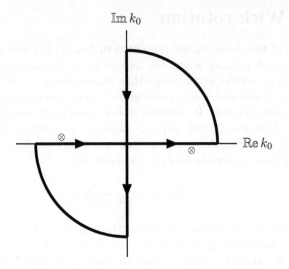

Figure G.1: The contour for performing k_0 integration is shown with thick lines. The crosses indicate the poles of the integrand.

$+\infty$ for k_0. This is not so. In fact, the limits on the $\underline{k}_0$-integration are also $-\infty$ to $+\infty$, for reasons that we describe now.

The transformation of variable introduced in Eq. (G.15) can also be seen as the result of a contour integration in the complex plane if we treat k_0 as a complex variable. We show the complex k_0 plane in Fig. G.1. We write the integral given in Eq. (G.14) in the form

$$I_D(r,s) = \int dk_0 \int \frac{d^{D-1}k}{(2\pi)^D} \frac{(k^2)^r}{(k_0^2 - \boldsymbol{k}^2 - B^2 + i\epsilon)^s} . \tag{G.18}$$

The limits of the k_0 integration are $-\infty$ and $+\infty$, which means that the integration should be performed along the horizontal line shown in Fig. G.1, in the limit that the radius of the curved portion becomes arbitrarily large.

Now suppose we consider the integration of the same integrand around the closed contour shown in the figure. The locations of the poles of the integrand are shown qualitatively in the figure, so that the contour does not include any of the poles. Theory of complex integration then tells us that the integral around the contour should vanish. We can therefore write

$$0 = \int_{-\infty}^{+\infty} dk_0 \, F(k_0) + \int_{C_1} dk_0 \, F(k_0) + \int_{+i\infty}^{-i\infty} dk_0 \, F(k_0) + \int_{C_2} dk_0 \, F(k_0) , \tag{G.19}$$

where $F(k_0)$ is the shorthand for the integrand, and C_1, C_2 represent the curved portions of the contour. One can show that for integrals of the form that we have here, the integrals from the curved portions vanish as we take the radius of these portions to infinity. Thus we can write

$$\int_{-\infty}^{+\infty} dk_0 \, F(k_0) = \int_{-i\infty}^{+i\infty} dk_0 \, F(k_0) = i \int_{-\infty}^{+\infty} d\underline{k}_0 \, F(\underline{k}_0) , \tag{G.20}$$

using Eq. (G.15) in the last step. Thus, the limits $-\infty$ to $+\infty$ of the variable k_0 translates to the same limits of the variable $\underline{k}_0$. In this sense, we can think of the integration path to be rotated from the real axis to the imaginary axis of the complex k_0 plane. This trick is called *Wick rotation*.

G.4 Angular integration

Since the integral $I_D(r, s)$ depends only on the magnitude of the integration variable k, the angular integrations should be trivial. In other words, we can write

$$\int d^D \underline{k} \, F(\underline{k}^2) = \int d\underline{k} \, \underline{k}^{D-1} \, F(\underline{k}^2) \int d\Omega_D \,, \qquad (G.21)$$

where

$$\underline{k} = \sqrt{\underline{k}^2} \qquad (G.22)$$

is the magnitude, and $d\Omega_D$ denotes the infinitesimal element of all angular variables in a D-dimensional space. The point is that we can determine the angular integration by taking any $F(\underline{k}^2)$ that we like, and use it for any other function of k^2. We take $F(\underline{k}^2) = \exp(-\underline{k}^2)$. This is convenient because we can use a Cartesian system and express the integral as a product of D integrals of a single exponential:

$$\int d^D \underline{k} \, \exp(-\underline{k}^2) = \prod_i \int_{-\infty}^{\infty} d\underline{k}_i \, \exp(-\underline{k}_i^2) = \pi^{D/2} \,, \qquad (G.23)$$

using the fact that

$$\int_{-\infty}^{\infty} dx \, e^{-x^2} = \sqrt{\pi} \,. \qquad (G.24)$$

On the other hand, the same integral can be expressed as in Eq. (G.21). Noting that

$$\int d\underline{k} \, \underline{k}^{D-1} \, e^{-\underline{k}^2} = \frac{1}{2} \int d\underline{k}^2 \left(\underline{k}^2 \right)^{(D/2)-1} e^{-\underline{k}^2} = \frac{1}{2} \Gamma(D/2) \,, \qquad (G.25)$$

we obtain

$$\int d\Omega_D = \frac{2\pi^{D/2}}{\Gamma(D/2)} \,. \qquad (G.26)$$

Therefore the integral of Eq. (G.16) can be written as

$$I_D(r, s) = \frac{2i(-1)^{r+s}}{(4\pi)^{D/2} \, \Gamma(D/2)} \int_0^{\infty} d\underline{k} \, \underline{k}^{D-1} \, \frac{(\underline{k}^2)^r}{(\underline{k}^2 + B^2)^s} \,. \qquad (G.27)$$

G.5 Integration over the magnitude

The only integration remaining is that over the magnitude of $\underline{k}$. Introducing a new integration variable

$$\xi = \underline{k}^2 / B^2 \,, \qquad (G.28)$$

the integral can be reduced to the form

$$I_D(r,s) = \frac{i(-1)^{r+s}}{(4\pi)^{D/2}\,\Gamma(D/2)}\,\frac{1}{(B^2)^{s-r-D/2}}\int_0^\infty d\xi\,\frac{\xi^{r+D/2-1}}{(\xi+1)^s}\,. \tag{G.29}$$

In the remaining integral, make a further change of variable:

$$y = \frac{1}{\xi+1}\,. \tag{G.30}$$

Then we find that

$$\int_0^\infty d\xi\,\frac{\xi^{r+D/2-1}}{(\xi+1)^s} = \int_0^1 dy\,y^{s-r-D/2-1}\,(1-y)^{r+D/2-1}\,. \tag{G.31}$$

This integral is a well-known representation of the beta function in mathematical analysis:

$$\int_0^1 dy\,y^{m-1}\,(1-y)^{n-1} \equiv B(m,n)\,. \tag{G.32}$$

Further, this function can be related to the gamma functions through the relation

$$B(m,n) = \frac{\Gamma(m)\Gamma(n)}{\Gamma(m+n)}\,. \tag{G.33}$$

Organizing these results then, we obtain

$$I_D(r,s) = \frac{(-1)^{r+s}}{(4\pi)^{D/2}}\,\frac{\Gamma(r+D/2)\Gamma(s-r-D/2)}{\Gamma(D/2)\Gamma(s)}\,\frac{i}{(B^2)^{s-r-D/2}}\,. \tag{G.34}$$

This is the general result. For the physical spacetime dimension $D = 4$, the result reads

$$\begin{aligned}
I(r,s) &\equiv \int \frac{d^4k}{(2\pi)^4}\,\frac{(k^2)^r}{[k^2-B^2]^s}\\
&= \frac{(-1)^{r+s}}{(4\pi)^2}\,\frac{\Gamma(r+2)\Gamma(s-r-2)}{\Gamma(s)}\,\frac{i}{(B^2)^{s-r-2}}\,,
\end{aligned} \tag{G.35}$$

where we do not put any subscript on I on the left hand side.

G.6   Divergent integrals

The gamma function, $\Gamma(z)$, is well-defined for all $z > 0$, but diverges when z equals zero or any negative integer. The loop integrals that we encounter in practical calculations are of the form $I(r,s)$ with positive integral values of both r and s. If we obtain a 4-dimensional integral with $s - r \leq 2$, we see that the result of Eq. (G.35) is divergent.

The divergence can be anticipated from the expression of the integral in Eq. (G.16). The integral can be divided into two regions, $\underline{k} < R$ and $\underline{k} > R$, where R is chosen such that $R^2 \gg B^2$. The first part will yield some finite result, depending on the values of R and B. In the second part we can neglect B in the denominator. Then, apart from a factor coming from the angular integration, the integral can be written as

$$\int_R^\infty d\underline{k}\, \underline{k}^{D-1}(\underline{k}^2)^{r-s} = \frac{1}{2(r-s)+D}\underline{k}^{2(r-s)+D}\bigg|_R^\infty . \tag{G.36}$$

For the integral to converge to a finite value, we therefore need $D + 2(r - s) < 0$, i.e., $s - r > 2$ for $D = 4$. Otherwise, the integral diverges. Since the divergence comes from high values of the loop momentum, it is called *ultraviolet divergence*.

We remarked in §12.2.1 that one has to introduce counterterms to cancel the infinities that arise from such divergent integrals. Let us see how that can be done if we interpret the integral to be an integral in D spacetime dimensions. To be specific, we consider the case $s - r = 2$. From Eq. (G.34), we find

$$I_D(s-2,s) = \frac{1}{(4\pi)^{D/2}} \frac{\Gamma(s-2+D/2)\Gamma(2-D/2)}{\Gamma(D/2)\Gamma(s)} \frac{i}{(B^2)^{2-D/2}} . \tag{G.37}$$

We now put $D = 4$ in all terms that do not have any problem for that value of D, obtaining

$$I_D(s-2,s) = \frac{i}{(4\pi)^2} \frac{\Gamma(\varepsilon)}{(B^2)^\varepsilon} . \tag{G.38}$$

where, for the sake of brevity, we have used the notation

$$\varepsilon = 2 - \frac{D}{2} . \tag{G.39}$$

The divergence now shows in the form of $\Gamma(0)$. However, we can separate out the divergence in a well-defined manner. Let us note that the gamma function has the property

$$z\Gamma(z) = \Gamma(z+1) . \tag{G.40}$$

Thus we can write

$$\Gamma(\varepsilon) = \frac{1}{\varepsilon}\Gamma(1+\varepsilon) = \frac{1}{\varepsilon}\left(1 + \varepsilon\Gamma'(1) + \mathcal{O}\left(\varepsilon^2\right)\right) . \tag{G.41}$$

The derivative of the gamma function at the value 1 is given by

$$\Gamma'(1) = \int_0^\infty dx\, \ln x\, e^{-x} . \tag{G.42}$$

The result of the integral is negative. The absolute value is called the *Euler–Mascheroni constant* and denoted by γ_E. Thus,

$$\Gamma(\varepsilon) = \frac{1}{\varepsilon} - \gamma_E + \mathcal{O}(\varepsilon). \tag{G.43}$$

Putting it back into Eq. (G.38), we can write

$$I_D(s-2,s) = \frac{i}{(4\pi)^2}\left(\frac{1}{\varepsilon} - \gamma_E + \mathcal{O}(\varepsilon)\right)\left(1 - \varepsilon\ln B^2 + \mathcal{O}(\varepsilon^2)\right), \tag{G.44}$$

where the last factor comes from the series expansion of $(B^2)^{-\varepsilon} = \exp(-\varepsilon\ln B^2)$. Hence

$$I_D(s-2,s) = \frac{i}{(4\pi)^2}\left(\frac{1}{\varepsilon'} - \ln B^2 + \mathcal{O}(\varepsilon)\right), \tag{G.45}$$

where ε' is defined through the relation

$$\frac{1}{\varepsilon'} = \frac{1}{\varepsilon} - \gamma_E. \tag{G.46}$$

The $\mathcal{O}(\varepsilon)$ terms in Eq. (G.45) vanish when we take the limit $\varepsilon \to 0$, i.e., to $D \to 4$. So we need not consider these terms further. Among the other terms, clearly the infinity resides in the part coming from the $\Gamma(\varepsilon)$, so we can make a separation of the infinite part and the finite part:

$$I_D(s-2,s)\Big|_{\text{divergent}} = \frac{i}{(4\pi)^2}\frac{1}{\varepsilon'}, \tag{G.47a}$$

$$I_D(s-2,s)\Big|_{\text{finite}} = -\frac{i}{(4\pi)^2}\ln B^2. \tag{G.47b}$$

Note that this is the reason we could set $D = 4$ to most terms before considering the powers of ε. For example, consider the powers of 4π in the denominator. If we had kept $(4\pi)^{D/2}$, as suggested in Eq. (G.37), instead of putting $D = 4$ in this factor, it would have amounted to an extra factor of $(4\pi)^\varepsilon$ in the denominator. Expanding this factor in powers of ε as well, we would have still obtained the expressions in Eq. (G.47), except that now ε' would have been defined through the relation

$$\frac{1}{\varepsilon'} = \frac{1}{\varepsilon} - \gamma_E - \ln(4\pi), \tag{G.48}$$

and the expression for the finite part would have been exactly the same as what appears in Eq. (G.47b).

□ **Exercise G.3** Following similar steps, show that

$$I_D(s-1,s) = \frac{is}{(4\pi)^2}B^2\left(\frac{1}{\varepsilon'} - \ln B^2\right). \tag{G.49}$$

There seems to be something strange in the expression in Eq. (G.47b). The quantity B^2 was defined in Eq. (G.12), from where it is clear that it should have the same dimensions as k^2. So it seems that Eq. (G.47b) contains the logarithm of an object carrying mass dimensions, which does not make any sense.

Agreed. The point is that in practical calculations, the integral shown in Eq. (G.12) is multiplied with some coupling constants. The dimensions of these coupling constants depend on the spacetime dimensions. For a definite example, consider the coupling constant of QED. Since the action must be dimensionless in the natural units, we want

$$\dim \mathscr{L} = D \tag{G.50}$$

for spacetime dimension equal to D. The free Lagrangians of the fermion field and the photon field dictate the following mass dimensions for these fields:

$$\dim \psi = \frac{D-1}{2}\,, \qquad \dim A_\mu = \frac{D}{2} - 1\,. \tag{G.51}$$

Thus the field combination in the interaction term has the dimension

$$\dim \overline{\psi}\gamma^\mu \psi A_\mu = \frac{3D}{2} - 2\,. \tag{G.52}$$

This means that the dimension of the coupling constant should be $2 - \frac{D}{2} = \varepsilon$. We can therefore denote the coupling constant by $e\mu^\varepsilon$, where e is the dimensionless coupling constant of a 4-dimensional theory and μ is an arbitrary parameter which has the dimension of mass. The quantity of real interest is therefore not the integral defined in Eq. (G.12), but something of the sort

$$J_D(r, s) = (g\mu^\varepsilon)^2 \int \frac{d^D k}{(2\pi)^D} \frac{(k^2)^r}{(k^2 - B^2)^s}\,, \tag{G.53}$$

where g is some dimensionless coupling constant. Performing the same exercise, we would obtain

$$J_D(s-2, s) = \frac{ig^2}{(4\pi)^2} \left(\frac{1}{\varepsilon'} - \ln(B^2/\mu^2) + \mathcal{O}\left(\varepsilon\right) \right) \tag{G.54}$$

instead of Eq. (G.45). This expression is free from the trouble of dimensions. However, it contains an unphysical parameter μ. Therefore, in calculations of observable quantities, this parameter must drop out.

Appendix H

Feynman rules for standard model

H.1 External lines

Type of particle	Feynman rule for	
	incoming	outgoing
Scalar	1	1
Dirac fermion	$u(\boldsymbol{p})$	$\bar{u}(\boldsymbol{p})$
Antifermion	$\bar{v}(\boldsymbol{p})$	$v(\boldsymbol{p})$
Vector boson	$\epsilon_\mu(\boldsymbol{p})$	$\epsilon_\mu^*(\boldsymbol{p})$

$$(\text{H.1})$$

Feynman rules for external Weyl or Majorana fermions are a bit more involved, and are given in Table 14.1 *(p 420)* and Table 22.1 *(p 683)* respectively.

H.2 Propagators

The general method for finding propagators has been described in §4.10.2. Here, we give the results only. As mentioned at the beginning of §G.3, the denominator of each propagator has an additional term $+i\epsilon$. We omit this term here.

a) Gauge bosons

$$g_\mu^a(k) \rightarrow g_\nu^b(k) \; \Box \; \frac{-i\delta^{ab}}{k^2}\left(g_{\mu\nu} - (1-\xi_g)\frac{k_\mu k_\nu}{k^2}\right), \qquad (\text{H.2a})$$

$$W_\mu^\pm(k) \rightarrow W_\nu^\pm(k) \; \Box \; \frac{-i}{k^2 - M_W^2}\left(g_{\mu\nu} - \frac{(1-\xi_W)k_\mu k_\nu}{k^2 - \xi_W M_W^2}\right), \qquad (\text{H.2b})$$

$$Z_\mu(k) \rightarrow Z_\nu(k) \; \Box \; \frac{-i}{k^2 - M_Z^2}\left(g_{\mu\nu} - \frac{(1-\xi_Z)k_\mu k_\nu}{k^2 - \xi_Z M_Z^2}\right), \qquad (\text{H.2c})$$

$$A_\mu(k) \rightarrow A_\nu(k) \; \Box \; \frac{-i}{k^2}\left(g_{\mu\nu} - (1-\xi_A)\frac{k_\mu k_\nu}{k^2}\right). \qquad (\text{H.2d})$$

In principle, ξ_g, ξ_W, ξ_Z and ξ_A can be different from one another, although in practice this freedom is rarely used. Most often, all these gauge parameters are set equal to unity.

b) Ghosts

$$c_a(k) \to c_b(k) \; \Box \; -\frac{\delta_{ab}}{k^2} \, . \tag{H.3}$$

c) Spinless bosons

$$w^\pm(k) \to w^\pm(k) \; \Box \; \frac{i}{k^2 - \xi_W M_W^2} \, , \tag{H.4a}$$

$$z(k) \to z(k) \; \Box \; \frac{i}{k^2 - \xi_Z M_Z^2} \, , \tag{H.4b}$$

$$H(k) \to H(k) \; \Box \; \frac{i}{k^2 - M_H^2} \, . \tag{H.4c}$$

d) Fermions

For any fermion of mass m, the propagator is as follows:

$$f(p) \to f(p) : i\,\frac{\not{p} + m}{p^2 - m^2} \, . \tag{H.5}$$

H.3   Vertices

Some couplings are given in terms of the incoming and outgoing particles, separated by an arrow. Others have been given in terms of field operators that generate them. Thus, e.g., the coupling given below in Eq. (H.10a) is appropriate for any of the following 16 combinations of events at a vertex:

$$\begin{bmatrix} W_\mu^+ \text{ in} \\ \text{or} \\ W_\mu^- \text{ out} \end{bmatrix} \text{ and } \begin{bmatrix} W_\nu^+ \text{ in} \\ \text{or} \\ W_\nu^- \text{ out} \end{bmatrix} \text{ and } \begin{bmatrix} W_\lambda^- \text{ in} \\ \text{or} \\ W_\lambda^+ \text{ out} \end{bmatrix} \text{ and } \begin{bmatrix} W_\rho^- \text{ in} \\ \text{or} \\ W_\rho^+ \text{ out} \end{bmatrix} . \tag{H.6}$$

H.3.1   Vertices in gauge sector

a) Three gauge bosons

To denote the Feynman rules for vertices involving three gauge bosons, we use a notation like $W \xrightarrow{V} W$. It means that at the vertex there are two W bosons and one V boson, where V can be either the photon or the Z. The 4-momenta of the W bosons are indicated in the notation, and that of the

V boson determined by momentum conservation. With this notation, the Feynman rules for vertices involving the electroweak gauge bosons are:

$$W_\mu^+(p) \xrightarrow{A_\lambda} W_\nu^+(q) \ \square \ ie S_{\mu\nu\lambda}(p,q)\,, \tag{H.7a}$$

$$W_\mu^+(p) \xrightarrow{Z_\lambda} W_\nu^+(q) \ \square \ ig\cos\theta_W S_{\mu\nu\lambda}(p,q)\,, \tag{H.7b}$$

$$W_\mu^-(p) \xrightarrow{A_\lambda} W_\nu^-(q) \ \square \ -ie S_{\mu\nu\lambda}(p,q)\,, \tag{H.7c}$$

$$W_\mu^-(p) \xrightarrow{Z_\lambda} W_\nu^-(q) \ \square \ -ig\cos\theta_W S_{\mu\nu\lambda}(p,q)\,, \tag{H.7d}$$

where

$$S_{\mu\nu\lambda}(p,q) = (p+q)_\lambda g_{\mu\nu} + (q-2p)_\nu g_{\mu\lambda} + (p-2q)_\mu g_{\lambda\nu}\,. \tag{H.8}$$

There is no other cubic coupling of electroweak gauge bosons. The cubic vertex involving gluons has the following Feynman rule:

$$g_\mu^a(p) \xrightarrow{g_\lambda^c} g_\nu^b(q) \ \square \ ig_3 f_{ab\bar{c}} S_{\mu\nu\lambda}(p,q)\,. \tag{H.9}$$

□ **Exercise H.1** Verify that the Feynman rule given in Eq. (H.9) is identical to that given in Fig. 11.2 (p 313).

b) Four gauge bosons

Possible vertices with electroweak gauge bosons are as follows:

$$W_\mu^+ W_\nu^+ W_\lambda^- W_\rho^- \ \square \ ig^2 S_{\mu\nu\lambda\rho}\,, \tag{H.10a}$$

$$A_\mu A_\nu W_\lambda^+ W_\rho^- \ \square \ -ie^2 S_{\mu\nu\lambda\rho}\,, \tag{H.10b}$$

$$Z_\mu Z_\nu W_\lambda^+ W_\rho^- \ \square \ -ig^2\cos^2\theta_W S_{\mu\nu\lambda\rho}\,, \tag{H.10c}$$

$$A_\mu Z_\nu W_\lambda^+ W_\rho^- \ \square \ -ieg\cos\theta_W S_{\mu\nu\lambda\rho}\,, \tag{H.10d}$$

where

$$S_{\mu\nu\lambda\rho} = 2g_{\mu\nu}g_{\lambda\rho} - g_{\mu\lambda}g_{\nu\rho} - g_{\mu\rho}g_{\nu\lambda}\,. \tag{H.11}$$

The quartic gluon vertex is given by

$$g_\mu^a g_\nu^b g_\lambda^c g_\rho^d \ \square \ -ig^2 \Big[f_{ab\bar{e}} f_{cde} E_{\mu\nu\lambda\rho} + f_{ac\bar{e}} f_{dbe} E_{\mu\lambda\rho\nu} + f_{ad\bar{e}} f_{bce} E_{\mu\rho\nu\lambda} \Big]\,, \tag{H.12}$$

where

$$E_{\mu\nu\lambda\rho} \equiv g_{\mu\lambda}g_{\nu\rho} - g_{\mu\rho}g_{\nu\lambda}\,. \tag{H.13}$$

c) Gauge bosons and ghosts

$$c_b(p') \xrightarrow{A_\mu^c} c_a(p) \ \square \ -gp_\mu f_{bca}\,. \tag{H.14}$$

H.3.2 Gauge bosons and Higgs bosons

By the term *Higgs bosons*, we here imply both the physical spinless boson of the standard model as well as the unphysical would-be Goldstone modes that are eaten up by the $W^{\pm}$ and the Z gauge bosons.

a) One gauge boson and two Higgs bosons

$$H(p) \xrightarrow{W_\mu} w^{\pm}(p') \;\square\; \mp \frac{ig}{2}(p+p')_\mu \,, \tag{H.15a}$$

$$z(p) \xrightarrow{W_\mu} w^{\pm}(p') \;\square\; \frac{g}{2}(p+p')_\mu \,, \tag{H.15b}$$

$$H(p) \xrightarrow{Z_\mu} z(p') \;\square\; \mp \frac{ig}{2\cos\theta_W}(p+p')_\mu \,, \tag{H.15c}$$

$$w^{\pm}(p) \xrightarrow{Z_\mu} w^{\pm}(p') \;\square\; \mp \frac{ig\cos 2\theta_W}{2\cos\theta_W}(p+p')_\mu \,, \tag{H.15d}$$

$$w^{\pm}(p) \xrightarrow{A_\mu} w^{\pm}(p') \;\square\; \mp ie(p+p')_\mu \,. \tag{H.15e}$$

b) Two gauge bosons and one Higgs boson

$$W_\mu^+ W_\nu^- z \;\square\; 0 \,, \tag{H.16a}$$

$$W_\mu^+ W_\nu^- H \;\square\; igM_W g_{\mu\nu} \,, \tag{H.16b}$$

$$Z_\mu Z_\nu H \;\square\; ig\frac{M_Z}{\cos\theta_W}g_{\mu\nu} \,, \tag{H.16c}$$

$$Z_\mu Z_\nu z \;\square\; 0 \,, \tag{H.16d}$$

$$W_\mu^+ A_\nu w^- \;\square\; ieM_W g_{\mu\nu} \,, \tag{H.16e}$$

$$W_\mu^+ Z_\nu w^- \;\square\; -igM_W g_{\mu\nu}\frac{\sin^2\theta_W}{\cos\theta_W} \,. \tag{H.16f}$$

c) Two gauge bosons and two Higgs bosons

$$W_\mu^+ W_\nu^- w^+ w^- \;\square\; \frac{ig^2}{2}g_{\mu\nu} \,, \tag{H.17a}$$

$$Z_\mu Z_\nu w^+ w^- \;\square\; \frac{ig^2 \cos^2 2\theta_W}{2\cos^2\theta_W}g_{\mu\nu} \,, \tag{H.17b}$$

$$Z_\mu A_\nu w^+ w^- \;\square\; \frac{ige\cos 2\theta_W}{\cos\theta_W}g_{\mu\nu} \,, \tag{H.17c}$$

$$A_\mu A_\nu w^+ w^- \;\square\; 2ie^2 g_{\mu\nu} \,, \tag{H.17d}$$

$$W_\mu^+ W_\nu^- zz \;\square\; \frac{ig^2}{2}g_{\mu\nu} \,, \tag{H.17e}$$

$$Z_\mu Z_\nu zz \;\square\; \frac{ig^2}{2\cos^2\theta_W}g_{\mu\nu} \,, \tag{H.17f}$$

$$W_\mu^+ W_\nu^- H H \ \square \ \frac{ig^2}{2} g_{\mu\nu} , \tag{H.17g}$$

$$Z_\mu Z_\nu H H \ \square \ \frac{ig^2}{2 \cos^2 \theta_W} g_{\mu\nu} . \tag{H.17h}$$

H.3.3   Scalar self-interactions

a)   Three Higgs bosons

$$w^+ w^- H \ \square \ -\frac{ig M_H^2}{2 M_W} , \tag{H.18a}$$

$$z z H \ \square \ -\frac{ig M_H^2}{2 M_W} , \tag{H.18b}$$

$$H H H \ \square \ -\frac{3 ig M_H^2}{2 M_W} . \tag{H.18c}$$

b)   Four Higgs bosons

$$w^+ w^- w^+ w^- \ \square \ -\frac{ig^2 M_H^2}{2 M_W^2} , \tag{H.19a}$$

$$w^+ w^- H H \ \square \ -\frac{ig^2 M_H^2}{4 M_W^2} , \tag{H.19b}$$

$$w^+ w^- z z \ \square \ -\frac{ig^2 M_H^2}{4 M_W^2} , \tag{H.19c}$$

$$z z H H \ \square \ -\frac{ig^2 M_H^2}{4 M_W^2} , \tag{H.19d}$$

$$H H H H \ \square \ -\frac{3 ig^2 M_H^2}{4 M_W^2} , \tag{H.19e}$$

$$z z z z \ \square \ -\frac{3 ig^2 M_H^2}{4 M_W^2} . \tag{H.19f}$$

There is no other quartic vertex. In particular, there is no vertex with an odd number of z or H lines.

H.3.4   Vertices involving fermions

a)   Fermions and a gauge boson

In the following, T_3 is the eigenvalue of the left-handed component of the fermion under the neutral generator of $SU(2)_L$, and Q is the electric charge in units of e.

$$\nu_l \xrightarrow{W_\mu} l \ \square \ \frac{-ig}{\sqrt{2}} \gamma_\mu \mathbb{L} , \tag{H.20a}$$

$$d_B \xrightarrow{W_\mu} u_A \ \square \ \frac{-ig}{\sqrt{2}} V_{AB} \gamma_\mu \mathbb{L} , \tag{H.20b}$$

$$f \xrightarrow{Z_\mu} f \;\square\; \frac{-ig}{\cos\theta_W}\gamma_\mu\left[T_3\mathbb{L} - \sin^2\theta_W Q\right], \qquad \text{(H.20c)}$$

$$f \xrightarrow{A_\mu} f \;\square\; -ieQ\gamma_\mu, \qquad \text{(H.20d)}$$

$$q \xrightarrow{g^a_\mu} q \;\square\; -ig_3\left(\lambda_a/2\right)\gamma_\mu. \qquad \text{(H.20e)}$$

b) Fermions and a Higgs boson

$$f_1 \xrightarrow{w^+} f_2 \;\square\; \frac{ig}{\sqrt{2}M_W}\left(m_1\mathbb{R} - m_2\mathbb{L}\right), \qquad \text{(H.21a)}$$

$$f \xrightarrow{z} f \;\square\; \frac{gm_f}{2M_Z\cos\theta_W}\gamma_5, \qquad \text{(H.21b)}$$

$$f \xrightarrow{H} f \;\square\; \frac{-igm_f}{2M_Z\cos\theta_W}. \qquad \text{(H.21c)}$$

Appendix I

Books and other reviews

This is not an exhaustive list of reference materials on the subject matter presented in the book. This is only a declaration that the following books and articles were around me at the time of writing this book. So, it is possible that I have been influenced by these books or articles. I also had some of my class notes, most importantly the notes that Professor Lincoln Wolfenstein circulated among the students when I took my first course of particle physics in 1979. However, I surely have been influenced by many other books, papers and review articles that I have read through my student life and research career, but it would be impossible for me to produce a complete list of everything I have read that is relevant to this book.

General textbooks on particle physics or quantum field theory

1) Francis Halzen, Alan D. Martin • *Quarks and leptons* (John Wiley & Sons, 1984)

2) T. P. Cheng, L. F. Li • *Gauge theory of elementary particle physics* (Oxford University Press, 1988)

3) Otto Nachtmann • *Elementary particle physics* (Springer-Verlag, 1989)

4) John F. Donoghue, Eugene Golowich, Barry R. Holstein • *Dyanmics of the standard model* (Cambridge University Press, 1992)

5) Michael E. Peskin, Daniel V. Schroeder • *An introduction to quantum field theory* (Westview Press, 1995)

6) Elliot Leader, Enrico Predazzi • *An introduction to gauge theories and modern particle physics* (Cambridge University Press, 1996, in two volumes)

7) B. R. Martin, G. Shaw • *Particle physics* (John Wiley & Sons, 2nd edition, 1997)

8) Quang Ho-Kim, Xuan-Yem Pham • *Elementary particles and their interactions* (Springer, 1998)

9) Amitabha Lahiri, Palash B. Pal • *A first book of quantum field theory* (Narosa Publishing House, 2nd edition, 2005)

10) Steven Weinberg • *The quantum theory of fields (Vols. 1 and 2)* (Cambridge University Press, 2005)

11) C. P. Burgess, G. D. Moore • *Standard model : a primer* (Cambridge University Press, 2007)

12) Alessandro Bettini • *Introduction to elementary particle physics* (Cambridge University Press, 2008)

Popular or semi-popular expositions

13) Yuval Ne'eman, Yoram Kirsh • *The particle hunters* (Cambridge University Press, 2nd edition, 1996)

14) Peter Woit • *Not even wrong* (Vintage books, 2007)

15) Lincoln Wolfenstein, João P. Silva • *Exploring fundamental particles* (CRC Press, 2011)

Monographs and review articles related to one or a few chapters of this book

16) M. K. Gaillard, M. Nikolic (eds) • *Weak interactions* (Institut National de Physique nucléaire et de physique des particules, Paris, publication date not mentioned)

17) Eugene D. Commins • *Weak interactions* (McGraw-Hill, 1973)

18) Richard Fernow • *Introduction to experimental particle physics* (Cambridge University Press, 1986)

19) Heinz J. Rothe • *Lattice gauge theories: an introduction* (World Scientific, 1992)

20) T. Muta • *Foundations of quantum chromodynamics* (World Scientific, 2nd edition, 1998)

21) I. I. Bigi, A. I. Sanda • *CP violation* (Cambridge University Press, 1999)

22) Aneesh V. Manohar, Mark B. Wise • *Heavy quark physics* (Cambridge University Press, 2000)

23) Andrei Smilga • *Lectures on Quantum chromodynamics* (World Scientific, Singapore, 2001)

24) G. Dissertori, I, Knowles, M. Schmelling • *Quantum chromodynamics* (Clarendon Press, 2003)

25) Jiri Horejsi • *Fundamentals of electroweak theory* (Karolinum Press, Charles University, 2003)

26) K. Zuber • *Neutrino physics* (Institute of Physics Publishing, 2004)

27) Rabindra N. Mohapatra, Palash B. Pal • *Massive neutrinos in physics and astrophysics* (World Scientific, 3rd edition 2004)

28) G. C. Branco, L. Lavoura, J. P. Silva • *CP violation* (Clarendon Press, 2007)

29) Robert Cahn, Gerson Goldhaber • *The experimental foundations of particle physics* (Cambridge University Press, 2nd edition, 2009)

30) R. Gupta , *"Introduction to lattice QCD"*, Les Houches Session LXVIII, *Probing the standard model of particle interactions*, edited by R. Gupta, A. Morel, E. de Rafael, F. David, pp 83–219 [arXiv:hep-lat/9807028v1]

31) Abdelhak Djouadi, *"The anatomy of electro-weak symmetry breaking I: The Higgs boson in the standard model"*, Phys. Rept. **457** (2008) 1–216 [hep-ph/0503172]

32) G. Bhattacharyya , *"A pedagogical review of electroweak symmetry breaking scenarios"*, Rept. Prog. Phys. **74** (2011) 026201 [arXiv:0910.5095[hep-ph]]

33) P. B. Pal , *"Representation-independent manipulations with Dirac matrices and spinors"*, http://arxiv.org/abs/physics/0703214

34) J. Beringer et al. (Particle Data Group), *"The Review of Particle Physics"*, Phys. Rev. D86, 010001 (2012)

Appendix J

Answers to selected exercises

Ex. 1.6 (p. 10): $0.9987c$.

Ex. 1.7 (p. 13): $M_P = \sqrt{\hbar c / G_N} = 2.17 \times 10^{-5}$ g.

Ex. 2.19 (p. 34): 0.

Ex. 3.1 (p. 37): $0, -2,$ 'yes'.

Ex. 3.10 (p. 44): $f_{+0+} = i$, $f_{-0-} = -i$, $f_{+-0} = -2i$.

Ex. 4.5 (p. 72): $u_{0,s} = \sqrt{2m}\,\xi_s$, $v_{0,s} = \sqrt{2m}\,\chi_{-s}$.

Ex. 5.9 (p. 129): 0.67×10^{-24} cm^2.

Ex. 6.21 (p. 175): A table entry J signifies that the particular combination of initial and final state is allowed by angular momentum conservation. Similarly, P and C mean allowed by parity and charge conjugation respectively.

	Final state			
Initial state	$^1S_0 + \gamma$	$^3S_1 + \gamma$	$\gamma + \gamma$	$\gamma + \gamma + \gamma$
$^3P_0$	P	JPC	JPC	J
$^3P_1$	JP	JPC	PC	J
$^3P_2$	P	JPC	JPC	J
$^1P_1$	JPC	JP	P	JC

Ex. 7.6 (p. 198): CP : $R_0^+ = R_0^-$, $A^- = -A^+$, $B^- = -B^+$, $C^- = C^+$, $D^- = -D^+$;
T : $D^+ = D^- = 0$;
CPT : $R_0^+ = R_0^-$, $A^- = -A^+$, $B^- = -B^+$, $C^- = C^+$, $D^- = D^+$.

Ex. 8.6 (p. 218): See Eq. (10.112, p 287) for the answer in a different notation.

Ex. 8.7 (p. 218): $S_z = +1 : \quad \frac{1}{\sqrt{2}}(p{\blacktriangle}n{\blacktriangle} - n{\blacktriangle}p{\blacktriangle})$,
$S_z = 0 \quad : \quad \frac{1}{2}(p{\blacktriangle}n{\blacktriangledown} - n{\blacktriangle}p{\blacktriangledown} - p{\blacktriangledown}n{\blacktriangle} + n{\blacktriangledown}p{\blacktriangle})$.

Ex. 8.9 (p. 224): $\pi^\pm = (\pi_1 \mp i\pi_2)/\sqrt{2}$, $\pi^0 = \pi_3$.

Ex. 9.3 (p. 246): $\sqrt{(E_1 + E_2)^2 - (\boldsymbol{p}_1 + \boldsymbol{p}_2)^2}$, i.e., $\sqrt{m_1^2 + m_2^2 + 2(E_1 E_2 - \boldsymbol{p}_1 \cdot \boldsymbol{p}_2)}$.

Ex. 10.9 (p. 267): Dimensions of the representations: 10, 15, 27.

Ex. 10.13 (p. 287): All members with charge eQ have magnetic moment equal to $-Q\mu_-$, where μ_- is the magnetic moment of the Δ^-.

Ex. 11.12 (p. 310): The factor 2 in the last term would have to be replaced by $1/C^{(f)}$.

Ex. 12.7 (p. 342): 4 and 2.

Ex. 12.22 (p. 368): $\sin^2 \frac{\theta}{2} \sin \alpha \, d\frac{\theta}{2} d\alpha d\beta$.

Ex. 13.2 (p. 391): M^{-2} and M^{-1}.

Ex. 14.9 (p. 429): About 11 GeV.

Ex. 15.1 (p. 449): $\sqrt{2\lambda v^2}$.

Ex. 17.4 (p. 494): M^1.

Ex. 17.9 (p. 497): 1.5×10^{-5}.

Ex. 18.1 (p. 525): $d_R u_R q_L \ell_L$, $q_L q_L u_R \ell_R$, $d_R u_R u_R \ell_R$, $u_R u_R d_R \ell_R$.

Ex. 18.2 (p. 529): $\theta_1 = \alpha - \beta$, $\theta_2 = \alpha + \beta$.

Ex. 18.6 (p. 532): $\sqrt{2\lambda v^2}$.

Ex. 18.7 (p. 532): No.

Ex. 18.13 (p. 547): The indefinite integral is:

$$\int dx \, [\tanh(x + a) - \tanh x] = \ln(1 + \tanh x \tanh a).$$

Ex. 20.1 (p. 597): The representations 6 and 15 are real.

Ex. 20.6 (p. 605): About 1 eV.

Ex. 21.1 (p. 622): $\text{Im}(a^*b)$, or equivalently, the phase of a^*b.

Ex. 21.12 (p. 633): $a = -d = \sqrt{2/3}$, $b = c = \sqrt{1/3}$.

Ex. 21.21 (p. 650): $\frac{1}{2} e^{-\gamma t}(1 + |\lambda_f|^2)$.

Ex. 22.16 (p. 699): -1.

Ex. 22.17 (p. 699): A Majorana mass term contains the combination $\nu_L \nu_L$, which has $T_3 = +1$. A singlet of $SU(2)_L$ contains only $T_3 = 0$.

Ex. 23.1 (p. 703): (2,2,0).

Index

* Capitalization has been disregarded in the alphabetization of the index. Word boundaries, even those involving hyphens and apostrophes, have also been disregarded. Thus, for the following entries, the order is as follows: *D-dimensional* comes before *de Broglie*, and *D mesons* comes later.

* For entries involving a symbol, the alphabetical order has been defined by the way the symbol is pronounced. For example, Λ appears at the alphabetical place appropriate for *Lambda*, and D^* comes where *Dstar* should appear.

* Names of scientists have not been included in the index, except where a name appears as an inseparable part of the name of a physical variable or phenomenon, like *Wick rotation* or *Dalitz plot*.

* Some entries have been marked with a bullet sign (•), implying that we have not listed all occurrences of that word or phrase in the index. The index only refers to the first occurrence or definition of the word or phrase.

For Product Safety Concerns and Information please contact our
EU representative GPSR@taylorandfrancis.com Taylor & Francis
Verlag GmbH, Kaufingerstraße 24, 80331 München, Germany